# Lecture Notes in Artificial Intelligence 11734

Subseries of Lecture Notes in Computer Science

## Series Editors

Randy Goebel
*University of Alberta, Edmonton, Canada*
Yuzuru Tanaka
*Hokkaido University, Sapporo, Japan*
Wolfgang Wahlster
*DFKI and Saarland University, Saarbrücken, Germany*

## Founding Editor

Jörg Siekmann
*DFKI and Saarland University, Saarbrücken, Germany*

More information about this series at http://www.springer.com/series/1244

Hilde Pérez García · Lidia Sánchez González ·
Manuel Castejón Limas ·
Héctor Quintián Pardo ·
Emilio Corchado Rodríguez (Eds.)

# Hybrid Artificial Intelligent Systems

14th International Conference, HAIS 2019
León, Spain, September 4–6, 2019
Proceedings

Springer

Editors
Hilde Pérez García (iD)
University of León
León, Spain

Lidia Sánchez González (iD)
University of León
León, Spain

Manuel Castejón Limas (iD)
University of León
León, Spain

Héctor Quintián Pardo (iD)
University of A Coruña
Ferrol, Spain

Emilio Corchado Rodríguez (iD)
University of Salamanca
Salamanca, Spain

ISSN 0302-9743          ISSN 1611-3349   (electronic)
Lecture Notes in Artificial Intelligence
ISBN 978-3-030-29858-6          ISBN 978-3-030-29859-3   (eBook)
https://doi.org/10.1007/978-3-030-29859-3

LNCS Sublibrary: SL7 – Artificial Intelligence

This Springer imprint is published by the registered company Springer Nature Switzerland AG
The registered company address is: Gewerbestrasse 11, 6330 Cham, Switzerland

# Preface

This volume of *Lecture Notes on Artificial Intelligence* (LNAI) includes accepted papers presented at the 14th International Conference on Hybrid Artificial Intelligence Systems (HAIS 2019), held in the beautiful and historic city of León, Spain, September 2019.

HAIS has become a unique, established, and broad interdisciplinary forum for researchers and practitioners who are involved in developing and applying symbolic and sub-symbolic techniques aimed at the construction of highly robust and reliable problem-solving techniques, as well as being responsible for the most relevant achievements in this field.

Hybridization of intelligent techniques, coming from different computational intelligence areas, has become popular because of the growing awareness that such combinations frequently perform better than the individual techniques such as neurocomputing, fuzzy systems, rough sets, evolutionary algorithms, agents and multiagent Systems, etc.

Practical experience has indicated that hybrid intelligence techniques might be helpful to solve some of the challenging real-world problems. In a hybrid intelligence system, a synergistic combination of multiple techniques is used to build an efficient solution to deal with a particular problem. This is, thus, the setting of the HAIS conference series, and its increasing success is the proof of the vitality of this exciting field.

The HAIS 2019 international Program Committee selected 64 papers that are published in this conference proceedings, with an acceptance rate of about 48% of the submissions.

The selection of papers was extremely rigorous in order to maintain the high quality of the conference and we would like to thank the Program Committee for their hard work in the reviewing process. This process is very important to the creation of a conference of high standard and the HAIS conference would not exist without their help.

The large number of submissions is certainly not only a testimony of the vitality and attractiveness of the field but an indicator of the interest in the HAIS conferences themselves.

HAIS 2019 enjoyed outstanding keynote speeches by distinguished guest speakers: Alberto Hernández General Chair Incibe, León, Spain; Dr. Pedro Larrañaga Múgica, Polytechnic University of Madrid, Spain; Dr. André C. P. L. F. de Carvalho, University of São Paulo, Brasil.

HAIS 2019 teamed up with *Neurocomputing*, Elsevier and the *Logic Journal of the IGPL* Oxford Journals for a suite of special issues including selected papers from HAIS 2019.

Particular thanks go as well to the main sponsors of the conference, Startup OLE, University of León, and University of Salamanca, who jointly contributed in an active and constructive manner to the success of this initiative.

We would like to thank Alfred Hoffman and Anna Kramer from Springer for their help and collaboration during this demanding publication project.

September 2019

Hilde Pérez García
Lidia Sánchez González
Manuel Castejón Limas
Héctor Quintián
Emilio Corchado

# Organization

## General Chair

Emilio Corchado — University of Salamanca, Spain

## Local Chairs

Hilde Pérez García — University of León, Spain
Lidia Sánchez González — University of León, Spain
Manuel Castejón Limas — University of León, Spain

## International Advisory Committee

Ajith Abraham — Machine Intelligence Research Labs, Europe
Antonio Bahamonde — University of Oviedo, Spain
Andre de Carvalho — University of São Paulo, Brazil
Sung-Bae Cho — Yonsei University, South Korea
Juan M. Corchado — University of Salamanca, Spain
José R. Dorronsoro — Autonomous University of Madrid, Spain
Michael Gabbay — Kings College London, UK
Ali A. Ghorbani — UNB, Canada
Mark Λ. Girolami — University of Glasgow, UK
Manuel Graña — University of País Vasco, Spain
Petro Gopych — Universal Power Systems USA-Ukraine LLC, Ukraine
Jon G. Hall — The Open University, UK
Francisco Herrera — University of Granada, Spain
César Hervás-Martínez — University of Córdoba, Spain
Tom Heskes — Radboud University Nijmegen, The Netherlands
Dusan Husek — Academy of Sciences of the Czech Republic, Czech Republic
Lakhmi Jain — University of South Australia, Australia
Samuel Kaski — Helsinki University of Technology, Finland
Daniel A. Keim — University Konstanz, Germany
Marios Polycarpou — University of Cyprus, Cyprus
Witold Pedrycz — University of Alberta, Canada
Xin Yao — University of Birmingham, UK
Hujun Yin — University of Manchester, UK
Michał Woźniak — Wroclaw University of Technology, Poland
Aditya Ghose — University of Wollongong, Australia
Ashraf Saad — Armstrong Atlantic State University, USA
Fanny Klett — German Workforce Advanced Distributed Learning Partnership Laboratory, Germany

| | |
|---|---|
| Paulo Novais | Universidade do Minho, Portugal |
| Rajkumar Roy | The EPSRC Centre for Innovative Manufacturing in Through-life Engineering Services, UK |
| Amy Neustein | Linguistic Technology Systems, USA |
| Jaydip Sen | Innovation Lab, Tata Consultancy Services Ltd., India |

## Program Committee

| | |
|---|---|
| Emilio Corchado (PC Chair) | University of Salamanca, Spain |
| Hilde Pérez García | University of León, Spain |
| Lidia Sánchez González | University of León, Spain |
| Manuel Castejón Limas | University of León, Spain |
| Héctor Quintián | University of A Coruña, Spain |
| Alberto Cano | Virginia Commonwealth University, USA |
| Amelia Zafra Gómez | University of Córdoba, Spain |
| Anca Andreica | Babes-Bolyai University, Romania |
| Andreea Vescan | Babes-Bolyai University, Romania |
| Ángel Arroyo | University of Burgos, Spain |
| Ángel Manuel Guerrero Higueras | University of León, Spain |
| Antonio Dourado | University of Coimbra, Portugal |
| Antonio Jesús Díaz Honrubia | Universidad Politécnica de Madrid, Spain |
| Arkadiusz Kowalski | Wroclaw University of Technology, Poland |
| Beatriz Remeseiro | University of Oviedo, Spain |
| Bruno Baruque | University of Burgos, Spain |
| Camelia Pintea | Technical University of Cluj-Napoca, Romania |
| Camelia Serban | Babes-Bolyai University, Romania |
| Carlos Cambra | University of Burgos, Spain |
| Carlos Carrascosa | GTI-IA DSIC Universidad Politecnica de Valencia, Spain |
| Carlos Mencía | University of Oviedo, Spain |
| Carlos Pereira | ISEC, Portugal |
| Cezary Grabowik | Silesian Technical University, Poland |
| Damian Krenczyk | Silesian University of Technology, Poland |
| Dario Landa-Silva | The University of Nottingham, UK |
| David Iclanzan | Sapientia – Hungarian Science University of Transylvania, Romania |
| Diego P. Ruiz | University of Granada, Spain |
| Dragan Simic | University of Novi Sad, Serbia |
| Eiji Uchino | Yamaguchi University, Japan |
| Eneko Osaba | University of Deusto, Spain |
| Enrique Onieva | University of Deusto, Spain |
| Esteban Jove Pérez | University of A Coruña, Spain |
| Eva Volna | University of Ostrava, Czech Republic |
| Federico Divina | Pablo de Olavide University, Spain |

| | |
|---|---|
| Fermin Segovia | University of Granada, Spain |
| Fidel Aznar | University of Alicante, Spain |
| Francisco Javier Martínez de Pisón Ascacíbar | University of La Rioja, Spain |
| Francisco Martínez-Álvarez | University Pablo de Olavide, Spain |
| George Papakostas | EMT Institute of Technology, Greece |
| Georgios Dounias | University of the Aegean, Greece |
| Giorgio Fumera | University of Cagliari, Italy |
| Gloria Cerasela Crisan | Vasile Alecsandri University of Bacau, Romania |
| Gonzalo A. Aranda-Corral | University of Huelva, Spain |
| Gualberto Asencio-Cortés | Pablo de Olavide University, Spain |
| Henrietta Toman | University of Debrecen, Hungary |
| Ignacio J. Turias Domínguez | University of Cádiz, Spain |
| Ioana Zelina | Technical University of Cluj Napoca, Romania |
| Ioannis Hatzilygeroudis | University of Patras, Greece |
| Irene Díaz | University of Oviedo, Spain |
| Isabel Barbancho | University of Málaga, Spain |
| Javier De Lope | Polytechnic University of Madrid, Spain |
| Jorge García-Gutiérrez | University of Seville, Spain |
| José Alfredo Ferreira | Federal University, Brazil |
| José Dorronsoro | Universidad Autónoma de Madrid, Spain |
| José García-Rodriguez | University of Alicante, Spain |
| José Luis Calvo-Rolle | University of A Coruña, Spain |
| José Luis Casteleiro-Roca | University of A Coruña, Spain |
| José Luis Verdegay | University of Granada, Spain |
| José M. Molina | University Carlos III of Madrid, Spain |
| José Manuel Lopez-Guede | Basque Country University, Spain |
| José María Armingol | University Carlos III of Madrid, Spain |
| José Ramón Villar | University of Oviedo, Spain |
| Juan Humberto Sossa Azuela | National Polytechnic Institute, Mexico |
| Juan J. Flores | Universidad Michoacana de San Nicolás de Hidalgo, Mexico |
| Juan Pavón | Complutense University of Madrid, Spain |
| Julio Ponce | Universidad Autónoma de Aguascalientes, Mexico |
| Krzysztof Kalinowski | Silesian University of Technology, Poland |
| Lauro Snidaro | University of Udine, Italy |
| Leocadio G. Casado | University of Almeria, Spain |
| Luis Alfonso Fernández Serantes | FH JOANNEU, University of Applied Sciences, Austria |
| Manuel Graña | University of Basque Country, Spain |
| Michal Wozniak | Wroclaw University of Science and Technology, Poland |
| Mohammed Chadli | University of Picardie Jules Verne, France |
| Oscar Fontenla-Romero | University of A Coruña, Spain |

| Ozgur Koray Sahingoz | Istanbul Kultur University, Turkey |
| Paula M. Castro | University of A Coruña, Spain |
| Paulo Novais | University of Minho, Portugal |
| Pavel Brandstetter | VSB-Technical University of Ostrava, Czech Republic |
| Peter Rockett | University of Sheffield, UK |
| Petrica Claudi Pop | Technical University of Cluj-Napoca, Romania |
| Ricardo Del Olmo | University of Burgos, Spain |
| Robert Burduk | Wroclaw University of Technology, Poland |
| Roman Senkerik | TBU in Zlin, Czech Republic |
| Rubén Fuentes-Fernández | Complutense University of Madrid, Spain |
| Sabo Cosmin | Technical University of Cluj-Napoca, Romania |
| Sean Holden | University of Cambridge, UK |
| Sebastián Ventura | University of Córdoba, Spain |
| Theodore Pachidis | Kavala Institute of Technology, Greece |
| Urszula Stanczyk | Silesian University of Technology, Poland |

## Organizing Committee

| Hilde Pérez García | University of León, Spain |
| Lidia Sánchez González | University of León, Spain |
| Manuel Castejón Limas | University of León, Spain |
| Laura Fernández Robles | University of León, Spain |
| Javier Díez González | University of León, Spain |
| Héctor Quintian | University of A Coruña, Spain |
| Emilio Corchado | University of Salamanca, Spain |

# Contents

## Learning Algorithms

## Visual Analysis and Advanced Data Processing Techniques

**Data Mining Applications**

## Hybrid Intelligent Applications

# Data Mining, Knowledge Discovery and Big Data

# Testing Modified Confusion Entropy as Split Criterion for Decision Trees

J. David Nuñez-Gonzalez[1,3,4], Alexander Gonzalo de Sá[4],
and Manuel Graña[2,3,4(✉)]

[1] Department of Applied Mathematics, University of the Basque Country
(UPV/EHU), San Sebastian, Spain
[2] Department of Computer Science and Artificial Intelligence,
University of the Basque Country (UPV/EHU), San Sebastian, Spain
[3] Computational Intelligence Group, University of the Basque Country (UPV/EHU),
San Sebastian, Spain
[4] University of the Basque Country (UPV/EHU), San Sebastian, Spain
manuel.grana@ehu.es

**Abstract.** Confusion Entropy (CEN) has been proposed as a performance measure for classification showing a better discrimination against
other metrics. Many works use CEN for other purposes. Recently, an
improvement in the definition of CEN has been proposed, a modified
CEN (MCEN). The aim of this work is to review a previous work based
on a classification tree that uses CEN as a pruning criterion, replacing
this criterion with the newly defined MCEN metric.

## 1 Introduction

During the data mining process different steps have to be covered with the aim of
extracting knowledge from raw data. Those steps are: business/research understanding phase, data understanding phase, data preparation phase, modeling
phase, evaluation phase, and deployment phase.

This work is focused on the evaluation phase, where a evaluation of the
performance of the machine learning models is done. Thus, different metrics
are classically used, like Accuracy, Precision, Area Under ROC Curve score,
Matthews correlation coefficient (MCC), or more recently, Confusion Entropy
(CEN), a metric based on Shannon's Entropy proposed by [11].

This new metric was proposed as a more discriminant than Accuracy or MCC.
CEN does not only have the percentage of well classified objects in consideration,
like Accuracy, but also considers the distribution of wrong classified objects
across the classes. In this way, CEN calculates how well the samples of different
classes are separated from each other, and gets better results in cases where
the wrongly classified objects are more concentrated in a class than in the cases
where the wrong classified objects have a uniform distribution across all classes.
The values of CEN fall in a range from 0 to 1, where 0 is the result of a perfect
classification.

© Springer Nature Switzerland AG 2019
H. Pérez García et al. (Eds.): HAIS 2019, LNAI 11734, pp. 3–13, 2019.
https://doi.org/10.1007/978-3-030-29859-3_1

Recently, proposed Modified Confusion Entropy (MCEN), a modification on CEN, which corrects an anomaly that happens in the calculation of CEN for binary confusion matrices.

Several works in the literature have used CEN in their experimental proposals [6–10], for instance proposes a Decision Tree classifier using CEN as a pruning method [3]. In this paper we refine the Decision Tree replacing CEN for MCEN.

The rest of the paper is organized as follows: Sect. 2 gives an overview on CEN and MCEN. Section 3 gives the Decision Tree classifier which used CEN as a pruning method and proposes the modification on the algorithm. Section 4 shows experimental results. Section 5 concludes this paper.

# 2 Related Works on Confusion Entropy

## 2.1 Confusion Entropy (CEN)

As mentioned before, [11] proposed a Confusion Entropy (CEN) as a classification evaluation metric based on Shannon's Entropy based on the confusion matrix, being CEN = 0 the best classification (all values fall in the main diagonal of the confusion matrix), and being 1 the worst possible misclassification (all values out of the main diagonal of the confusion matrix and uniformly distributed). Let it be a confusion matrix $C = (C_{i,j})_{i,j=1,...,n}$, which collects the results of predictions of a classifier.

As described in the original publication [11], a multi-class classifier learned from a training data-set, with $N \geq 2$ classes labelled as $\{1, 2, ..., N\}$, is used to classify items of a test data-set, or in other words, predict the class of these items with the data of their attributes. Since this work is focused in supervised classification, classes of these test data-set entries are known so that a $N \times N$ confusion matrix $C = (C_{i,j})_{i,j=1,...,N}$ can be built, collecting the results of the classification of the test data-set. $C_{i,j}$ is the number of items of class $i$ wrongly classified as class $j$. $S$ denotes the sum of values of the confusion matrix, which corresponds to the total number of entries in the test data-set, $S = \sum_{i=1}^{N} \sum_{j=1}^{N} C_{i,j}$.

The probability for an element of class $i$ to be classified as class $j$ subjected to class $j$, $P_{i,j}^{j}$, is shown as:

$$P_{i,j}^{j} = \frac{C_{i,j}}{\sum_{k=1}^{N} (C_{j,k} + C_{k,j})}, \; i,j = 1,...,N, \; i \neq j \tag{1}$$

Therefore, $P_{i,j}^{j}$ should be the relative frequency of cases of class $i$ classified as class $j$ among all cases that are of class $j$ or have been classified as class $j$. However, the metric is biased because the correctly classified cases of class $j$, $C_{j,j}$ are counted twice in the denominator [1,2]. In the same way, the probability for an elements of class $i$ to be classified as class $j$, subjected to class $i$, $P_{i,j}^{i}$, is defined as:

$$P_{i,j}^{i} = \frac{C_{i,j}}{\sum_{k=1}^{N} (C_{i,k} + C_{k,i})}, \; i,j = 1,...,N, \; i \neq j \tag{2}$$

Thus, the value of $CEN$ associated to class $j$ is defined in the following way:

$$\mathrm{CEN}_j = -\sum_{k=1,k\neq j}^{N} \left( P_{j,k}^j \log_{2(N-1)}(P_{j,k}^j) + P_{k,j}^j \log_{2(N-1)}(P_{k,j}^j) \right) \quad (3)$$

Last, $CEN$ associated to the confusion matrix is the combination of the $CEN$ associated to each class:

$$\mathrm{CEN} = \sum_{j=1}^{N} P_j \, \mathrm{CEN}_j \quad (4)$$

Where the non-negative weights $P_j$, summing 1, are:

$$P_j = \frac{\sum_{k=1}^{N} (C_{j,k} + C_{k,j})}{2\sum_{k,\ell=1}^{N} C_{k,\ell}}. \quad (5)$$

## 2.2   Modified Confusion Entropy (MCEN)

An enhanced version of $CEN$ named $MCEN$ introducing a correction that solves CEN erratic behavior in cases of binary classification is proposed in [1], where the formula to calculate the probability of classifying elements of class $i$ as elements of class $j$ subjected to class $j$ is replaced with the following formula:

$$\widetilde{P}_{i,j}^j = \frac{C_{i,j}}{\sum_{k=1}^{N} (C_{j,k} + C_{k,j}) - C_{j,j}}, \; i,j = 1, ..., N, \; i \neq j \quad (6)$$

This way the problem of counting twice the correctly classified cases in the denominator is solved, reflecting in a correct way the relative frequency of cases of class $i$ classified as class $j$ among all cases that are of class $j$ or that have been wrongly classified as class $j$.

The formula in Eq. (2) is modified in the same way:

$$\widetilde{P}_{i,j}^i = \frac{C_{i,j}}{\sum_{k=1}^{N} (C_{i,k} + C_{k,i}) - C_{i,i}}, \, , i,j = 1, ..., N, \; i \neq j \quad (7)$$

Hence, $\widetilde{P}_{i,j}^i$ correctly represents the relative frequency of cases of class $i$ classified as class $j$ among all cases that are of class $i$ or that have been wrongly classified as class $i$. The formula for the weights in Eq. (5) is also modified:

$$\widetilde{P}_j = \frac{\sum_{k=1}^{N} (C_{j,k} + C_{k,j}) - C_{j,j}}{2\sum_{k,\ell=1}^{N} C_{k,\ell} - \alpha \sum_{k=1}^{N} C_{k,k}} \quad (8)$$

where

$$\alpha = \begin{cases} 1/2 & \text{if } N = 2 \\ 1 & \text{if } N > 2. \end{cases}$$

Consequently, the MCEN associated to class $j$ (3) is redefined as follows:

$$\mathrm{MCEN}_j = - \sum_{k=1,k\neq j}^{N} \left( \widetilde{P}^j_{j,k} \log_{2(N-1)}(\widetilde{P}^j_{j,k}) + \widetilde{P}^j_{k,j} \log_{2(N-1)}(\widetilde{P}^j_{k,j}) \right), \quad (9)$$

Finally, the formula for CEN in (4) is modified as follows:

$$\mathrm{MCEN} = \sum_{j=1}^{N} \widetilde{P}_j \, \mathrm{MCEN}_j . \qquad (10)$$

In both measures *CEN* of Eq. (4) and *MCEN* of Eq. (10), classes can be separated, making it easy to check the effect of modifications in the classifier that affect single classes.

The differences in the values of *CEN* and *MCEN* for some example binary confusion matrices can be seen in Table 1. This table shows some simple examples of symmetric and balanced confusion matrices. The inconsistent nature of *CEN* is reflected in the cases where its value surpass *1*.

**Table 1.** CEN and MCEN values for several example confusion matrices in the binary case [2, 4]

| | $\begin{pmatrix} 6\ 0 \\ 0\ 6 \end{pmatrix}$ | $\begin{pmatrix} 5\ 1 \\ 1\ 5 \end{pmatrix}$ | $\begin{pmatrix} 4\ 2 \\ 2\ 4 \end{pmatrix}$ | $\begin{pmatrix} 3\ 3 \\ 3\ 3 \end{pmatrix}$ | $\begin{pmatrix} 2\ 4 \\ 4\ 2 \end{pmatrix}$ | $\begin{pmatrix} 1\ 5 \\ 5\ 1 \end{pmatrix}$ | $\begin{pmatrix} 0\ 6 \\ 6\ 0 \end{pmatrix}$ |
|---|---|---|---|---|---|---|---|
| CEN | 0 | 0.5975 | 0.8617 | 1 | 1.0566 | 1.0525 | 1 |
| MCEN | 0 | 0.4678 | 0.6667 | 0.7925 | 0.8813 | 0.9479 | 1 |

## 3   Decision Tree Algorithm

According to [3], one particular application of *CEN* must be emphasized. They did not only use *CEN* for measuring the performance of a classifier but also for constructing the classifier itself. They propose to apply CEN in decision tree construction as follows.

A decision tree consists of a set of decision nodes (interior) and response nodes (leaves): A decision node is associated with one of the attributes and has 2 or more branches that leave it, each of them representing the possible values that the associated attribute can take. In some way, a decision node is like a question that is asked to the analyzed example, and depending on the answer that of, the flow will take one of the outgoing branches. A node-response is associated with the classification to be provided, and returns the decision of the tree with respect to the input example.

In mentioned work, authors present the *CENTree*, a decision tree classifier based on the C4.5 algorithm [5]. To build a decision tree algorithm, a splitting criterion must be choosen to determine how the data is divided, from an initial impure node representing the entire data-set to multiple and purer nodes, in a branches and leaves(nodes) tree-like structure. Algorithm 1 specifies the decision tree construction with MCEN pruning method as an improvement of the one in [3].

---

**Algorithm 1.** New Decision Tree with MCEN pruning method based on [3]

---

```
MCENTree (Samples, Target_attribute, Attribute):
1.- Create a Root node.
2.- If (all Samples are belonging to one class)
{return Root whose label = classname}
3.- If (Attributes == empty)
{return Root, with label =
most common value Target_attribute in Samples}
4.- Otherwise Begin
4.1.- A<- best(Attribute)
4.2.- Decision attribute for Root <- A
4.3.- For each possible value v(i) of A
{
4.3.1.- Add a new tree branch below Root where test(A=v(i))
4.3.2.- Take subset Samples(v(i)) from Samples for A
4.3.3.- If (Samples(v(i))==empty)
{Add below this new branch a node with label =
min MCEN optimal labeling}
else
{add below this new branch the subtree:
MCENTree (Samples(v(i)), Target_attribute, Attributes-{A}}
}
5.- return Root
```

---

## 4    Empirical Experimental Analysis

In this work the experiments made in [3] with *CENTree* have been replied to the extent possible, focusing in the binary class datasets used in the mentioned article, the ones that are susceptible to have been wrongly classified in the original paper. In addition, classification made in four more datasets has been studied. Datasets are available from the UCI repository[1]. In order to reproduce the C4.5 algorithm used as basis of the *CENTree*, a python code created by Baris Can Esmer and found in his GitHub repository[2] is used to create the model. Changing the splitting method of the maximum information gain, used by the C4.5, for a function to calculate and choose the minimum *CEN*, the code for constructing the *CENTree* is created.

---

[1] https://archive.ics.uci.edu/.
[2] https://github.com/barisesmer/C4.5.

With the *CENTree* model constructor codified, a classifier has been programmed to classify test data. For this purpose, the data is splitted in the same percentage as in [3], a 40% of the instances of each data set is used as train data to construct the *CENTree* model, and the remaining 60% of the instances is used as test data to be classified through the model and measure the effectiveness of the classifier. However, different factors have made the aforementioned experiments not reproducible, and therefore, the results obtained in this work do not match exactly those reported in [3]. On the one hand the lack of the original code of a non-fully described modified algorithm, and on the other hand, the lack of information about the random number generator seed used in the experiments.

For these reasons the basic classical version of the C4.5 algorithm has been implemented to construct the model, and we have repeated 100 times the training and test on each data-set using 100 different random number generator seeds, from seed number 0 to number 99. The results shown in this paper are the outcome of calculating the mean of those 100 different results. Furthermore, experiments have been made adding noise to the data-sets to study the behaviour of CEN and MCEN in more adverse situations. The analyzed noise levels are of 0%, 20% and 40% and have been applied using operators from Weka. Figure 1 presents the pipeline of the experiment.

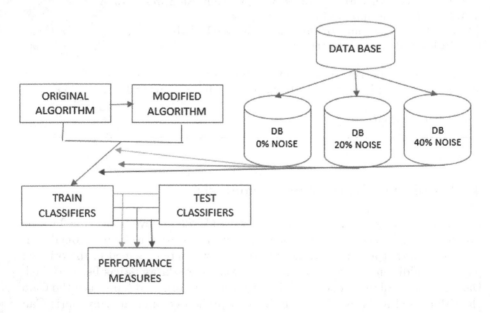

**Fig. 1.** Pipeline of the experiment

We have made a preprocessing of the data from open databases in the aforementioned repository. Specifically, we have added 20% and 40% noise to the data. In this way, for each database we will have: the original database, the database with 20% added noise and the database with 40% added noise. It should be noted that only databases with 2 classes have been selected for experimentation.

**Table 2.** Average values of performance measures for each model for 0% noise

| DB | Model | Acc 0% | CEN 0% | MCEN 0% |
|---|---|---|---|---|
| Tic-tac-toe | CENTree | 0.8098 | 0.6256 | 0.6103 |
| | MCENTree | 0.8190 | 0.6072 | 0.5948 |
| Breast-cancer | CENTree | 0.6247 | 0.7932 | 0.7482 |
| | MCENTree | 0.6231 | 0.7877 | 0.7434 |
| Balloons | CENTree | 0.6682 | 0.7993 | 0.7339 |
| | MCENTree | 0.6666 | 0.7992 | 0.7333 |
| Monks-2 | CENTree | 0.5849 | 0.8994 | 0.8337 |
| | MCENTree | 0.5815 | 0.9023 | 0.8361 |
| Spect | CENTree | 0.7695 | 0.6386 | 0.6198 |
| | MCENTree | 0.7778 | 0.6279 | 0.6119 |
| Vote | CENTree | 0.9285 | 0.3307 | 0.3476 |
| | MCENTree | 0.9299 | 0.3281 | 0.3458 |
| Chess | CENTree | 0.8055 | 0.6502 | 0.6335 |
| | MCENTree | 0.7897 | 0.6792 | 0.6569 |
| Mushroom | CENTree | 0.9971 | 0.0254 | 0.0301 |
| | MCENTree | 0.9977 | 0.0198 | 0.0234 |

**Table 3.** Average values of performance measures for each model for 20% noise

| DB | Model | Acc 20% | CEN 20% | MCEN 20% |
|---|---|---|---|---|
| Tic-tac-toe | CENTree | 0.5918 | 0.9185 | 0.8382 |
| | MCENTree | 0.5936 | 0.9159 | 0.8356 |
| Breast-cancer | CENTree | 0.5169 | 0.9193 | 0.8444 |
| | MCENTree | 0.5159 | 0.9152 | 0.8414 |
| Balloons | CENTree | 0.5219 | 0.9182 | 0.8274 |
| | MCENTree | 0.5242 | 0.9181 | 0.8276 |
| Monks-2 | CENTree | 0.5467 | 0.9457 | 0.8652 |
| | MCENTree | 0.5445 | 0.9468 | 0.5467 |
| Spect | CENTree | 0.5501 | 0.9387 | 0.8537 |
| | MCENTree | 0.5500 | 0.9387 | 0.8533 |
| Vote | CENTree | 0.6732 | 0.8347 | 0.7758 |
| | MCENTree | 0.6708 | 0.8374 | 0.7777 |
| Chess | CENTree | 0.6117 | 0.9148 | 0.8392 |
| | MCENTree | 0.6095 | 0.9168 | 0.8406 |
| Mushroom | CENTree | 0.6661 | 0.8581 | 0.7959 |
| | MCENTree | 0.6685 | 0.8565 | 0.7951 |

**Table 4.** Average values of performance measures for each model for 40% noise

| DB | Model | Acc 40% | CEN 40% | MCEN 40% |
|---|---|---|---|---|
| Tic-tac-toe | CENTree | 0.5033 | 0.9824 | 0.8869 |
| | MCENTree | 0.5023 | 0.9802 | 0.8843 |
| Breast-cancer | CENTree | 0.4776 | 0.9818 | 0.8884 |
| | MCENTree | 0.4801 | 0.9748 | 0.8821 |
| Balloons | CENTree | 0.4579 | 0.9636 | 0.8679 |
| | MCENTree | 0.4573 | 0.9659 | 0.8703 |
| Monks-2 | CENTree | 0.5038 | 0.9834 | 0.8912 |
| | MCENTree | 0.5050 | 0.9834 | 0.8913 |
| Spect | CENTree | 0.5036 | 0.9758 | 0.8825 |
| | MCENTree | 0.5001 | 0.9785 | 0.8847 |
| Vote | CENTree | 0.5205 | 0.9698 | 0.8782 |
| | MCENTree | 0.5168 | 0.9733 | 0.8813 |
| Chess | CENTree | 0.5116 | 0.9898 | 0.8968 |
| | MCENTree | 0.5125 | 0.9894 | 0.8965 |
| Mushroom | CENTree | 0.5125 | 0.9866 | 0.8936 |
| | MCENTree | 0.5123 | 0.9889 | 0.8960 |

The three databases have been trained with the previous algorithm (C4.5 with CEN as a pruning method) and with the modified algorithm (C4.5 with MCEN as a pruning method) obtaining their respective values of CEN and MCEN performance measures. As mentioned at the beginning, the experiment has been repeated 100 times. So that the average values do not blur the particular cases that we think are of interest.

**Table 5.** Confusion Entropy values for the worst Confusion Matrices when no noise is added for both algorithms

| Dataset | CENTree CEN | CENTree MCEN | MCENTree MCEN | Dataset | CENTree CEN | CENTree MCEN | MCENTree MCEN |
|---|---|---|---|---|---|---|---|
| Tic-tac-toe | 0.7286 | 0.6972 | 0.6686 | Spect | 0.7281 | 0.6935 | 0.6215 |
| | 0.7090 | 0.6783 | 0.6015 | | 0.7107 | 0.6832 | 0.6566 |
| | 0.7034 | 0.6745 | 0.6609 | | 0.7087 | 0.6843 | 0.6421 |
| Breast-cancer | 0.9412 | 0.8702 | 0.8728 | Vote | 0.4449 | 0.4441 | 0.3749 |
| | 0.9055 | 0.8450 | 0.8365 | | 0.4256 | 0.4391 | 0.3931 |
| | 0.8894 | 0.8289 | 0.8213 | | 0.4226 | 0.4319 | 0.3854 |
| Balloons | 0.9513 | 0.8625 | 0.8431 | Chess | 0.6923 | 0.6678 | 0.6765 |
| | 0.9411 | 0.8605 | 0.8605 | | 0.6854 | 0.6630 | 0.6894 |
| | 0.9260 | 0.8397 | 0.8397 | | 0.6842 | 0.6617 | 0.6664 |
| Monks-2 | 0.9438 | 0.8584 | 0.8650 | Mushroom | 0.0511 | 0.0599 | 0.0528 |
| | 0.9385 | 0.8643 | 0.8653 | | 0.0490 | 0.0574 | 0.0494 |
| | 0.9363 | 0.8663 | 0.8745 | | 0.0467 | 0.0549 | 0.0549 |

Tables 2, 3, and 4 give the average accuracy, CEN an MCEN values when we build the decision tree using CEN and MCEN for the datasets without noise, with 20% noise, and 40% noise, respectively. These results show that the MCEN-tree has systematically better accuracy, better MCEN, and, even, better CEN performance metrics.

In the Tables 5, 6 and 7 we make a more detailed comparison the CEN and MCEN results for the worst cases of confusion matrices, without noise, with 20% noise, and 40% noise, respectively.

As the noise increases, the CEN values become higher than 1 while the MCEN values are within their range. In any case, most of the time the new algorithm presents less entropy than the previous algorithm. That is, the MCEN values are closer to 0.

**Table 6.** Confusion Entropy values for some of worst Confusion Matrixes when 20% of noise is added for both algorithms

| Dataset | CENTree CEN | CENTree MCEN | MCENTree MCEN | Dataset | CENTree CEN | CENTree MCEN | MCENTree MCEN |
|---------|-------------|--------------|---------------|---------|-------------|--------------|---------------|
| Tic-tac-toe | 0.9573 | 0.8736 | 0.8677 | Spect | 1.0172 | 0.9200 | 0.9164 |
| | 0.9541 | 0.8702 | 0.8606 | | 1.0020 | 0.9091 | 0.9139 |
| | 0.9531 | 0.8684 | 0.8366 | | 0.9993 | 0.9010 | 0.8856 |
| Breast-cancer | 0.9958 | 0.9088 | 0.9129 | Vote | 0.8407 | 0.8407 | 0.8398 |
| | 0.9874 | 0.9074 | 0.9066 | | 0.8976 | 0.8242 | 0.8488 |
| | 0.9862 | 0.8978 | 0.9017 | | 0.8904 | 0.8194 | 0.8272 |
| Balloons | 1.0577 | 0.9654 | 0.9654 | Chess | 0.9411 | 0.8598 | 0.8598 |
| | 1.0402 | 0.9450 | 0.9450 | | 0.9338 | 0.8540 | 0.8538 |
| | 1.0369 | 0.9389 | 0.9389 | | 0.9317 | 0.8522 | 0.8576 |
| Monks-2 | 1.0006 | 0.9076 | 0.9118 | Mushroom | 0.0511 | 0.8097 | 0.8060 |
| | 0.9853 | 0.8972 | 0.8900 | | 0.0490 | 0.8055 | 0.8036 |
| | 0.9836 | 0.8966 | 0.8984 | | 0.0467 | 0.8046 | 0.8093 |

**Table 7.** Confusion Entropy values for some of worst Confusion Matrixes when 40% noise is added for both algorithms

| Dataset | CENTree CEN | CENTree MCEN | MCENTree MCEN | Dataset | CENTree CEN | CENTree MCEN | MCENTree MCEN |
|---------|-------------|--------------|---------------|---------|-------------|--------------|---------------|
| Tic-tac-toe | 1.0123 | 0.9158 | 0.9231 | Spect | 1.0389 | 0.9414 | 0.9446 |
| | 1.0096 | 0.9131 | 0.9160 | | 1.0275 | 0.9288 | 0.9316 |
| | 1.0080 | 0.9117 | 0.9211 | | 1.0243 | 0.9263 | 0.9189 |
| Breast-cancer | 1.0331 | 0.9341 | 0.9231 | Vote | 1.0158 | 0.9191 | 0.9207 |
| | 1.0290 | 0.9319 | 0.9357 | | 1.0024 | 0.9084 | 0.9155 |
| | 1.0264 | 0.9310 | 0.9360 | | 1.0001 | 0.9036 | 0.9074 |
| Balloons | 1.0606 | 0.9708 | 0.9708 | Chess | 1.0084 | 0.9124 | 0.9086 |
| | 1.0479 | 0.9498 | 0.9498 | | 1.0038 | 0.9085 | 0.9044 |
| | 1.0396 | 0.9470 | 0.9470 | | 0.9991 | 0.9042 | 0.9031 |
| Monks-2 | 1.0179 | 0.9205 | 0.9206 | Mushroom | 0.9988 | 0.9047 | 0.9084 |
| | 1.0136 | 0.9178 | 0.9108 | | 0.9968 | 0.9021 | 0.8996 |
| | 1.0129 | 0.9177 | 0.9161 | | 0.9948 | 0.9011 | 0.9006 |

## 5   Conclusion

As mentioned in the very firsts Sections, Confusion Entropy, as a performance measure, has been used in several works. Also as a pruning method as in this work. A redefinition of CEN provides the occasion to review all the works that have used CEN.

This time we chose a work to replicate it but replacing the ancient definition of CEN and using the new one, MCEN. While at the beginning the results may seem similar, this responds to a mere coincidence, because what underlies is that CEN was not correctly defined. In fact, results become more and more noticeable when we introduce noise in the model and therefore we are overloading the confusion matrix outside the main diagonal.

We conclude that, with the results obtained, the substitution of CEN by MCEN yields better results than the previous work.

As a future research lines, our proposal will be to continue with the replication of works from the literature in which the concept of Entropy Confusion has been used. Thus, we will replace CEN with MCEN and we can make comparisons in the results between these and other performance measures. In this way, we could conclude the improvement presented by MCEN with respect to CEN by empirical experimentation.

**Acknowledgments.** The work in this paper has been partially supported by FEDER funds for the MINECO project TIN2017-85827-P, and projects KK-2018/00071 and KK-2018/00082 of the Elkartek 2018 funding program of the Basque Government.

## References

1. Delgado, R., Núñez-González, J.D.: Enhancing confusion entropy as measure for evaluating classifiers. In: Graña, M., López-Guede, J.M., Etxaniz, O., Herrero, Á., Sáez, J.A., Quintián, H., Corchado, E. (eds.) SOCO'18-CISIS'18-ICEUTE'18 2018. AISC, vol. 771, pp. 79–89. Springer, Cham (2019). https://doi.org/10.1007/978-3-319-94120-2_8
2. Delgado, R., Núñez-González, J.D.: Enhancing confusion entropy (CEN) for binary and multiclass classification. PLoS ONE **14**(1), e0210264 (2019)
3. Jin, H., Wang, X.-N., Gao, F., Li, J., Wei, J.-M.: Learning decision trees using confusion entropy. In: 2013 International Conference on Machine Learning and Cybernetics (ICMLC), vol. 2, pp. 560–564. IEEE (2013)
4. Jurman, G., Furlanello, C.: A unifying view for performance measures in multiclass prediction. arXiv preprint arXiv:1008.2908 (2010)
5. Quinlan, J.R.: C4. 5: Programs for Machine Learning. Elsevier (2014)
6. Roumani, Y.F., May, J.H., Strum, D.P., Vargas, L.G.: Classifying highly imbalanced ICU data. Health Care Manage. Sci. **16**(2), 119–128 (2013)
7. Roumani, Y.F., Roumani, Y., Nwankpa, J.K., Tanniru, M.: Classifying readmissions to a cardiac intensive care unit. Ann. Oper. Res. **263**(1–2), 429–451 (2018)
8. Salari, N., Shohaimi, S., Najafi, F., Nallappan, M., Karishnarajah, I.: A novel hybrid classification model of genetic algorithms, modified k-nearest neighbor and developed backpropagation neural network. PLoS ONE **9**(11), e112987 (2014)

9. Sigdel, M., Aygün, R.S.: Pacc - a discriminative and accuracy correlated measure for assessment of classification results. In: Perner, P. (ed.) MLDM 2013. LNCS (LNAI), vol. 7988, pp. 281–295. Springer, Heidelberg (2013). https://doi.org/10.1007/978-3-642-39712-7_22

10. Sublime, J., Matei, B., Cabanes, G., Grozavu, N., Bennani, Y., Cornuéjols, A.: Entropy based probabilistic collaborative clustering. Pattern Recogn. **72**, 144–157 (2017)

11. Wei, J.-M., Yuan, X.-J., Qing-Hua, H., Wang, S.-Q.: A novel measure for evaluating classifiers. Expert Syst. Appl. **37**(5), 3799–3809 (2010)

# Generating a Question Answering System from Text Causal Relations

E. C. Garrido Merchán[1], C. Puente[2(✉)], and J. A. Olivas[3]

[1] Universidad Autónoma de Madrid,
Francisco Tomas y Valiente 11, 28049 Madrid, Spain
eduardo.garrido@uam.es
[2] Escuela Técnica Superior de Ingeniería ICAI,
Universidad Pontificia Comillas, Madrid, Spain
cristina.puente@icai.comillas.edu
[3] Departamento de Tecnologías y Sistemas, Universidad de Castilla-La Mancha,
Ciudad Real, Spain
joseangel.olivas@uclm.es

**Abstract.** The aim of this paper is to present a methodology for creating expert systems by processing texts in order to respond to the queries of a question answering system. In previous work, we have shown several algorithms that were able to extract causal information from text documents and to summarize it. These approaches extracted knowledge from unstructured information, but the performed representation could not be processed automatically to infer new knowledge. Generated summaries only present the information in natural language, and hence cannot be processed in order to generate complex implications. In this paper, we introduce a procedure capable of using this knowledge in order to infer new causal relations between concepts automatically by creating expert systems from the processed texts. These expert systems will contain the causal relations presented in the processed texts. In this representation, by using logic programming, we can infer new concepts that are implied by causal relations. We describe the methodology, technical details of the implementation of our question answering system and a full example where its usefulness is described.

**Keywords:** Causality · Question answering system · Causal texts · Causal detection · Causal summary

## 1 Introduction

Unstructured information contains latent knowledge that can be useful to retrieve, discarding the rest of information that is useless for a predefined task [1]. An example of unstructured information where knowledge can be retrieved are texts. In this context, text mining systems have traditionally dealt with the task of extracting knowledge from texts [2]. It is necessary that the extracted knowledge is represented in a way that allow computers to process the retrieved knowledge for making decisions. By representing it in that format, and not for example in an unstructured format as images or natural language, we can infer new knowledge using an inference engine, as the one of Swi-Prolog [3], for example. These new inferences can help in a decision making process. It

© Springer Nature Switzerland AG 2019
H. Pérez García et al. (Eds.): HAIS 2019, LNAI 11734, pp. 14–25, 2019.
https://doi.org/10.1007/978-3-030-29859-3_2

is also useful if that knowledge can be visualized clearly, which, for humans, can support the decision making process with visual evidence [4]. The automatic decision making process by computers becomes critical in scenarios where unstructured information is the only source of knowledge and it comes in a massive amount [5]. Social networks, for instance, are an example of an unstructured information flow that must be processed automatically, and ideally, visually studied. In this work, we are going to focus on texts that are an example of the described unstructured information.

In scenarios as the one described before, if we are targeting causal knowledge from enormous amounts of texts, we need a system that process automatically this information, discarding the rest, as in [6], were Sobrino et al., created a procedure to extract, filter and process causal sentences from medical texts.

This can be done by an expert system [7]. In the past decades, expert systems were a trend as they represented knowledge from unstructured information, but they had to be written by experts on the field [8]. They were very precise, but this accuracy is static, in the sense that if the information stream or the context vary, the encoded knowledge cannot be adapted to the new scenario, and the expert system will need to be fully maintained or rewritten again [9]. Not only they lack of adaptation, but also coverage. In very broad problems, where lots of patterns need to be captured, expert systems written by humans need to be enormous, and patterns are easily missed. Machine learning [10] emerged as an alternative approach to solve inference problems due to the described issues. Unfortunately, machine learning sometimes suffers from the issue of not having the best precision. We present an alternative methodology to adapt classic expert systems that are processed by logic inference engines. By using this methodology, we try to combine the best of the two worlds, adaptability to new knowledge, precision and coverage, in the domain of causal relations.

We manage to automatically build causal relation expert systems by using an automatic extractor and summarizer of causal relations from texts [11]. We describe a methodology that adapts to the content of new texts, changing the degree of uncertainty between causes and effects in function of the new texts analyzed. These texts can be fed into the system continuously, in a reinforcement or active learning fashion [12]. Hence, we adapt the new information coming into the system, using several linguistic resources to take into account different meanings of the same concept, making our system to have a great coverage, as machine learning systems. Our representation of the knowledge can also be revised by experts, as can be easily visualized by using graphs or diving into the logical clauses, so it can be refined, gaining the precision of classic expert systems. As seen, our methodology tries to tackle both, the issues of classic expert systems and machine learning systems.

The rest of the article is organized as follows: We briefly describe in the following section our previous work on this topic and how we retain causal relation knowledge, creating a natural text summary. This summary represents the causal relations obtained, but cannot work on them. In the third section, we describe the procedure that create expert systems that uses logic programming to infer new concepts. We show all the architecture of our proposed system to answer queries from users in the fourth and fifth section. We then show a full example and conclude the article with some conclusions and further work.

## 2 Generating Automatic Summaries from Causal Sentences

In previous work, we have developed a system that generates natural language summaries from a causal graph [11, 13]. This causal graph was generated by means of an automatic parser that contains a set of rules that filter causal relations presented in a text. Through this process, we generate a causal directed graph, an ordered pair $G = (V, E)$ where $V$ is a set of nodes resembling concepts of the causal relations, and $E$ is a set of edges. Every edge $e_{ij} \in E$ connects the nodes $n_i$ in $V$ with $n_j$ in $V$. We define that every edge $e \in E$ has an associated degree of uncertainty $c \in [0, 1]$ that represents that concept $n_i$ implies concept $n_j$ through $e_{ij}$ with an associated uncertainty degree or with probability $c_{ij}$.

The summary process consists in a mixture of algorithms and linguistic resources as Wordnet [14], used to process every node in the causal graph. The process contains a set of algorithms and procedures that use linguistic resources that we define as $A$, let us also define every algorithm or process as $a \in A$. Every algorithm has an associated weight $w_i \in R$ such as $\sum_{i=1}^{N} w_i = 1$.

Every algorithm outputs a score $p \in [0, 1]$ for every concept. Hence, every concept is given the final score $f = \sum_{j=1}^{M} w_j a_j$, where $M$ is the total number of algorithms presented in the summary process.

This procedure generates a ranking of the concepts which scores are given by the previous expression. This ranking creates an order of the concepts that are the most relevant and less redundant. Given a compression rate $g$ in [0, 1] by the user, the procedure deletes from the summary all the concepts which position in the ranking does not fit in the compression rate. That is, if the compression rate is 0.6, we will only retain 40% of the most relevant concepts ordered by the algorithm. The other concepts will be deleted from the summary. In the example of Fig. 1, there are concepts with a degree of similitude as *smoking* and *tobacco use*. With this ranking of concepts, the most relevant would be selected for the next steps:

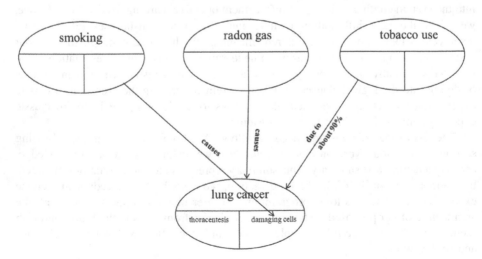

**Fig. 1.** Example of similitude between concepts in an automated extracted graph.

Once we have the compressed graph, the natural language is created through a generative grammar that contains a set of rules that process every causal relation and incorporates natural text content. It generates different adverbs between the causal sentences depending on the uncertainty that connects the causes and effects.

By the end of this procedure we have a visual representation of the text by means of a graph and a natural text summary. But, unfortunately, we cannot make inference in the graph nor the summary, wasting useful implications and relations. For example, suppose that the concept A implies concept B with some probability degree, and then, concept B implies concept C and so on. If we reach concept F, for example, in this chain of relations, we may lose this relation that can be useful by only reading the summary or visualizing the graph. We would like to answer a query as, is concept A related with concept F, and in which degree? In the next section we propose a procedure that creates an expert system able to answer this kind of queries automatically. We show a flow diagram with all the described process in Fig. 2.

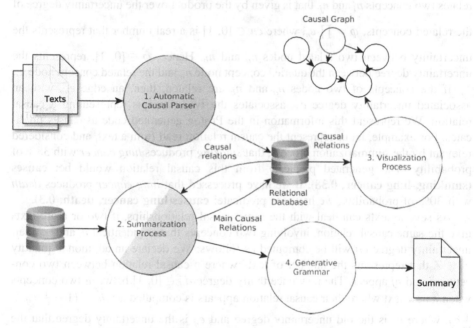

**Fig. 2.** Flow diagram of the causal extractor system.

## 3   Creating Expert Systems from Causal Relations

In this section we describe the generation of expert systems from texts. As we have described in the previous section, the natural language and graph representations make not possible to infer new knowledge automatically or to answer queries from related concepts. In order to solve this, we have used logic programming and create a Prolog [15] file automatically for every summarized causal graph. The objective is to search for the consequences or causes of a given concept $ni$. We may as well want to know whether two concepts $n_i$ and $n_j$ are related and to which degree $c_{ij}$.

In order to answer to these requirements, the content of the generated Prolog code is the following one. First, we need to create a mechanism to traverse the graph automatically, performing a search for the consequences or the causes of a concept. In order to do that, we print the following clauses in the Prolog file:

At the beginning of the Prolog file:

**causes(X, Y) :- causes(X, Y, ).**

At the end of the Prolog file we print:

**implication(X, Y, Z) :- causes(X, Y, Z).**
**implication(X, Z, A) :- causes(X, Y, C), implication(Y, Z, B), A is C * B.**

These clauses make the predicate implication perform a recursive search over the causal graph represented by causal relations, taking into account or not the uncertainty degree $c_{ij}$. Moreover, we compute in this recursive search the uncertainty degree that relates two concepts $n_i$ and $n_j$, that is given by the product over the uncertainty degree of the related concepts, $\varphi = \prod_{n=1}^{N} c_n$, where $cn \in [0, 1]$ is a real number that represents the uncertainty between two related nodes $n_a$ and $n_b$. Hence, $\varphi \in [0, 1]$, represents the uncertainty degree between the queried concept node $n_q$ and the related concept node $n_r$.

If the concepts of two nodes $n_a$ and $n_b$ are related, then, an edge $e_{ab}$ with an associated uncertainty degree $c_{ab}$ associates the two concepts representing a causal relation. We represent this information in the Prolog generated code as causes predicates. For example, let us represent the causal relation read from a text, and considered relevant by the summarization process, that *smoking* produces *lung cancer* with 38% of probability. The generated predicate from this causal relation would be: **causes (smoking, lung cancer, 0.38)**. If we have processed that *lung cancer* produces *death* with 30% of probability, we have the predicate: **causes(lung cancer, death, 0.3)**.

As several texts can deal with the same causal relationships, if two or more texts give the same causal relation, involving two concepts that are identical $n_i$ and $n_j$, their uncertainty degree $cij$ will be computed as follows: We declare an additional quantity $v_{ij} \in Z$ that represents the number of texts where a causal relation between two concepts $ni$ and $nj$ appear. The new uncertainty degree $d_{ij} \in [0, 1]$ between two concepts when a new text where their causal relation appears is computed as: $d_{ij} = (1 - \frac{1}{v_{ij}})c_{ij} + \frac{1}{v_{ij}}e_{ij}$, where $c_{ij}$ is the old uncertainty degree and $e_{ij}$ is the uncertainty degree that the new text present for the causal relation. We update the Prolog predicate with the new uncertainty degree, updating the expert system with new knowledge every time that a new text is processed.

This representation of the processed knowledge by the causal parser and summarization procedure makes able to perform logical inferences. In the next section, we briefly introduce the role that causality plays in question answering systems. With those fundamentals, we describe how we have created a question answering system by using Prolog as inference engine and the rest of procedures described in Sect. 5.

## 4 Causality and Question Answering Systems

Causality is an ancient topic that came from Aristotle [16, 17]. The Aristotelian view of causality traditionally offered a frame for providing answers to causal questions, as what-q or why-q. In effect, Aristotle's typology serves to answer what and for what questions. For example, in the presence of a statue, we can ask for the following queries: 'What is it made from?' It is made of metal (material cause); 'What is its form?' A man in a praying attitude (formal cause); 'What produced it?' The sculptor (efficient cause); 'For what purpose?' To pay tribute to a virtuous person (final cause).

More in depth, and referring to a medical context, we can argue that the efficient cause of a diagnosis it is not a doctor, but his medical knowledge. As we previously said, Aristotle pointed out that efficient causes are the primary source of change. The transmission of such changes in a causal network is known as a mechanism. So, efficient causes are related to mechanisms. The following pyramid arranges interrogative particles depending on the potential complexity of their answers [18] (Fig. 3).

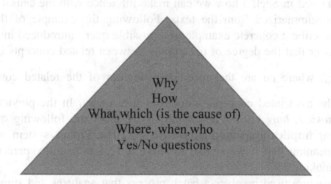

**Fig. 3.** Pyramid of questions' complexity.

Ascending in the pyramid means the use of causal interrogatives, demanding complex answers instead of yes, no replies to questions, stimulating reflective and deepening thinking. At the top, why-questions ask for some kind of explanation.

An explanation is better understood showing a causal mechanism, describing the interrelations among multiple factors involved in its origin and development.

In Medicine, diseases like schizophrenia, bulimia, or anorexia, frequently show fuzzy causal boundaries, as they present similar symptoms. Thus, causes frequently are complex and vague. That is, in medicine we should shift from the classic scientific paradigm of 'theory' to the more evasive of 'mechanism'. Causes are not single, but complex [19] and are not crisp, but imprecise [20].

Medical question answering systems aim to provide the users with direct answers to the posed questions; instead of furnishing them with a large amount of relevant potentially documents. For seeking direct answers, Q/A systems need to go beyond the surface-level analysis of texts, providing lexical-syntactic and semantic resources, as well as reasoning capabilities, providing inference mechanisms to obtain more adequate answers, as we propose in this paper. Traditional Q/A systems accept questions as inputs; those questions trigger a search engine providing relevant documents and, finally, the system operates over those documents in order to reach the direct answer to the question. Although most Q/A systems have exploited syntactic or semantic resources, few approaches have scanned the utility of inference mechanisms. Our approach uses syntactic and semantic resources to perform an inference-based one, as aims to extract causal semantic relations from inferential mechanisms.

## 5   Creating a Question Answering System

We have described in Sect. 3 how we can make inference with the causal knowledge extracted and summarized from the texts. Following the example of the previous section, we describe a concrete example of a possible query introduced in the system. Let us remember that the degree of uncertainty between related concepts is computed as: $\varphi = \prod_{n=1}^{N} c_n$, where $cn$ are the uncertainty degrees of the related concept nodes $ni$ between the associated concepts introduced in a query. In the previous example involving *smoking*, *lung cancer* and *death*, if we enter the following query to the Prolog system **implication(smoking, death, X).**, the Prolog system will answer X = 0.114, meaning that if we smoke we have a 11.4 probability percent of dying because of smoking.

The architecture used involves a batch process that analyzes and summarizes the texts and an online process that serves as the question and answering system. The batch process is formed by the *C* parser of texts, that stores the relations in a relational database, in our case we have used *mysql* for the database, the summarization java tool that access this database via a driver and persist in the same database the summarized causal relations and the Prolog persistence manager, that are included in a.*pl* file with the described Prolog predicates. We show in Fig. 4 a diagram flow of all the processes that are involved in the system.

The online question answering system is a *JavaEE* application that access the Prolog predicates with the *jpl* library. We implemented these accesses with java servlets and services that manage the possible queries that users can ask to the system. The user interface is implemented through *jsps*, giving several services to the users as introducing new texts, summarizing them and accessing to the question answering system. The visualization of the graph has been generated with Javascript in the *jsps*.

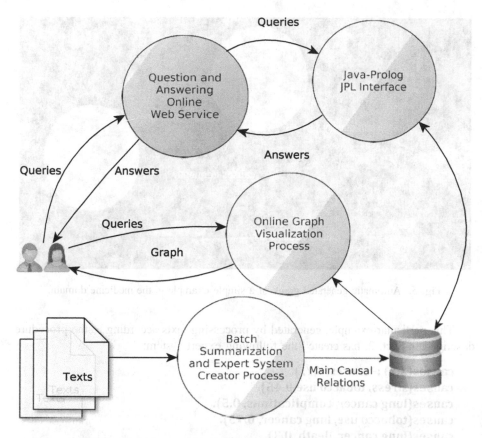

**Fig. 4.** Flow diagram of all the processes involved in the system.

## 6   Example Processed by the System

We present in this section a full example of the possible actions that the described procedure can perform. We develop the medicine example drawing the relation between smoking and lung cancer shown in the previous section. In this scenario, we have the following graph generated by the visualization process in Javascript, that is the result of the summarization and extraction of the batch procedures. We show the graph in Fig. 5. The graph circles can be moved arbitrarily by the user, in order to generate the graph as the user considers. The size of the circles, which are the concepts, represent the number of texts that we have processed with that concepts. The bigger the circles, the higher the number of texts that contain that concept.

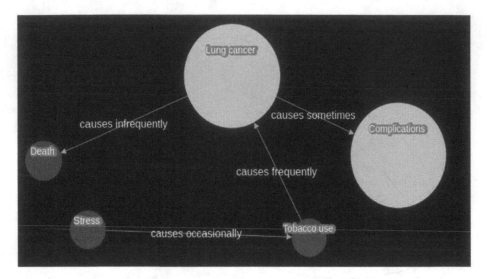

**Fig. 5.** Automatic generated graph of a simple example in the medicine domain.

This particular example, generated by processing texts according to the procedure described in Sect. 2, has created the following expert system:

**causes(X,Y) :- causes(X,Y, ).**
**causes(stress, tobacco use,0.45).**
**causes(lung cancer, complications, 0.5).**
**causes(tobacco use, lung cancer, 0.75).**
**causes(lung cancer, death, 0.3).**
**implication(X,Y,Z) :- causes(X,Y,Z).**
**implication(X,Z,A) :- causes(X,Y,C), implication(Y,Z,B), A is C\*B.**

This expert system has been created from several medicine texts associated to lung cancer. An example of a summary that those texts have generated is the following one: *"In this text, the author argues that lung cancer is frequently implied by tobacco use. It is important to end knowing that lung cancer sometimes causes severe complications".* We can observe that the time adverbs written by the generative grammar are associated to the uncertainty degrees. Synonyms have been used extracted from Wordnet synsets to associate different written representations of the same concept, as for example, *smoking* and *tobacco use*.

The question answering system solves the queries given by the user based in the previous automatically generated expert system. These would be some examples of possible queries and their answers. We see that stress can cause several effects, that are indirectly linked. If we ask the system for the causes of stress, the system would output: *Tobacco use, Lung Cancer, Complications* and *Death* and the respective degrees of uncertainty about the effect. The query performed and the answer in Prolog, is the next one:

**implication(stress,X,Y).**
**X = tobacco use,**
**Y = 0.45 ;**
**X = lung cancer,**
**Y = 0.3375 ;**
**X = complications,**
**Y = 0.16875 ;**
**X = death,**
**Y = 0.10124999999999999;**
**false.**

We can also ask for the possible causes of a concept. For example, let us consider the concept *death*, the last one in the chain of implications of this example. We can also ask the system for possible causes of *death*. In this case, the result would be the following one:

**implication(X,death, ). X = lung cancer ;**
**X = stress ;**
**X = tobacco use ; false.**

We see that every possible cause of the concept *death* is listed and retrieved by the system. This representation of the extracted knowledge and summarized by the described procedures is very concise and uses all the inference of any Prolog interpreter. By using this Prolog interpreter we can automatically infer new facts about concepts that can be used for another application or for human users.

## 7 Conclusions and Future Work

In this work, we have presented a question answering system that uses a Prolog expert system to solve causal queries performed by users. The novelty of this system is that the Prolog expert systems are automatically generated by the system using a summarization and causal extraction process that retrieves causal relations from texts. We have, hence, created a system that adapts to new knowledge, has coverage due to its linguistic resources and can be easily refined by human experts due to its visualization features to gain accuracy. By an online and batch process, we have presented a system that can automatically build an expert system reuniting all the causal relations presented in different texts and solving queries from users.

Our system can perform logical inferences through Prolog, but as we use an uncertainty degree, we could use the Fuzzy Prolog [21] interpreter and use fuzzy logic to answer questions. There is also another fascinating research branch that we would like to develop and that would make the system to produce more accurate inferences. The summarization is a weighted mixture of algorithms but needs to be configured by the user. Ideally, depending on the context of the texts, we would like to have the optimum weights of these algorithms in order to produce good summaries. This is a subjective task which evaluation does not have gradients and it is very expensive to compute. In this context and having several objectives and constraints into account as the compression rate, the evaluation of the text according to gold standards and subjective evaluations; Multi-Objective Constrained Bayesian Optimization [22] emerges as an ideal tool to solve this problem. Bayesian Optimization has been used with success in other complicated previous work such as the optimization of Bayesian Networks constructions [23] or real tasks as ocean wave features prediction [24].

**Acknowledgments.** Work supported by the Spanish Ministry for Economy and Innovation and by the European Regional Development Fund (ERDF/FEDER) under grant TIN2016-76843-C4-2-R TIN2014-56633-C3-1-R and TIN2017-84796-C2-1-R, The authors acknowledge the use of the facilities of Centro de Computación Científica (CCC) at Universidad Autonoma de Madrid, and financial support from the Spanish Plan Nacional I+D+i, Grants TIN2016-76406-P and TEC2016-81900-REDT, and from Comunidad de Madrid, Grant S2013/ICE- 2845 CASI-CAM-CM.

# References

1. Rao, R.: From unstructured data to actionable intelligence. IT Prof. **5**(6), 29–35 (2003)
2. Tan, A.-H., et al.: Text mining: the state of the art and the challenges. In: Proceedings of the PAKDD 1999 Workshop on Knowledge Disocovery from Advanced Databases, vol. 8, pp. 65–70, sn (1999)
3. Wielemaker, J., Schrijvers, T., Triska, M., Lager, T.: Swi-prolog. Theor. Pract. Log. Program. **12**(1–2), 67–96 (2012)
4. Lurie, N.H., Mason, C.H.: Visual representation: implications for decision making. J. Mark. **71**(1), 160–177 (2007)
5. Lohr, S.: The age of big data. New York Times, **11**(2012) (2012)
6. Puente, C., Sobrino, A., Olivas, J.A., Merlo, R.: Extraction, analysis and representation of imperfect conditional and causal sentences by means of a semi- automatic process. In: Proceedings IEEE International Conference on Fuzzy Systems (FUZZ-IEEE 2010), Barcelona, Spain, pp. 1423–1430 (2010)
7. Waterman, D.: A Guide to Expert Systems. Addison-Wesley, Boston (1986)
8. Hayes-Roth, F., Waterman, D.A., Lenat, D.B. (eds.): Building Expert System. Addison-Wesley, Boston (1983)
9. Compton, P., Jansen, R.: Knowledge in context: a strategy for expert system maintenance. In: Barter, C.J., Brooks, M.J. (eds.) AI 1988. LNCS, vol. 406, pp. 292–306. Springer, Heidelberg (1990). https://doi.org/10.1007/3-540-52062-7_86
10. Nasrabadi, N.M.: Pattern recognition and machine learning. J. Electron. Imaging **16**(4), 049901 (2007)
11. Puente, C., Sobrino, A., Garrido, E., Olivas, J.A.: Summarizing information by means of causal sentences through causal graphs. J. Appl. Logic **24**(Part B), 3–14 (2017)
12. Sutton, R.S., Barto, A.G.: Introduction to Reinforcement Learning, vol. 135. MIT Press, Cambridge (1998)
13. Puente, C., Olivas, J., Garrido, E., Seisdedos, R.: Creating a natural language summary from a compressed causal graph. In: IFSA World Congress and NAFIPS Annual Meeting (IFSA/NAFIPS), 2013 Joint, pp. 513–518. IEEE (2013)
14. Miller, G.A.: Wordnet: a lexical database for English. Commun. ACM **38**(11), 39–41 (1995)
15. Sterling, L., Shapiro, E.Y.: The Art of Prolog: Advanced Programming Techniques. MIT Press, Cambridge (1994)
16. Waterfield, R.: Aristotle, Physics. Oxford University Press, Oxford (1996)
17. Lawson-Tancred, H.: Aristotle, The Metaphysics. Penguin Books, London (1998)

18. Vogt, E., Brown, J., Isaacs, D.: The Art of Powerful Questions. Whole Systems Associates (2003)
19. Mackie, J.L.: The Cement of the Universe: A Study in Causation. Clarendon Press, Oxford (1988)
20. Pechsiri, C., Kawtrakul, A.: Mining causality from texts for question answering system. IEICE Trans. Inf. Syst. **E90-D**(10), 1523–1533 (2007)
21. Martin, T.P., Baldwin, J.F., Pilsworth, B.W.: The implementation of fprolog a fuzzy prolog interpreter. Fuzzy Sets Syst. **23**(1), 119–129 (1987)
22. Garrido-Merchan, E.C., Hernandez-Lobato, D.: Predictive entropy search for multi-objective bayesian optimization with constraints. arXiv preprint arXiv:1609.01051 (2016)
23. Córdoba, I., Garrido-Merchán, E.C., Hernández-Lobato, D., Bielza, C., Larrañaga, P.: Bayesian optimization of the PC algorithm for learning Gaussian Bayesian networks. In: Herrera, F., et al. (eds.) CAEPIA 2018. LNCS (LNAI), vol. 11160, pp. 44–54. Springer, Cham (2018). https://doi.org/10.1007/978-3-030-00374-6_5
24. Cornejo-Bueno, L., Garrido-Merchan, E.C., Hernandez-Lobato, D., Salcedo-Sanz, S.: Bayesian optimization of a hybrid system for robust ocean wave features pre- diction. Neurocomputing **275**, 818–828 (2018)

# Creation of a Distributed NoSQL Database with Distributed Hash Tables

Agustín San Román Guzmán(ID), Diego Valdeolmillos(ID), Alberto Rivas(ID),
Angélica González Arrieta, and Pablo Chamoso(✉)(ID)

BISITE Research Group, University of Salamanca,
Calle Espejo 2, 37007 Salamanca, Spain
{id00681813,dval,rivis,angelica,chamoso}@usal.es

**Abstract.** Databases are an essential tool in the real world. Traditionally, the relational model and centralized architectures have been used mostly. However, with the growth of the Internet in recent decades, both in the number of users and in the amount of information, the use of decentralized architectures and alternative database models to the relational model has been extended, which receive the name of NoSQL (Not only Structured Query Language) databases. With the present end of degree work, the development of a distributed NoSQL database is proposed, which will try to achieve high availability and high scalability through a decentralized architecture based on DHT (Distributed Hash Tables).

**Keywords:** Database · NoSQL · Hash table · Distributed · Decentralized · Availability · Scalability

## 1 Introduction

This paper describes a database system designed to achieve high availability and high scalability. Nowadays these two concepts are essential due both the huge amount of generated data that needs to be saved in a database and the increasing amount of users, witch increases the number of transactions and requires the system to be available and with good response times.

Over time, new algorithms and architectures have been developed in order to achieve high availability and high scalability. Generally the rely on new data models different from the traditional relational model (NoSQL models) and use distribution and replication strategies to achieve high scalability and high availability respectively [1].

The proposed solution consists on a database system implemented using a NoSQL data model, which we will discuss later in this paper, with a distributed and completely decentralized architecture, governed by a DHT, specifically the rendezvous hashing algorithm, and with a CRUD (Create, Read, Update, Delete) operations model.

© Springer Nature Switzerland AG 2019
H. Pérez García et al. (Eds.): HAIS 2019, LNAI 11734, pp. 26–37, 2019.
https://doi.org/10.1007/978-3-030-29859-3_3

The resulting database system was called *DistrDocDB* as the abbreviations of distributed (distributed network architecture), documents (documents data model) and database.

The content of this paper consists, in the first place, of an exposition of the theoretical concepts in those referring to NoSQL databases and DHT. Below is a brief summary of the state of the art prior to the completion of the work. Next, the most relevant aspects of the system are explained, which include the data and persistence model, the network architecture, the distribution and replication strategies, the consistency and synchronization model and the operations model. Finally, a series of conclusions are presented, as well as possible future lines of work.

## 2    Theoretical Concepts

The two most relevant theoretical concepts covered by this project are NoSQL database models and DHT. These theoretical concepts are explained in more detail in the sections below.

### 2.1    NoSQL Databases Models

Relational databases have been used extensively for many applications. Due to their general purpose design and their high number of functionalities, including indexes, type restrictions, use of primary and foreign keys, among others, they have been widely used [2].

However, with the expansion of the Internet in recent decades it has become increasingly necessary to process large volumes of data and types of data with variable structure. In addition, for systems that require support for a large number of users that access continuously, systems with high availability and very high scalability are necessary [3].

Relational databases do not scale well when it comes to distributed systems. In addition, they present problems in handling flexible or variable structure data [2]. This is why NoSQL data models arise, among which are grouped all data models other than the traditional relational model, distributed models and horizontally scalable database systems [4].

One of the main characteristics shared by almost all NoSQL databases, especially when dealing with distributed systems, is the fact that ACID properties (atomicity, consistency, isolation, and durability) are not guaranteed, but rather an approach known as BASE properties (basically available, soft state and eventual consistency) is used [5]. Availability is given priority over consistency, which can be temporarily lost to recover within a reasonable period of time.

One of the most relevant problems faced by NoSQL database models in distributed environments is data version control. Because, if they temporarily lose consistency, they will have different versions of the data, there is the need to determine which version of the data is the most recent. A variety of mechanisms are available, including timestamps, the use of logical times, the storage of historical data, time vectors, and multi-version storage [5].

There are multiple models that are grouped within NoSQL databases, the most relevant of which include documentary databases, key-value databases, graph-based databases, multi-value databases, and object-oriented databases.

Document databases associate each record with a separate document, which has a flexible structure. These documents are very similar to the objects of some programming languages, which facilitates the development of applications and reduces the cost of altering the stature of data in a relational database [6].

Key-value databases are based on the data model of a dictionary, a structure formed by key-value pairs. In general, priority is given to obtaining the value from the key, leaving aside union and aggregation operations [5].

Graph-oriented databases use nodes and relationships between them to carry out the same function as relational databases. This is why, in many cases, graph-oriented databases are considered the new generation of relational databases, fitting much better than these into distributed models [7].

Multi-Value databases are an alternative to relational databases based on discarding the first normal form, allowing multiple values to be stored in each attribute [8].

Object-oriented databases present an alternative to most other database models, which store data of different types: strings, integers, real numbers, dates, etc. In contrast to this, object-oriented databases store "objects", such as those used in object-oriented programming languages [9].

## 2.2    Distributed Hash Tables

A DHT is a decentralized distributed system that distributes data among nodes using a hash table as the fundamental structure. DHT store pairs key-value, being able to obtain the given value of the key efficiently [10].

The two most relevant approaches for the implementation of DHT are consistent hashing and rendezvous hashing [11].

Consistent hashing consists of dividing the space of possible outcomes of a hash function (e.g., SHA or MD5) into ranges for each node. These ranges can be the same size or different depending on the capabilities of each node [12].

Rendezvous hashing pursues a completely random distribution of the keys using the properties of the hash functions [13]. Unlike consistent hashing, instead of dividing the range of results, each node is assigned a score for each key, which is calculated by the hash function of the node identifier concatenated with the key itself. From among all the nodes, the one with the best score is selected [13].

## 3    State of the Art

In this section, a brief summary of the state of the art regarding NoSQL databases will be made. Due to the great amount of NoSQL databases that exist in the market [4], only the most relevant ones will be treated taking into account the requirements of the project, which are Apache Cassandra, MongoDB and, Redis.

Apache Cassandra is an open source NoSQL database, written in Java, based on a key-value model, with great scalability capabilities, since it allows storing huge amounts of data in multiple nodes, increasing performance and reducing response times [14].

Apache Cassandra uses a node discovery protocol called Gossip. This protocol consists of sending every second to a maximum of three nodes status information about themselves and about the nodes they know [14].

As for data storage, Apache Cassandra uses a data storage model based on three fundamental structures: The commit log (operations log file that allows the recovery of information losses and avoids inconsistencies), the mem-tables (used as a buffer that accelerates data access) and the SS-Tables (sorted string tables, which store data persistently and implement binary search to speed up queries) [14].

With regard to replication, Apache Cassandra allows the use of two replication strategies depending on the specific needs of the application: A simple strategy based on consistent hashing and a strategy based on network topology, which consists on keeping one or more copies of the data in each data center [14].

MongoDB is an open source NoSQL database that follows a document model, handling documents in JSON (Javascript Object Notation) format and allowing a flexible data structure. In addition, MongoDB is a database built to follow a distributed model, achieving high scalability with distribution strategies and data replication [6].

MongoDB uses horizontal scaling through its data distribution method: Sharding, which consists of dividing the data into fragments or shards and distributing them according to a fragments distribution strategy [6]. There are two strategies available: (1) hashed sharding (a hash function is applied to the key, the results range of the hash function is divided into parts for each of the nodes) and (2) range sharding (simply divides the key range between the nodes) [6].

Replication in MongoDB is based on replica sets. A replica set is a group of MongoDB instances containing the same data set. Each replica set presents an internal organization, under which the nodes take differentiated roles: primary node (in charge of the read and write operations), secondary nodes (replicating the information of the primary node) and arbiter node (it does not store data but is in charge of monitoring the rest of nodes and choosing the primary and secondary nodes) [6].

Redis is an open source NoSQL database system for Linux, based on structured storage in memory and can be used as a database, cache or message broker. [15] Redis uses a key-value model, and its main attraction lies in the enormous performance it achieves when used in non-persistent scenarios, i.e., using only RAM [16].

Redis supports master-slave replication. It consists in the distinction between the master nodes, in charge of processing the transactions, and the slave nodes, in charge of replicating the data of the master nodes [15].

# 4    Data Model and Persistence

The implemented data model consists of a document-based data model, similar to those of other document databases, such as MongoDB, CouchDB or ArangoDB.

Within this data model, we can distinguish four fundamental structures: databases, collections, documents, and fields.

Databases contain collections, each of which is identified by a unique name. Collections are containers of documents. The documents are the basic storage units and are identified by a unique key within the collection to which they belong. These documents are formed by fields, which are name-content pairs. The content of the fields is a non-typed binary content, designed for the storage of files interpretable by the clients. A schema of this model is shown in Fig. 1.

The main advantage of this data model is the support of unstructured data. Also, they do not require costly and time-consuming migrations when application requirements change [6].

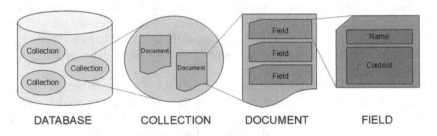

DATABASE            COLLECTION        DOCUMENT            FIELD

**Fig. 1.** Data model schema

Based on this data model, a persistent storage system has been implemented, which has the following characteristics:

- It treats collections independently so that increasing the size of one collection does not affect the performance of others.
- With the implementation of a commit log it ensures the atomicity of writing operations, so that, if a writing operation is interrupted, any changes made to the document are reverted. In addition, isolation is also ensured, so that the read operations will always get a consistent version of the data (before or after the write operation).
- Thanks to the implementation of a backup system, it avoids data corruption in the event of uncontrolled interruptions (system failures, voltage drops, etc.).
- It presents a good performance in the searches of the documents by its key and fields by its name thanks to the use of indexes.

Indexes are used to improve the performance of operations [17], considerably reducing response time thanks to the use of the binary search algorithm, which has the lowest complexity in terms of search algorithms [18].

The commit log is a record of the writing operations carried out on the documents. Each write operation is isolated, writing its changes temporarily in the commit log. Once the operation is committed, the changes are applied to the document.

The backup system consists of making a backup copy of the data before making changes on an archive, indicating with files as indicators the state of the writing. Thanks to these indicators, in the event of system failure, during the restart it returns to a consistent state, choosing to keep the pre-writing data if it was interrupted before being completed.

## 5   Network Architecture

The implemented database system presents a distributed and decentralized architecture. It consists of a set of nodes or database servers with identical functionalities and coordinated in a decentralized manner by means of a DHT (see Fig. 2). Thanks to this decentralized architecture, it is possible to access the database from any of the nodes that form it, allowing a distribution of the load of requests, thus increasing the scalability of the system.

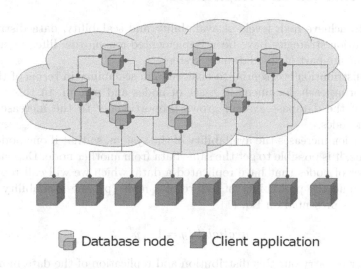

Database node       Client application

**Fig. 2.** Database distributed network architecture

Being a distributed system, exposed to the Internet, security is a fundamental point. That is why a security system based on asymmetric encryption (RSA) has been implemented, which ensures the confidentiality of communications by encrypting messages, as well as the authentication of the parties, ensuring at all times that the node with which a communication is being established is an authorized node of the database and its messages have not been altered by third parties during transmission.

In order to carry out the coordination, it is necessary that the nodes have knowledge of each other. The gossip protocol has been used for this purpose.

The Gossip protocol is a paradigm of communication based on the phenomena of dissemination of information between people observed on social networks. This paradigm allows the dissemination of information and the formation of network topologies in environments where scale and dynamism are crucial [19].

The gossip protocol is based on the dissemination of information in an epidemic manner. When an element of the system wishes to make certain information known, it communicates this information to only a few elements. The elements that receive the information carry out a "gossip" process, from which the protocol is named, spreading the information to other elements they know. Performing this process in a repetitive manner, the information is disseminated throughout the network and, eventually, all the elements receive it.

The dissemination of information in this way allows the number of elements received by the information to increase exponentially over time, thus ensuring that the information is disseminated to all elements in a logarithmic time order respecting the number of elements in the network [19].

## 6   Distribution and Replication

In order to achieve high levels of availability and scalability, data distribution and replication strategies have been implemented among the different nodes or servers that are part of the database system.

Data distribution is essential to achieve high scalability in terms of the data set [3], storing each document in a set of nodes and not all. In this way, the capacity of the database system grows proportionally to the increase in the number of nodes.

Replication increases the availability of the system, so that if one node in the system fails, it is possible to get the same data from another node. Depending on the number of nodes that have replicated a data, which we will call replication factor (f), and the probability of failure of a node (p), the availability can be calculated as presented on Eq. (1).

$$avaliability = 1 - p^f \tag{1}$$

In order to carry out this distribution and replication of the data in a decentralized way, a DHT has been implemented using the rendezvous hashing algorithm, which presents the following characteristics that have led to its use:

- It is a fairly efficient algorithm, as the hash function used is very cheap in terms of CPU time. Modern computers can perform millions of these operations per second [13].
- Due to the dispersion property of the hash function used, good load balancing is achieved regardless of the keys used [13].
- It allows the selection of multiple replicas at no considerable additional cost.

– It presents a minimum cost of key redistribution. When a node disappears, only the documents assigned to that node need to be redistributed [13].

The implementation of the rendezvous hashing algorithm makes it possible to select a set of replicas in which to store a given document based on its key [20].

The algorithm consists in assigning to each node a score relative to the key, calculated by applying the hash function SHA-256 to the identifier of the node concatenated with the key of the document. The nodes that obtain a lower score will be the nodes assigned as replicas. The number of replicates is determined by the replication factor, which is configurable by the database administrator.

In order to improve the performance of the algorithm, since it is an algorithm that is executed very frequently, it has been implemented using a maximum heap, data structure which always maintains as first element the element with maximum value, being able to insert and eliminate elements with a logarithmic complexity with respect to the number of elements.

Thanks to the maximum heap, instead of ordering a complete list of values of a size equal to the number of nodes, which means a complexity $nlogn$, we only focus on the F elements with the best score, F being the replication factor. When working with a heap of size F, being F constant, the complexity of the algorithm is linear.

---

**Algorithm 1.** Rendezvous Hashing

---

1: **procedure** CHOOSEREPLICAS(*Nodes, Key, ReplicationFactor*)
2:     *heap* ← **Heap**(*ReplicationFactor*)
3:     **for all** *Node* ∈ *Nodes* **do**
4:         *score* ← SHA256(*Node.ID* + *Key*)
5:         **if** *heap.length* < *ReplicationFactor* **or** *heap.max().score* > *score* **then**
6:             **if** *heap.length* >= *ReplicationFactor* **then**
7:                 heap.removeMax()
8:             **end if**
9:             *heap.add*({*ID* : *Node.ID, score* : *score*})
10:         **end if**
11:     **end for**
12:     *chosen* ← **List**()
13:     **while** *heap.length* > 0 **do**
14:         *chosen.addAtBeginning*(*heap.max()*)
15:         heap.removeMax()
16:     **end while**
17:     **return** *chosen*
18: **end procedure**

---

Making use of the maximum heap, the resulting algorithm is the one presented by the pseudocode in Algorithm 1.

# 7   Consistency and Synchronization

The use of document replication to increase availability presents the problem that it is necessary to maintain the consistency of the replicas in order for such replication to be effective.

As it is a NoSQL database oriented to documents, it is not necessary a permanent consistency, being able to give intervals of times in which some document is in an inconsistent state, as long as it eventually reaches a state of consistency [5].

In order to achieve this consistency, a system of fully sorted versions has been implemented, as well as a document synchronization protocol to resolve inconsistencies between replicas caused by node failures or network changes.

Each document and each of its fields are marked with a version number, which is a logical time that allows the system to determine, given two versions of a field or a document, which of them is the most recent, which has a major version.

The versions are determined using a total order consensus algorithm, which consists of three phases:

1. A node requires a new version to make a modification in a document, requesting this new version to all the replicas assigned to the document according to the rendezvous hashing algorithm.
2. The replicas increase the version of the document they have stored in a unit and return the result to the node that requested it.
3. The highest version is chosen, which is diffused to the mirrors to update their version. In case of equal versions, the tie is broken by the identifier of the node that provided the version.

The document synchronization protocol allows, from two replicas with different versions of the same data, to reach a consistent state. This is achieved by choosing the highest version and, in the case of equal versions, breaking the tie with the node identifier that provided the version.

To determine if two replicas need synchronization, the hashes of your documents, calculated with the hash function SHA-256, are compared. If the hashes match, it is concluded that the document is in a consistent state. Otherwise, the synchronization protocol is carried out. The use of hashes comparison avoids checking the content of the document, which can be considerably large.

The synchronization protocol is carried out when detecting failures in the replicas during write operations, as well as changes in the network (inputs and outputs of nodes).

# 8   Operations Model

The operation model implemented is a CRUD model, which fits the document-oriented data model and is applicable to most REST (Representational State Transfer) applications [5]. The Table 1 summarizes the main creation, reading, updating, and deletion operations applicable to documents and their fields.

The table also includes performance measures for a simulated network with 8 virtual nodes in a host computer with a 2.80 GHz Intel(R) Core(TM) 7700HQ CPU and 16 GB of RAM. The connections between the nodes were also emulated with a bandwidth of 100 MB/s. The goals of this test were to measure the performance of each operation in a high load situation to be able to compare them.

**Table 1.** Overview of implemented operation model

| Operation name | Category | Description | Performance |
|---|---|---|---|
| EXISTS | Reading | Checks if a document exists within a collection and, if it exists, returns its hash | ~200 OP/s |
| DESCRIBE | Reading | Obtains the status (whether it exists or not), the hash and the list of fields of a document, being able to apply a filter to the fields | ~123 OP/s |
| READ | Reading | read the contents of the fields in a document, and a filter can be applied to select the fields to be read | ~92 OP/s |
| CREATE | Creation | Creates a new document within a collection | ~43 OP/s |
| UPDATE | Updating | Updates an existing document by creating new fields or by overwriting or deleting existing fields | ~39 OP/s |
| DELETE | Deletion | Deletes a document | ~104 OP/s |

Collections are created implicitly by creating documents within them, in the same way that MongoDB does [6]. This greatly facilitates the use of queries and is consistent with a flexible data structure.

The three writing operations which allow the creation, updating and deletion of documents act on all the replicas associated to the document according to the algorithm rendezvous hashing, applying the total order for the consensus as for versions to use.

The reading operations consult all the replicas for the version of the document to be read and, choosing the larger version, only read the content from one replica, thus increasing availability.

When a client carries out any type of operation, the node with which he establishes the connection will be the coordinating node of that operation, responsible for searching for replicas using the rendezvous hashing algorithm and responding to the client based on the results of the operation.

# 9   Conclusions and Future Work

Relational databases, despite having very broad and general purpose functionality, present difficulties in adapting to highly flexible data models and distributed environments. This is why the NoSQL database model arises, encompassing all those databases with a data model different from the relational one or which presents a distributed architecture.

High scalability and high availability are two of the most crucial objectives of a distributed database. To achieve this, the most convenient is the design of a decentralized architecture [3], carrying out a distribution of information to achieve a large storage capacity and applying a replication strategy to increase availability and fault tolerance, which can be achieved using a DHT.

Specifically, the gossip algorithm has been implemented to coordinate the nodes in a decentralized architecture, as well as the rendezvous hashing algorithm, which allows carrying out the distribution and replication of data in a balanced and efficient way [13].

As future work is proposed the extension of the operations model, improve the security system and system performance, as well as the implementation of libraries that allow interaction with the database from different programming languages.

**Acknowledgments.** This work has been developed as part of the project "Virtual-Ledgers-Tecnologías DLT/Blockchain y Cripto-IOT sobre organizaciones virtuales de agentes ligeros y su aplicación en la eficiencia en el transporte de última milla" (ID SA267P18), financed by Junta Castilla y León, Consejería de Educación and ERDF funds.

# References

1. Hannan, T.: Replication is the Key for Scalability & High Availability. http://basho.com/posts/technical/replication-is-the-key-for-scalability-high-availability/. Accessed 12 Feb 2019
2. Mohamed, M., Altrafi, G.O., Ismail, M.O.: Relational vs. NoSQL databases: a survey. Int. J. Comput. Inf. Technol. (IJCIT) **3**, 598 (2014)
3. Roehm, B., et al.: IBM websphere v5. 0 performance, scalability, and high availability websphere handbook series. IBM Corp. (2003)
4. NoSQL Databases. http://nosql-database.org/. Accessed 12 Feb 2019
5. Strauch, C.: NoSQL Databases (2012). http://www.christof-strauch.de/nosqldbs.pdf. Accessed 12 Feb 2019
6. MongoDB official website. https://www.mongodb.com. Accessed 12 Feb 2019
7. Neo4j. From Relational to Graph Databases. https://neo4j.com/developer/graph-db-vs-rdbms/. Accessed 13 Feb 2019
8. Zumasys Inc.: What is MultiValue? http://www.openqm-zumasys.com/openqm/what-is-multivalue/. Accessed 13 Feb 2019
9. CompTechDoc.org: Object Oriented Databases. http://www.comptechdoc.org/independent/database/basicdb/dataobject.html. Accessed 13 Feb 2019
10. Wiley, B.: Distributed Hash Tables, Part I (2013). http://www.linuxjournal.com/article/6797. Accessed 13 Feb 2019

11. Gryski, D.: Consistent Hashing: Algorithmic Tradeoffs. https://medium.com/@dgryski/consistent-hashing-algorithmic-tradeoffs-ef6b8e2fcae8. Accessed 13 Feb 2019
12. Nielsen, M.: Consistent hashing (2009). http://michaelnielsen.org/blog/consistent-hashing/. Accessed 13 Feb 2019
13. Resch, J.: New Hashing Algorithms for Data Storage (2015). http://www.snia.org/sites/default/files/SDC15_presentations/dist_sys/Jason_Resch_New_Consistent_Hashings_Rev.pdf. Accessed 13 Feb 2019
14. DataStax website - Apache Cassandra documentation. https://docs.datastax.com/. Accessed 13 Feb 2019
15. Official Redis project page. https://redis.io/. Accessed 14 Feb 2019
16. Zawodny, J.: Redis: lightweight key/value Store That Goes the Extra Mile (2009). http://www.linux-mag.com/id/7496/. Accessed 14 Feb 2019
17. Schmid, S., Galicz, E., et al.: Performance investigation of selected SQL and NoSQL databases (2015)
18. Kumari, A., Tripathi, R., et al.: Linear search versus binary search: a statistical comparison for binomial inputs. Int. J. Comput. Sci. Eng. Appl. **2**, 29 (2012)
19. Montresor, A.: Gossip and Epidemic Protocols (2017)
20. Khuong, P.: Rendezvous Hashing: My Baseline "Consistent" Distribution Method (2017). https://www.pvk.ca/Blog/2017/09/24/rendezvous-hashing-my-baseline-consistent-distribution-method/. Accessed 18 Feb 2019

# Study of Data Pre-processing for Short-Term Prediction of Heat Exchanger Behaviour Using Time Series

Bruno Baruque[1(✉)], Esteban Jove[2], Santiago Porras[3],
and José Luis Calvo-Rolle[2]

[1] Departamento de Ingeniería Civil, University of Burgos,
C/Francisco de Vitoria, s/n, 09006 Burgos, Burgos, Spain
bbaruque@ubu.es
[2] Departamento de Ingeniería Industrial, University of A Coruña,
Avda. 19 de febrero s/n, 15495 Ferrol, A Coruña, Spain
[3] Departamento de Economía Aplicada, University of Burgos,
Plaza Infanta Doña Elena, s/n, 09001 Burgos, Burgos, Spain

**Abstract.** Geothermal exchangers are among the most interesting solutions to equip a modern house with a renewable energy heating installation. The present study shows the computational modelling of an instance of an installation of such type, aiming to predict the behaviour of the system in the short term, basing on registered data in previous time instants. A correct prediction could potentially be of interesting use in the design of smart power grids. In this study, several models and configurations have been compared to determine the best and most economical setup needed for registering data of the prediction. The study includes comparisons of several ways of arranging the temporal data and pre-processing it with unsupervised techniques and several regression models. The novel approach has been tested empirically with a real dataset of measurements registered along a complete year; obtaining good results in all the operating condition ranges.

**Keywords:** Heat exchanger · Time series modelling ·
Time series clustering

## 1 Introduction

Nowadays, an increase in the use of renewable energies can be seen in most developed countries. Causes like the growth of energy prices or newly enacted laws, for instance, are reasons encouraging the use of alternative energies [12]. Therefore, a clear trend of contributions in this field is oriented to develop or optimize emergent methods in the alternative energy field, with the purpose of increasing the different installations' performance [8]. Still considering the essential collective objective of the environment preservation, it is safe to claim that consumers would also demand that the facilities based on alternative energies

© Springer Nature Switzerland AG 2019
H. Pérez García et al. (Eds.): HAIS 2019, LNAI 11734, pp. 38–49, 2019.
https://doi.org/10.1007/978-3-030-29859-3_4

have, at least, the same cost as non-renewable solutions to contemplate its use. The requirement also affects the installations based on heat pump with geothermal heat exchangers, such as the one presented here.

A Heat Pump can be defined as a device used to provide a regular housing building the energy produced by a natural source [4]. One common source used to obtain energy is the ground. The heat exchangers which take the energy can be placed in two main configurations: horizontal and vertical. Despite the vertical topology is more efficient than the horizontal topology, the horizontal one is usually more economical [9]. This configuration is frequently placed deeper in the ground in order to increase the efficiency.

Intelligent Systems are being used to model or optimize several problems on engineering fields nowadays [3]. Some studies have been conducted with the aim of improving a given system performance, in which classical methods, like the applications based on PID controller [6], are not capable to solve this task. In other studies, new intelligent techniques are developed to ensure a right system operation [1,15].

Usually, when the time component is registered along the rest of sensor information while these systems are working, it enables the use of other techniques, specifically devised to incorporate the temporal dimension into the analysis to further improve the results obtained by generalist techniques. In the specialized literature, this family of techniques is called time series modelling.

This study shows the experimentation performed with the bioclimatic house heat exchanger retrieved data. This analysis aims to determine what would be the best set up for a complete system that would enable the prediction of short-term behaviour of the observed system. This set up configuration is defined in terms of the amount and position of needed sensors, the time intervals whose data should be registered, or the pre-processing and organization of the data before being treated by a given regression system.

## 2 Case of Study

### 2.1 Sotavento Bioclimatic House

Sotavento bioclimatic house is a project of a demonstrative bioclimatic house belonging to the Sotavento Galicia Foundation, located within the Sotavento Experimental Wind Farm. Thermal and electrical installations of the bioclimatic house have some renewable energy systems for a better efficiency. The thermal installation has 3 different sources (solar, biomass and geothermal), which provide energy to the domestic hot water and the heating systems. The electrical installation includes two renewable energy systems (wind and photovoltaic), and one connection to the power grid, to supply the lighting and power systems. The thermal installation can be divided into 3 functional groups: generation, energy storage and consumption.

## 2.2  Geothermal System Under Study

This section gives a detailed description real geothermal system's operation and its components.

– System description - The Heat Pump has two different circuits; the primary one provides the heat from the ground (the geothermal exchanger) to the Heat Pump unit. The secondary one connects the unit and the inertial accumulator.
– Geothermal exchanger - The horizontal exchanger has five different circuits. To study the ground temperature while the system is running, the installation has several temperature sensors installed along the heat exchanger, distributed in four different loops. Figure 1 shows the geothermal exchanger sensors layout.

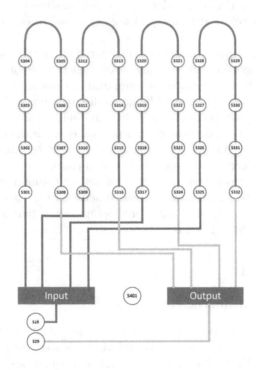

**Fig. 1.** Geothermal exchanger sensors layout

# 3  Regression Schemas

The main objective of the study performed is to predict future states of the geothermal heating system in short-term. Considering the sensor located at point S28 of the system as the geothermal installation's output to the rest of the heating system, the aim of the study is the accurate prediction of the temperature registered on sensor S28, considering different possible subsets or distributions of the data previously registered on the system.

As an interesting step of a more general study on progress, the experimentation process shown in this contribution tries to determine if there is a relatively straightforward pre-processing step that would enable any of the algorithms most commonly used for regression purposes to have its results clearly improved.

In the approach presented in this work, a comparison has been performed between the following schemas to organise data before it is fed to any of the regressors for its training:

– Using the readings obtained by a subset of all sensors in the geothermal installation
– Using only the readings of the same sensor to predict in previous states

Both approaches have been tested in combination with other two schemas:

– Using all data available to train a single regressor
– Performing an initial step of clustering of similar samples and then training a different regressor with the samples included in each of the clusters, using a hybrid non supervised and supervised approach. A depiction of this configuration can be seen on Fig. 2.

This set up, implies the testing of four pre-processing options by the combination of the mentioned options, which are the objective of the tests and results presented.

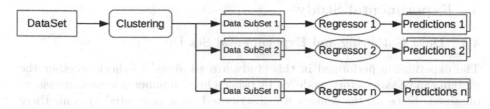

**Fig. 2.** Clustering and regression schema used for the regression of the temporal data regression.

## 3.1 Regression Models

In the first step of the study, several of the most common regression algorithms used in time series literature were initially selected to be tested. Within the scope of this first experiment, we chose five different models: k-Nearest Neighbour Regressor, Ridge Regression (Linear regression) [7], CART Decision Trees [10], Support Vector Regressor, and Multi-Layer Perceptron [13]. They were chosen according both to their widespread use in these kind of tasks and their relative simplicity in the training process. Thus, the testing did not require too much time and the differences in performance could be more clearly related with the data pre-processing procedure than the training of the models. All of them have been used previously in a wide range of regression problems, including dynamic systems modelling, weather forecasting and other related tasks.

So as to present the most interesting results of the study, we selected those that were performing better in the initial tests (described in Sect. 4.2): Ridge Regressor, Decision Tree and Multi-Layer Perceptron.

## 3.2 Clustering Models

In order to perform an initial step of sample clustering to obtain the sub-sets each of different regressors are going to be trained with, the data configuration must be taken into account. This implies that, for the experiments in which the dataset is considered as a set of measurements obtained from different points of the system, a common clustering model such as the K-Means has been selected. In the case of this study, a slightly optimized version of the original algorithm, known as Mini Batch k-Means [2] has been used to reduce the computing times of this step.

On its hand, the setup in which the model uses previous measurements of the same sensor, has several advantages if it is interpreted as a time series. It is reasonable to think that the information about how the conditions changed along time in previous occasions would provide interesting information about the way it will change in a given situation. The most common approach to cluster time series, is to use more advanced models that take into account the shape of the series, rather than the measures and their concrete positions. The well-known Dynamic Time Warping (DTW) [5] technique has been used in this study to determine which are the most similar series according to the internal structure or shape of its components.

## 4 Experimental Study

### 4.1 Data Gathering and Experimental Set Up

The experiments performed in this study are all aimed to check whether the system can be satisfactory modelled by using a lower number of sensors measures obtained. Since all the sensors are distributed on a connected system, there is a strong correlation between them, and therefore, it is reasonable to think that there are few dimensions that are really needed to capture the system's behaviour.

The real data used for testing consists in the temperature readings for each of the sensors distributed along the described circuit, plus the ground temperature sensor. Readings were taken along a complete year (2012), with a time frequency of 10 min. Therefore, the total amount of data samples included 52,645 samples of 29 features. Few of them were clearly identifiable as errors and were discarded, so the final dataset counted 52,639 samples.

An exploratory analysis of data was performed as an initial step to the presented work. A conclusion brought out from the initial study of the dataset is that there are quite long periods of time where the heat pump is inactive. That is, the heat system does not need of the use the geothermal pump and the water in the conduction is not circulating. These periods can be identified quite easily. To do this, samples in which the temperature difference between sensors S28 and sensor S29 (considered as the output and input sensors respectively) are equals or lower than 1 °C, can be considered belonging to time instants where the pump is idle.

It is crucial to note this circumstance because in this moments, the best prediction is always the same temperature as the present instant. This situation would hinder some of the results characteristics, as there are many time instants where the prediction is very accurate by not doing a prediction and those instances lower considerably the mean errors in a somehow artificial way.

So as to obtain better error approximations in the time instants when the system is functioning (and are more relevant to our study), even if all algorithms were trained in the complete dataset (not including major time gaps), the error results shown have been calculated over only time instants where the pump is determined to be working. This means that the errors are calculated over a dataset that was reduced to the 6,735 samples where the system exhibits changing behaviour.

## 4.2 Experiments and Results

**Parameters Setting.** The initial tests both served to select the best performing models and to determine the best set of parameters for each of the models selected and for each of the four combinations for preparing data. To keep this phase of experiments as simple as possible, the complete dataset was divided into 2/3 of samples (January to September) for training and 1/3 (September to December) for testing. A randomized search among a reasonable range of values for different parameters for every regressor, including 100 different sets of parameters for each regressor was conducted, obtaining the results shown in Table 1. All implementations of the regression and clustering algorithms used in the study have been obtained from the Scikit-learn python library [11]. For all parameters that are not included in the table, the default values were used.

**Table 1.** Parameter settings for the different regression models used in the study, according to each of the input data configuration

| Model | Parameters | |
|---|---|---|
| | 8 sensors/13 dimensions | 1 sensor/40 dimensions |
| Ridge Regression | Alpha: 107.51 | Alpha: 17.83 |
| Decision Tree (CART 4.5) | Quality of a split: MSE | Quality of a split: MSE |
| | Max depth: 6 | Max depth: 9 |
| | Max features: 10 | Max features: 21 |
| | Min samples on a leaf: 99 | Min samples on a leaf: 96 |
| | Min samples to split: 16 | Min samples to split: 9 |
| Multi-Layer Perceptron | Num hidden layers: 1 | Num hidden layers: 1 |
| | Hidden layer size: 10 | Hidden layer size: 25 |
| | Activation function: relu | Activation function: relu |
| | Alpha: 1.0 | Alpha: 1.0 |

As a means of comparison for all schemas, a test has been completed using the same readings of the S28 sensor to predict the value of the S28 in 3 different future time horizons. That is, assuming that within the time of 1, 3 or 6 h, the temperature would be the same as in the reference time instant. The results presented are the well known error measures calculated for regression problems: Mean Absolute Error (MAE), Root Mean Squared Error (MSE) and Median Absolute Error (MDAE). Their common characteristic is that all of them provide a measure of the error in the same magnitude and units as the variable that is predicted; that is, in this case can be interpreted as error in Celsius degrees. For a deep revision of forecasting errors, the work by Shcherbakov et al. [14] is recommended.

**Table 2.** Regression errors obtained when using the same S28 time series data to try to predict the value of the series in different time horizons

| Time Horz. | All data | | | Working pump | | |
|---|---|---|---|---|---|---|
| | MAE | RMSE | MDAE | MAE | RMSE | MDAE |
| 1 h | 1.06 | 2.059 | 0.25 | 1.64 | 1.997 | 1.47 |
| 3 h | 1.74 | 3.029 | 0.49 | 2.95 | 3.620 | 2.75 |
| 6 h | 2.29 | 3.686 | 0.81 | 3.91 | 4.906 | 3.37 |

These error results are included in Table 2. As expected, results are not very satisfactory and can be improved with the use of regression models. In this, and the rest of experiments shown in the study a time series cross validation schema has been adopted: the dataset has been divided into 12 subsets, determined sequentially according to the time stamp of the measure, corresponding roughly to each month. Then, all models have been trained in all data available up to a given moment and tested by calculating the regression values of samples along the following month, so results shown are the mean value of the 12 tests for each setting.

**Regression with Several Sensors.** The first batch of tests have been made using a dataset composed by the readings of 8 different sensors along the complete circuit. Since all sensors are disposed along the same circuit, a subset has been selected in order to reduce computational requirements. The sensors included in the subset were S29, S301, S308, S309, S316, S317, S324. Additionally, each sample of the dataset includes the reading of the reference temperature of the ground sensor (S401) and the month and day of the week for every sample, formatted using the polar coordinates to preserve its periodic nature. This makes a dataset with a total of 13 dimensions. As explained before, this dataset regression has been tested both for the direct calculations of each fold of a sequential cross-validation and for the schema including a step of clustering prior to the regression on each of the folds.

**Table 3.** Error measures obtained when using simultaneous data from 8 sensors along the circuit. One regressor was trained with all data available.

| Time Horz. | Ridge | | | Decision Tree | | | MLP | | |
|---|---|---|---|---|---|---|---|---|---|
| | MAE | RMSE | MDAE | MAE | RMSE | MDAE | MAE | RMSE | MDAE |
| 1 h | 1.238 | 1.880 | 0.793 | 1.696 | 2.129 | 1.465 | 1.432 | 2.009 | 1.048 |
| 3 h | 2.543 | 3.114 | 2.203 | 3.174 | 3.763 | 2.932 | 2.696 | 3.199 | 2.470 |
| 6 h | 3.314 | 4.012 | 2.891 | 3.689 | 4.461 | 3.229 | 3.284 | 3.947 | 3.054 |

**Table 4.** Error measures obtained when using simultaneous data from 8 sensors along the circuit. 10 different regressors were trained in data corresponding to each of the clusters identified in the data.

| Time Horz. | Ridge | | | Decision Tree | | | MLP | | |
|---|---|---|---|---|---|---|---|---|---|
| | MAE | RMSE | MDAE | MAE | RMSE | MDAE | MAE | RMSE | MDAE |
| 1 h | 2.146 | 2.910 | 1.573 | 3.450 | 4.519 | 2.479 | 3.712 | 5.520 | 2.526 |
| 3 h | 3.725 | 4.743 | 3.126 | 3.620 | 4.431 | 3.072 | 4.840 | 6.564 | 3.729 |
| 6 h | 4.208 | 5.326 | 3.652 | 4.352 | 5.344 | 3.664 | 6.750 | 8.711 | 5.630 |

From Tables 2 and 3 can be seen that common regressors can offer a better prediction than the naïve prediction of invariant temperature with a quite simple schema (Table 2). In this case, the more complex model (Multi-Layer Perceptron) obtains the best results.

Comparing Tables 3 and 4, the use of simple regressor models seems to be more effective. While using a single classifier for all samples appears to capture better the temporal relationships between different samples, the clustering of the samples, seems to break that relationships, including samples from different time periods into the same cluster. Even though their static characteristics can be quite similar, in this case the dynamics of the system could be more important to determine the output of future states. What is more, reducing the number of samples that each of the models is using seems to degrade the performance of the models that use them more intensively, such as the MLP; in a considerable way. While MLP it is the best performing model in Table 3, it is the worst one in Table 4.

A visual comparison of the regression performance obtained with this set up is shown in Fig. 3. It depicts the ground truth of sensor S28 (in blue colour) and the two best performing models. Both belong to the simplest approach, in which no previous clustering is performed.

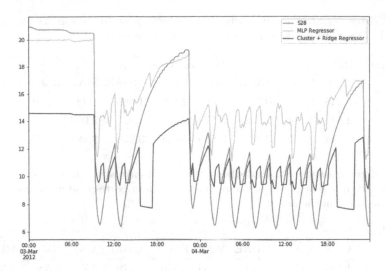

**Fig. 3.** Regression obtained from samples composed by readings from 8 sensors during days $3^{rd}$ and $4^{th}$ of March, 2012 (Color figure online)

**Regression with One Sensor.** On the second set of experiments, the objective was to test whether similar results to those obtained with 8 measures along the conduction could be obtained using only the accumulated measures taken from sensor S28. That is, predicting the value of S28 in a given time horizon with the measures of several previous time instants to construct each data sample. In the case of this study, the data from the 6 previous hours was considered as input to each of the models (36 samples) obtained by sweeping a time window of 36 samples along the complete dataset. In each sample, the information of the month and day of the week is also added, meaning that each sample fed to the regression models is composed by 40 dimensions or features.

Then, two set of experiments have been devised to test this approach in a similar way to the ones using different sensors: the first set consists in training one instance of all different regressors directly on each of the samples of the dataset. In the second one, firstly the data is clustered attending to its similar characteristics, by using the classical KNN clustering algorithm, considering the DTW (see Sect. 3.2) as the distance measure between two samples. Then, a different instance of each of the classifiers is trained exclusively with the samples belonging to a given cluster. In order to obtain the regression prediction for a new sample, first this sample is compared to each of the centroids obtained in the clustering process, determining then the cluster to which it should belong. The regression is requested then to the regressor that was trained with the subset of data corresponding to that cluster.

Data samples are clustered in this case attending to the tendency of 6 consecutive hours (36 samples), giving as a result centroids or "template tendencies".

**Table 5.** Error measures obtained when using measures of only sensor S28 from the previous 6 h. One regressor was trained with all data available.

| Time Horz. | Ridge | | | Decision Tree | | | MLP | | |
|---|---|---|---|---|---|---|---|---|---|
| | MAE | RMSE | MDAE | MAE | RMSE | MDAE | MAE | RMSE | MDAE |
| 1 h | 1.535 | 2.040 | 1.229 | 1.577 | 2.210 | 1.142 | 1.48 | 2.069 | 1.123 |
| 3 h | 2.826 | 3.361 | 2.779 | 2.957 | 3.670 | 2.442 | 2.765 | 3.319 | 2.622 |
| 6 h | 3.375 | 4.083 | 3.121 | 3.792 | 4.611 | 3.361 | 3.353 | 4.050 | 3.112 |

Comparing Tables 5 and 3, it can be concluded that the composition of the second dataset, using only accumulated data from a single sensor obtains results that are quite close to the ones obtained requiring 8 sensors. Results are slightly worse in the second case (Table 3), but the difference in their errors is overall less than 0.1 °C.

Contrasting Tables 4 and 6 it is clear than in the case of this problem, the idea of clustering data based on its time series nature, better captures the nature of the dataset; since it considers the similar evolution of the system in previous short periods of time, rather than the similarity of several measures taken in the same time instants.

From the comparison between Tables 5 and 6 it can be concluded that, also in this case the single model appears to capture better the complexities of the system than the clustering of samples; although in this case, the differences between the single and clustered approach are much less significant.

**Table 6.** Error measures obtained when using measures of only sensor S28 from the previous 6 h. 10 different regressors were trained in data corresponding to each of the clusters identified in the data.

| Time Horz. | Ridge | | | Decision Tree | | | MLP | | |
|---|---|---|---|---|---|---|---|---|---|
| | MAE | RMSE | MDAE | MAE | RMSE | MDAE | MAE | RMSE | MDAE |
| 1 h | 1.847 | 2.884 | 1.273 | 2.300 | 3.143 | 1.655 | 1.825 | 2.701 | 1.117 |
| 3 h | 2.799 | 3.299 | 2.541 | 3.103 | 3.650 | 2.928 | 2.936 | 3.561 | 2.625 |
| 6 h | 3.571 | 4.191 | 3.365 | 3.667 | 4.416 | 3.297 | 3.638 | 4.366 | 3.296 |

A visual comparison of the regression performance obtained with the time series approach set up is shown in Fig. 4. It depicts the ground truth of sensor S28 (in blue colour) and the two best performing models. In this case, they belong to each of the variants: simple algorithm (MLP) and clustering and regression (Ridge).

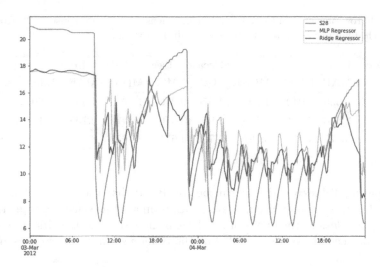

**Fig. 4.** Regression obtained from samples composed by readings from one sensor along 6 h during days $3^{rd}$ and $4^{th}$ of March, 2012 (Color figure online)

## 5   Conclusions and Future Work

The contribution presents a heat exchanger designed to help regulate the temperature of a bio-climatic installation. The research presented focuses its effort in proving that the behaviour exhibited by this type of installation can be accurately modelled for later purposes, such as expected energy consumption.

The experiments presented in this contribution, are focused on determining whether the system can be modelled trustfully without the need of an extensive array of sensors along its installation. As result of the analysis, one of the first conclusions is that the modelling of the system, can be achieved using very few of the sensors readings, as long as they are logging the measurements with a reasonable enough frequency: a 10 min frequency is used in our experiments. Another conclusion achieved is that the schema of clustering and regression that has been useful in other applications did not achieve significant better results in this case.

A path to pursue the advances of the research would be testing of more advanced regression models, such as the state of the art deep learning architectures, like the Long-Short Term Memory (LSTM) Recurrent Neural Networks; which have reported to achieved good results in time modelling problems.

**Acknowledgements.** We would like to thank the 'Instituto Enerxético de Galicia' (INEGA) and 'Parque Eólico Experimental de Sotavento' (Sotavento Foundation) for their technical support on this work.

# References

1. Baruque, B., Porras, S., Jove, E., Calvo-Rolle, J.L.: Geothermal heat exchanger energy prediction based on time series and monitoring sensors optimization. Energy **171**, 49–60 (2019)
2. Béjar, J.: K-means vs mini batch k-means: a comparison. In: 2014 International Symposium on Biometrics and Security Technologies (ISBAST) (2014)
3. Casteleiro-Roca, J.L., Pérez, J.A.M., Piñón-Pazos, A.J., Calvo-Rolle, J.L., Corchado, E.: Modeling the electromyogram (EMG) of patients undergoing anesthesia during surgery. In: Herrero, Á., Sedano, J., Baruque, B., Quintián, H., Corchado, E. (eds.) 10th International Conference on Soft Computing Models in Industrial and Environmental Applications, vol. 368, pp. 273–283. Springer, Cham (2015). https://doi.org/10.1007/978-3-319-19719-7_24
4. Cui, P., Li, X., Man, Y., Fang, Z.: Heat transfer analysis of pile geothermal heat exchangers with spiral coils. Appl. Energy **88**(11), 4113–4119 (2011)
5. Dürrenmatt, D.J., Del Giudice, D., Rieckermann, J.: Dynamic time warping improves sewer flow monitoring. Water Res. **47**(11), 3803–3816 (2013)
6. García, R.F., Rolle, J.L.C., Castelo, J.P., Gomez, M.R.: On the monitoring task of solar thermal fluid transfer systems using NN based models and rule based techniques. Eng. Appl. Artif. Intell. **27**, 129–136 (2014)
7. Hoerl, A.E., Kennard, R.W.: Ridge regression: applications to nonorthogonal problems. Technometrics **12**(1), 69–82 (1970)
8. Jenssen, T.: Glances at Renewable and Sustainable Energy. Springer, London (2013). https://doi.org/10.1007/978-1-4471-5137-1
9. Kakaç, S., Liu, H., Pramuanjaroenkij, A.: Heat Exchangers: Selection, Rating, and Thermal Design, 2nd edn. Taylor & Francis (2002). Designing for heat transfer
10. Loh, W.Y.: Classification and regression trees. Wiley Interdisc. Rev. Data Min. Knowl. Discov. **1**(1), 14–23 (2011)
11. Pedregosa, F., et al.: Scikit-learn: machine learning in Python. J. Mach. Learn. Res. **12**, 2825–2830 (2011)
12. Porter, D.: Comprehensive Renewable Energy. Elsevier, Amsterdam (2012)
13. Rumelhart, D.E., Hinton, G.E., Williams, R.J.: Learning representations by back-propagating errors. Nature **323**(533) (1986)
14. Shcherbakov, M., Brebels, A., Shcherbakova, N., Tyukov, A., Janovsky, T., Kamaev, V.: A survey of forecast error measures. World Appl. Sci. J. **24**, 171–176 (2013). https://doi.org/10.5829/idosi.wasj.2013.24.itmies.80032
15. Vilar-Martínez, X.M., Montero-Sousa, J.A., Calvo-Rolle, J.L., Casteleiro-Roca, J.L.: Expert system development to assist on the verification of TACAN system performance. DYNA **89**(1), 112–121 (2014)

# Can Automated Smoothing Significantly Improve Benchmark Time Series Classification Algorithms?

James Large, Paul Southam, and Anthony Bagnall[✉]

University of East Anglia, Norwich Research Park, Norwich, UK
{James.Large,Paul.Southam,ajb}@uea.ac.uk
http://www.timeseriesclassification.com

**Abstract.** tl;dr: no, it cannot, at least not on average on the standard archive problems. We assess whether using six smoothing algorithms (moving average, exponential smoothing, Gaussian filter, Savitzky-Golay filter, Fourier approximation and a recursive median sieve) could be automatically applied to time series classification problems as a preprocessing step to improve the performance of three benchmark classifiers (1-Nearest Neighbour with Euclidean and Dynamic Time Warping distances, and Rotation Forest). We found no significant improvement over unsmoothed data even when we set the smoothing parameter through cross validation. We are not claiming smoothing has no worth. It has an important role in exploratory analysis and helps with specific classification problems where domain knowledge can be exploited. What we observe is that the automatic application does not help to improve classification performance and that we cannot explain the improvement of other time series classification algorithms over the baseline classifiers simply as a function of the absence of smoothing.

**Keywords:** Time series · Classification · Smoothing · Benchmark

## 1 Introduction

Time Series Classification (TSC) is differentiated from standard classification by the fact that the ordering of the attributes may be important in finding discriminatory features. Standard vector classifiers such as rotation forest and standard time dependent approaches such dynamic time warping with 1-NN are strong benchmark algorithms to compare against the range of bespoke TSC algorithms that have been proposed in recent years. Some of these achieve impressive performance and are significantly better than the benchmarks. Nevertheless, there has always been a suspicion that sensible standard preprocessing of the data would perhaps increase the accuracy of benchmark classifiers and that would make at least some of the bespoke algorithms redundant [1]. Broadly speaking, there are four types of preprocessing that may improve classifier performance: normalisation; smoothing; dimensionality reduction; and discretization. We address the

© Springer Nature Switzerland AG 2019
H. Pérez García et al. (Eds.): HAIS 2019, LNAI 11734, pp. 50–60, 2019.
https://doi.org/10.1007/978-3-030-29859-3_5

question of whether smoothing series can significantly improve the accuracy of benchmark classifiers. Smoothing is the process of reducing the noise in the series to make patterns in the data more apparent and is generally used as part of an exploratory analysis.

It is important to stress we are only concerned with class independent noise, since class dependent noise is possibly useful as a discriminatory feature. This is where we diverge from the majority of signal processing research into noise modeling and reduction. We are not necessarily trying to "clean up" a signal. Instead, we are trying to remove artifacts that may confound the classifier.

We test whether six smoothing algorithms improve three base classifiers. These are described in detail in Sect. 2. It is clearly important to set the parameters of the algorithm when smoothing so that it is relevant to a specific problem. Because we are attempting to smooth to improve classification, we set parameters through cross validation on the train data using the base classifier we are testing. The experimental design is described in Sect. 3. Our experiments address the following two questions.

1. Does smoothing with default parameters increase the accuracy of benchmark classifiers?
2. Can we learn smoothing parameters on the train data to significantly improve benchmark TSC algorithms?

A priori, we believed it unlikely that systematic smoothing would improve accuracy over the diverse data sets in the archive, since many of the series have very little noise. However, we thought that supervised smoothing, where no smoothing was an option, would improve performance albeit at the large computational cost of the parameter search. Our results, presented in Sect. 4 show that in fact smoothing makes very little difference, even when supervised. We discuss these results in Sect. 5 and conclude in Sect. 6.

## 2   Background

### 2.1   Time Series Classification

A large number of new classification problems have been proposed in the last ten years. While not exhaustive by itself, it is important to evaluate new algorithms against sensible benchmark classifiers on standard test problems in order to ascertain the usefulness of new research. The UCR archive is a widely used archive of test problems [8]. The archive is a continually growing collection of real valued TSC datasets[1] which come from a range of different domains and have a range of characteristics, in terms of size, number of classes, imbalances, etc. Most TSC publications benchmark against a 1-NN classifier using either Euclidean distance (ED) or Dynamic Time Warping (DTW) distance. DTW compensates for potential misalignments amongst series of the same class. DTW

---

[1] http://www.timeseriesclassification.com.

has a single parameter, the maximum warping window, and DTW performs significantly better when this parameter is set through cross validation. A recent comparative study [2] found that the classifier rotation forest [16] (RotF) was also a strong benchmark. It is able to discover relationships in time through the internal principle component transformation it uses, and is not significantly worse than DTW with window set through cross validation (DTWCV henceforth). The same study compared 22 TSC algorithms on 85 of the UCR archive data and found that just nine out of twenty two TSC algorithms were significantly more accurate than both a rotation forest and DTW classifier. Some of these TSC algorithms are highly complex and both memory and computationally expensive. A case was made that the superior algorithms achieved higher accuracy because the representation they use allows for the detection of discriminatory features that the benchmarks cannot find. This was further demonstrated on the archive and through data simulation [14]. We wish to test whether simple preprocessing can significantly improve the benchmarks and hence narrow the gap between DTW and rotation forest and the nine significantly better TSC algorithms.

## 2.2   Time Series Smoothing

Given a time series $T =< t_0, \ldots, t_{m-1} >$, a smoothing function produces a new series $S =< s_0, \ldots, s_{p-1} >$, where $p \leq m$ (we index from zero to make the equations simpler). Most algorithms employ a sliding window, of length $w$, along the series, resulting in a series of length $p = m - w$. The simplest form of smoothing is to take the **moving average (MA)** [9], often called the Simple Moving Average,

$$s_j = \frac{\sum_{i=j-w}^{j} t_i}{w} \quad \text{for } j = w \ldots m - 1,$$

where $w$ is the single parameter, window size. **Exponential smoothing (EXP)** [9] is a generalisation of moving average smoothing that assigns a decaying weight to each element rather than averaging over a window.

$$s_0 = t_0 \quad \text{and} \quad s_j = \alpha \cdot t_j + (1 - \alpha) \cdot t_{j-1}$$

where $0 \leq \alpha \leq 1$. For consistency with other smoothing algorithms, EXP is often given a window size $w$, then the decay weight is set as $\alpha = \frac{2}{w+1}$.

A **Gaussian filter (GF)** [9] applies a fixed convolution over a window

$$s_j = \sum_{i=j-w}^{j} t_i \cdot c_i,$$

where the convolution values $c_i$ are derived from a standard normal distribution over the window $w$, the single parameter.

Like GF, the **Savitzky-Golay (SG)** filtering method is a convolutional method of smoothing. Instead of using a fixed convolution, it estimates a different convolution on each window based on local least-squares polynomial approximation.

$$s_j = \sum_{i=j-w}^{j} t_i \cdot c_{i,j}$$

Since its initial introduction [17], it has been used successfully and pervasively across many signal processing domains for different purposes, particularly in chemometrics [7,12]. SG has two parameters, window size $w$ and polynomial order $n$. For accessible explanations of how the polynomial coefficients are calculated, we refer the reader to [18].

**Discrete Fourier Approximation (DFT)** [10] smooths the series by first transforming into the frequency domain, discarding the high frequency terms, then transforming back to the time domain. DFT has a single parameter, $r$, the proportion of Fourier terms to retain.

The **Recursive Median Sieve (SIV)** is a one-dimensional recursive median filter [3] that filters the data by removing extrema of specific scales. The sieve uses morphological scale-space operations, specifically openings and closings, or combinations of them, to filter an input signal. It does this by applying flat structuring elements to an input signal, which unlike conventional morphological operators such as those used in granulometries, have a fixed size but variable shape. They were introduced as a one-dimensional non-linear scale-space decomposition algorithm in [5], but can be extend to $n$-dimensions by adopting techniques from graph morphology [4].

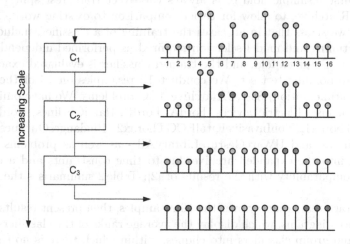

**Fig. 1.** An example sieve decomposition of a 1D signal. Green vertices are the vertices affected at each scale level. (Color figure online)

The sieve performs a decomposition removing extrema (both maxima and minima) at different scales as shown in Fig. 1. At the scale $c_1$ the maxima and minima at points 6,7 and 10 are smoothed to equal the nearest value of the neighbours. At scale $c_2$ the pairs at (4,5) and (12,13) are smoothed. At the highest scale, the series is uniform. The sieve takes in a single parameter, $c$, which is the scale to smooth the signal to.

## 3  Experimental Setup

**Table 1.** The parameter spaces searched for each filtering method over the course of our experiments (default value in bold). $m$ is the series length. For Savitzky-Golay (SG), all combinations of $w, n$ are searched where $w > 2n$.

| Method | Parameters and default values in bold |
|---|---|
| Moving Average (MA) | $w \in \{2, 3, \mathbf{5}, 10, 25, 50, 100, \sqrt{m}, \log_2(m)\}$ |
| Exponential Smoothing (EXP) | $w \in \{2, 3, \mathbf{5}, 10, 25, 50, 100, \sqrt{m}, \log_2(m)\}$ |
| Gaussian Filtering (GF) | $w \in \{2, 3, \mathbf{5}, 10, 25, 50, 100, \sqrt{m}, \log_2(m)\}$ |
| Savitzky-Golay (SG) | $w \in \{\mathbf{5}, 9, 17, 33, 65\}$ |
| | $n \in \{\mathbf{2}, 3, 4, 8, 16, 32\}$ |
| Fourier Approximation (DFT) | $r \in \{0.01, 0.05, \mathbf{0.1}, 0.25, 0.5, \log_2(m)/m\}$ |
| Sieve (SIV) | $c \in \{\frac{1}{15} \cdot \log_{10}(m), \ldots, \mathbf{\frac{5}{15} \cdot \log_{10}(m)}, \ldots, \log_{10}(m)\}$ |

For each of pair of filter+classifier combination, we perform 10 stratified random resamples of each data set and report the average results across those resamples. The first resample, fold 0, is always the exact train/test split published on the UCR archive, to allow for easier comparison to existing work. To avoid ambiguity, we stress that in all cases the training of a classifier, including any parameter tuning and model selection required, is performed independently on the train set of a given fold, and the trained classifier is evaluated exactly once on the corresponding test set. We conduct 10 resamples on 76 of the (at the time of experimentation) 85 UCR archive TSC problems. We have omitted the largest problems - ElectricDevices, FordA, FordB, HandOutlines, NonInvasive-FatalECGThorax1, NonInvasiveFatalECGThorax2, PhalangesOutlinesCorrect, StarlightCurves, and UWaveGestureLibraryAll - as well as problems recently introduced into the expanded archive due to time constraints and a desire to maintain comparability with the results of [2]. Table 2 summarises the datasets used.

We average test accuracy over the 10 resamples, then present results in critical difference diagrams, which display the average ranks of the classifiers over all problems and group classifiers into cliques, within which there is no significant difference. For each resample, we perform a 10 fold cross validation (CV) on that resamples' train data to find smoothing parameters, such that no transfer learning of optimal parameters is performed.

**Table 2.** The 76 UCR time series classification problems used in the experiments.

| Dataset | Atts | Classes | Train | Test | Dataset | Atts | Classes | Train | Test |
|---|---|---|---|---|---|---|---|---|---|
| Adiac | 176 | 37 | 390 | 391 | Meat | 448 | 3 | 60 | 60 |
| ArrowHead | 251 | 3 | 36 | 175 | MedicalImages | 99 | 10 | 381 | 760 |
| Beef | 470 | 5 | 30 | 30 | MidPhalOutAgeGroup | 80 | 3 | 400 | 154 |
| BeetleFly | 512 | 2 | 20 | 20 | MidPhalOutCorrect | 80 | 2 | 600 | 291 |
| BirdChicken | 512 | 2 | 20 | 20 | MiddlePhalanxTW | 80 | 6 | 399 | 154 |
| Car | 577 | 4 | 60 | 60 | MoteStrain | 84 | 2 | 20 | 1252 |
| CBF | 128 | 3 | 30 | 900 | OliveOil | 570 | 4 | 30 | 30 |
| ChlorineConcentration | 166 | 3 | 467 | 3840 | OSULeaf | 427 | 6 | 200 | 242 |
| CinCECGtorso | 1639 | 4 | 40 | 1380 | Phoneme | 1024 | 39 | 214 | 1896 |
| Coffee | 286 | 2 | 28 | 28 | Plane | 144 | 7 | 105 | 105 |
| Computers | 720 | 2 | 250 | 250 | ProxPhalOutAgeGroup | 80 | 3 | 400 | 205 |
| CricketX | 300 | 12 | 390 | 390 | ProxPhalOutCorrect | 80 | 2 | 600 | 291 |
| CricketY | 300 | 12 | 390 | 390 | ProximalPhalanxTW | 80 | 6 | 400 | 205 |
| CricketZ | 300 | 12 | 390 | 390 | RefrigerationDevices | 720 | 3 | 375 | 375 |
| DiatomSizeReduction | 345 | 4 | 16 | 306 | ScreenType | 720 | 3 | 375 | 375 |
| DisPhalOutAgeGroup | 80 | 3 | 400 | 139 | ShapeletSim | 500 | 2 | 20 | 180 |
| DisPhalOutCor | 80 | 2 | 600 | 276 | ShapesAll | 512 | 60 | 600 | 600 |
| DislPhalTW | 80 | 6 | 400 | 139 | SmallKitchApps | 720 | 3 | 375 | 375 |
| Earthquakes | 512 | 2 | 322 | 139 | SonyAIBORSurface1 | 70 | 2 | 20 | 601 |
| ECG200 | 96 | 2 | 100 | 100 | SonyAIBORSurface2 | 65 | 2 | 27 | 953 |
| ECG5000 | 140 | 5 | 500 | 4500 | Strawberry | 235 | 2 | 613 | 370 |
| ECGFiveDays | 136 | 2 | 23 | 861 | SwedishLeaf | 128 | 15 | 500 | 625 |
| FaceAll | 131 | 14 | 560 | 1690 | Symbols | 398 | 6 | 25 | 995 |
| FaceFour | 350 | 4 | 24 | 88 | SyntheticControl | 60 | 6 | 300 | 300 |
| FacesUCR | 131 | 14 | 200 | 2050 | ToeSegmentation1 | 277 | 2 | 40 | 228 |
| FiftyWords | 270 | 50 | 450 | 455 | ToeSegmentation2 | 343 | 2 | 36 | 130 |
| Fish | 463 | 7 | 175 | 175 | Trace | 275 | 4 | 100 | 100 |
| GunPoint | 150 | 2 | 50 | 150 | TwoLeadECG | 82 | 2 | 23 | 1139 |
| Ham | 431 | 2 | 109 | 105 | TwoPatterns | 128 | 4 | 1000 | 4000 |
| Haptics | 1092 | 5 | 155 | 308 | UWaveX | 315 | 8 | 896 | 3582 |
| Herring | 512 | 2 | 64 | 64 | UWaveY | 315 | 8 | 896 | 3582 |
| InlineSkate | 1882 | 7 | 100 | 550 | UWaveZ | 315 | 8 | 896 | 3582 |
| InsectWingbeatSound | 256 | 11 | 220 | 1980 | Wafer | 152 | 2 | 1000 | 6164 |
| ItalyPowerDemand | 24 | 2 | 67 | 1029 | Wine | 234 | 2 | 57 | 54 |
| LargeKitchApps | 720 | 3 | 375 | 375 | WordSynonyms | 270 | 25 | 267 | 638 |
| Lightning2 | 637 | 2 | 60 | 61 | Worms | 900 | 5 | 181 | 77 |
| Lightning7 | 319 | 7 | 70 | 73 | WormsTwoClass | 900 | 2 | 181 | 77 |
| Mallat | 1024 | 8 | 55 | 2345 | Yoga | 426 | 2 | 300 | 3000 |

For comparing multiple classifiers on multiple datasets, we follow the recommendation of Demšar [11] and use the Friedmann test to determine if there are any statistically significant differences in the rankings of the classifiers. However, following recommendations in [6] and [13], we have abandoned the Nemenyi post-hoc test originally used by [11] to form cliques (groups of classifiers within which there is no significant difference in ranks). Instead, we compare all classifiers with pairwise Wilcoxon signed-rank tests, and form cliques using the Holm correction (which adjusts family-wise error less conservatively than a Bonferonni adjustment).

Our code[2] reproduces the splits used in this evaluation exactly, and full, reproducible results are available[3]. For all smoothing algorithms except the sieve, we used the standard MATLAB implementations and performed the smoothing and classification in separate stages. The default parameters given in Table 1 are those of the Matlab implementations. The sieve is implemented in C and was similarly isolated from the classification stage.

We use three baseline classifiers. 1-NN with Euclidean distance is a weak baseline in a TSC context, but it is still frequently used in research. 1-NN with DTW is the most common benchmark, although it is important to set the window through cross validation [15]. This is computationally expensive, although we use the DTW version described in [19] which speeds up the calculation by orders of magnitude. All the UCR data are normalised. For consistency, we renormalise each series after smoothing.

## 4   Results

We present results for three baseline classifiers with both default smoothing and tuned smoothing through critical difference diagrams in Fig. 2, and average

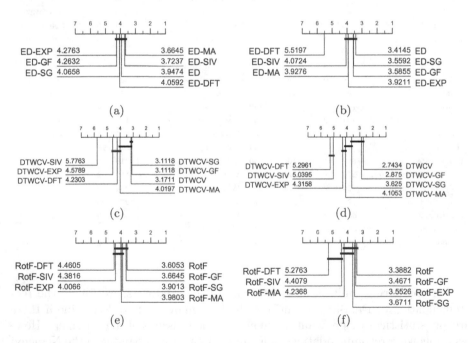

(a)                              (b)

(c)                              (d)

(e)                              (f)

**Fig. 2.** Average ranks on 76 UCR problems while smoothing with the six methods described in Sect. 2, with default (left) and tuned (right) parameters using the parameters given in Table 1, for Euclidean distance ((a) and (b)), dynamic time warping ((c) and (d)) and rotation forest ((e) and (f)).

---

[2] https://github.com/TonyBagnall/uea-tsc.
[3] http://www.timeseriesclassification.com/Smoothing.php.

**Table 3.** Average accuracies across the 76 datasets of the default (top) and tuned (bottom) smoothing methods when using the three benchmark classifiers. Within each group, classifier are ordered by rank to mirror Fig. 2.

| Untuned smoothing filters | | | | | |
|---|---|---|---|---|---|
| | Accuracy | | Accuracy | | Accuracy |
| ED-MA | 0.718 | DTWCV-SG | 0.77 | RotF | 0.769 |
| ED-SIV | 0.714 | DTWCV-GF | 0.769 | RotF-GF | 0.768 |
| ED | 0.714 | DTWCV | 0.771 | RotF-SG | 0.768 |
| ED-DFT | 0.716 | DTWCV-MA | 0.765 | RotF-MA | 0.767 |
| ED-SG | 0.716 | DTWCV-DFT | 0.764 | RotF-EXP | 0.768 |
| ED-GF | 0.716 | DTWCV-EXP | 0.763 | RotF-SIV | 0.761 |
| ED-EXP | 0.717 | DTWCV-SIV | 0.74 | RotF-DFT | 0.761 |
| Tuned smoothing filters | | | | | |
| | Accuracy | | Accuracy | | Accuracy |
| ED | 0.714 | DTWCV | 0.771 | RotF | 0.769 |
| ED-SG | 0.717 | DTWCV-GF | 0.771 | RotF-GF | 0.769 |
| ED-GF | 0.716 | DTWCV-SG | 0.769 | RotF-EXP | 0.77 |
| ED-MA | 0.718 | DTWCV-MA | 0.763 | RotF-SG | 0.769 |
| ED-EXP | 0.719 | DTWCV-EXP | 0.764 | RotF-MA | 0.768 |
| ED-SIV | 0.716 | DTWCV-SIV | 0.748 | RotF-SIV | 0.76 |
| ED-DFT | 0.689 | DTWCV-DFT | 0.735 | RotF-DFT | 0.753 |

accuracies are summarised in Table 3. For all three classifiers, smoothing of any kind provides no benefit. Tuning provides no benefit over using default values, and in many cases makes things worse due to overfitting.

For all six experiments, the classifiers built on unsmoothed data are in the top clique. For four of the experiments, the unsmoothed classifier is the highest ranked. Setting the parameter through cross validation is if anything worse than using a default parameter. Given the order of magnitude more computation required to tune these parameters, this is surprising, particularly as *no smoothing* was one of the options. Further analysis shows that *no smoothing* was selected approximately 25% of the time. This could be an indication that the archive data are simply not suited to smoothing, however even in that case this suggests that the improvements being found by complex TSC classifiers cannot be easily explained away by simple, or even costly, attempts to smooth the data.

## 5  Analysis

We examine whether there are any characteristics of the data that could help determine whether any of the six types of smoothing would improve performance. We would expect that smoothing might be more useful for longer series. Figures 3 and 4 show the scatter plot of length against classifier rank for DTWCV and rotation forest. We find no obvious relationship between the performance of the unsmoothed classifier and series length. We repeated a similar analysis for number of instances and classes, but again found no correlation as one should expect when considering smoothing. Further, we find no significant areas where particular smoothing methods improve over the others.

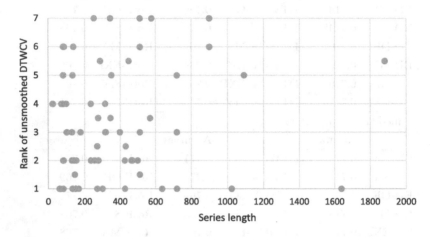

**Fig. 3.** Ranks per dataset on 76 UCR problems of unsmoothed DTWCV compared to six *untuned* smoothed versions plotted against series length. No obvious correlation is found.

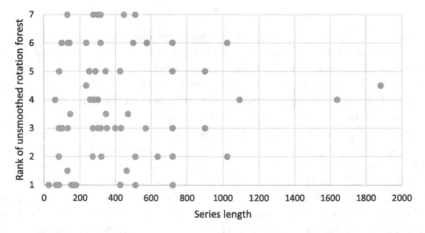

**Fig. 4.** Ranks per dataset on 76 UCR problems of unsmoothed rotation forest compared to six *untuned* smoothed versions plotted against series length. No obvious correlation is found.

# 6   Conclusion

It has long been a suspicion of many researchers in this field that much of the improvement seen in complex TSC algorithms could equally be achieved with comparatively simple preprocessing. Our experiments indicate for the case of smoothing, this is not true. We have taken six very popular smoothing algorithms and applied them using sensible default parameters and using extensive extra computation to discover optimal parameters through cross validation. We have found no significant difference between smoothed and unsmoothed classification with three benchmarks. The nature of the UCR data may explain this to a degree: the data from problems such as image processing will have less noise than, for example, financial data. We are not claiming that smoothing has no role to play in the analysis of time series data, merely that the automated application of smoothing without domain expertise does not on average improve the performance of baseline classifiers and that the absence of smoothing cannot explain the performance of algorithms that outperform the baselines.

**Acknowledgement.** This work is supported by the UK Engineering and Physical Sciences Research Council (EPSRC) [grant number EP/M015807/1] and Biotechnology and Biological Sciences Research Council [grant number BB/M011216/1]. The experiments were carried out on the High Performance Computing Cluster supported by the Research and Specialist Computing Support service at the University of East Anglia and using a Titan X Pascal donated by the NVIDIA Corporation.

# References

1. Chen, Y., Hu, B., Keogh, E.: Time series classification under more realistic assumption. In Proceedings 13th SIAM International Conference on Data Mining (2013)
2. Bagnall, A., Lines, J., Bostrom, A., Large, J., Keogh, E.: The great time series classification bake off: a review and experimental evaluation of recent algorithmic advances. Data Min. Knowl. Disc. **31**(3), 606–660 (2017)
3. Bangham, J.: Data-sieving hydrophobicity plots. Anal. Biochem. **174**(1), 142–145 (1988)
4. Bangham, J., Harvey, R., Ling, P., Aldridge, R.: Morphological scale-space preserving transforms in many dimensions. J. Electron. Imaging **5**, 283–299 (1996)
5. Bangham, J., Ling, P., Harvey, R.: Scale-space from nonlinear filters. IEEE Trans. Pattern Anal. Mach. Intell. **18**(5), 520–528 (1996)
6. Benavoli, A., Corani, G., Mangili, F.: Should we really use post-hoc tests based on mean-ranks? J. Mach. Learn. Res. **17**, 1–10 (2016)
7. Betta, G., Capriglione, D., Cerro, G., Ferrigno, L., Miele, G.: The effectiveness of Savitzky-Golay smoothing method for spectrum sensing in cognitive radios. In: Proceedings of the 2015 18th AISEM Annual Conference, pp. 1–4 (2015)
8. Chen, Y., et al.: The UEA-UCR time series classification archive (2015). http://www.cs.ucr.edu/~eamonn/time_series_data/
9. Chou, Y.: Statistical Analysis. Holt International (1975)
10. Cooley, J., Lewis, P., Welch, P.: The fast fourier transform and its applications. IEEE Trans. Educ. **12**(1), 27–34 (1969)

11. Demšar, J.: Statistical comparisons of classifiers over multiple data sets. J. Mach. Learn. Res. **7**, 1–30 (2006)
12. Fernandes, B., Colletta, G., Ferreira, L., Dutra, O.: Utilization of Savitzky-Golay filter for power line interference cancellation in an embedded electrocardiographic monitoring platform. In: Proceedings IEEE International Symposium on Medical Measurements and Applications, pp. 7–12 (2017)
13. García, S., Herrera, F.: An extension on "statistical comparisons of classifiers over multiple data sets" for all pairwise comparisons. J. Mach. Learn. Res. **9**, 2677–2694 (2008)
14. Lines, J., Taylor, S., Bagnall, A.: HIVE-COTE: the hierarchical vote collective of transformation-based ensembles for time series classification. In Proceedings IEEE International Conference on Data Mining (2016)
15. Ratanamahatana, C., Keogh, E.: Three myths about dynamic time warping data mining. In: Proceedings 5th SIAM International Conference on Data Mining (2005)
16. Rodriguez, J., Kuncheva, L., Alonso, C.: Rotation forest: a new classifier ensemble method. IEEE Trans. Pattern Anal. Mach. Intell. **28**(10), 1619–1630 (2006)
17. Savitzky, A., Golay, M.: Smoothing and differentiation of data by simplified least squares procedures. Anal. Chem. **36**(8), 1627–1639 (1964)
18. Schafer, R.: What is a Savitzky-Golay filter? IEEE Signal Process. Mag. **28**(4), 111–117 (2011)
19. Tan, C., Herrman, C., Forestier, G., Webb, G., Petitjean, F.: Efficient search of the best warping window for dynamic time warping. In: Proceedings 18th SIAM International Conference on Data Mining (2018)

# Dataset Weighting via Intrinsic Data Characteristics for Pairwise Statistical Comparisons in Classification

José A. Sáez[1(✉)], Pablo Villacorta[2], and Emilio Corchado[1]

[1] Department of Computer Science and Automatics, University of Salamanca,
Plaza de los Caídos s/n, 37008 Salamanca, Spain
{joseasaezm,escorchado}@usal.es
[2] Department of Computer Science and Artificial Intelligence, CITIC-UGR,
University of Granada, 18071 Granada, Spain
pjvi@decsai.ugr.es

**Abstract.** In supervised learning, some data characteristics (e.g. presence of errors, overlapping degree, etc.) may negatively influence classifier performance. Many methods are designed to overcome the undesirable effects of the aforementioned issues. When comparing one of those techniques with existing ones, a proper selection of datasets must be made, based on how well each dataset reflects the characteristic being specifically addressed by the proposed algorithm. In this setting, statistical tests are necessary to check the significance of the differences found in the comparison of different methods. Wilcoxon's signed-ranks test is one of the most well-known statistical tests for pairwise comparisons between classifiers. However, it gives the same importance to every dataset, disregarding how representative each of them is in relation to the concrete issue addressed by the methods compared. This research proposes a hybrid approach which combines techniques of measurement for data characterization with statistical tests for decision making in data mining. Thus, each dataset is weighted according to its representativeness of the property of interest before using Wilcoxon's test. Our proposal has been successfully compared with the standard Wilcoxon's test in two scenarios related to the noisy data problem. As a result, this approach stands out properties of the algorithms easier, which may otherwise remain hidden if data characteristics are not considered in the comparison.

## 1 Introduction

Classification tasks aim to create a model, called a classifier, from labeled examples of the problem. The classifier is then used to predict the class label of new examples from the value of their attributes. Thus, the characteristics of the data used to build the classifier directly influence it, affecting its complexity and classification performance. Moreover, the presence of some undesirable properties

© Springer Nature Switzerland AG 2019
H. Pérez García et al. (Eds.): HAIS 2019, LNAI 11734, pp. 61–72, 2019.
https://doi.org/10.1007/978-3-030-29859-3_6

in the data, which are usually quantified using well-known metrics, may negatively affect classifier learning, causing problems broadly studied in the literature [6,9,12]. For instance, data that are characterized by an unequal distribution of the examples among the classes (which is usually measured using the *imbalance ratio*) results in the *imbalanced classification* problem [6]. Data that are characterized by a large quantity of unknown attribute values (which is usually measured as the percentage of these values in the data) results in the *missing values* problem in classification [9].

In order to build a classifier from data heavily characterized by these undesirable characteristics, many techniques have been proposed in the literature. In these works, new algorithms are typically compared against existing ones considering a set of datasets, which should be representative of the issues being addressed [14] (degree of imbalance between classes, presence of missing values, errors, etc.). In such a way, it is possible to analyze the effects of these properties on the classifiers built by all these methods. However, the selection of the datasets must be done carefully, since some of them may be much more representative of the problematic characteristic than others.

In the last decade, the analysis of results has been closely related to the usage of statistical tests [3,15], which are needed to confirm whether a new method provides a significant improvement with respect to existing ones. Among them, *Wilcoxon's signed-ranks* test [3] (hereafter called *Wilcoxon*'s test) is commonly used in the machine learning literature to perform pairwise comparisons. It is a non-parametric statistical test that, in the context of classification algorithms, compares the performance of two methods when they are applied to a set of problems to find differences between them. In this test, each problem (i.e. dataset) has the same relevance. However, the relevance of each dataset when determining the output of the test should be different, according to the degree to which a dataset reflects the characteristic of interest. Datasets in which the presence of the characteristic being addressed is stronger should have a stronger influence on the result of the test, as they are more representative of such characteristic.

This study proposes an hybrid system which weights each one of the datasets used in the comparison performed by *Wilcoxon*'s test according to its representativeness with respect to the characteristic under study. To the best of our knowledge, this approach has not been previously used in the context of pairwise statistical comparisons between classifiers. Weighting schemes do have been applied to statistical comparisons between multiple classifiers (not pairwise). *Quade*'s test [11] considers that some problems are more difficult than others, and proposes scaling each problem depending on the differences observed in the algorithms' performances. However, there are two main differences between *Quade*'s test and our proposal. First, our weighting scheme is for pairwise comparisons, whereas *Quade*'s test is aimed at detecting differences between more than two classifiers and requires the usage of post-hoc procedures to characterize these differences. Second, *Quade*'s test only allows a weighting scheme based on the performance of the classifiers, whereas our proposal can be used with the performance of the methods, but also with any other metric computable from the data, as explained in Sect. 3.

The novel approach proposed implies a modification of *Wilcoxon*'s test in order to include information about the relevance of each dataset in the comparison, giving more importance to those problems that are interesting from the point of view of the metric under study. When comparing supervised classification algorithms, experiments are usually conducted to show that one of the algorithms is better than some others. However, it is well-known that no algorithm can be better than any other for any condition (*no free lunch theorem* [17]). Thus, instead of this kind of comparisons trying to find the best overall algorithm, experiments should be conducted to show which characteristics of the problem (such as the class imbalance, missing values, noisy data, etc.) lead to a better or worse performance of a classification algorithm. The weighting scheme enables us to easily focus on the exploration of such conditions, which is recommended when proposing new methods [14]. For this reason, the modification of *Wilcoxon*'s test presented in this study is an interesting analysis tool that allows to include additional information about the characteristics of each classification problem in order to deal with the comparison of two classification algorithms and better identify the conditions that are most favorable for each of them.

To assess the feasibility of our proposal, *Wilcoxon*'s test and its weighted version will be experimentally compared in different scenarios in the framework of the *noisy data* in classification [12]–note that the proposal is not exclusively applicable to this task, but can be used in any comparison of classification methods. In them, different metrics that can be computed from the data will be studied and the performance of several classifiers will be compared using the unweighted and weighted *Wilcoxon*'s test. The output of both tests will be compared with findings already published in the literature about noisy data.

The rest of this research is organized as follows. Section 2 provides an overview of metrics to characterize datasets and the description of *Wilcoxon*'s test. Section 3 introduces the hybrid version of *Wilcoxon*'s test for classifier comparison, including data weighting. Then, Sect. 4 includes the comparison of the unweighted and weighted versions of *Wilcoxon*'s test. Finally, Sect. 5 points out some concluding remarks.

## 2   Background

### 2.1   Measuring Data Characteristics

Each dataset has particular characteristics that define it, such as its size, the generality of the data and the inter-relationships among the variables. These properties are generally quantified defining numerical metrics directly computable from the data, with the aim of increasing the knowledge about the problem and determining the best way to deal with the data.

Classification datasets are composed by examples that are described by several attributes (numerical or nominal) and a class label (always nominal). This formation enables one to compute any numerical measure that summarizes the full data or a concrete part of these, such as relative or absolute frequencies of concrete values, means and medians, correlations among variables and so on.

Among the simplest metrics that can be computed from a dataset are the number of examples, attributes and classes. These are traditionally used to quantify, in a simple way, data characteristics such as the size and complexity of a dataset. Thus, higher values of these metrics usually represent a higher size and complexity of the corresponding dataset, resulting in some of the most studied problems within the framework of classification tasks - a clear parallelism among the characteristics of the data, the metrics employed to quantify them and the type of problems studied can be established. For example, some works focus on datasets with a high number of attributes, a fact that results in the *curse of dimensionality* problem [1]. The presence of unknown attribute values (which can be quantified by the ratio of missing values) for some examples results in *missing values* problems [9]. *Big data* problems [2] are characterized, among other factors, by the presence of a high number of attributes and examples simultaneously. Datasets with a highly unequal number of examples in each class (quantified using the imbalance ratio) lead to *imbalanced data* problems [6].

Another recent trend proposes more sophisticated *data complexity measures* [5] to quantify more subtle characteristics of the data which are considered difficult in classification tasks, such as the overlapping among classes, their separability or the linearity of the decision boundaries. Among the metrics estimating the overlapping existing in a dataset, one can find the F1 metric, which computes the maximum *Fisher's discriminant ratio*; the F2 metric, which estimates the volume of the overlapping region; and the F3 metric, representing the maximum feature efficiency, which is the maximum fraction of examples within the overlapping region distinguishable with only one attribute. Class separability measures include the N2 metric, which is used to estimate whether the examples of the same class lie close in the feature space, and the N3 metric, which denotes how close the examples of different classes are. Other metrics, such as L3 or N4, are used to estimate the linearity of the decision boundaries.

Finally, other artificial mechanisms have been proposed in the literature to build synthetic datasets from the modification of real-world ones, controlling the presence of a particular characteristic of the data [9,12]. These schemes enable one to extract conclusions based on the properties of the data which are modified. Examples of these mechanisms are, for example, the introduction of errors into the data (resulting in *noisy data* problems [12]) or the introduction of missing values [9]. In all these cases, the metric to study usually represents the amount (commonly, a percentage) of the property modified in the data. In this research, we consider both the computation of data complexity metrics and the use of artificial mechanisms to modify the datasets in two different experiments related to the problem of noisy data in classification.

## 2.2  Wilcoxon's Signed-Ranks Test

*Wilcoxon's signed-ranks* test [3] is a simple, yet safe and robust, nonparametric procedure that aims to detect whether two related samples come from two different populations. When applied to classification algorithms comparison, it performs an statistical comparison between the performance of two techniques

$X$ and $Y$ when they are applied to a common set of $n$ problems. Let $x_1, \ldots, x_n$ and $y_1, \ldots, y_n$ be the performance results of $X$ and $Y$ in the $n$ datasets considered in the comparison. *Wilcoxon*'s test proceeds as follows:

1. Compute the difference between the performance results of $X$ and $Y$ for each one of the $n$ datasets, that is, $d_i = x_i - y_i, i = 1, \ldots, n$.
2. Rank the differences from the lowest value of $|d_i|$ (with $rank(d_i) = 1$) up to the highest one (with $rank(d_i) = n$), by increments of 1. If there are $t$ tied differences $|d_i|, \ldots, |d_{i+t}|$, they are sorted in any order, but the final rank of all of them is computed as the average rank $(rank(d_i) + \ldots + rank(d_{i+t}))/t$.
3. Let $R^+$ be the sum of ranks for the datasets in which $X$ outperforms $Y$ $(d_i > 0)$, and $R^-$ the sum of ranks for the opposite $(d_i < 0)$. Those $rank(d_i)$ | $d_i = 0$ are evenly splitted between $R^+$ and $R^-$ and, if there is an odd number of them, one is ignored –note that $R^+ + R^- = n \cdot (n+1)/2$:

$$R^+ = \sum_{d_i > 0} rank(d_i) + \frac{1}{2} \sum_{d_i = 0} rank(d_i) \qquad R^- = \sum_{d_i < 0} rank(d_i) + \frac{1}{2} \sum_{d_i = 0} rank(d_i)$$

4. Let $W = min(R^+, R^-)$ be the *Wilcoxon* statistic. If this value is less than or equal to the theoretical value of the specific distribution this statistic is known to follow (Table B.12 in [18]), the null hypothesis of mean equality is rejected. This fact implies that a given method outperforms the other one, with the corresponding $p$-value associated.

Note that *Wilcoxon*'s test is analogous to the paired $t$-test, but it does not require any parametric assumption. It is more sensitive; thus, higher differences $|d_i|$ have a greater influence than in the $t$-test when determining the final result of the comparison, which is probably desired, although the absolute magnitudes of these differences are ignored. This means that two differences $d_i, d_j$ can be equal, but we lose the information about the magnitudes of the performance values of the classifiers $x_i, y_i, x_j, y_j$ they come from. This fact is one of the reasons that motivates the proposal of a weighting scheme for the data considered.

## 3 Hybridizing Wilcoxon's Signed-Rank Test with Dataset Weighting Based on Data Characteristics

When applying *Wilcoxon*'s test to compare two methods over a set of problems, only the performance results of each algorithm in each dataset are considered. The intrinsic properties of these datasets are not usually taken into account in the statistical comparison, and each one of the problems receives the same importance for determining the result of the test. As mentioned previously, sometimes we are interested in comparing methods which explicitly address a specific problematic characteristic of the data, such as noise, class imbalance, missing values, etc. However, the datasets considered may present this characteristic to different degrees, which can be quantified by an appropriate metric. Consequently, each dataset should have a different influence on the test result.

Our novel hybrid proposal in this scenario is to weight each one of the datasets of the comparison depending on its relevance with respect to the characteristic being studied when applying *Wilcoxon*'s test: the higher the value of the metric in the dataset, the higher the weight, that is, the importance of the dataset when determining the test result. Therefore, the test conclusions are based on the representativeness of each dataset with respect to the metric used.

The weighting proposed for *Wilcoxon*'s test is based on the following steps:

1. **Computation of the metric to study on each dataset.** A metric of interest is computed over each one of the $n$ datasets considered in the comparison of the methods $X$ and $Y$, resulting in the values $m_i$, $i = 1, \ldots, n$.
2. **Computation of the weight associated to each dataset.** The values $m_i$ are normalized to obtain a weight $w_i$ for each dataset as follows:

$$w_i = m_i / \sum_{j=1}^{n} m_j \qquad (1)$$

Note that the aforementioned equation is used to maximize the weight of those datasets with higher values $m_i$. In case we want to maximize the weight of the datasets having lower values, the equation of the weights is $w_i = (M - m_i)/\sum_{j=1}^{n} (M - m_j)$, being $M = \max_i\{m_i\}$.

Due to the large variety of metrics that can be computed from the data, this paper proposes the calculation of weights linearly distributed across the domain of the metric (as shown by Eq. 1). However, other schemes are also possible, such as logarithmic or exponential. The choice of a concrete scheme should be made by the data analyst based on the problems involved in the comparison and the distribution of values $m_i$. In any case, it is important to remark that we do not manipulate the weight of each dataset arbitrarily, since the weights are determined by the values $m_i$ of the metric.

3. **Modification of the performance values of the methods to compare.** The performance results $x_i$ and $y_i$ of the two algorithms $X$ and $Y$ involved in the comparison are replaced by their weighted versions $x_i^w$ and $y_i^w$ as follows:

$$x_i^w = x_i \cdot w_i \qquad\qquad y_i^w = y_i \cdot w_i$$

4. **Application of *Wilcoxon*'s test over the weighted results.** Finally, *Wilcoxon*'s test is applied to the weighted data $x_i^w$ and $y_i^w$, $i = 1, \ldots, n$.

Note this weighting scheme is not a new statistical test; it requires a modification of the original data samples based on the relevance of each dataset with respect to the characteristic of interest. It neither intends to be a substitute of *Wilcoxon*'s test, but a complementary tool to be used when analyzing the results obtained in experiments. Even though a classifier should show differences against another one when using unweighted *Wilcoxon*'s test, the use of its weighted version enhances this issue. Unlike the unweighted *Wilcoxon*'s test, the weighting can capture the relevance of those data more representative with respect to the property of interest. Our novel proposal modifies (considering the property of the data) the differences in performance on the two methods, which determine the rankings assigned to each method and, thus, the final test result.

Although in some works the only goal of new proposals is to improve the classification performance of existing methods (without focusing on particular data characteristics), the use of the weighting scheme for statistical comparisons could be also interesting. Several metrics could be computed from the data and applied to the statistical comparisons to obtain another batch of results. With them, we could check whether giving more importance to some property of the data makes a particular algorithm stand out. Thus, the weighting approach can be useful to uncover properties of the algorithms, that is, their better behavior in datasets presenting a particular characteristic to a high degree, even when they were not specifically designed for dealing with this type of data.

Finally, even if we are not studying any specific characteristic of the data, a relevance metric based on the accuracy itself can still be defined. The weighting procedure can be applied to assign more importance to those datasets that are intrinsically more difficult for constructing a model (and therefore every classifier performs poorly on them in terms of accuracy), since those are more interesting from a classification point of view. This would partially remedy the fact that *Wilcoxon's signed-ranks* test does not consider absolute magnitudes of the samples to reach the final result, which in our scenarios may often represent an omission of valuable information. In these cases, datasets could be weighted, for example, depending on the intrinsic difficulty of the problem to be learnt by any classifier, which can be quantified, for instance, as the minimum theoretical error achievable in the dataset by the *Bayes* classifier. Other option could be the minimum error obtained experimentally by any known classifier in that dataset.

## 4 Combining Data Weighting and Wilcoxon's Test in the Framework of Noisy Data in Classification

In this section we assess the differences of applying *Wilcoxon*'s test with or without the weighting scheme proposed in Sect. 3. Two scenarios are studied focusing on a common problem in classification, namely, the existence of errors or *noise* in the data [12]. These scenarios are not aimed at showing that the weighted version allows to reach statistically significant results, neither if it is better or worse than the unweighted version, but showing that giving more importance to more representative datasets for the comparison of algorithms leads to different conclusions that are closer to those claimed in the literature.

### 4.1 Robustness of Different Classifiers to Noise

This scenario considers the results of two classification algorithms, C4.5 [8] and SVM [16], over a collection of 30 datasets taken from the UCI repository [4], which are affected by different amounts of noise (added artificially).

These algorithms are known to behave differently dealing with noisy data [10]. On the one hand, C4.5 is a well-known robust learner, which is less affected by the presence of noise thanks to the pruning mechanisms which reduce the chances that the trees overfit the data [8]. On the other hand, SVM usually obtains better results when data have little or no noise at all, but performs poorly in presence of severe noise [10].

Ideally, *Wilcoxon*'s test without weights should be able to find statistically significant differences in this setting, as already noted in existing literature. Furthermore, since we are focusing on noisy data, the amount of noise present in each dataset should be considered when drawing a conclusion.

The performance of the methods is calculated as the average of the accuracy in test sets of five independent runs of 5-fold stratified cross validation (SCV).

To control the presence of noise, different noise levels $l\%$ are introduced into each training set. To this end, the *uniform class noise scheme* [12] is used, in which $l\%$ of the examples are corrupted randomly changing their class labels by other randomly chosen one. For each one of the 30 datasets used we introduce a different random noise level $l \in [5, 50]\%$, in steps of 5%. The procedure to introduce a noise level $l\%$ in the original dataset is the following:

1. A noise level $l\%$ is introduced into a copy of the full original dataset.
2. The original dataset and the noisy copy are partitioned into 5 folds, maintaining the same examples in each one.
3. The training sets are built from the noisy copy, whereas the test sets are built from the original dataset.

The algorithms have been executed with the following configurations:

- C4.5: *confidence* = 0.25, *instances per leaf* = 2, *prune* after the tree building.
- SVM: $C = 100$, *tolerance* = 0.001, $\epsilon = 10^{-12}$, *kernel* = PUK ($\sigma = 1$, $\omega = 1$).

Table 1 shows the datasets considered in the experiments, their noise levels and the performance of C4.5 and SVM. The usage of unweighted *Wilcoxon*'s test over the results shown above, comparing C4.5 versus SVM in the 30 datasets, provides a total sum of ranks $R^+ = 289$ in favor of C4.5 and a sum of ranks $R^- = 176$ for SVM. These results show that C4.5 presents, as expected, a better behavior than SVM dealing with this type of data (since it obtains a sum of ranks higher than that of SVM). However, this difference is not statistically significant to a significance level of 0.05 since the $p$-value obtained is 0.241.

In order to include information about the noise level of each dataset, the weighting scheme of Eq. 1 is used. In this case, we want to give more importance to those datasets with higher amount of noise when determining the outcome of the test. The weight $w_i$ in this problem is computed by dividing each noise level $l_i$ in Table 1 by the sum of noise levels, $S = 10 + 25 + \ldots + 15 + 15 = 635$. The weighted version of *Wilcoxon*'s test provides $R^+ = 333$ for C4.5 and $R^- = 132$ for SVM, with an associated $p$-value = 0.038420.

As these results show (Table 2), C4.5 obtains a higher sum of ranks than SVM when using the weighting scheme (like in the first comparison without using weights), but in this case the $p$-value obtained is much lower and now determines statistical significance, which is consistent with the results published in the literature [10]. Thus, by weighting each dataset based on the noise level, we were able to uncover a particular property of C4.5 versus SVM, that is, a better behavior when dealing with datasets affected by severe noise.

**Table 1.** Datasets, noise levels and accuracy of C4.5 and SVM.

| Dataset | Noise | C4.5 | SVM | Dataset | Noise | C4.5 | SVM | Dataset | Noise | C4.5 | SVM |
|---|---|---|---|---|---|---|---|---|---|---|---|
| autos | 10 | 73.56 | 64.5 | heart | 40 | 70.81 | 69.33 | satimage | 5 | 85.4 | 89.25 |
| banana | 25 | 88.03 | 90.04 | ionosp. | 40 | 80 | 75.05 | segment | 20 | 93.09 | 87.77 |
| cleveland | 20 | 49.96 | 36.82 | iris | 25 | 90.27 | 86.8 | sonar | 20 | 68.76 | 83.46 |
| contracep. | 10 | 50.52 | 46.8 | led7digit | 35 | 68.72 | 66.88 | twonorm | 5 | 84.61 | 96.96 |
| dermat. | 5 | 93.07 | 96.53 | lymph. | 10 | 77.31 | 80.96 | vehicle | 45 | 52.84 | 53.12 |
| ecoli | 15 | 77.73 | 63.28 | magic | 45 | 83.36 | 83.73 | vowel | 10 | 74.91 | 87.11 |
| flare | 30 | 73.81 | 70.71 | pageblocks | 30 | 95.59 | 95.77 | wdbc | 25 | 88.61 | 85.77 |
| german | 30 | 69.68 | 64.46 | penbased | 20 | 93.28 | 84.57 | wine | 5 | 89.55 | 94.93 |
| glass | 5 | 67.28 | 67.75 | phoneme | 10 | 84.76 | 86.66 | yeast | 15 | 51.97 | 54.3 |
| hayes | 15 | 81.24 | 73.15 | pima | 50 | 70.94 | 60.21 | zoo | 15 | 92.26 | 72.85 |

**Table 2.** *Wilcoxon*'s test with and without weights using the results of Table 1.

| Wilcoxon | C4.5 ($R^+$) | SVM ($R^-$) | $p$-value |
|---|---|---|---|
| Unweighted | 289 | 176 | 0.241038 |
| Weighted | 333 | 132 | 0.038420 |

## 4.2 Noise Filtering Efficacy

This scenario focuses on using *noise filters* in classification problems [12]. These are preprocessing methods to identify and remove noisy data before building a classifier. The removal of noisy examples has shown to be beneficial in many cases, improving the performance of the classifiers used later [12]. However, examples containing valuable information may also be removed, which implies that filters do not always provide an improvement in performance [13].

The work of Sáez et al. [13] show that the efficacy of noise filters, i.e., whether their usage causes an improvement in classifier performance, is somehow related to the characteristics of the data. The authors show that the overlapping among the classes, measured with the F2 metric, is important to determine whether filters will improve classifier performance. Thus, when the amount of overlapping is high enough, filters usually improve classifier performance. The F2 metric computes the volume of the overlapping region among the examples of two different classes $C_1$ and $C_2$, by means of the following equation:

$$F2 = \prod_{i=1}^{d} \frac{minmax_i - maxmin_i}{max(f_i, C_1 \cup C_2) - min(f_i, C_1 \cup C_2)}, \qquad (2)$$

being $d$ the number of attributes, $max(f_i, C_j)$ and $min(f_i, C_j)$ the maximum and minimum values of the feature $f_i$ in the set of examples of class $C_j$, $minmax_i$ the minimum of $max(f_i, C_j)$ and $maxmin_i$ the maximum of $min(f_i, C_j)$.

In this case, we compare a well-known noise filter, *Edited Nearest Neighbor* (ENN) [7], versus not considering any preprocessing of the data. As in [13], the classifier used in both cases is the *Nearest Neighbor* (NN) rule [8], known to be sensitive to noise. ENN is run considering $k = 3$, and both ENN and NN consider the HVDM distance, which is valid for nominal and numerical attributes.

These two methods are compared over 20 datasets taken from the UCI repository [4]. The performance is measured using AUC, which is an evaluation metric less sensitive to class imbalance, an issue posed by some of the data used. The AUC shown in Table 3 is the average of 5 independent runs of a 5-fold SCV, when no preprocessing is done before NN (None) and when ENN is used prior to NN.

**Table 3.** Datasets, F2 metric and performance of ENN and None.

| Dataset | F2 | None | ENN | Dataset | F2 | None | ENN |
|---|---|---|---|---|---|---|---|
| appendicitis | 4.50E-02 | 0.7551 | 0.7511 | monk-2 | 6.67E-01 | 0.7531 | 0.7601 |
| australian | 3.00E-03 | 0.8236 | 0.8120 | phoneme | 2.71E-01 | 0.8683 | 0.8440 |
| banana | 6.26E-01 | 0.8710 | 0.8917 | pima | 2.52E-01 | 0.6487 | 0.6601 |
| breast | 1.88E-01 | 0.5570 | 0.5978 | sonar | 1.00E-06 | 0.8614 | 0.7964 |
| bupa | 7.30E-02 | 0.6218 | 0.6116 | spambase | 2.53E-33 | 0.8965 | 0.8749 |
| crx | 3.00E-03 | 0.8233 | 0.8172 | spectfheart | 3.60E-19 | 0.6299 | 0.6203 |
| haberman | 7.18E-01 | 0.5519 | 0.5658 | tic-tac-toe | 1.00E+00 | 0.9088 | 0.8970 |
| heart | 1.96E-01 | 0.7663 | 0.8043 | twonorm | 4.12E-03 | 0.9424 | 0.9518 |
| housevotes | 1.00E+00 | 0.9484 | 0.9550 | wdbc | 5.90E-11 | 0.9507 | 0.9469 |
| mammograp. | 7.44E-01 | 0.7494 | 0.7968 | wisconsin | 2.17E-01 | 0.9547 | 0.9691 |

The results of the tests are summarized in Table 4. Regarding the unweighted version, the test slightly favors ENN as it has a larger $R^- = 117$ than None, $R^+ = 93$, but no statistically significant differences are found ($p$-value $= 0.640744$). To sum up, no interesting conclusions can be drawn about the efficacy of ENN when the only information is the AUC.

When the data are weighted by their F2 metric, the output of the test is the same, but it is closer to statistical significance. Thanks to such additional information, ENN is now clearly favored, with $R^- = 141$ versus a much smaller value $R^+ = 69$ obtained when no preprocessing is done. The $p$-value corresponding to this comparison is 0.189340, closer to the significance threshold.

This fact shows that, giving more importance to datasets with higher degrees of overlapping, the preprocessing can work better than not considering it, which is in concordance with the results claimed in [13]. This conclusion can be drawn from the great decrease in the $p$-value when the weighted version of *Wilcoxon's*

**Table 4.** *Wilcoxon*'s test with and without weights using the results of Table 3.

| *Wilcoxon* | None ($R^+$) | ENN ($R^-$) | $p$-value |
|---|---|---|---|
| Unweighted | 93 | 117 | 0.640744 |
| Weighted | 69 | 141 | 0.189340 |

test is employed instead of the conventional one, even though the new $p$-value is still larger than the significance threshold. It is important to note that the conclusions claimed in [13] on the data properties that determine in which cases the filtering is statistically beneficial are based on a combination of several data complexity metrics considered simultaneously, among which F2 is included. This fact can explain why the weighting scheme does not show significant differences in the comparison, since we only consider the isolated metric F2 in our study.

## 5    Concluding Remarks

This research proposes a hybrid approach to weight data before using *Wilcoxon*'s test and give more or less importance to the different data in a comparison. The weights of the datasets are computed using characteristics of the datasets used in the comparison. The conclusions reached by the statistical test consider the property that determines the weighting, which constitutes additional information not exploited by the unweighted version of *Wilcoxon*'s test.

We have evaluated our proposal in two scenarios related to the problem of noisy data. In the first scenario, we have compared the C4.5 robust learner and the noise-sensitive SVM classifier when they are trained over data with different noise levels. The results revealed that the weighting scheme based on the noise ratio of each dataset leads to statistically significant differences that the unweighted *Wilcoxon*'s test could not find. Such differences support the claims done in the existing literature about the superiority of C4.5 over SVM on noisy data, particularly when the amount of noise is high enough [10].

In the second scenario, we have compared the efficacy of the ENN filter versus not-preprocessing. In the literature, it is claimed that noise filters are usually useful when the overlapping among the classes is noticeable. Neither of the tests were able to detect such differences, but the weighted version showed a clear advantage towards the use of ENN, supported by a large decrease of the $p$-value. This can be considered additional information which the unweighted version could not uncover, in accordance with the existing literature [13].

As a final note, it is clear that the information returned by the weighted *Wilcoxon*'s test is a revenue for the weights we have computed at the input of the test, but this constitutes a desirable approach: we are orienting our analysis and conclusions to be based on the properties of the data we are interested in.

**Acknowledgment.** José A. Sáez holds a *Juan de la Cierva-formación* fellowship (*Ref. FJCI-2015-25547*) from the Spanish Ministry of Economy, Industry and Competitiveness.

# References

1. Bach, F.: Breaking the curse of dimensionality with convex neural networks. J. Mach. Learn. Res. **18**, 1–53 (2017)
2. Bello-Orgaz, G., Jung, J., Camacho, D.: Social big data: recent achievements and new challenges. Inf. Fusion **28**, 45–59 (2016)
3. Demšar, J.: Statistical comparisons of classifiers over multiple data sets. J. Mach. Learn. Res. **7**, 1–30 (2006)
4. Dua, D., Karra Taniskidou, E.: UCI machine learning repository (2017). http://archive.ics.uci.edu/ml
5. Jain, S., Shukla, S., Wadhvani, R.: Dynamic selection of normalization techniques using data complexity measures. Expert Syst. Appl. **106**, 252–262 (2018)
6. Khalilpour Darzi, M., Niaki, S., Khedmati, M.: Binary classification of imbalanced datasets: the case of coil challenge 2000. Expert Syst. Appl. **128**, 169–186 (2019)
7. Kuncheva, L., Galar, M.: Theoretical and empirical criteria for the edited nearest neighbour classifier, vol. January, pp. 817–822 (2016)
8. Larose, D.T., Larose, C.D.: Data Mining and Predictive Analytics, 2nd edn. Wiley Publishing, Hoboken (2015)
9. Luengo, J., García, S., Herrera, F.: A study on the use of imputation methods for experimentation with radial basis function network classifiers handling missing attribute values: the good synergy between RBFs and eventcovering method. Neural Networks **23**(3), 406–418 (2010)
10. Nettleton, D., Orriols-Puig, A., Fornells, A.: A study of the effect of different types of noise on the precision of supervised learning techniques. Artif. Intell. Rev. **33**, 275–306 (2010)
11. Quade, D.: Using weighted rankings in the analysis of complete blocks with additive block effects. J. Am. Stat. Assoc. **74**, 680–683 (1979)
12. Sáez, J.A., Galar, M., Luengo, J., Herrera, F.: INFFC: an iterative class noise filter based on the fusion of classifiers with noise sensitivity control. Inf. Fusion **27**, 19–32 (2016)
13. Sáez, J.A., Luengo, J., Herrera, F.: Predicting noise filtering efficacy with data complexity measures for nearest neighbor classification. Pattern Recogn. **46**(1), 355–364 (2013)
14. Santafe, G., Inza, I., Lozano, J.: Dealing with the evaluation of supervised classification algorithms. Artif. Intell. Rev. **44**(4), 467–508 (2015)
15. Singh, P., Sarkar, R., Nasipuri, M.: Significance of non-parametric statistical tests for comparison of classifiers over multiple datasets. Int. J. Comput. Sci. Math. **7**(5), 410–442 (2016)
16. Vapnik, V.: Statistical Learning Theory. Wiley, New York (1998)
17. Wolpert, D.H., Macready, W.G.: No free lunch theorems for optimization. IEEE Trans. Evol. Comput. **1**(1), 67–82 (1997)
18. Zar, J.: Biostatistical Analysis. Prentice Hall, Upper Saddle River (2009)

# Mining Network Motif Discovery by Learning Techniques

Bogdan-Eduard-Mădălin Mursa(✉) [iD], Anca Andreica [iD], and Laura Dioşan [iD]

Faculty of Mathematics and Computer Science, Babeş-Bolyai University,
Cluj Napoca, Romania
{bmursa,anca,lauras}@cs.ubbcluj.ro

**Abstract.** Properties of complex networks represent a powerful set of tools that can be used to study the complex behaviour of these systems of interconnections. They can vary from properties represented as simplistic metrics (number of edges and nodes) to properties that reflect complex information of the connection between entities part of the network (assortativity degree, density or clustering coefficient). Such a topological property that has valuable implications on the study of the networks dynamics are network motifs - patterns of interconnections found in real-world networks. Knowing that one of the biggest issue with network motifs discovery is its algorithmic NP-complete nature, this paper intends to present a method to detect if a network is prone or not to generate motifs by making use of its topological properties while training various classification models. This approach wants to serve as a time saving pre-processing step for the state-of-the-art solutions used to detect motifs in Complex networks.

**Keywords:** Complex networks · Network motifs · Network topology · Classification

## 1 Introduction

From micro to macro, during our lifetime we get in contact with signals emitted by an immeasurable number of different sources. All these signals and the entities they interact with, can be represented as topological data structures named Complex Networks. Being part of various systems that affect us directly or indirectly, we are in a continuous modeling during our lifetime. Metaphorically we can say that every human being is a product of the interconnections' fingerprints from an infinite Complex Network - our Universe.

The interconnections found in a network are one of the structural key component in understanding individual behaviour of network's entities or the behaviour of the network by itself. Complex Network and Graph Theory present various topological properties with a focus on the study of entities interconnections from which we can remind: assortativity degree [16], density [18], local global clustering coefficient [22], network motifs [12] and many others.

© Springer Nature Switzerland AG 2019
H. Pérez García et al. (Eds.): HAIS 2019, LNAI 11734, pp. 73–84, 2019.
https://doi.org/10.1007/978-3-030-29859-3_7

The focus of this paper will be on the property named network motifs, which are patterns of interconnection which appear in a higher frequency in real-world networks, than in random networks with the same topology as the original network. Motifs can be observed in various natural systems from fields like biology (ecosystem relationships of predator-prey or symbiosis), sociology (social networks [8, 20]), cosmology (cosmic webs [4]).

From a topological perspective, networks motifs are frequent patterns of nodes and interconnections with a contextualised cause of their appearance, which context is given by the real-world origin of the network. In order to confirm that a given configuration of nodes and interconnection is indeed a motif, the current state-of-the-art presents a series of algorithms that follow a common paradigm which unfortunately is ranked as a NP-complete problem [7]. While implemented in a real-world application, all proposed solutions reach time or memory bottlenecks as long as the size of the motifs grows.

Although given its NP-complete nature, the problem was not solved yet by a perfect algorithm, the literature presents various approaches to partially improve the time or memory bottlenecks. Part of the algorithms are using a statistical approach that process only samples of the original network, but having as a trade-off an approximate result. The most recent trend is to make use of computational power existing nowadays, while re-modelling existing algorithms as parallel solutions able to run on cluster of computers [13].

Encouraged by the need of more solutions that could improve the time spent while analysing networks that are suitable to contain network motifs, this paper's goal is a method of deciding if a network with a given topological configuration is prone to contain network motifs or not. While making use of our previous studies that revealed important statistical correlation between network motifs and other network properties from which we can remind the ones with the strongest correlation: articulation points, local clustering coefficient or assortativity degree, our goal is to train a series of classification models that can be further used to predict the existence of motifs in a network. A such solution could drastically cut the time spent by researches while analysing networks, while using this solution as a pre-processing step in their process, approach which from our knowledge is not currently presented by literature.

The structure of the paper contains a series of sections that are willing to introduce the reader into the context of Complex Network theory while presenting the most relevant topological properties that will be later used into the classification training in Sect. 2, followed by a theoretical description of the classification methods that will be used for modeling (Sect. 3). In Sect. 4 it will be presented the data set used in the analysis and the steps followed while training a model. This section will also contain the final results accompanied by a discussion over the comparison between these results. The final section of the paper is dedicated to the conclusions which will reiterate over the presented results and will describe future plans of research.

## 2    Complex Networks Theoretical Insights

This section will present a series of properties from Complex Network Theory that will be used in the following analysis sections. The reader will have the chance to understand the theoretical insights of these properties, and how they vary from simple concepts to more complicated ones which can have a deep impact in the appearance of motifs in the topological space of a network (Table 1).

**Table 1.** Studied topological properties

| Property | Annotation | Property | Annotation |
|---|---|---|---|
| Articulation point | $AP$ | k-core | $k_{core}$ |
| Assortivity degree | $\rho$ | Local clustering coefficient | $CC_L$ |
| Average degree | $d_{avg}$ | Maximum degree | $d_{max}$ |
| Average triangles | $tr_{avg}$ | Maximum triangles number | $tr_{max}$ |
| Density | $D$ | Minimum degree | $d_{min}$ |
| Relative edge distribution entropy | $H_{er}$ | Number of nodes | $N$ |
| Number of edges | $E$ | Power law exponent | $\gamma$ |
| Gini coefficient | $g$ | Triangles number | $tr_{no}$ |
| Global clustering coefficient | $CC_G$ | Network motif | $M$ |

The following subsections represent a more in-depth description of network motifs and properties that can be considered favourable measurements in correlation with the existence of network motifs. The reason of choosing this particular subset of properties to explicitly define the network motifs and describe their application in our modelling problem is set out in our previous studies [14,15], which revealed a high correlation factor between the properties and the appearance of network motifs. Building on the results offered by this previous research, the following subsections will present theoretical descriptions of the properties and discuss why and how these topological measurements may be good indicators of networks prone to contain motifs, even before applying motifs discovery algorithms.

### 2.1    Network Motifs

Although at the first glance the systems that surround us tend to have a chaotic behaviour, human mind is designed to find patterns in the large set of observable connections as a way to abstract the complex dynamics of the systems it is part of. While learning from these patterns the human mind can easily adapt to new similar stimuli that might appear.

These patterns tend to appear in lots of domains, where various system can be modelled as complex networks: from clusters of astronomic bodies named galaxies to micro interactions between proteins (Fig. 1).

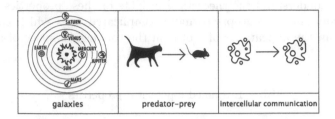

| galaxies | predator–prey | intercellular communication |

**Fig. 1.** Network motifs examples

Initially studied in a biological context [19], one of the first network motifs definition was that they represent patterns of interconnections that appear in a higher concentration in real-world networks than in randomly generated networks with the same topology as the original network [12]. A motif can be associated with a functional behaviour of the context it is part of, function which can be an important factor in system's controllability (E.G: a motif that is a structure of with a functional behaviour of predator-prey).

As topological structures, network motifs represent sub-components of a given size $N$ from the original network ($N$ being the number of nodes from a motif). Figure 2 shows all possible motifs of size $N = 3$ from an undirected network.

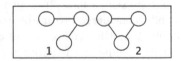

**Fig. 2.** Network motifs of size 3 examples

The state-of-the art presents various algorithms [6, 9, 10, 12, 17, 23] for network motifs detection following the next steps:

– enumerate all sub-components of a given size $N$
– clusterise the enumerated sub-components using isomorphic principles

In order to gain performance improvements, part of the existing algorithms [17, 23] perform a statistical trick while executing the steps described above with a trade-off in the accuracy of the results obtained. These algorithms do not apply the steps on the entire network, but on an arbitrary sample of a given size from the original network such that less computations are required. As the sample size gets bigger, the accuracy of the results will be better.

To consider a sub-component $G'$ to be a network motif, firstly it is required to compute its concentration in the initial network $G$ ($F_G(G')$). If this concentration is significantly higher than in $n$ random networks ($R(G)$) that hold the same degree distribution as the original network, then we can consider the sub-component $G'$ to be a network motif. Performing a statistical approach to compute the significance of a motif, it is defined the Z-score function (1) which result is directly proportional with the significance of a network motif.

$$Z(G') = \frac{(F_G(G') - \mu_R(G'))}{\sigma_R(G')} \tag{1}$$

where $\mu_R(G')$ and $\sigma_R(G')$ are the mean and standard deviation frequency in set $R(G)$ [12].

## 2.2 Clustering Coefficient

The topological space created when a system is modelled as a complex network, encodes properties that studied can quantify real-world contextual behaviours of these systems. Such example are social networks on which their members tend to follow specific behaviours studied by fields like sociology. Societies, religions or communities are the result of a social network behaviours where theirs members tend to clusterise themselves into groups that share the same goals.

Topologically it is possible to quantity these behaviours by making use of different measurements like global clustering coefficient ($CC_G$) and local clustering coefficient ($CC_L$) [22]. Although both measurements describe a clustering behaviour of the network, the difference between them is in the level of localisation where they are focused. $CC_L$ is computed while considering how close are the neighbors of a node to create a clique together, hence a local behaviour, when $CC_G$ is studying the network as whole by computing a ratio between number of closed triangles and both open and closed triangles.

Intuitively and empirically observed, the patterns of interconnections tend to appear in the topology of the networks with clusters that can evolve specific behaviours among the nodes from such a cluster. As it was proven in a our previous research [14], a network with noticeable values for $CC_G$ and especially $CC_L$ are prone to evolve motifs.

## 2.3 Assortativity Degree

Another measurement that reveals a natural behaviour of the networks is assortativity degree $\rho$ [16] which quantifies if the nodes tend to have neighbors with the same number of connections (degree). While performing a real-world analogy, we can consider a member of a society popular if it has a considerably higher number of connections than the average. These members of societies will indeed tend to surround themselves with other members that have the same level of popularity, hence they are prone to evolve specific behaviours in their cluster that can be encoded in network's topology as network motifs.

## 2.4   Relative Edge Distribution Entropy

A measurement that denotes the uniformity of the connections between nodes is the relative edge distribution entropy $H_{er}$ [11]. Although in general an equal distribution would be ideally, such a topology would be too tedious to favour the appearance of motifs, hence we are presuming that networks with a lower value of $H_{er}$ are more probably to contain motifs.

## 2.5   Articulation Points

Some nodes of a complex network can have a more significant importance while considering different factors. One of them can be the number of connections that a given node have, which can says if a node is a popular or not (Subsect. 2.3).

Similar to popular nodes, a topology can have the so called articulation points ($AP$), which are nodes that removed from network's topologies split the original network into two or more connected components (Fig. 3).

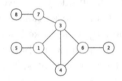

**Fig. 3.** Example of articulation points (in red) (Color figure online)

We can see these nodes as communication bridges between two or more isolated communities. The isolation factors gives to articulation points a great power of controllability of the behaviours of these isolated communities. One of our previous work [15] stated and proved a hypothesis for a higher frequency of network motifs in the proximity of the articulation points than to any other part of a network topology. This can be explained by the controllability factor mentioned earlier, than can be used by the articulation point to inject ideologies to its followers.

## 2.6   Density

A property that reveals the complexity of the topology interconnections is density ($D$) [18] which measures how close is a network to reach the maximal number of edges. Density might not be a relevant property in the topological modeling for network motifs, but together with other properties it should reflect a more visible context of the maturity of the topology.

## 2.7  Gini Coefficient

Gini coefficient $g$ is a measure widely applied in the analysis of distributions. In the context of complex networks, $g$ is used to compute the distribution of the edges among the nodes with values between 0 and 1, a 0 value being a perfect distribution of the edges among the nodes and 1 opposite of a perfect distribution. Frankly it is unlikely to find balanced networks in real-world systems, which means a better chance for the network participants to evolve specific behaviours materialised as patterns of interconnections.

## 2.8  Power Law Exponent

Properties that indicate preferential attachment between nodes receive a great interest as they are candidates that might explain the dynamics of different topologies modelled from real-world systems. A such example are scale-free networks [1] which are networks that hold a very specific degree distribution between nodes that is a correlation between network's nodes with a given degree $d$ and a power $d^{-\gamma}$, where $\gamma$ is the power law exponent. Explained in a more natural way, scale-free networks are networks with a Paretto distribution of the node degrees, where large portion of degrees is held by a small fraction of the nodes. Following a similar analytic approach as for the other properties, we can say that rare nodes with a large fraction of connection can also favour the appearance the motifs on their proximity.

# 3  Classification Methods

Our approach is one of using a set of independent variables, which are the properties presented in Sect. 2, to predict if a network contains network motifs or not (dependent variable).

Following the "no free lunch" theorem for supervised learning which states that there is no perfect algorithm for classification or regression problems, only models that perform better in specific circumstances [5], our goal is to test most promising state-of-the-art statistical models for our classification problem. These models will be evaluated by using the same test data set, such that in the end it would be possible a fair comparison between their results.

Our analysis will be performed while using 4 different statistical classification techniques, different by their paradigm: Gradient boosting (XGBoost) as boosting algorithm [3], Decision Tree Classifier and Random Forest Classifier as tree-based models and last but not least SVM.

Knowing that each of these models are known to perform better under specific data distribution, in the next section it will be described the analysis approach it was applied starting from pre-processing steps of data normalisation and model training to the validation of the results obtained while performing a comparison of their performance.

# 4   Classification Modelling

## 4.1   Data

While performing the classification modelling, we used a consistent data set of undirected networks of different sizes coming from various fields. For each of these networks we extracted 17 topological properties that will be considered independent variables in the classification training that was performed - those properties considered to be the most important from a correlation factor perspective in our previous research [14,15] described in depth in Sect. 2. For each network, the frequency of motifs of size $N = 3$ was already extracted. Table 2 shows a numerical description of the networks used, by considering the mentioned properties.

**Table 2.** Average values for the studied topological properties

| Property | Mean | Property | Mean |
|---|---|---|---|
| $AP$ | 340.85 | $k_{core}$ | 14.39 |
| $\rho$ | 0.19 | $CC_L$ | 0.3 |
| $d_{avg}$ | 16.8 | $d_{max}$ | 117.86 |
| $tr_{avg}$ | 372.2 | $tr_{max}$ | 611.79 |
| $D$ | 0.05 | $d_{min}$ | 8.85 |
| $H_{er}$ | 0.9 | $N$ | 5047.73 |
| $E$ | 19371.89 | $\gamma$ | 3.72 |
| $g$ | 0.22 | $tr_{no}$ | 143000.86 |
| $CC_G$ | 0.28 | | |

The training set was labeled with two possible outcomes, 1 being for networks that contain at least one motif in their topology and 0 being for network missing motifs. A question that may derive from the applied labeling is why it was not considered the frequency of motifs also during this process. Our approach does not consider the frequency of motifs as being relevant in this case, as by its definition (Subsect. 2.1) a topology sub-component that is considered to be a motif it is already an important validation.

From a total of 386 networks, 184 have at least one motif in their structure, hence we can consider the data set as being a balanced one for the intended binary classification training. Although networks which contain motifs are rare in a real-world context, in line with literature studies that indicate a possible biased accuracy between unbalanced classes [21] we kept a balance between the classes found in the training set using manual intervention as a first iteration and use complex fitness functions to evolve unbiased classifiers [2] if the training set became unbalanced in the next iteration of the model.

## 4.2 Model Training and Results Comparison

Knowing that properties range scale is different, before training the classifiers it was applied a pre-processing step to normalise the data using a scaler robust to outliers.

The training set, respectively testing data set was obtained by splitting the initial data using a 80/20 proportion.

The models were trained using *sklearn* Python package and there were not required any adjustments to the configurations of their default implementation.

**Table 3.** Validation test results

|          | XGBoost | Decision tree classifier | Random forest classifier | SVM  |
|----------|---------|--------------------------|--------------------------|------|
| Accuracy | **0.95** | 0.88                   | 0.91                     | 0.72 |
| Precision | **0.92** | 0.84                  | 0.89                     | 0.67 |
| Recall   | **0.97** | 0.92                    | 0.92                     | 0.78 |
| F1-score | **0.95** | 0.88                    | 0.91                     | 0.72 |

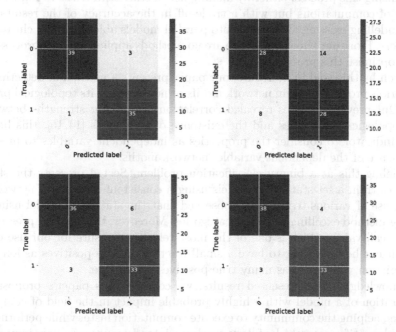

**Fig. 4.** Quality evaluation of the trained model using confusion matrices (Top-Left: XGBoost, Top-Right: SVM, Bottom-Left: Decision Tree Classifier, Bottom-Right: Random Forest Classifier). 0 label - network without motifs, 1 label - network with motifs

The validation tests revealed that the boosting method is the most qualitative one in each testing categories (Table 3). By doing a deeper interpretation analysis of these values, our use case is to look for models that are that are able to predict as many true positives as possible, even if there is a small fraction of false positives in the prediction (Fig. 4). The reasoning behind this expectation is that using in a pre-processing phase a classification model to say if the network is prone to contain motifs or not, it will not be bothering to know that there is a small chance to process a network that will not contain motifs as long as most of the predictions will indicate expected values. In this worst case scenario, the network will be processed without any expected motifs.

Using these results we can conclude that using our previous study on the correlation between the topological properties and appearance of the network motifs it is possible to train a statistical model to classify given networks as networks prone to contain motifs or not. Moreover this analysis was performed on various models such that it was possible to propose the boosting method as the best solution for this use case.

## 5   Conclusions

The discovery of network motifs in complex networks is a very intensive process being ranked as an NP-complete problem. Literature presents various solutions of optimising this process, either by making use of statistical tricks to reduce the number of computations but with a trade-off in the accuracy of the results, or by re-modelling existing algorithms into parallel models able to run on cluster of computers. From our knowledge there are no methods applied as a pre-processing step to optimise this process.

Driven by this void in literature, our paper presents a method of detecting if a network is prone to contain network motifs or not by using its topological properties. Our previous studies revealed correlation of different strengths between these topological properties and the existence of the motifs [14,15], this being a good indicator to consider the properties as independent variables to predict the presence of the dependent variable, network motifs.

Modeling this as a binary classification problem, Sect. 4 presents the steps applied to train a set statistical models using a consistent data set of networks. Making use of various tests, it was observed that the statistical model using a boosting method excelling in all test categories. Moreover, this method presented a high recall value, which is one of the most relevant measure for our use case as it will not be bothering to have a small fraction of false positives as long as the prediction will return as many true positives as possible.

Acknowledging the presented results, we consider this paper's proposal a first iteration of a model with a highly probable impact in the field of complex networks, helping the community to execute computation faster while performing a network motifs research. In future work we intend to propose new improved versions of the model that will take into account more detailed information such as the frequency of motifs. A perspective of transforming the model from a binary classifier into a regression model that is be able to predict the frequency of the motifs could be an important addition to the literature.

Moreover, we will continue our research implied for our ultimate goal of developing a generator model able to generate networks having a certain concentration of motifs by using as input topological measurements as ones discussed. The results presented by this paper together with our previous results will represent a solid statistical confirmation for when the network generator will be proposed.

## References

1. Barabási, A.L., Albert, R.: Emergence of scaling in random networks. Science **286**(5439), 509–512 (1999). https://doi.org/10.1126/science.286.5439.509
2. Bhowan, U., Johnston, M., Zhang, M.: Developing new fitness functions in genetic programming for classification with unbalanced data. IEEE Trans. Syst. Man Cybern. Part B Cybern. **42**, 406–421 (2012). https://doi.org/10.1109/TSMCB.2011.2167144
3. Chen, T., Guestrin, C.: XGBoost: a scalable tree boosting system. In: Proceedings of the 22nd ACM SIGKDD International Conference on Knowledge Discovery and Data Mining, KDD 2016, pp. 785–794. ACM, New York (2016). https://doi.org/10.1145/2939672.2939785
4. Coutinho, B., et al.: The Network Behind the Cosmic Web. http://cosmicweb.kimalbrecht.com
5. Fernandez-Delgado, M., Cernadas, E., Barro, S., Amorim, D.: Do we need hundreds of classifiers to solve real world classification problems? J. Mach. Learn. Res. **15**, 3133–3181 (2014)
6. Grochow, J.A., Kellis, M.: Network motif discovery using subgraph enumeration and symmetry-breaking. In: Speed, T., Huang, H. (eds.) RECOMB 2007. LNCS, vol. 4453, pp. 92–106. Springer, Heidelberg (2007). https://doi.org/10.1007/978-3-540-71681-5_7
7. Hu, J., Shang, X.: Detection of network motif based on a novel graph canonization algorithm from transcriptional regulation networks. Molecules **22**, 2194 (2017). https://doi.org/10.3390/molecules22122194
8. Jin, X., Li, J., Zhang, L.: Online social networks based on complex network theory and simulation analysis. In: Wong, W.E. (ed.) Proceedings of the 4th International Conference on Computer Engineering and Networks, pp. 1129–1138. Springer, Cham (2015). https://doi.org/10.1007/978-3-319-11104-9_130
9. Kashani, Z.R.M., et al.: Kavosh: a new algorithm for finding network motifs. BMC Bioinform. **10**, 318 (2009)
10. Kashtan, N., Itzkovitz, S., Milo, R., Alon, U.: Efficient sampling algorithm for estimating subgraph concentrations and detecting network motifs. Bioinformatics (Oxford, England) **20**, 1746–1758 (2004)
11. Kunegis, J., Preusse, J.: Fairness on the web: alternatives to the power law. In: Proceedings of the 3rd Annual ACM Web Science Conference, WebSci 2012, pp. 175–184, June 2012. https://doi.org/10.1145/2380718.2380741
12. Milo, R., Shen-Orr, S., Itzkovitz, S., Kashtan, N., Chklovskii, D., Alon, U.: Network motifs: simple building blocks of complex networks. Science **298**, 824–827 (2002)
13. Mursa, B.E.M., Andreica, A., Laura, D.: Parallel acceleration of subgraph enumeration in the process of network motif detection. In: 20th International Symposium on Symbolic and Numeric Algorithms for Scientific Computing (2018)
14. Mursa, B.E.M., Andreica, A., Laura, D.: An empirical analysis of the correlation between the motifs frequency and the topological properties of complex networks. In: 23rd International Conference on Knowledge-Based and Intelligent Information & Engineering Systems (2019)

15. Mursa, B.E.M., Andreica, A., Laura, D.: Study of connection between articulation points and network motifs in complex networks. In: Proceedings of the 27th European Conference on Information Systems (ECIS) (2019)
16. Noldus, R., Van Mieghem, P.: Assortativity in complex networks. J. Complex Netw. **2015** (2015). https://doi.org/10.1093/comnet/cnv005
17. Omidi, S., Schreiber, F., Masoudi-Nejad, A.: MODA: an efficient algorithm for network motif discovery in biological networks. Genes Genet. Syst. **84**, 385–395 (2009)
18. Krishna Raj, P.M., Mohan, A., Srinivasa, K.G.: Basics of graph theory. Practical Social Network Analysis with Python. CCN, pp. 1–23. Springer, Cham (2018). https://doi.org/10.1007/978-3-319-96746-2_1
19. Shen-Orr, S., Milo, R., Mangan, S., Alon, U.: Network motifs in the transcriptional regulation network of escherichiacoli. Nat. Genet. **31**, 1061–4036 (2002)
20. Svenson, P.: Complex networks and social network analysis in information fusion. In: 2006 9th International Conference on Information Fusion, pp. 1–7, August 2006. https://doi.org/10.1109/ICIF.2006.301554
21. Van Hulse, J., Khoshgoftaar, T., Napolitano, A.: Experimental perspectives on learning from imbalanced data, vol. 227, pp. 935–942, January 2007. https://doi.org/10.1145/1273496.1273614
22. Watts, D.H. Strogatz, S.: Collective dynamics of 'small-world' networks. In: The Structure and Dynamics of Networks, December 2011. https://doi.org/10.1515/9781400841356.301
23. Wernicke, S.: A faster algorithm for detecting network motifs. Algorithms Bioniform. Proc. **3692**, 165–177 (2005)

# Deep Structured Semantic Model for Recommendations in E-commerce

Anna Larionova[1], Polina Kazakova[2(✉)], and Nikita Nikitinsky[3]

[1] Picturer LLC, Petrovka Street 20 Building 1, 127051 Moscow, Russia
`tarelo4ka76@mail.ru`
[2] Integrated Systems, Vorontsovskaya Street 35B Building 3,
Room 413, 109147 Moscow, Russia
`kazakova1537@gmail.com`
[3] National University of Science and Technology MISIS,
Leninsky Avenue 4, 119049 Moscow, Russia
`torselllo@yandex.ru`

**Abstract.** This paper presents an approach for building a recommender system that makes use of heterogeneous side information based on a modification of Deep Structured Semantic Model (DSSM). The core idea is to unite all side-information features into two subnetworks of user-related and item-related features and learn the similarity between their latent representations using neural matrix factorization. We tested the proposed model in the task of products recommendation on the dataset provided by a Russian online-marketing company. To deal with the sparsity of the data, we suggest recommending categories of items first and then use any other algorithm to rank items inside categories. We compared the performance of the proposed model to several traditional methods and demonstrated that DSSM with heterogeneous input significantly increases overall recommendation quality which makes it suitable for recommendations with rich side information about items and users.

**Keywords:** Recommender systems ·
Deep structured similarity model · E-commerce

## 1 Introduction

The goal of a recommender system is to suggest relevant items to users (items that users are expected to like). E-commerce recommender systems recently receive slightly less attention in scientific research than, for example, music, news or movies recommendations. For instance, the deep learning papers that we cite in Sect. 2 used data on news, movies, images, web pages, and apps downloads [8,14,25] to evaluate the performance of their proposed architectures. We argue that the reason for this fact is that there are fewer open datasets applicable to this task since commercial companies seem not to will to share their private purchase data. Among large existing recommendation datasets in e-commerce one could name perhaps only Amazon product data [1,13] and Instacart [18]. At the

© Springer Nature Switzerland AG 2019
H. Pérez García et al. (Eds.): HAIS 2019, LNAI 11734, pp. 85–96, 2019.
https://doi.org/10.1007/978-3-030-29859-3_8

same time, commercial recommender systems deserve attention as a high-quality recommender system could help a company to improve its sales performance and increase customer loyalty. The present paper attempts to contribute to the topic of recommendations in e-commerce.

The model we describe here is designed for a rather specific recommendation case as the data contains catalogs of many different stores, and the system has to be capable of recommending items independently of the store. Since the data was excessively large, we first focused on suggesting categories of items instead of items themselves to deal with the high sparsity of the data. After the relevant categories are found one could use any other algorithm to recommend products. This two-step recommendation approach allows one to reduce the dimensionality of the data and apply a model to the product catalogs of different stores that share the same categories but sell different items.

The present paper is structured as follows. The next section provides a brief description of conventional recommendation methods as well as an overview of using DSSM in the task of recommendation. Section 3 provides a detailed description of the data we worked on. Section 4 describes the neural network architecture that we introduce and evaluate in the present research and Sect. 4.1 demonstrates how the proposed architecture could be used to find similar items and users. Section 5 presents the results of comparing our proposed approach to several conventional baselines and Sect. 6 summarizes the results of this study and formulate the plans for the future research.

## 2 Related Work

One of the basic approaches to building a recommender system include latent factor models and neighborhood methods. The former group of methods aims at projecting both users and items into the same low-dimensional vector space. It is usually done by matrix factorization that allows one to approximate a user-item matrix with a product of two matrices that can be interpreted as matrices of the user and item features (or latent factors). Complex MF models also take into account user-item biases (individual differences between user behavior trends and item features that influence interaction values systematically) as well as incorporate additional information on users and items (for instance, gender, age, demographics, price, description) and temporal features [19]. This factorization approach could also be generalized to tensors [16].

The neighborhood-based approach uses $k$ Nearest Neighbors algorithm to predict ratings exploiting the similarity between users and items. In the user-user setting, unknown ratings are generated using existing ratings of similar users [15]. Accordingly, in the item-item setting, ratings are predicted based on existing ratings that the same users made on similar items [24].

Though factor models and neighborhood methods are still used in recommender systems rather widely, more recent models usually involve deep neural nets. Companies such as YouTube [7], Google [6], Yahoo [23] has recently deployed deep learning recommender systems. The key advantage of using deep learning in recommender systems, apart from its ability to model non-linearity and reduction of time and effort for feature engineering, is that it is capable of integrating heterogeneous side information into the model more easily and naturally than in case of conventional algorithms. In fact, information heterogeneity is what one usually has to deal with when building a recommender system. For a detailed overview of deep neural networks in recommendations see [26].

One of the ways to incorporate information of different types in neural networks is using Deep Structured Semantic Model. DSSM is a deep neural network architecture that projects various heterogeneous entities into a common low-dimensional semantic space and outputs similarity scores between them. It consists of several sub-networks that process each input entity separately. DSSM was first introduced in the information retrieval task [17] and then adapted for generating recommendations [8].

The authors of [8] used a modification of DSSM to handle users query history (user features) and item information from multiple heterogeneous domains simultaneously as a multi-view architecture which compares each domain of item features with user features and then calculate the overall similarity score. In [25] the authors proposed to modify DSSM with RNN to account for multi-rate temporal factors. Their DSSM is composed of two sub-networks (MLP) for the static item and user features and three RNN (LSTM) that model user behavior at different time rates: daily, weekly and globally.

In our model, we exploit the concept of DSSM but instead of computing the relevance of an item to a user as cosine similarity between their latent representations we use neural matrix factorization. We suppose that neural matrix factorization could capture more complex user-item interactions because of involving non-linearity (see [14] for a more detailed discussion).

## 3   Dataset Description

In our experiments, we used the private data provided by a Russian online-marketing company that was gathered from 211 Russian online retail stores. It contains information about user interactions with products for a period of one week and a full catalog of products. Each item in the catalog has category information (id, name, description, parent category) and product information (id, name, description, characteristics, manufacturer, price, store that sells this product). The user-item interactions data has the following information: date-time of a click, event (interaction) type, session id, item id, IP address, user-agent string, URL of the event.

User data is anonymous: it does not contain any personal details like gender or age that might be helpful to form a better user representation. The data was additionally enriched with information about device (device type, family name, OS family name, browser family name; extracted from user-agent string), location (continent, country, region, city, time zone, postal code, longitude and latitude; extracted from IP address), and domain (name and zone; extracted from URL).

The dataset has only implicit feedback from users (users had never rated any item directly) in the form of user interactions type names. These names are mapped to numbers as follows:

- 0 - other, unknown, login, error in records, etc: any user interaction not connected directly to the item
- 1 - main, index: an item was displayed to a user somewhere on the main page, and he or she might have seen it;
- 2 - list, lists: an item was displayed to a user on the search or catalog page when a user was searching or browsing the category items;
- 3 - card: user opened a particular item card contained detailed information;
- 4 - basket: user added a particular item to the basket;
- 5 - thanks: user purchased an item.

The higher the value is, the stronger positive signal about user preferences and interaction with items is sent to the recommender model. To help a recommender system detect positive and negative feedback we set the threshold equal to 3 to binarize the target variable for models that can only work with binary targets (less or equal than 3 - negative; 4 and 5 - positive). Note that this division is rather conventional since the feedback is implicit.

The dataset has a large amount of raw data (several gigabytes of text logs per day), and the product catalog contains several millions of items. Therefore it is necessary to reduce its size to fit into the memory. As the dataset has information about products from multiple stores and some of the information could be outdated or incomplete, some products could be highly popular while others have almost no interactions (high sparsity of user-item interaction matrix), we decided to focus on recommending categories at the first step. Additionally, the catalog structure appeared to be more stable and up-to-date, which allows one to build more accurate models. After recommending the category, one could sample products based on additional criteria, including availability, store, and price. The catalog has a hierarchical structure, and we focus on recommending child categories since categories at this level are more specific and contain the smallest number of items. We will later refer to a category as an item.

The data was grouped by users and categories so that the resulting user-item interaction is equal to the maximum interaction value that a corresponding user had with a corresponding category. This aggregation step allowed us to significantly reduce the number of training examples (though at the cost of losing information on sequences of actions). All users that had interacted with less than 10 categories were excluded from the dataset. Table 1 describes the statistics for both the original dataset and the data obtained by applying the aforementioned transformations.

To prepare the data for training, we used the following preprocessing techniques:

- The textual category information (name and description) was merged into a single text. We generated a vector representation for each text by inferring

**Table 1.** Dataset statistics.

| Dataset type | # users | # items | # user-item pairs | Avg. interactions per user | Sparsity |
|---|---|---|---|---|---|
| Raw (initial) data | 981862 | 8134826 | 3958806 | 4 | 99.7% |
| Processed (final) data | 10049 | 1383 | 76127 | 10 | 99.4% |

FastText embeddings [4] from their tokens and averaging them (FastText model is pretrained on the Russian language subset of the Common Crawl corpus [10]).

– Categorical features (device type, family name, os family name, browser family name, continent, country, region, city, etc.) and identificators (store id, category id, parent category, user id) were mapped to a number from 0 to N, where N is a number of unique values. Missed values were replaced with zeros.
– We transformed dates to the POSIX timestamp format; we also extracted numbers of a day of the week and day of the year.
– We applied a logarithmic transformation to the price.
– The other numeric features (longitude, latitude, time zone) were left as is.

User features were split into two types: short-term features describing a specific interaction and long-term preferences based on analysis of their previous history and activity (note that 'long-term' here is a convention since the data contains information only about one week). Short-term features include datetime, days of the week and year, IP address, domain name, device information, device browser families, longitude, latitude, time zone, continent, city, country and whether user browsed from a mobile device. Long-term preferences include average user purchase price, the average price of all viewed items, popular items (most frequent item and its frequency).

Following the common empirical rule, we did 80/20 train-test split so that the subsets share the same users and the testing data contain approximately 20% of user *last* interactions. We did not train our model to make predictions for new users (warm start scenario).

## 4 Deep Structured Semantic Model for Recommendation

In the present research, we compare two different neural architectures. The first one concatenates all input features into one layer. The second one concatenates input into two subnetworks: for user and item features separately (Fig. 1).

Each feature has a separate input layer. If the feature is categorical, its input layer is followed by an embedding layer with an embedding size approximately equal to N/10 and a dense layer with 64 neurons. If it is text data we apply

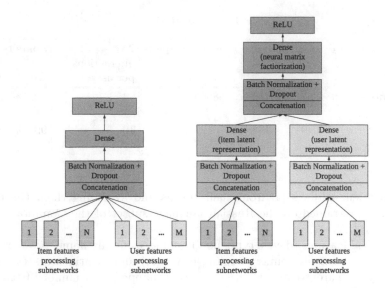

**Fig. 1.** Neural recommender architectures with one concatenation of all input features (on the left) and with two separate subnetworks of the user and item features (on the right).

GRU RNN with 100 hidden units layer after embedding and then add a dense layer to capture the sequences of words (Fig. 2). For numeric features, we do not add any additional layers.

We experimented with different number (from 16 to 128) of hidden neurons in a dense layer. 64 neurons gave almost as high quality as 128 neurons, and we chose to use 64 neurons to prevent overfitting. The dimensionality of text embeddings usually varies from 100 to 300 [22], and we decided to set the number of hidden units in GRU to 100 as the textual descriptions of products are rather short (usually less than 200–300 words).

In the first model, all inputs are passed through feature processing subnets and their outputs are concatenated in a single layer, followed by a dense layer and ReLU [20] activation function. We will refer to it as S-DSSM (for simple DSSM).

In the second model, the outputs of the feature processing subnets are concatenated into two groups (user features and item features) each followed by a dense layer which allows the model to learn internal representations for users and items. Then the user and item subnetworks are concatenated into one layer followed by a dense layer of the same size to simulate user-item matrix factorization, i.e. to learn the interactions between user and item internal representations [14]. The output layer is a single neuron with ReLU activation and its output score can be interpreted as the 'similarity' between user and item. We will refer to this architecture as 2B-DSSM (for two-branches DSSM).

For implementing the networks we used python 3.6 and the Keras library with a Tensorflow backend. We trained both models with Adam optimizer with

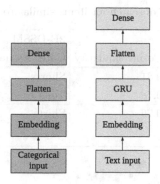

**Fig. 2.** Feature processing subnetworks architectures: for categorical features (on the left) and texts (on the right).

learning rate 0.0001 during 1000 epochs of training. We used ReLU as activation in all hidden layers as well. Although we tried to use binary cross entropy for classification, our experiments showed that using MSE loss function and regression of the target variable provided better results. The overall task of the network is to measure semantic similarity between user and item representations and output it in the form of real number equal or larger than 0.

## 4.1 Similarity Search

The architecture of the second network allows one to get the hidden representation of users and items (from concatenation layers) so that one could use them in other tasks, for example, in the similarity search. The latter could be crucial in a cold start scenario when there is little information on a new user or item. Despite S-DSSM in fact produces the latent representation of a user-item pair and is less suitable for exploring similarity between items or users, one could use the following hack: fix user features and variate only item features to get vectors that mainly describe items.

However, to operate on these vector representations one needs to calculate them for each item and user respectively and then store. To look for similar entities one could further use approximate nearest neighbor search [5]. Moving one step forward, Table 2 represents an example output of similar items search using Annoy [3].

To make sure that the 2B-DSSM network is capable of finding similar items we projected item representations into two-dimensional space using t-SNE [21] and applied DBSCAN clustering [9] to the resulting projections. Figure 3 demonstrates that the projections are clustered in several rather distinct blobs which means that the item embedding is capable to represent similar categories closer to each other.

**Table 2.** The result of search for items similar to 'suits and coats'.

| | S-DSSM | | 2B-DSSM | |
|---|---|---|---|---|
| Rank | Category name | Cosine similarity | Category name | Cosine similarity |
| 1 | Socks | 0.81 | Shorts | 0.94 |
| 2 | Shirts | 0.78 | Tops | 0.93 |
| 3 | Swimming suits | 0.76 | Shirts | 0.92 |

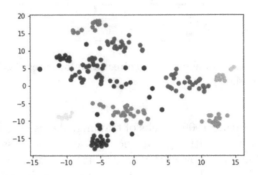

**Fig. 3.** T-SNE visualization of categories representations from 2B-DSSM.

## 5  Experiments

To evaluate and compare the proposed DSSM-based recommenders with existing well-known[1] approaches in recommendation we selected several models from traditional machine learning algorithms without and with side information as baselines:

- **Random model** – generates recommendations randomly for each user.
- **Most popular** – always recommends top $n$ items with the highest ratings (independently of particular user preferences and ratings).
- **Item-to-item** – neighborhood model that generates recommendations for a user based on similarity score between items [24]. The model suggests items that are similar to what the user has already bought.
- **PureSVD** – simple low-rank matrix approximation via singular value decomposition. We used the implementation from the Polara framework [11].
- **CoFFee** (Collaborative Full Feedback model) – a novel algorithm [12] based on tensor factorization, implemented in the recommender systems framework Polara [11]. The key feature of this model compared to other matrix and tensor factorization techniques is that it makes use of negative feedback too.
- **TFM** (without ranking) – a Factorization Machine model which main advantage over conventional matrix factorization is that they model latent factors not only for users and items but also for all side features and interactions

---

[1] In fact, we also tested a novel collaborative model named *CoFFee* [12].

between variables. Therefore, they are more suitable to learn complex interactions from the data. We used the implementation from the Turicreate framework [2].

- **TFMR** (with ranking) – a Factorization Machine that additionally uses ranking (by modifying the loss and regularization) to improve the recommendation
quality by placing user-relevant items on top of recommendation list (improving accuracy and recall).
- **TFMR+** – extends TFMR model by tuning its regularisation parameters to
improve the quality of recommendations.

Both proposed models could be used for selecting the best matching candidate to be recommended. To provide a more user-relevant output we additionally apply ranking to the predictions of a neural network model. For ranking we
used gradient boosting trees on the same set of features with additional similarity score (output of the neural network) and information about previous user
interaction ('seen', 'purchased' or 'unseen') with category (to rank unseen items
higher than seen and seen higher than purchased) with lambda rank objective
where target was constructed from previous user history interactions as a recommendation list of size $N$ ($N = 10$ in our case). This allowed the system to
move more relevant items up and improve overall model quality.

We selected the following metrics to evaluate the performance of the models:
accuracy, precision, recall, fallout, miss, and nDCG@10, where $k = 10$; values
are averaged over all users. Additionally, we compute serendipity and catalog
coverage. The experimental results are presented in Table 3.

**Table 3.** Recommendation quality of the models.

| Type | Model | Acc | Prec | Recall | Fallout | Miss | nDCG@10 | Serendipity | Coverage |
|------|-------|-----|------|--------|---------|------|---------|-------------|----------|
| No side inf. | Random | 0.4706 | 0.0640 | 0.0120 | 0.0119 | 0.9360 | 0.0092 | 0.9000 | **1.0000** |
| | MP | 0.4826 | 0.5040 | 0.2429 | 0.2584 | 0.6464 | 0.2285 | 0.0000 | 0.0881 |
| | Item-item | 0.4801 | 0.5098 | 0.2616 | 0.2649 | 0.6863 | 0.2372 | 0.0024 | 0.1165 |
| | Coffee | 0.4693 | 0.4644 | 0.1973 | 0.1799 | 0.7506 | 0.1641 | 0.0647 | 0.1658 |
| | PureSVD | 0.4701 | 0.4880 | 0.2325 | 0.2164 | 0.7154 | 0.2062 | 0.0774 | 0.1865 |
| With side inf. | TCF[a] | 0.4496 | 0.0500 | 0.0098 | **0.0062** | 0.9445 | 0.0079 | **0.9971** | 0.0352 |
| | TCFR | 0.4846 | 0.5223 | 0.1865 | 0.1331 | 0.7678 | 0.1596 | 0.0315 | 0.1036 |
| | TCFR+ | 0.4795 | 0.5302 | 0.1943 | 0.1373 | 0.7593 | 0.1722 | 0.0801 | 0.1721 |
| | S-DSSM | 0.6561 | 0.6577 | 0.6539 | 0.3529 | 0.3601 | 0.6521 | 0.2760 | 0.2130 |
| | 2B-DSSM | **0.6968** | **0.6964** | **0.6965** | 0.0389 | **0.3246** | **0.6629** | 0.3810 | 0.3420 |

[a]This model was unable to converge as the result it had almost random behavior, but unlike
Random model it recommends mostly rare elements which is the reason for serendipity to be
that high.

Note that the metrics values are rather large (relative to those reported in
scientific papers on average). The reason is that we recommend categories instead
of specific products.

Among the simplest baselines which do not utilize side features Most Popular and Item-item models appear to be the best in terms of accuracy, precision

and recall. At the same time, CoFFee and PureSVD give lower values of fallout indicating that these models are more sensitive to specific user preferences. Simple Factorization Machine performs even worse that the random model despite integrating side information. Nevertheless, Factorization Machines *with* ranking do outperform other baselines by accuracy, precision and recall.

Finally, 2B-DSSM outperforms all the other models, including S-DSSN. The architecture with two subnetworks for user and item features provides better feature representation learning that helps to capture similarity between user-item pairs and therefore generate more relevant predictions.

## 6   Conclusion

Our research demonstrated that recommender systems can not only be used for recommending particular products or services but also groups of similar items (categories). Grouping products into categories allows one to reduce the overall amount of information and improve recommendation quality due to the smaller number of different items which describe general user preferences. To recommend a particular item one could later pick up it from a group it belongs to by any other algorithm. This reduces the search area for a particular product recommendation which could be done by any other appropriate algorithm.

The experiments proved that utilizing side information even in non-deep learning algorithms of collaborative filtering gives an additional gain in accuracy and precision. As side information can be heterogeneous (different sources, types and so on), manual feature engineering might be a rather difficult task. This issue could be overcome by using feature extraction subnetworks. The experiments showed that DSSM with separate item and user feature concatenation provides better recommendations and is appropriate for searching for similar items due to learning item and user embeddings before the neural factorization layer.

The future work of this research is thereby to investigate the second step of the proposed two-step approach, i.e. recommending products inside categories, and compare its joint quality in recommending items to the quality of a standard approach of recommending items directly.

**Acknowledgments.** We are very grateful to Anton Lozhkov and Diana Khakimova for the efforts in revising the present paper in the very last minutes before the deadline.

This research was supported by the Russian Research Foundation grant no. 19-11-00281.

## References

1. Amazon: Amazon product data. http://jmcauley.ucsd.edu/data/amazon/
2. Apple Inc.: Turi create. https://github.com/apple/turicreate
3. Bernhardsson, E.: Annoy: Approximate nearest neighbors in c++/python. https://github.com/spotify/annoy
4. Bojanowski, P., Grave, E., Joulin, A., Mikolov, T.: Enriching word vectors with subword information. Trans. Assoc. Comput. Linguist. **5**, 135–146 (2017)

5. Cayton, L.: Fast nearest neighbor retrieval for Bregman divergences. In: Proceedings of the 25th International Conference on Machine Learning. ACM (2018)
6. Cheng, H.T., et al.: Wide & deep learning for recommender systems. In: Proceedings of the 1st Workshop on Deep Learning for Recommender Systems, pp. 7–10. ACM (2016)
7. Covington, P., Adams, J., Sargin, E.: Deep neural networks for Youtube recommendations. In: Proceedings of the 10th ACM Conference on Recommender Systems, pp. 191–198. ACM (2016)
8. Elkahky, A.M., Song, Y., He, X.: A multi-view deep learning approach for cross domain user modeling in recommendation systems. In: Proceedings of the 24th International Conference on World Wide Web, pp. 278–288. International World Wide Web Conferences Steering Committee (2015)
9. Ester, M., Kriegel, H.P., Sander, J., Xu, X., et al.: A density-based algorithm for discovering clusters in large spatial databases with noise. In: KDD, vol. 96, pp. 226–231 (1996)
10. Facebook Research: fastText: Word vectors for 157 languages. https://fasttext.cc/docs/en/crawl-vectors.html
11. Frolov, E.: Polara: Recommender system and evaluation framework for top-n recommendations tasks that respects polarity of feedbacks. https://github.com/Evfro/polara
12. Frolov, E., Oseledets, I.: Fifty shades of ratings: how to benefit from a negative feedback in top-n recommendations tasks. In: Proceedings of the 10th ACM Conference on Recommender Systems, pp. 91–98. ACM (2016)
13. He, R., McAuley, J.: Ups and downs: modeling the visual evolution of fashion trends with one-class collaborative filtering. In: Proceedings of the 25th International Conference on World Wide Web, pp. 507–517. International World Wide Web Conferences Steering Committee (2016)
14. He, X., Liao, L., Zhang, H., Nie, L., Hu, X., Chua, T.S.: Neural collaborative filtering. In: Proceedings of the 26th International Conference on World Wide Web, pp. 173–182. International World Wide Web Conferences Steering Committee (2017)
15. Herlocker, J.L., Konstan, J.A., Borchers, A., Riedl, J.: An algorithmic framework for performing collaborative filtering. In: 22nd Annual International ACM SIGIR Conference on Research and Development in Information Retrieval, SIGIR 1999, pp. 230–237. Association for Computing Machinery, Inc. (1999)
16. Hidasi, B., Tikk, D.: Fast ALS-based tensor factorization for context-aware recommendation from implicit feedback. In: Flach, P.A., De Bie, T., Cristianini, N. (eds.) ECML PKDD 2012. LNCS (LNAI), vol. 7524, pp. 67–82. Springer, Heidelberg (2012). https://doi.org/10.1007/978-3-642-33486-3_5
17. Huang, P.S., He, X., Gao, J., Deng, L., Acero, A., Heck, L.: Learning deep structured semantic models for web search using clickthrough data. In: Proceedings of the 22nd ACM International Conference on Information & Knowledge Management, pp. 2333–2338. ACM (2013)
18. Instacart: The instacart online grocery shopping dataset 2017. https://www.instacart.com/datasets/grocery-shopping-2017
19. Koren, Y., Bell, R., Volinsky, C.: Matrix factorization techniques for recommender systems. Computer **8**, 30–37 (2009)
20. Maas, A.L., Hannun, A.Y., Ng, A.Y.: Rectifier nonlinearities improve neural network acoustic models. In: Proceedings ICML, vol. 30, p. 3 (2013)
21. Maaten, L.V.D., Hinton, G.: Visualizing data using t-SNE. J. Mach. Learn. Res. **9**(Nov), 2579–2605 (2008)

22. Mikolov, T., Chen, K., Corrado, G., Dean, J.: Efficient estimation of word representations in vector space. arXiv preprint arXiv:1301.3781 (2013)
23. Okura, S., Tagami, Y., Ono, S., Tajima, A.: Embedding-based news recommendation for millions of users. In: Proceedings of the 23rd ACM SIGKDD International Conference on Knowledge Discovery and Data Mining, pp. 1933–1942. ACM (2017)
24. Sarwar, B.M., Karypis, G., Konstan, J.A., Riedl, J., et al.: Item-based collaborative filtering recommendation algorithms. In: WWW, vol. 1, pp. 285–295 (2001)
25. Song, Y., Elkahky, A.M., He, X.: Multi-rate deep learning for temporal recommendation. In: Proceedings of the 39th International ACM SIGIR Conference on Research and Development in Information Retrieval, pp. 909–912. ACM (2016)
26. Zhang, S., Yao, L., Sun, A., Tay, Y.: Deep learning based recommender system: a survey and new perspectives. ACM Comput. Surv. (CSUR) **52**(1), 5 (2019)

# Bio-inspired Models and Evolutionary Computation

# Improving Comparative Radiography by Multi-resolution 3D-2D Evolutionary Image Registration

Oscar Gómez[1,2](✉), Oscar Ibáñez[1,2,3], Andrea Valsecchi[1,2,3],
and Oscar Cordón[1,2]

[1] Department of Computer Science and Artificial Intelligence,
University of Granada, Granada, Spain
valsecchi.andrea@gmail.com
[2] Instituto Andaluz Interuniversitario DaSCI, University of Granada, Granada, Spain
{ogomez,ocordon}@decsai.ugr.es
[3] Panacea Cooperative Research S. Coop., Ponferrada, Spain
oscar.ibanez@panacea-coop.com

**Abstract.** Comparative radiography has a crucial role in the forensic identification endeavor. A proposal to automate the comparison of ante-mortem and post-mortem radiographs has been recently proposed based on an evolutionary image registration method. It considers the use of differential evolution to estimate the parameters of a 3D-2D registration transformation that automatically superimposes a bone surface model over a radiograph of the same bone. The main drawback of this proposal is the high computational cost. This contribution tackled this high computational cost by incorporating multi-resolution and multi-start strategies into its optimization process. We have studied the accuracy, robustness and computation time of the different configurations of the proposed method with synthetic images of patellae, clavicles and frontal sinuses. A significant improvement has been obtained in comparison to the state-of-the-art method in term of the robustness of the optimization method and computational cost with a drop in accuracy smaller than the 0.5% of the pixels of the silhouette of the bone or cavity.

**Keywords:** Comparative radiography · Evolutionary computation · 3D-2D image registration

## 1 Introduction

Comparative radiography (CR) [1] is a forensic identification technique based on the comparison of skeletal structures in ante-mortem (AM) and post-mortem (PM) radiographs. In the literature, several bones and cavities have been reported as useful for candidate short listing or positive identification based on their individuality and uniqueness [2]. Furthermore, CR techniques have a lower costs and time requirements in comparison to DNA analysis, which are crucial

© Springer Nature Switzerland AG 2019
H. Pérez García et al. (Eds.): HAIS 2019, LNAI 11734, pp. 99–110, 2019.
https://doi.org/10.1007/978-3-030-29859-3_9

factors in mass disasters victim identification scenarios. However, the application of CR requires a difficult task: the superimposition of the AM and PM data for their visual comparison by producing PM radiographs simulating the AM ones in scope and projection. This is a time consuming trial and error process, that relies completely on the skills and experience of the analyst. Furthermore, the utility of the method is reduced because of the time required and the errors related to analyst's fatigue and subjectivity.

The automation of the CR's superimposition process is complex and computationally expensive (see Sect. 2 for further details). It is due to several reasons such as the unknown set-up of the AM radiograph or the fact that image intensities are not reliable or even not captured, etc. These reasons make classic 2D-3D image registration (IR) techniques [3] not suitable for CR, and more sophisticated techniques should be considered in order to solve it as evolutionary algorithms (EAs). In particular, the state-of-the-art approach for CR follows and evolutionary 3D-2D IR methodology based on an EA called Differential Evolution (DE) [4] with a superb performance.

The aim of this work is to improve the robustness and reduce the time required by the state-of-the-art approach for CR [4] while keeping its accuracy. We will approach this goal by incorporating multi-resolution and multi-start strategies [5] (see Sect. 3) into the evolutionary process of the classic DE optimizer. This goal is studied with synthetic images of two bones (clavicles and patellae) and one cavity (frontal sinuses) as in [4] in order to allow a fair comparison.

This paper is structured as follows. Section 2 briefly reviews the current state of the art in IR for CR. Section 3 describes our proposal to tackle the 3D bone scan-2D radiograph superimposition problem with a multi-resolution IR approach. Section 4 presents the experiments and the results obtained. The main conclusions are detailed in Sect. 5.

## 2    Background and Related Works

IR [3] is the process of aligning two or more images of the same or different dimensionalities into one coordinate system. 3D-2D IR approaches (i.e. Computed Tomographies (CTs)-radiograph or 3D surface mode-radiograph in CR) are classified into intensity-based and feature-based. Intensity-based methods compare intensities [3] of a 2D projection of the volumetric image with a fixed 2D image. However, one crucial consideration in any automatic method for CR is that the intensity level could have changed between the AM and the PM images as the bone density changes within the individual through time (due to factors as aging, osteoporosis, and the PM interval). Additionally, the intensity information cannot be acquired by 3D surface scans, which are being increasingly used by forensic labs [6] to get accurate scans of PM "clean" bones due to their great availability and low cost, while just a few of them can afford a CT scan.

The previous limitations lead to feature-based approaches. Feature-based methods minimize the distance between geometrical features to be segmented in both images. Forensic anthropologists consider bone morphology (silhouette) a reliable marker for performing the CR-based identification to compare radiographs of frontal sinus [7], clavicles [8], and patellae [9]. However, these works cannot be considered as IR method since they only compare silhouettes using elliptical Fourier analysis [10]. The former one is based on the comparison of AM radiograph with a PM radiograph. The latter two, in contrast, are base on the comparison of the AM radiograph with a set of a predefined 2D projected images obtained through the rotation of 3D surface models acquired with a 3D laser range scanner. However, none of these completely automate the search for the best possible 2D projection of the PM 3D surface model of the bone.

Another important consideration about 3D-2D IR methods for medical domains [11] is that most of them are designed for a controllable set-up. Therefore, they can assume a calibrated case where the parameters related to the perspective distortions are known, and only considering the parameters related to translations and rotations (6 degrees of freedom, a.k.a. DoF), and with a initialization pose close to the ground truth (GT) pose (i.e. a maximum target registration error [12] of 16 mm in [11], etc). However, these assumptions are not suitable for CR since the AM radiograph was taken in an unknown pose with an unknown radiograph device. Therefore, the search for the optimal solution in the CR scenario is more complex. Of course, there are a few exception such as Feldman et al. [13] that proposed a 3D-2D IR method based on the silhouette that does not relies on assumptions about the initial pose by using free-form curves and surfaces, but it is only applicable in the calibrated case (6 DoF).

IR methods based on EAs, a.k.a. evolutionary IR methods, have demonstrated to overcome some of these drawbacks in others IR problems [14]. In particular, Gómez et al. [4] proposed an evolutionary 3D-2D IR approach for CR based on the bone or cavity silhouette. It automatizes the search of the best possible 2D projection of the PM 3D surface model of the bone (either obtained using a 3D scanner or segmented from a PM CT). It does not consider any assumption on the initialization or the main parameter related to the perspective distortions in radiographs (i.e. the source to image distance, a.k.a. SID). This proposal is based on the use of an EA called DE [15], a modification of the DICE metric [16] that considers occlusion regions (which are regions hard to segment either because of the fuzzy borders of the bone or occlusions caused by other overlapped structures), and a simple perspective transformation (with 7 parameters: 3 translations; 3 rotations; and the SID). This method was tested with frontal sinuses, clavicles and patellae obtaining a superb performance.

The best results were obtained when using frontal sinuses that have been used in several works for positive identification [17], probably due to the singularity of the visible region, followed by clavicles and patellae, which have been mainly used for candidate short listing [9, 18]. However, the method showed the following drawbacks: (1) the robustness of the DE algorithm, especially with clavicles and patellae, that in some runs leaded to bad superimpositions due to the stochastic nature of DE and the highly multimodal search space tackled; and (2) the large amount of time required to obtain a superimposition with DE (on average, 1800 s).

This long time is motivated by the high computational time required by each evaluation (on average, it takes 0.250 s for a projection of 1290×1050 pixels in a standard computer [4]), uncovering the computationally expensive optimization nature of the CR problem, and the high number of evaluations required by the optimizer to converge to a solution.

## 3    Methodology

### 3.1    Problem Statement

The evolutionary IR method requires the five following components (these are further detailed in [4] and depicted in Fig. 1): (1) the model (PM 3D surface model of the bone/cavity) and the scene image (AM radiograph, where the silhouette of the bone/cavity is segmented as well as the occlusion region, a.k.a. the region where the segmentation expert cannot distinguish if there is bone/cavity or not); (2) the projective transformation responsible of generating a 2D image from a 3D object; (3) the expert knowledge of the problem that delimits the target transformation (i.e. radiographs acquisition protocols [19]); (4) a similarity metric which measures the resemblance of a 2D projection with the original 2D image (overlapping); and (5) an EA, which looks for the best parameters for the transformation to minimize the error of the similarity metric.

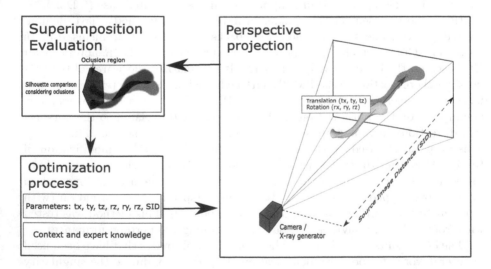

**Fig. 1.** Scheme of the proposal of 3D-2D IR for CR. Three main interconnected blocks are represented: (Right) the projective transformation to obtain a projection of the 3D model with 7 parameters: translation ($t_x$, $t_y$, and $t_z$), rotation ($r_x$, $r_y$, and $r_z$), and perspective distortions ($SID$); (Top left) The similarity metric that compares the PM projection and the AM segmentation considering an occlusion region; (Button left) the optimization process of the 7 projective transformation parameters, that are only weakly limited by the expert knowledge from the x-ray acquisition protocol.

The projective transformation [20] behind a radiograph image is a simple perspective transformation [21] with 7 parameters (6 extrinsic parameters: 3 translations; 3 rotation; and 1 intrinsic parameter). Notice that in a radiograph the perspective distortion is related to the source to image receptor distance (SID) [21] (see Fig. 1) instead of the focal distance. Most works consider a calibrated scenario (only 6 parameters) and the SID is assumed as known which is not the case for the CR problem [11]. Although the perspective distortion can be small in many radiographs because of the large distance between the x-ray generator and receptor (as in chest radiographs), its consideration has shown to be crucial in the IR endeavour. This has been shown in [4], where better results were obtained using the perspective transformation than the orthographic transformation, despite the more challenging optimization problem involved.

The goodness of a projection of the PM 3D model with respect to fixed 2D segmentation of the radiograph is measured using a similarity metric, used as fitness function. This metric measures the resemblance of a 2D projection with the original 2D image (overlapping). The most utilized metric to measure the overlap of silhouettes is the DICE metric [16]. However, this metric is not robust against occlusion and does not allow partial matching. These drawbacks are overcome for the CR problem using the Masked DICE metric [4] (see Eq. 1), designed ad-hoc for the CR problem. This metric incorporates the information of an occlusion region segmented by the expert into the DICE metric [16] increasing the robustness to occlusions and partial matching.

$$\text{Masked DICE} = \frac{2 \cdot |(I_A \setminus M) \cap (I_B \setminus M)|}{|I_A \setminus M| + |I_B \setminus M|} \tag{1}$$

where $I_A$ is the set of pixels of object A (segmented bone) silhouette, $I_B$ is the set of pixels of object B (PM project bone) silhouette, and M is the occlusion region.

## 3.2 Differential Evolution

The search space of the perspective transformation is complex and highly multimodal [4]. Therefore, classic numerical optimization methods are not sufficient, and more sophisticated techniques should be considered in order to solve it satisfactorily as EAs [14].

DE is a variant of an EA proposed by Storn and Price [15]. DE iteratively tries to improve a population of $n$ candidate solutions (each of them called chromosomes) within a repetitive loop of $m$ generations by combining the genes (the set of problem parameters to be found) of candidate solutions from the previous generation, and also preserving the best candidate solution in term of a fitness function (e.g. the Masked DICE metric) until the moment. DE has the following parameters that modulates its optimization behaviour: the number of individuals in the population $p$, the differential weigh $F$, the crossover probability $P_c$, and the number of generations to be performed $m$ (or number of evaluations which is equal to the population's size multiplied by the number of generations).

### 3.3  Multi-resolution and Multi-start Strategies

The high computation time required to obtain a projection of the 3D model for a given candidate solution is strongly related with the resolution of the input 2D image. For instance, the time required to perform a projection of $1000 \times 1000$ pixels is four times the time required for a projection of $500 \times 500$ pixels. Meanwhile, the resolution of the input 3D image does not have such a significant effect when using an AABB tree ray-tracing approach [22]. However, reducing the 2D image resolution also results in losing details that can be crucial to obtain a good superimposition and leads to local minima. A tradeoff between quality and time required can be found by using a multi-resolution IR strategy [5,23], that performs the optimization process in multiple stages with increasing complexity in the image resolution. The change between stages is performed after a set numbers of generations of the EA (in our case DE). The first resolutions are tackled using down-sampled 2D images of increasing resolutions, and thus being computationally cheaper. Furthermore, only a few operations are required to change from one resolution to the next. The candidate solutions' parameters are not affected since they are encoded in physical units. Furthermore, the projective solution does not depends on the image's resolution but instead on the image receptor resolution (set by the radiograph acquisition protocol [19] and encoded in physical units, see [4] for further details). However, when changing from one resolution to the next, new details are visible in the images, and therefore the fitness (i.e. the similarity metric) of all the candidate solutions have to be computed again. Finally, the last stage tackles the original image resolution, thus being the most computationally expensive, and is mostly focused on the fine-tuning of the solutions achieved in the previous optimization stages.

Another important consideration is that when a low quality solution is carried over to the second stage/resolution, the optimization process is unlikely to recover and produce a good final solution. Therefore it is more appropriate to perform again the search for a suitable initial registration by restarting the optimization process and repeating the first stage. However, it is not easy to determinate a threshold in the fitness value to distinguish between good and bad solutions since it would depend on several factors of the input images (such as the bone/cavity considered, the particular projection, and the occlusion region) and it is hard to predict. Therefore, instead of following a restart strategy [5], we follow a multi-start strategy. A multi-start strategy consists of performing the fist resolution $n$ times, independently of their outcome, and continuing the optimization process in the posterior resolutions only with the best of them. The computational cost of performing multiple starts at the first resolution is quite low as the resolution of the images involved is still small and it helps to improve the robustness and quality of the superimposition obtained.

Lastly, a distinctive feature of the multi-resolution IR approach is the inclusion of a search space adaptation mechanism (a.k.a. a dynamic boundary mechanism). Dynamic boundary [5] allows us to further take advantage of the use of the multiple resolutions approach by reducing the parameter ranges by a shrinking factor $sf$ (i.e. a shrinking factor of 2 halves the parameter range of all the parameters) in each change of resolution around the values in the best solution. The initial range of the parameters are quite wide since they are only delimited by the acquisition protocol. However, when we change from one stage/resolution to the next one, it can be expected that the optimal solution lies in an area nearby the best solution found at the current stage/resolution. Thus, the optimizer's intensification can be improved by reducing the parameters' range inside this area.

The incorporation of these strategies into the classic CR-DE approach results in the multi-resolution and multi-start CR-DE approach. The parameters to be tuned for this approach are those from the CR-DE approach (detailed in Subsect. 3.2) and the parameters related to the multi-resolution and multi-start strategies which are: the number of resolutions $r$, the down-sampling ratio in each resolution with respect to the original resolution $dr$ (coded as an array of size $r$), the number of generations to be devoted to each resolution $n$ (coded as an array of size $r$), the number of starts $s$, and the shrinking factor $sf$.

## 4    Experimental Study

The experimental study is devoted to analyze the performance, robustness and computational time of the proposed multi-resolution and multi-start CR-DE approach against the classic CR-DE approach, proposed in the state-of-the-art work for solving the CR problem. Furthermore, it also aims to fine tune the parameters that modulates the optimization behaviour of the multi-resolution and multi-start CR-DE approach. For this experiment, simulated CR problems (positive cases, i.e. the AM and PM data belong to the same person) of frontal sinuses, clavicles and patellae are considered.

All the experiments have been performed on the high performance computing server Alhambra from the University of Granada composed of Fujitsu PRIMERGY CX250/RX350/RX500 nodes running Red Hat Enterprise 6.4 with a total of 1808 cores, although on average only 50 cores were available for this experimentation.

### 4.1    Data Set

The dataset utilised for this experimentation is composed by 450 positive simulated CR problems (150 frontal sinuses, 150 clavicles, and 150 patellae) with different degrees of occlusions (0%, 20% and 40%) in the target bone/cavity's silhouette, and their corresponding GT. A simulated CR problem is composed of a 3D surface model and a 2D perspective projection of the 3D model with a random transformation (within a given parameter range showed in Table 1, and with a resolution of 2 pixels per mm) to be superimposed[1]. Simulated CR problems were generated to objectively measure the optimization performance even if there are occlusions present (see [4] for further information about the dataset generation protocol).

---

[1] The materials for generating this dataset were provided by Physical Anthropology Lab at the University of Granada and the Hospital de Castilla la Mancha.

**Table 1.** Parameter range of each bone/cavity for the perspective transformation constrained by international acquisition protocols [19] and expert knowledge.

| Parameter | Bone/Cavity | | |
|---|---|---|---|
| | Frontal sinuses | Patellae | Clavicles |
| Image receptor dimension (mm) | $240 \times 300$ | | $430 \times 350$ |
| $t_x$ (mm) | $[-125, 125]$ | $[-125, 125]$ | $[-210, 210]$ |
| $t_y$ (mm) | $[-150, 150]$ | | $[-175, 175]$ |
| $t_z$ (mm) | $[900 - 200, 900 + 200]$ | | $[900 - 200, 1700 + 200]$ |
| $r_x$, $r_y$, and $r_z$ (degrees) | $[-40°, 40°]$ | | |
| SID (mm) | $[1000 - 100, 1000 + 100]$ | | $[1800 - 100, 1800 + 100]$ |

## 4.2    Performance Metrics

Two metrics are employed to evaluate the quality of the results quantitatively: GT DICE [16] and mean reprojection distance error (mRPD) [12]. The GT DICE metric measures the percentage of not superimposed pixels of the 2D projection obtained by the optimizer and the GT projection (equal to the AM 2D projection used by the optimizer, but without considering occlusions). Notice that, in cases without occlusion, the Masked DICE metric being used by the EAs as fitness function (see Sect. 3) is equal to GT DICE. To avoid any possible bias caused by the high correlation between the two DICE metrics, a second independent metric is also utilized. mRPD [12] measures the average distance from each 3D point of the 3D surface model to a reprojected line, which is the line of all the 3D points whose projection under a certain transformation results in the same 2D point (see [4] for further details).

## 4.3    Experimental Set-Up

This experimentation involves the experimental study of different configurations of the multi-resolution and multi-start CR approach based on DE to all the 450 CR cases to determine the best configuration for the automatic CR method, and also to validate if the new approach improves the results obtained for basic DE in terms of robustness and computational time while keeping its accuracy.

The parameters of the classic DE optimizers were set based on the state-of-the-art method for CR [4] (100 individuals, 500 generations/50.000 evaluations, a crossover probability $P_c$ of 0.5, and F set to 0.5). Meanwhile, the parameters related to the multi-resolution and multi-start strategies have to be studied and fine tuned. We have utilized a grid search where the parameter values are chosen based on preliminary experiments and expert knowledge about its behaviour. The possible values of the parameters are the following:

- Number of resolutions $r$: 2.
- Number of generations devoted to the first resolution $n$: 250 generations (50%), and 375 generations (75%).

- Resolution of the down-sample 2D image in the first resolution $dr$: 25%, and 50%.
- Number of starts $s$: 1 (i.e. the multi-start strategy is not utilized) and 3.
- Shrinking factor $sf$: 1 (i.e. the dynamic bounding strategy is not utilized) and 2 (i.e. the ranges of the parameters are halved).

Since the DE approach is based on a stochastic process, 16 independent runs were performed for each problem instance to compare the robustness of the methods and to avoid any possible bias. The initialization of each run is random in the whole parameter range for all the degrees of freedom of its corresponding projective transformation. A closer initialization would be unrealistic in a real identification scenarios where the AM radiograph was taken in an unknown conditions. Lastly, a stop criteria is set when the optimizer reaches an error lower than 0.01%, which means that the 99.99% of the pixels outside the occlusion region are superimposed correctly with respect to the GT or when 10.000 evaluations have been performed without improving the fitness value of the best solution. Notices that this stop criteria is only applicable in the last optimization stage, i.e. when full resolution images are utilized.

### 4.4    Results

The results obtained are shown in Table 2 according to the Masked DICE, GT DICE and the mRPD metrics together with their computational time requirements. Notice that the results of the DE optimizer are obtained using the state-of-the-art proposal [4] without any modification to allow a fair comparison. The accuracy does not vary significantly depending on the approach and its configuration in terms of mean results. Furthermore, there is neither a significant difference among the approaches behaviour depending on the bone or cavity. The best approaches, in terms of mean of the GT metrics (i.e. GT DICE and mRPD), are the classic CR-DE approach and CR-DE_dr=50%_n=50%_s=3 approach (independently of the value of shrinking factor $sf$), although the are closely tied. Nevertheless, the proposed CR-DE_dr=50%_n=50%_s=3 approach clearly outperforms the classic CR-DE approaches in terms of the standard deviation, probably thanks to the multi-start strategy. However, it requires a higher computational time, whose impact is decreased thanks to the multi-resolution strategy but is still greater than the time of the classic CR-DE approach. Thus, CR-DE_dr=50%_n=50%_s=3 approach only solves the problem of the robustness but not the problem related to the computational cost.

Therefore, in order to overcome the two drawbacks of the classic CR-DE approach for solving the CR problem (see Sect. 2), we have to look for a trade-off between performance and computational time. The best balance is found by CR-DE_dr=25%_n=50%_s=3 approach (independently of the value of shrinking factor $sf$). It shows a performance's mean slowly worse than the CR-DE_dr=50%_n=50%_s=3 and CR-DE approaches and a performance's standard deviation slightly worse than the CR-DE_dr=50%_n=50%_s=3 approach but significantly better than the one of the CR-DE approach, while having a (significantly in some cases) lower computational time.

**Table 2.** Summary of the Masked DICE, GT DICE, and mRPD metrics results and computational time according to bone/cavity type, and optimization approach.

| Bone | Optimizer | Masked DICE | | GT DICE | | mRPD | | Time (s) | |
|---|---|---|---|---|---|---|---|---|---|
| | | Mean | Std | Mean | Std | Mean | Std | Mean | Std |
| Clavicle | DE (state-of-the-art method [4]) | **0.005** | 0.021 | **0.010** | 0.036 | 0.46 | 3.12 | 1515 | 652 |
| | DE_dr=25%_n=50%_s=1_sf=1 | 0.012 | 0.032 | 0.022 | 0.051 | 1.02 | 4.70 | 1268 | 597 |
| | DE_dr=25%_n=50%_s=1_sf=2 | 0.012 | 0.032 | 0.022 | 0.051 | 1.02 | 4.69 | 1276 | 622 |
| | *DE_dr=25%_n=50%_s=3_sf=1* | 0.008 | 0.011 | 0.014 | 0.022 | 0.51 | 0.79 | 1340 | 583 |
| | *DE_dr=25%_n=50%_s=3_sf=2* | 0.008 | 0.011 | 0.014 | 0.022 | 0.50 | 0.79 | 1390 | 610 |
| | DE_dr=25%_n=75%_s=1_sf=1 | 0.021 | 0.033 | 0.034 | 0.054 | 1.44 | 4.76 | 827 | 380 |
| | DE_dr=25%_n=75%_s=1_sf=2 | 0.021 | 0.033 | 0.034 | 0.054 | 1.45 | 4.76 | **822** | **345** |
| | DE_dr=25%_n=75%_s=3_sf=1 | 0.016 | 0.015 | 0.025 | 0.025 | 0.84 | 0.91 | 1009 | 404 |
| | DE_dr=25%_n=75%_s=3_sf=2 | 0.016 | 0.015 | 0.025 | 0.025 | 0.84 | 0.91 | 997 | 397 |
| | DE_dr=50%_n=50%_s=1_sf=1 | 0.012 | 0.022 | 0.020 | 0.039 | 0.83 | 3.24 | 1674 | 805 |
| | DE_dr=50%_n=50%_s=1_sf=2 | 0.012 | 0.022 | 0.020 | 0.039 | 0.83 | 3.24 | 1715 | 837 |
| | DE_dr=50%_n=50%_s=3_sf=1 | 0.007 | **0.009** | 0.012 | **0.018** | **0.44** | **0.61** | 2082 | 753 |
| | DE_dr=50%_n=50%_s=3_sf=2 | 0.007 | **0.009** | 0.012 | **0.018** | **0.44** | 0.61 | 2136 | 88 |
| | DE_dr=50%_n=75%_s=1_sf=1 | 0.024 | 0.022 | 0.036 | 0.040 | 1.29 | 3.21 | 1276 | 578 |
| | DE_dr=50%_n=75%_s=1_sf=2 | 0.024 | 0.022 | 0.036 | 0.040 | 1.29 | 3.21 | 1257 | 581 |
| | DE_dr=50%_n=75%_s=3_sf=1 | 0.019 | 0.011 | 0.027 | 0.021 | 0.83 | 0.72 | 2172 | 855 |
| | DE_dr=50%_n=75%_s=3_sf=2 | 0.019 | 0.011 | 0.027 | 0.021 | 0.83 | 0.72 | 2159 | 770 |
| Patella | DE (state-of-the-art method [4]) | 0.014 | 0.022 | 0.045 | 0.057 | 7.06 | 16.32 | 814 | 369 |
| | DE_dr=25%_n=50%_s=1_sf=1 | 0.019 | 0.026 | 0.051 | 0.060 | 8.08 | 16.98 | 656 | 350 |
| | DE_dr=25%_n=50%_s=1_sf=2 | 0.019 | 0.026 | 0.051 | 0.060 | 8.09 | 16.98 | 696 | 387 |
| | *DE_dr=25%_n=50%_s=3_sf=1* | 0.013 | 0.019 | 0.040 | 0.052 | 6.11 | 14.87 | 664 | 333 |
| | *DE_dr=25%_n=50%_s=3_sf=2* | 0.013 | 0.019 | 0.040 | 0.052 | 6.11 | 14.87 | 654 | 327 |
| | DE_dr=25%_n=75%_s=1_sf=1 | 0.021 | 0.027 | 0.052 | 0.060 | 8.26 | 17.09 | 421 | 202 |
| | DE_dr=25%_n=75%_s=1_sf=2 | 0.021 | 0.027 | 0.053 | 0.060 | 8.26 | 17.09 | 407 | 185 |
| | DE_dr=25%_n=75%_s=3_sf=1 | 0.015 | 0.020 | 0.042 | 0.053 | 6.30 | 14.88 | 477 | **171** |
| | DE_dr=25%_n=75%_s=3_sf=2 | 0.015 | 0.020 | 0.042 | 0.053 | 6.30 | 14.88 | 480 | 196 |
| | DE_dr=50%_n=50%_s=1_sf=1 | 0.018 | 0.024 | 0.051 | 0.061 | 7.87 | 17.77 | 834 | 432 |
| | DE_dr=50%_n=50%_s=1_sf=2 | 0.018 | 0.024 | 0.051 | 0.061 | 7.87 | 17.77 | 815 | 419 |
| | DE_dr=50%_n=50%_s=3_sf=1 | **0.010** | **0.015** | **0.033** | **0.045** | **4.95** | **13.41** | 996 | 428 |
| | DE_dr=50%_n=50%_s=3_sf=2 | **0.010** | **0.015** | **0.033** | **0.045** | **4.95** | **13.41** | 1009 | 443 |
| | DE_dr=50%_n=75%_s=1_sf=1 | 0.020 | 0.025 | 0.054 | 0.061 | 8.07 | 17.93 | 593 | 258 |
| | DE_dr=50%_n=75%_s=1_sf=2 | 0.020 | 0.025 | 0.054 | 0.061 | 8.07 | 17.93 | 592 | 271 |
| | DE_dr=50%_n=75%_s=3_sf=1 | 0.012 | 0.016 | 0.036 | 0.045 | 5.03 | 13.39 | 1035 | 427 |
| | DE_dr=50%_n=75%_s=3_sf=2 | 0.012 | 0.016 | 0.036 | 0.045 | 5.03 | 13.39 | 1066 | 422 |
| Frontal Sinus | DE (state-of-the-art method [4]) | **0.008** | 0.034 | **0.015** | 0.048 | 0.31 | 1.40 | 3429 | 22889 |
| | DE_dr=25%_n=50%_s=1_sf=1 | 0.016 | 0.043 | 0.027 | 0.063 | 0.53 | 1.64 | 1213 | 668 |
| | DE_dr=25%_n=50%_s=1_sf=2 | 0.016 | 0.043 | 0.027 | 0.063 | 0.53 | 1.64 | 1203 | 680 |
| | *DE_dr=25%_n=50%_s=3_sf=1* | 0.010 | 0.028 | 0.018 | 0.042 | 0.30 | 0.98 | 1216 | 654 |
| | *DE_dr=25%_n=50%_s=3_sf=2* | 0.010 | 0.028 | 0.018 | 0.042 | 0.30 | 0.99 | 1266 | 707 |
| | DE_dr=25%_n=75%_s=1_sf=1 | 0.022 | 0.044 | 0.035 | 0.063 | 0.63 | 1.62 | 826 | **400** |
| | DE_dr=25%_n=75%_s=1_sf=2 | 0.022 | 0.044 | 0.035 | 0.063 | 0.63 | 1.61 | **800** | 433 |
| | DE_dr=25%_n=75%_s=3_sf=1 | 0.015 | 0.029 | 0.025 | 0.043 | 0.39 | **0.96** | 997 | 533 |
| | DE_dr=25%_n=75%_s=3_sf=2 | 0.016 | 0.029 | 0.025 | 0.043 | 0.39 | **0.96** | 1010 | 534 |
| | DE_dr=50%_n=50%_s=1_sf=1 | 0.016 | 0.041 | 0.028 | 0.059 | 0.50 | 1.64 | 1604 | 825 |
| | DE_dr=50%_n=50%_s=1_sf=2 | 0.016 | 0.041 | 0.028 | 0.059 | 0.50 | 1.64 | 1543 | 850 |
| | DE_dr=50%_n=50%_s=3_sf=1 | **0.008** | **0.024** | 0.016 | **0.036** | **0.25** | 1.04 | 2086 | 967 |
| | DE_dr=50%_n=50%_s=3_sf=2 | **0.008** | **0.024** | 0.016 | **0.036** | **0.25** | 1.04 | 2107 | 1040 |
| | DE_dr=50%_n=75%_s=1_sf=1 | 0.025 | 0.043 | 0.040 | 0.061 | 0.63 | 1.63 | 1263 | 654 |
| | DE_dr=50%_n=75%_s=1_sf=2 | 0.025 | 0.043 | 0.040 | 0.061 | 0.63 | 1.63 | 1308 | 690 |
| | DE_dr=50%_n=75%_s=3_sf=1 | 0.016 | 0.025 | 0.025 | 0.037 | 0.35 | 1.03 | 2196 | 1012 |
| | DE_dr=50%_n=75%_s=3_sf=2 | 0.016 | 0.025 | 0.025 | 0.037 | 0.35 | 1.03 | 2195 | 1014 |

# 5   Discussion and Future Developments

This contribution has tackled the complex and computation expensive optimization scenario of the CR problem with a multi-resolution and multi-start evolutionary 3D-2D IR approach based on the silhouette of the bone or cavity. It considers a more complex version of the CR problem than the one studied in [4]: occlusions up to the 40% of their silhouettes, and rotations of up to 80° ($[-40°, 40°]$) in the three axis. Our aim was to overcome the drawbacks of the lack of robustness and computational cost to obtain a good superimposition of the state-of-the-art work [4] based on the DE optimizer.

Promising results have been obtained. The best optimization configuration to solve the CR problem, in terms of accuracy and robustness, is the CR-DE_dr=50%_n=50%_s=3 approach. However, it requires a higher computational time. Nevertheless, another configuration of the multi-resolution and multi-start strategies, CR-DE_dr=25%_n=50%_s=3 approach, has managed to improve the results obtained by the CR-DE approach in terms of both robustness and computational time with only small drop in accuracy (lower than the 0.5% of the pixels of the silhouette according to the GT DICE metric).

In conclusion, we have managed to improve the robustness and computational time of the state-of-the-art proposal for the CR problem while keeping a similar accuracy, obtaining high quality superimpositions with any bone or cavity even in the most complex version of the problem.

Future research is planned to model radiographies where the x-ray generator is not perpendicular to the image receptor (e.g. in the Water's projection of radiographs of frontal sinuses). Lastly, we will perform several reliability studies for the validity of different bones and cavities for the CR task [24] through a collaboration with the Israel National Centre of Forensic Medicine and the Hebrew University of Jerusalem.

**Acknowledgements.** Mr. Gómez's work was supported by Spanish MECD FPU grant [grant number FPU14/02380].

# References

1. Thali, M.J., Brogdon, B., Viner, M.D.: Forensic Radiology. CRC Press, Boca Raton (2002)
2. Kahana, T., Hiss, J.: Identification of human remains: forensic radiology. J. Clin. Forensic Med. **4**(1), 7–15 (1997)
3. Markelj, P., Tomaževič, D., Likar, B., Pernuš, F.: A review of 3D/2D registration methods for image-guided interventions. Med. Image Anal. **16**(3), 642–661 (2012)
4. Gómez, O., Ibáñez, O., Valsecchi, A., Cordón, O., Kahana, T.: 3D–2D silhouette-based image registration for comparative radiography-based forensic identification. Pattern Recogn. **83**, 469–480 (2018)
5. Valsecchi, A., Damas, S., Santamaria, J.: Evolutionary intensity-based medical image registration: a review. Curr. Med. Imaging Rev. **9**(4), 283–297 (2013)
6. Damas, S., et al.: Forensic identification by computer-aided craniofacial superimposition: a survey. ACM Comput. Surv. **43**(4), 1–27 (2011)

7. Christensen, A.M.: Testing the reliability of frontal sinuses in positive identification. J. Forensic Sci. **50**(1), 18–22 (2005)
8. Stephan, C.N., Amidan, B., Trease, H., Guyomarc'h, P., Pulsipher, T., Byrd, J.E.: Morphometric comparison of clavicle outlines from 3D bone scans and 2D chest radiographs: a shortlisting tool to assist radiographic identification of human skeletons. J. Forensic Sci. **59**(2), 306–313 (2014)
9. Niespodziewanski, E., Stephan, C.N., Guyomarc'h, P., Fenton, T.W.: Human identification via lateral patella radiographs: a validation study. J. Forensic Sci. **61**(1), 134–140 (2016)
10. Caple, J., Byrd, J., Stephan, C.N.: Elliptical fourier analysis: fundamentals, applications, and value for forensic anthropology. Int. J. Legal Med. **131**(6), 1675–1690 (2017)
11. Russakoff, D.B., et al.: Fast generation of digitally reconstructed radiographs using attenuation fields with application to 2D–3D image registration. IEEE Trans. Med. Imaging **24**(11), 1441–1454 (2005)
12. van de Kraats, E.B., Penney, G.P., Tomazevic, D., Van Walsum, T., Niessen, W.J.: Standardized evaluation methodology for 2-D-3-D registration. IEEE Trans. Med. Imaging **24**(9), 1177–1189 (2005)
13. Feldmar, J., Ayache, N., Betting, F.: 3D–2D projective registration of free-form curves and surfaces. In: Fifth International Conference on Computer Vision, Proceedings, pp. 549–556. IEEE (1995)
14. Damas, S., Cordón, O., Santamaría, J.: Medical image registration using evolutionary computation: an experimental survey. IEEE Comput. Intell. Mag. **6**(4), 26–42 (2011)
15. Storn, R., Price, K.: Differential evolution-a simple and efficient heuristic for global optimization over continuous spaces. J. Global Optim. **11**(4), 341–359 (1997)
16. Sørensen, T.: A method of establishing groups of equal amplitude in plant sociology based on similarity of species and its application to analyses of the vegetation on danish commons. Kongelige Danske Videnskabernes Selskab **5**, 1–34 (1948)
17. Quatrehomme, G., Fronty, P., Sapanet, M., Grévin, G., Bailet, P., Ollier, A.: Identification by frontal sinus pattern in forensic anthropology. Forensic Sci. Int. **83**(2), 147–153 (1996)
18. Stephan, C., Winburn, A., Christensen, A., Tyrrell, A.: Skeletal identification by radiographic comparison: blind tests of a morphoscopic method using antemortem chest radiographs. J. Forensic Sci. **56**(2), 320–332 (2011)
19. Bontrager, K.L., Lampignano, J.: Textbook of Radiographic Positioning and Related Anatomy. Elsevier Health Sciences (2013)
20. Hartley, R., Zisserman, A.: Multiple View Geometry in Computer Vision. Cambridge University Press, Cambridge (2003)
21. Mery, D.: Computer Vision for X-Ray Testing. Springer, Cham (2015)
22. The CGAL Project, CGAL User and Reference Manual, 4.9.1 Edition, CGAL Editorial Board (2017)
23. Valsecchi, A., Damas, S., Santamaria, J., Marrakchi-Kacem, L.: Genetic algorithms for voxel-based medical image registration. In: IEEE Fourth International Workshop on Computational Intelligence in Medical Imaging (CIMI), pp. 22–29. IEEE (2013)
24. Page, M., Taylor, J., Blenkin, M.: Uniqueness in the forensic identification sciences-fact or fiction? Forensic Sci. Int. **206**(1), 12–18 (2011)

# Evolutionary Algorithm for Pathways Detection in GWAS Studies

Fidel Díez Díaz[1], Fernando Sánchez Lasheras[2(✉)],
Francisco Javier de Cos Juez[3], and Vicente Martín Sánchez[4,5]

[1] CTIC Centro Tecnológico, Gijón, Spain
[2] Department of Mathematics, University of Oviedo, Oviedo, Spain
sanchezfernando@uniovi.es
[3] Department of Exploitation and Exploration of Mines,
University of Oviedo, Oviedo, Spain
[4] Biomedicine Institute (IBIOMED),
Research Group of Gene-Environment-Health Interactions,
University of Leon, Leon, Spain
[5] Consortium for Biomedical Research in Epidemiology and Public Health
(CIBERESP), Carlos III Institute of Health, Madrid, Spain

**Abstract.** In genetics, a genome-wide association study (GWAs) involves an analysis of the single-nucleotide polymorphisms (SNPs) that constitute the genome. This analysis is performed on a large set of individuals usually classified as cases and controls. The study of differences in the SNP chains of both groups is known as pathway analysis. The analysis alluded to allows the researcher to go beyond univariate results like those offered by the p-value analysis and its representation by Manhattan plots. Pathway analysis makes it possible to detect weaker single-variant signals and is also helpful in order to understand molecular mechanisms linked to certain diseases and phenotypes. The present research proposes a new algorithm based on evolutionary computation, capable of finding significant pathways in GWA studies. Its performance has been tested with the help of synthetic data sets created with an *ad hoc* developed genomic data simulator.

**Keywords:** Genome wide association studies (GWAS) ·
Single nucleotide polymorphism (SNP) · Pathway analysis ·
Genomic data simulator · Genetic algorithms

## 1  Introduction

In recent years, advances in genomics have provided researchers with an amount of information never available before. This information includes complete genome sequences of individuals from different species as one of its main sources. In most of the genome-wide association studies (GWAS) [1], single-nucleotide polymorphisms are considered as the basic information unit. In general, a SNP can be defined as a variation in a single nucleotide that occurs at a specific position in the genome. In order to detect relevant variations, only those present in a certain appreciable degree in the population are considered [2].

© Springer Nature Switzerland AG 2019
H. Pérez García et al. (Eds.): HAIS 2019, LNAI 11734, pp. 111–122, 2019.
https://doi.org/10.1007/978-3-030-29859-3_10

Despite the large amount of information available, previous studies [3–6] have demonstrated how difficult it is to find genetically relevant explanations to complex phenotypes. Therefore, the development of methodologies able to find relevant relationship is still ongoing. Pathway Analysis [7] can be defined as a methodology whose main purpose is the detection of relevant groups of related genes that are altered in case samples in comparison with a control.

Nowadays, algorithms based on Evolutive Computation are considered to be a complementary approach for GWAS studies [8]. One of the main advantages of this kind of algorithms is their ability to manage the computational complexity of the search for pathways in GWAS studies. Under the name of algorithms based on Evolutive Computation, we refer to a vast class of generic optimization algorithms that will be able to find the relationships between genetic variants and phenotypes [9–12]. The present research develops an algorithm based on an Evolutive Computation methodology called Genetic Algorithms (GAs) that can find relevant pathways in GWAS studies. Although in some contexts GAs are considered as an optimization methodology out of the machine learning scope, in our understanding they should be considered as another machine learning methodology, one very efficient at selecting features and with a high performance when coupled with other machine learning algorithms. In general, over the last few decades machine learning techniques have demonstrated their problem-solving efficacy in the field of health sciences [13–17].

One of the main problems found in GWAS analysis is the large amount of spurious relationships that can occur due to the high dimensionality of the information employed, and the continual presence of many more variables (SNPs) than individuals (patients and controls) that makes GWAS analysis problematic. Also, and due to the general lack of *a priori* knowledge of the possible modes of inheritance, a genomic data simulator has been developed in order to check the input data employed in the research. A check of this kind would be impossible with a real genomic data set. The optimization capabilities of GA allow us to find relevant pathways. Also, permutations of cases and controls in the data set are performed in order to avoid spurious relationships. In the Materials and Methods section of this paper can be found a description of the genomic data simulator as well as information about the database generated and the proposed algorithm based on GA.

## 2   Materials and Methods

In this section, the foundations of the genomic simulator created and the algorithm developed are detailed. Both were programmed in the statistical programming language R [18]. The GA library for genetic algorithms [19] was also employed for the development of the proposed algorithm. All the calculations were performed in a server with 64 GB RAM memory, Intel® Xeon® CPU-1650 v3 @3.50 GHz and 500 GB of solid-state hard disk.

## 2.1   The Genomic Data Simulator

The genomic data simulator developed is based on the epidemiological concept of relative risk. Relative risk is the ratio of the event occurrence in the risk group divided by the risk in the control group. Table 1 shows a two by two odds ratio frequency table, where cases and controls are divided into groups considering the presence or absence of the trait under analysis.

**Table 1.** Two by two odds ratio frequency table.

|          | Cases   | Controls |
|----------|---------|----------|
| Presence | $a$     | $b$      |
| Absence  | $c$     | $d$      |
| Total    | $a+c$   | $b+d$    |

The odds ratio (OR) equation is as follows:

$$OR = \frac{\frac{a}{b}}{\frac{c}{d}} = \frac{a \cdot d}{b \cdot c} \tag{1}$$

Where $a$, $b$, $c$ and $d$ correspond to the variables represented in Table 1. The 95% confidence interval of the OR is calculated considering:

$$\log(95\%CI) = \log(OR) \pm 1.96 \cdot SE(\log(OR)) \tag{2}$$

Please note that the standard error of the logarithm of the odd ratio is calculated as follows [20]:

$$SE(\log(OR)) = \sqrt{\frac{1}{a} + \frac{1}{b} + \frac{1}{c} + \frac{1}{d}} \tag{3}$$

This genomic data simulator makes use of the equation above and allows us to create SNPs with different frequencies of each allele in cases and controls. For this research, relevant SNPs, i.e. those with different frequencies in cases and controls, are created in such a way that their 95% OR confidence intervals (CI) do not contain 1, as this would mean the same risk probability in cases and controls, while those SNPs that are not considered as relevant to finding differences in cases and controls have an OR value of 1 (central value of their CI). While for the relevant SNPs the simulator creates the SNP values considering the OR defined and the width of the CI, for those SNPs that are not relevant, the OR is always equal to 1 and it is only possible to define different widths of the confidence interval.

After the selection of the OR and the width of the CI, the algorithm calculates the number of individuals that present the characteristic in the case and control groups. Next it is decided which alleles are to be linked to the characteristic and which not. In our case, we have programmed four possible scenarios (called types), presented in

Table 2. For example, Type I means that alleles 1 and 2 are linked to the presence of the trait in cases, while for Type II, the presence of the trait is associated with allele 2.

**Table 2.** Alleles allocated to cases and controls in the four types.

| Type I | Alleles |
|---|---|
| Cases (1) | 1, 2 |
| Controls (0) | 0 |
| **Type II** | **Alleles** |
| Cases (1) | 2 |
| Controls (0) | 0, 1 |
| **Type III** | **Alleles** |
| Cases (1) | 0 |
| Controls (0) | 1, 2 |
| **Type IV** | **Alleles** |
| Cases (1) | 1, 0 |
| Controls (0) | 2 |

## 2.2    The Database

The algorithm developed was tested with the help of synthetic data sets, each one formed by 5,000 cases, 5,000 controls and a total of 11,000 SNPs belonging to 10 different chromosomes. These sets were developed with the help of the genomic data simulator described earlier. In this database, only 8 SNPs were created as being significant. The odds ratio selected for the referred SNPs was 1.15 with a 95% confidence interval of $(1.01, 1.29)$. As four different types of alleles allocation with cases and controls were available, two SNPs were created using each one. The rest of the SNPs were created with an odds ratio from 0.95 to 1.05 and a confidence interval width randomly chosen from 0.3 to 0.5.

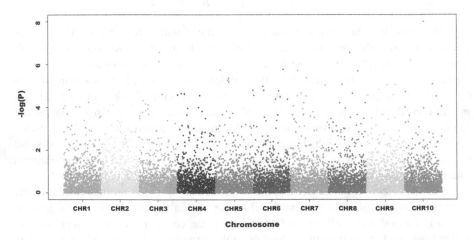

**Fig. 1.** Manhattan plot of the database.

Figure 1 shows the Manhattan plot of the database [21]. This chart represents the $-log10(p)$ where $p$ is the p-value of the Chi-squared test [22] when the occurrence of each allele in cases and controls for each SNP is tested. In the case of this Figure, SNPs are classified in 10 different chromosomes. the $-log10(p)$ of these SNPs ranges from 0 to 3.46711. Please note that the $-log10(0.05)$ is equal to 1.30103. A total of 152 SNPs (1.38% of the total) has a value above this value, including the eight SNPs with an OR of 1.15. Significant SPNs were randomly located in the chromosomes and in this case, two of them, numbers 2 and 8, belong to chromosome number one, another (number 5) to chromosome number 3, another two to chromosome 6 (numbers 1 and 6) another one to chromosome 9 (number 3) and the last two (numbers 4 and 7) to chromosome 10.

## 2.3    Genetic Algorithms

Over the last few decades, the use of algorithms that mimic the behavior of nature has become more widespread. Genetic Algorithms (GA) are a kind of Evolutionary Algorithms inspired by the evolution of organisms [9, 23]. Most of these organisms make use of natural selection and reproduction for its evolution.

The theory of natural selection and evolution is based on the following principles [23]:

- Nature is in constant evolution. Species are in a state of continuous change. Some of them appear while others disappear.
- Changes in species are gradual and continuous.
- Organisms that are similar are usually linked.
- The evolutionary change is the result of a natural selection process where those individuals that are more efficient in adapting to their habitat and able to create offspring survive.

The development of the mathematical theory of GA started with the work of Holland [23]. GAs are a global and robust method for the search of problems solutions. Some of the advantages of GA are as follows [11]:

- They do not require an *a priori* knowledge of the problem to be solved.
- They make use of a subset of the total solutions space in order to find information about the global solutions space by means of the optimization function to be evaluated.
- They make use of probabilistic operators.
- They are easy to program and run in modern computers.
- When used in optimization problems, they are less affected by local maximums than traditional methodologies.

In a GA, the population, initially created either at random or considering certain characteristics, evolves with the help of three main rules that are based on the theory of natural selection cited above. These rules are as follows:

- Selection of the best-performing individuals, whose crossing probabilities increases.
- Crossover: creation of the offspring of the next generation using the genetics of the best individuals in the population.

– Mutation: with a low probability, the characteristics of some of the individuals are changed at random in order to obtain greater diversity in the solutions population. Please note that these probability values are low in order to guarantee convergence.

In the case of the present research, each member of the GA population is formed by a list of 0 s and 1 s. Each 0 represents a non-active SNP and each 1 an active SNP. Active SNPs are those SNPs that are considered to be forming a pathway. For example, the following GA population member '0001011000...000' represents a pathway formed by the SNPs in the positions 4, 6 and 7 of the total of 11,000 SNPs. Table 3 shows the pathways formed with the alleles of the SNPs referred to above.

In our case, the initial population is created with a total of 11,000 members each one with only one active SNP. This population evolves by trying to increase the value of the objective function.

The minor allele frequency (MAF) is defined as the frequency of the least often occurring allele at a specific location. Due to a possible lack of power in order to detect association of the MAF and following a well-known rule [24] generally applied to SNPs in GWAS studies, those pathways that were present in less than 5% of the total of the individuals are discarded. Also, when the total pathways discarded has more than 20% of the total of individuals in the database, the value returned by the objective function is 0. In the case of alleles in Table 3, only those sequences in bold ('222', '121', '221', '021', '122' and '022') are over 5%; as all these pathways contains 92.70% of the individuals, the return value of the objective function is calculated by means of the minus decimal logarithm of the p-value of the Chi-squared test applied to these pathways in cases versus controls.

**Table 3.** Pathways formed with SNPs in the positions 4, 6 and 7.

| Pathway | Cases | Controls | All | Percentage |
|---------|-------|----------|-----|------------|
| 210 | 13 | 13 | 26 | 1.30% |
| 000 | 22 | 5 | 27 | 1.35% |
| 110 | 17 | 12 | 29 | 1.45% |
| 010 | 14 | 16 | 30 | 1.50% |
| 100 | 14 | 20 | 34 | 1.70% |
| **222** | **66** | **65** | **131** | **6.55%** |
| **121** | **72** | **66** | **138** | **6.90%** |
| **221** | **72** | **69** | **141** | **7.05%** |
| **021** | **158** | **94** | **252** | **12.60%** |
| **122** | **169** | **176** | **345** | **17.25%** |
| **022** | **383** | **464** | **847** | **42.35%** |
| **All** | **1000** | **1000** | **2000** | 100.00% |

In the case of the present research, the GA is programmed in such a way that at least 600 iterations are performed in each run. The algorithm stops if after 200 iterations there is no improvement in the value of the objective function of the best solution found. The initial mutation probability is set to 0.1 and divided by 2 after each 100 iterations, in order to accelerate the convergence.

## 2.4 The Proposed Algorithm

Figure 2 shows the flowchart of the proposed algorithm. The first step consists of using the GA in order to find all possible relevant pathways in the database. Afterwards, the same algorithm is applied 1,000 times to the database permutating the cases and controls categories in search of pathways. Therefore, the pathways found in this second step will be spurious. Each one of the algorithms executions took an average time of 27.51 s, with a standard deviation of 3.02 s. Considering the p-values obtained by means of the Chi-squared test and represented by $p$, a cutting point is defined as follows: $max\{-log10(p)\}$. In other words, the cutting point is the maximum of the $-log10(p)$ of the p-values of the pathways obtained by means of permutations that would be considered as spurious associations. Finally, all the pathways that come from the GA application to the original data set and are over the cutting point are considered as relevant.

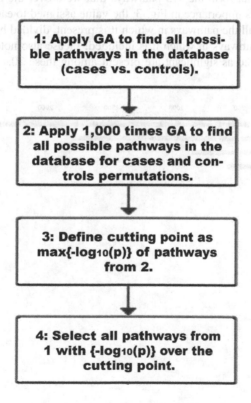

Fig. 2. Flowchart of the proposed algorithm.

## 3  Results and Discussion

The algorithm was applied 1,000 times to the original data set in order to find pathways that would present different alleles in cases and controls. This means that a total of 1,000 replications for cases and controls and 1,000,000 for permutations were performed. An average of 798 pathways were presented for each result, with very different p-values among the pathways in the same solution.

One of the 1,000 executions of the algorithm has been randomly chosen in order for its results to be presented in this paper. Please note that a complete execution of the algorithm involves a total of 1,000 executions of the GA with permutated data and another without data permutation. The p-value considered as a threshold is the maximum p-value obtained from the permutations which minus logarithm in base ten is equal to 6.257662. Results obtained in the 1,000 tests were all in the range from 5 to 7. In the case of the solution randomly chosen to be presented in this research, the average number of SNPs of these pathways was 5.528, with a standard deviation of 0.4993. These pathways are considered to be the significant ones.

Figure 3 shows the scores obtained for the 20 most relevant SNPs from a total of 229 different SNPs that form the 473 pathways that were over the threshold value. In order to classify SNPs importance in Fig. 3, the value assigned to each SNP is the sum of the $-log10(p)$ of all the pathways in which it is present, divided by the total number of SNPs of all the pathways in which it is included. Please also note that 6 out of the total of 8 SNPs created as significant are present among these 20.

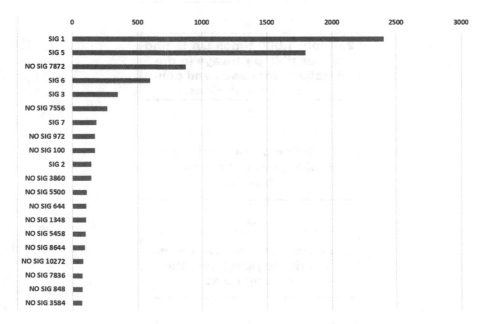

**Fig. 3.** List of the top 20 most relevant SNPs included in significative pathways.

Please remember that the results obtained employed a p-value calculated by means of Chi-squared test in the objective function, but that is not the only way to find statistically-significant differences in GWAS studies. For future research, authors will also consider other methodologies for getting a p-value that are already employed in GWAS analysis, such as the normal approximation to the Fisher's Exact test or the Cochran-Armitage Trend test. In the case of the Cochran-Armitage test, criticized for it being necessary to know the mode of inheritance (recessive, additive or dominant) beforehand, from our point of view the test with all possible modes of inheritance would be also implemented in the GA. Also, for those cases where the large number of SNPs considered greatly increases the computation times, the use of the allele test would be interesting. In this test, a $2 \times 2$ association is performed, classifying cases and controls as carriers or not of the minor risk-carrying allele and afterwards performing a Chi-squared test. Finally, we also consider it to be of interest to test the performance of the GA employed in the present study with other models such as Logistic Regression Linear and Categorial Model tests.

Also, we would like to remark on one of the main limitations of the GA. Due to its evolutive characteristics, we cannot be sure of having found the most relevant pathway in the data base under analysis, as it would be possible for us to get trapped in a local optimum of the search space. This problem is partially solved by repeating the algorithm several times (1,000 in our case). Please note that in an analysis of the pathways proposed over the cutting point in the 1,000 different runs of the algorithm, it was found that a total of 1,071 different SNPs out of the 11,000 were involved in all of them. The 8 significant SNPs were involved in most of the solutions proposed and the relevant SNP with the fewest appearances was number 7 and it was present in 67.6% of the solutions. Only 117 different SNPs were involved in at least 50% of the solutions proposed.

Finally, as can be observed, in our case we have used the maximum value of the $-log10(p)$ of the p-values of the pathways obtained by means of permutations as a cut-off point, so it is possible to change the criteria in order to avoid too many spurious results by simply selecting another cutting value such as the 99th percentile of all the results. In our research, this has not been necessary, but it would be convenient to avoid this issue in future research.

The pathways created by the present method are in general short, with an average of 5.528 SNPs. The reason for this is two-fold: on the one hand, the clear significance of the 8 SNPs created with an OR of 1.15 and also the constraint that does not make use of those pathways with less than 5% of information in SNPs combinations that total more than 20% of cases and controls. These values defined by the researchers would be changed in future. In the permutation analysis, the average number of SNPs present in each pathway was 5.184. Please note that it is also possible to execute the algorithm again after removing one or more of the most significant SNPs in order to find new pathways.

# 4 Conclusions

In this research we present a novel methodology based on GA which is able to detect relevant pathways in GWAS analysis. The algorithm has been tested on synthetic databases that allow the researcher to control the performance of the algorithm with promising results.

The next step of this research will consist of carrying out intensive tests in order to assess the performance of the algorithm under different circumstances and then applying it to real data sets.

One of the advantages of this method over most of the methodologies currently applied for GWAS studies [1] is that it does not require any prior knowledge of the SNPs that are candidates for being included in pathways. Despite this, in a future approach those SNPs candidates may be employed as active variables in the individuals of the initial population.

From our point of view, for future research, the hybridization of Pathways of Distinction Analysis (PoDA) [25] with GA would be of interest in such a way that the performance of different pathways created with the help of a GA be employed as the result of the objective function of the GA. In the case of the database of the present research, PoDA was applied in order to record the times required for performing 1,000 permutations in pathways from 3 to 8 significant SNPs. The average time for a 3 SNPs pathway was 329.98 s. Considering that in the first step it is necessary to create all three possible SNP pathways and to evaluate them in order to determine the best three, in the case of the database employed in this study with 11,000 SNPs, this means more than $221 \cdot 10^9$ different pathways to assess, which makes it impossible an exhaustively apply PoDA to all the candidate pathways.

# References

1. Gonzalez-Donquiles, C., et al.: PoDA algorithm: predictive pathways in colorectal cancer. In: Pérez García, H., Alfonso-Cendón, J., Sánchez González, L., Quintián, H., Corchado, E. (eds.) SOCO/CISIS/ICEUTE -2017. AISC, vol. 649, pp. 419–427. Springer, Cham (2018). https://doi.org/10.1007/978-3-319-67180-2_41
2. Gutiérrez, D.Á., et al.: A multiregressive approach for SNPs identification in prostate cancer. In: Pérez García, H., Alfonso-Cendón, J., Sánchez González, L., Quintián, H., Corchado, E. (eds.) SOCO/CISIS/ICEUTE -2017. AISC, vol. 649, pp. 400–409. Springer, Cham (2018). https://doi.org/10.1007/978-3-319-67180-2_39
3. McCarthy, M., Abecasis, G., Cardon, L., et al.: Genome-wide association studies for complex traits: consensus, uncertainty and challenges. Nat. Rev. Genet. 9, 356–369 (2008). https://doi.org/10.1038/nrg2344
4. Moore, J., Asselbergs, F., Williams, S.: Bioinformatics challenges for genome-wide association studies. Bioinformatics 26, 445–455 (2010). https://doi.org/10.1093/bioinformatics/btp713
5. Visscher, P., Brown, M., McCarthy, M., Yang, J.: Five years of GWAS discovery. Am. J. Hum. Genet. 90, 7–24 (2012). https://doi.org/10.1016/j.ajhg.2011.11.029

6. Fan, Y., Song, Y.: Finding the missing heritability of genome-wide association study using genotype imputation. Matters **2**, e201604000013 (2016). https://doi.org/10.19185/matters. 201604000013

7. García-Campos, M., Espinal-Enríquez, J., Hernández-Lemus, E.: Pathway analysis: state of the art. Front. Physiol. **6**, 383 (2015). https://doi.org/10.3389/fphys.2015.00383

8. Marees, A., de Kluiver, H., Stringer, S., et al.: A tutorial on conducting genome-wide association studies: quality control and statistical analysis. Int. J. Methods Psychiatr. Res. **27**, e1608 (2018). https://doi.org/10.1002/mpr.1608

9. Alonso Fernández, J., Díaz Muñiz, C., Garcia Nieto, P., de Cos, J.F., Sánchez Lasheras, F., Roqueñí, M.: Forecasting the cyanotoxins presence in fresh waters: a new model based on genetic algorithms combined with the MARS technique. Ecol. Eng. **53**, 68–78 (2013). https://doi.org/10.1016/j.ecoleng.2012.12.015

10. Moore, J.H., White, B.: Genome-wide genetic analysis using genetic programming: the critical need for expert knowledge. In: Riolo, R., Soule, T., Worzel, B. (eds.) Genetic Programming Theory and Practice IV. Genetic and Evolutionary Computation, pp. 11–28. Springer, Boston (2007). https://doi.org/10.1007/978-0-387-49650-4_2

11. Ordóñez Galán, C., Sánchez Lasheras, F., de Cos, J.F., Bernardo Sánchez, A.: Missing data imputation of questionnaires by means of genetic algorithms with different fitness functions. J. Comput. Appl. Math. **311**, 704–717 (2017). https://doi.org/10.1016/j.cam.2016.08.012

12. Sánchez Lasheras, J.E., et al.: Classification of prostate cancer patients and healthy individuals by means of a hybrid algorithm combining SVM and evolutionary algorithms. In: de Cos Juez, F.J., et al. (eds.) HAIS 2018. LNCS, pp. 547–557. Springer, Heidelberg (2018). https://doi.org/10.1007/978-3-319-92639-1_46

13. Suárez Sánchez, A., Riesgo Fernández, P., Sánchez Lasheras, F., et al.: Prediction of work-related accidents according to working conditions using support vector machines. Appl. Math. Comput. **218**, 3539–3552 (2011). https://doi.org/10.1016/j.amc.2011.08.100

14. García Nieto, P., Alonso Fernández, J., Sánchez Lasheras, F., de Cos, J.F., Díaz Muñiz, C.: A new improved study of cyanotoxins presence from experimental cyanobacteria concentrations in the Trasona reservoir (Northern Spain) using the MARS technique. Sci. Total Environ. **430**, 88–92 (2012). https://doi.org/10.1016/j.scitotenv.2012.04.068

15. Rosado, P., Lequerica-Fernández, P., Villallaín, L., et al.: Survival model in oral squamous cell carcinoma based on clinicopathological parameters, molecular markers and support vector machines. Expert Syst. Appl. **40**, 4770–4776 (2013). https://doi.org/10.1016/j.eswa. 2013.02.032

16. Vilán Vilán, J., Alonso Fernández, J., García Nieto, P., et al.: Support vector machines and multilayer perceptron networks used to evaluate the cyanotoxins presence from experimental cyanobacteria concentrations in the Trasona reservoir (Northern Spain). Water Resour. Manage. **27**, 3457–3476 (2013). https://doi.org/10.1007/s11269-013-0358-4

17. García Nieto, P., Sánchez Lasheras, F., García-Gonzalo, E., de Cos, J.F.: PM10 concentration forecasting in the metropolitan area of Oviedo (Northern Spain) using models based on SVM, MLP, VARMA and ARIMA: a case study. Sci. Total Environ. **621**, 753–761 (2018). https://doi.org/10.1016/j.scitotenv.2017.11.291

18. R Core Team: A language and environment for statistical computing. R Foundation for Statistical Computing, Vienna, Austria (2018). https://www.R-project.org/

19. Scrucca, L.: GA: a package for genetic algorithms in R. J. Stat. Softw. **53**(4), 1–37 (2013). https://www.jstatsoft.org/v53/i04/

20. Szumilas, M.: Explaining odds ratios. J. Can. Acad. Child Adolesc. Psychiatry **19**(3), 227–229 (2010)

21. Turner, S.D.: qqman: an R package for visualizing GWAS results using Q-Q and Manhattan plots. J. Open Source Softw. **3**, 731 (2018). https://doi.org/10.21105/joss.00731

22. Satagopan, J., Smith, A.: Statistical methods in genomics research. Heart Drug **3**, 48–60 (2003). https://doi.org/10.1159/000070907
23. Holland, J.: Adaptation in Natural and Artificial Systems. University of Michigan Press, Ann Arbor (1975)
24. Östensson, M.: Statistical methods for genome wide association studies. Chalmers University of Technology and the University of Gothenburg, Göteborg (2012)
25. Braun, R., Buetow, K.: Pathways of distinction analysis: a new technique for multi-SNP analysis of GWAS data. PLoS Genet. **7**(6), e1002101 (2011). https://doi.org/10.1371/journal.pgen.1002101

# Particle Swarm Optimization-Based CNN-LSTM Networks for Anomalous Query Access Control in RBAC-Administered Model

Tae-Young Kim and Sung-Bae Cho[✉]

Department of Computer Science, Yonsei University, Seoul, South Korea
{taeyoungkim, sbcho}@yonsei.ac.kr

**Abstract.** As most organizations and companies depend on the database to process confidential information, database security has received considerable attention in recent years. In the database security category, access control is the selective restriction of access to the system or information only by the authorized user. However, access control is difficult to prevent information leakage by structured query language (SQL) statements created by internal attackers. In this paper, we propose a hybrid anomalous query access control system to extract the features of the access behavior by parsing the query log with the assumption that the DBA has role-based access control (RBAC) and to detect the database access anomalies in the features using the particle swarm optimization (PSO)-based CNN-LSTM network. The CNN hierachy can extract important features for role classification in the vector of elements that have converted the SQL queries, and the LSTM model is suitable for representing the sequential relationship of SQL query statements. The PSO automatically finds the optimal CNN-LSTM hyperparameters for access control. Our CNN-LSTM method achieves nearly perfect access control performance for very similar roles that were previously difficult to classify and explains important variables that influence the role classification. Finally, the PSO-based CNN-LSTM networks outperform other state-of-the-art machine learning techniques in the TPC-E scenario-based virtual query dataset.

## 1 Introduction

A relational database management system (RDBMS) is based on a relational model and often used to store financial records, personnel data, manufacturing and logistics information [1]. Organizations and enterprises keep critical information in the database for a long period of time and securely manage data using the consistency, integrity, and security of the RDBMS [2]. However, as the network has been developed recently, the risk of exploiting the vulnerability of the database is increasing, and the damage caused by the leakage of information lowers the productivity of the enterprise [3]. It is also important to protect the database from unauthorized users, as database corruption can seriously affect the operation of an organization that owns the infrastructure. An intrusion detection system (IDS) generally detects abnormal operations in a system. IDS has been widely applied for the security of host and network systems, but there are not many IDS studies targeting RDBMS security. Database security should also detect insider threats as

© Springer Nature Switzerland AG 2019
H. Pérez García et al. (Eds.): HAIS 2019, LNAI 11734, pp. 123–132, 2019.
https://doi.org/10.1007/978-3-030-29859-3_11

well as outsider threats. External threats can be solved with existing security technologies, but insider threats are not clear targets and it is difficult to maintain security due to various exceptions [4]. It is necessary to develop an IDS that can maintain a high level of security for RDBMS. The role-based access control (RBAC) is a method of controlling system access to authorized users in computer system security.

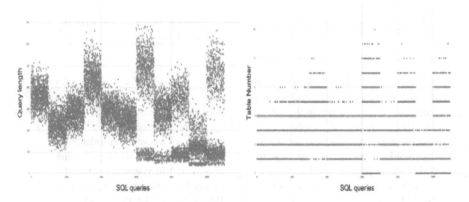

**Fig. 1.** Distribution of TPC-E query data

Figure 1 shows the distribution of query length and table number according to the role of TPC-E query data. We can see the nested distribution between each role through Fig. 1. We classify and analyze access control roles using benchmark database queries provided by TPC-E. The Transaction Processing Performance Council (TPC) is a non-profit corporation founded to define database benchmarks and transaction processing. TPC-E is one of the TPC benchmarks for simulating brokerage OLTP workloads. This dataset has 11 roles in total and contains 33 tables and 191 attributes.

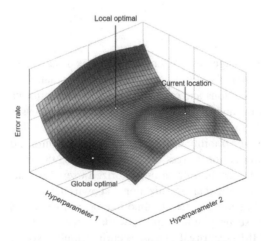

**Fig. 2.** Search space according to hyperparameters

The CNN-LSTM model combining CNN and LSTM for access control increases the complexity of the neural network. In particular, to improve the performance of access control, administrators must adjust large hyperparameters. The size of the hyperparameter space that must be adjusted to classify roles in access control is known to be at least a billion. Figure 2 conceptually shows the search space according to the hyper parameters of a general neural network.

It is the first time when PSO-based CNN-LSTM has been designed and trained for RBAC-based IDS. PSO-based CNN-LSTM detects normal and abnormal behaviors by learning whether the feature extracted from the SQL query belongs to a specific role or not. The remainder of this paper is organized as follows. We provide empirical rules and principles for designing PSO-based CNN-LSTM architectures to better perform abnormal query classifications that deviate from roles. In Sect. 2, we discuss the related work on intrusion detection system. Section 3 details the proposed hybrid PSO-CNN-LSTM network architecture. Section 4 represents experimental results and Sect. 5 concludes the paper.

## 2   Related Works

In Table 1, many researchers have attempted to extract the features of usage data and perform anomaly detection for an intrusion detection system, which can be categorized into four parts: Statistical modeling, rule based modeling, neural network based modeling, and genetic algorithm based modeling.

**Table 1.** Related works on intrusion detection system

| Category | Author | Year | Data | Method | Description |
|---|---|---|---|---|---|
| Statistical modeling | Chen et al. [6] | 2016 | Database log | Hidden Markov model | State-based HMM classification |
| | Islam et al. [7] | 2015 | Database query | Hidden Markov model | Intrusion detection using query syntax |
| Rule based modeling | Ronao et al. [8] | 2016 | Database query | Random forest | Using weighted voting and PCA |
| | Puthran et al. [9] | 2016 | TCP/IP packet | Decision tree | Using IDTBS to improve detection rate |
| Neural network based modeling | Dias et al. [10] | 2017 | TCP/IP packet | Artificial neural network | Signature based IDS using ANN |
| | Devikrishna [11] | 2013 | TCP/IP packet | Multi-layer perceptron | Classification of six attack types using MLP |
| Genetic algorithm based modeling | Ali et al. [12] | 2018 | TCP/IP packet | ANN and GA | Intrusion detection using PSO-FLN |
| | Aslahi et al. [13] | 2016 | TCP/IP packet | GA and SVM | Select feature using GA algorithm |

Chen *et al.* performed intrusion detection in the collected database logs using the hidden Markov model [6]. They applied a statistical method of ordering to detect advanced attacks in a series of intrusion steps. The proposed anomaly detection method is suitable for reporting attack phases, predicting ongoing multistep attacks and preventing further damage to the database. However, since it is based on probability, it is difficult to improve the generalization performance because each event is affected only in the immediately preceding stage and is structurally easy to converge to the local optimum.

Ronao and Cho performed intrusion detection using PCA and random forest with weighted voting [8]. This method uses a weighted voting technique to minimize false alarm and is fast in terms of access control. This approach has the advantage that the user can interpret what query patterns affect the intrusion of the relational database. However, since the continuous variable is treated as a discontinuous value, the prediction error may be large in the vicinity of the boundary point of the separation.

Devikrishna and Ramakrishna performed intrusion detection in TCP/IP communication using multi-layer perceptron [11]. However, the learning time is slow because the learning progresses without extracting the characteristics of data, and the overfitting problem occurs because it is difficult to find the optimal parameters.

Ali *et al.* performed intrusion detection by adjusting the artificial neural network using GA [12]. GA finds the optimal weights when the structure of the neural network is simple [14]. But the optimization performance deteriorates as the neural network becomes complicated.

## 3    The Proposed Method

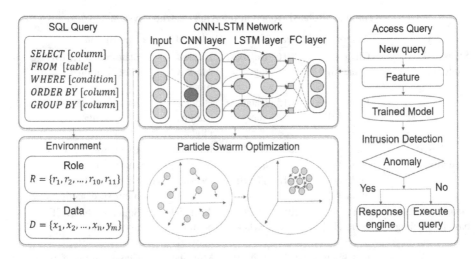

**Fig. 3.** The proposed PSO-based CNN-LSTM intrusion detection structure

The proposed PSO-based CNN-LSTM is composed of CNN, LSTM, and DNN layers and connected in a linear structure. Figure 3 shows the architecture for database access control using the proposed method. CNN-LSTM uses the feature vector extracted by separating the SQL query. The complex feature vectors of each role are extracted by the convolution filters [5]. Features of SQL queries are modeled continuously by the LSTM layer. The trained model uses the softmax classifier to perform intrusion detection on test data. We extract feature by separating input SQL query into clause for database intrusion detection. Features extracted from the query are composed of elements (VALUE-CTR[], GRPBY-ATTR-DEC[], ORDBY-ATTR-DEC[], SEL-ATTR-DEC[], PROJ-ATTR-DEC[], PROJ-REL-DEC[], SQL-CMD[]). The PSO finds the optimal hyperparameter structure while moving the encoded particles. In the parsed feature, we preprocess a total of 277 function vectors by encoding the number and location.

### 3.1 Query Log Parsing and Feature Extraction

In the parsed feature, we preprocess a total of 277 function vectors by encoding the number and location of specific elements in decimal numbers. The generated parsed query input vector is used as input to the proposed CNN-LSTM model. Figure 4 represents extracting parsed features from SQL queries.

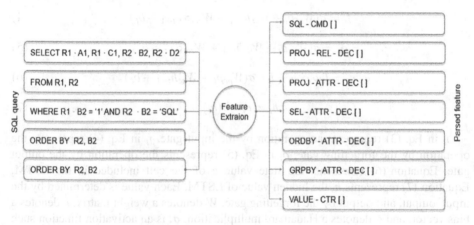

**Fig. 4.** Extracting parsed features from SQL queries

### 3.2 CNN-LSTM Network

Assume that $x = (x_1, x_2, \cdots, x_n)$ be a parsed query log input vector, $n$ is the number of SQL query feature vectors. $x_i$ is normalized feature values. $i$ is the index of the SQL query feature, and $j$ is the index of convolution feature map for each role of RBAC. We use the Eq. (1) to automatically extract the features of the input query. The input value $x_{ij}^0$ represents the parsed query log of the TPC-E dataset. $y_{ij}^1$ represents the feature vector extracted by the convolution layer. The convolution layer is adjusted by the back-propagation algorithm, the learning process, of the convolution layer. Also, $M$ represents the size of the filter for the convolution operation and $b_j^0$ represents the bias for the $j^{th}$ feature map. We use the max pooling layer to reduce the size of the feature vector $y_{ij}^1$ of the SQL query. This layer reduces the complexity of the feature vector $y_{ij}^1$

to reduce the amount of computation. It also effectively reduces the feature size independently for each feature map. Equation (2) represents the operation of the max pooling layer.

$$y_{ij}^l = \sigma(b_j^{l-1} + \sum_{m=1}^{M} W_{m,j}^{l-1} x_{i+m-1,j}^{l-1}) \tag{1}$$

$$p_{ij}^l = \max_{r \in R} y_{i \times T + r, j}^{l-1} \tag{2}$$

In CNN-LSTM, the LSTM layer uses recursive memory cells to store sequential information about SQL feature vectors. By learning SQL feature vectors using the LSTM layer, we can use sequential information to improve performance. The LSTM unit consists of input, output, and forget gates. The LSTM cell updates $h_t$, which is a hidden value according to the input feature, at each time step $t$.

$$i_t = \sigma(W_{pi}p_t + W_{hi}h_{t-1} + W_{ci} \circ c_{t-1} + b_i) \tag{3}$$

$$f_t = \sigma(W_{pf}p_t + W_{hf}h_{t-1} + W_{cf} \circ c_{t-1} + b_f) \tag{4}$$

$$o_t = \sigma(W_{po}p_t + W_{ho}h_{t-1} + W_{co} \circ c_t + b_o) \tag{5}$$

$$c_t = f_t \circ c_{t-1} + i_t \circ \sigma(W_{pc}p_t + W_{hc}h_{t-1} + b_c) \tag{6}$$

$$h_t = o_t \circ \sigma(c_t) \tag{7}$$

$i_t$ in Eq. (3) represents an operation by the input gate. $f_t$ in Eq. (4) represents the operation by the forgetting gate. $o_t$ in Eq. (5) represents the operation by the output gate. Equation (6) represents the state value $c$ of the cell included in the LSTM. Equation (7) represents $h$, the hidden value of LSTM. Each value is determined by the input, output, and output by the forgetting gate. $W$ denotes a weight matrix, $b$ denotes a bias vector, and $\circ$ denotes a Hadamard multiplication. $\sigma$. is an activation function such as tanh for nonlinear decision boundary. Modeling sequential information of SQL queries using LSTM cells provides outperformance and provides superior classification results in access control.

### 3.3 Particle Swarm Optimization

In this paper, we apply the PSO algorithm that automatically searches hyper parameter space of complex CNN-LSTM network. Applying evolutionary design to machine learning and neural networks has great impact as an optimization tool [15]. We create chromosomes by 8-bits binary encoding each hyperparameter. Each chromosome is represented as a particle by the PSO algorithm. The PSO algorithm updates the motion of the particle. We created a 32-bits chromosome by setting the kernel size, units of LSTM, and the output size of the fully connected layer as hyperparameter. We move particles in a direction that enhances access control performance. The movement of particles is affected by the recording of past particles. The velocity of the particles at each iteration is updated by the operation of Eq. (8).

$$v_i(t+1) = (c_1 \times rand() \times (p_i^{best} - p_i(t)))$$
$$+ c_2 \times rand() \times (p_{gbest} - p_i(t))) \quad (8)$$
$$+ v_i(t)$$

In Eq. (8), $v_i(t+1)$ represents the new velocity updated by the PSO algorithm for the ith particle. $c_1$ and $c_2$ represent weighting coefficients for the individual maximum and global optimal positions in the group of particles, respectively. $p_i(t)$ represents the position of the $i^{th}$ particle at time t. $p_i^{best}$ represents the optimal position of the ith particle. $p_g^{best}$ represents the global optimal location. The position of the particle is updated using Eq. (9). The update Eq. (9) takes into account the most suitable location at time $t$ near the region of the particle (Fig. 5).

$$p_i(t+1) = p_i(t) + v_i(t). \quad (9)$$

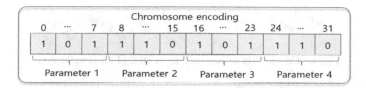

**Fig. 5.** Chromosome representation of hyperparameters

# 4 Experiments

## 4.1 Dataset

In this paper, we have experimented the TPC-E database benchmark dataset. It is generated using the RBAC schema, which simulates the online transaction processing (OLTP) workload of the brokerage company provided by the TPC [8]. To verify the proposed CNN-LSTM intrusion detection system, we generated 1000 data for each of the eleven roles. Each role is composed of customer, broker, market and so on. The TPC-E dataset consists of read/write and read-only transactions. We separate the query by clause to extract the feature vector from the SQL query. Each feature value represents the number of specific elements in the query clause, and encodes the position of the element in decimal number to generate a total of 277 features. To test the proposed method, 11 labels were added to 11,000 data. Since the input values of the PSN-based CNN-LSTM are between 0 and 1, the feature vector is preprocessed to classify roles in access control. $X$ is the decimal encoding value before preprocessing, and $X'$ represents the normalized value.

## 4.2    Performance Comparison with Other Machine Learning Systems

We experimented the proposed PSO-based CNN-LSTM and the conventional machine learning method using 10-fold cross validation. We optimized the existing machine learning method for performance comparison. The proposed PSO-based CNN-LSTM achieved the highest performance in the role classification for access control. The proposed access control system has a performance improvement of about 10% compared to the conventional machine learning method (Fig. 6).

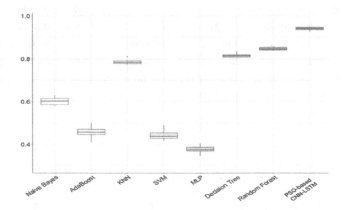

**Fig. 6.**  Comparison of accuracy from 10-fold cross-validation

## 4.3    PSO-Based CNN-LSTM Performance Evaluation by Generation

We measure the fitness value of each generation to see if the PSO technique influences the optimization of the CNN-LSTM method. We performed about 100 iterations for the experiment. As a result of the experiment, the chromosome encoding the hyperparameter of CNN-LSTM was optimized by the PSO technique. Although the fitness value tended to decrease in the initial generation, the fitness value increases with the generation (Fig. 7).

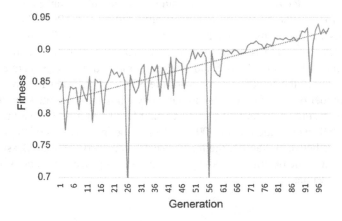

**Fig. 7.**  Fitness value per generation

### 4.4  Cluster Analysis Through t-SNE

Figure 8 shows the output by t-SNE, a technique to visualize each query log data of TPC-E through dimension reduction. t-SNE shows the result of classification of the last fully connected layer of PSO-based CNN-LSTM. We show the last complete connection layer of CNN and PSO-based CNN-LSTM. We can confirm that the intrusion detection of PSO-based CNN-LSTM is performed better than CNN through Fig. 8. CNN and PSO-based CNN-LSTM tend to misclassify role 8 as role 6 in common. While CNN has classified the roles of 11 and 10 as inappropriate, PSO-based CNN-LSTM shows a tendency to misclassify role 6 as 8.

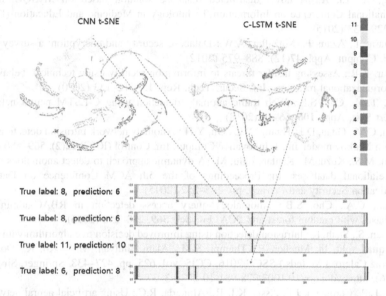

**Fig. 8.** The visualization from the last fully connected layer by t-SNE

## 5  Conclusion

We propose PSO-based CNN-LSTM model for robust and efficient intrusion detection for RBAC-administered relational database. Our model allows a DBMS to parse a large number of queries to quickly and accurately classify unauthorized queries to protect against insider threats. We optimized the hyperparameters of CNN-LSTM using PSO. We have also shown its usefulness by comparing the proposed method with conventional machine learning methods. The proposed method models the complex patterns of database queries and the sequence of features. It has achieved high classification performance for TPC-E datasets with eleven roles, and also handles uneven database query access well. Through the t-SNE analysis, we have identified the characteristics of the data that the proposed method failed to properly classify. In order to verify the practical usefulness of the proposed system, the actual SQL query should be collected from the data server. We also need to collect large amounts of data to demonstrate the effectiveness of intrusion detection systems.

**Acknowledgement.** This work was supported by an Electronics and Telecommunications Research Institute (ETRI) grant funded by the Korean government (19ZS1110, Development of self-improving and human augmenting cognitive computing technology).

# References

1. Shmueli, E., Vaisenberg, R., Elovici, Y., Glezer, C.: Database encryption: an overview of contemporary challenges and design considerations. ACM SIGMOD Rec. **38**(3), 29–34 (2010)
2. Dong, X., Li, X.: A novel distributed database solution based on MySQL. In: 7th International Conference on Information Technology in Medicine and Education (ITME), pp. 329–333 (2015)
3. Basharat, I., Azam, F., Muzaffar, A.W.: Database security and encryption: a survey study. Int. J. Comput. Appl. **47**(12), 888–975 (2012)
4. Sarkar, K.R.: Assessing insider threats to information security using technical, behavioural and organizational measures. Inf. Secur. Tech. Rep. **15**, 112–133 (2010)
5. Kim, T.-Y., Cho, S.B.: Web traffic anomaly detection using C-LSTM neural networks. Expert Syst. Appl. **106**, 66–76 (2018)
6. Chen, C.M., Guan, D.J., Huang, Y.Z., Ou, Y.H.: Anomaly network intrusion detection using hidden Markov model. Int. J. Innovative Comput. Inf. Control (ICIC) **12**(2), 569–580 (2016)
7. Islam, M.S., Kuzu, M., Kantarcioglu, M.: A dynamic approach to detect anomalous queries on relational databases. In: Proceedings of the 5th ACM Conference on Data and Application Security and Privacy, pp. 245–252 (2015)
8. Ronao, C.A., Cho, S.B.: Anomalous query access detection in RBAC-administered databases with random forest and PCA. Inf. Sci. **369**, 238–250 (2016)
9. Puthran, S., Shah, K.: Intrusion detection using improved decision tree algorithm with binary and quad split. In: Mueller, P., Thampi, S.M., Alam Bhuiyan, M.Z., Ko, R., Doss, R., Alcaraz Calero, J.M. (eds.) SSCC 2016. CCIS, vol. 625, pp. 427–438. Springer, Singapore (2016). https://doi.org/10.1007/978-981-10-2738-3_37
10. Dias, L.P., Cerqueira, J.J., Assis, K.D.R., Almeida, R.C.: Using artificial neural network in intrusion detection systems to computer networks. In: Computer Science and Electronic Engineering (CEEC), pp. 145–150 (2017)
11. Devikrishna, K.S., Ramakrishna, B.B.: An artificial neural network based intrusion detection system and classification of attacks. Int. J. Eng. Res. Appl. (IJERA) **3**(4), 1959–1964 (2013)
12. Aslahi-Shahri, B.M., et al.: A hybrid method consisting of GA and SVM for intrusion detection system. Neural Comput. Appl. **27**(6), 1669–1676 (2016)
13. Ali, M.H., Mohammed, B.A.D., Ismail, A., Zolkipli, M.F.: A new intrusion detection system based on fast learning network and particle swarm optimization. IEEE Access **6**, 20255–20261 (2018)
14. Seo, Y.-G., Cho, S.-B., Yao, X.: The impact of payoff function and local interaction on the N-player iterated prisoner's dilemma. Knowl. Inf. Syst. **2**(4), 461–478 (2000)
15. Cho, S.-B., Shimohara, K.: Evolutionary learning of modular neural networks with genetic programming. Appl. Intell. **9**(3), 191–200 (1998)

# d(Tree)-by-dx: Automatic and Exact Differentiation of Genetic Programming Trees

Peter Rockett[1][(✉)] [ID], Yuri Kaszubowski Lopes[1], Tiantian Dou[1], and Elizabeth A. Hathway[2]

[1] Department of Electronic and Electrical Engineering, University of Sheffield, Portobello Centre, Pitt Street, Sheffield S1 4ET, UK
p.rockett@sheffield.ac.uk
[2] Department of Civil and Structural Engineering, University of Sheffield, Mappin Street, Sheffield S1 3JD, UK

**Abstract.** Genetic programming (GP) has developed to the point where it is a credible candidate for the 'black box' modeling of real systems. Wider application, however, could greatly benefit from its seamless embedding in conventional optimization schemes, which are most efficiently carried out using gradient-based methods. This paper describes the development of a method to automatically differentiate GP trees using a series of tree transformation rules; the resulting method can be applied an unlimited number of times to obtain higher derivatives of the function approximated by the original, trained GP tree. We demonstrate the utility of our method using a number of illustrative gradient-based optimizations that embed GP models.

**Keywords:** Genetic programming · Automatic differentiation · Optimization · Real-world applications

## 1 Introduction

It is now widely accepted that genetic programming (GP) is capable of competitive application across a range of sciences and engineering [5]. One area that has not hitherto received much attention is the important topic of integration of GP into conventional optimization-based applications, such as control, that typically require the computation of derivatives for fast solution. Izzo et al. [8] have recently listed a range of diverse applications of GP that require derivatives for effective solution.

In particular, our specific interest is model predictive control (MPC) [3] where a dynamic predictive model of system behavior is used to optimize future inputs over a so-called *prediction horizon* extending many time steps into the future. MPC is especially suited to systems where there is a significant time lag between application of an input and an observable response, and has been widely used in

© Springer Nature Switzerland AG 2019
H. Pérez García et al. (Eds.): HAIS 2019, LNAI 11734, pp. 133–144, 2019.
https://doi.org/10.1007/978-3-030-29859-3_12

the process industries although interest is growing for the indoor environmental control of buildings [14]. One of the major problems—and indeed costs—of MPC is the economic acquisition of a suitable predictive model, and GP, as a machine learning technique, is particularly attractive for this purpose. Hitherto, however, the ability to integrate GP predictive models into the fast, gradient-based optimizers conventionally used in MPC has been a significant barrier to the deployment of GP.

Potentially, so-called derivative-free conventional optimizers [4] can avoid the need for explicit derivatives although the solution times are invariably longer than for techniques explicitly based on derivatives; further, derivative-free solvers often internally approximate derivatives, a limitation we will expand upon later in this paper. In essence, the derivatives of the objective function to be optimized usually have to implicitly exist, at least up to second order.

Mousavi Astarabadi and Ebadzadeh [12] have calculated tree derivatives using finite difference approximations, but there is a well-known trade-off here between making the perturbation on the variable of differentiation too small and the accuracy being dominated by round-off errors, and making it too big and suffering large truncation errors. Unfortunately, the optimal value of perturbation typically varies across the domain.

Finally, stochastic search methods, such as differential evolution, particle swarm optimization, etc. strictly require no gradient information but tend to be too slow for real-time applications, such as control [6].

To expand the range of applications of GP, a means of exploiting fast, gradient-based optimizers is therefore highly desirable. This, in turn, requires a straightforward and reliable method of calculating the derivatives of GP trees, and this is the key contribution of the present paper.

Izzo et al. [8] have recently reported the application of truncated Taylor polynomials in the context of Cartesian genetic programming to calculate derivatives. These authors concede, however, that the "necessary algebraic manipulations" are "non trivial". The approach we present here, on the other hand, is much simpler. Indeed, we believe all of the concepts will be very familiar to the GP community.

In this paper, we describe a set of tree transformations that generate the partial derivative (with respect to a given variable) of a real function described by conventional tree-based GP. In Sect. 2, we describe the problem formulation, and we derive the necessary tree transformations in Sect. 3. We give practical implementation details in Sect. 4. Section 5 describes two examples of the application of our tree-differentiation technique for embedding GP within conventional, gradient-based optimizations. Possible future work, including extensions, is discussed in Sect. 6; Sect. 7 concludes the paper.

## 2    Problem Statement

A very general requirement for a real function—be it implemented by a GP or otherwise—to be differentiable is for it to be *analytic*. (More pedantically, an analytic function is infinitely differentiable [15]). Since GP tree mappings ($\mathbb{R}^N \to \mathbb{R}$)

are simply compositions of the internal function nodes, and a composition of analytic functions is itself analytic [15], it follows that a GP tree will be analytic, and therefore (infinitely) differentiable, if each of the function nodes is analytic. At this stage, the principal challenge becomes apparent. The commonly-used set of function nodes in GP comprises: addition ('+'), subtraction ('−'), multiplication ('×') and some version of protected division (PD) defined by, for example:

$$\frac{f}{g} \begin{cases} \frac{f}{g} & g \neq 0 \\ 1 & g = 0 \end{cases} \tag{1}$$

where $f, g$ are the values returned by the two subtrees of the PD operator.

Although '+', '−' and '×' are everywhere analytic, protected division is not. In particular, when $g = 0$ the limit defining the derivative does not exist. While GP systems using only a function set of $\{+, -, \times\}$ have been explored, the ability to divide some quantity into 'parts' appears to give greater expressiveness [10]. (More generally, Keijzer [10] criticizes the use of 'protected' operators in GP.) Fortunately, an elegant solution to the above problem has already been reported by Ni et al. [13] who replaced the protected division operator with an analytic quotient defined by

$$AQ(f, g) = \frac{f}{\sqrt{1 + g^2}}$$

that tends asymptotically to the quotient value $f/|g|$ when $|g| \gg 1$, but retains analyticity at $g = 0$. Although the principal motivation in [13] was to avoid problems with the PD operator returning huge values when $|g|$ was very small but strictly $> 0$, the differentiability of the operator was remarked upon in [13]. In consequence, in this work we replace the conventional protected division operator with the analytic quotient (AQ) operator.

## 3   Theory

Since the fundamental requirement for all function nodes to be analytic can now be met with a function set of $\{+, -, \times, AQ\}$ supplemented by a leaf node set of $\{x_i \in \mathbb{R} \, \forall i \in [1 \ldots n], \text{ and constants} \in \mathbb{R}\}$ where $n$ is the dimensionality of the input vector $\mathbf{x}$, an arbitrary GP tree can be differentiated using repeated, recursive application of the standard chain rule of calculus for the differentiation of a composition of functions. Thus, where $F(x) = h(k(x))$:

$$F'(x) = h'(k(x)).k'(x) \tag{2}$$

### 3.1   Internal Function Nodes

Differentiation of each of the internal function nodes follows straightforwardly from (2), and the chain rule and can be conveniently formulated as a transformation of the elements of the GP tree.

*Addition/Subtraction.* The derivative can be written as:

$$\frac{\partial(f \pm g)}{\partial x_i} = f' \pm g' \tag{3}$$

which can be represented by the simple tree transformation in Fig. 1.

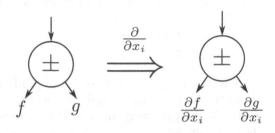

**Fig. 1.** Addition/subtraction transformation

*Multiplication.* Similarly, the derivative can be written as:

$$\frac{\partial(fg)}{\partial x_i} = f'g + fg' \tag{4}$$

which is represented by the tree transformation shown in Fig. 2.

*Analytic Quotient.* Finally, the derivative of the analytic quotient can be written as (5), which can be rearranged as (6) in the form of other analytic quotient functions:

$$\frac{\partial AQ(f,g)}{\partial x_i} = f'(1 + g^2)^{-1/2} - fgg'(1 + g^2)^{-3/2} \tag{5}$$

$$= \frac{f'}{\sqrt{1 + g^2}} - \frac{f}{\sqrt{1 + g^2}} \cdot \frac{g}{\sqrt{1 + g^2}} \cdot \frac{g'}{\sqrt{1 + g^2}} \tag{6}$$

The practical choice between (5) and the less compact (6) is discussed further in Sect. 4, but for the present, the tree transformation corresponding to (6) is shown in Fig. 3.

### 3.2 Terminal Nodes

Any given GP tree can be differentiated by repeated, recursive application of the transformations shown above. Finally, however, the inputs to the transformed (differentiated) trees will reach the leaves of the tree. For completeness, the derivative of a variable is given by (7), and for a constant leaf node by (8).

$$\frac{\partial x_i}{\partial x_j} = \begin{cases} 1 & i = j \\ 0 & i \neq j \end{cases} \tag{7}$$

$$\frac{\partial c}{\partial x_i} = 0 \quad \forall i \in [1 \dots n] \tag{8}$$

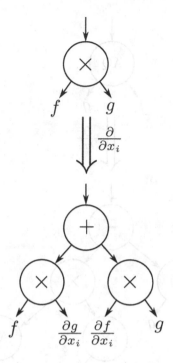

**Fig. 2.** Multiplication transformation

## 4    Implementation

Since trees are differentiated by application of the chain rule of calculus, our implementation is correct by construction.

With reference to Sect. 3, it is clear that multiplication, and especially the AQ operator, result in significant growth in tree size compared to the original, undifferentiated tree. One AQ node in the original tree transforms to (at least) seven nodes in the differentiated tree. This tree growth is a direct consequence of the chain rule of calculus. (For reference, the derivative of a regular (unprotected) quotient operation would transform to five nodes.) The increased complexity of derivatives is well-recognized in numerical analysis, and many derivative-based solvers include either automatic differentiation libraries to generate derivatives from instrumented source code [1], or additional utilities to roughly check hand-generated derivatives using finite difference approximations.

In the context of the present work, the tree growth from differentiation poses a design choice: should we use the same function set and express the derivative of the AQ operator using the existing function set at the inevitable cost of some tree growth? Or should we augment the function set with a new type of 4-ary function node that embeds the derivative of AQ as a single node and takes $f$, $g$, $\partial f/\partial x_i$ and $\partial g/\partial x_i$ as inputs? This latter option would lead to less tree growth since the differentiated AQ node would be replaced by a single derivative node.

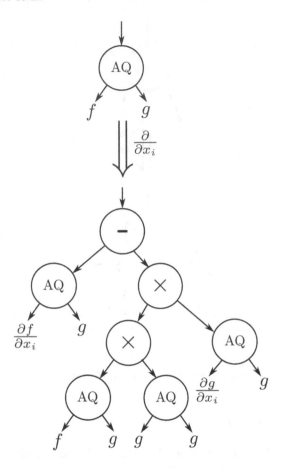

**Fig. 3.** Analytic quotient transformation

We have chosen the former route since expressing the derivative tree using only the original function set allows repeated application of the tree differentiation operation without limit to form higher derivatives; for example, the Hessian matrix, the elements of which comprise second-derivatives such as $\partial^2 f/\partial x_i \partial x_j$, is often invaluable in optimization problems. The alternative implementation of expressing an AQ derivative within the tree as a new 4-ary node would require the addition of further, special nodes to handle the derivatives of the derivative of an AQ operator. Generating third and higher order derivatives would cause further significant complications.

While we give an example later of the direct use of second-order derivatives in optimization, at the present time we can suggest no immediate application for third and higher-order derivatives; nonetheless, such quantities are frequently used in mathematical analysis [15]. The ability to generate higher derivatives, or to assure the analyticity of a GP model thus creates an enabling platform for possible future research.

Detailed implementation was in the form of a function that takes the GP tree as an argument together with the index in **x** of the variable of differentiation, and returns a tree generated by application of the tree transformations in Sect. 3. This newly created tree is completely independent of the original, undifferentiated tree and can be recursively evaluated in the normal manner to obtain the value of the derivative of the original tree at some arbitrary **x**. Our initial implementation is in C++ and we reference both the original and differentiated trees using C(++) pointers to the trees.

## 5   Experiments

In order to fully demonstrate the efficacy of automatically-calculated tree gradients, we present two example use cases of embedding of GP within conventional optimization frameworks. Sect. 5.1 reports the use of first-derivative information only while Sect. 5.2 describes the inclusion of explicit Hessian information. Both cases would be typical of control or related applications of GP.

### 5.1   Gradient-Based Minimization - I

As an initial demonstration, we have used the tutorial example in the documentation of the highly-regarded NLopt[1] nonlinear optimization library [9], and given in (9) and (10).

$$\min_{\mathbf{x}\in\mathbb{R}^2}\sqrt{x_2} \tag{9}$$

$$\text{s.t.}\ \ 2x_1^3 - x_2 \leq 0\ \ \text{and}\ (-x_1 + 1)^3 - x_2 \leq 0 \tag{10}$$

We selected ten evenly spaced points in both $x_1$ and $x_2$ to train three separate GP approximations to the objective function (9), and the two constraint functions (10). We have used a fairly standard GP framework with a population of 100 individuals, a hard depth limit of 8, and a fixed number of 10,000 function evaluations per function. (The exact details are unimportant—we have deliberately made no attempt to accurately learn the functions in (9, 10) since our objective was not to try to reproduce the optimization results for this test problem, but rather demonstrate the utility of our tree differentiation approach.) Having (approximately) learned each of the three functions in (9, 10), these trained trees were incorporated into the NLopt example program. We used the sequential least-squares quadratic programming (SLSQP) algorithm from NLopt [11] to solve the optimization problem, which requires derivatives of both the objective function and of the two constraint functions. Six separate trees implementing the derivatives in $x_{1,2}$ of each of the three functions described above were generated *at runtime*, and evaluated as demanded by the SLSQP algorithm to return the values of the derivatives. The optimization was set to terminate when the relative

---

[1] We have used NLopt version 2.5 downloadable from https://github.com/stevengj/nlopt/archive/v2.5.0.tar.gz.

error of the optimized parameters fell below $10^{-4}$, which required 27 function evaluations; the optimal solution was obtained at the point $(0.300489, 0.282026)$ and an objective function minimum of $0.256948$.

(In contrast, the original problem in (9, 10) had an optimum at $(0.3333, 0.2963)$, a function minimum of $0.5443$, and required 53 iterations with the same termination criterion. The differences can, of course, be explained by the fact that we have solved an *approximation* of the original problem in (9, 10) using GP trees.)

To judge whether our GP-generated solution was indeed a minimum, we generated 1 million random points in a 2D square of 0.1 on a side and centered on the solution point. We were unable to find any point in the neighborhood of the solution that had a lower objective function and that satisfied the constraints. We thus infer that the identified solution point is indeed a minimum (of that modified problem).

## 5.2   Gradient-Based Minimization - II – Using Hessian Information

As a second example, we have used our GP differentiation framework within the Ipopt [16] large-scale interior-point algorithm for nonlinear programming that combines line-search and trust region methods [17]. We have used Ipopt version 3.12.12 freely downloadable from http://www.coin-or.org/download/source/Ipopt. We have used the example problem from Ipopt's documentation:

$$\min_{x_1,x_2,x_3,x_4 \in \mathbb{R}} x_1 x_4 (x_1 + x_2 + x_3) + x_3 \tag{11}$$

$$\text{s.t. } x_1 x_2 x_3 x_4 \geq 25 \tag{12}$$

$$x_1^2 + x_2^2 + x_3^2 + x_4^2 = 40 \tag{13}$$

$$1 \leq x_{1,2,3,4} \leq 5 \tag{14}$$

which, in turn, is a standard, constrained nonlinear benchmark problem from the collection published by Hock and Schittkowski [7]. The problem has an optimal solution given by: $\mathbf{x}^* = (1.00000000, 4.74299963, 3.82114998, 1.37940829)$ with a minimum objective value of $17.014017140224134$.

Ipopt is an interesting application of our method because, as well as needing gradient values, it requires the Hessian of the Lagrangian functional, $\nabla^2 f + \lambda_1 \nabla^2 g_1 + \lambda_2 \nabla^2 g_2$, where $f$ is the objective (11), $g_{1,2}$ are the two constraint functions (13) and (14), and $\lambda_{1,2}$ are the Lagrange multipliers. If the Hessian information is not explicitly available, Ipopt approximates the Hessian of Lagrangian internally with some loss of accuracy. We are thus able to explore if our approach can produce comparable results to conventional coding of the objective and constraint functions, and their first and second derivatives.

Since both the objective and constraint functions in (11)–(14) are exactly representable with addition and multiplication operators, we have hand-constructed GP trees for the objective function (11) and the two constraint functions (13) and (14) only, and verified that these trees returned identical numerical results compared to direct C coding of the functions within floating-point round-off error. We have then embedded these hand-coded trees within the example Ipopt C code. Rather than hand-evaluate the necessary gradients and Hessians (as was done in the Ipopt example code), we have automatically generated these quantities by application of our tree differentiation procedures. Since the gradient vectors $\nabla f, g_{1,2}$ are each composed of four partial derivatives, each differentiation with respect to $x_{1-4}$, we generated—at *runtime*—a total of $4 \times 3 = 12$ trees implementing $\nabla f$ and $\nabla g_{1,2}$.

Similarly, we generated the $4 \times 4$ Hessian matrices $\nabla^2 f$ and $\nabla^2 g_{1,2}$ with a second round of applications of the procedure to $\nabla f$ and $\nabla g_{1,2}$, again at runtime. That is, we have differentiated the derivatives. Each element of each Hessian matrix required a separate GP tree making a total of $4 \times 4 \times 3$ trees to produce the second-order information. The generation of none of the first or second-order derivative information required any manual intervention (other than coding the invocations of the tree differentiations).

We then compared the GP-based optimization against the use of the hard-coded implementation of the objective, constraints, gradient, Lagrangian, and Hessian functions. Since manually calculating Hessian matrices is usually tedious and error-prone, Ipopt offers the facility to internally approximate this quantity albeit with some loss of accuracy. We thus also used this method to provide the second-order information within our GP-based implementation of the optimization to compare with exact, automatic generation of the Hessian by tree differentiation. We tried various combinations of calculating the necessary functions and their performances are shown in Table 1.

**Table 1.** Results for the gradient-based minimization - II. Comparison between the use of hand-coded (CD) functions, the use of GP trees, and the use Ipopt's Hessian approximation (HAP). The table shows the method of calculating: the objective function ($f$), constraints ($g$), the gradient of the objective function ($\nabla f$), the Jacobian of the constraints ($J$), and the Hessian ($H$). The final column shows the difference in the objective value compared to the fully hand-coded solution.

| id | $f$ | $g$ | $\nabla f$ | $J$ | $H$ | Objective | Difference |
|----|-----|-----|-----|-----|-----|-----------|------------|
| 01 | CD | CD | CD | CD | CD | 17.014017140224134 | n/a |
| 02 | GP | CD | CD | CD | CD | 17.014017140224134 | 0 |
| 03 | GP | GP | CD | CD | CD | 17.014017140224137 | $+3 \times 10^{-15}$ |
| 04 | GP | GP | GP | CD | CD | 17.014017140224134 | 0 |
| 05 | GP | GP | GP | GP | CD | 17.014017140224137 | $+3 \times 10^{-15}$ |
| 06 | GP | GP | GP | GP | GP | 17.014017140224137 | $+3 \times 10^{-15}$ |
| 07 | CD | CD | CD | CD | HAP | 17.014017140224176 | $+4.2 \times 10^{-14}$ |
| 08 | GP | CD | CD | CD | HAP | 17.014017140224176 | $+4.2 \times 10^{-14}$ |
| 09 | GP | GP | CD | CD | HAP | 17.014017140224176 | $+4.2 \times 10^{-14}$ |
| 10 | GP | GP | GP | CD | HAP | 17.014017140224180 | $+4.6 \times 10^{-14}$ |
| 11 | GP | GP | GP | GP | HAP | 17.014017140224176 | $+4.2 \times 10^{-14}$ |

The basic comparison to be made is between the first and sixth rows of Table 1 since these show the results of implementing the objectives and constraint functions, their derivatives and Hessians using hand-coded C (id = 01), and implementing the objective and constraint function as GP trees and then automatically generating the first and second-order derivative information at runtime by tree differentiation (id = 06). The differences between the two objective values obtained (using standard IEEE 754:1985 double-precision floating point arithmetic with roughly 16–17 decimal digits) occurs in the least-significant digit and is almost certainly due to rounding error. To summarize this sub-section, we have demonstrated that our automatic tree differentiation procedure is able to produce accuracies that differ from the results of hand-coded evaluation by what we believe to be rounding error alone.

The first six rows of Table 1 show various combinations of using hand-coded functions, derivatives and Hessians and the corresponding quantities evaluated using GP trees. Any differences that do exist also appear due to rounding errors rather than issues with the accuracy of the GP derivative trees.

Rows id = 07 to id = 11 show the results of using Ipopt's internal approximation of the Hessian information. The differences compared to row id = 01 are all very similar and would thus all appear dominated by the errors due to Ipopt's internal Hessian approximation. A noteworthy point here is that hand-coding Hessian information is well-known to be a tedious and error-prone process due to the increasing complexity of the derivative-of-derivative expressions. Using automatic tree differentiation, however, there is a negligible cost to exact evaluation of the second-derivative information.

To summarize this sub-section, we have demonstrated that our automatic tree differentiation procedure is able to produce accuracies that differ from the results of hand-coded evaluation by what we believe to be rounding error alone.

## 6  Discussion and Future Work

Although the analytic quotient operator (AQ) [13] allows us to produce the transformation in Sect. 3.1, it is almost certainly not the only suitable function. Other analytic, quotient-type operators may be preferable, but the property of AQ of being able to construct derivatives of any order without extending the function set is very attractive for easily generating higher-order derivatives. Consideration of other quotient options is an area for future work.

Another area for future work is extension to additional function nodes (e.g. circular and other transcendental functions). Extension to sine and cosine would appear straightforward since these generate complementary derivatives. That is, the derivative of a sine function is a cosine, and vice versa. Similarly, the derivative of an exponential function is another exponential. The consequence is that forming higher derivatives by repeated transformation of already-calculated derivative trees is straightforward, and would not require extension of the GP function set. Other functions commonly used in GP might be more problematic.

We have framed the present paper in terms of tree transformations, but automatic differentiation (AD) of computer code has received considerable attention—see [1], for example, for a review. AD is, of course, also based on the chain rule of calculus but allows users to generate derivatives of expressions in conventional computer code by instrumenting that code, and typically using a pre-processor system to substitute compilable code implementing the required derivatives. The discriminating factor between AD and the present work is that GP trees are not usually manually programmed in conventional languages, but rather exist as trained tree models in the computer's memory. Our automatic differentiation operates directly on the in-memory data structures. Nonetheless, there may be opportunities for cross fertilization with the AD literature. In this context, Baydin et al. [2] have recently reviewed the links between AD and machine learning.

One major area of our future work will be to exploit GP tree derivatives in model predictive control (MPC) [3], as set out in Sect. 1. Results will be published elsewhere.

## 7   Conclusions

In this paper, we have introduced a series of tree transformations that can be applied recursively to generate independent trees for the evaluation of the derivative of the function implemented by the original tree. Since the differentiated tree can be expressed in terms of the original function set, we can apply the tree differentiation procedure any numbers of times to produce, again independent, GP trees, that implement higher-order derivatives.

We have demonstrated the application of our tree differentiation procedure on two representative constrained, non-linear optimization problems. These demonstrators involved conventional optimization in which both the objective and constraint functions were implemented with genetic programming trees.

The present paper thus makes an important contribution to extending the application of genetic programming to novel, real-world problems, especially control.

**Acknowledgements.** We gratefully acknowledge supported by the UK Engineering and Physical Sciences Research Council (EPSRC) under grant EP/N022351/1.

## References

1. Bartholomew-Biggs, M.C., Brown, S., Christianson, B., Dixon, L.C.W.: Automatic differentiation of algorithms. J. Comput. Appl. Math. **124**, 171–190 (2000). https://doi.org/10.1016/S0377-0427(00)00422-2
2. Baydin, A.G., Pearlmutter, B.A., Radul, A.A., Siskind, J.M.: Automatic differentiation in machine learning: a survey. J. Mach. Learn. Res. **18**(1), 5595–5637 (2017). http://jmlr.org/papers/v18/17-468.html
3. Camacho, E.F., Bordons, C.: Model Predictive Control, 2nd edn. Springer, London (2004)

4. Conn, A.R., Scheinberg, K., Vicente, L.N.: Introduction to Derivative-Free Optimization. Society for Industrial and Applied Mathematics (2009). https://doi.org/10.1137/1.9780898718768
5. Gandomi, A.H., Alavi, A.H., Ryan, C. (eds.): Handbook of Genetic Programming Applications. Springer, Cham (2015). https://doi.org/10.1007/978-3-319-20883-1
6. Goldberg, D.E.: Genetic Algorithms in Search, Optimization and Machine Learning. Addison Wesley, Reading (1989)
7. Hock, W., Schittkowski, K.: Test Examples for Nonlinear Programming Codes. Lecture Notes in Economics and Mathematical Systems, vol. 187. Springer, Heidelberg (1981). https://doi.org/10.1007/978-3-642-48320-2
8. Izzo, D., Biscani, F., Mereta, A.: Differentiable genetic programming. In: 20th European Conference (EuroGP 2017), Amsterdam, The Netherlands, pp. 35–51 (2017). https://doi.org/10.1007/978-3-319-55696-3_3
9. Johnson, S.G.: The NLopt nonlinear-optimization package (2019). http://ab-initio.mit.edu/nlopt
10. Keijzer, M.: Improving symbolic regression with interval arithmetic and linear scaling. In: Ryan, C., Soule, T., Keijzer, M., Tsang, E., Poli, R., Costa, E. (eds.) EuroGP 2003. LNCS, vol. 2610, pp. 70–82. Springer, Heidelberg (2003). https://doi.org/10.1007/3-540-36599-0_7
11. Kraft, D.: Algorithm 733: TOMP-Fortran modules for optimal control calculations. ACM Trans. Math. Softw. **20**(3), 262–281 (1994). https://doi.org/10.1145/192115.192124
12. Mousavi Astarabadi, S.S., Ebadzadeh, M.M.: Avoiding overfitting in symbolic regression using the first order derivative of GP trees. In: Genetic and Evolutionary Computation Conference (GECCO Companion 2015), Madrid, Spain, pp. 1441–1442, 11–15 July 2015. https://doi.org/10.1145/2739482.2764662
13. Ni, J., Drieberg, R.H., Rockett, P.I.: The use of an analytic quotient operator in genetic programming. IEEE Trans. Evol. Comput. **17**(1), 146–152 (2013). https://doi.org/10.1109/TEVC.2012.2195319
14. Rockett, P., Hathway, E.A.: Model-predictive control for non-domestic buildings: critical review and prospects. Build. Res. Inform. **45**(5), 556–571 (2017). https://doi.org/10.1080/09613218.2016.1139885
15. Rudin, W.: Principles of Mathematical Analysis. McGraw-Hill, New York (1976)
16. Wächter, A., Biegler, L.T.: On the implementation of an interior-point filter line-search algorithm for large-scale nonlinear programming. Math. Program. **106**(1), 25–57 (2006). https://doi.org/10.1007/s10107-004-0559-y
17. Waltz, R.A., Morales, J.L., Nocedal, J., Orban, D.: An interior algorithm for nonlinear optimization that combines line search and trust region steps. Math. Program. **107**(3), 391–408 (2006). https://doi.org/10.1007/s10107-004-0560-5

# Genetic Algorithm-Based Deep Learning Ensemble for Detecting Database Intrusion via Insider Attack

Seok-Jun Bu and Sung-Bae Cho[✉]

Department of Computer Science, Yonsei University,
Seoul, Republic of Korea
{sjbuhan, sbcho}@yonsei.ac.kr

**Abstract.** A database Intrusion Detection System (IDS) based on Role-based Access Control (RBAC) mechanism that has capability of learning and adaptation learns SQL transaction patterns represented by roles to detect insider attacks. In this paper, we parameterize the rules for partitioning the entire query set into multiple areas with simple chromosomes and propose an ensemble of multiple deep learning models that can effectively model the tree structural characteristics of SQL transactions. Experimental results on a large synthetic query dataset verify that it quantitatively outperforms other ensemble methods and machine learning methods including deep learning models, in terms of 10-fold cross validation and chi-square validation.

**Keywords:** Deep learning ensemble · Genetic algorithms · Database intrusion detection · Role-based access control

## 1 Introduction

Although various methods of securing database against policy violations and malicious activity have been proposed for many years, it is still a difficult problem to provide a parametric Intrusion Detection System (IDS) that learns and detects the intrusion patterns inductively from the SQL transactions. The attacks on relational database management system can be categorized as insider and outsider attacks. An attack from the outside may be able to gain unauthorized access to data by sending carefully crafted queries to the back-end database of web applications [1]. However, an insider attack that requests unauthorized privileges is more likely to gain abnormal database access beyond their privileges and causes critical financial loss.

The primary method of securing database from internal attacks is to limit the access to database based on user roles [2]. Under the Role-based Access Control (RBAC) mechanism which provides a user-level abstraction, the machine learning approach that has the capability of learning and adaptation can be addressed. An IDS based on RBAC mechanism that has capability of learning and adaptation, especially machine learning based IDS, learns transaction patterns represented by roles.

© Springer Nature Switzerland AG 2019
H. Pérez García et al. (Eds.): HAIS 2019, LNAI 11734, pp. 145–156, 2019.
https://doi.org/10.1007/978-3-030-29859-3_13

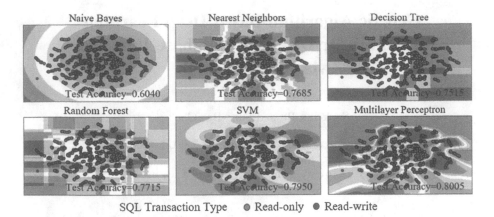

SQL Transaction Type    ● Read-only  ● Read-write

**Fig. 1.** Decision boundaries of SQL transaction roles by conventional learning methods.

Figure 1 shows the decision boundaries of machine learning-based IDS that classifies the role of SQL transactions. Each red or blue dot represents a two-dimensional embedding of read-only and read-write SQL transactions using t-SNE [3]. The misclassified cases within the model represented by the red and blue areas imply that it is difficult to classify the role of SQL transactions that hold various cases even with the MLP model known to effectively perform complex mappings.

In order to develop an IDS that effectively performs the complex mapping between these SQL transactions and the roles, we focused on the fact that the SQL queries are in a tree-structure [4, 5] and the entire data cannot be modeled at one time. In this paper, we parameterize the rules for partitioning the entire dataset into multiple areas with simple chromosomes and propose an ensemble of multiple deep learning models that can effectively model the SQL transactions within each area. The partitioning rules and the ensemble of deep learning models are learned not exclusively but in end-to-end fashion.

The combination of a Genetic Algorithm (GA) that is well-known for their global exploration and optimization ability [6] and Convolutional Neural Network (CNN) that models the complex hidden relations between the transaction features and roles outperforms other machine learning models including deep learning models on a synthetic query dataset. We have conducted 10-fold cross validation tests and analysis through chi-squared test to prove the validity of the results.

The remainder of this paper is organized as follows. In Sect. 2, we review existing machine learning-based IDS's and highlight the contributions. Section 3 explains how the evolving partitioning rules and CNN are combined in end-to-end manner. The performance of our model is evaluated in Sect. 4 through various experiments, including 10-fold cross validation and chi-square test as well as the visualization of partitioning rules. Finally, some conclusions and a discussion of the future directions for this study are presented in Sect. 5 (Table 1).

**Table 1.** Related works on IDS using machine learning algorithms

| Author | Method | Description |
|---|---|---|
| Valeur (2005) [7] | Naive Bayes | Propose the database IDS based on inductive learning of SQL transaction patterns |
| Ramasubramanian (2006) [8] | GA, ANN | Improve the training process of an ANN by modified GA |
| Pinzon (2013) [9] | SVM Ensemble | Propose a multi-agent based on SVM for hierarchical query modeling |
| Kim (2014) [10] | Query Tree, SVM | Extract the syntactic features of SQL using its tree-structural characteristics |
| Alom (2015) [11] | Deep Neural Network | Model the distribution of intrusion patterns using generative neural networks that stack restricted Boltzmann machines |
| Ronao (2016) [12] | PCA, Random Forest | Select the feature using PCA prior to classify the SQL transactions using an ensemble of decision trees |
| Bu (2017) [13] | GA, CNN | Extract the best subset of SQL transaction features using GA that maximize classification performance of CNN |

## 2 Related Works

In the mid to late 2000s, machine learning-based database IDS pioneers focused on accuracy, efficiency and automatic feature extraction capabilities derived from inductive learning [7]. Ramasubramanian et al. showed that the combination of Artificial Neural Network (ANN)-based exploitation and GA-based exploration was statistically significant in the IDS [8].

Ensemble strategies including SVM and DNN were shown to be effective in the mid-2010s. In Pinzon et al., the existing IDS performance is significantly increased through ensemble of multiple SVMs [9], and especially Kim et al. developed IDS that learn SVM by concentrating on the characteristics of SQL query with tree structure [10]. As noted above, the SVM or deep neural network is intuitively suitable because the mapping function between SQL transactions and roles must consider complex nonlinear techniques as well as the nature of SQL queries.

Alom et al. proposed the IDS based on stacked Restricted Boltzmann Machine capable of modeling the probability distributions among layers in a neural network model [11]. In return for high computational complexity of modeling the probability distributions of SQL transaction features in unsupervised manner, the developed IDS is robust to new data instance and guarantees convergence of learning.

Bu and Cho proposed the combination of modified Pittsburgh-style Learning Classifier Systems (LCS) for the optimization of feature selection rules and one-dimensional Convolutional Neural Network (CNN) for modeling SQL transactions [13]. In this line of research, we propose the combination of partial solutions that divides the input space and ensembles CNNs. An ensemble approach based on GA is more intuitive in terms of modeling tree-structured SQL transactions. In Sect. 4, we validated GA-based ensemble approach by 10 fold cross validation and chi-squared test compared to existing machine learning method.

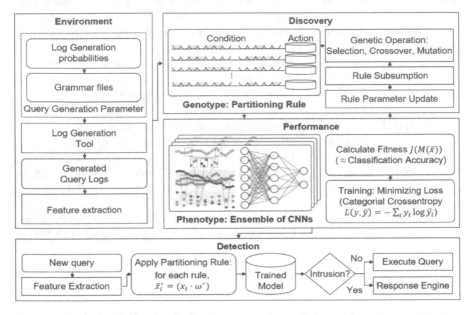

**Fig. 2.** Four components of GA-based CNN ensemble and its architecture.

## 3   Proposed Method

### 3.1   Overview

We can exploit Genetic Algorithm (GA) that is well-known for its flexibility and adaptability to environmental changes [14]. The overall intrusion detection system consists of four subcomponents described in Fig. 2: an environment that preprocesses and supplies data, a discovery component that generates and optimizes the partitioning rules based on GA [15], a performance component that builds and learns a CNN for each dataset partition, and a detection component that detects intrusions based on the learned model for the input query.

In this figure, for each subset of data entered from the environment, the discovery component iteratively creates new partitioning rules, or enhances and deletes existing rules after the performance component trains and validates the CNN ensemble along the set of rules indicates.

In order to find an efficient partitioning rule among $N_f$ features, the $2^{N_f}$ sized rule space to which discovery component must search makes it difficult to exhaustively build and train deep models with large computational complexity. GAs are commonly used to generate high-quality solutions to optimization and search problems by relying on biology-inspired operations, such as mutation, crossover and selection [16].

The implicit parallelism inherent in the evolutionary metaphor has its advantage in finding an optimal solution in a large searching space of ensemble strategies. This advantage comes from the exploration ability of the GA-based process and the exploitation ability gained by modeling the latent correlations between features.

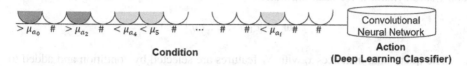

$> \mu_{a_0}$  #  $> \mu_{a_2}$  #  $< \mu_{a_4} < \mu_5$  #  ...  #  #  $< \mu_{a_i}$  #  #

Convolutional Neural Network

**Condition**

**Action (Deep Learning Classifier)**

**Fig. 3.** Encoding of partitioning rule consisting of condition and action.

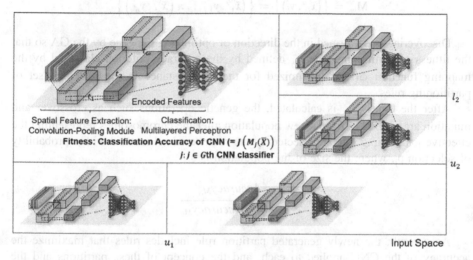

Encoded Features

Spatial Feature Extraction: Convolution-Pooling Module

Classification: Multilayered Perceptron

**Fitness: Classification Accuracy of CNN** $(= J\left(M_j(\bar{X})\right))$

$j: j \in G$th **CNN classifier**

$l_2$

$u_2$

$u_1$

Input Space

**Fig. 4.** A conceptual diagram of the input space partitioning and the ensemble of CNNs

## 3.2  GA-Based CNN Ensemble

The main purpose of the GA used in the proposed IDS is to find the best rules that satisfy an optimal ensemble performance criterion $J(\cdot)$, which parameterizes the partitioning rule of input space. One of the most important issues in evolving strategies is their representation [17]. In terms of modeling the normal behavior of the database, it is advantageous to select effective information with a focus on a local area, rather than the entire input space.

One of the most critical factors affecting the performance in rule-based machine learning models is rule encoding. Figure 3 shows the chromosome encoding in the GA framework, which consists of condition and action. Data instances that match the conditions of orange, blue and empty semicircles are modeled by CNN, the phenotype of chromosome.

Given a training dataset D that consists of a tuple of $(x_i, y_i)$,

$$D = \{(x_1, y_1), \ldots, (x_n, y_n)\},$$

we define the ternary condition $c = \{ > \mu_{a_0}, < \mu_{a_1}, \# \}$ for the $j$ th chromosome $\omega_j$ with don't care symbol # for each attribute.

$$\omega_j = [c_1, c_2, \ldots, c_{N_f}]$$

The query data instances $x_i$ with $N_f$ features are selected by condition and added to the matching set $M_{\omega_j}$ assigned to each $j$ th partitioning rule $\omega_j$ in generation G:

$$M_{\omega_j} = \left\{ \left( \bar{x}_i^{\omega_j}, y_i \right) \right\} = \left\{ \left( \bar{x}_1^{\omega_j}, y_1 \right), \ldots, \left( \bar{x}_{n'}^{\omega_j}, y_{n'} \right) \right\}$$

Discovering is performed in the direction of optimizing the gene by the GA so that the fitness of partitioning rule $J_{\omega_j}$ defined by the accuracy criterion classified by the mapping function $\phi(\cdot)$ is maximized for the data instance in the matching set of partitioning rules.

After the fitness $J_{\omega_j}$ is calculated, the genetic operations such as crossover and mutation are applied to form a new population after the chromosomes representing the effective partitioning rule are selected by the roulette wheel method with a probability of selection $p_j$, where the size of the population is s:

$$p_j = \frac{Accuracy_{M_{\omega_j}}}{\sum_{k=1}^{s} Accuracy_{M_k}}$$

As a result, the newly generated partition rule includes rules that maximize the accuracy of the CNN applied to each, and the concept of these partitions and the ensemble of the CNNs are shown in Fig. 4.

The convolution operation $\phi_c(\cdot)$ and pooling operation $\phi_p(\cdot)$ in CNN are well-known methods based on local-connectivity and shared-weights mechanism [18]. The convolution operation, which preserves the spatial relationships between features by learning filters that extract correlations, is known to reduce the translational variance between features [19]. The latent correlations between features from SQL transaction are modeled as a feature-map:

$$\phi_{c_i}^l = \sum_{k=0}^{f-1} w_i y_{i+k}^{l-1}$$

where the output of convolution $\phi_{c_i}$ from the $i$ th node of the $l$ th convolutional layer performs the convolution operation on $y^{l-1}$, using $1 \times f$ sized filter F.

Since the feature vectors distorted by the convolution operation are stacked by the number of the convolution filters, a pooling operation $\phi_{p_i}$ is applied to lower the computational complexity while preserving effective features:

$$\phi_{p_i}^l = \max \phi_{c_i \times \tau}^{l-1}$$

where the output of pooling $\phi_{p_i}$ from the $l$ th max-pooling layer selects the maximum value from a $1 \times k$ sized area, where $\tau$ is the pooling stride.

The output vector from the CNN represents the probabilities of the $t$ th roles associated with an input query sequence, where a softmax activation function is used so that the output vector is encoded as a probability in the range [0, 1]:

$$p\left(\hat{y}_i | \bar{x}_i^{\omega_j}\right) = argmax \frac{\exp(\phi_p(\phi_c(\bar{x}_i^{\omega_j})) + b_t)}{\sum_{k=1}^{11} \exp(\phi_p(\phi_c(\bar{x}_i^{\omega_j})) + b_k)}$$

**Table 2.** Roles and specifications for the generated queries based on TPC-E

| Transactions | Specifications | Transactions | Specifications |
|---|---|---|---|
| Read-only transactions | | Read/Write transactions | |
| Broker-volume (0) | SELECT only 714 kb | Trade-order (6) | SELECT, INSERT 759 kb |
| Customer-position (1) | SELECT only 566 kb | Trade-update (7) | SELECT, UPDATE 499 kb |
| Market-watch (2) | SELECT only 863 kb | Data-maintenance (8) | SELECT, UPDATE 246 kb |
| Security-detail (3) | SELECT only 571 kb | Market-feed (9) | SELECT, INSERT, UPDATE, DELETE 456 kb |
| Trade-status (4) | SELECT only 571 kb | Trade-result (10) | SELECT, INSERT, UPDATE, DELETE 419 kb |
| Trade-lookup (5) | SELECT only 490 kb | **Total 11,000 queries generated** | |

## 4  Experimental Results

Synthetic queries are generated in order to inductively model the SQL transactions based on its role and remedy the class imbalance problem as shown in Table 2 [6, 12]. We refer the transaction database footprint and pseudocode from the TPC-E benchmark schema for our role implementations [20]. We used only the commands specified in each entry for each transaction type in the table and noted the total size generated. 10,000 queries for each of the 11 roles are generated based on the schema.

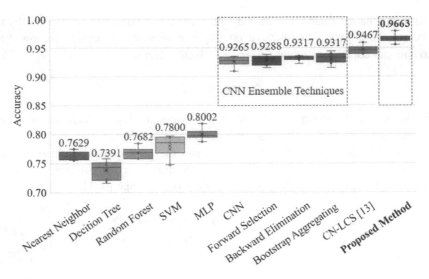

**Fig. 5.** Comparison of 10-fold cross validation with other machine learning-based IDS including ensemble-based IDS

The feature extraction process consists of two steps: parsing and extraction steps. The parsing step reshapes and reformats the input for the extraction step. This step is implemented by simply dividing the clauses from the queries. In the extraction step, we generate a feature vector which contains the seven fields: Command, projection relation, projection attribute, selection attribute, ORDER-BY features, GROUP-BY features and value counter.

Figure 5 compares the accuracy of the proposed method with the other machine learning-based IDS's. The proposed GA-based CNN ensemble has achieved the highest accuracy of 0.9663, while the combination of Pittsburgh-style LCS and CNN in Bu and Cho [13] achieved 0.9467. Compared with Bu and Cho in which feature selection masks were parameterized using GA and the whole dataset was learned by CNN, parameterization of partitioning rules to reflect the tree structural characteristics of SQL transactions are considered as the main factors for performance improvement.

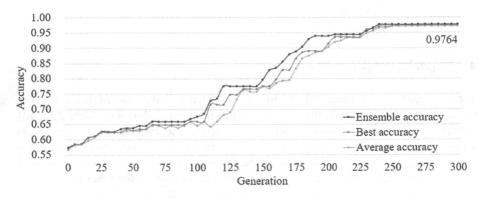

**Fig. 6.** The ensemble, best and average classification accuracies for each generation

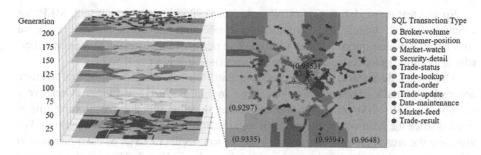

**Fig. 7.** Visualization of partitioning rules encoded by GA per generation

**Table 3.** The chi-square values of GA-based CNN ensemble and CNN

| | | Predicted | | | | | | | | | | |
|---|---|---|---|---|---|---|---|---|---|---|---|---|
| | | 0 | 1 | 2 | 3 | 4 | 5 | 6 | 7 | 8 | 9 | 10 |
| **Actual** | 0 | 0.03 | 0.00 | 0.00 | 0.00 | 0.00 | 0.00 | 0.00 | 0.00 | 0.00 | 0.00 | 0.00 |
| | 1 | 0.00 | 0.04 | 0.00 | 0.00 | 0.00 | 2.50 | 0.00 | 0.00 | 0.00 | 0.00 | 0.00 |
| | 2 | 0.00 | 0.00 | 0.02 | 0.00 | 0.00 | 0.00 | 0.00 | 0.00 | 0.00 | 0.00 | 0.00 |
| | 3 | 0.00 | 0.00 | 0.40 | 0.00 | 0.00 | 0.00 | 0.00 | 0.00 | 0.00 | 0.00 | 0.00 |
| | 4 | 0.00 | 0.00 | 0.00 | 0.00 | 0.00 | 2.13 | 0.00 | 0.00 | 0.00 | 0.00 | 0.00 |
| | 5 | 0.00 | 0.00 | 0.00 | 0.00 | 0.00 | 2.04 | 0.00 | 10.80 | 10.00 | 10.00 | 3.60 |
| | 6 | 0.00 | 0.00 | 0.40 | 0.00 | 0.10 | 20.00 | 42.63 | 50.00 | 20.00 | 30.00 | 37.38 |
| | 7 | 0.00 | 0.00 | 0.40 | 0.00 | 0.10 | 0.05 | 20.00 | 2.91 | 0.00 | 4.05 | 18.05 |
| | 8 | 0.00 | 0.00 | 0.00 | 0.00 | 4.90 | 0.00 | 28.90 | 0.00 | 9.13 | 20.00 | 32.40 |
| | 9 | 0.00 | 0.00 | 0.00 | 0.00 | 0.00 | 3.60 | 6.05 | 8.10 | 0.00 | 1.73 | 8.45 |
| | 10 | 0.00 | 0.00 | 0.00 | 0.00 | 5.00 | 0.00 | 22.82 | 24.03 | 25.60 | 12.03 | 21.33 |

**Table 4.** Computational complexity and performance per population

| Population | Training time [h] | | | Testing time [s] | Test accuracy |
|---|---|---|---|---|---|
| | Per population | Per generation | Total | | |
| 10 | 0.0029 | 0.0292 | 5.84 | 58.29 | 0.9328 |
| 20 | 0.0027 | 0.0566 | 11.35 | 116.28 | 0.9287 |
| 30 | 0.0027 | 0.0814 | 14.84 | 145.24 | 0.9313 |
| 40 | 0.0025 | 0.1023 | 20.47 | 213.86 | 0.9455 |
| **50** | **0.0026** | **0.1342** | **26.84** | **258.33** | **0.9659** |
| 100 | 0.0027 | 0.2766 | 54.76 | 484.95 | 0.9657 |
| 150 | 0.0024 | 0.0362 | 73.84 | 784.34 | 0.9645 |
| 200 | 0.0024 | 0.0586 | 113.71 | 942.25 | 0.9558 |

Spatial feature modeling using CNN achieves 0.9264 in comparison with MLP that achieves 0.8002, which is an empirical evidence that the consideration of spatial correlation between query features is valid. The performance of naïve Bayes classifier, which assumes independence between features, is the lowest at 0.5877.

The ensemble accuracy of the proposed partition rules and the best and average accuracies of each generation are visualized in Fig. 6. Average fitness is relatively unstable in 100 – 200 generations, and it seems that the effective partitioning rule is created since the accuracy of the model is significantly increased in this generation. We visualize the areas that have dominant classification results of CNN of each partition in color with respect to the five highest partitioning rules in Fig. 7. The partitioning rules that maximize classification accuracy have nonlinear boundaries with each other, and the ensemble accuracy is higher than the accuracy of each CNN. The implicit parallelism of GA has found more accurate partial solutions for ensemble purposes.

Moreover, we conduct chi-square test with the standard CNN in Table 3. Given the classification result of the proposed method, we show that CNN frequently misclassifies the read-write transactions as indicated at the bottom right of the chi-square table. We have noted that trade-order and trade-update transactions are often misclassified by standard CNN, and SELECT and INSERT are common to both SQL queries. On the other hand, the proposed method has an advantage in the case because it applies the partitioning rule to improve classification accuracy. The overall superiority is statistically significant at the degree of freedom given as 521.70 ($p \leq 0.001$).

Meanwhile, the computational complexity required to training multiple CNNs to find optimal ensemble rule for classification should be addressed in terms of practicality. In Table 4, the time complexity and performance were observed while controlling the population which is a critical hyperparameter of GA. The training time of 200 times that of standard CNN is clearly an issue to overcome, and we noted that the time to load the CNN into memory is the biggest factor.

## 5  Conclusions

We have proposed a GA-based deep learning ensemble method for database IDS that models the roles associated with queries by optimizing the input space partitioning rules of CNNs. Since the partitioning rule is optimized based on GA prior to learning process of CNN, the proposed method reflects the tree structural characteristic of the SQL transaction and outperforms the conventional query role classification accuracy. We have verified the statistical significance of the proposed method through 10-fold cross-validation and chi-square validation.

Future work will include comparing the GA-based deep learning ensemble with latest deep learning models in other domains, especially where complex mappings should be considered or characteristics of the data structure are already known. We also plan to improve the performance by considering the weighted ensemble by reliability of partitioning rules. The time complexity caused by CNN ensemble during the performance process is another issue to work out.

**Acknowledgements.** This research was supported by Korea Electric Power Corporation. (Grant number: R18XA05).

# References

1. Mathew, S., Petropoulos, M., Ngo, Hung Q., Upadhyaya, S.: A data-centric approach to insider attack detection in database systems. In: Jha, S., Sommer, R., Kreibich, C. (eds.) RAID 2010. LNCS, vol. 6307, pp. 382–401. Springer, Heidelberg (2010). https://doi.org/10.1007/978-3-642-15512-3_20
2. Murray, M.C.: Database security: what students need to know. J. Inf. Technol. Educ. Innovates Pract. **9**, 44–61 (2010)
3. Maaten, L.V.D., Hinton, G.: Visualizing data using t-SNE. J. Mach. Learn. Res. **9**, 2579–2605 (2008)
4. Buehrer, G., Weide, B.W., Sivilotti, P.A.: Using parse tree validation to prevent SQL injection attacks, In: Proceedings of the 5th International Workshop on Software Engineering and Middleware, pp. 106–113 (2005)
5. Bockermann, C., Apel, M., Meier, M.: Learning SQL for database intrusion detection using context-sensitive modelling, In: International Conference on Detection of Intrusions and Malware, and Vulnerability Assessment, pp. 196–205 (2009)
6. Xue, B., Zhang, M., Browne, W.N., Yao, X.: A survey on evolutionary computation approaches to feature selection. IEEE Trans. Evol. Comput. **4**, 606–626 (2016)
7. Valeur, F., Mutz, D., Vigna, G.: A learning-based approach to the detection of SQL attacks. In: International Conference on Detection of Intrusions and Malware, and Vulnerability Assessment, pp. 123–140, (2005)
8. Ramasubramanian, P., Kannan, A.: A genetic-algorithm based neural network short-term forecasting framework for database intrusion prediction system. Soft Comput. **10**, 699–714 (2006)
9. Pinzon, C.I., De Paz, J.F., Herrero, A., Corchado, E., Bajo, J., Corchado, J.M.: idMAS-SQL: Intrusion detection based on MAS to detect and block SQL injection through data mining. Inf. Sci. **231**, 15–31 (2013)
10. Kim, M.Y., Lee, D.H.: Data-mining based SQL injection attack detection using internal query trees. Expert Syst. Appl. **41**, 5416–5430 (2014)
11. Alom, Z., Bontupalli, V.R., Taha, T.M.: Intrusion detection using deep belief network and extreme learning machine. Int. J. Monitoring Surveillance Technol. Res. **3**, 35–56 (2015)
12. Ronao, C.A., Cho, S.B.: Anomalous query access detection in RBAC-administered databases with random forest and PCA. Inf. Sci. **369**, 238–250 (2016)
13. Bu, S.J., Cho, S.B.: A hybrid system of deep learning and learning classifier systems for database intrusion detection, In: International Conference on Hybrid Artificial Intelligence Systems, pp. 615–625 (2017)
14. Cho, S.B., Shimohara, K.: Evolutionary learning of modular neural networks with genetic programming. Appl. Intell. **9**, 191–200 (1998)
15. Urbanowicz, R.J., Moore, J.H.: Learning classifier systems: a complete introduction, review and roadmap. J. Artif. Evol. Appl. **1**, 1–25 (2009)
16. Holland, John H.: What is a learning classifier system? In: Lanzi, P.L., Stolzmann, W., Wilson, Stewart W. (eds.) IWLCS 1999. LNCS (LNAI), vol. 1813, pp. 3–32. Springer, Heidelberg (2000). https://doi.org/10.1007/3-540-45027-0_1
17. Seo, Y.G., Cho, S.B., Yao, X.: The impact of payoff function and local interaction on the N-player iterated prisoner's dilemma. Knowl. Inf. Syst. **2**, 461–478 (2000)

18. Rippel, O., Snoek, J., Adams, R.P.: Spectral representations for convolutional neural networks. In: Advances in Neural Information Processing Systems, pp. 2449–2457 (2015)
19. Sainath, T.N., Parada, C.: Convolutional neural networks for small-footprint keyword spotting, In: 16th Annual Conference of International Speech Communication Association (2015)
20. Transaction Process Performance Council (TPC), TPC Benchmark E, Standard Specification Ver. 1.0 (2014)

# An Efficient Hybrid Genetic Algorithm for Solving a Particular Two-Stage Fixed-Charge Transportation Problem

Ovidiu Cosma, Petrica C. Pop[(✉)], and Cosmin Sabo

Department of Mathematics and Computer Science,
Technical University of Cluj-Napoca, North University Center at Baia Mare,
Baia Mare, Romania
{ovidiu.cosma,petrica.pop}@cunbm.utcluj.ro, sabo.cosmin@gmail.com

**Abstract.** In this paper we address a particular capacitated two-stage fixed-charge transportation problem using an efficient hybrid genetic algorithm. The proposed approach is designed to fit the challenges of the investigated optimization problem and is obtained by incorporating an linear programming (LP) optimization problem within the framework of a genetic algorithm. We evaluated our proposed solution approach on two sets of instances often used in the literature. The experimental results that we achieved show the efficiency of our hybrid algorithm in yielding high-quality solutions within reasonable running-times, besides the superiority of our approach against other existing competing methods.

## 1 Introduction

We consider a particular transportation problem, i.e. a fixed charge transportation problem which occurs in a two-stage supply chain network, consisting of a manufacturer, a set of distribution centers and a set of customers. The objective of the investigated problem is to determine and pick out the distribution centers (DC's), connect them with the manufacturer and satisfy the requests of the customers such that the total distribution cost is minimized. The main characteristic of the investigated problem is that a fixed charge is associated with each route connecting distribution centers to customers that may be opened in addition to the variable transportation cost which is proportional to the amount of goods distributed and an opening cost for each potential DC.

The existing literature regarding the considered two-stage transportation problem with fixed charges associated to the routes is rather scarce. Molla et al. [3] introduced this variant of the two-stage transportation problem with only one manufacturer. They described a mathematical model based on integer programming and developed a spanning tree-based genetic algorithm with a Prüfer number representation, as well as an artificial immune algorithm meant for solving the problem. El-Sherbiny [9] pinpointed some inaccuracies regarding the mathematical formulation proposed by Molla et al. [3] and provided a valid mixed integer programming mathematical model of the problem. Pintea et al. [4–7]

© Springer Nature Switzerland AG 2019
H. Pérez García et al. (Eds.): HAIS 2019, LNAI 11734, pp. 157–167, 2019.
https://doi.org/10.1007/978-3-030-29859-3_14

developed some classical solution approaches and came up with an improved hybrid algorithm in which the Nearest Neighbor search heuristic is combined with a local search procedure. A novel and improved hybrid heuristic approach was described by Pop et al. [8] obtained by combining a genetic algorithm based on a hash table coding of the individuals with a powerful local search procedure. Recently, Cosma et al. [1] proposed an efficient hybrid Iterated Local Search (HILS) procedure that constructs an initial solution and uses a local search procedure in order to increase the exploration.

We organized the remainder of the paper as follows: in Sect. 2, the considered particular two-stage fixed-cost transportation problem is defined and a mathematical model is presented; in Sect. 3, the developed hybrid genetic algorithm is described, while in Sect. 4 we present and analyze the computational experiments and the achieved results. Finally, we conclude our work and discuss our plans for future work in Sect. 5.

## 2   Definition of the Problem

In order to define the considered particular two-stage fixed-charge transportation problem we start by providing the following notations:

$m$ is the number of DC's, $i$ is the DC identifier, $i \in \{1, ..., m\}$
$n$ is the number of customers, $j$ is the customer identifier, $j \in \{1, ..., n\}$
$D_j$ is the demand of customer $j$, $S_i$ is the stocking capacity of DC $i$
$f_i$ is the opening cost of the DC $i$
$g_{ij}$ is the opening cost for the link from DC $i$ to customer $j$
$b_i$ is the unit cost of transportation from manufacturer to DC $i$
$c_{ij}$ is the unit cost of transportation from DC $i$ to customer $j$
$Z$ the total cost of the distribution solution.

Given a manufacturer, a set of $m$ potential DC's and a set of $n$ customers satisfying the following properties:

- the manufacturer may ship to any DC at a transportation cost $b_i$, $i \in \{1, ..., m\}$,
- each DC may ship to any customer at a transportation cost $c_{ij}$ from DC $i \in \{1, ..., m\}$ to customer $j \in \{1, ..., n\}$, plus a fixed charge $g_{ij}$ for operating the route,
- the opening cost for a potential DC $i$ is denoted by $f_i$, $i \in \{1, ..., m\}$,
- each DC $i$, $i \in \{1, ..., m\}$ has a capacity $S_i$ and each customer $j$, $j \in \{1, ..., n\}$ has a demand $D_j$,

the aim of the considered two-stage fixed-charge transportation problem is to determine the routes to be opened and corresponding shipment quantities on these routes, such that the customer demands are fulfilled, all shipment constraints are satisfied, and the total distribution costs are minimized.

An illustration of the particular investigated two-stage fixed-charge transportation problem is presented in the next figure (Fig. 1).

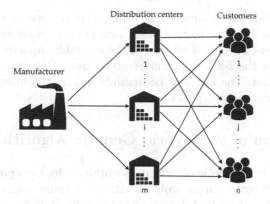

**Fig. 1.** Illustration of the particular two-stage fixed-charge transportation problem

By introducing the linear variables:

$x_i$ representing the number of units transported from manufacturer to DC $i$,
$y_{ij}$ representing the number of units transported from DC $i$ to customer $j$,

and the binary variables:

$w_{ij}$ is 1 if the route from DC $i$ to customer $j$ is used and 0 otherwise,
$v_i$ is 1 if the DC $i$ is open and 0 otherwise,

then the particular two-stage fixed charge transportation problem can be modeled as the following mixed integer problem described by Molla et al. [3] and improved by El-Sherbiny [9]:

$$\min \sum_{i=1}^{m} \left( b_i x_i + f_i v_i \right) + \sum_{i=1}^{m} \sum_{j=1}^{n} \left( c_{ij} y_{ij} + g_{ij} w_{ij} \right)$$

$$s.t. \sum_{j=1}^{n} y_{ij} \leq S_i, \ \forall \, i \in \{1, ..., m\} \tag{1}$$

$$\sum_{i=1}^{m} y_{ij} = D_j, \ \forall \, j \in \{1, ..., n\} \tag{2}$$

$$x_i \geq 0, \ \ \forall \, i \in \{1, ..., m\} \tag{3}$$

$$y_{ij} \geq 0, \ \forall \, i \in \{1, ..., m\}, \ \forall \, j \in \{1, ..., n\} \tag{4}$$

$$w_{ij} = \begin{cases} 1, & \text{if } y_{ij} > 0 \\ 0, & \text{if } y_{ij} = 0 \end{cases} \ \forall \, i \in \{1, ..., m\}, \ \forall \, j \in \{1, ..., n\} \tag{5}$$

$$v_i = \begin{cases} 1, & \text{if } \sum_{j=1}^{n} y_{ij} > 0 \\ 0, & \text{if } \sum_{j=1}^{n} y_{ij} = 0 \end{cases} \quad \forall \, i \in \{1, ..., m\} \tag{6}$$

The objective function minimizes the total distribution cost: the fixed charges and transportation per-unit costs. Constraints (1) guarantee that the quantity shipped out from each DC does not exceed the available capacity and constraints (2) guarantee that the total shipment received from DCs by each customer is equal to its demand. The last four constraints ensure the integrality and non-negativity of the decision variables.

## 3   Description of the Hybrid Genetic Algorithm

Our developed hybrid genetic algorithm is obtained by incorporating an linear programming (LP) optimization problem within the framework of a genetic algorithm. Genetic algorithms (GAs) were introduced by Holland [2] and are search heuristic methods inspired from the theory of natural evolution depeloped by Charles Darwin based on the "survival of the fittest". GAs have the ability to deliver a "good-enough" solution "fast-enough", making them very attractive in solving optimization problems.

In our proposed genetic algorithm, each chromosome contains $m$ genes associated to the links from the manufacturer to the $m$ distribution centers denoted by $\tilde{x}_i$ and $m \times n$ genes associated to the links from the m distribution centers to the $n$ customers denoted by $\tilde{y}_{ij}$. The value of each gene represents an estimate of the number of units transported on the corresponding link.

Based on the genes of a chromosome $(\tilde{x}_i, \tilde{y}_{ij})$, we define the following linear programming optimization problem:

$$\min \sum_{i=1}^{m} \tilde{b}_i x_i + \sum_{i=1}^{m} \sum_{j=1}^{n} \tilde{c}_{ij} y_{ij}$$

$$s.t. \sum_{j=1}^{n} y_{ij} \leq S_i, \ \forall \, i \in \{1, ..., m\}$$

$$\sum_{i=1}^{m} y_{ij} = D_j, \ \forall \, j \in \{1, ..., n\}$$

$$x_i \geq 0, \quad \forall \, i \in \{1, ..., m\}$$

$$y_{ij} \geq 0, \quad \forall \, i \in \{1, ..., m\}, \ \forall \, j \in \{1, ..., n\}$$

$$\tilde{b}_i = \begin{cases} b_i + \dfrac{f_i}{\tilde{x}_i}, & \text{if } \tilde{x}_i > 0 \\ b_i + f_i, & \text{if } \tilde{x}_i = 0 \end{cases} \quad \forall \, i \in \{1, ..., m\}$$

$$\tilde{c}_{ij} = \begin{cases} c_{ij} + \dfrac{g_{ij}}{\tilde{y}_{ij}}, & \text{if } \tilde{y}_{ij} > 0 \\ c_{ij} + g_{ij}, & \text{if } \tilde{y}_{ij} = 0 \end{cases} \quad \forall \, i \in \{1, ..., m\}, \ \forall \, j \in \{1, ..., n\}$$

This is a well-known optimization problem (the minimum cost flow problem), for which there are available several algorithms that solve it efficiently. In our proposed solution approach, we used the Network Simplex algorithm. By solving the described optimization problem, we effectively determine the quantities $x_i$ and $y_{ij}$ that correspond to the optimal solution.

The genes of a chromosome do not introduce any restrictions in the mathematical model of the considered fixed charge transportation problem. Their only effect is limited to the $\tilde{b}_i$ and $\tilde{c}_{ij}$ costs that occur in the described linear programming optimization problem. Consequently, if the constraints are satisfied, then any chromosome, even a randomly generated one, leads to a feasible solution of the transportation problem. It is unlikely, however, that a good quality solution will be obtained based on a randomly generated chromosome. For improving the quality of the chromosomes, we have developed the following optimization algorithm:

**Algorithm Chromosome Optimization**
**input:** chromosome $(\tilde{x}_i, \tilde{y}_{ij})$
**output:** feasible solution
    1.  $\tilde{Z} \leftarrow \infty$
    2.  Solve the linear programming optimization problem.
    3.  Calculate the corresponding objective value Z.
    4.  **If** $Z > \tilde{Z}$ or the solution is a duplicate, **then**
        4.1.  Result ← saved solution
        4.2.  **STOP**
    5.  **Else**
        5.1.  $\tilde{Z} \leftarrow Z$
        5.2.  $\tilde{x}_i \leftarrow x_i$
        5.3.  $\tilde{y}_{ij} \leftarrow y_{ij}$
        5.4.  Save the solution.
        5.5.  Continue with step 2.

Step 1 initializes the estimated cost $\tilde{Z}$ corresponding to the optimal solution of the considered transportation problem. The previously defined linear programming optimization problem is solved in step 2, using the Network Simplex algorithm. Thus, the optimum solution of this problem is obtained efficiently. Step 3 calculates the total distribution cost based on the $x_i$ and $y_{ij}$ amounts determined in step 2. Steps 2 and 3 may be repeated several times in the loop created by the jump performed in step 5.5. However, there is no guarantee that each iteration of the loop will improve the solution. Step 4 takes the decision whether to continue or end the loop. The loop ends if the last iteration has not improved the solution, or if a duplicate that has been generated in previous iterations, has been obtained. Various estimates $\tilde{x}_i$ and $\tilde{y}_{ij}$ can lead to the same solution. Two chromosomes are considered identical if the amounts $x_i$ and $y_{ij}$ obtained by solving the linear programming optimzation problem are identical, without considering the $\tilde{x}_i$ and $\tilde{y}_{ij}$ estimates. When the loop ends, the last saved solution represents the result of the algorithm. Step 5 is reached only if the last iteration has improved the solution. In this case, the estimates $\tilde{x}_i$, $\tilde{y}_{ij}$ and $\tilde{Z}$ are updated, the solution is saved, and the optimization process continues with a jump to step 2.

The initial population in our genetic algorithm consists of $P$ randomly generated individuals. The chromosomes of each individual in the initial population are randomly chosen such that: $0 < \tilde{x}_i < 2S_i$ and $0 < \tilde{y}_{ij} < 2D_j$, where $i \in \{1, ..., m\}$ and $j \in \{1, ..., n\}$. Each random generated chromosome is passed through the optimization algorithm described above.

The individuals in the current population are combined to form the next generation, hopping that the new born individuals will be better. The individuals that are combined to form the new generations are selected by an elitist strategy, taking into consideration their fitness values. Better chromosomes have better chances to be selected in the process.

The crossover is one of the most important elements in our genetic algorithm. For the creation of each new chromosome in the next generation, two parents are randomly chosen from the current population. The first one is always chosen from the fittest 20% chromosomes and the second one is selected randomly from the entire population. This selection rule was chosen in order to increase the convergence speed of our genetic algorithm. The genes that make up the chromosome of the offspring are taken from the two parents with equal probabilities. Thus, the descendant will inherit an equal amount of genetic information from both parents. Each new chromosome is processed using the optimization algorithm, which ultimately determines its fitness value. If the achieved fitness is weaker than that of the last individual in the current population, then the new individual is considered unfit and it is destroyed. For the creation of each new generation, at least $3P$ crossover operations are performed. If the $3P$ attempts could create at least $P$ fit individuals, then the number of attempts is extended to maximum $10P$, or until $P$ fit individuals are created.

When the crossover process creates a new individual, before its chromosome is optimized, some of its genes may suffer a mutation. In our proposed GA, the mutation probability is not constant in the process of creating a new generation. In the first $3P$ crossover operations, mutations cannot occur. If the number of crossover operations exceeds $3P$, then the probability $p$ of mutation occurrence is given by the following relation: $p = \dfrac{1}{10 - \lfloor \frac{NC}{P} \rfloor}$, where $NC$ is the number of the current crossover operation.

Three mutation variants have been defined. Every time a chromosome suffers a mutation, one of the three variants is randomly chosen.

In the first variant, a client $j$ and a set $DCS_1$ consisting of maximum 5 distribution centers are randomly chosen. All the estimates $\tilde{y}_{ij}$ are replaced with random values in the interval $[0, 2D_j]$, where $i \in DCS_1$. Next a set $DCS_2$ of maximum 2 distribution centers are randomly selected. The estimates $\tilde{x}_i$ are replaced with random values in the interval $[0, 5S/m]$, where $i \in DCS_2$ and $S$ is the requierd production capacity of the manufacturer (the total demand of the customers).

In the second variant, a random number r in the interval $[1, 2(m + n + 1)]$ is selected. Next, $r$ distribution links are randomly picked from the distribution system, and their estimated amounts $\tilde{x}_i$ or $\tilde{y}_{ij}$ are modified as follows: if the selected link connects the manufacturer with DC $i$, then the $\tilde{x}_i$ estimate is replaced with a random value in the interval $[0, 5S/m]$ and if the link connects DC $i$ with customer $j$, then its estimate $\tilde{y}_{ij}$ is replaced with a random number in the interval $[0, 2D_j]$.

The third variant processes all the customers that are supplied from multiple distribution centers. For each such customer $j$, all estimates $\tilde{y}_{ij}$ are cleared, and then a set $DCS_3$ consisting of maximum 3 DCs is randomly selected. Next all the estimates $\tilde{y}_{ij}$ are replaced with random numbers in the interval $[0, D_j]$.

After mutation, the chromosome follows the normal optimization process. If after optimization the chromosome is found unfit, it is saved in a separate list of unfit mutants.

Both the initial population generator and the crossover operator which are fundamental components of our genetic algorithm, are based on the chromosome optimization algorithm, which solves a linear optimization problem, using the well-known Network Simplex algorithm. Therefore the linear programming method is deeply embedded in our genetic algorithm.

A selection process follows the creation of each new generation. In the selection process, the current population is merged with the offspring, all duplicates are eliminated, the chromosomes are sorted by their fitness value, and only the first $2P$ individuals are retained. The result forms the new population. The best 5% of the chromosomes in the unfit mutants list are also injected in the current population, and then the list is cleared.

During the experiments, the running time of our GA was limited as follows: 3 s for the first 12 instances in Table 1, 5 min for the last 3 instances in Table 1 and 60 min for the instances in Table 2.

## 4   Computational Results

In order to evaluate the performance of our described hybrid algorithm, we conducted our computational experiments on two sets of instances containing in total 24 instances. The first set of instances contains 15 instances used in the computational experiments of Pintea and Pop [6] and Pop et al. [8]. The second set of instances contains 9 new randomly instances of larger sizes generated according to Molla et al. strategy [3]. All the instances used in our computational experiments are available at the address: https://sites.google.com/view/tstp-instances/.

Our genetic algorithm was coded in Java 8. We used the C++ Network Simplex implementation in the Library for Efficient Modeling and Optimization in Networks template library [10], which was integrated in our application through the Java Native Interface (JNI). We performed 10 independent runs for each instance on a Procesor Intel Core i5-4590 3.3 GHz, 4 GB RAM, Windows 10 Education 64 bit.

Genetic parameters are very important for the success of a GA. In our proposed algorithm, based on preliminary computational experiments, the genetic

parameters were set as follows: the initial population contains $P = \dfrac{n + nm}{5}$ individuals for the first 12 instances in Table 1 and 500 individuals for the remaining
3 instances in Table 1 and for all the instances in Table 2. The size of the following generations was fixed at maximum 2P individuals. The mutation probability
$p \in [0, 1]$. It depends on the number of crossover operations performed previously
for creating the individuals in the next generation.

Table 1 displays our achieved computational results in comparison with the
hybrid heuristic algorithm described by Pintea et al. [4], denoted by HA, the GA
and hybrid based GAs introduced by Pop et al. [8], denoted by GA and HGA
and the hybrid Iterated Local Search (HILS) algorithm developed by Cosma
et al. [1] The first column provides the type of the instances followed by two
columns that contain the number of distribution centers $(m)$ and the number
of customers $(n)$. The next columns provide the best solution achieved by the
hybrid heuristic algorithm described by Pintea et al. [4], the best and average
solutions obtained by the genetic algorithm and hybrid based genetic algorithm
introduced by Pop et al. [8], the HILS heuristic algorithm [1] and the achieved
results by our hybrid algorithm. The results written in bold represent cases for
which the obtained solution is the best existing from the literature.

**Table 1.** Computational results achieved by our proposed hybrid GA compared to
existing approaches

| Type | m | n | HA [4] | GA [8] | | HGA [8] | | HILS [1] | | Our approach | |
|---|---|---|---|---|---|---|---|---|---|---|---|
| | | | | Best sol. | Avg. sol. | Best sol. | Avg. sol. | Best sol. | Avg. sol. | Best sol. | Avg. sol. |
| 1 | 10 | 10 | 21980 | 20450 | 21430 | 20400 | 21320 | **20395** | **20395** | **20395** | **20395** |
| 2 | 10 | 10 | 12160 | 11240 | 11850 | 11220 | 11740 | **11214** | **11214** | **11214** | **11214** |
| 3 | 10 | 10 | 14000 | 14100 | 14620 | 14040 | 14520 | **13999** | **13999** | **13999** | **13999** |
| 1 | 10 | 20 | 36000 | 35400 | 36200 | 35380 | 35860 | **35371** | **35371** | **35371** | **35371** |
| 2 | 10 | 20 | 39660 | 37840 | 38470 | **37800** | 38250 | **37800** | **37800** | **37800** | **37800** |
| 3 | 10 | 20 | 36060 | 36000 | 36110 | 36000 | 36000 | 35988 | 35988 | **35583** | **35583** |
| 1 | 10 | 30 | 55660 | 52700 | 54880 | 52650 | 53700 | **52643** | **52643** | **52643** | **52643** |
| 2 | 10 | 30 | 55380 | 54650 | 55640 | 54540 | 54880 | **54539** | **54539** | **54539** | **54539** |
| 3 | 10 | 30 | 49860 | 48580 | 49470 | 48540 | 49240 | **48535** | **48535** | **48535** | **48535** |
| 1 | 15 | 15 | 26680 | 25420 | 27640 | 25420 | 26710 | **25417** | **25417** | **25417** | **25417** |
| 2 | 15 | 15 | 29100 | 28600 | 29230 | 28600 | 28940 | **28598** | **28598** | **28598** | **28598** |
| 3 | 15 | 15 | 29200 | 28840 | 29470 | 28750 | 29120 | **28411** | **28411** | **28411** | **28411** |
| 1 | 50 | 50 | 92400 | 91550 | 92410 | 91500 | 92104 | **91455** | **91455** | **91455** | **91455** |
| 2 | 50 | 50 | 116500 | 114660 | 117440 | 114150 | 115420 | **114111** | 114114.2 | **114111** | 114124.6 |
| 3 | 50 | 50 | 105000 | 105000 | 107400 | 105000 | 106480 | 104904 | 104920.8 | **104884** | **104884** |

Analyzing the computational results reported in Table 1, we can observe that
our hybrid GA has a better computational performance compared to the hybrid
heuristic algorithm described by Pintea et al. [4], the genetic algorithm and
hybrid based genetic algorithm introduced by Pop et al. [8] and the hybrid iterated local search HILS algorithm [1]. For all the instances in Table 1, excepting
the second last instance, the best and the average solutions are the same. In the
case of the sixth and fifteenth instances, our hybrid GA has obtained at every

run better solutions than those reported so far. We would like to emphasize the fact that each achieved solution was obtained in under 25 s of running time and because our hybrid GA obtains the same solution in each of the 10 runs for each instance, excepting the second last instance, proves the robustness of our proposed algorithm.

In Table 2, we present the computational results achieved by our proposed hybrid GA in comparison to the HILS algorithm developed by Cosma et al. [1] in the case of the 9 randomly generated instances of larger sizes proposed by the same authors. The first column of Table 2 gives the type of the instances followed by two columns that contain the number of distribution centers ($m$) and the number of customers ($n$) and the next six columns contain the following results: the best solution, the average solution and the corresponding average computational times necessary to obtain the provided solutions obtained by the HILS algorithm and our novel solution approach. The execution times are reported in minutes (min).

**Table 2.** Computational results achieved by our proposed hybrid GA in the case of large size instances

| Type | m | n | HILS [1] | | | Our approach | | |
|---|---|---|---|---|---|---|---|---|
| | | | Best sol. | Avg. sol. | Avg. time | Best sol. | Avg. sol | Avg. time |
| A | 50 | 50 | **361171** | 361187.6 | 133.55 | **361171** | 361189.6 | **20.77** |
| B | 50 | 50 | **366870** | 366904.6 | 133.55 | **366870** | **366894.0** | **9.30** |
| C | 50 | 50 | **374285** | 374285.0 | 48.84 | **374285** | 374285.0 | **0.76** |
| A | 30 | 100 | **242099** | 242099.0 | 276.08 | **242099** | 242101.0 | **2.76** |
| B | 30 | 100 | 233279 | 233379.5 | 329.92 | **233214** | **233293.2** | **9.04** |
| C | 30 | 100 | 277087 | 277327.7 | 440.55 | **277074** | **277229.2** | **22.261** |
| A | 50 | 100 | 236916 | 237187.0 | 326.84 | **236417** | **236504.4** | **18.74** |
| B | 50 | 100 | 245628 | 245809.3 | 99.88 | **245219** | **245341.6** | 44.16 |
| C | 50 | 100 | 288768 | 289229.5 | 93.84 | **287898** | **288210.6** | **25.79** |

Analyzing the results displayed in Table 2, we can observe that, regarding the best solutions, our hybrid GA performs slightly better in comparison to the HILS algorithm [1]. For the last five instances, our proposed hybrid GA provided better values. For the first four instances, the achieved best solutions coincide with those found by the HILS algorithm. Concerning the average solutions, we remark that in one out of nine instances they coincide, in six out of nine our achieved average solutions are better and in the case of the first and fourth instances the average solution delivered by HILS algorithm [1] is slightly better. Regarding the average computational times necessary to obtain the provided solutions, we can observe that the proposed hybrid GA is much faster in comparison to the HILS algorithm. On average, the hybrid GA is 29.4 times faster than the HILS algorithm.

# 5    Conclusions

Our main goal is to design an efficient hybrid genetic algorithm for solving a particular two-stage fixed-charge transportation problem, which models an important transportation application in a supply chain, from a given manufacturer to a number of customers through distribution centers. The proposed hybrid genetic algorithm is obtained by incorporating an linear programming (LP) optimization problem within the framework of a genetic algorithm. We evaluate our approach using two sets of benchmark instances used in the literature. The computational results show that our hybrid genetic algorithm is robust and yields high-quality solutions within reasonable running-times.

In future, we plan to improve the developed hybrid GA by combining with local search methods and to evaluate the generality and scalability of the proposed solution approach by testing it on larger instances.

# References

1. Cosma, O., Pop, P.C., Matei, O., Zelina, I.: A hybrid iterated local search for solving a particular two-stage fixed-charge transportation problem. In: de Cos Juez, F., et al. (eds.) HAIS 2018. LNCS, vol. 10870, pp. 684–693. Springer, Cham (2018). https://doi.org/10.1007/978-3-319-92639-1_57
2. Holland, J.H.: Adaptation in Natural and Artificial Systems: An Introductory Analysis with Applications to Biology, Control and Artificial Intelligence. MIT Press, Cambridge (1992)
3. Molla-Alizadeh-Zavardehi, S., Hajiaghaei-Kesteli, M., Tavakkoli-Moghaddam, R.: Solving a capacitated fixed-cost transportation problem by artificial immune and genetic algorithms with a Prüfer number representation. Expert Syst. Appl. **38**, 10462–10474 (2011)
4. Pintea, C.-M., Sitar, C.P., Hajdu-Macelaru, M., Petrica, P.: A hybrid classical approach to a fixed-charged transportation problem. In: Corchado, E., Snášel, V., Abraham, A., Woźniak, M., Graña, M., Cho, S.-B. (eds.) HAIS 2012. LNCS (LNAI), vol. 7208, pp. 557–566. Springer, Heidelberg (2012). https://doi.org/10.1007/978-3-642-28942-2_50
5. Pintea, C.-M., Pop, P.C., Hajdu-Măcelaru, M.: Classical hybrid approaches on a transportation problem with gas emissions constraints. In: Snášel, V., Abraham, A., Corchado, E. (eds.) Soft Computing Models in Industrial and Environmental Applications. Advances in Intelligent and Soft Computing, vol. 188, pp. 449–458. Springer, Heidelberg (2013). https://doi.org/10.1007/978-3-642-32922-7_46
6. Pintea, C.M., Pop, P.C.: An improved hybrid algorithm for capacitated fixed-charge transportation problem. Log. J. IJPL **23**(3), 369–378 (2015)
7. Pop, P.C., Pintea, C.-M., Pop Sitar, C., Hajdu-Macelaru, M.: An efficient reverse distribution system for solving sustainable supply chain network design problem. J. Appl. Log. **13**(2), 105–113 (2015)
8. Pop, P.C., Matei, O., Pop Sitar, C., Zelina, I.: A hybrid based genetic algorithm for solving a capacitated fixed-charge transportation problem. Carpath. J. Math. **32**(2), 225–232 (2016)

9. El-Sherbiny, M.M.: Solving a capacitated fixed-cost transportation problem by artificial immune and genetic algorithms with a Prüfer number representation. Expert Syst. Appl. **39**, 11321–11322 (2011). Molla-Alizadeh-Zavardehi, S.: Expert Syst. Appl. (2012)
10. Egerváry Research Group on Combinatorial Optimization: Library for Efficient Modeling and Optimization in Networks. https://lemon.cs.elte.hu/trac/lemon

# Analysis of MOEA/D Approaches for Inferring Ancestral Relationships

Sergio Santander-Jiménez[1(✉)], Miguel A. Vega-Rodríguez[2], and Leonel Sousa[1]

[1] Instituto de Engenharia de Sistemas e Computadores - Investigação e
Desenvolvimento em Lisboa (INESC-ID), Instituto Superior Técnico,
Universidade de Lisboa, 1000-029 Lisbon, Portugal
sergio.jimenez@tecnico.ulisboa.pt, leonel.sousa@ist.utl.pt
[2] Instituto de Investigación en Tecnologías Informáticas Aplicadas de Extremadura
(INTIA), Univ. de Extremadura, Avda. de la Universidad s/n, 10003 Caceres, Spain
mavega@unex.es

**Abstract.** Throughout the years, decomposition approaches have been
gaining major research attraction as a promising way to solve complex multiobjective optimization problems. This work investigates the
application of decomposition-based optimization techniques to address a
challenging problem from the bioinformatics domain: the reconstruction
of ancestral relationships from protein data. A comparative analysis of
different design alternatives for the Multiobjective Evolutionary Algorithm based on Decomposition (MOEA/D) is undertaken. Particularly,
MOEA/D variants integrating genetic operators (MOEA/D-GA) and
differential evolution (MOEA/D-DE) are studied. Hybrid search mechanisms are included to improve the accuracy of these methods, combining evolutionary strategies with problem-specific heuristics. Experimental results on four real-world problem instances give account of the
significance of these techniques, especially when differential evolution
approaches are used to conduct the search. As a result, significant multiobjective performance and biological solution quality are accomplished
when compared with other methods from the literature.

**Keywords:** Evolutionary algorithms · Multiobjective optimization ·
Decomposition · Bioinformatics

## 1 Introduction

Multiobjective optimization problems play a major role in a wide variety of
research domains. These problems are characterized by mathematical formulations that involve multiple objective functions to be simultaneously optimized.
As a result, the goal of the search in a multiobjective context is not a single solution but a set of trade-off Pareto solutions [4]. The attainment of high-quality
solutions in such scenarios often implies dealing with large, multi-dimensional

© Springer Nature Switzerland AG 2019
H. Pérez García et al. (Eds.): HAIS 2019, LNAI 11734, pp. 168–180, 2019.
https://doi.org/10.1007/978-3-030-29859-3_15

search spaces and complex Pareto front shapes. These hardness factors are particularly difficult to address in real-world scenarios, so accurate search strategies hybridized with problem-specific, low-level heuristics are consequently required.

Complex multiobjective problems are a constant in current bioinformatics. One of the most challenging problems in this field is the reconstruction of phylogenetic relationships from molecular sequence data [23]. Throughout the years, multiobjective optimization techniques have been applied to this problem in order to address biological incongruence issues. For example, Poladian and Jermiin studied the application of multiobjective evolutionary algorithms to consistently conduct phylogenetic analyses over conflicting datasets [13]. Genetic algorithm approaches were examined in [8] to perform single and multiobjective reconstructions of the primate evolutionary tree. Cancino and Delbem applied the Nondominated Sorting Genetic Algorithm II (NSGA-II) to conduct phylogeny reconstructions according to two widely-used phylogenetic functions, parsimony and likelihood [2]. Multiobjective immune-inspired approaches were also reported in this problem [3], along with other swarm intelligence methods [17,19]. However, these previous approaches are oriented towards the analysis of DNA sequence data, thus not supporting more complex data types, such as protein data.

This paper tackles phylogenetic reconstructions from protein alignments by using decomposition [24], one of the most relevant multiobjective design trends. Under decomposition frameworks, a multiobjective problem with $n$ objectives can be split into $m$ scalar subproblems e.g. by using weighted aggregation vectors:

$$minimize\ g_i(y|\lambda^i, z^*) = max_{1 \leq j \leq n} \left\{ \lambda_j^i |f_j(y) - z_j^*| \right\}, \tag{1}$$

where $y$ is a candidate solution for the $i$-th subproblem, $f_j(y)$ the score of $y$ at the $j$-th objective function, $\lambda^i$ the aggregation vector for $g_i$, and $z^*$ a reference point based on the best scores found for the original objectives. The focus of this work lies in the evaluation of different design alternatives for the reference method in decomposition: the Multiobjective Evolutionary Algorithm based on Decomposition (MOEA/D). Particularly, we compare the multiobjective performance of MOEA/D approaches based on genetic operators (MOEA/D-GA [24]) and differential evolution (MOEA/D-DE [11]). Due to the complexity of the problem, these approaches are hybridized with problem-aware heuristics and optimization techniques to boost the quality of the solutions. Four real-world protein datasets will be used to experimentally assess the performance of each approach, also conducting comparisons of solution quality with reference biological tools.

This paper is organized as follows. The next section introduces the main features of the tackled problem, while Sect. 3 details the MOEA/D designs considered in this work. The evaluation of experimental results is conducted in Sect. 4. Finally, Sect. 5 draws conclusions and future research lines.

## 2  Formulation of the Problem

Given a multiple sequence alignment containing $N$ amino acid sequences of length $M$, the phylogeny reconstruction problem is aimed at inferring hypotheses about the evolutionary history of the organisms characterized in the alignment. Tree-shaped structures, named as phylogenetic trees $T = (V, E)$, are used to describe evolutionary events and ancestor-descendant relationships. In a phylogenetic topology, the node set $V$ defines the extant organisms from the alignment as terminal nodes, while internal nodes are used to represent hypothetical ancestral species. Ancestral relationships are then described according to the branching patterns established by the branch set $E$.

The inference of optimal phylogenetic trees can be formulated as an optimization problem that involves the exploration of a phylogeny space under the guidance of some phylogenetic objective functions. The main complexity issue in this context is given by the relationship between the number of sequences $N$ and the size of the phylogeny space, as the number of candidate solutions exponentially grows following the double factorial $(2N - 5)!!$ [23]. Furthermore, the processing of protein alignments introduces additional complexity into phylogenetic analyses, due to the consideration of 20 possible amino acids instead of the four nucleotides from the DNA case [15]. Consequently, efficient and accurate optimizers are required to handle the NP-hard nature of this problem.

This work addresses phylogenetic reconstruction under a multiobjective formulation based on two widely-used phylogenetic optimality criteria. The first criterion is given by the parsimony function, which gives priority to the phylogenetic trees that represent the simplest evolutionary hypotheses. Given a phylogenetic tree $T = (V, E)$, the parsimony score $P(T)$ measures the number of changes observed at the sequence level between related nodes:

$$P(T) = \sum_{i=1}^{M} \sum_{(u,v) \in E} C(u_i, v_i), \tag{2}$$

where $(u, v) \in E$ is the branch that links the nodes $u, v \in V$, $u_i$ and $v_i$ the state values at the $i$-th character of the sequences for $u$ and $v$, and $C(u_i, v_i)$ a function quantifying the substitutions (mutations) between $u$ and $v$ ($C(u_i, v_i) = 1$ iff $u_i \neq v_i$, $C(u_i, v_i) = 0$ otherwise). Candidate solutions with lower $P(T)$ values are preferred under the parsimony approach.

The second criterion used in the multiobjective formulation is the likelihood function. This approach employs models of protein evolution to calculate a probabilistic measurement of how likely a phylogenetic hypothesis gave rise to the data observed in the input alignment. The likelihood score $L(T)$ for a phylogenetic tree $T = (V, E)$ can be calculated as follows:

$$L(T) = \prod_{i=1}^{M} \sum_{x,y \in \Lambda} \pi_x \left[ P_{xy}(t_{ru}) L_p(u_i = y) \right] \times \left[ P_{xy}(t_{rv}) L_p(v_i = y) \right], \tag{3}$$

where $\Lambda$ is the amino acid state alphabet, $\pi_x$ the stationary probability of observing the state value $x \in \Lambda$, $r \in V$ the root node of $T$ with child nodes $u, v \in V$, $P_{xy}(t)$ the probability of mutation from $x$ to $y$ within a time interval $t$ (given by the branch lengths $t_{ru}$, $t_{rv}$), and $L_p(u_i = y)$, $L_p(v_i = y)$ the partial likelihoods of observing $y$ at $u_i$ and $v_i$. Candidate solutions with higher $L(T)$ values are preferred from a likelihood perspective.

# 3   MOEA/D Approaches

This work examines the performance of MOEA/D when tackling phylogenetic reconstructions on protein sequence data. This section introduces the general features of the algorithm and describes the two design variants herein considered: MOEA/D-GA and MOEA/D-DE.

## 3.1   Individual Representation and Search Methodology

In order to adapt MOEA/D to the tackled problem, an indirect solution encoding is employed. More specifically, a solution is represented by a $N \times N$ matrix-shaped structure $\Delta$ named distance matrix [23]. Each element $\Delta[u, v]$ in the matrix contains a floating-point number that represents the evolutionary distance between the organisms $u$ and $v$. The search for new candidate solutions is therefore conducted over a distance matrix decision space, thus allowing the consideration of more diverse, matrix-adapted evolutionary operators.

Prior to the calculation of objective functions, the resulting distance matrices are mapped to the phylogeny space by using a tree-building method i.e. the neighbour-joining approach BIONJ [23]. Local searches are also included in the designs to boost solution quality, applying problem-specific heuristics aimed at refining the topology and branch lengths. Particularly, two topological rearrangement operators, nearest neighbour interchange and subtree pruning and regrafting [18], are included, along with gradient-descent branch length optimization. In this way, the matrix-based evolutionary approach is hybridized with a tree-based local search procedure to improve the accuracy of the method.

## 3.2   MOEA/D Framework

MOEA/D is based on the idea of evolving a population $P$ of *popSize* individuals, where each individual $P_i$ represents the best solution found for the $i$-th scalar subproblem. The objective function that governs each subproblem is given by a weighted aggregation of the original objectives, in accordance with the chosen decomposition mechanism (Tchebycheff's approach in our case, as shown in Eq. 1). Each subproblem $g_i$ is then associated to an aggregation vector $\lambda^i$, whose components define the weight of each original objective in $g_i$.

Two algorithmic strategies are included for addressing complex optimization problems [11]. Firstly, a selection probability is included to determine the sources of information for the generation of new candidate solutions. With probability $\delta$, these sources are restricted to the $T$ closest subproblems in the neighbourhood of the currently processed one. On the other hand, information from the whole population can also be considered, with probability $1 - \delta$. The second strategy refers to the way the newly generated solutions are integrated in the population.

---

**Algorithm 1.** MOEA/D Framework

---

**Input:** *popSize* (number of subproblems = population size), $T$ (neighbourhood size), $\delta$ (probability of selecting parent solutions from the neighbourhood), $n_r$ (maximum number of solutions replaced by each child solution), *maxEval* (stop criterion, maximum number of evaluations).

**Output:** *PS* (Pareto approximation set).

1: $P$, $z^*$ ← Initialize Subproblems and Reference Point (*popSize*)
2: $\lambda$ ← Initialize Weight Vectors (*popSize*)
3: $PS$ ← $\emptyset$
4: **while** ! Stop Criterion Reached (*maxEval*) **do**
5:     **for** $i = 1$ to *popSize* **do**
6:         *Index* ← Select Neighbourhood Indexes ($P_i$, $T$, *popSize*, $\delta$)
7:         $y$ ← Generate Candidate Solution ($P$, *Index*)
8:         $y$ ← Perform Problem-Specific Local Search ($y$)
9:         $z^*$ ← Update Reference Point ($y$, $z^*$)
10:        $c$ ← 0
11:        **while** $c \neq n_r$ && *Index* $\neq \emptyset$ **do**
12:            $j$ ← Get Index Value (*Index*)
13:            **if** $g_j(y|\lambda^j, z^*)$ is better than $g_j(P_j|\lambda^j, z^*)$ **then**
14:                $P_j$ ← $y$
15:                $c$ ← $c + 1$
16:            **end if**
17:            *Index* ← Remove Index Value ($j$, *Index*)
18:        **end while**
19:    **end for**
20:    $PS$ ← Update Pareto Approximation Set ($P$, *PS*)
21: **end while**
22: **return** *PS*

---

Particularly, a maximum number of $n_r$ subproblems can be updated for each generated solution, with the idea of promoting diversity in $P$.

Algorithm 1 illustrates the pseudocode of MOEA/D. The algorithm undertakes first the definition of initial distance matrices from starter phylogenies generated via bootstrapping [23], as well as the initialization of the reference point $z^*$ with the best starting objective scores (line 1 in Algorithm 1). Moreover, the weight vectors $\lambda$ are defined by assigning to each separated weight a value in $\{0/H, 1/H, ..., H/H\}$, where $H = popSize\text{-}1$ in the case of bi-objective optimization problems (line 2). After the initializations, the algorithm proceeds with the evolutionary mechanisms until the stop criterion is satisfied.

The processing of each subproblem $P_i$ is conducted as follows. The indexes of the individuals (from the neighbourhood or from the whole $P$) to be considered in the candidate solution computations are first retrieved (line 6). A new solution $y$ is generated by using this information, combining matrix-based operators and tree-based heuristics (lines 7 and 8). The algorithm then proceeds to integrate $y$ into the population, in case this new solution improves any of the subproblems considered in the indexes. With this purpose in mind, the reference point $z^*$ is updated provided that $y$ has, in any of the original objectives, a score better than the one currently stored in $z^*$ (line 9). Those individuals $P_j$ showing a score worse than $y$ at the corresponding scalar function $g_j$ are updated with $y$ (lines 10–18, up to a maximum of $n_r$ subproblems). Upon termination of this procedure, the algorithm repeats these steps over the next subproblem $P_{i+1}$. An iteration of the algorithm finishes with the update of the Pareto approximation set $PS$ (line 20), which is returned once the stop criterion is reached (line 22).

On the basis of this general framework, we describe next the main features of the considered design variants, MOEA/D-GA and MOEA/D-DE.

- **MOEA/D-GA**: This design implements the generation of new candidate solutions via genetic operators (selection, crossover, and mutation). More specifically, the selection operator randomly chooses two individuals, $P_{r1}$ and $P_{r2}$, from the indexes of $P_i$. These individuals take the role of parents, being their distances matrices subject of processing by the crossover operator. The new offspring individual $y$ is obtained from $P_{r1}.\Delta$ and $P_{r2}.\Delta$ by using uniform crossover, alternatively interchanging rows from the parent matrices. Mutation is applied over the resulting offspring matrix, updating randomly selected entries $y.\Delta[u, v]$ according to the following expression:

$$y.\Delta[u, v] = y.\Delta[u, v] + \Phi(\gamma * y.\Delta[u, v]), \tag{4}$$

where $\Phi$ is a uniformly distributed random number in the interval $[-1, 1]$, and $\gamma$ a gamma-distributed factor of genetic rate variation [23].
- **MOEA/D-DE**: This version employs a differential evolution (DE) approach and mutations to generate new candidate solutions. In this design, we have included the DE expressions from Li and Zhang's work [11], adapting them to our distance matrix encoding. Hence, the new offspring solution $y$ is calculated by processing the matrix entries $y.\Delta[u, v]$ through DE as follows:

$$y.\Delta[u, v] = \begin{cases} P_i.\Delta[u, v] + F(P_{r1}.\Delta[u, v] - P_{r2}.\Delta[u, v]) & \text{with prob. } CR \\ P_i.\Delta[u, v] & \text{with prob. } 1 - CR, \end{cases} \tag{5}$$

where $P_i$ is the individual under processing, $P_{r1}$, $P_{r2}$ two randomly selected solutions from the indexes of $P_i$, and $F$, $CR$ the DE control parameters. Additional genetic variability is introduced by using the mutation operator from Eq. 4, which is based on the standard Lewis' proposal [10].

## 4   Experimental Evaluation

This section undertakes the experimental assessment of the MOEA/D variants under analysis. This experimental evaluation has been carried out on a hardware platform composed of AMD Opteron 6174 processors at 2.2 GHz, with 12 MB L3 cache, and 32 GB DDR3 RAM. The software employed in this setup includes Ubuntu 14.04LTS as operating system and the compiler GCC 5.2.1. Four real-world protein datasets have been used for experimentation purposes, each one

**Table 1.** Description of protein datasets used in the experimentation

| Dataset | Description |
|---------|-------------|
| M67x11333 | 67 sequences, 11333 characters of bacterial ancestry euBac proteins [7] |
| M88x3329 | 88 sequences, 3329 characters of Thermophilic fungi proteins [12] |
| M187x814 | 187 sequences, 814 characters of Mycorrhiza fungi ABC-B transporters [9] |
| M260x1781 | 260 sequences, 1781 characters of Beta vulgaris proto-oncogene proteins [21] |

**Table 2.** Median hypervolume $I_H$ results

| Dataset | $I_H$(MOEA/D-GA) | $I_H$(MOEA/D-DE) |
|---------|------------------|------------------|
| M67x11333 | 79.031% | **80.518%** |
| M88x3329 | 56.302% | **58.083%** |
| M187x814 | 54.298% | **61.819%** |
| M260x1781 | 70.367% | **73.429%** |

**Table 3.** $I_H$ statistical testing

| Dataset | P-value | Stat. Sign.? |
|---------|---------|--------------|
| M67x11333 | 5.07E-08 | ✓ |
| M88x3329 | 1.64E-08 | ✓ |
| M187x814 | 1.33E-11 | ✓ |
| M260x1781 | 2.71E-09 | ✓ |

representing different problem sizes with a variant number of sequences and sequence lengths. The description of these datasets is provided in Table 1. The probabilistic models of protein evolution used in the calculation of the likelihood objective function are the Le-Gascuel $LG+\Gamma$ model (for M67x11333, M88x3329, and M187x814) and the Jones-Taylor-Thornton $JTT+\Gamma$ model (for M260x1781) [1]. Moreover, the input parameters of MOEA/D were configured by checking uniformly distributed values in the range of each parameter. In accordance with these parametric studies, the final parameter values were set to $popSize = 96$, $T = 6$, $\delta = 25\%$, and $n_r = 1$. The crossover probability in MOEA/D-GA was set to 70%, while the DE control parameters in MOEA/D-DE took the values F = 0.5 and CR = 75%. The mutation probability in both versions was established to 5%, and the stop criterion was set to $maxEval = 12,000$ evaluations.

### 4.1   Multiobjective Performance

The experimentation involved 31 independent runs per dataset and method. The outcomes reported by the MOEA/D designs under evaluation were assessed by using two multiobjective performance metrics: hypervolume $I_H$ and the coverage relation $SC$. While hypervolume $I_H(X)$ measures the area of the objective space covered by the solutions reported by an algorithm $X$, the coverage relation $SC(X,Y)$ compares the outcomes of two optimizers $X$ and $Y$ by calculating the fraction of solutions from $Y$ that are weakly-dominated by $X$. Higher $I_H$ and $SC$ scores denote better multiobjective quality attending to these two performance metrics. Hypervolume calculations were performed by using the following ideal and nadir points (in format $P(T), L(T)$) for normalization purposes:

- **Ideal**: (171540, −473023.58) in M67x11333, (33456, −149020.30) in M88x3329, (29832, −133804.97) in M187x814, and (43507, −163813.35) in M260x1781.

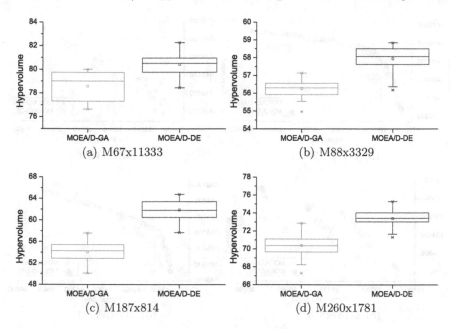

**Fig. 1.** Representation of hypervolume box plots for MOEA/D-GA and MOEA/D-DE

**Table 4.** Coverage relation $SC$ results

|  | M67x11333 | M88x3329 | M187x814 | M260x1781 |
|---|---|---|---|---|
| $SC$(MOEA/D-GA, DE) | 19.318% | 40.000% | 40.541% | 38.211% |
| $SC$(MOEA/D-DE, GA) | **69.737%** | **63.636%** | **58.065%** | **57.955%** |

– *Nadir*: (196869, −491215.09) in M67x11333, (33668, −149450.07) in M88x3329, (30213, −134944.68) in M187x814, and (44519, −165660.07) in M260x1781.

Table 2 reports the median hypervolume scores achieved by MOEA/D-GA and MOEA/D-DE in the experiments. The box plots for the observed hypervolume samples are graphically depicted in Fig. 1. According to these results, the design variant MOEA/D-DE attains more satisfying multiobjective performance with regard to MOEA/D-GA in all the datasets under analysis. MOEA/D-DE obtains hypervolume scores in the interval 58.1%–80.5%, while MOEA/D-GA lies in the range 54.3%–79.0%. The most noticeable difference is observed in the dataset M187x814, where MOEA/D-DE leads to an improvement of 7.5% in hypervolume over MOEA/D-GA. The box plots in Fig. 1 also shed light on the improved behaviour observed when using MOEA/D-DE. In fact, the statistical analysis of hypervolume samples under the Wilcoxon-Mann-Whitney test [20] confirms the attainment of statistically significant improvements in all the datasets. Table 3 shows the P-values reported by the statistical testing procedure, suggesting significant differences with a confidence level of 95%.

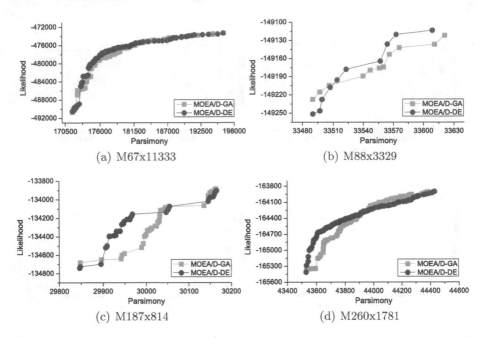

Fig. 2. Median-hypervolume Pareto fronts for MOEA/D-GA and MOEA/D-DE

Regarding the second metric, Table 4 includes the coverage results attained by MOEA/D-GA over MOEA/D-DE ($SC(MOEA/D$-$GA,DE)$) and the ones achieved by MOEA/D-DE over MOEA/D-GA ($SC(MOEA/D$-$DE,\ GA)$). For illustration purposes, the Pareto fronts obtained by each design variant in the median-hypervolume executions are shown in Fig. 2. The coverage metric also points out MOEA/D-DE as the design variant that accomplishes the most successful results from a multiobjective perspective, covering up to a 69.7% of the solutions generated by MOEA/D-GA. On the other hand, MOEA/D-GA obtains $SC$ scores lower than 50% (from 19.3% to 40.5%). These results denote the improved quality of the solutions returned by MOEA/D-DE, which cover noticeable percentages of the fronts obtained by MOEA/D-GA. Additionally, the distances to the ideal point (L2 metric) further confirm the significance of MOEA/D-DE (0.317 –DE– vs. 0.370 –GA– on M67x11333, 0.533 vs. 0.647 on M88x3329, 0.495 vs. 0.645 on M187x814, and 0.460 vs. 0.507 on M260x1781).

These metrics therefore suggest that the design variant MOEA/D-DE leads to improved multiobjective performance in comparison to MOEA/D-GA. Consequently, the tackled problem benefits from the consideration of multiple sources of information (individuals) to generate new candidate solutions, thus being the DE strategy more consistent than the genetic approach within MOEA/D.

**Table 5.** Comparisons of phylogenetic quality with other methods

|  | Parsimony $P(T)$ | | Likelihood $L(T)$ | | |
|---|---|---|---|---|---|
| Dataset | MOEA/D-DE | TNT | MOEA/D-DE | FastTree | MrBayes |
| M67x11333 | **171623** | **171623** | **−473311.888** | −473546.618 | −473989.471 |
| M88x3329 | **33490** | **33490** | **−149113.272** | −149306.304 | −149827.652 |
| M187x814 | **29847** | **29847** | **−133899.987** | −133908.553 | −134141.657 |
| M260x1781 | **43529** | **43529** | **−163899.882** | −164614.024 | −165221.774 |

## 4.2 Comparisons with Other Methods

In order to put into context the significance of the decomposition-based design, comparisons with other approaches for protein-based phylogeny reconstruction have been conducted. With this purpose in mind, we have used as reference the results from the best MOEA/D design variant, that is, MOEA/D-DE. We first introduce a comparative evaluation of multiobjective results with one of the most widely-used multiobjective evolutionary algorithms, NSGA-II [5]. 31 independent runs per dataset were carried out under the same experimental conditions as MOEA/D-DE, reporting the following results:

- *Hypervolume*: NSGA-II returned median hypervolume scores of 78.8% for M67x11333, 55.9% for M88x3329, 58.1% for M187x814, and 65.2% for M260x-1781. According to Table 2, MOEA/D-DE manages to achieve significant improvements in hypervolume over NSGA-II in all the evaluation scenarios.
- *Coverage Relation*: NSGA-II covered limited front percentages of 15.9% (M67x11333), 20% (M88x3329), 37.8% (M187x814), and 22.8% (M260x1781) over MOEA/D-DE. On the other hand, MOEA/D-DE reported an average coverage percentage of 60% over NSGA-II.

Therefore, both multiobjective metrics agree in pointing out the satisfying performance obtained by MOEA/D-DE with regard to the standard NSGA-II.

The assessment of phylogenetic quality (in terms of parsimony and likelihood) is presented in Table 5. This table introduces comparisons between MOEA/D-DE and three single-objective biological tools: TNT [6], FastTree [14], and MrBayes [16]. Particularly, we compare the extreme points of the median-hypervolume Pareto fronts from MOEA/D-DE with the median results from each tool, after 31 independent runs per experiment. Firstly, the analysis of parsimony scores highlights that MOEA/D-DE is able to reconstruct parsimony topologies with the same quality as the ones inferred by the reference method TNT in all the datasets. Secondly, MOEA/D also reports significant solutions from the likelihood perspective, improving the results obtained by FastTree and MrBayes in all the problem instances under analysis. Consequently, the comparative evaluation of biological quality also suggests the relevance of applying decomposition with robust optimization strategies in this hard-to-solve problem.

# 5  Conclusions

This work examined decomposition-based multiobjective optimizers to tackle the reconstruction of ancestral relationships from protein data. The focus of the research was the evaluation of different design variants for the reference multiobjective algorithm based on decomposition, MOEA/D. Particularly, we have compared the integration of different evolutionary strategies to generate new candidate solutions, namely genetic operators (MOEA/D-GA) and differential evolution (MOEA/D-DE). These alternative versions were adapted to the addressed problem by using a hybrid search methodology, combining matrix-based evolutionary operators with tree-based heuristics to boost solution's quality.

MOEA/D-GA and MOEA/D-DE were experimentally assessed by using four real-world protein datasets, which served as an appropriate representation of problem sizes in terms of number of sequences and sequence length. A comparative evaluation of multiobjective performance was carried out by means of two well-known multiobjective quality metrics: hypervolume and coverage relation. The attained results pointed out the benefits provided by the MOEA/D-DE design variant. More specifically, statistically significant improvements in hypervolume were observed over MOEA/D-GA in all the problem instances, along with relevant coverage scores in the reported Pareto fronts. The relevance of MOEA/D-DE results was examined by conducting comparisons with four alternative methods for phylogeny inference. The comparisons performed at the multiobjective and biological levels confirmed the significant performance of the decomposition approach and the accuracy of the designed search strategies.

Future research directions are aimed at analyzing other decomposition-based multiobjective designs to handle computationally expensive, large-scale phylogeny reconstructions. We will put emphasis on extending the evaluation of MOEA/D approaches with different design alternatives integrating computational resource allocation strategies and other reproduction operators [22]. Furthermore, we will address the question on how to effectively parallelize MOEA/D in state-of-the-art high-performance architectures, dealing with the intrinsic data dependencies of its algorithmic design.

**Acknowledgments.** This work was partially funded by the AEI (State Research Agency, Spain) and the ERDF (European Regional Development Fund, EU), under the contract TIN2016-76259-P (PROTEIN project), as well as Portuguese national funds through FCT (Fundação para a Ciência e a Tecnologia, Portugal) projects UID/CEC/50021/2019 and PTDC/CCI-COM/31901/2017 (HiPErBio). Sergio Santander-Jiménez is supported by the Post-Doctoral Fellowship from FCT under Grant SFRH/BPD/119220/2016.

# References

1. Arenas, M.: Trends in substitution models of molecular evolution. Frontiers Genet. **6**(319), 1–9 (2015)

2. Cancino, W., Delbem, A.C.B.: A multi-criterion evolutionary approach applied to Phylogenetic reconstruction. In: New Achievements in Evolutionary Computation, pp. 135–156. InTech (2010)
3. Coelho, G.P., Silva, A.E.A., Zuben, F.J.V.: An immune-inspired multi-objective approach to the reconstruction of phylogenetic trees. Neural Comput. Appl. **19**(8), 1103–1132 (2010)
4. Deb, K.: Multi-objective evolutionary algorithms. In: Kacprzyk, J., Pedrycz, W. (eds.) Springer Handbook of Computational Intelligence, pp. 995–1015. Springer, Heidelberg (2015). https://doi.org/10.1007/978-3-662-43505-2_49
5. Deb, K., Pratap, A., Agarwal, S., Meyarivan, T.: A fast and elitist multi-objective genetic algorithm: NSGA-II. IEEE Trans. Evol. Comput. **6**(2), 182–197 (2002)
6. Goloboff, P.A., Catalano, S.A.: TNT version 1.5, including a full implementation of phylogenetic morphometrics. Cladistics **32**(3), 221–238 (2016)
7. He, D., Fiz-Palacios, O., Fu, C., Fehling, J., Tsai, C., Baldauf, S.L.: An alternative root for the eukaryote tree of life. Curr. Biol. **24**(4), 465–470 (2014)
8. Jayaswal, V., Poladian, L., Jermiin, L.S.: Single- and multi-objective phylogenetic analysis of primate evolution using a genetic algorithm. In: Proceedings of IEEE CEC 2007, pp. 4146–4153. IEEE (2007)
9. Kovalchuk, A., Kohler, A., Martin, F., Asiegbu, F.O.: Diversity and evolution of ABC proteins in mycorrhiza-forming fungi. BMC Evol. Biol. **15**(249), 1–19 (2015)
10. Lewis, P.O.: Phylogenetic systematics turns over a new leaf. Trends Ecol. Evol. **16**(1), 30–37 (2001)
11. Li, H., Zhang, Q.: Multiobjective optimization problems with complicated pareto sets, MOEA/D and NSGA-II. IEEE Trans. Evol. Comput. **13**(2), 284–302 (2009)
12. Morgenstern, I., et al.: A molecular phylogeny of thermophilic fungi. Fungal Biol. **116**(4), 489–502 (2012)
13. Poladian, L., Jermiin, L.: Multi-objective evolutionary algorithms and phylogenetic inference with multiple data sets. Soft. Comput. **10**(4), 359–368 (2006)
14. Price, M.N., Dehal, P.S., Arkin, A.P.: FastTree 2 - approximately maximum-likelihood trees for large alignments. PLoS ONE **5**(3), 1–10 (2010). (e9490)
15. Rokas, A.: Phylogenetic analysis of protein sequence data using the Randomized Axelerated Maximum Likelihood (RAxML) program. Curr. Protoc. Mol. Biol. **96**, 1–14 (2011). 19.11
16. Ronquist, F., et al.: MrBayes 3.2: efficient bayesian phylogenetic inference and model choice across a large model space. Syst. Biol. **61**(3), 539–542 (2012)
17. Santander-Jiménez, S., Vega-Rodríguez, M.A.: On the design of shared memory approaches to parallelize a multiobjective bee-inspired proposal for phylogenetic reconstruction. Inf. Sci. **324**, 163–185 (2015)
18. Santander-Jiménez, S., Vega-Rodríguez, M.A., Gómez-Pulido, J.A., Sánchez-Pérez, J.M.: Comparing different operators and models to improve a multiobjective artificial bee colony algorithm for inferring phylogenies. In: Dediu, A.-H., Martín-Vide, C., Truthe, B. (eds.) TPNC 2012. LNCS, vol. 7505, pp. 187–200. Springer, Heidelberg (2012). https://doi.org/10.1007/978-3-642-33860-1_16
19. Santander-Jiménez, S., Vega-Rodríguez, M.A.: Applying a multiobjective meta-heuristic inspired by honey bees to phylogenetic inference. BioSyst. **114**(1), 39–55 (2013)
20. Sheskin, D.J.: Handbook of Parametric and Nonparametric Statistical Procedures, 5th edn. Chapman & Hall/CRC, New York (2011)
21. Stracke, R., Holtgräwe, D., Schneider, J., Pucker, B., Sörensen, T.R., Weisshaar, B.: Genome-wide identification and characterisation of R2R3-MYB genes in sugar beet (Beta vulgaris). BMC Plant Biol. **14**(249), 1–17 (2014)

22. Trivedi, A., Srinivasan, D., Sanyal, K., Ghosh, A.: A survey of multiobjective evolutionary algorithms based on decomposition. IEEE Trans. Evol. Comput. **21**(3), 440–462 (2017)
23. Warnow, T.: Computational Phylogenetics: An Introduction to Designing Methods for Phylogeny Estimation. Cambridge University Press, Cambridge (2017)
24. Zhang, Q., Li, H.: MOEA/D: a multi-objective evolutionary algorithm based on decomposition. IEEE Trans. Evol. Comput. **11**(6), 712–731 (2007)

# Parsimonious Modeling for Estimating Hospital Cooling Demand to Reduce Maintenance Costs and Power Consumption

Eduardo Dulce and Francisco Javier Martinez-de-Pison$^{(\boxtimes)}$

EDMANS Group, Department of Mechanical Engineering,
University of La Rioja, 26004 Logroño, Spain
edmans@dim.unirioja.es, fjmartin@unirioja.es

**Abstract.** Hospitals are massive consumers of energy, and their cooling systems for HVAC and sanitary uses are particularly energy-intensive. Forecasting the thermal cooling demand of a hospital facility is a remarkable method for its potential to improve the energy efficiency of these buildings. A predictive model can help forecast the activity of water-cooled generators and improve the overall efficiency of the whole system. Therefore, power generation can be adapted to the real demand expected and adjusted accordingly. In addition, the maintenance costs related to power-generator breakdowns or ineffective starts and stops can be reduced. This article details the steps taken to develop an optimal and efficient model based on a genetic methodology that searches for low-complexity models through feature selection, parameter tuning and parsimonious model selection. The methodology, called GAparsimony, has been tested with neural networks, support vector machines and gradient boosting techniques. This new operational method employed herein can be replicated in similar buildings with comparable water-cooled generators, regardless of whether the buildings are new or existing structures.

**Keywords:** GAparsimony · Parsimonious modeling ·
Hybrid forecasting · Thermal demand forecasting ·
Cooling demand forecasting · Building energy management system

## 1 Introduction

European buildings are responsible for 38% of CO2 emissions [1]. On November 28th 2018, the European Commission presented its new strategic long-term vision with the target of limiting the global temperature increase to well below 2 °C and pursuing efforts to maintaining it at 1.5 °C. Hence, by 2050, the EU should reduce its emissions by 80%, to below 1990 levels.

Hospitals are buildings that require vast amounts of energy. Those that use chilled water, for air conditioning (HVAC) or other essential health care services

© Springer Nature Switzerland AG 2019
H. Pérez García et al. (Eds.): HAIS 2019, LNAI 11734, pp. 181–192, 2019.
https://doi.org/10.1007/978-3-030-29859-3_16

and activities, are among the hospital facilities that consume the most energy. Past studies have shown that energy used to generate chilled water exceeds 45% of the total energy demanded by a building [2].

Building Management Systems (BMS) can contribute to energy efficiency and economic savings [3,4]. In addition, cooling demand forecasting can be a useful function to implement in a BMS, as it also represents a common problem and a key factor in thermal generation.

The BMS used in this study was implemented during the construction of the hospital in January 2008. The existing BMS, like most systems installed in buildings, is based on real-time control that uses information captured by sensors. But the control system generated more starts and stops than necessary in the liquid-cooled generators. This led to premature ageing in the generators, higher cooling demand than necessary, frequent breakdowns, and unnecessary thermal variations that did not correspond to the demand. Thus, **a predictive model of the thermal cooling demand can help forecast the water-cooled generators' activity, controlled by the BMS**, and improve the building's overall efficiency. Thus, the generation of cooling water can be adapted to meet the real demand expected for the day ahead, while the maintenance costs related to power generator breakdowns or ineffective starts and stops can also be reduced.

Several studies have been conducted about efficiency in hospitals [5], and forecasting cooling demand [6,7], short-term electrical load forecasting [8,9], often using Gaussian Processes [10], SVR methods [11,12], artificial neural networks (ANN) [13–15], and Hybrid methods [16].

## 1.1  The Search for Parsimonious Solutions

Many forecasting applications in 'Industry 4.0' (I4.0) are based on regression models that are constructed with small databases obtained from a short period of time. In this case, the information has been collected over more than three years, but this period has included dramatic changes, improvements and different measurement conditions in the facilities. Therefore, the prediction model obtained is not as accurate as in the case of studies based on shorter time frames.

This system recorded data every second, but the pre-processing strategy adopted to optimize the model averaged data by the hour, considerably reducing the size of the training dataset. In this kind of problem, the search for low-complexity models (more parsimonious), among different accurate solutions, is usually a good strategy for finding models that are robust against perturbations or noise. These kinds of models are also easier to maintain and understand [17,18].

In recent years, there is an increasing tendency to create methods to automate modeling processes with hyperparameter optimization (HO), and feature selection (FS), in order to reduce the human effort involved in these time-consuming tasks [19,20]. Among the currently available methods, GAparsimony [21] is a genetic algorithm (GA) methodology for searching for parsimonious models. It is designed specifically to work with small datasets. GAparsimony optimizes HO and FS by executing a parsimonious model selection (PMS), which is based on criteria that considers complexity and accuracy separately. Although GAparsimony performs quite well with HO, model selection with a complexity measurement based on the number of selected features has proven to be useful for obtaining more parsimonious solutions as compared to previous experiments [22].

This methodology has been successfully applied in a variety of real applications such as steel industrial processes [23], hotel room booking forecasting [24], mechanical design [25] and solar radiation forecasting [26]. This method has been extremely useful with classical machine learning methods, such as extreme gradient boosting machines (XGBoost), support vector regression (SVR), random forest (RF) or artificial neural networks (ANNs) [27]. What's more, the GAparsimony package for R is available since July 2017 [28].

This study presents promising initial research to create parsimonious predictive models of a building's cooling demand that can be useful to reducing the electrical consumption of hospital cooling systems, improve energy efficiency, and reduce the number of starts and stops of the water-cooled generators.

## 2   Case Study Description

The San Pedro Hospital is located in the city of Logroño (SPAIN), and is the top hospital in the autonomous community of La Rioja, and is part of the Spanish public healthcare system.

The San Pedro Hospital building covers an area of about $125.000\,\text{m}^2$. Most of the thermal generation, gas and high voltage installations are located in a separate building. Let us make note of the medical services offered by this hospital that are most energy-intensive: over 600 beds for hospitalization, a diagnostic imaging area, 23 operating rooms, emergency and consultation area with 21 boxes, hemodialysis, an intensive care unit, endoscopy, rehabilitation, laboratories, pharmacy, sterilization, and other general services.

### 2.1   Description of the Installations

The hospital has a central cold-water production system for cooling the building and its dependencies, which consists of 4 chillers EF1, EF2, EF3 and EF4: 3 centrifugal units of 3.51 MW (Trane CVFG model), and 1 screw machine of 1 MW (Trane RTHD model) of cooling capacity. The electrical consumption data of the system is described in Table 1. The BMS of the hospital is comprised primarily by controllers belonging to the Sauter EY3600 family and which communicate with each other through the novaNet bus. The Building Management System is a SCADA application, whose environment is novaPro Open 4.1. The server is located in the hospital data center.

Chilled water in a hospital has essential applications not only for human welfare, but for industrial and sanitary needs as well: air conditioning operating rooms, outpatient surgery, intensive care, delivery rooms, and emergency rooms, for example. It is also utilized in radiology and diagnostic imaging equipment, scanners, mammography, etc.; and for refrigeration storage such as a blood bank, kitchen, or pharmacy; for Kardex, cooling kitchen trolleys, pathological anatomy, the morgue, laboratories, data center racks, etc.

This article focuses on the study of a prediction model for a chilled-water system, given in its significance for hospital services and its high total electrical consumption.

**Table 1.** Chilled Water production data

|  | Electric power | Flow |
|---|---|---|
| **Cooling unit with 3.5 MW of cooling power (per unit)** | **754.60 kW** | —— |
| Centrifugal Chiller (EF1, EF2, EF3) | 574.60 kW | |
| Group of evaporation pumps | 45.00 kW | 615.60 m3/h |
| Group of condensation pumps | 90.00 kW | 770.40 m3/h |
| Fans (3 units) | 45.00 kW | |
| **Cooling unit with 1 MW of cooling power (total)** | **317.50 kW** | —— |
| Screw Chiller (EF4) | 280.00 kW | |
| Group of evaporation pumps | 7.50 kW | **205.00 m3/h** |
| Group of condensation pumps | 15.00 kW | 253.00 m3/h |
| Fan | 15.00 kW | |
| **Chilled Water circuit** | —— | 2019.60 m3/h |
| Group of drive pumps (4 pumps) | **37.00 kW** | 673.20 m3/h |

## 2.2 Description of Existing Problems and Proposed Solutions

Four main problems were detected and need to be addressed: uncontrolled starts and stops, breakdowns, sub-cooling of ring water and higher temperature of the ring than set-points.

Therefore, as a result of the previous Exploratory Data Analysis (EDA) and the review of installations, some actions were implemented to improve the efficiency of the system:

- Installation of frequency inverter systems in EF4. The frequency inverter (AFD) can regulate the speed of the compressor motor with a partial load.
- In EF1, EF2, EF3, which are centrifugal chillers, AFDs installation is not possible, since they still have a modulation with the refrigerant charge. Communication hardware cards have been installed that improve integration with the control system.
- Improved calculation of cold water ring temperature setpoint in order to reduce the number of starts and stops in the chillers.

These actions affected the modeling process from the moment they were implemented, as can be observed in Fig. 2.

# 3  Dataset

## 3.1  Extraction of Data

The BMS installed in the San Pedro Hospital has two ways of recording data: a measurement logger in the generation system (the BMS Sauter novaPro Open); and a system to record consumption parameters such as electricity, water, and gas consumption, etc. (the Energy Management System, Sauter EMS). The cooling energy was not measured by the EMS system, so for this study only the generation system data was used. However, in order to create a useful database, it was necessary to calculate cooling energy, and preprocess the data.

The data list of variables extracted from the BMS generation system to perform this study is included in Table 2.

**Table 2.** Control system variables

| Short name | Description |
| --- | --- |
| EF1 | EF1 - Status |
| EF2 | EF2 - Status |
| EF3 | EF3 - Status |
| EF4 | EF4 - Status |
| TIMP | Cold Ring Drive Temperature [°C] |
| TEXT | Exterior temperature of Facilities Building [°C] |
| TCONSIG | Calculated Setpoint of the regulation for Cold Production Drive [°C] |
| TENEF1 to 4 | Water temperature at the inlet of the EF1 to EF4 [°C] |
| TSALEF1 to 4 | Water temperature at the outlet of the EF1 to EF4 [°C] |

## 3.2  Data Preprocessing

During the process, the following actions were taken:

1. Creating variables by calculating them as thermal power generated.
2. Grouping data every hour. The system logs when a variable alters its state or changes its measurement. The time difference between measurements can be seconds. Therefore it was necessary to establish an algorithm that allows for similar temporal measurements.
3. NA data resolution.
4. Filtering of the predicted variable, ENERGIAKWHPOST.

The current BMS lacks a measurement function for the thermal energy generated, but both the instantaneous thermal power and the thermal energy generated can be calculated. Thanks to the other variables available in the measurement system and the fact that in this system the pump flow has a set value; thermal power can be calculated by the following formula:

$$Thermal\ Power = Flow * Thermal\ jump * Ce \tag{1}$$

The thermal power is expressed in watts [W]. Flow rate in l/h. Thermal jump in the chiller is expressed in degrees Celsius [°C]. The specific heat of the water is 1.16 Wh/kg°C. The specific weight is 1 kg/l.

The time differences between thermal power measurements is a known value, so thermal energy can be calculated. Considering that the minimum working time of generators is one hour, the chosen prediction variable was energy, ENERGIAKWHPOST [kWh], rather than instantaneous power [kW].

Due to the previous adjusting of incorrect starts/stops and setpoints in the generators, the generated variable of Thermal Energy, ENERGIAKWHPOST, exhibits a sawtooth graph, see Fig. 2. This could later lead to an incorrect learning process, so thermal energy needed to be filtered in order to soften the data.

**Table 3.** Data filtering of Prediction variable

| Filter: | ENERGY Sept.-2017 | RMS | MAE |
|---|---|---|---|
| ENERGIAKWHPOST | **886726.7** | 0 | 0 |
| ENE_GAUSSFILT3 | 886191.2 | **400.5** | **345.0** |
| ENE_GAUSSFILT5 | 885854.9 | 630.0 | 543.4 |
| ENE_GAUSSFILT7 | 885745.9 | **666.6** | **574.9** |

Different functions were tested, as shown in Table 3. The Gaussian method was the method selected to filter the variable of thermal energy, ENE_GAUSSFILT7, represented by the blue line. This method was chosen for its low RMS and MAE errors, and because the accumulated energy in the tested month was similar to the real amount of accumulated energy. In addition, compared to the red line of ENE_GAUSSFILT3, ENE_GAUSSFILT7 displays a much smoother curve:

## 3.3   Final Dataset

The attributes selected to predict the energy demanded are the following (Fig. 1):

- **ENE_GAUSSFILT7**, variable to predict.
- month, month of measurement.
- day_of_week, day of the week.
- Is_festivity, boolean variable for holiday.
- TIMP, instant impulsion temperature.
- TEXT, instant exterior temperature.
- TMEAN, average daily temperature.
- TMAX, maximum daily temperature.
- TMIN, minimum daily temperature.

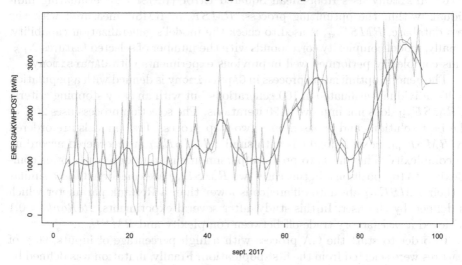

**Fig. 1.** Filtering ENERGIAKWHPOST with different Gaussian steps.

## 4 Parsimonious Modeling

The search for parsimonious models is performed with the GA-PARSIMONY methodology. For this purpose, three popular algorithms are used: artificial neural networks (ANN), support vector machines (SVR) with RBF kernel, and extreme gradient boosting machines (XGB).

All the experiments were implemented with the GAparsimony [29] package in R programming language.

### 4.1 GAparsimony Settings

GAparsimony optimization extracts from $\lambda_g^i$ chromosome and for each individual $i$ of the generation $g$, the algorithm's parameters and the selected input features. Chromosome $\lambda_g^i$ is defined for each method as:

$$ANN(\lambda_g^i) = [size, \ decay, \ num\_epochs, \ Q]$$
$$SVR(\lambda_g^i) = [cost, \ gamma, \ epsilon, \ Q]$$
$$XGB(\lambda_g^i) = [subsample, \ colsample\_bytree,$$
$$max\_depth, \ alpha, \ lambda, \ Q]$$

(2)

Where the values correspond to the algorithm's parameters except the last one, $Q$, that is a vector of probabilities for selecting each input feature $j$ when $Q_j \geqslant 0.5$.

Models were trained with the dataset from period between 01-01-2017 and 28-02-2018 dates. Validation database corresponds with the even weeks between 01-03-2018 and 28-02-2019, and testing with the odd weeks of the same period.

GAparsimony uses Root Mean Squared Error (RMSE) for evaluating individuals within the optimizing process, $RMSE_{val}$. RMSE measured with the test database, $RMSE_{tst}$, is used to check the model's generalization capability. Finally, model complexity corresponds with the number of selected features $N_{FS}$. This complexity performed well in previous experiments with GAparsimony.

The genetic optimization process in GAparsimony is defined with a population of 40 individuals evaluated in 100 generations but with an early stopping criteria if $RMSE_{val}$ does not improve in 20 iterations. The selection process uses 20% of the best solutions and is based on a two-step process: first, models are ordered by $RMSE_{val}$, next, individuals with similar $RMSE_{val}$ are reordered according to complexity. The aim is to promote parsimonious solutions (with lower complexity) to top positions. In this case, two $RMSE_{val}$ are considered to be similar if their $RMSE_{val}$ absolute difference is lower than a ReRank parameter which is defined by the user. In this study, after several experiments, $ReRank = 0.1$ achieved a satisfactory trade-off between complexity and $RMSE_{val}$.

In order to start the GA process with a high percentage of inputs, 90% of features were selected from the first population. Finally, mutation was defined by the number of most elite individuals that were not mutated (2), the probability of mutation in the model's parameter in the chromosome (10%), and the probability of a feature having the value of 1 if the feature is selected to be mutated (10%). This parameter was fixed to a low value of 10% to ease the reduction of input features in following generations.

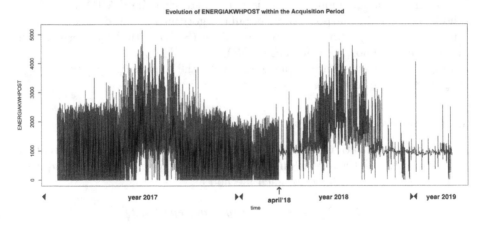

**Fig. 2.** Evolution of ENERGIAKWHPOST within the acquisition period.

## 5   Results and Discussion

Figure 3 shows $RMSE_{val}$ and $RMSE_{tst}$ evolution, white and gray box-plots respectively, for the elite population of the best GAparsimony iteration with XGB algorithm. In this case, GAparsimony converges in 7 generations to a solution with only 4 features.

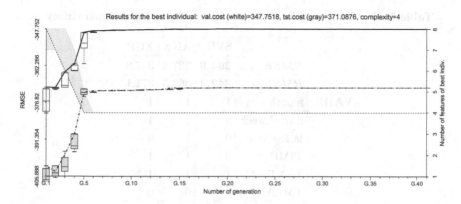

**Fig. 3.** Evolution of the elitist solutions' errors with XGB algorithm. White and gray box-plots represent $RMSE_{val}$ and $RMSE_{tst}$ evolution respectively and, continuous and shaded lines the best individual of each population. The gray area covers the maximum and minimum number of features $N_{FS}$ (right axis).

The SVR algorithm obtained the best validation and testing error with only three attributes: the month (month), and, the external (TEXT) and minimum temperatures (TMIN). ANN came in second with 7 features and, finally, XGB selected only 4.

Table 5 shows validation and testing errors, and the final selected features of the best model from the last generation with SVR, ANN, and XGB, respectively. In addition, Table 4 presents the models parameters.

**Table 4.** Best individuals' parameters

| SVR parameters | | ANN parameters | | XGB parameters | |
|---|---|---|---|---|---|
| cost | 0.173 | size | 27.00 | subsample | 0.999 |
| gamma | 0.161 | decay | 844.35 | colsample_bytree | 0.371 |
| epsilon | 1.303 | maxit | 28.00 | max_depth | 5.338 |
| | | | | alpha | 0.774 |
| | | | | lambda | 0.008 |

Surprisingly, `GAparsimony` with SVR is capable of obtaining a parsimonious model with only 3 attributes and acceptable validation and testing errors. To some degree, an explication for this can be found in the improvements realized in the control process at the half-acquisition period. Figure 2 shows the ENERGIAKWHPOST evolution during the two-year acquisition period. Due to the Exploratory Data Analysis (EDA) done at the beginning and applied after April 2018, SVR obtained a better model with a low-complexity solution that averages 'the noise' and reduces the differences between the training database created with the first year and the validation/testing data built within the last 12 months.

**Table 5.** Best individual for each algorithm obtained with GAparsimony

|  |  | SVR | ANN | XGB |
|---|---|---|---|---|
|  | $RMSE_{val}$ | **294.9** | 327.4 | 347.8 |
|  | $RMSE_{tst}$ | **342.4** | 363.3 | 371.1 |
| **VARS:** | month | 1 | 1 | 1 |
|  | day_of_week | 0 | 1 | 1 |
|  | Is_festivity | 0 | 1 | 0 |
|  | TIMP | 0 | 1 | 1 |
|  | TEXT | 1 | 1 | 1 |
|  | TMEAN | 0 | 0 | 0 |
|  | TMAX | 0 | 1 | 0 |
|  | TMIN | 1 | 1 | 0 |
|  | complexity | **3** | 7 | 4 |

# 6    Conclusions

This study has demonstrated that GAparsimony can serve as an advanced method for selecting the best model among different forecasting methodologies, and to adjust internal parameters as well.

The final step of this study is to implement the model in the BMS decision software, and then track the real response and make the corresponding adjustments.

Probably, with more new data, and by eliminating the first 12 months, the search for the best models will improve by including more features and the differences between the algorithms will be smaller.

# References

1. OECD/IEA: International energy agency (2014)
2. IDAE, Fenercom: Guía de ahorro y eficiencia energética en hospitales. Fenercom (2010)
3. Yoon, S.H., Kim, S.Y., Park, G.H., Kim, Y.K., Cho, C.H., Park, B.H.: Multiple power-based building energy management system for efficient management of building energy. Sustain. Cities Soc. **42**, 462–470 (2018)
4. Missaoui, R., Joumaa, H., Ploix, S., Bacha, S.: Managing energy smart homes according to energy prices: analysis of a building energy management system. Energy Build. **71**, 155–167 (2014)
5. Palme, M.: The possible shift between heating and cooling demand of buildings under climate change conditions: are some mitigation policies wrongly understood? In: Sayigh, A. (ed.) Mediterranean Green Buildings & Renewable Energy, pp. 417–422. Springer International Publishing, Cham (2017). https://doi.org/10.1007/978-3-319-30746-6_30
6. Saeedi, M., Moradi, M., Hosseini, M., Emamifar, A., Ghadimi, N.: Robust optimization based optimal chiller loading under cooling demand uncertainty. Appl. Therm. Eng. **148**, 1081–1091 (2019)

7. Wang, L., Lee, E.W., Yuen, R.K.: Novel dynamic forecasting model for building cooling loads combining an artificial neural network and an ensemble approach. Appl. Energy **228**, 1740–1753 (2018)
8. Abdel-Aal, R.: Modeling and forecasting electric daily peak loads using abductive networks. Int. J. Electr. Power Energy Syst. **28**(2), 133–141 (2006)
9. Chitsaz, H., Shaker, H., Zareipour, H., Wood, D., Amjady, N.: Short-term electricity load forecasting of buildings in microgrids. Energy Build. **99**, 50–60 (2015)
10. Shepero, M., van der Meer, D., Munkhammar, J., Widén, J.: Residential probabilistic load forecasting: a method using Gaussian process designed for electric load data. Appl. Energy **218**, 159–172 (2018)
11. Li, Y., Che, J., Yang, Y.: Subsampled support vector regression ensemble for short term electric load forecasting. Energy **164**, 160–170 (2018)
12. Yang, Y., Che, J., Deng, C., Li, L.: Sequential grid approach based support vector regression for short-term electric load forecasting. Appl. Energy **238**, 1010–1021 (2019)
13. Bagnasco, A., Fresi, F., Saviozzi, M., Silvestro, F., Vinci, A.: Electrical consumption forecasting in hospital facilities: an application case. Energy Buildings **103**(Complete), 261–270 (2015)
14. Jetcheva, J.G., Majidpour, M., Chen, W.P.: Neural network model ensembles for building-level electricity load forecasts. Energy Build. **84**, 214–223 (2014)
15. Hsu, Y.Y., Tung, T.T., Yeh, H.C., Lu, C.N.: Two-stage artificial neural network model for short-term load forecasting. IFAC-PapersOnLine **51**(28), 678–683 (2018). 10th IFAC Symposium on Control of Power and Energy Systems CPES 2018
16. Singh, P., Dwivedi, P., Kant, V.: A hybrid method based on neural network and improved environmental adaptation method using controlled gaussian mutation with real parameter for short-term load forecasting. Energy **174**, 460–477 (2019)
17. Avalos, M., Grandvalet, Y., Ambroise, C.: Parsimonious additive models. Comput. Stat. Data Anal. **51**(6), 2851–2870 (2007)
18. Li, H., Shu, D., Zhang, Y., Yi, G.Y.: Simultaneous variable selection and estimation for multivariate multilevel longitudinal data with both continuous and binary responses. Comput. Stat. Data Anal. **118**, 126–137 (2018)
19. Husain, H., Handel, N.: Automated machine learning. A paradigm shift that accelerates data scientist productivity, May 2017
20. Feurer, M., Klein, A., Eggensperger, K., Springenberg, J., Blum, M., Hutter, F.: Efficient and robust automated machine learning. In: Cortes, C., Lawrence, N.D., Lee, D.D., Sugiyama, M., Garnett, R. (eds.) Advances in Neural Information Processing Systems vol. 28, pp. 2962–2970. Curran Associates, Inc. (2015)
21. Sanz-Garcia, A., Fernandez-Ceniceros, J., Antonanzas-Torres, F., Pernia-Espinoza, A., Martinez-de Pison, F.J.: GA-PARSIMONY: A GA-SVR approach with feature selection and parameter optimization to obtain parsimonious solutions for predicting temperature settings in a continuous annealing furnace. Appl. Soft Comput. **35**, 13–28 (2015)
22. Urraca, R., Sodupe-Ortega, E., Antonanzas, J., Antonanzas-Torres, F., de Pison, F.M.: Evaluation of a novel GA-based methodology for model structure selection: the GA-PARSIMONY. Neurocomputing **271**(Supplement C), 9–17 (2018)
23. Sanz-García, A., Fernández-Ceniceros, J., Antoñanzas-Torres, F., Martínez-de Pisón, F.J.: Parsimonious support vector machines modelling for set points in industrial processes based on genetic algorithm optimization. In: Herrero, Á., et al. (eds.) International Joint Conference SOCO13-CISIS13-ICEUTE13. Advances in Intelligent Systems and Computing, vol. 239, pp. 1–10. Springer, Cham (2014). https://doi.org/10.1007/978-3-319-01854-6_1

24. Urraca, R., Sanz-Garcia, A., Fernandez-Ceniceros, J., Sodupe-Ortega, E., Martinez-de-Pison, F.J.: Improving hotel room demand forecasting with a hybrid GA-SVR methodology based on skewed data transformation, feature selection and parsimony tuning. In: Onieva, E., Santos, I., Osaba, E., Quintián, H., Corchado, E. (eds.) HAIS 2015. LNCS (LNAI), vol. 9121, pp. 632–643. Springer, Cham (2015). https://doi.org/10.1007/978-3-319-19644-2_52
25. Fernandez-Ceniceros, J., Sanz-Garcia, A., Antonanzas-Torres, F., de Pison, F.M.: A numerical-informational approach for characterising the ductile behaviour of the T-stub component. Part 2: Parsimonious soft-computing-based metamodel. Eng. Struct. **82**, 249–260 (2015)
26. Antonanzas-Torres, F., Urraca, R., Antonanzas, J., Fernandez-Ceniceros, J., de Pison, F.M.: Generation of daily global solar irradiation with support vector machines for regression. Energy Convers. Manag. **96**, 277–286 (2015)
27. Martinez-de-Pison, F.J., Fraile-Garcia, E., Ferreiro-Cabello, J., Gonzalez, R., Pernia, A.: Searching parsimonious solutions with GA-PARSIMONY and XGBoost in high-dimensional databases. In: Graña, M., López-Guede, J.M., Etxaniz, O., Herrero, Á., Quintián, H., Corchado, E. (eds.) ICEUTE/SOCO/CISIS -2016. AISC, vol. 527, pp. 201–210. Springer, Cham (2017). https://doi.org/10.1007/978-3-319-47364-2_20
28. Sanz-Garcia, A., Fernandez-Ceniceros, J., Antonanzas-Torres, F., Pernia-Espinoza, A., Martinez-de Pison, F.: GA-parsimony. Appl. Soft Comput. **35**(C), 13–28 (2015)
29. Martínez-De-Pisón, F.J.: GAparsimony: GA-based optimization R package for searching accurate parsimonious models (2017). R package version 0.9-1

# Haploid Versus Diploid Genetic Algorithms. A Comparative Study

Adrian Petrovan$^{(\boxtimes)}$, Petrica Pop-Sitar, and Oliviu Matei

Technical University of Cluj-Napoca, North University Center of Baia Mare,
Cluj-Napoca, Romania
adrian.petrovan@cunbm.utcluj.ro

**Abstract.** Genetic algorithms (GAs) are powerful tools for solving complex optimization problems, usually using a haploid representation. In the past decades, there has been a growing interest concerning the diploid genetic algorithms. Even though this area seems to be attractive, it lacks wider coverage and research in the Evolutionary Computation community. The scope of this paper is to provide some reasons why this situation happens and in order to fulfill this aim, we present experimental results using a conventional haploid GA and a developed diploid GA tested on some major benchmark functions used for performance evaluation of genetic algorithms. The obtained results show the superiority of the diploid GA over the conventional haploid GA in the case of the considered benchmark functions.

**Keywords:** Haploid and diploid genetic algorithms ·
Benchmark functions · Comparative study

## 1 Introduction

Genetic algorithms (GAs) were introduced by Holland [6] and are search heuristic methods inspired from the theory of natural evolution depeloped by Charles Darwin based on the "survival of the fittest". GAs have the ability to deliver a "good-enough" solution "fast-enough", making them very attractive in solving optimization problems. The main difficulties encountered while solving optimization problems are the occurrence of the local optima and finding a good starting point of the search.

The idea behind GA is to model the natural evolution by using genetic inheritance together with Darwin's theory. In GAs, we have a population that consists of a set of feasible solutions or individuals instead of chromosomes. A selection procedure, simulating the natural selection, selects a certain number of parent solutions that undergo recombination and mutation (like in natural genetics), producing new solutions also called offspring, and the process is repeated over various generations. At the end of each iteration the offspring together with the solutions from the previous generation form a new generation. Each candidate

© Springer Nature Switzerland AG 2019
H. Pérez García et al. (Eds.): HAIS 2019, LNAI 11734, pp. 193–205, 2019.
https://doi.org/10.1007/978-3-030-29859-3_17

solution is evaluated in terms of its fitness value (based on its objective function value) and the fitter individuals are given a chance to mate and yield more "fitter" individuals.

In most of the cases the considered GAs make use of haploid representation of individuals, but the idea of using diploid representation (an individual comprising of two chromosomes) in GAs is not new and has proven as an important feature of the natural evolution. Goldberg and Smith [5] suggested diploids in the case of dynamic optimization and reported the first experimental and theoretical results of using dominance for boolean chromosomal representations, Yukiko and Nobue [14] described a diploid GA (DGA) for preserving the population diversity using the idea of meiosis to convert the genotype to phenotype, also based on discrete representation of alleles. Lieckens et al. [7] introduced the diploid simple GA that limits the GA mainly in its discrete time, non-overlapping populations setup and its representation of genotypes and as well provided formal methods to be used to study finite population models of diploid genetic algorithms. In another approach, Bull [3] presents a new variant of evolutionary algorithm that harnesses the haploid-diploid cycle present in eukaryotic organisms.

Recently, Bhasin et al. [2] described in the case of dynamic traveling salesman person a diploid GA that outperforms a greedy approach and a simple genetic algorithm and discussed the performance of diploid GAs in comparison to other approaches in dynamic environments and Pop et al. [10–12] used successfully diploid GAs in order to solve the generalized traveling salesman problem, the generalized minimum spanning tree problem and the family traveling salesman problem.

The aim of this article is to compare diploid and haploid GAs, respectively to found the diploids on a sound scientific basis, by testing them on some major benchmark functions used for performance evaluation of genetic algorithms. The rest of the paper is organized as follows: in Sect. 2 we describe the proposed conventional haploid GA (HGA) and the diploid GA (DGA) with the means of measuring the corresponding fitness and the other genetic operators used in our algorithms. Section 3 details the experiments performed and analyses the obtained results. Finally, the conclusions and future research directions are presented in the last section.

## 2 Haploid Versus Diploid Genetic Algorithms

In superior life forms, chromosomes contain two sets of genes, known as diploids, and this suggest that a more complex structure can yield a more complex phenotype [1]. Should there appear a conflict between two values within the same couple of genes at a particular locus, the phenotype will be determined by the dominant one, while the one called recessive, whilst still present, can be passed on to the offspring. Therefore, the diploids allow a wider diversity of alleles.

However, most GAs concentrate on haploid representation in particular as they are much easier to build. In haploid representation only one set of each gene is stored, thus one avoids the entire process of determining which allele should be dominant and which one should be recessive is avoided.

## 2.1   The Haploid Genetic Algorithm

In the considered conventional haploid GA, the chromosomes, representing possible solutions of the given optimization problem, consist of a number of genes having real values. The algorithm starts with a set of solutions (represented by chromosomes) called population, which are generated randomly. Solutions from one population are taken and used to form a new population, motivated by a hope, that the new population will be better than the old one. Solutions which are chosen to form new solutions (offspring) are selected according to the roulette wheel selection strategy based on their fitness. The better the chromosomes are, the more chances to be selected they have. Crossover is one of the most significant phases in a GA. In our case, for each pair of parents to be mated, two crossover points are chosen at random from the parent chromosomes and the genes in between the two points are swapped between the parent chromosomes. This results in two offspring, each carrying some genetic information from both parents. In certain new offspring formed, one of their genes can be subjected to a mutation with a low random probability. This implies that the value of a gene is altered.

## 2.2   The Diploid Genetic Algorithm

In the case of the diploid GA, the individuals consist of two (haploid) coupled chromosomes. We call this kind of representation the diploid (bi-chromosomal) representation.

Consequently, unlike the situation of classical individuals, which were considered synonymous to chromosomes, in this situation we deal with individuals that are made up of a pair of chromosomes. Thus the natural diploid individuals are mimicked [8]. The superiority of this representation resides in the fact that twice as much information is borne by each individual as compared to the classical haploid approach, thus a higher diversity being ensured in terms of potential feasible solutions [5].

The ultimate features of a phenotype of an individual are decided by the dominance schemes which have an important role on the performance of the algorithm. It is important to design a good dominance scheme in order to guarantee the performance of the diploid GA in comparison to the conventional GA. Some of the most important dominance scheme described in the literature have been proposed by Ng and Wong [9] in which the dominant allele will always be part of the phenotype and should there be a conflict between two dominant or two recessive alleles, one is randomly selected and by Yang and Yao [13] called dominance learning scheme in which a dominant probability vector is defined, where each element is a dominance probability that represents the probability that a genotypic allele can be expressed in the phenotype in the corresponding locus. An individual is composed by a pair chromosomes and is represented as follows:

$$I = (C_1, C_2) \tag{1}$$

where $C_i$ are the chromosomes, with $i \in \{1,2\}$. In fact, both chromosomes are potential feasible solutions belonging to the space of solutions of the given optimization problem.

Further, $C_d \in \{C_1, C_2\}$ is defined as the dominant chromosome, should it be the best of the two chromosomes, and $C_r \in \{C_1, C_2\}$ as the recessive chromosome, should it be the worst of the two. Obviously, $\{C_d, C_r\} = \{C_1, C_2\}$.

As mentioned in the description of the haploid GA, in our approach we used real-valued chromosomes which consists of pre-defined number of cells having real values.

Let us denote the fitness of the individual by $f(I)$ and subsequently the fitness of the two chromosomes by $f(C_i)$, with $i \in \{1,2\}$. Of course, according to the definition provided in this section, we have $f(C_d) \leq f(C_r)$.

As it was pointed out by Pop et al. [10], there exist various options of defining an individual's fitness:

1. the fitness of the individual is the fitness of the dominant chromosome, that is: $f(I) = f(C_d)$;
2. the fitness of the individual is the fitness of the recessive chromosomes: $f(I) = f(C_r)$;
3. the fitness of the individual is the sum (or average) of the fitness values of the two chromosomes:
$$f(I) = f(C_1) + f(C_2) \tag{2}$$
4. the fitness of the individual is a weighted average of the fitness values of the two chromosomes:
$$f(I) = w_1 \cdot f(C_1) + w_2 \cdot f(C_2) \tag{3}$$
where $w_i = \frac{f(C_i)}{f(C_1)+f(C_2)}$, with $i \in \{1,2\}$.

In our developed diploid GA, we opted for the first possibility, namely the fitness of an individual is given by the fitness value of the dominant chromosome.

Also in the case of diploid GA we used roulette wheel as a selection procedure, using the above defined fitness values. Once the individuals have been selected for producing the next generation, the crossover genetic operator is used to combine the genetic information of two individuals in order to generate new offspring.

Having two individuals $I_1 = (C_1^1, C_2^1)$ and $I_2 = (C_1^2, C_2^2)$, we can define the crossover operator.

Two randomly selected diploid parents yield two offspring. The two chromosomes of each individual recombine with each chromosome of the other individual, based on 2-cutting points discrete crossover principle.

$$O_1 = (C_1^{11}, C_2^{11}) \text{ and } O_2 = (C_1^{22}, C_2^{22}), \tag{4}$$

where $C_i^{kk}$, $k \in \{1,2\}$ are obtained using the 2-cutting points discrete crossover principle at the haploid level.

Figure 1 depicts the two diploid parents, with the two chromosomes (gametes in the case of recombination). After crossover, 8 offspring are created (an example being shown in Fig. 2).

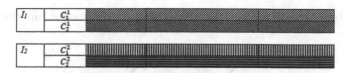

**Fig. 1.** The diploid parents

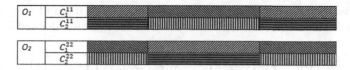

**Fig. 2.** The diploid offspring

In our diploid GA, the mutation is defined in the same way with the one used in the conventional GA, each gene in both chromosomes forming the individual has a small probability of exchanging the state.

In both haploid and diploid GAs, after mutation the offspring and the parents form a common pool and the best ones are selected to form the new generation. This approach assures a faster convergence, but a higher probability to stick the population in local optima due to the possible better parents which do not allow the new generation to explore other regions of the solution space. In this way, the ability of the proposed GAs to avoid local optima is totally challenged and prove the advantages of the developed diploid GA.

## 3   Experimental Study

The experiments consisted of running a straightforward haploid, respectively diploid genetic algorithm on several benchmark functions. The target was to determine the behaviour and the improvements brought in by one algorithm or the other (especially the diploid representation itself), not tunings or tweaks applied to them. That is the reason why the results are not compared with the results of other techniques reported in other articles.

Our haploid and diploid GAs have been implemented on Java 8 and we have performed 20 independent tests for each considered benchmark function. The experiments have been conducted on a machine with CPU Intel Core i5, 2.4 GHz, 8 GB RAM, running JDK 8 on macOS High Siera 10.13.

The two developed algorithms, haploid, respectively diploid, have been tested on the following benchmark functions described in what it follows:

1. **Sphere function** is a smooth unimodal function defined as follows:

$$f(\mathbf{x}) = f(x_1, x_2, ..., x_n) = \sum_{i=1}^{n} x_i^2 \tag{5}$$

where $-5.12 \leq x_i \leq 5.12$.

2. **Ackley function** is a continuous, non-convex multimodal function defined as follows:

$$f(\mathbf{x}) = -a.exp\left(-b\sqrt{\frac{1}{n}\sum_{i=1}^{n}x_i^2}\right) - exp\left(\frac{1}{n}\sum_{i=1}^{n}\cos(cx_i)\right) + a + exp(1) \quad (6)$$

where usually $-32.768 \le x_i \le 32.768$ and $a = 20$, $b = 0.2$ and $c = 2\pi$.

3. **Griewank function** is a non-linear multimodal function defined as follows:

$$f(\mathbf{x}) = f(x_1, ..., x_n) = 1 + \sum_{i=1}^{n}\frac{x_i^2}{4000} - \prod_{i=1}^{n}\cos\left(\frac{x_i}{\sqrt{i}}\right) \quad (7)$$

where usually $-600 \le x_i \le 600$.

4. **Rastrigin function** which is a non-linear multimodal function containing millions of local optima and defined as follows:

$$f(\mathbf{x}) = 20A + \sum_{i=1}^{n}\left(x_i^2 - 10\cos(2\pi x_i)\right) \quad (8)$$

where usually $A = 10$ and $-5.12 \le x_i \le 5.12$.

5. **Schwefel function** which is a non-linear multimodal function defined as follows:

$$f(\mathbf{x}) = f(x_1, x_2, ..., x_n) = 10V - \sum_{i=1}^{n}x_i \sin(\sqrt{|x_i|}) \quad (9)$$

where usually $V = 4189,829101$ and $-500 \le x_i \le 500$.

All the considered test functions are dealing with minimization problems. For more information regarding the properties of the considered benchmark functions and some other functions used in order to test the performance of GAs we refer to [4].

The initial population was generated randomly and for each benchmark function, the tests have been made on two dimensions: 25, respectively 50. As turns out from the experimental results, the conclusions are similar in both cases, therefore no further dimensional searches were needed. This assures an extremely large solution space along with complex individuals. The size of the population makes use of the same number of chromosomes, meaning 2000 haploid individuals, respectively 1000 diploids and the intermediary size of the population is 8000. The best 1000 individuals are selected to form the next generation, the percentage of elitist population representing 12.5%. The algorithms have been run 20 times for each benchmark function for maximum 100 epochs and for each epoch, the best, the worst and the average individual have been recorded, along with the standard deviation of the population. All the experiments conducted use a mutation rate of 5%.

For the envisaged benchmark functions with dimensions 25 respectively 50, we compared the proposed haploid GA (HGA), and the diploid GA (DGA), with the purpose of studying the performance of the considered GAs. As the purpose of this article is to make a sound and fair comparison between HGA and DGA, no advanced configurations or tuning's have been performed on the two approaches. All parameters considered here are quite straightforward and the global outcomes do not change. In other words, the relative improvements of the DGA comparing with HGA remain however the parameters are set.

The obtained computational results are summarized in Table 1. The first column of the table presents the used benchmark functions, the next two columns contain the average values obtained using the haploid GA in the two cases considered one with dimension 25 and the other with dimension 50, the next two columns contain the average values obtained using the diploid GA in the considered cases and the last two columns present the improvement of the diploid GA over the haploid GA calculated as a percentage.

**Table 1.** The computational results of comparing the haploid and diploid GAs on five benchmark functions

| Function | Haploid GA | | Diploid GA | | Improvement | |
|---|---|---|---|---|---|---|
| | Case 1 | Case 2 | Case 1 | Case 2 | Case 1 | Case 2 |
| Sphere | 0.314 | 2.511 | 0.090 | 1.182 | 71,34% | 52,93% |
| Ackley | 3.740 | 5.704 | 1.7640 | 5.374 | 52,83% | 5,79% |
| Griewank | 1.348 | 8.880 | 1.222 | 3.535 | 9,35% | 60,19% |
| Rastrigin | 10.454 | 27.417 | 4.835 | 26.349 | 53,75% | 3,90% |
| Schwefel | 67.029 | 524.344 | 23.162 | 202.135 | 65,44% | 61,45% |

Analyzing the obtained results presented in Table 1, we can observe that the diploid GA significantly outperforms the conventional haploid GA in terms of the quality of the achieved solutions.

The corresponding execution times per epoch are summarized in Table 2.

**Table 2.** Execution times per epoch for haploid and diploid GAs

| Function | Haploid GA | | Diploid GA | | Improvement | |
|---|---|---|---|---|---|---|
| | Case 1 | Case 2 | Case 1 | Case 2 | Case 1 | Case 2 |
| Sphere | 0.22455 | 0.25434 | 0.07977 | 0.115 | 64.48% | 54.78% |
| Ackley | 0.24597 | 0.30863 | 0.11022 | 0.18553 | 55.19% | 39.89% |
| Griewank | 0.22202 | 0.31087 | 0.10899 | 0.17589 | 50.91% | 43.42% |
| Rastrigin | 0.19826 | 0.31038 | 0.10081 | 0.1693 | 49.15% | 45.45% |
| Schwefel | 0.24597 | 0.31896 | 0.11022 | 0.18206 | 55.19% | 42.92% |

We can observe that the execution time per epoch in the case of diploid GA is smaller compare to the corresponding execution time of the haploid GA because the population is lower (1,000 comparing with 2,000), although the information carried by a population is the same (2,000 chromosomes, meaning 2,000 possible solutions) in both representations. The times are measured in seconds. The improvements are more than 40%. The convergence times for haploid and diploid GAs are summarized in Table 3. The early stopping criterion used is that there are no improvements of the solution over 10 consecutive generations.

**Table 3.** Convergence times for haploid and diploid GAs

| Function | Haploid GA | | Diploid GA | | Improvement | |
|----------|-----------|---------|-----------|---------|-------------|---------|
|          | Case 1    | Case 2  | Case 1    | Case 2  | Case 1      | Case 2  |
| Sphere   | 8,982     | 15,2604 | 4,7862    | 9,2     | 46.71%      | 39.71%  |
| Ackley   | 9,8388    | 15,4315 | 6,6132    | 11,1318 | 32.78%      | 27.86%  |
| Griewank | 8,8808    | 15,5435 | 5,4495    | 12,3123 | 38.63%      | 20.78%  |
| Rastrigin| 7,9304    | 18,6228 | 5,0405    | 13,544  | 36.44%      | 27.27%  |
| Schwefel | 9,8388    | 15,948  | 5,511     | 12,7442 | 43.98%      | 20.08%  |

As can be seen from Table 3, the convergence times are significantly better in the case of diploid GA compared to those of haploid GA, with and an improvement of more than 20%. As we expected, we can observe that the times to convergence of the constructed GAs increased with the dimension of the considered benchmark functions.

The standard deviations of the haploid and diploid GAs in the case of the considered benchmark function with dimensions 25 and 50 are summarized in Tables 4 and 5.

In Tables 4 and 5 the first column represent the epoch and the next ten columns provide the achieved standard deviation at a certain epoch by the haploid GA and diploid GA in the case of the considered benchmark functions.

**Table 4.** Standard deviations for the benchmark functions with dimension 25

| Epoch | Sphere | | Ackley | | Griewank | | Rastrigin | | Schwefel | |
|-------|--------|-------|--------|-------|----------|-------|-----------|-------|----------|---------|
|       | HGA    | DGA   | HGA    | DGA   | HGA      | DGA   | HGA       | DGA   | HGA      | DGA     |
| 10    | 0.868  | 2.284 | 0.551  | 0.460 | 5.268    | 8.242 | 5.268     | 8.206 | 143.557  | 195.821 |
| 20    | 0.101  | 0.244 | 0.129  | 0.276 | 0.398    | 1.005 | 0.398     | 1.882 | 15.644   | 55.325  |
| 30    | 0.009  | 0.035 | 0.024  | 0.075 | 0.026    | 0.105 | 0.026     | 0.452 | 0.490    | 6.554   |
| 40    | 0      | 0.003 | 0      | 0.027 | 0        | 0.009 | 0         | 0.139 | 0        | 0.940   |
| 50    | 0      | 0     | 0      | 0.001 | 0        | 0     | 0         | 0     | 0        | 0.006   |
| 60    | 0      | 0     | 0      | 0     | 0        | 0     | 0         | 0     | 0        | 0       |
| 70    | 0      | 0     | 0      | 0     | 0        | 0     | 0         | 0     | 0        | 0       |
| 80    | 0      | 0     | 0      | 0     | 0        | 0     | 0         | 0     | 0        | 0       |
| 90    | 0      | 0     | 0      | 0     | 0        | 0     | 0         | 0     | 0        | 0       |
| 100   | 0      | 0     | 0      | 0     | 0        | 0     | 0         | 0     | 0        | 0       |

**Table 5.** Standard deviations for the benchmark functions with dimension 50

| Epoch | Sphere | | Ackley | | Griewank | | Rastrigin | | Schwefel | |
|---|---|---|---|---|---|---|---|---|---|---|
| | HGA | DGA | HGA | DGA | HGA | DGA | HGA | DGA | HGA | DGA |
| 10 | 4,485 | 6,832 | 0,200 | 0,192 | 16,389 | 21,936 | 12,331 | 13,625 | 299,486 | 326,272 |
| 20 | 1,132 | 1,879 | 0,152 | 0,294 | 2,842 | 5,495 | 2,883 | 6,751 | 144,991 | 179,397 |
| 30 | 0,164 | 0,404 | 0,079 | 0,167 | 0,563 | 1,245 | 1,204 | 2,725 | 35,931 | 77,251 |
| 40 | 0,040 | 0,122 | 0,027 | 0,063 | 0,066 | 0,337 | 0,344 | 1,199 | 3,756 | 22,634 |
| 50 | 0,001 | 0,029 | 0 | 0,024 | 0,005 | 0,067 | 0 | 0,341 | 0 | 4,886 |
| 60 | 0 | 0,007 | 0 | 0 | 0 | 0,017 | 0 | 0,003 | 0 | 0,149 |
| 70 | 0 | 0 | 0 | 0 | 0 | 0,002 | 0 | 0 | 0 | 0 |
| 80 | 0 | 0 | 0 | 0 | 0 | 0 | 0 | 0 | 0 | 0 |
| 90 | 0 | 0 | 0 | 0 | 0 | 0 | 0 | 0 | 0 | 0 |
| 100 | 0 | 0 | 0 | 0 | 0 | 0 | 0 | 0 | 0 | 0 |

We can observe that the initial standard deviations vary, as the first population is created randomly, but after 20 epochs, when the population starts converging and the random individuals are eliminated, the standard deviation of the diploid algorithms is always larger than the corresponding one in the case of haploid GA. And yet, the convergence is slower in the case of diploid GA than in the case of conventional haploid GA.

For all the considered benchmark functions the convergence rates of the haploid GA and diploid GA are different. The diploid GA has a slower convergence, as we can see from Figs. 4 and 6, which means that it avoids local minima better. This fact is also obvious from Figs. 3 and 5, where both the best and the worst individuals achieved by the diploid GA become significantly better over epochs than their haploid GA counterparts, however the initial starting conditions are.

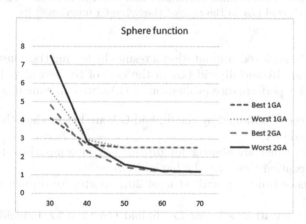

**Fig. 3.** Comparative performance of the best and worst individuals evolved using the haploid GA and diploid GA in the case of the Sphere function

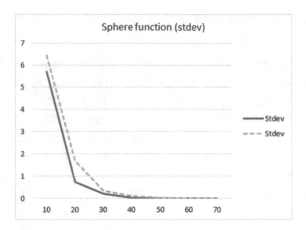

**Fig. 4.** The standard deviation for both haploid GA and diploid GA in the case of the Sphere function

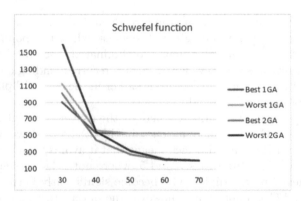

**Fig. 5.** Comparative performance of the best and worst individuals evolved using the haploid GA and diploid GA in the case of the Schwefel function

Analyzing the achieved computational results by testing the constructed conventional haploid GA and diploid GA in the case of the considered benchmark functions used for performance evaluation of GAs, we can conclude that:

- the overall results obtained by the diploid GA are better than those achieved by the haploid GA;
- the execution times are shorter for diploid GA in comparison to the corresponding execution times for the haploid GA;
- the convergence times are with at least 20% shorter for diploid GA than for haploid GA;
- the standard deviation is larger for diploid GA than for haploid GA, which means larger diversification and which avoid the local optima better, thus the results are better.

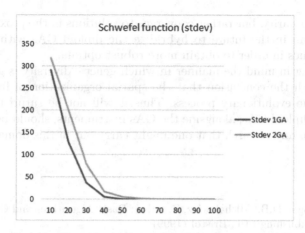

**Fig. 6.** The standard deviation for both haploid GA and diploid GA in the case of the Schwefel function

The diploid GA has proven better performance than their haploid counterpart due to the reasons shown so far. However, the underlying characteristic to all these is that each individual carries two solutions, a better one along with a worse one. The worse chromosomes have the potential of driving the generated offspring from local optima imposed by the better chromosomes, therefore the diversity of the population persists along the evolutionary process.

It is worth mentioning that in the scientific literature there is well founded comparison between haploid GA and diploid GA. Moreover, the diploid approach has been run either on specific optimization problems or on discrete search spaces, which makes a potential comparison between our results and the ones reported in the related work impossible and senseless.

## 4    Conclusions

In this paper we made a comparative study between a conventional haploid GA and a diploid GA. The constructed GAs have been tested on five benchmark functions used for performance evaluation of genetic algorithms and the achieved results show that the diploid GA found more robust optimum in comparison to haploid GA.

The results obtained through the use of the diploid GA are very promising, thus providing a reason to apply these GAs to other optimization problems, with the aim of assessing the real practicality of the method.

Future research will focus on defining, detailing and adapting the genetic operators (crossover, mutation and selection) for diploid representation. In addition, we plan to test our developed diploid GA on some more benchmarking functions and as well in the case of complex combinatorial optimization problems. As it happens with all GA's, the diploid GA does not find the global optima with regularity, but rather converges to solutions in the proximity of the optima. We plan in the future to hybridize the diploid GAs with some local search approaches in order to obtain more robust optima.

When having in mind the manner in which genetic diversity is generated in nature, we reach the conclusion that the spatial organization of living species is critical in the evolutionary process. Thus it will not be surprising that the integration of diploid GAs, alongside the GAs instruments, should be a topic for consideration in any research that one might carry out in this domain.

# References

1. Back, T., Fogel, D.B., Michalewicz, Z. (eds.): Basic Algorithms and Operators, 1st edn. IOP Publishing Ltd., Bristol (1999)
2. Bhasin, H., Behal, G., Aggarwal, N., Saini, R.K., Choudhary, S.: On the applicability of diploid genetic algorithms in dynamic environments. Soft. Comput. **20**(9), 3403–3410 (2016). https://doi.org/10.1007/s00500-015-1803-5
3. Bull, L.: Haploid-diploid evolutionary algorithms: the Baldwin effect and recombination nature's way. In: AISB (2017)
4. Digalakis, J., Margaritis, K.: On benchmarking functions for genetic algorithms. Int. J. Comput. Math. **77**(4), 481–506 (2001). https://doi.org/10.1080/00207160108805080
5. Goldberg, D., Smith, R.: Nonstationary function optimization using genetic algorithms with dominance and diploidy. In: Proceedings of Second International Conference on Genetic Algorithms and Their Application, pp. 59–68 (1987)
6. Holland, J.H.: Adaptation in Natural and Artificial Systems. University of Michigan Press, Ann Arbor (1975). Second edition 1992
7. Liekens, A., Eikelder, H., Hilbers, P.: Modeling and simulating diploid simple genetic algorithms. In: Proceedings Foundations of Genetic Algorithms VII, pp. 151–168. FOGA VII (2003)
8. Mitchell, M.: An Introduction to Genetic Algorithms. MIT Press, Cambridge (1998)
9. Ng, K.P., Wong, K.C.: A new diploid scheme and dominance change mechanism for non-stationary function optimization. In: Proceedings of the 6th International Conference on Genetic Algorithms, pp. 159–166. Morgan Kaufmann Publishers Inc., San Francisco (1995). http://dl.acm.org/citation.cfm?id=645514.657904
10. Pop, P., Oliviu, M., Sabo, C.: A hybrid diploid genetic based algorithm for solving the generalized traveling salesman problem. In: Martínez de Pisón, F.J., Urraca, R., Quintián, H., Corchado, E. (eds.) HAIS 2017. LNCS (LNAI), vol. 10334, pp. 149–160. Springer, Cham (2017). https://doi.org/10.1007/978-3-319-59650-1_13
11. Pop, P., Matei, O., Sabo, C., Petrovan, A.: A two-level solution approach for solving the generalized minimum spanning tree problem. Eur. J. Oper. Res. **265**(2), 478–487 (2018)
12. Pop, P., Matei, O., Pintea, C.: A two-level diploid genetic based algorithm for solving the family traveling salesman problem. In: Proceedings of the Genetic and Evolutionary Computation Conference, GECCO 2018, pp. 340–346. ACM, New York (2018). https://doi.org/10.1145/3205455.3205545

13. Yang, S., Yao, X.: Experimental study on population-based incremental learning algorithms for dynamic optimization problems. Soft Comput. **9**(11), 815–834 (2005). https://doi.org/10.1007/s00500-004-0422-3
14. Yukiko, Y., Nobue, A.: A diploid genetic algorithm for preserving population diversity — Pseudo-Meiosis GA. In: Davidor, Y., Schwefel, H.-P., Männer, R. (eds.) PPSN 1994. LNCS, vol. 866, pp. 36–45. Springer, Heidelberg (1994). https://doi.org/10.1007/3-540-58484-6_248

# Entropy and Organizational Performance

José Neves[1(✉)] ⓘ, Nuno Maia[1] ⓘ, Goreti Marreiros[2] ⓘ,
Mariana Neves[3] ⓘ, Ana Fernandes[4] ⓘ, Jorge Ribeiro[5] ⓘ,
Isabel Araújo[6] ⓘ, Nuno Araújo[6] ⓘ, Liliana Ávidos[6] ⓘ,
Filipa Ferraz[1] ⓘ, António Capita[7] ⓘ, Nicolás Lori[1,8] ⓘ,
Victor Alves[1] ⓘ, and Henrique Vicente[1,4] ⓘ

[1] Centro Algoritmi, Universidade do Minho, Braga, Portugal
{jneves, valves}@di.uminho.pt,
nuno.maia@dialogue-diversity.pt,
filipatferraz@gmail.com, nicolas.lori@gmail.com
[2] Departamento de Engenharia Informática,
Instituto Superior de Engenharia do Porto, Porto, Portugal
goreti@dei.isep.ipp.pt
[3] Deloitte, London, UK
maneves@deloitte.co.uk
[4] Departamento de Química, Universidade de Évora, Évora, Portugal
anavilafernandes@gmail.com, hvicente@uevora.pt
[5] Escola Superior de Tecnologia e Gestão,
Instituto Politécnico de Viana do Castelo, Viana do Castelo, Portugal
jribeiro@estg.ipvc.pt
[6] CESPU, Instituto Universitário de Ciências da Saúde, Gandra, Portugal
{isabel.araujo, nuno.araujo,
liliana.avidos}@ipsn.cespu.pt
[7] Instituto Superior Técnico Militar, Luanda, Angola
antoniojorgecapita@gmail.com
[8] ICVS, Escola de Medicina, Universidade do Minho, Braga, Portugal

**Abstract.** The main purpose of this article is to analyze the impact of the workers' behavior in terms of their emotions and feelings in system's performance, i.e., one is looking at issues concerned with Organizational Sustainability. Indeed, one's aim is to define a process that motivates and inspires managers and personnel to act upon the limit, i.e., to achieve the organizational goals through an effective and efficient implementation of operational and behavioral strategies. The focus will be on the importance of specific psychosocial variables that may affect collective pro-organizational attitudes. Data that is increasing exponentially, and somehow being out of control, i.e., the question is to know the correct value of the information that may be behind these numbers.

**Keywords:** Entropy · Organizational Sustainability · Logic Programming · Knowledge Representation and Reasoning

© Springer Nature Switzerland AG 2019
H. Pérez García et al. (Eds.): HAIS 2019, LNAI 11734, pp. 206–217, 2019.
https://doi.org/10.1007/978-3-030-29859-3_18

# 1 Introduction

*Efficiency* refers to the understanding *of the phenomenon of improving organizational sustainability*, considering the inputs used and the achievable results. When the focus is on the process of improving *work effectiveness*, one of the dimensions that affect the process depends on the interaction among people and their behavior in the organizational environment, as well as their impact on the process of achieving improved outcomes. People, while a piece of the engine to improve efficiency at the organizational level act in order to support the enterprise sustainability. In 1973, David McClelland in his article *"Testing for Competence Rather than Intelligence"* [1] challenged this paradigm by arguing that academic aptitude and literacy could not consistently anticipate a person's good performance. He argued that there was a *set of competencies* that played an important role on the *organizational setting* which were centered on a set of abilities that supported the *professional capacity for success*, such as *empathy, self-discipline*, and *initiative*. These were *good predictors for the leaders' performance* in the conduction of their teams. A *characteristic of these people* was associated with the *ability to read, interpret, and act on the evaluation of other people's emotions and feelings*. This is the attitude that lies behind this work, which develops through the stretches, i.e., following the introduction, the fundamentals adopted in the article are set, namely the use of *Logic Programming* for *Knowledge Representation and Reasoning*, whose sceneries are understood as a process of energy devaluation [2, 3]. Next, a case study on organizational efficiency will be presented and evaluated, considering the emotions and feelings of employees. It uses an eclectic mix of characteristics that indicates the extent of *collective self-esteem, gratitude*, and *prosocialness* predicates, i.e., how employees respond to specific questionnaires that, once translated into logical programs, and taken in isolation, form the company's knowledge base. Then conclusions are drawn and future work is outlined.

# 2 Fundamentals

*Knowledge Representation and Reasoning (KRR)* practices may be understood as a process of energy devaluation [2]. A data item is to be understood as being in a given moment at a particular entropic state as untainted energy which is in the interval [0,1] and, according to the *First Law of Thermodynamics* is a quantity well-preserved that cannot be consumed in the sense of destruction, but may be consumed in the sense of devaluation. It may be introduced by dividing a certain amount of energy in terms of:

- *exergy*, sometimes called available energy or more precisely available work, is the part of the energy which can be arbitrarily used after a transfer operation or, in other words, the *entropy* generated by it. In Fig. 1 it is given by the dark colored areas;
- *vagueness*, i.e., the corresponding *energy values* that *may or may not have been transferred and consumed*. In Fig. 1 are given by the gray colored areas; and

- *anergy*, that stands for an *energetic potential* that was not yet transferred and consumed, being therefore available, i.e., *all of energy that is not exergy*. In Fig. 1 it is given by the white colored areas.

These terms denote all possible energy's operations as pure energy transfer and consume practices. In order to make the process comprehensible, it will be presented in a graphical form. Taking as an example a group of 6 (six) questions that make the *Collective Self-Esteem Questionnaire-Six-Item (CSEQ – 6)* [4]:

*Q1 – I am a worthy member of the social groups I belong to;*
*Q2 – Overall, my social groups are considered good by others;*
*Q3 – The social groups I belong to are an important reflection of who I am;*
*Q4 – I am a cooperative participant in the social groups I belong to;*
*Q5 – In general, others respect the social groups that I am a member of; and*
*Q6 – In general, belonging to social groups is an important part of my self-image.*

This questionnaire is designed to assess the workers' general feelings about their corporation, on the assumption that high *collective self-esteem* will cause positive outcomes and benefits. The scale used was made upon the terms [5, 6]:

*strongly agree (4), agree (3), disagree (2), strongly disagree (1), disagree (2), agree (3), strongly agree (4)*

Moreover, it is included a neutral term, *neither agree nor disagree*, which stands for *uncertain* or *vague*. The reason for the individual's answers is in relation to the query:

*On the other hand, as an individual, how much would you agree with each one of CSEQ – 6* referred to above? (Table 1)

**Table 1.** *CSEQ – 6* single worker answer

| Questions | Scale | | | | | | | |
|---|---|---|---|---|---|---|---|---|
| | (4) | (3) | (2) | (1) | (2) | (3) | (4) | *vagueness* |
| Q1 | × | × | | | | | | |
| Q2 | | | | | × | × | | |
| Q3 | | | | | × | | × | |
| Q4 | | | | | | | | × |
| Q5 | | × | | | | | | |
| Q6 | | | | | | × | | |

**Leading to** → **Fig. 1** ← **Leading to**

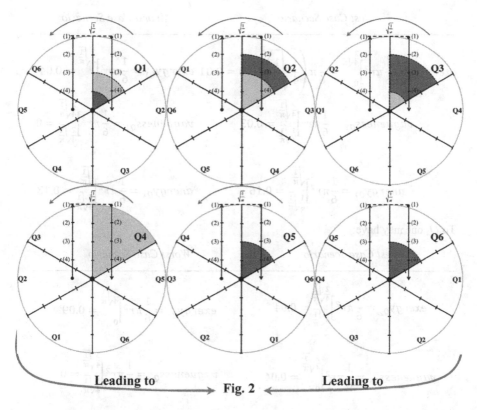

**Fig. 1.**    An assessment of the attained energy with respect to a single worker answer to *CSEQ-6*.

Where the input for *Q1* means that he/she *strongly agrees* (*4*) but does not rule out that he/she will *agree* (*3*) in certain situations. The inputs are be read from left to right, from *strongly agree* (*4*) to *strongly disagree* (*1*) (with increasing entropy), or from *strongly disagree* (*1*) to *strongly agree* (*4*) (with decreasing entropy), i.e., the markers on the axis correspond to any of the possible scale options, which may be used from *bottom → top* (from *strongly agree* (*4*) to *strongly disagree* (*1*)), indicating that the performance of the system decreases as entropy increases, or is used from *top → bottom* (from *strongly disagree* (*1*) to *strongly agree* (*4*)), indicating that the performance of the system increases as entropy decreases). The contribution of each individual to the system entropic state as untainted energy is evaluated as follows (Fig. 2):

| Best Case Scenario | Worst Case Scenario |
|---|---|

$$exergy_{Q1} = \frac{1}{6}\pi r^2 \Big]_0^{\frac{1}{4}\sqrt{\frac{1}{\pi}}} = \frac{1}{6}\pi \left(\frac{1}{4}\sqrt{\frac{1}{\pi}}\right)^2 - 0 = 0.01 \quad exergy_{Q1} = \frac{1}{6}\pi r^2 \Big]_0^{\frac{2}{4}\sqrt{\frac{1}{\pi}}} = 0.04$$

$$vagueness_{Q1} = \frac{1}{6}\pi r^2 \Big]_{\frac{1}{4}\sqrt{\frac{1}{\pi}}}^{\frac{2}{4}\sqrt{\frac{1}{\pi}}} = 0.03 \qquad vagueness_{Q1} = \frac{1}{6}\pi r^2 \Big]_{\frac{2}{4}\sqrt{\frac{1}{\pi}}}^{\frac{2}{4}\sqrt{\frac{1}{\pi}}} = 0$$

$$anergy_{Q1} = \frac{1}{6}\pi r^2 \Big]_{\frac{1}{4}\sqrt{\frac{1}{\pi}}}^{\sqrt{\frac{1}{\pi}}} = 0.16 \qquad anergy_{Q1} = \frac{1}{6}\pi r^2 \Big]_{\frac{2}{4}\sqrt{\frac{1}{\pi}}}^{\sqrt{\frac{1}{\pi}}} = 0.13$$

To $Q_2$ one may have:

| Best Case Scenario | Worst Case Scenario |
|---|---|

$$exergy_{Q2} = \frac{1}{6}\pi r^2 \Big]_{0\sqrt{\frac{1}{\pi}}}^{\frac{2}{4}\sqrt{\frac{1}{\pi}}} = 0.04 \qquad exergy_{Q2} = \frac{1}{6}\pi r^2 \Big]_0^{\frac{3}{4}\sqrt{\frac{1}{\pi}}} = 0.09$$

$$vagueness_{Q2} = \frac{1}{6}\pi r^2 \Big]_0^{\frac{2}{4}\sqrt{\frac{1}{\pi}}} = 0.04 \qquad vagueness_{Q2} = \frac{1}{6}\pi r^2 \Big]_{\frac{3}{4}\sqrt{\frac{1}{\pi}}}^{\frac{3}{4}\sqrt{\frac{1}{\pi}}} = 0$$

$$anergy_{Q2} = \frac{1}{6}\pi r^2 \Big]_{\frac{2}{4}\sqrt{\frac{1}{\pi}}}^{\sqrt{\frac{1}{\pi}}} = 0.13 \qquad anergy_{Q2} = \frac{1}{6}\pi r^2 \Big]_{\frac{3}{4}\sqrt{\frac{1}{\pi}}}^{\sqrt{\frac{1}{\pi}}} = 0.08$$

$Q_3$ is similar to $Q_2$. Therefore, one may have:

| Best Case Scenario | Worst Case Scenario |
|---|---|

$$exergy_{Q3} = \frac{1}{6}\pi r^2 \Big]_0^{\frac{1}{4}\sqrt{\frac{1}{\pi}}} = 0.01 \qquad exergy_{Q3} = \frac{1}{6}\pi r^2 \Big]_0^{\frac{3}{4}\sqrt{\frac{1}{\pi}}} = 0.09$$

$$vagueness_{Q3} = \frac{1}{6}\pi r^2 \Big]_0^{\frac{1}{4}\sqrt{\frac{1}{\pi}}} = 0.01 \qquad vagueness_{Q3} = \frac{1}{6}\pi r^2 \Big]_{\frac{1}{4}\sqrt{\frac{1}{\pi}}}^{\frac{1}{4}\sqrt{\frac{1}{\pi}}} = 0$$

$$anergy_{Q3} = \frac{1}{6}\pi r^2 \Big]_{\frac{1}{4}\sqrt{\frac{1}{\pi}}}^{\sqrt{\frac{1}{\pi}}} = 0.16 \qquad anergy_{Q3} = \frac{1}{6}\pi r^2 \Big]_{\frac{3}{4}\sqrt{\frac{1}{\pi}}}^{\sqrt{\frac{1}{\pi}}} = 0.08$$

To $Q_4$ one may have:

**Fig. 2.** *System entropic states' best* and *worst-case scenarios.*

| Best Case Scenario | Worst Case Scenario |
|---|---|

$$exergy_{Q_4} = \frac{1}{6}\pi r^2 \Big]_0^0 = 0 \qquad exergy_{Q_4} = \frac{1}{6}\pi r^2 \Big]_0^{\sqrt{\frac{1}{\pi}}} = 0.17$$

$$vagueness_{Q_4} = \frac{1}{6}\pi r^2 \Big]_0^{\sqrt{\frac{1}{\pi}}} = 0.17 \qquad vagueness_{Q_4} = \frac{1}{6}\pi r^2 \Big]_{\sqrt{\frac{1}{\pi}}}^{\sqrt{\frac{1}{\pi}}} = 0$$

$$anergy_{Q_4} = \frac{1}{6}\pi r^2 \Big]_0^{\sqrt{\frac{1}{\pi}}} = 0.17 \qquad anergy_{Q_4} = \frac{1}{6}\pi r^2 \Big]_{\sqrt{\frac{1}{\pi}}}^{\sqrt{\frac{1}{\pi}}} = 0$$

To $Q_5$ one may have:

| Best Case Scenario | Worst Case Scenario |
|---|---|

$$exergy_{Q_5} = \frac{1}{6}\pi r^2 \Big]_0^{\frac{2}{4}\sqrt{\frac{1}{\pi}}} = 0.04 \qquad exergy_{Q_5} = \frac{1}{6}\pi r^2 \Big]_0^{\frac{2}{4}\sqrt{\frac{1}{\pi}}} = 0.04$$

$$vagueness_{Q_5} = \frac{1}{6}\pi r^2 \Big]_{\frac{2}{4}\sqrt{\frac{1}{\pi}}}^{\frac{2}{4}\sqrt{\frac{1}{\pi}}} = 0 \qquad vagueness_{Q_5} = \frac{1}{6}\pi r^2 \Big]_{\frac{2}{4}\sqrt{\frac{1}{\pi}}}^{\frac{2}{4}\sqrt{\frac{1}{\pi}}} = 0$$

$$anergy_{Q_5} = \frac{1}{6}\pi r^2 \Big]_{\frac{2}{4}\sqrt{\frac{1}{\pi}}}^{\sqrt{\frac{1}{\pi}}} = 0.13 \qquad anergy_{Q_5} = \frac{1}{6}\pi r^2 \Big]_{\frac{2}{4}\sqrt{\frac{1}{\pi}}}^{\sqrt{\frac{1}{\pi}}} = 0.13$$

$Q_6$ is similar to $Q_5$. Therefore, one may have:

| Best Case Scenario | Worst Case Scenario |
|---|---|

$$exergy_{Q_6} = \frac{1}{6}\pi r^2 \Big]_0^{\frac{2}{4}\sqrt{\frac{1}{\pi}}} = 0.04 \qquad exergy_{Q_6} = \frac{1}{6}\pi r^2 \Big]_0^{\frac{2}{4}\sqrt{\frac{1}{\pi}}} = 0.04$$

$$vagueness_{Q_6} = \frac{1}{6}\pi r^2 \Big]_{\frac{2}{4}\sqrt{\frac{1}{\pi}}}^{\frac{2}{4}\sqrt{\frac{1}{\pi}}} = 0 \qquad vagueness_{Q_6} = \frac{1}{6}\pi r^2 \Big]_{\frac{2}{4}\sqrt{\frac{1}{\pi}}}^{\frac{2}{4}\sqrt{\frac{1}{\pi}}} = 0$$

$$anergy_{Q_6} = \frac{1}{6}\pi r^2 \Big]_{\frac{2}{4}\sqrt{\frac{1}{\pi}}}^{\sqrt{\frac{1}{\pi}}} = 0.13 \qquad anergy_{Q_6} = \frac{1}{6}\pi r^2 \Big]_{\frac{2}{4}\sqrt{\frac{1}{\pi}}}^{\sqrt{\frac{1}{\pi}}} = 0.13$$

**Fig. 2.** (*continued*)

However, once the performance of a system depends on its entropic state, the data collected above can be structured in terms of the extent of predicate *collective self-esteem (cse)*:

$$cse : EXergy, VAgueness, System's\,Performance,$$
$$Quality-of-Information \rightarrow \{True, False\}$$

whose graphical and formal description are given below (Fig. 3) (Table 2):

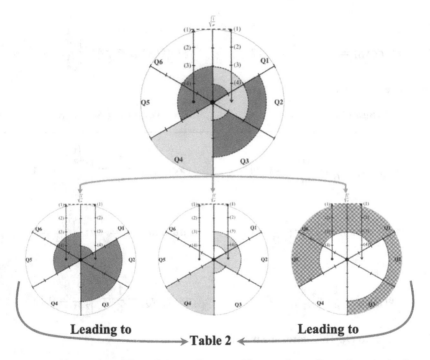

**Fig. 3.** A graphic representation of the *collective self-esteem's* predicate that results from the responses of a single employee to *CSEQ-6*.

**Table 2.** The *collective self-esteem* predicate's extent obtained according to the answer of a single worker to the *CSEQ-6*.

| Exergy$_{BCS}$ | Vague$_{BCS}$ | SP$_{BCS}$ | QoI$_{BCS}$ | Exergy$_{WCS}$ | Vague$_{WCS}$ | SP$_{WCS}$ | QoI$_{WCS}$ |
|---|---|---|---|---|---|---|---|
| 0.14 | 0.25 | 0.92 | 0.36 | 0.47 | 0 | 0.88 | 1 |

Which may now be depicted as the logical program (it was considered the worst-case scenario) [3, 4]:

$$\neg \, cse \, (EX, \, VA, \, SP, \, QoI) \leftarrow not \; cse(EX, \, VA, \, SP, \, QoI),$$

$$not \; exception_{cse}(EX, \, VA, \, SP, \, QoI)$$

$$cse \, (0.47, 0, 0.88, 1).$$

**Program 1.** The extent of the *collective self-esteem* (*cse*) predicate for the *worst-case scenario*.

where the evaluation of *SP* and *QoI* for the different items that make the *CSEQ – 6* are given in the form:

- *SP* is figured out using $SP = \sqrt{1 - ES^2}$, where *ES* stands for an *exergy's* value in the worst-case scenario (i.e., $ES = exergy + vagueness$), a value that ranges in the interval [0, 1] (Fig. 4).

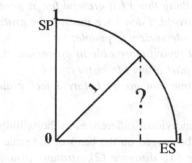

**Fig. 4.** *SP* evaluation.

$$SP = \sqrt{1 - (0.47 + 0)^2} = 0.88$$

- *QoI* is evaluated in the form:

$$QoI = exergy/(exergy + vagueness)$$

$$QoI = 0.47/(0.47 + 0) = 1$$

# 3   Organizational Efficiency

*Organizational Efficiency (OE)* is a complex theme. *In order to evaluate it*, it is necessary to introduce *a system analysis* where technology, organizing methods and people interact in order to perform the maximum efficiency. In this process, the inter-relational interactions between people have an important role in system performance

[7]. For this purpose, we look at principles that govern the *Emotional* development that underlies this process and particularly *empathy*, as a way to develop people involvement and participation in organizational challenges, all this supported by the dynamics of social relations. This social relations support involvement and participation, and they denote a collective dimension. The former one is associated with the ability to include compromise and involve members of a particular group; the last one is linked with the ability to inform, to transmit, to communicate, and to share. The principles that govern this dimension that support the relationship between peers are associated with the phenomenon of Identity [7, 8]. It is the Identity that creates the common vision among people as members of a working group. This psychological process associated with Social Identity [4, 8–10] is in itself responsible for generating group behavior, in which solidarity among members and conformity to norms form the basis of behaviors and attitudes [8], and will be addressed here in terms of the extensions of predicates collective self-esteem (cse), gratitude (gra) and prosocialness (pro). The *Collective Self-Esteem Questionnaire-Six-Item (CSEQ – 6)* [4], was already attended above. To the *Gratitude Questionnaire-Six-Item* (GQ – 6) one may have:

*Q1 – I have so much in life to be thankful for;*
*Q2 – If I had to list everything that I felt grateful for, it would be a very long list;*
*Q3 – When I look at the world, I don't see much to be grateful for;*
*Q4 – I am grateful to a wide variety of people;*
*Q5 – As I get older I find myself more able to appreciate the people, events, and situations that have been part of my life history; and*
*Q6 – Long amounts of time can go by before I feel grateful to something or someone.*

With this questionnaire, individual differences in the willingness to feel comfortable in everyday life should be judged on the basis of the scale (Tables 3 and 4):
*strongly agree (4), agree (3), disagree (2), strongly disagree (1), disagree (2), agree (3), strongly agree (4)*

**Table 3.** *GQ – 6* single worker answer

| Questions | Scale | | | | | | | |
|---|---|---|---|---|---|---|---|---|
| | (4) | (3) | (2) | (1) | (2) | (3) | (4) | *vagueness* |
| Q1 | | × | | | | | | |
| Q2 | | × | | | × | | | |
| Q3 | | | | | | × | | |
| Q4 | | | | × | × | | | |
| Q5 | × | | | | | | | |
| Q6 | | | | | | | | × |

**Leading to** →Table 4← **Leading to**

**Table 4.** The *gratitude* predicate's extent obtained according to a single worker answer to *GQ – 6*.

| Exergy$_{BCS}$ | Vague$_{BCS}$ | SP$_{BCS}$ | QoI$_{BCS}$ | Exergy$_{WCS}$ | Vague$_{WCS}$ | SP$_{WCS}$ | QoI$_{WCS}$ |
|---|---|---|---|---|---|---|---|
| 0.21 | 0.39 | 0.80 | 0.35 | 0.69 | 0 | 0.72 | 1 |

For the *Prosocialness Questionnaire-Six-Item (PQ – 6)* one has:

*Q1 – I am pleased to help my colleagues in their activities;*
*Q2 – I try to help others;*
*Q3 – I am emphatic with those who are in need;*
*Q4 – I do what I can to help others avoid getting into trouble;*
*Q5 – I am willing to make my knowledge and abilities available to others; and*
*Q6 – I easily share with friends any good opportunity that comes to me.*

This questionnaire aims to assess the extent to which individuals engage in sharing, helping, caring for the needs of others and empathizing with their feelings. The scale used is given below (Tables 5 and 6):

*strongly agree (4), agree (3), disagree (2), strongly disagree (1), disagree (2), agree (3), strongly agree (4)*

**Table 5.** *PQ – 6* single worker answer

| Questions | Scale | | | | | | | |
|---|---|---|---|---|---|---|---|---|
| | (4) | (3) | (2) | (1) | (2) | (3) | (4) | vagueness |
| Q1 | | | | | | | | × |
| Q2 | | | × | | | | | |
| Q3 | | | | | | × | × | |
| Q4 | | × | | × | | | | |
| Q5 | | | | | | | | × |
| Q6 | | | | × | | × | | |

Leading to → Table 6 ← Leading to

**Table 6.** The *prosocialness* predicate's extent obtained according to the single worker answer to the *PQ – 6*.

| Exergy$_{BCS}$ | Vague$_{BCS}$ | SP$_{BCS}$ | QoI$_{BCS}$ | Exergy$_{WCS}$ | Vague$_{WCS}$ | SP$_{WCS}$ | QoI$_{WCS}$ |
|---|---|---|---|---|---|---|---|
| 0.22 | 0.57 | 0.61 | 0.28 | 0.86 | 0 | 0.51 | 1 |

# 4  Computational Make-up

The following describes nothing less than a mathematical logic program that, through insights that are subject to formal proof (i.e., through the use of theorem proofs), allows one to understand and even adapt the actions and attitudes of individuals or groups, and towards them, the organization as a whole.

*Assessing Organizational Performance for the Worst-case Scenario*

$\neg cse\ (EX,\ VA,\ SP,\ QoI) \leftarrow not\ cse\ (EX,\ VA,\ SP,\ QoI),$

$not\ exception_{cse}\ (EX,\ VA,\ SP,\ QoI).$

$cse\ (0.47, 0, 0.88, 1).$

$\neg gra\ (EX,\ VA,\ SP,\ QoI) \leftarrow not\ gra\ (EX,\ VA,\ SP,\ QoI),$

$not\ exception_{gra}\ (EX,\ VA,\ SP,\ QoI).$

$gra\ (0.69,\ 0,\ 0.72,\ 1).$

$\neg pro\ (EX,\ VA,\ SP,\ QoI) \leftarrow not\ pro\ (EX,\ VA,\ SP,\ QoI),$

$not\ exception_{pro}\ (EX,\ VA,\ SP,\ QoI).$

$pro\ (0.86,\ 0,\ 0.51,\ 1).$

**Program 2.** The make-up of the company's knowledge base for a single worker

An evolutionary knowledge base that is at the heart of symbolic Artificial Intelligence. A knowledge-based system that decides how to act by running formal reasoning procedures over a body of explicitly represented knowledge. A system that is not programmed for specific tasks; rather, it is told what it needs to know and is expected to infer the rest.

# 5  Conclusions and Future Work

It showed how perceptions of *Psychology, Management, Computer Science* and *Mathematical Logic* may be used to promote sustainable decisions and behaviors, i.e., a new interdisciplinary field of research and practice aimed at investigating and solving complex computer problems that are unique in scope, impact and complexity, that exist in very dynamic and uncertain environments. An example of such an event is presented in this paper, which focuses on issues such as how to deal with a diverse workforce. Effects of individual differences in attitude, job satisfaction and engagement, including their impact on performance and management, were examined in terms of the dimensions of *collective self-esteem, gratitude* and *pro-socialness*. It will allow one to monitor the sustainability of the organization in terms of the attitudes of its various actors and to predict its evolution by varying its entropic states. Future work will focus on topics such as personality, including the impact of different cultures. Perception and its impact on decision-making; employee's values; emotions, including emotional intelligence, emotional work and the effects of positive and negative impact on decision-making and creativity; and motivation, including the impact of rewards and goals and their impact on management.

**Acknowledgments.** This work has been supported by FCT – Fundação para a Ciência e Tecnologia within the Project Scope: UID/CEC/00319/2019.

# References

1. McClelland, D.: Testing for competence rather than intelligence. Am. Psychol. **28**, 1–14 (1973)
2. Wenterodt, T., Herwig, H.: The entropic potential concept: a new way to look at energy transfer operations. Entropy **16**, 2071–2084 (2014)
3. Neves, J.: A logic interpreter to handle time and negation in logic databases. In: Muller, R., Pottmyer, J. (eds.) Proceedings of the 1984 Annual Conference of the ACM on the 5th Generation Challenge, pp. 50–54. ACM, New York (1984)
4. Luhtanen, R., Crocker, J.: A collective self-esteem scale: self-evaluation of one's social identity. Pers. Soc. Psychol. Bull. **18**, 302–318 (1992)
5. Rosenberg, M.: Society and the Adolescent Self-Image. Princeton University Press, Princeton (1965)
6. Baumeister, R.F., Campbell, J.D., Krueger, J.I., Vohs, K.D.: Does high self-esteem cause better performance, interpersonal success, happiness, or healthier lifestyles? Psychol. Sci. Public Interest **4**, 1–44 (2003)
7. Haslam, S.: Psychology in Organizations – The Social Identity Approach. SAGE Publications, London (2004)
8. Tajfel, H.: Human Groups and Social Categories – Studies in Social Psychology. Cambridge University Press, Cambridge (1981)
9. Tajfel, H., Turner, J.C.: The social identity theory of inter-group behavior. In: Worchel, S., Austin, W.G. (eds.) Psychology of Inter-Group Relations, pp. 7–24. Nelson-Hall Publishers, Chicago (1986)
10. Hogg, M., Abrams, D.: Social Identifications – A Social Psychology of Intergroup Relations and Group Processes. Routledge, New York (1998)

# A Semi-supervised Method to Classify Educational Videos

Alexandru Stefan Stoica[1], Stella Heras[2(✉)], Javier Palanca[2], Vicente Julian[2], and Marian Cristian Mihaescu[1]

[1] Faculty of Automation, Computers and Electronics, University of Craiova, Craiova, Romania
stoicaastefan@gmail.com, mihaescu@software.ucv.ro
[2] D. Sistemas Informáticos y Computación, Universitat Politècnica de València, Valencia, Spain
{sheras,jpalanca,vinglada}@dsic.upv.es

**Abstract.** Currently, topic modelling has regained interest in the world of e-learning, where it is necessary to search through an extensive database of online learning objects, mainly in the form of educational videos. The main problem is the retrieval of those learning objects that are best suited to students' keyword searches. Today this problem is still an open topic. According to this, this paper aims to provide a more sophisticated method to improve the search of educational videos and thus show those that best fit the learning objectives of students. To do this, a semi-supervised method to cluster and classify a large data-set of educational videos from the *Universitat Politècnica de València* has been developed. The proposed method employs open content resources from Wikipedia as labelled data to train the model.

**Keywords:** Topic modelling · E-learning · Clustering · Classification

## 1 Introduction

Over the last 20 years, topic modelling has been a critical and challenging issue for information retrieval, natural language processing and other related research communities. Here, different approaches to discover topics and hidden semantic structures in text have been proposed, such *n-gram* statistical and probabilistic models [7,8], *bag-of-words* based models and other hybrid approaches [10].

At the same time, we have experienced a boom of online learning, fostered by the increasing availability of online learning objects (LO) [3] and Massive Online Learning Courses (MOOCs). The success of *Flipped Classroom* [9] (a teaching methodology where the idea is to flip the common instructional approach: the teacher creates online LOs -videos and interactive lessons-, which students visualize at home to devote classes to work through problems and engage in collaborative learning). This boom of online learning has also brought an information overload problem since students have many more learning resources

© Springer Nature Switzerland AG 2019
H. Pérez García et al. (Eds.): HAIS 2019, LNAI 11734, pp. 218–228, 2019.
https://doi.org/10.1007/978-3-030-29859-3_19

available than they can locate and consume efficiently. Therefore, topic modelling is again a trending topic for the e-learning research community, in which searching through a big database of online learning resources and retrieving those that best-fit keyword searches of students is still an open issue.

The *Universitat Politècnica de València* (UPV, Spain) has also echoed this trend, launching its MOOC platform https://www.upvx.es/, powered by the edX[1] MOOC platform of MIT and Hardvard University, and its video lectures sharing website https://media.upv.es. Figures 1 and 2 show a snapshot of their websites, respectively.

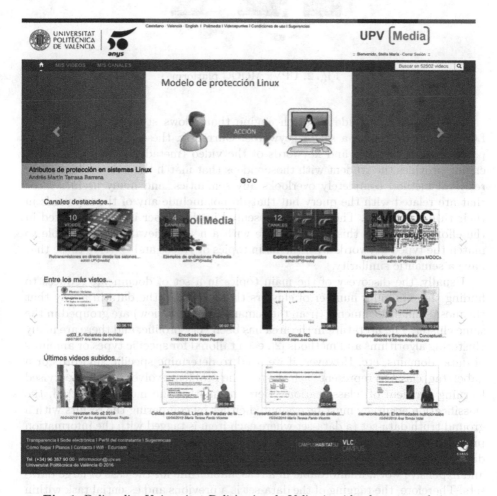

**Fig. 1.** Polimedia: Universitat Politècnica de València video lectures website

---

[1] edX MOOC platform: https://www.edx.org/.

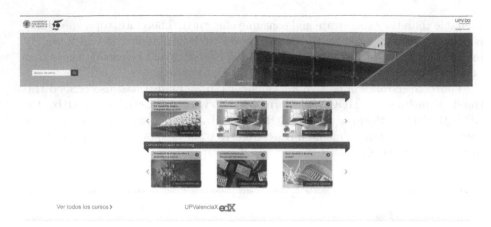

**Fig. 2.** UPVx MOOC platform

Both websites include a search engine that allows students to search for LO (videos) by typing a set of keywords. Currently, these keywords are compared against the title and keywords of the video (metadata), and the search engine provides the student with those videos that match the query. This simple retrieval method completely overlooks any semantics, and many useful videos that are related with the query but that do not include any of the keywords in their title are missed. The research presented in this paper is contextualized in the effort to improve this search engine with a new retrieval algorithm able to match the typed keywords with the main topics of videos and retrieve those that have a semantic similarity.

Usually, the discovery of the main topics in a set of documents reduces to finding the minimum number of clusters that separates the data in a way that the most similar documents (from the semantic point of view) are grouped in the same cluster. The literature in the area has been very prolific, and there are many clustering algorithms and methods [2], each tailored for specific types or amounts of data, domains, etc. However, if we need to determine specific keywords or a 'label/tag/class' to represent each cluster, the problem evolves to the supervised learning problem of 'classification'. Here, no matter the method selected, the classification accuracy must be evaluated either by comparing the results with a ground truth data-set (a data-set where each item is tagged with the information provided by direct observation, i.e. empirical evidence). However, ground truth data-sets must be manually created and/or curated by experts, which is hard and expensive work, unaffordable in a reasonable time for the case of large data-sets. Therefore, the tagging of the data-set is a previous and essential task within the area of topic modelling, where several approaches have been investigated.

Among them, semi-supervised learning approaches have been successfully applied to improve the learning accuracy of clustering methods (where no data is labelled) by the usage of unlabeled data together with a small amount of labelled data. For instance, in work presented in [12], authors proposed a method of categorising Flickr tags as WordNet semantic categories by first categorising Wikipedia articles and then mapping Flickr tags onto these categorised articles. In [5], the authors face the challenge of identifying the outcome and prerequisite concepts within a piece of educational content and present a semi-supervised approach that leverages textbooks as a source of indirect supervision. With this approach, they can learn a model that can generalise to arbitrary web documents, but without requiring expert annotation. Also, in [4], a framework to process MOOCs data to be structured for further knowledge management and mining is proposed. The framework uses human knowledge through labelling data and proposes a method for concept extraction based on machine learning (Conditional Random Fields). Results demonstrate that only 10% of labelled data can lead to acceptable performance.

As the first step to improve the search engine of the media websites of the *Universitat Politècnica de València*, this work presents a semi-supervised method to cluster and classify the LO data-set of the university, by using open content resources from Wikipedia as labelled data to train the model [6]. The paper is structured as follows: Sect. 2 presents the actual UPV-Media data-set; Sect. 3 explains the semi-supervised method proposed in this work; Sect. 4 provides the evaluation of the approach; and finally, Sect. 5 summarizes the paper and points out future work.

## 2 UPV-Media Data-set

In our work, we make use of the LO database of the Universitat Politècnica de València (Spain), which contains around 50.000 educational videos (UPV-Media data-set). The videos cover different subjects which are taught in the university and are usually presented by a lecturer in Spanish. Each video has several slides (screen-shots from the video) which are chosen by the presenter as a summary for the video, and sometimes, a transcription. This information can be found in the metadata, as is presented in the example below:

```
1   {'_id': '00054a38-5a32-4db2-ae9c-85c296015c3b',
2    'hidden': False,
3    'title': 'Programa Mathematica gratis y online',
4    'source': {'type': 'polimedia',
5               'videos': [{'mimetype': 'video/mp4',
6                          'width': 640, 'height': 480
7                          'src': 'politube.mp4',
8                          }],
9               ...},
10   'slides': [
11       {'mimetype':'image/jpg','url':'frame.0.jpg','time':'0'},
12       {'mimetype':'image/jpg','url':'frame.48.jpg','time':'48'
             },
13       ...
14       {'mimetype':'image/jpg','url':'frame.432.jpg','time':'43
             2'}
15   ],
```

```
16    "metadata": {
17              "keywords": [
18                "croma"
19          ]
20        },
21    'duration': 564.523537,
22    'transcription': '...very long text... (or empty)'
23  }
```

As we can see, there is no parameter which specifies from what field the videos are (ex. Biology, Engineering, Humanities, etc.). The only parameter which could be useful in finding the class to which the video belongs to is the transcription (by, for instance, analyzing keywords), but this is not available for all the videos (just 15.387 videos have a transcription and keywords attached).

Even when transcriptions are available, there are problems regarding their quality. Some transcriptions seem to have no punctuation since most of them were auto-generated by a speech recognition engine. This practical scenario makes the task of separating a transcription in sentences a challenging one.

For this work, we processed the original data-set and created a new data-set with each video's transcription and the keywords. Figure 3 shows the statistics for the number of observed words in each transcription.

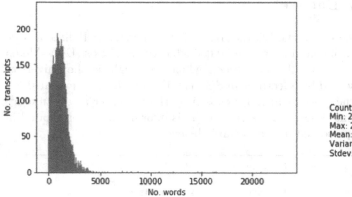

**Fig. 3.** Statistics of the number of words in each transcription

The statistics helped in understanding the actual structure of the available data-set. Several consequences that follow up with this analysis regard defining outliers in our practical context. For example, transcripts with less than 50 words or with more than 5000 words may be considered outliers and may not be taken into consideration within the data analysis pipeline. In fact, a transcript with less than 50 words may contain too few information such that it may be reliably used for training or testing.

# 3   Proposed Approach

Our objective is to separate the educational data-set of videos mentioned above into a set of distinct topics by using an external open content educational data-set, gathered from Wikipedia articles, as ground truth. We want to show that, what it seems as an unsupervised problem at first glance, can be approached in a semi-supervised way without any manual labelling of the original data-set.

The problem which must be addressed is that of finding the number of topics in our transcripts. Luckily, in our case, we can approximate the number of topics by using an observation which implies that the videos are educational and besides that, they come from a specific university, so we could say that the number of topics is smaller than the number of bachelor degrees in the university. Based on the main field of these degrees, we can deduce that there are three topics *Biology* (e.g. bacteria, diseases, bioengineering, biomedicine, etc.), *Engineering* (e.g. computer, electrical, architecture, civil, aerospace, etc.), *Humanities* (e.g. laws, arts, social, economy, etc.).

Our solution proposes a method in which we use a semi-supervised approach to pseudo label the transcripts, by using a Support Vector Machine (SVM) [11] model complemented by a mechanism which tries to correct some of the errors using the keywords in the data-set. To train and create the model, we did not directly use our data-set, since the number of examples for each class is not balanced (for instance, the number of engineering related videos and their attached transcripts is enormous compared with those related with humanities). We chose to use instead Wikipedia articles, consisting of pages from different subcategories of biology, engineering and humanities which were extracted using the Wikipedia API and the Wikipedia python library https://pypi.org/project/wikipedia/. These articles were used as a training set because they are more general and more natural to acquire.

The pipeline for the approach is described in Fig. 4:

- **Preprocessing:** For each Wikipedia article, and each video from the UPV-Media data-set (represented by its keywords and transcript), we lowercased the characters and removed anything which did not represent a word. We chose not to do any stemming or removing the stop words because the used embedding method requires that the paragraphs keep their semantic value.
- **Embeddings:** Since we are processing text, we need a method which maps them to a vector of numerical values and takes into account their semantic value. Thus, we used a pre-built text embedding method based on feed-forward Neural-Net Language Models [1] with pre-built Out Of Vocabulary terms (OOV) from *tensorflow* hubs feed-forward Neural-Net Language Models[2] that maps the text to a 128-dimensional embedding vector[3].

---

[2] https://tfhub.dev/google/nnlm-es-dim128-with-normalization/1.

[3] We decided to use this embeddings method since its suitability is well demonstrated by state of the art in this field. Furthermore, it was already trained on a considerable amount of data.

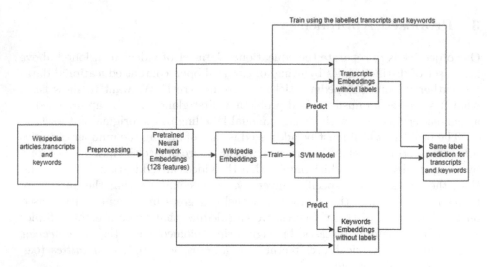

**Fig. 4.** Pipeline for determining the labels

– **Training and prediction:** We trained the Wikipedia embeddings using an SVM model with a Radial Basis Function (RBF) kernel[4] from the *sklearn* library (https://scikit-learn.org/stable/modules/svm.html) and used cross-validation to determine the accuracy of the model. We used this model classify each video of the UPV-Media data-set as belonging to one of the three classes (Biology, Engineering or Humanities). As the UPV-Media data-set is not labelled, we do not know the actual class of each video, and thus, we followed a two-fold validation approach. Thus, we used the model to predict the classes by using as input the transcripts and the keywords of each video separately. The resulting predictions were separated into two sets, A and B, where A represents those videos for which the model predicted the same class for both the transcripts and the keywords, and B the ones for which the model predicted different classes. Then, we added the A set to the Wikipedia articles data-set but removing the keywords. Afterwards, we run a new iteration of the process trying to use this new model (formed by the original Wikipedia data-set plus the transcripts added from set A) to predict the classes of the remaining videos (those of set B). The process has been repeated until we could not add any more transcripts to the training set.

## 4    Experimental Results and Evaluation

We have pre-trained the language model on Wikipedia articles which we obtained by using Wikipedia's API and Wikipedia python library. We have to build a

---

[4] This design decision was supported by the fact that it behaved better than other algorithms and it required a smaller data-set than the commonly used neural networks.

custom tool that extracts all the pages from a specific category on Wikipedia. For example, the page https://es.wikipedia.org/wiki/Categora:Biologia provides the following nine categories that we labeled as **Biology**: *Biologia, Anatomia, Bioinformatica, Biologia celular, Bioquimica, Biotecnologia, Botanica, Microbiologia* and *Genetica*. For **Engineering** Wikipedia provides in the same way eight categories: *Ingenieria, Materiales en ingenieria, Bases de datos, Computacion distribuida, Computacin grafica, Geomatica, Ingenieria de software,Seguridad informatica*. For **Humanities** the obtained labels are: *Arte, Tecnicas de arte, Antropologia, Simbolos, Ciencias Historicas, Ciencias sociales, Economia, Sociologia, Comunicacion, Terminos juridicos, Justicia, Derecho, Principios del derecho*.

For a total number of thirty categories the tool obtained a data-set of 3747 pages from which 1219 are for the categories belonging to *Biology*, 1060 pages are from categories belonging to *Engineering* and the rest of 1468 pages are for the categories belonging to *Humanities*.

To validate our model, we split the original set of Wikipedia articles in three disjoint data-sets that were used for training, testing and validation (70%/15%/15%). The training set was used to build the initial model, as described in the pipeline of Fig. 4. Later, in each iteration, we used the currently trained model on the test data-set by computing the precision/recall/f1 scores.

The motivation for following this approach regards the need to observe how much does the cross-validation accuracy changes after each iteration, or more precisely, to see how sensible is the model to the transcripts that are being added. This approach is not used in any way for validating the model, but to check the stability. Monitoring the accuracy levels regards the internal logging and debugging of the training process as situations of high accuracy drop (i.e., from 90% to 70%) need to identified and investigated in detail.

The items that were correctly labelled (i.e., the ones from set A) were appended to the training data-set after which a new model is obtained. This process was stopped after 10 iterations since, after this threshold, we were not able to add transcripts anymore. Thus, the stopping criteria regards the lack of improvement of the accuracy of the model and the small number of transcripts that are being added at each iteration. Table 1, shows the results of each iteration of the training process, with the number of transcripts added to the model in each iteration (labeled in column head as *tr. same classes*) and the classification accuracy obtained for each label. The classification accuracy metrics are computed on the validation data-set, which contains only unseen data during training.

After that, we evaluated the resulting model by using it to predict the classes of the transcripts from validation data-set and computed the prediction scores. Table 2 presents the evaluation results obtained. The model obtained very good results in labeling unseen data into Biology, Engineering or Humanities classes. The greatest achievement is that the model has been created without having a ground truth data-set with labeled data or manually labeling a percentage of the

**Table 1.** Validation scores for each iteration in the pipeline

| Iteration (tr. same classes/tr. no classes) | Accuracy (cross-validation) | Class | Precision | Recall | F1-score |
|---|---|---|---|---|---|
| 1(9132/15387) | 0.90 | Biology | 0.90 | 0.92 | 0.91 |
| | | Engineering | 0.90 | 0.85 | 0.87 |
| | | Humanities | 0.93 | 0.95 | 0.94 |
| 2(1740/6255) | 0.96 | Biology | 0.90 | 0.92 | 0.91 |
| | | Engineering | 0.90 | 0.87 | 0.88 |
| | | Humanities | 0.93 | 0.94 | 0.94 |
| 3(568/4515) | 0.96 | Biology | 0.90 | 0.91 | 0.91 |
| | | Engineering | 0.90 | 0.86 | 0.88 |
| | | Humanities | 0.92 | 0.93 | 0.93 |
| 4(226/3947) | 0.95 | Biology | 0.90 | 0.92 | 0.91 |
| | | Engineering | 0.90 | 0.87 | 0.88 |
| | | Humanities | 0.93 | 0.93 | 0.93 |
| 5(137/3721) | 0.95 | Biology | 0.90 | 0.92 | 0.91 |
| | | Engineering | 0.89 | 0.86 | 0.88 |
| | | Humanities | 0.93 | 0.93 | 0.93 |
| 6(83/3584) | 0.95 | Biology | 0.90 | 0.92 | 0.91 |
| | | Engineering | 0.89 | 0.87 | 0.88 |
| | | Humanities | 0.94 | 0.93 | 0.93 |
| 7(47/3501) | 0.94 | Biology | 0.90 | 0.92 | 0.91 |
| | | Engineering | 0.89 | 0.86 | 0.88 |
| | | Humanities | 0.94 | 0.93 | 0.93 |
| 8(41/3454) | 0.94 | Biology | 0.92 | 0.92 | 0.91 |
| | | Engineering | 0.89 | 0.87 | 0.88 |
| | | Humanities | 0.94 | 0.93 | 0.93 |
| 9(46/3413) | 0.94 | Biology | 0.90 | 0.92 | 0.91 |
| | | Engineering | 0.89 | 0.86 | 0.88 |
| | | Humanities | 0.94 | 0.93 | 0.93 |
| 10(18/3367) | 0.94 | Biology | 0.90 | 0.92 | 0.91 |
| | | Engineering | 0.89 | 0.86 | 0.88 |
| | | Humanities | 0.94 | 0.93 | 0.93 |

available data-set. This approach paves the way for future developments of more intelligent algorithms that allow building a search or recommendation engine to retrieve videos that fit students' queries or profiles in terms of their learning style.

**Table 2.** Validation scores for the labels of the transcripts

| Accuracy (cross-validation) | Class | Precision | Recall | F1-score |
|---|---|---|---|---|
| 0.96 | Biology | 0.92 | 0.89 | 0.90 |
| | Engineering | 0.83 | 0.87 | 0.85 |
| | Humanities | 0.86 | 0.87 | 0.87 |

# 5  Conclusions

This paper has presented a semi-supervised method to cluster and classify a large data-set of educational videos. The data-set has been obtained from the offer of online educational videos available at the *Universitat Politècnica de València*. To train and create the model, we have employed open content resources from Wikipedia as labelled data. Experiments have shown the accuracy of the model to classify video lectures. Therefore, the model allows obtaining a labelled data-set without the need of performing a manual data curation.

As a next step, we have prepared the integration of the proposed model to improve the search process of the global service of educational videos offered by the *Universitat Politècnica de València*. The use of the proposed approach in combination with a more sophisticated procedure for indexing and retrieving video transcripts will improve the final ranked list of videos obtained as the output of the searching process.

Further improvements may take into consideration other input data-sets for pre-training the language model such as scientific articles, open book corpus or a domain ontology.

**Acknowledgements.** This work was partially supported by the project TIN2017-89156-R of the Spanish government, and by the grant program for the recruitment of doctors for the Spanish system of science and technology (PAID- 10-14) of the Universitat Politècnica de València.

# References

1. Bengio, Y., Ducharme, R., Vincent, P., Jauvin, C.: A neural probabilistic language model. J. Mach. Learn. Res. **3**(Feb), 1137–1155 (2003)
2. Berkhin, P.: A survey of clustering data mining techniques. In: Kogan, J., Nicholas, C., Teboulle, M. (eds.) Grouping Multidimensional Data, pp. 25–71. Springer, Heidelberg (2006). https://doi.org/10.1007/3-540-28349-8_2
3. Downes, S.: Learning objects: resources for distance education worldwide. Int. Rev. Res. Open Distrib. Learn. **2**(1) (2001)
4. Jiang, Z., Zhang, Y., Li, X.: MOOCon: a framework for semi-supervised concept extraction from MOOC content. In: Bao, Z., Trajcevski, G., Chang, L., Hua, W. (eds.) DASFAA 2017. LNCS, vol. 10179, pp. 303–315. Springer, Cham (2017). https://doi.org/10.1007/978-3-319-55705-2_24

5. Labutov, I., Huang, Y., Brusilovsky, P., He, D.: Semi-supervised techniques for mining learning outcomes and prerequisites. In: Proceedings of the 23rd ACM SIGKDD International Conference on Knowledge Discovery and Data Mining, pp. 907–915. ACM (2017)
6. Overell, S., Sigurbjörnsson, B., Van Zwol, R.: Classifying tags using open content resources. In: Proceedings of the Second ACM International Conference on Web Search and Data Mining, pp. 64–73. ACM (2009)
7. Papadimitriou, C.H., Raghavan, P., Tamaki, H., Vempala, S.: Latent semantic indexing: a probabilistic analysis. J. Comput. Syst. Sci. **61**(2), 217–235 (2000)
8. Steyvers, M., Griffiths, T.: Probabilistic topic models. Handb. Latent Semant. Anal. **427**(7), 424–440 (2007)
9. Tucker, B.: The flipped classroom. Educ. Next **12**(1), 82–83 (2012)
10. Wallach, H.M.: Topic modeling: beyond bag-of-words. In: Proceedings of the 23rd International Conference on Machine learning, pp. 977–984. ACM (2006)
11. Wang, L.: Support Vector Machines: Theory and Applications, vol. 177. Springer, Heidelberg (2005). https://doi.org/10.1007/b95439
12. Zhu, X.J.: Semi-supervised learning literature survey. Technical report, University of Wisconsin-Madison, Department of Computer Sciences (2005)

# New Approach for the Aesthetic Improvement of Images Through the Combination of Convolutional Neural Networks and Evolutionary Algorithms

Juan Abascal, Miguel A. Patricio$^{(\boxtimes)}$ ![ORCID], Antonio Berlanga ![ORCID], and José M. Molina ![ORCID]

Grupo de Inteligencia Artificial Aplicada, Universidad Carlos III de Madrid, Madrid, Spain
juanabascalsanchez@gmail.com, mpatrici@inf.uc3m.es,
{aberlan,molina}@ia.uc3m.es

**Abstract.** Programs for aesthetic improvements of the images have been one of the applications more widely in the last years so much from the commercial point of view like the private one. The improvement of images has been made through the application of different filters that transform the original image into another whose aesthetics have been improved. In this work a new approach for the automatic improvement of the aesthetics of images is presented. This approach uses a Convolutional Neural Network (CNN) network trained with the AVA photography data set, which contains around 255,000 images that are valued by amateur photographers. Once trained, we will have the ability to assess an image in terms of its aesthetic characteristics. Through an evolutionary differential algorithm, an optimization process will be carried out in order to find the parameters of a set of filters that improve the aesthetics of the original image. As a fitness function the trained CNN will be used. At the end of the experimentation, the viability of this methodology is presented, analyzing the convergence capacity and some visual results.

**Keywords:** Convolution Neural Network ·
Differential evolutionary algorithm · Aesthetic improvement of images

## 1 Introduction

Since 2012, when AlexNet [5] was created to compete in the ImageNet Large Scale Visual Recognition Challenge, Convolutional Neural Networks (CNN) have changed the landscape of artificial intelligence. The application area of this type of networks is very wide, and they are used to recognize handwritten texts,

This work was funded by public research projects of Spanish Ministry of Economy and Competitivity (MINECO), references TEC2017-88048-C2-2-R, RTC-2016-5595-2, RTC-2016-5191-8 and RTC-2016-5059-8.

© Springer Nature Switzerland AG 2019
H. Pérez García et al. (Eds.): HAIS 2019, LNAI 11734, pp. 229–240, 2019.
https://doi.org/10.1007/978-3-030-29859-3_20

semantic analysis of photographs and recognition of people. In some cases they have exceeded human performance when performing some of the tasks mentioned above. One of the biggest drawbacks of implementing solutions based on convolutional neural networks is that they are very computational and data intensive [11], so a lot of training time is required to achieve the desired accuracy. The training of a deep neural network like GoogleLeNet that has $6,797,700$ parameters and $1,502,000,000$ operations can be done in a week using several GPUs in parallel [12]. Other studies have taken neural networks to other more subjective fields such as paintings and music. Nowadays, you can find networks that have been trained to draw pictures with the style of Vicent van Gogh [2] or to compose songs that resemble the melodies composed by the German composer Johann Sebastian Bach [4]. However, in other less defined fields such as photography, machine learning has not been always as successful. Different works have tried to solve the problem of quality evaluation of images. These works have provided different alternatives for the resolution of this question. Many of these works have tried to solve the problem by performing a binary classification to determine if an image is good or of poor quality. Although this approach could be valid, it does not provide a metric to compare two different images. There are also works that offer another approach such as NIMA (Neural IMage Assessment) [13], in which a measure of image quality is achieved by simulating scores in the range of one to ten that an image would receive in a photo contest. In this case, the output of the neural network is a distribution from which the mean and the standard deviation can be extracted to offer a quantitative assessment of the aesthetic quality of the image.

The aim of this paper is to present a novel method for improving the aesthetics of images based on the ability of a Convolution Neural Network, similar to NIMA network, to determine the aesthetic value of a certain image. As a more relevant contribution, the paper presents a methodology that uses an evolutionary algorithm (more specifically, differential evolutionary algorithm [10]) to optimize a set of photographic filter that are applied to enhance an specific image. From a specific image, a set of individuals is selected, each of which represents the value of the parameters of the filters to be applied on the input image. For each individual, the resulting image is obtained by applying the filters with the determined parameters. The output of the NIMA network will serve as a fitness function for each of the individuals. Throughout the generations what will be obtained will be a set of images resulting from applying each of the filters encoded in each individual. The objective is to obtain the filtered image with the best aesthetic evaluation.

## 2    Deep Convolutional Neural Network for the Assessment of Image Aesthetics

The convolutional neuronal network used to solve this problem is based on the paper Neural Image Assesment [13]. For the training of this network, the AVA photography data set [7] is used. AVA contains around 255,000 images and are valued by amateur photographers. Each image is associated with the frequency with which the users have selected each of the scores from 1 to 10. In addition, they present metadata related to the type of contest for which the image was taken. In addition, the architecture used has been MobileNet [3], a lightweight architecture designed to be executed on mobile devices.

MobileNet [3] convolutional neuronal network is formed by a type of convolutional layer, "Depthwise Separable Convolution", which allows a substantial improvement in both the number of operations and the number of parameters. The last two layers, which are non-convolutional layers, have been removed from MobileNet architecture. These last two layers are a classifier, which receives the characteristics extracted by the convolutional part of the neural network and results in the probability of belonging to a class.

This behavior needs to be modified to resolve the problem presented in this work. For this, we create a score distribution generator by adding two fully connected layers. The output of the neural network is an array of ten positions one of each representing the percentage of the votes received for each score from 1 to 10. The value each position takes is the percentage of total votes given to that specific score. This array is a distribution of percentages that can be summarized doing a weighted average to get an overall score.

The error function in this work measures the distance between two score distributions, the predicted distribution and the ground truth input. To compare two distributions, the Earth Mover's Distance (EMD) error [9] is used, which measures the minimum cost of transforming one distribution into the other. This metric, therefore, gives a value for the Neural Network model's accuracy. $\hat{p}$ in Eq. 1 represents the predicted distribution and $p$ is the ground truth input distribution. In this case $r$ is equal to 2. Earth Mover's distance is represented in the following equation.

$$EMD(p, \hat{p}) = \left( \frac{1}{N} \sum_{k=1}^{N} |CDF_p(k) - CDF_{\hat{p}}(k)|^r \right)^{\frac{1}{r}} \tag{1}$$

where $CDF_p(k)$ is the cumulative probability function for an image $k$.

As a starting point, MobileNet has been used previously trained to solve ImageNet, since it is considered that this distribution of the weights is closer to the distribution that minimizes the objective function. The training hyperparameters are explained in Table 1. The training has been carried out in two stages in which different hyperparameters have been used. In the first place, only the dense layer of the model has been trained, that is, the part that has been added to the trained network, while the convolutional layers remain constant. Since the weights of this layer have been randomly initialized, they will produce a high error, so that the weights of the convolutional part would suffer a large modification. As the objective of this first phase is to approximate the weights of the last layer to the possible solution, a high learning rate is used that allows a broader search in the hypothesis space. The weights of the neural network have been obtained from this GitHub project [6] and have resulted in days of time reduction.

Table 1. Parameters used for neuronal network training process.

| Parameter | Value |
|---|---|
| Batch size | 96 |
| Learning rate I | 0.001 |
| Learning rate II | 0.00003 |
| Learning rate decay III | 0.000023 |
| Dropout | 0.75 |

In total to train the model, 5 complete epochs have been made with the first of the configurations. And another 9 epochs with the second configuration. At the end of the training an error of less than 0.07 points has been achieved, in both the training set and the test set [6].

## 3    Differential Evolutionary Algorithm

There are problems in which you have to optimize the properties of a system by adjusting different parameters of it. To begin to solve this type of problem computationally, it is necessary to define an objective function. In the case of the present work, the objective function will not be formally defined, but it will be the result of the processing of a neural network. The proposed optimization problem is a minimization problem, since we try to find the values of the set of parameters that minimize the result of the neural network. The algorithm chosen to carry out the optimization task is a differential evolutionary algorithm [10]. It is a direct search that does not need information about the gradient of the objective function. In addition, it uses a "greedy criterion", ie a new set of parameters is accepted when it reduces the value of the objective function. Optimization algorithms that use this type of criterion tend to find local minima instead of global minimum, so the evolutionary differential optimization method executes several sets of parameters simultaneously. The evolutionary differential algorithm is an evolutionary stochastic process that presents four different stages as exposed in Fig. 1:

- Initialization: The population of individuals is created. Each individual is a random instance of the parameter set. In addition, each individual is evaluated with the objective function to save computational costs in the future. This represents steps 1 and 2 of Fig. 1.
- Mutation and recombination: In our case, we use binomial crossover, which is the most used in applications. For each individual of the population, three other random individuals are selected, which are combined to generate a mutant individual. The original individual and the mutant individual are recombined to create the candidate individual who will be evaluated to see their performance compared to the original individual. This represents steps 3 and 4 of Fig. 1.

- Evaluation: The result obtained by the candidate individual in the objective function is compared with the result obtained by the original individual. This represents step 5 of Fig. 1.
- Replacement: If the result of the candidate is better than the original, it will be substituted in the population.

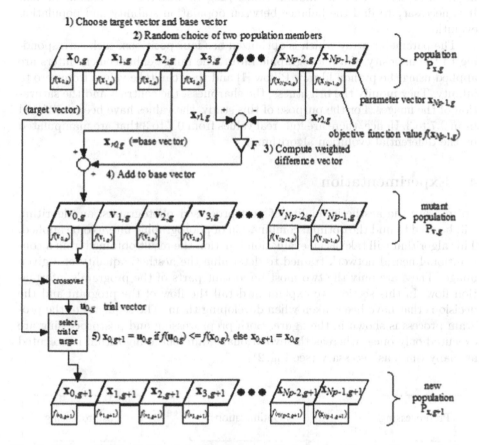

**Fig. 1.** Differential Evolution algorithm diagram. It shows how the starting population become the new population after doing the process explained before [8].

The objective function of the evolutionary differential algorithm is given by the score that a photograph receives when applying the parameters of an individual of the population. The image score is a number between 0 and 1, and the goal is to maximize that value. As the evolutionary differential algorithm, in this work, has been modeled to perform a minimization, the result produced by the neural network must be transformed:

$$min(1 - score) \tag{2}$$

Therefore, the evolutionary differential algorithm manages to obtain the set of parameters that optimizes the score of the image. Figure 1 shows this process with the specific data of the problem. It is necessary to choose the value for two different constants. One of them is the mutation constant and it uses dithering technique, between 0.5 and 1. This technique helps with convergence speed. The other parameter is the recombination constant that takes a value of 0.7. Setting a higher value would let more mutant values progress into the next recombination. It is necessary to find the balance between population stability and population evolution.

The parameter array which is optimized has four positions, each corresponding to the intensity of a photographic filter. The image enhancement filters are applied using the python library Pillow [1] and they can take values from zero to infinity. They modify the brightness, the sharpness, the contrast and the saturation of the images. For the purpose of this study, the values have been restricted from 0.7 to 3. Hence, there are four real values from 0.7 to 3 that are manipulated by the differential evolution algorithm.

## 4   Experimentation

To improve the aesthetic quality of post-processed photographs, an algorithm will be used to find the optimal combination of photographic filters to be applied. This algorithm will take as an evaluation function the result obtained from a convolutional neural network trained to determine the aesthetic quality of a given image. These are only the two most important parts of the program's information flow. In this section we explain in detail the flow of the program and the decisions that have been taken when developing them. The sequence of the program process is shown in the figure. Both preprocessing and postprocessing are executed only once, whereas the evolutionary differential algorithm is executed as many times as necessary (see Fig. 2).

**Fig. 2.** Process that follows the proposed solution to make the improvement in the aesthetic quality of an image.

In the preprocessing stage, the image is prepared so that it can be used by the evolutionary algorithm. Due to the characteristics of the neural network the image must have dimensions of $224 \times 224$. There are different ways to perform this preprocessing, trimming or resizing. Both techniques have advantages and disadvantages. On the one hand, the clipping removes information from the photograph that may affect the composition of the photograph, the information that is discarded could be crucial when determining the score of the photograph.

On the other hand, the resizing also affects the composition of the image, but this time instead of losing information, the image is deformed by altering its aspect ratio. For the optimization of the parameters, a differential evolutionary algorithm will be used. Once decided which is the best combination of filters that can be applied to the image, it is processed. The processed image is stored in the path specified by the user.

A subset of 1,000 images from the AVA dataset has been used in order to evaluate the optimization algorithm. The selected images have been processed by the algorithm while storing data related with the score distribution and the DE algorithm.

In the first place, the average time the DE algorithm takes has been obtained. The distribution of time convergence are shown in Fig. 3. At first glance, it can be observed that the distribution is similar to a normal distribution, so the values are concentrated towards the mean. Fifty percent of the experiments took between 33.4 and 49.3 s to execute. Table 2 shows the quartiles, mean, median, min and max of the distribution.

**Table 2.** Summarized figures of the execution time distribution.

| Summary | Value |
| --- | --- |
| Mean | 42.5 |
| Standard deviation | 14.7 |
| First quartile | 31.4 |
| Median | 39.6 |
| Second quartile | 50.5 |
| Min | 11.6 |
| Max | 147 |

The scores obtained before and after applying the aesthetic filters have also been studied. All the images have improved after applying the enhance parameter. The average improvement was 0.398 points with a standard deviation of 0.237 points. The improvement of the images has been substantial, however it has been limited by the neural network model's bias when assigning ratings to photographs. The model, due to an unbalanced training set, tends to give very centered scores around the average. Figure 4 compares a distribution of the scores before applying the optimization algorithm and after applying it.

Finally, it has been observed which parameters have been modified the most by the optimization process. The boxplot of Fig. 5 shows the distribution of changes made by the algorithm on the different parameters. It can be observed that although the parameters gather around a mean, they have enough variability to show that the algorithm applies different filter intensities depending on the image. Hence, the applied filters depend to a large extent on the original image and its score, and it gets better results than applying the same filter for all the images.

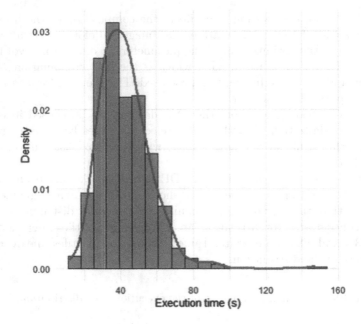

**Fig. 3.** Execution time distribution.

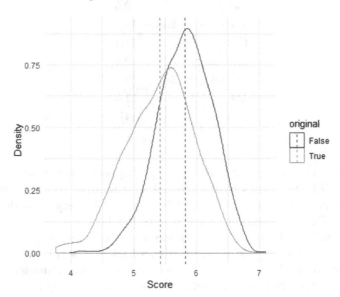

**Fig. 4.** Scores densities. The red line is the density of the scores after applying the optimization whereas the blue line the original score density. (Color figure online)

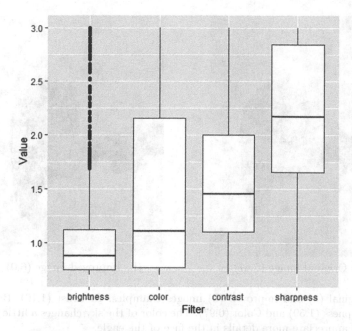

**Fig. 5.** Distribution of the factors applied by each filter.

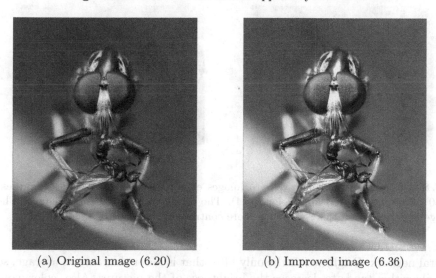

(a) Original image (6.20)                    (b) Improved image (6.36)

**Fig. 6.** Original (a) and transformed (b) images examples. Contrast (0.98), Brightness (1.49), Sharpness (1.97) and Color (1.16). It can be appreciated brighter colors in the leaf and the body of the insect. Also, the details of the hair in the nose are more defined.

Moreover, the parameters let us know more about the neural network preferences. On average, sharpness and contrast are the parameters that increase the most, so probably images with high sharpness and contrast are preferred by the

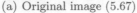

(a) Original image (5.67)　　　　　　　(b) Improved image (6.01)

**Fig. 7.** Original (a) and improved (b) images examples. Contrast (1.16), Brightness (1.14), Sharpness (1.59) and Color (0.97). The color of the sky changes a little and also the you can appreciate more details in the face of the eagle.

(a) Original image (3.88)　　　　　　　(b) Improved image (4.71)

**Fig. 8.** Original (a) and improved (b) images examples. Contrast (2.99), Brightness (0.80), Sharpness (2.85) and Color (0.74). The change is clear in this example. The image is dehazed and there are much more contrast in the picture.

neural network. Brightness is the only filter that is lower than one on average, so the algorithm tends to decrease the brightness of the pictures. Also, color, contrast an sharpness are the filters with higher variance. This assumptions must be taken carefully because the filters are not completely independent to each other.

In Figs. 6, 8, 7 and 9 original and improved images are depicted. It also shows the filters applied to the image. A value of 1 means that the filter does not have an effect on the image. A value of zero means a lack of that characteristic in the picture, for example an image with brightness value of 0 would be black.

(a) Original image (7.05)                    (b) Improved image (7.09)

**Fig. 9.** Original (a) and improved (b) images examples. Contrast (1.06), Brightness (0.86), Sharpness (1.46) and Color (2.23). The colors of the improved picture are purer and more saturated.

## 5    Conclusion

This study shows how differential evolution algorithms and convolutional neural networks can be applied to find a set of filters that improve image aesthetics. This approach is imagewise, hence the filters that are applied are not fixed but they change depending of the photography necessities. The key for getting this result is to be able to generate score distribution functions so a continuous score value can be obtained. This allow us to know when the image aesthetic quality has improved after applying the filters, which could not be possible using a binary classifier.

Although good results have been found, there is still room for improvement. The convergence speed is not fast enough to be implemented in devices that need real time inferences like a smartphone's camera. Using an inference hardware accelerator such as Google's Tensor Processing Units, which are 30X faster than a CPU, would speed up one of the most expensive tasks of the algorithm. Another option would be doing multithreading for evaluating the population in parallel. However, the improvement this method can get is relatively small.

Sometimes the optimization algorithm tends to over-saturate certain images. An improved neural network model could be used to avoid this. Altough AVA dataset is powerful to generate score distributions the quality of the majority of the images is usually quite poor. The lack of very good scores, over 8 points, and very bad, under 3 points, generates an unbalanced dataset that is clearly biased. A better dataset would improve the predictions, hence the filters would end in a better result.

As shown above, although more research needs to be done, this method provides a start point to a variable photo editing that helps improving the aesthetic of any given picture.

## References

1. Clark, A., Contributors: Pillow is the friendly PIL fork. PIL is a python imaging library (2001). https://pillow.readthedocs.io/

2. Gatys, L.A., Ecker, A.S., Bethge, M.: A neural algorithm of artistic style (2015). http://arxiv.org/abs/1508.06576
3. Howard, A.G., et al.: MobileNets: efficient convolutional neural networks for mobile vision applications. CoRR abs/1704.04861 (2017). http://arxiv.org/abs/1704.04861
4. Huang, A., Wu, R.: Deep learning for music. CoRR abs/1606.04930 (2016). http://arxiv.org/abs/1606.04930
5. Krizhevsky, A., Sutskever, I., Hinton, G.E.: ImageNet classification with deep convolutional neural networks. In: Pereira, F., Burges, C.J.C., Bottou, L., Weinberger, K.Q. (eds.) Advances in Neural Information Processing Systems 25, pp. 1097–1105. Curran Associates, Inc. (2012). http://papers.nips.cc/paper/4824-imagenet-classification-with-deep-convolutional-neural-networks.pdf
6. Lennan, C., Nguyen, H., Tran, D.: Image quality assessment. https://github.com/idealo/image-quality-assessment (2018)
7. Murray, N., Marchesotti, L., Perronnin, F.: AVA: a large-scale database for aesthetic visual analysis. In: 2012 IEEE Conference on Computer Vision and Pattern Recognition, pp. 2408–2415, June 2012. https://doi.org/10.1109/CVPR.2012.6247954
8. Price, K., Storn, R.M., Lampinen, J.A.: Differential Evolution: A Practical Approach to Global Optimization (Natural Computing Series). Springer, Heidelberg (2005). https://doi.org/10.1007/3-540-31306-0
9. Rubner, Y., Tomasi, C., Guibas, L.J.: The earth mover's distance as a metric for image retrieval. Int. J. Comput. Vision **40**(2), 99–121 (2000). https://doi.org/10.1023/A:1026543900054
10. Storn, R., Price, K.: Differential evolution – a simple and efficient heuristic for global optimization over continuous spaces. J. Global Optim. **11**(4), 341–359 (1997). https://doi.org/10.1023/A:1008202821328
11. Strigl, D., Kofler, K., Podlipnig, S.: Performance and scalability of GPU-based convolutional neural networks. In: 2010 18th Euromicro Conference on Parallel, Distributed and Network-based Processing, pp. 317–324, February 2010. https://doi.org/10.1109/PDP.2010.43
12. Szegedy, C., et al.: Going deeper with convolutions. In: Computer Vision and Pattern Recognition (CVPR) (2015). http://arxiv.org/abs/1409.4842
13. Talebi, H., Milanfar, P.: NIMA: neural image assessment. IEEE Trans. Image Process. **27**(8), 3998–4011 (2018). https://doi.org/10.1109/TIP.2018.2831899

# Learning Algorithms

# Evaluating Strategies for Selecting Test Datasets in Recommender Systems

Francisco Pajuelo-Holguera[1], Juan A. Gómez-Pulido[1(✉)],
and Fernando Ortega[2]

[1] Department Technologies of Computers and Communications,
Universidad de Extremadura, 10003 Caceres, Spain
jangomez@unex.es
[2] Department Sistemas Informáticos, ETSI Sistemas Informáticos,
Universidad Politécnica de Madrid, Madrid, Spain

**Abstract.** Recommender systems based on collaborative filtering are widely used to predict users' behaviour in large databases, where users rate items. The prediction model is built from a training dataset according to matrix factorization method and validated using a test dataset in order to measure the prediction error. Random selection is the most simple and instinctive way to build test datasets. Nevertheless, we could think about other deterministic methods to select test ratings uniformly along the database, in order to obtain a balanced contribution from all the users and items. In this paper, we perform several experiments of validating recommender systems using random and deterministic strategies to select test datasets. We considered a zigzag deterministic strategy that selects ratings uniformly across the rows and columns of the ratings matrix, following a diagonal path. After analysing the statistical results, we conclude that there are no particular advantages in considering the deterministic strategy.

**Keywords:** Recommender systems · Collaborative filtering ·
Matrix factorization · Test datasets · Prediction

## 1 Introduction

Nowadays, many databases collect the behaviour of thousands users rating items. From this information, *recommender systems* (RS) [12] allow making personalized recommendations according to user behavior and preferences when requesting information and rating items [2], disregarding the non-relevant information. Popular fields like movies databases consider RS for recommending purposes [3].

A popular implementation of RS is *collaborative filtering* (CF) [1,18]. This implementation revolves around the fact that the future behaviour of users is similar to the past behaviour when rating items [7]. For example, if two users have rated the same items with similar scores, the new items that one user rates with the same score might be rated by the another user with similar score.

© Springer Nature Switzerland AG 2019
H. Pérez García et al. (Eds.): HAIS 2019, LNAI 11734, pp. 243–253, 2019.
https://doi.org/10.1007/978-3-030-29859-3_21

The CF model is built from a rating matrix $R$ that relates users with items. This matrix stores the ratings of the users to the items without additional information. Usually, $R$ is very sparsed since not all the users rate all the items. This is the reason why CF tries to fill the gaps of $R$ [10]. This goal is tackled by *matrix factorization* (MF) algorithm [11,16]. Matrix factorization builds a mathematical model based on two matrices composed of two matrices $P$ and $Q$ [17] that makes the prediction of a rating as a linear combination of factors.

In the MF model, the items are described and conditioned by $K$ latent factors. For example, for a movies database, the genres of the movies condition the ratings scored by each user. For example, if a user likes movies of genre A and dislikes movies of genre B, this user will probably rate any movie of genre A with high score, whereas will rate with low score any movie of genre B. Therefore, MF tries to find these hidden factors through $R$, considering a particular number $K$.

The MF model is built, or trained, from known rating data. As $R$ usually includes few ratings in comparison with the matrix size, the *training dataset* $D_{train}$ considered to build the RS model tends to include as many known data as be possible, having in mind the available computational resources. On the other hand, the obtained model is validated by means of the *test dataset* $D_{test}$. There are different methods for selecting training and test datasets [19]. Depending on the nature of the ratings database, the way to select these datasets can influence strongly on the performance results [6]. In this paper, we focus on the test data since they validate the RS model, whereas the training data are usually provided according to a particular strategy that we cannot modify. Therefore, the questions to answer are: How to select the data to test the model? Can the way to select $D_{test}$ influence on the performance results? To do a first approach, we consider two methods or strategies for selecting test data: random or deterministic strategies. The deterministic strategy selects the test data going across the rows and columns of the ratings matrix uniformly following a zigzag path. Therefore, we would like to know if this method provides less prediction error than a simple random selection.

## 2   Basic Algorithm for the Recommender System

The rating matrix $R$ includes the ratings from $U$ users (rows) to $I$ items (columns). The rating $r_{u,i}$ corresponds to user $u$ rating item $i$. The MF model approaches $R$ by its factorization into two dense matrices $P$ and $Q$ [13]:

- $P$: It contains the suitability of each latent factor $k$ with each user $u$.
- $Q$: It contains the suitability of each latent factor $k$ with each item $i$.

The values of $P$ and $Q$ are learned from $R$ in order to satisfy Eq. (1).

$$R \approx P \cdot Q^T \tag{1}$$

Matrix factorization provides good performance and accuracy in comparison with other collaborative filtering implementations [14,15]. Once the model ($P$,$Q$)

is obtained (learned), predictions can be computed by a simple dot product of two $K$-dimensional vectors. However, the model should be re-trained if new ratings, users or items update $R$.

We consider that the model has been trained when we have found optimal values of $P$ and $Q$ that minimize the error in the predictions provided by the model and the training dataset $D_{train}$. To learn the factor vectors the system minimizes the regularized squared error of the known rating set, as Eq. (2) shows, where $\kappa$ is the set of $(u, i)$ pairs for with the rating $r_{u,i}$ is known, $p_u$ is the latent factors vector of the user $u$, $q_i$ is the latent factors vector of the item $i$ and $\lambda$ is a regularization hyper-parameter to avoid overfitting.

$$\min_{p_u, q_i} \sum_{(u,i) \in \kappa} (r_{u,i} - q_i^T \cdot p_u)^2 + \lambda(||q_i||^2 + ||p_u||^2) \tag{2}$$

The model can be trained using Stochastic Gradient Descent (SGD) [8]. SGD loops over all the existing ratings $(r_{u,i})$ until convergence. For each training case (i.e. each known $r_{u,i}$) SGD updates the values of $P$ and $Q$ according to Eq. (3) and Eq. (4) respectively.

$$p_u \leftarrow p_u + \gamma \cdot (e_{u,i} \cdot p_u - \lambda \cdot q_i) \tag{3}$$

$$q_i \leftarrow q_i + \gamma \cdot (e_{u,i} \cdot q_i - \lambda \cdot p_u) \tag{4}$$

The hyper-parameter $\gamma$ controls the learning speed. The error $e_{u,i}$ in the prediction of the rating of the user $u$ to the item $i$ is defined in Eq. (5).

$$e_{u,i} = r_{u,i} - q_i^T \cdot p_u \tag{5}$$

*Alternating Least Squares* (ALS) [5] is an alternative technique to SGD that allows to speed up the training process by parallelizing the updates of the factors vectors of each user and item, since SGD does not allow it using Eqs. (3) and (4) (to update each $p_u$ the value of each $q_i$ is required, and vice versa). ALS rotates between fixing the $q_i$ and fixing the $p_u$; therefore, the system computes each $q_i$ independently of the other item factors and computes each $p_u$ independently of the other user factors.

Once the model is trained, predictions can be performed computing the dot product of the factors vector of the target user and the factors vector of the target item. The predicted rating $\hat{r}_{u,i}$ of the user $u$ to the item $i$ is calculated according to Eq. (6). Recommendations to each user can be obtained from the set of $T$ unrated items with the highest predictions $(\hat{r}_{u,i})$.

$$\hat{r}_{u,i} = q_i^T \cdot p_u \tag{6}$$

The basic MF pseudo code is shown in Algorithm 1. The algorithm reads the inputs from $R$, the number of latent factors $K$, and the hyper-parameters to control the learning process $\lambda$ and $\gamma$. It returns the latent factors matrices $P$ and $Q$ learned from the $R$. Convergence criteria is usually defined as a fixed number of iterations.

---

**Algorithm 1.** Matrix Factorization

---

**Input:**
*numIters* (number of iterations of the algorithm)
*numItems* (number of items)
*numUsers* (number of users)
*lambda; gamma*
$K$ (number of latent factors)
$p_{uk}$ and $q_{ik}$ (matrices of sizes $(numUsers \times K)$ and $(numItems \times K)$ that contains initial factors)
*ratings$_{ui}$* (matrix $(numUsers \times numItems)$ that contains ratings)
**Output:**
$p_{uk}$ and $q_{ik}$: matrices of sizes $(numUsers \times K)$ and $(numItems \times K)$ that contains final factors

---

```
 1: for iter ← 1 to numIters do
 2:     for u ← 1 to numUsers do
 3:         for i ← 1 to numItems do
 4:             if ratings_ui ≠ −1 then
 5:                 error ← ratings_ui - (p_u · q_i)
 6:             end if
 7:             for k ← 1 to K do
 8:                 p_uk ← p_uk + gamma * ( error * q_ik - lambda * p_uk )
 9:             end for
10:         end for
11:     end for
12:     for i ← 1 to numItems do
13:         for u ← 1 to numUsers do
14:             if ratings_ui ≠ −1 then
15:                 error ← ratings_ui - (p_u · q_i)
16:             end if
17:             for k ← 1 to K do
18:                 q_ik ← q_ik + gamma * ( error * p_uk - lambda * q_ik )
19:             end for
20:         end for
21:     end for
22: end for
```

---

# 3   Two Strategies for Selecting Test Datasets

Once we decided a particular number $NT$ of known ratings for building a test dataset, we considered two different methods to select these ratings among the available ones:

- *Random strategy.* This method chooses the test data randomly. Therefore, it presents a non-deterministic nature that suggests us to tackle any test by performing several runs in order to obtain statistical results.
- *Diagonal strategy.* This method chooses the test data following a zigzag path that goes across the rows and columns of the ratings matrix uniformly.

The reason of proposing this method is that we guarantee that the test data represent most of users and items with a similar weight. There is no need to do any statistical study since diagonal strategy is deterministic.

Basically, the diagonal strategy chooses one rating value by each user (row) following consecutive items (columns). Figure 1 shows a simple example of the diagonal strategy: when the rating to be selected falls on an unknown value, we skip to the next column of the same row (A), unless the end of the row has been reached, in which case we select the first column of the next row (B).

Each test dataset, independently of the method from which it was generated, is applied on a RS model in order to obtain the *root mean squared error* (RMSE) value as error quality metric. This metric is calculated according to Eq. (7).

$$RMSE = \sqrt{\frac{\sum_{u,i \in D^{test}} (r_{u,i} - \hat{r}_{u,i})^2}{|D^{test}|}} \tag{7}$$

## 4  Experimental Results

This section shows the experimental results after applying both strategies on three test datasets. Since some parts of the algorithms are non-deterministic, we have followed an experimental procedure that performs different runs and collects enough statistical information in order to obtain accurate conclusions.

### 4.1  Datasets

The diagonal and random strategies for selecting test datasets were tested using three state-of-the-art datasets: Movielens-100K and Movielens-1M [9], and The Movies Dataset (Kaggle) [4]. These datasets collect the ratings of movies from many users, where the rating is scored from 1 to 5, and each user rates at least 20 movies. The main features of these datasets are shown in Table 1.

### 4.2  Matrix Factorization Parameters

Table 2 shows the values chosen for the main parameters involved in MF, as they are described in Algorithm 1. These values remained unchanged in the different experiments.

### 4.3  Experimental Procedure

We considered different issues in order to design the experimental procedure:

1. It is advisable to consider more than one ratings dataset.
2. Independently of the test validation, each time that a RS model is trained, the matrices $P$ and $Q$ are randomly initialized. Consequently, each training process must be repeated several times in order to obtain statistical validity. In this sense, we considered 31 runs for each model training (an odd value is needed to calculated the median value).

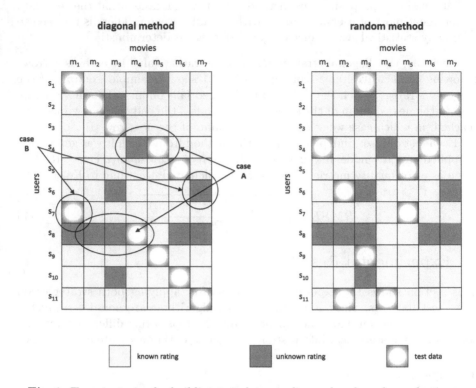

**Fig. 1.** Two strategies for building test dataset: diagonal and random selection.

**Table 1.** Datasets

| Dataset | Movielens-100K | Movielens-1M | Kaggle |
|---------|---------------|--------------|--------|
| Ratings | 100,000 | 1,000,000 | 100,000 |
| Users | 943 | 6,000 | 700 |
| Items | 1,682 | 4,000 | 9,000 |

**Table 2.** Parameters for matrix factorization algorithm

| Parameter | Name in Algorithm 1 | Value |
|-----------|---------------------|-------|
| Iterations | $numIters$ | 150 |
| Number of latent factors (K) | $K$ | 15 |
| Regularization factor ($\lambda$) | $lambda$ | 0.055 |
| Learning speed ($\gamma$) | $gamma$ | 0.01 |

3. Once a RS model is obtained after the training process, the accuracy provided by matrices $P$ and $Q$ is validated by applying test datasets built from the two different strategies. Then, we consider:
   (a) A test dataset obtained from the diagonal strategy just needs one run to be applied on the RS model, since this strategy is deterministic. The RMSE value obtained is named $RMSE\_diag$.
   (b) Different test datasets obtained from the random strategy should be applied on the RS model, since this strategy has a stochastic nature. For this purpose, we generated 31 test datasets randomly, in order to perform 31 runs of the RS model validation. After these runs, we take the mean of the RMSE values obtained and named it $RMSE\_rand\_mean$.
4. Both $RMSE\_diag$ and $RMSE\_rand\_mean$ are compared in order to obtain the accuracy for a particular RS model.
5. Finally, as we generated 31 RS models, it is needed to calculate the mean of the 31 $RMSE\_diag$ values ($RMSE\_diag\_mean$) and the mean of the 31 $RMSE\_rand\_mean$ values ($RMSE\_rand\_mean\_mean$), and then compare them in order to obtain a final conclusion.

Algorithm 2 reflects the experimental procedure described in the above items.

## 4.4 Results

Table 3 shows the results obtained according to the experimental procedure described in Algorithm 2. This table lists the RMSE results from the two strategies for the three datasets considered. Each row is a run, which means: (a) initial P and Q are randomly generated; (b) the RS model is obtained (final P and Q); (c) 1 test dataset is generated with diagonal strategy, and the corresponding RMSE value (RMSE_diag) is calculated; (d) 31 test datasets are randomly generated, and the mean of their corresponding RMSE values (RMSE_rand-mean) is calculated. Last row shows the global results for each dataset, as the means of the 31 runs: RMSE_diag-mean and RMSE_rand-mean-mean for diagonal and random strategies respectively.

The results by run and global, shown in Table 3, are graphically represented in Fig. 2 and Fig. 3 respectively.

We can extract to main conclusions from the experimental results:

1. The random strategy provides much better result (lower RMSE) than the diagonal strategy in all the runs of two datasets (Movielens-100k and Kaggle), whereas this improvement is very slight for the third dataset (Movielens-1M), where even one of the 31 runs reports best result for diagonal strategy.
2. Observing the global results, we can conclude that diagonal strategy does not provide advantages in accuracy terms with regard to a simple random method to select test datasets.

The diagonal approach tries to represent almost every user and item with a similar weight in the test data. This deterministic sampling does not consider

**Algorithm 2.** Experimental Procedure

```
 1: for  dataset ← Movielens-100k to Kaggle  do
 2:        Selects train data from rating matrix R
 3:        for run ← 1 to 31 do
 4:              Initial Pᵢ and Qᵢ randomly generated
 5:              RS algorithm applied
 6:              Final Pf and Qf obtained (RS model)
 7:              Test dataset D_test generated with diagonal strategy
 8:              RMSE_diag ← RS model applied to D_test
 9:              for testrand ← 1 to 31 do
10:                    Test dataset D_test generated with random values
11:                    RMSE_diag ← RS model applied to D_test
12:              end for
13:              RMSE_rand-mean ← mean of the above 31 RMSE_rand values
14:              if  RMSE_rand-mean < RMSE_diag  then
15:                    Random better than Diagonal for this RS model
16:              else
17:                    Diagonal better than Random for this RS model
18:              end if
19:        end for
20:        RMSE_diag-mean ← mean of the above 31 RMSE_diag values
21:        RMSE_rand-mean-mean ← mean of the above 31 RMSE_rand-mean values
22:        if  RMSE_rand-mean-mean < RMSE_diag-mean  then
23:              Random better than Diagonal for this ratings dataset
24:        else
25:              Diagonal better than Random for this ratings dataset
26:        end if
27: end for
```

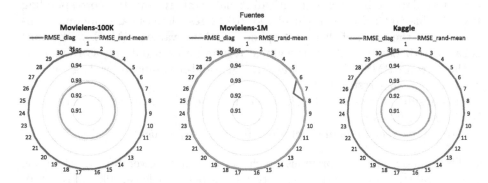

**Fig. 2.** Results by run, for the three datasets.

the possible cases where some users rate order of magnitude more items than the others and some items are much more popular and thus much more frequently rated as others. Consequently, this might explain the worse RMSE results for diagonal strategy.

**Table 3.** RMSE results from the two strategies for the three datasets. Each row run includes: (a) initial P and Q are randomly generated; (b) the RS model is obtained (final P and Q); (c) 1 test dataset is generated with diagonal strategy, and the corresponding RMSE value (RMSE_diag) is calculated; (d) 31 test datasets are randomly generated, and the mean of their corresponding RMSE values (RMSE_rand-mean) is calculated

| | Movielens-100k dataset | | | | | Movielens-1M dataset | | | | | Kaggle dataset | | | | |
| | | RMSE_rand | | | | | RMSE_rand | | | | | RMSE_rand | | | |
| run | RMSE_diag | min | max | mean | stdev | RMSE_diag | min | max | mean | stdev | RMSE_diag | min | max | mean | stdev |
|---|---|---|---|---|---|---|---|---|---|---|---|---|---|---|---|
| 1 | .94927 | .92835 | .92884 | .92858 | 1.1E-4 | .94935 | .94910 | .94918 | .94912 | 1.6E-5 | .94957 | .92623 | .92669 | .92655 | 7.7E-5 |
| 2 | .94927 | .92835 | .92874 | .92856 | 1.1E-4 | .94916 | .94908 | .94919 | .94914 | 3.5E-5 | .94942 | .92650 | .92670 | .92660 | 6.1E-5 |
| 3 | .94916 | .92830 | .92879 | .92853 | 1.6E-4 | .94936 | .94909 | .94919 | .94913 | 3.0E-5 | .94946 | .92651 | .92669 | .92660 | 5.5E-5 |
| 4 | .94918 | .92831 | .92881 | .92854 | 1.7E-4 | .94938 | .94908 | .94919 | .94914 | 3.5E-5 | .94942 | .92650 | .92669 | .92660 | 6.1E-5 |
| 5 | .94950 | .92829 | .92874 | .92850 | 1.5E-4 | .94930 | .94908 | .94919 | .94913 | 3.2E-5 | .94941 | .92650 | .92662 | .92662 | 6.0E-5 |
| 6 | .94932 | .92827 | .92877 | .92850 | 1.7E-4 | .94931 | .94908 | .94919 | .94914 | 3.2E-5 | .94944 | .92650 | .92668 | .92658 | 5.5E-5 |
| 7 | .94932 | .92825 | .92879 | .92849 | 1.8E-4 | .94315 | .94908 | .94919 | .94914 | 3.1E-5 | .94947 | .92651 | .92670 | .92660 | 5.2E-5 |
| 8 | .94950 | .92823 | .92879 | .92849 | 1.9E-4 | .94937 | .94908 | .94919 | .94913 | 3.0E-5 | .94942 | .92650 | .92670 | .92659 | 5.9E-5 |
| 9 | .94952 | .92824 | .92877 | .92848 | 1.8E-4 | .94932 | .94909 | .94919 | .94914 | 3.8E-5 | .94942 | .92650 | .92669 | .92660 | 6.7E-5 |
| 10 | .94941 | .92826 | .92877 | .92849 | 1.7E-4 | .94932 | .94908 | .94919 | .94914 | 3.7E-5 | .94950 | .92650 | .92669 | .92659 | 5.8E-5 |
| 11 | .94942 | .92831 | .92880 | .92853 | 1.6E-4 | .94940 | .94908 | .94919 | .94914 | 3.0E-5 | .94951 | .92650 | .92670 | .92658 | 6.8E-5 |
| 12 | .94940 | .92825 | .92877 | .92851 | 1.7E-4 | .94941 | .94908 | .94919 | .94914 | 3.4E-5 | .94949 | .92651 | .92670 | .92661 | 5.8E-5 |
| 13 | .94911 | .92825 | .92877 | .92852 | 1.6E-4 | .94939 | .94908 | .94919 | .94913 | 3.4E-5 | .94945 | .92650 | .92669 | .92661 | 6.5E-5 |
| 14 | .94916 | .92833 | .92887 | .92858 | 1.8E-4 | .94935 | .94908 | .94919 | .94914 | 3.3E-5 | .94957 | .92650 | .92669 | .92659 | 6.6E-5 |
| 15 | .94918 | .92838 | .92886 | .92860 | 1.6E-4 | .94947 | .94909 | .94919 | .94914 | 3.3E-5 | .94959 | .92650 | .92669 | .92659 | 5.6E-5 |
| 16 | .94922 | .92838 | .92886 | .92860 | 1.6E-4 | .94949 | .94909 | .94919 | .94913 | 3.3E-5 | .94951 | .92650 | .92667 | .92658 | 5.4E-5 |
| 17 | .94929 | .92839 | .92881 | .92862 | 1.3E-4 | .94941 | .94909 | .94919 | .94914 | 2.7E-5 | .94951 | .92650 | .92669 | .92659 | 5.4E-5 |
| 18 | .94919 | .92838 | .92885 | .92862 | 1.5E-4 | .94941 | .94908 | .94919 | .94913 | 3.0E-5 | .94941 | .92651 | .92669 | .92658 | 5.4E-5 |
| 19 | .94919 | .92839 | .92885 | .92866 | 1.4E-4 | .94931 | .94908 | .94918 | .94913 | 3.0E-5 | .94955 | .92651 | .92669 | .92660 | 6.0E-5 |
| 20 | .94933 | .92839 | .92885 | .92862 | 1.4E-4 | .94945 | .94909 | .94919 | .94914 | 3.2E-5 | .94940 | .92651 | .92668 | .92660 | 5.5E-5 |
| 21 | .94934 | .92839 | .92859 | .92859 | 1.4E-4 | .94930 | .94909 | .94919 | .94913 | 2.8E-5 | .94945 | .92650 | .92670 | .92659 | 5.9E-5 |
| 22 | .94932 | .92839 | .92886 | .92863 | 1.3E-4 | .94936 | .94908 | .94919 | .94913 | 3.1E-5 | .94946 | .92651 | .92668 | .92659 | 4.9E-5 |
| 23 | .94924 | .92839 | .92883 | .92860 | 1.4E-4 | .94948 | .94909 | .94919 | .94914 | 3.3E-5 | .94958 | .92651 | .92669 | .92661 | 6.1E-5 |
| 24 | .94944 | .92839 | .92885 | .92861 | 1.6E-4 | .94941 | .94908 | .94918 | .94913 | 3.1E-5 | .94951 | .92650 | .92670 | .92661 | 5.8E-5 |
| 25 | .94948 | .92840 | .92886 | .92862 | 1.3E-4 | .94948 | .94909 | .94919 | .94914 | 2.9E-5 | .94958 | .92651 | .92669 | .92660 | 6.3E-5 |
| 26 | .94935 | .92839 | .92886 | .92860 | 1.6E-4 | .94940 | .94908 | .94919 | .94914 | 3.4E-5 | .94950 | .92650 | .92670 | .92660 | 5.9E-5 |
| 27 | .94935 | .92839 | .92885 | .92864 | 1.5E-4 | .94946 | .94909 | .94918 | .94913 | 3.0E-5 | .94956 | .92651 | .92669 | .92660 | 5.9E-5 |
| 28 | .94918 | .92839 | .92885 | .92862 | 1.5E-4 | .94934 | .94908 | .94919 | .94913 | 3.4E-5 | .94944 | .92650 | .92670 | .92660 | 6.2E-5 |
| 29 | .94942 | .92840 | .92883 | .92860 | 1.5E-4 | .94948 | .94908 | .94919 | .94914 | 3.2E-5 | .94958 | .92652 | .92670 | .92661 | 6.2E-5 |
| 30 | .94932 | .92839 | .92883 | .92864 | 1.4E-4 | .94932 | .94908 | .94912 | .94912 | 3.1E-5 | .94951 | .92651 | .92670 | .92660 | 6.3E-5 |
| 31 | .94923 | .92839 | .92883 | .92861 | 1.4E-4 | .94936 | .94908 | .94918 | .94913 | 3.5E-5 | .94953 | .92650 | .92670 | .92660 | 5.9E-5 |
| | mean | | | mean | | mean | | | mean | | mean | | | mean | |
| | .94931 | | | .92857 | | .94918 | | | .94913 | | .94949 | | | .92660 | |

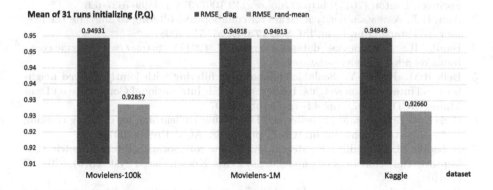

Fig. 3. Global results, for the three datasets.

# 5  Conclusions

In this work we have demonstrated that a deterministic method to select test data for validating recommender systems models, does not provide better results than a simple random selection. The reason to design a deterministic method was to build test datasets whose values collect ratings from users and items uniformly across the rating matrix. This way, a deterministic method like the one proposed in this work has the aim of obtaining a representative contribution of the users' behavior. Nevertheless, this method did not provide significant advantages.

A third possible strategy could tackle the cases where there is a strong ratings sparsity among the different users and items. This strategy should try to select representative pairs of users/items in the test data, after a quick initial scanning of the rating matrix. This idea is left for future research.

Other future research issues could be focused on designing methods that build test datasets by combining both random and deterministic strategies, with the aim of providing the advantages of good representation and fast selection. Moreover, a higher number of datasets should be considered in order to obtain more robust conclusions.

**Acknowledgments.** This work was partially funded by the Government of Extremadura under the project IB16002, and by the AEI (State Research Agency, Spain) and the ERDF (European Regional Development Fund, EU) under the contract TIN2016-76259-P.

# References

1. Adomavicius, G., Tuzhilin, A.: Toward the next generation of recommender systems: a survey of the state-of-the-art and possible extensions. IEEE Trans. Knowl. Data Eng. **17**(6), 734–749 (2005)
2. Adomavicius, G., Tuzhilin, A.: Context-aware recommender systems. In: Ricci, F., Rokach, L., Shapira, B. (eds.) Recommender Systems Handbook, pp. 191–226. Springer, Boston (2015). https://doi.org/10.1007/978-1-4899-7637-6_6
3. Ahn, H.J.: A new similarity measure for collaborative filtering to alleviate the new user cold-starting problem. Inf. Sci. **178**(1), 37–51 (2008)
4. Banik, R.: The movies dataset, version 7 (2017). https://www.kaggle.com/rounakbanik/the-movies-dataset
5. Bell, R.M., Koren, Y.: Scalable collaborative filtering with jointly derived neighborhood interpolation weights. In: Seventh IEEE International Conference on Data Mining (ICDM 2007), pp. 43–52, October 2007
6. Bickel, S., Brückner, M., Scheffer, T.: Discriminative learning for differing training and test distributions. In: In: ICML, pp. 81–88. ACM Press (2007)
7. Bobadilla, J., Serradilla, F., Bernal, J.: A new collaborative filtering metric that improves the behavior of recommender systems. Knowl. Based Syst. **23**(6), 520–528 (2010)
8. Bottou, L.: Large-scale machine learning with stochastic gradient descent. In: Lechevallier, Y., Saporta, G. (eds.) Proceedings of 19th International Conference on Computational Statistics, pp. 177–186. Springer, Heidelberg (2010). https://doi.org/10.1007/978-3-7908-2604-3_16
9. Harper, F.M., Konstan, J.A.: The MovieLens datasets: History and context. ACM Trans. Interact. Intell. Syst. **5**(4), 19:1–19:19 (2015)
10. Herlocker, J.L., Konstan, J.A., Terveen, L.G., Riedl, J.T.: Evaluating collaborative filtering recommender systems. ACM Trans. Inf. Syst. (TOIS) **22**(1), 5–53 (2004)
11. Hernando, A., Bobadilla, J., Ortega, F.: A non negative matrix factorization for collaborative filtering recommender systems based on a Bayesian probabilistic model. Knowl. Based Syst. **97**, 188–202 (2016)

12. Jannach, D., Zanker, M., Felfernig, A., Friedrich, G.: Recommender Systems. An Introduction. Cambridge University Press, New York (2011)
13. Koren, Y., Bell, R., Volinsky, C.: Matrix factorization techniques for recommender systems. Computer **42**(8), 30–37 (2009)
14. Mnih, A., Salakhutdinov, R.R.: Probabilistic matrix factorization. In: Advances in Neural Information Processing Systems, pp. 1257–1264 (2008)
15. Ortega, F., Hernando, A., Bobadilla, J., Kang, J.H.: Recommending items to group of users using matrix factorization based collaborative filtering. Inf. Sci. **345**, 313–324 (2016)
16. Paatero, P., Tapper, U.: Positive matrix factorization: a non-negative factor model with optimal utilization of error estimates of data values. Environmetrics **5**(2), 111–126 (1994)
17. Rendle, S., Schmidt-Thieme, L.: Online-updating regularized kernel matrix factorization models for large-scale recommender systems. In: Proceedings of the 2008 ACM Conference on Recommender Systems, pp. 251–258 (2008)
18. Ricci, F., Rokach, L., Shapira, B.: Introduction to recommender systems handbook. In: Ricci, F., Rokach, L., Shapira, B., Kantor, P.B. (eds.) Recommender Systems Handbook, pp. 1–35. Springer, Boston (2011). https://doi.org/10.1007/978-0-387-85820-3_1
19. Storkey, A.J.: When training and test sets are different: characterising learning transfer. In: Dataset Shift in Machine Learning, pp. 3–28. MIT Press (2009)

# Use of Natural Language Processing to Identify Inappropriate Content in Text

Sergio Merayo-Alba, Eduardo Fidalgo, Víctor González-Castro$^{(\boxtimes)}$,
Rocío Alaiz-Rodríguez, and Javier Velasco-Mata

Department of Electrical, Systems and Automation Engineering,
Universidad de León, León, Spain
sergiomerayo@gmail.com, {efidf,vgonc,rocio.alaiz,jvelm}@unileon.es

**Abstract.** The quick development of communication through new technology media such as social networks and mobile phones has improved our lives. However, this also produces collateral problems such as the presence of insults and abusive comments. In this work, we address the problem of detecting violent content on text documents using Natural Language Processing techniques. Following an approach based on Machine Learning techniques, we have trained six models resulting from the combinations of two text encoders, Term Frequency-Inverse Document Frequency and Bag of Words, together with three classifiers: Logistic Regression, Support Vector Machines and Naïve Bayes. We have also assessed StarSpace, a Deep Learning approach proposed by Facebook and configured to use a Hit@1 accuracy. We evaluated these seven alternatives in two publicly available datasets from the Wikipedia Detox Project: Attack and Aggression. StarSpace achieved an accuracy of 0.938 and 0.937 in these datasets, respectively, being the algorithm recommended to detect violent content on text documents among the alternatives evaluated.

## 1 Introduction

Due to the development of the Internet over the last years and the change in habits and behaviour of people regarding technology, Social Networks have gained more and more popularity and users, generating an impressive daily amount of comments about any topic. Unfortunately, this also implies that a significant amount of these comments may contain inappropriate content such as obscene, aggressive, rude, racist, sexist or violent sentences [1].

Nowadays, more and more children have access to the Internet [2], so it is imperative to prevent them to access inappropriate contents such as those mentioned before. When it comes to text, due to the large amount of material available, reviewing all of it is an unmanageable task to be efficiently accomplished by human inspectors. Therefore, it is necessary to develop automated solutions to filter inappropriate textual contents. Machine Learning methods applied to text classification can be used to build parental filters for online content and to detect illegal activities such as hate speeches [3].

© Springer Nature Switzerland AG 2019
H. Pérez García et al. (Eds.): HAIS 2019, LNAI 11734, pp. 254–263, 2019.
https://doi.org/10.1007/978-3-030-29859-3_22

In this work, we explored the use of some Natural Language Processing (NLP) and Machine Learning techniques to detect violent content in text. We have compared six combinations of encoder-plus-classifier, where the former is either Term frequency - Inverse document frequency (TF-IDF) or Bag of Words (BoW), and the latter is either Logistic Regression (LR), Support Vector Machines (SVM) or Naïve Bayes (NB). In addition, we have also assessed an alternative based on Deep Learning called StarSpace, an algorithm proposed by Facebook[1]. We evaluated these seven methods using two public datasets, with attacking aggressive comments respectively. Finally, we made a recommendation about the best approach to detect and classify inappropriate content on text documents. A scheme of this work is shown in Fig. 1.

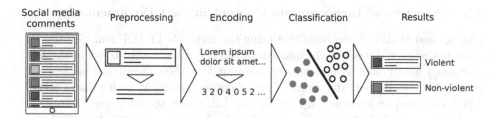

**Fig. 1.** Proposed pipeline to detect inappropriate content in text.

The rest of the work is presented as follows: in Sect. 2 we review the state-of-the-art in this field. Next, we describe the techniques used in this work in Sect. 3, and in Sect. 4 we explain the experimental settings, describe the datasets and discuss the achieved results. Finally, we summarize this work and discuss future lines of work in Sect. 5.

## 2   Related Work

### 2.1   Text Classification

According to the literature review carried out by Mironczuk and Protasiewicz in 2016 [4], the text classification process can be divided into six phases: (1) data acquisition, (2) data analysis and labelling, (3) feature construction and weighting, (4) feature selection and projection, (5) training of a classification model and, finally, (6) the solution evaluation. Different researches focus on different steps. For example, Bui et al. [5] focused on the data acquisition step on PDF documents where the relevant text was mixed with metadata or semi-structured text, while the work of Chen et al. [6] focused on the feature construction phase by comparing different features extractors for NB.

Rogati et al. [7] assessed the feature and selection phase comparing the performance of four classifiers, namely NB, Rocchio classifier, k-Nearest Neighbors

---

[1] https://research.fb.com/downloads/starspace/.

(kNN) and SVM, using different feature selection strategies describing the samples of two well known datasets: RCV1 and Reuters-21578. The highest scores were achieved using only 3% of the available features.

The model training and the solution evaluation phases are usually studied together. An example is the work of Diab and El Hindi [8], where they compared different techniques for fine-tuning the NB algorithm. They used 53 text-classification datasets obtained from the UCI repository[2]. Their research concluded that a Multi Parent Differential Evolution (MPDE) allowed NB to reach a peak performance comparing to other tuning method such as regular Differential Evolution (DE), Genetic Algorithms (GA) and Simulated Annealing (SA).

### 2.2  Detection of Inappropriate Content in Text Documentation

Chavan and Shylaja [9] encoded text using the methods TF-IDF and N-grams. It was applied on a dataset with comments from the Kaggle website[3]. The dataset contained about 4000 comments for training and 2500 comments for testing. Next, they tested the performance of two classifiers: an SVM with a linear kernel which obtained an accuracy of 77.65%, and LR whose accuracy was 73.76%.

Later, in 2017, Hammer [10] used a logistic LASSO regression to detect violent content in threads about minorities, immigrants and homosexuals in 24840 manually tagged sentences from YouTube comments. The classifier only showed an approximate rate of 10% of violent texts classified as non-violent, and 5% of non-violent text classified as violent.

Also in 2017, Eshan and Hasan [11] classified Bengali text from Facebook comments with abusive content using TF-IDF, BoW and CountVectorizer along with the classifiers Random Forest (RF), multinomial NB, and SVM with different kernels: Linear, Radial Basis Function (RBF), Polynomial and Sigmoid. A TF-IDF encoder combined with a SVM with linear kernel achieved the best performance.

Recently, Deep Learning techniques [12] have also been used for NLP. Chu et al. [13] compared two Recurrent Neural Networks, Long-Short Term Memory (LSTM) and Convolutional Neural Networks (CNN), against two datasets of 150,000 Wikipedia comments where the highest accuracy (94%) was achieved by CNN with character embedding. Moreover, Badjatiya et al. [14] achieved a F1 score of 93% detecting sexist and racist tweets over a dataset with 16,000 samples using a CNN with a random embedding.

## 3   Methodology

In this work, we compared two different schemes for text classification: one based on different Machine Learning methods, and the other one based on a

---

[2] http://archive.ics.uci.edu/ml/index.php.
[3] https://kaggle.com.

Deep Learning technique. On the one hand, we combined two text encoders (see Subsect. 3.1) and three different classifiers (see Subsect. 3.2) representing the Machine Learning methods. On the other hand, Deep Learning, we chose a deep neuronal network classifier.

For any of the two schemes, we need to perform a preprocessing on the text of the datasets. First, we eliminated start-line words and tags from markup languages like HTML. Next, we separated punctuation symbols such as dots or commas to facilitate the recognition of text elements. After that, we transformed the text elements such as words, spaces and punctuation symbols into numerical values that can be used by a learning algorithm. Finally, we encoded the text into a vector of numbers, i.e. the process of *vectorization* or *encoding*.

## 3.1 Encoding Techniques

In *Term Frequency - Inverse Document Frequency (TF-IDF)*, each term has a weight given by the product of two factors [15]: TF and IDF as shown in (1). TF refers to the frequency of appearance of a word in the text, and IDF is a measure of the amount of information provided by the word, i.e., how common is the word in the considered text.

$$\text{TF-IDF} = \text{TF} \times \text{IDF} \tag{1}$$

Within the framework of *Bag of Words (BoW)*, the word codification is based on the use of a dictionary previously generated with the documents used for training. Given a text, the algorithm expresses it in a form of a vector where each of its components correspond with a word of the dictionary, and it records the number of times that this word appears on the text [16] disregarding the grammar and the word order of the original text. This process can be visualized on Fig. 2.

## 3.2 Classifiers

*Logistic Regression (LR)* was proposed by Cox [17] and it is used to classify binary data, i.e., belonging to a two-class scheme, using a linear combination of the variables used to characterize the samples. The basic idea behind the model consists in fitting a logistic function with the samples of the training data expressed as points in a bi-dimensional plane where the X coordinate expresses the result of a linear combination of the characteristics and the Y coordinate a "0" or "1" depending on the class of the sample. After the training, for each sample to be classified the same linear combination of characteristics as in training is calculated, then the fitted logistic function is used to calculate the corresponding $y$ value from the linear combination and this $y$ value is considered as the probability of belonging to the class labelled as "1".

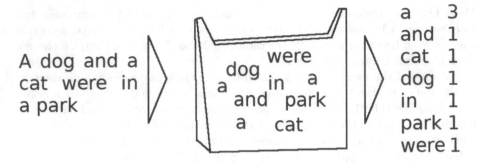

**Fig. 2.** Representation of the BoW encoding model

*Support Vector Machine (SVM)* classifier was proposed in 1995 by Cortes and Vapnik [18] and it consists on representing the data samples in a $n$-dimensional space where $n$ is the number of characteristics used to describe the data, and then find the hyperplane that separates the two classes of the dataset with the largest margin, as shown in Fig. 3.

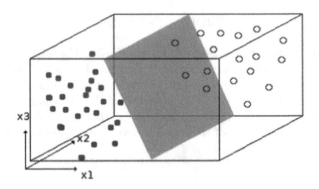

**Fig. 3.** Visualization of the hyperplane generated by a SVM model separating the samples of two classes represented as filled and empty dots, and assuming that their characteristics can be represented on a tridimensional space.

Finally, *Naïve Bayes (NB)* proposed by McCallum [19] is based on the Bayesian theorem: it considers that each of the characteristics of a sample contributes independently to the probability of the sample to belong to a class. Therefore, the presence or absence of a particular characteristic is not related to the presence or absence of any other characteristic.

### 3.3   StarSpace

StarSpace is a general-purpose deep neural model proposed by Wu et al. [20]. The model learns how to represent different encoded entities in a common linear space and then compares them against each other. Because this is a general scheme, it can be used in a wide variety of tasks[4]:

1. *PageSpace user/page embedding*: Used to recommend which Facebook pages may interest someone to follow based on the tracking of others.
2. *DocSpace document recommendation*: Recommendation of web pages based on the "like" history and clicks of a user.
3. *GraphSpace: Link Prediction in Knowledge Bases*: Map between entities and relationships in Freebase[5].
4. *SentenceSpace: Learning Sentence Embedding*: Given the coding of a sentence, it tries to find other semantically similar ones.
5. *ArticleSpace: Learning Sentence and Article Embedding*: Given the coding of a sentence, it tries to search relevant articles.
6. *ImageSpace: Learning Image and Label Embedding*: Learn the relationship between images and other entities.
7. *TagSpace word/tag embedding*: Classic classification of short texts.

In this work, we use the algorithm to embed the text and classify its content automatically. The algorithm can be configured with a Hit @ $n$ accuracy, i.e., in the classification of a sample the model indicates the $n$ most probable classes. In this work we use a Hit @ 1 accuracy, i.e., the model only indicates the most probable class in the classification of a sample.

## 4    Experimental Results

### 4.1    Experimental Settings

This work has been carried out in a computer environment with an Intel Core i5 CPU (1.3 GHz) and 8 GB of RAM. The software was developed on Python using the modules Anaconda[6], Scikit-Learn[7], Pandas[8] and Numpy[9]. The classic Machine Learning classifiers provided by the used modules were configured as follows:

1. The LR and SVM classifiers were configured to automatically adjust the weights of the classes as inversely proportional to class frequencies in the training data.
2. In SVM, we chose a linear kernel, as well as it was done in similar works of the related literature.
3. We chose a multinomial NB classifier since the input data contained discrete features, such as word counts.

---

[4] https://github.com/facebookresearch/StarSpace.
[5] https://developers.google.com/freebase/.
[6] https://www.anaconda.com/distribution/.
[7] https://scikit-learn.org/stable/.
[8] https://pandas.pydata.org/.
[9] http://www.numpy.org/.

## 4.2 Datasets

In our experiments we used two public datasets manually tagged by ten human collaborators from the Wikipedia Detox Project[10]: the "Aggression" content and the "Attack" content ones. Both datasets contain approximately 150,000 comments dual tagged by each operator with an integer value from 3 (i.e., peaceful comment) to −3 (i.e., very aggressive comment), and with a binary value indicating if the comment is inappropriate or not, as shown on Table 1. In the case of the "Attack" dataset, instead of the binary tag it presented a label indicating if the comment was classified as mentioned, received, attack to third parties or another attack; and we simplified this information into a binary tag informing whether the comment was an attack or not. Table 2 shows examples of violent and non-violent comments of both datasets.

**Table 1.** Description of the Attack and Aggression datasets. Amount of comments (violent vs. non violent.) in both datasets

| Dataset | Violent | Not violent |
|---|---|---|
| Attack | 13590 | 102274 |
| Aggression | 7498 | 58452 |

**Table 2.** Examples of violent and non-violent comments on the Attack and Aggression datasets

|  | Attack | Aggression |
|---|---|---|
| Violent | - People as stupid as you should not edit Wikipedia! | - Charles, you are a real fag aren't you? Siding with chinks? |
|  | - NO! Im not gunna sign my posts you ass! | - You are a raging faggot. Kill yourself. |
|  | - Fuck you and Fuck your mom. And her dog | - Please take time today to kill yourself. We would all enjoy it. |
| Non-violent | - Thank you for your contribution, you did a great job! | - Thank you for reminding me about my signature |
|  | - I think Mac mini is just a ordinary desktop in a small case | - I think you're a bit late with that last post. |
|  | - For your own safety, please do not post personal information | - Correct, and noted in article |

The datasets are already divided into a training subset and a testing subset with 75% and 25% of the data respectively, and we discarded the comments where the standard deviation of the operators was higher than 1.25 since we

---

[10] https://meta.wikimedia.org/wiki/Research:Detox.

consider that a larger deviation in a system of scoring between −3 and 3 indicates a misleading judgment of the character of the comment. After this, in our experiments, we only consider the binary tag, i.e., if the comment is appropriate or not.

## 4.3 Results

For each of the two Wikipedia detox datasets, we trained the seven evaluated methods with the training subset, i.e., 75% of the data, and we assessed them on the test subset. We presented the achieved accuracy of each model on Table 3.

**Table 3.** Accuracies achieved by the seven tested methods on the Attack and the Aggression datasets of the Wikipedia Detox Project

| Method | Attack | Aggression |
|---|---|---|
| TF-IDF + LR | 0.922 | 0.923 |
| TF-IDF + SVM | 0.907 | 0.905 |
| TF-IDF + NB | 0.931 | 0.931 |
| BoW + LR | 0.919 | 0.917 |
| BoW + SVM | 0.898 | 0.899 |
| BoW + NB | 0.926 | 0.927 |
| StarSpace | **0.938** | **0.937** |

The results show that the achieved accuracy of each classifier is similar in both datasets. StarSpace achieved the best scores on the Attack and Aggression datasets, with an accuracy of 0.938 and 0.937, respectively. Among the tested Machine Learning methods, the best performance was achieved by the combination of TF-IDF and NB with an accuracy of 0.931 on each of the two datasets, only behind of the Deep Learning classifier by less than 1% in both datasets.

We can also observe that considering the six combination methods, our results are higher when using TF-IDF than when using BoW for each of the three classifiers (LR, SVM and NB). We consider two reasons behind this event. First, BoW relies on a dictionary, which means that if there are offensive words present in the comments but not in the dictionary then this method will fail to encode those violent words. However, TF-IDF does not rely on a previous knowledge on the words and thus is capable to considerate new terms. Second, TF-IDF gives each word a score based on the information that it provides through the IDF term, unlike BoW which only accounts the number of times that each word in the predefined dictionary appears in the comment, and therefore cannot differentiate the importance of different words.

Moreover, either using TF-IDF or BoW, NB outperforms other approaches while the lowest performance results are achieved with the SVM classifier. The advantage of the NB algorithm is explainable due to the nature of the features provided by the encoders, such as the number of times that a word appears or the information that a single word provides, i.e., the features are mostly independent from each other which is the basic idea behind the NB classifier.

## 5    Conclusions and Future Work

The increase of inappropriate content on the Internet over the last years is forcing the development of new tools to filter it. In this work, we evaluated the accuracy of the combination of two encoders with three Machine Learning classifiers, and a Deep Learning model to detect violent content on two datasets from the Wikipedia Detox Project: the Attack and the Aggression ones. We tested six classic combination methods resulting from selection one of two encoders, i.e. TF-IDF and BoW, and one out of three classifiers, i.e. Logistic Regression, Support Vector Machine and Naive Bayes. Additionally, these six methods were compared with the Deep Learning classifier StarSpace, developed by Facebook.

The highest scores were achieved by StarSpace in both datasets, with an accuracy of 0.938 on the Attack dataset and of 0.937 on the Aggression one. However, these results are only less than 1% higher than the achieved by the combination of the simple algorithms TF-IDF and Naive Bayes, which obtained an accuracy of 0.931 in both datasets.

The experimental results achieved in this work suggest that we can apply these models to develop real filters for social networks which have minors as potential users, and other areas where this type of comments is not relevant for the reader, such as YouTube where we can find a large number of comments without regulation.

In future works, we will try to improve the obtained results by developing new Deep Learning models. In addition, we will explore the use of these techniques to the detection of other kinds of inappropriate contents in text, e.g. sexual, terrorism-related, hatred contents, etc.

## References

1. Hussainalsaid, A., Azami, B.Z., Abhari, A.: Automatic classification of the emotional content of URL documents using NLP algorithms. In: Proceedings of the 18th Symposium on Communications & Networking, pp. 56–59 (2015)
2. Chin, H., Kim, J., Kim, Y., Shin, J., Yi, M.Y.: Explicit content detection in music lyrics using machine learning. In: IEEE International Conference on Big Data and Smart Computing, pp. 517–521 (2018)
3. Duarte, N., Llanso, E., Loup, A.: Mixed Messages? The Limits of Automated Social Media Content Analysis. In: FAT, vol. 106 (2018)
4. Mironczuk, M., Protasiewicz, J.: A recent overview of the state-of-the-art elements of text classification. Expert Syst. Appl. **106** (2016)
5. Bui, D.D.A., Del Fiol, G., Jonnalagadda, S.: PDF text classification to leverage information extraction from publication reports. J. Biomed. Inform. **61**, 141–148 (2016)

6. Chen, J., Huang, H., Tian, S., Qu, Y.: Feature selection for text classification with Naïve Bayes. Expert Syst. Appl. **36**(3), 5432–5435 (2009)
7. Rogati, M., Yang, Y.: High-performing feature selection for text classification. In: Proceedings of the Eleventh International Conference on Information and Knowledge Management, pp. 659–661 (2002)
8. Diab, D.M., Hindi, K.: Using differential evolution for fine tuning Naïve Bayesian classifiers and its application for text classification. Appl. Soft Comput. **54** (2016)
9. Chavan, V., Shylaja, S.: Machine learning approach for detection of cyber-aggressive comments by peers on social media network, pp. 2354–2358 (2015)
10. Hammer, H.: Automatic detection of hateful comments in online discussion. Ind. Netw. Intell. Syst., 164–173 (2017)
11. Eshan, S., Hasan, M.: An application of machine learning to detect abusive Bengali text. In: International Conference of Computer and Information Technology, pp. 1–6 (2017)
12. LeCun, Y., Bengio, Y., Hinton, G.: Deep learning. Nature **521**, 436–444 (2015)
13. Chu, T., Jue, K., Wang, M.: Comment abuse classification with deep learning. Stanford University (2016)
14. Badjatiya, P., Gupta, S., Gupta, M., Varma, V.: Deep learning for hate speech detection in tweets. In: International Conference on World Wide Web Companion, pp. 759–760 (2017)
15. Aizawa, A.: An information-theoretic perspective of TF-IDF measures. Inf. Process. Manag. **39**(1), 45–65 (2003)
16. Harris, Z.: Distributional structure. Word **10**(2–3), 146–162 (1954)
17. Cox, D.: The regression analysis of binary sequences. J. Roy. Stat. Soc. B **20**(2), 215–232 (1958)
18. Cortes, C., Vapnik, V.: Support-vector networks. Mach. Learn. **20**(3), 273–297 (1995)
19. McCallum, A., Nigam, K.: A comparison of event models for naive Bayes text classification. In: AAAI-98 Workshop on Learning for Text Categorization, vol. 752, no. 1, pp. 41–48 (1998)
20. Wu, L., Fisch, A., Chopra, S., Adams, K., Bordes, A., Weston, J.: StarSpace: embed all the things!. In: AAAI Conference on Artificial Intelligence, pp. 5569–5577 (2018)

# Classification of Human Driving Behaviour Images Using Convolutional Neural Network Architecture

Emine Cengil$^{(\boxtimes)}$ (iD) and Ahmet Cinar (iD)

Firat University, 23800 Elazig, Turkey
ecengil@firat.edu.tr

**Abstract.** Traffic safety is a problem that concerns the worldwide. Many traffic accidents occur. There are many situations that cause these accidents. However, when we look at the relevant statistics, it is seen that the traffic accident is caused by the behavior of the driver. Drivers who exhibit careless behavior, cause an accident. Preliminary detection of such actions may prevent the accident. In this study, it is possible to recognize the behavior of the state farm distracted driver detection data, which includes nine situations and one normal state image, which may cause an accident. The images are preprocessed with the LOG (Laplasian of Gaussian) filter. The feature extraction process is carried out with googlenet, which is the convolutional neural network architecture. As a result, the classification process resulted in 97.7% accuracy.

**Keywords:** Convolutional neural networks · GoogleNet ·
Laplasian of Gaussian · Classification

## 1 Introduction

With the rapid development of the automobile industry and the spread of private cars, the number of vehicles in traffic is increasing. The increase in the number of vehicles in traffic brings traffic accidents together. Traffic accidents often result in loss of life and property. Based on the fact that a large part of the accidents are caused by the steepness of the drivers, it is possible to prevent accidents by detecting the drivers and giving warnings in advance.

The field of image classification is one of the areas where deep learning methods are used. There are many methods used to incorporate images into the relevant class. The methods used depend on the type of data to be classified. However, deep learning is mostly preferred in the field of image classification.

Deep learning is one of the machine learning techniques. The structure of deep learning comes from artificial neural networks. It is based on learning the levels of representations that match a hierarchy of properties [1].

In the literature, many classification techniques are recommended for recognizing driver behavior. Nowadays, machine learning techniques are widely used in the multi-image classification problem. They provide high classification accuracy.

© Springer Nature Switzerland AG 2019
H. Pérez García et al. (Eds.): HAIS 2019, LNAI 11734, pp. 264–274, 2019.
https://doi.org/10.1007/978-3-030-29859-3_23

Škrjanc et al. [2], provides a developing cloud-based algorithm for the recognition of drivers' operations. The general idea is to detect different maneuvers by operating the standard signals measured in a car such as speed, revolutions, angle of the steering wheel, position of the pedals and others without additional intelligent sensors. The main purpose of this research is to propose a concept that can be used to recognize various driver actions. All experiments are done on a realistic car simulator.

Wang et al. [3], using a sample of the car's case, the multi-source and dynamic data of the human-vehicle environment under the different emotional states of drivers, experiments based on emotions in this study, real driving experiments and virtual driving experiments were obtained. The main factors affecting the typical driving emotions were removed by the factor analysis method and a feeling definition model was formed based on the fuzzy evaluation and PAD emotional model.

Olabiyi et al. [4], in the system they propose, contain camera-based information about the driving environment and the driver themselves, as well as the traditional vehicle dynamics. It then uses an deeper bidirectional repetitive neural network (DBRNN) to learn the relationship between sensory inputs and driver behavior that predicts accurate and high horizon action. The proposed system can correctly predict basic driver actions, including acceleration, braking, lane change and 5-second turns before the driver handles operation.

Braunagel et al. [5], provide a new approach to automatic recognition of drive efficiency. The proposed approach is evaluated on the data recorded during a driving simulator operation, with 73 people performing different secondary tasks while driving in an autonomous environment. The proposed architecture shows promising results in the recognition of in-vehicle driver effectiveness.

Yan et al. [6], propose a vision-based solution to recognize the driver's behavior based on evolutionary neural networks. Specifically, in an image verild-spindle, skin-like regions are subtracted by the Gaussian Mixture Model passed to a deep model of evolutionary neural networks (R * CNN) to produce action labels. We tested the proposed methods on the Southeast University Driving Posture Data Set and 97.76% of the average precision was obtained in the dataset which proved that the proposed method was effective in the drivers' recognition process.

The existing studies in the literature are carried out using many different machine learning algorithms. Their performances vary according to the data set and problem used.

It has been pointed out that traffic accidents are widespread and can lead to serious loss of life and property. The cause of accidents is mostly due to the careless behavior of the driver. In this study, driver behavior classification is made by focusing on this point. The googlenet network architecture was used when using the Matlab tool during the implementation of the method. System training is shown using the State Farm Distracted Driver Detection dataset.

The structure of this study is as follows; Sect. 2 describes the architectural structure of convolutional neural networks. Section 3 is the proposed method which includes the preprocessing, feature extraction and classification steps. Finally, Sect. 4 describes the results of the study.

## 2    Theoretical Background

### 2.1    Laplasian of Gaussian

Laplasian is a 2-D isotropic measure of the second spatial derivative of an image. The Laplacian of an image highlights areas of rapid density change and is therefore often used for edge detection. Laplacian is often applied to an image that is smoothed with something that approaches the gaussian smoothing filter to reduce its sensitivity to noise. The operator receives a single gray level image as the normal input and produces another gray level image as output. Laplacian and Gaussian are combined to form a single filter applied to the image [7].

Since the input image is represented as a series of individual pixels, there is a separate convolutional kernel that can approach Laplayian's second derivatives in the definition. Using one of these nuclei, Laplasian can be calculated using standard convolution methods.

These kernels are very sensitive to interference as they approach a second derivative measurement on the image. To counteract this, the image is usually Gaussian before applying the Laplacian filter. This preprocessing step reduces the high-frequency noise components before the differentiation step [8].

$$LOG(x, y) = -\frac{1}{\pi\sigma^4}\left[1 - \frac{x^2 + y^2}{2\sigma^2}\right] e^{-\frac{x^2+y^2}{2\sigma^2}} \tag{1}$$

The 2-D LoG function is centered at zero and the gauss standard deviation is as in (1).

### 2.2    Convolutional Neural Network

Convolutional neural networks (CNN) are multi-layered architectures based on generalized neural networks as "Deep Learning" [9]. The architecture of CNN is similar to that of the connection model of neurons in the Human Brain, and has been inspired by the visual cortex's organization. Individual neurons respond to stimuli only in a limited region of the visual field known as the Receptive field. A collection of such areas overlaps to cover the entire visual field [10].

The role of CNN, which is widely preferred in the image processing field, is to download images into an easy-to-compile form without losing the critical features to get a good estimate.

Images are high-dimensional vectors. Requires a large amount of parameters to characterize the network. CNNs aim to reduce the number of parameters and to adapt the network architecture, especially for visual tasks. The CNN is generally composed of a series of layers that can be grouped by its functions [11].

#### Convolutional Layer
Convolutional neural networks are composed of several layers. These layers are designed to perform separate functions. The purpose of applying the convolution layer, the first layer of the network, is to extract the properties of the processed data. Convolutional neural networks, thanks to the convolutional layers contained in the feature provides very good performance.

Filtering process is applied to input image in convolution layer. In a convolutional neural network, there are usually multiple layers of convolution. In the first layers more general features of the image are extracted. In later layers of the network, more distinctive features are extracted. Filtration is the mathematical process of several matrices. Since the filters used are different, the results are not the same as they express the difference of their digital content. The network is able to learn the values mentioned in the training process [12].

The convolution process is carried out by entering the new value by multiplying mutual pixels by circulating on the input image of the selected filter. The filter is completely circulated on the image matrix and the corresponding values are assigned to the output pixel. The result image generated by filtering is known as a property map. Parameters that change the size of property maps are depth, pitch size, and zero-padding [13]. Figure 1 shows the application of convolution process.

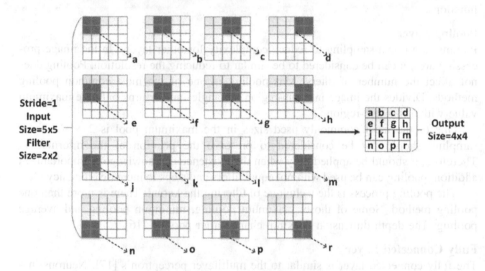

**Fig. 1.** Applying convolution operation

Depth refers to the number of filters used in convolution process. Step size is how many pixels we will jump when applying a mask to the image. In this case, if the size of the step is chosen large, the feature maps of smaller size are obtained. Another currency-meter that specifies the property map size, zero-padding, refers to filling the matrix representing the input image with zeroes. In this way, the filter can be applied to the edge pixels of the image. The dimension map of the feature map is also provided in the implementation of Zero-padding.

**Activation Layer**
Deep learning is generally used to solve non-linear problems. Deep learning is preferred because it is more successful in non-linear problems than other methods. The values obtained after the matrix multiplication in the convolution fold are linear. Activation functions are used to convert values to non-linearity. The hidden layer outputs of the network are normalized using the activation function.

**Table 1.** Activation functions and their formulas

| Activation function | Formula |
|---|---|
| Tanh | tanh(x) |
| ReLU | max(0, x) |
| LeakyReLU | max(0.1x, x) |
| ELU | x $\quad\quad$ x >= 0 <br> $\alpha(e^x - 1)$ $\quad$ x < 0 |
| Maxout | $\max(w_1^T + b1, \ w_2^T + b2)$ |

It is used to increase the non-linearity of the network without affecting the receiving areas of the convolution layers [14]. Tanh, Sigmoid, ReLU and LeakyReLU are some of the preferred activation functions in CNN models. Table 1 gives the activation functions.

**Pooling Layer**
Pooling is a down-sampling to reduce complexity for other layers. In the image processing area, it can be considered to be similar to reducing the resolution. Pooling does not affect the number of filters. Max-pooling is one of the most common pooling methods. Divides the image into sub-region rectangles and returns only the maximum value within that sub-region.

One of the most commonly used sizes in the maximum pool is $2 \times 2$. Down sampling should not be considered to preserve the position of the information. Therefore, it should be applied only when the existence of knowledge is important. In addition, pooling can be used with not equal filters and steps to increase efficiency [15].

The pooling process is the technique of filtering the kernel. There is more than one pooling method. Some of those; maximum pooling, minimum pooling and average pooling. The depth dimension does not change after pooling [16].

**Fully Connected Layer**
The fully connected layer is similar to the multilayer perceptron's [17]. Neurons in a fully connected layer have complete connections to all the activations in the previous layer, as seen in the Neural Networks. Thus activity can be calculated by a matrix multiplication followed by a bias offset. This layer depends on all neurons of the previous layer. The fully connected layer is the upper-level representation of the input image, the output obtained at the end of the convolution, activation and pooling layers previously applied. These layers are not expected to make classification estimates. The fully connected layer is used to classify the input image based on the training set by looking at the features.

# 3   Proposed Method

The proposed method is to perform the classification task of driver behavior. In the study, convolutional neural networks (CNN) were chosen based on the success of image processing. The training process was categorized with 10 classes of state farm distracted driver detection data set. Images were first preprocessed by laplasian gaussian filter. The proposed method is to perform the classification task of driver behavior. It then displays the pre-trained CNN architecture googlenet which was carried out by trained classification. The flowchart of the proposed method is as in Fig. 2.

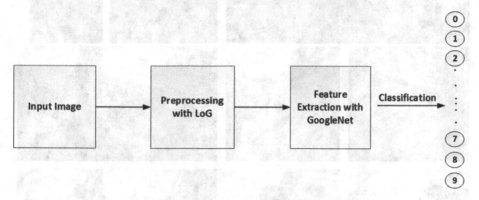

**Fig. 2.** Flowchart of proposed method

## 3.1 Dataset

Recognizing driver behavior is one of the ways to avoid accidents. State-related data is available. From this dataset, the state-farm distracted driver detection dataset was chosen in the training of the method (Table 2).

State farm distracted driver detection dataset contains ten different categories of driver behavior. These categories are; safe driving, texting – right, talking on the phone – right, texting – left, talking on the phone – left, operating the radio, drinking, reaching behind, hair and makeup and talking to passenger. Some images of the dataset are given in Fig. 3.

**Table 2.** Categories of dataset

| Class Name | Category |
| --- | --- |
| C0 | Safe driving |
| C1 | Texting – right |
| C2 | Talking on the phone-right |
| C3 | Texting – left |
| C4 | Talking on the phone – left |
| C5 | Operating the radio |
| C6 | Drinking |
| C7 | Reaching behind |
| C8 | Hair and makeup |
| C9 | Talking to passenger |

**Fig. 3.** Images from state farm distracted driver detection dataset

## 3.2    Preprocessing

Gaussian Laplacian is a filter that can be used to find a rapid change in pixel values in image data. It can smooth the Gaussian image in LoG and reduce its effect on noise. It can also compensate for the effect of increased noise caused by the second derivative of Laplacian. Gaussian's Laplacian can be used as appropriate properties of images by finding meaningful changes in pixels rather than using all pixel values that may contain meaningless data [18].

The raw images in the dataset were subjected to a preprocessing step with the laplasian Gaussian filter. Our aim is to reduce the area covered by the data and to perform the training process in a shorter time. Log_filter = fspecial ('log', [5, 5], 1.0) command with the sigma value 1 and hsize = [5, 5] is taken as. After filtering, the images are as shown in Fig. 4.

**Fig. 4.** Images from state farm distracted driver detection dataset after preprocessing

### 3.3    Experimental Results

GoogleNet, the first in the image classification competition, ILSVRC2014, was developed by researchers from the Google group, as its name suggests. Network architecture is quite different from VGGNet, ZFNet and AlexNet. In the middle of the network contains $1 \times 1$ Convolution. In addition, the overall average pooling is used at the end of the network instead of using fully connected layers. The network used a CNN inspired by LeNET-5 [19]. The architecture, consisting of 22 layers, reduced the number of parameters from 60 million to 4 million. In the study, googleNet was preferred. The accuracy function of the training performed in Matlab is shown in Fig. 5.

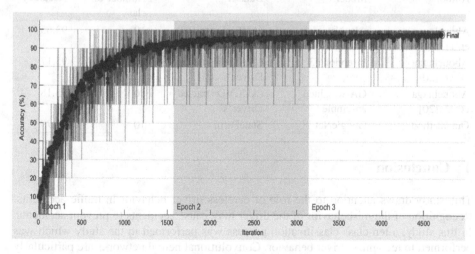

**Fig. 5.** Accuracy graphic of model

We trained our model using a NVIDIA Geforce GTX 950 M GPU. The system was completed in 1866 min with 4707 iterations by using single Gpu and its validation accuracy was 97.79%. The loss function of the system is as given in Fig. 6.

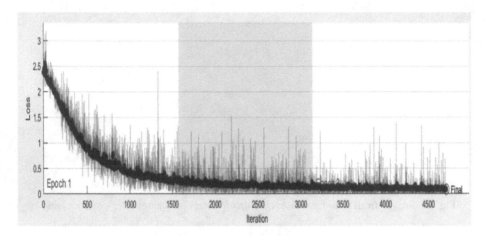

**Fig. 6.** Loss graphics of model

The proposed method was compared with the studies of Y. Abouelnaga et al. [20]. In the study, (AUC) Distracted Driver's dataset, which is a Statefarm dataset-like dataset was used. In addition to testing with the AlexNet and InseptionV3 models, a new method consisting of a genetic-weighted CNN population was also classified. The classification results are as in Table 3.

**Table 3.** Classification results

| Author | Model | Dataset | Number of class | Accuracy |
|--------|-------|---------|-----------------|----------|
| Abouelnaga et. al. [20] | AlexNet | (AUC) Distracted Driver's | 10 | 93.65 |
| Abouelnaga et. al. [20] | InseptionV3 | (AUC) Distracted Driver's | 10 | 95.17 |
| Abouelnaga et. al. [20] | GA-weighted ensemble | (AUC) Distracted Driver's | 10 | 95.18 |
| Our method | GoogleNet | StateFarm | 10 | 97.79 |

## 4   Conclusion

This study draws attention to the role of careless driver behavior in traffic accidents. Recognition of the misbehavior of the driver may cause accidents to prevent accidents. In this study, a ten-class classification process was performed in the study which was performed to recognize driver behavior. Convolutional neural networks are particularly well known for its success in image classification.

In the study, images were preprocessed with Laplacian Gaussian filter. Then googlenet architecture is used for classification. The proposed method showed 97.79% accuracy. In the study, the existing studies in the literature were examined and compared with the proposed method. Considering the large proportion of driver behavior in existing accidents, the study will be useful when used in accident prevention and autonomous vehicle production. The method can be developed and implemented in real time.

# References

1. Liu, T., Fang, S., Zhao, Y., Wang, P., Zhang, J.: Implementation of training convolutional neural networks. arXiv preprint arXiv:1506.01195 (2015)
2. Škrjanc, I., Andonovski, G., Ledezma, A., Sipele, O., Iglesias, J.A., Sanchis, A.: Evolving cloud-based system for the recognition of drivers' actions. Expert Syst. Appl. **99**, 231–238 (2018)
3. Wang, X., Liu, Y., Wang, F., Wang, J., Liu, L., Wang, J.: Feature extraction and dynamic identification of drivers' emotions. Transp. Res. Part F Traffic Psychol. Behav. **62**, 175–191 (2019)
4. Olabiyi, O., Martinson, E., Chintalapudi, V., Guo, R.: Driver action prediction using deep (bidirectional) recurrent neural network. arXiv preprint arXiv:1706.02257 (2017)
5. Braunagel, C., Kasneci, E., Stolzmann, W., Rosenstiel, W.: Driver-activity recognition in the context of conditionally autonomous driving. In: 2015 IEEE 18th International Conference on Intelligent Transportation Systems, pp. 1652–1657. IEEE (2015)
6. Yan, S., Teng, Y., Smith, J.S., Zhang, B.: Driver behavior recognition based on deep convolutional neural networks. In: 2016 12th International Conference on Natural Computation, Fuzzy Systems and Knowledge Discovery (ICNC-FSKD), pp. 636–641. IEEE (2016)
7. Kong, H., Akakin, H.C., Sarma, S.E.: A generalized Laplacian of Gaussian filter for blob detection and its applications. IEEE Trans. Cybern. **43**(6), 1719–1733 (2013)
8. Kong, H., Sarma, S.E., Tang, F.: Generalizing Laplacian of Gaussian filters for vanishing-point detection. IEEE Trans. Intell. Transp. Syst. **14**(1), 408–418 (2013)
9. Ma, W., Lu, J.: An equivalence of fully connected layer and convolutional layer. arXiv preprint arXiv:1712.01252 (2017)
10. Karpathy, A., Toderici, G., Shetty, S., Leung, T., Sukthankar, R., Fei-Fei, L.: Large-scale video classification with convolutional neural networks. In: Proceedings of the IEEE Conference on Computer Vision and Pattern Recognition, pp. 1725–1732 (2014)
11. Cengil, E., Çınar, A.: A new approach for image classification: convolutional neural network. Eur. J. Tech. **6**(2) (2016)
12. Accelerate Machine Learning with the cuDNN Deep Neural Network Library. https://devblogs.nvidia.com/accelerate-machine-learning-cudnn-deep-neural-network-library/
13. Han, S., Meng, Z., O'Reilly, J., Cai, J., Wang, X., Tong, Y.: Optimizing filter size in convolutional neural networks for facial action unit recognition. arXiv preprint arXiv:1707.08630, 26 July 2017
14. Agostinelli, F., Hoffman, M., Sadowski, P., Baldi, P.: Learning activation functions to improve deep neural networks. arXiv preprint arXiv:1412.6830 (2014)
15. Albawi, S., Mohammed, T.A., Al-Zawi, S.: Understanding of a convolutional neural network. In: 2017 International Conference on Engineering and Technology (ICET), pp. 1–6. IEEE, August 2017

16. Lee, C.Y., Gallagher, P.W., Tu, Z.: Generalizing pooling functions in convolutional neural networks: mixed, gated, and tree, In: Artificial Intelligence and Statistics, pp. 464–472 (2016)
17. An Intuitive Explanation of Convolutional Neural Networks. https://ujjwalkarn.me/2016/08/11/intuitive-explanation-convnets/
18. Uçar, M., Uçar, E.: Computer-aided detection of lung nodules in chest X-rays using deep convolutional neural networks. Sakarya Univ. J. Comput. Inf. Sci. 2(1), 41–52 (2019)
19. LeCun, Y.: LeNet-5, convolutional neural networks (2015). http://yannlecun.com/exdb/lenet
20. Abouelnaga, Y., Eraqi, H.M., Moustafa, M.N.: Real-time distracted driver posture classification. arXiv preprint arXiv:1706.09498 (2017)

# Building a Classification Model Using Affinity Propagation

Christopher Klecker[(✉)] and Ashraf Saad

Georgia Southern University, Savannah, GA 31419, USA
david.klecker@bcsav.net

**Abstract.** Regular classification of data includes a training set and test set. For example for Naïve Bayes, Artificial Neural Networks, and Support Vector Machines, each classifier employs the whole training set to train itself. This study will explore the possibility of using a condensed form of the training set in order to get a comparable classification accuracy. The technique we explored in this study will use a clustering algorithm to explore how the data can be compressed. For example, is it possible to represent 50 records as a single record? Can this single record train a classifier as similarly to using all 50 records? This thesis aims to explore the idea of how we can achieve data compression through clustering, what are the concepts that extract the qualities of a compressed dataset, and how to check the information gain to ensure the integrity and quality of the compression algorithm. This study will explore compression through Affinity Propagation using categorical data, exploring entropy within cluster sets to calculate integrity and quality, and testing the compressed dataset with a classifier using Cosine Similarity against the uncompressed dataset.

**Keywords:** Classification · Clustering · Clustering analysis ·
Condensed dataset · Prediction model · Affinity propagation · Exemplars ·
Damping factor · Preference value · Similarity matrix · Categorical data ·
Elbow method · Naïve-Bayes

## 1 Introduction and Objectives

Classification is a machine learning task which seeks to understand the features of the data to learn how to assign objects to a predefined category or class [1]. Classification seeks to split data into training and test sets where the training set is sent to a learning/decision function which builds the prediction model from which the test set can be applied and a classification measurement can be assessed [2]. The training set is a predefined series of records from the original dataset and is usually included as a whole to train the classifier.

This study seeks to find if classification accuracy is comparable when using the full training dataset versus a subset of the training dataset. For large datasets, this can be advantageous to help speed up the training process for classifiers as it is no longer necessary to have the whole training dataset to create a classification model. In creating this kind of data subset the following questions must be answered: how to define the data subset, if a data subset represents relevant records in a dataset, how to define these

© Springer Nature Switzerland AG 2019
H. Pérez García et al. (Eds.): HAIS 2019, LNAI 11734, pp. 275–286, 2019.
https://doi.org/10.1007/978-3-030-29859-3_24

relevant records? How can this relevancy be measured and proved? How far can the data be compressed before accuracy is reduced? This study aims to show the technique of this process of defining the relevant records of a dataset making up the subset of a training dataset used as a classification model, and will show the classification results are comparable to using the whole dataset.

# 2    Clustering to Find Relevant Records in a Dataset

## 2.1    What is Clustering?

Clustering is a methodology of data mining to learn meaningful patterns in a dataset. It is an exploratory technique which aims to discover an optimal grouping of data points [3]. Clustering seeks to find features of a dataset, and how those features can shape the entire dataset into meaningful groups which can represent patterns of interest. Because of this feature of clustering, it may be possible to discover relevant data records of a dataset by clustering the records and analyzing each cluster as a relevant feature.

Clustering algorithms can be centroid based, where the shape of the clusters surrounds a centroid or datapoint. They can be hierarchal based from which new clusters are descendants of a parent cluster. They can also be density based where clusters are determined based on the density of the data points mapped to a plane. For the purposes of this study, all datasets tested will be of mixed features, that is categorical and numerical values with an emphasis on categorical data and will be clustered using centroid based clustering algorithms as such algorithms when applied to categorical data offer unique results for returning relevant data records we seek.

## 2.2    Affinity Propagation – Clustering Through Message Passing

Affinity Propagation is a modern unsupervised clustering algorithm that "takes as input a collection of real-valued similarities between data points, where the similarity $s(i, k)$ indicates how well the data point $k$ is suited to be the exemplar for data point $i$" [4]. The centroids returned from Affinity Propagation are actual data points, and these data points are what will become the basis for what defines a relevant data point in a dataset. An example of a cluster returned with Affinity Propagation is shown in Fig. 1.

| age+(10.00) | menopause+(10.00) | tumor-size+(10.00) | inv-nodes+(10.00) | node-caps+(10.00) | deg-malig+(10.00) | breast+(10.00) | breast-quad+(10.00) | irradiat+(10.00) | recurrence+(10.00) |
|---|---|---|---|---|---|---|---|---|---|
| 40-49+(60.00) | premeno+(70.00) | 20-24+(30.00) | 0-2+(90.00) | no+(90.00) | 2+(90.00) | left+(90.00) | left_low+(40.00) | no+(80.00) | no-recurrence-events+(90.00) |
| 40-49+(60.00) | premeno+(70.00) | 15-19+(10.00) | 0-2+(90.00) | no+(90.00) | 2+(90.00) | left+(90.00) | left_low+(40.00) | no+(80.00) | no-recurrence-events+(90.00) |
| 30-39+(30.00) | premeno+(70.00) | 30-34+(30.00) | 0-2+(90.00) | no+(90.00) | 2+(90.00) | left+(90.00) | left_up+(30.00) | no+(80.00) | no-recurrence-events+(90.00) |
| 30-39+(30.00) | premeno+(70.00) | 20-24+(30.00) | 0-2+(90.00) | no+(90.00) | 2+(90.00) | left+(90.00) | left_low+(40.00) | no+(80.00) | no-recurrence-events+(90.00) |
| 30-39+(30.00) | premeno+(70.00) | 25-29+(20.00) | 0-2+(90.00) | no+(90.00) | 2+(90.00) | left+(90.00) | left_low+(40.00) | no+(80.00) | no-recurrence-events+(90.00) |
| 40-49+(60.00) | premeno+(70.00) | 20-24+(30.00) | 0-2+(90.00) | no+(90.00) | 2+(90.00) | left+(90.00) | left_up+(30.00) | no+(80.00) | no-recurrence-events+(90.00) |
| 40-49+(60.00) | premeno+(70.00) | 30-34+(30.00) | 0-2+(90.00) | no+(90.00) | 2+(90.00) | left+(90.00) | left_low+(40.00) | no+(80.00) | no-recurrence-events+(90.00) |
| 40-49+(60.00) | ge40+(20.00) | 25-29+(20.00) | 0-2+(90.00) | no+(90.00) | 2+(90.00) | left+(90.00) | left_low+(40.00) | no+(80.00) | no-recurrence-events+(90.00) |
| 40-49+(60.00) | ge40+(20.00) | 30-34+(30.00) | 0-2+(90.00) | no+(90.00) | 2+(90.00) | left+(90.00) | left_up+(30.00) | yes+(10.00) | no-recurrence-events+(90.00) |

Fig. 1.  Cluster example from affinity propagation

The cluster size is 9 records with the centroid or exemplar record listed as the first record shown. Each feature of this exemplar represents a majority characterized from all the records' features in each attribute. For example in Attribute "Age", the exemplar contains the feature "40–49". This feature is contained in 60% of the records in the cluster. Therefore "40–49" represents the majority of the features for this attribute in this cluster. This can be shown for all attribute features of the cluster. Therefore, it is possible to represent this entire cluster through the exemplar. What was 9 records can be represented in just one record, the exemplar.

The goal of Affinity Propagation is to discover these exemplars through a process of message passing. This message passing algorithm is based on the Sum Product algorithm and aims to find a maximal value of the responsibility and availability of each data point [4]. The responsibility is the measurement of a data point's ability to be assigned to a cluster, and the availability is the measurement of a data point's ability to be labeled as an exemplar for data points assigned to it. This procedure is illustrated in Fig. 2.

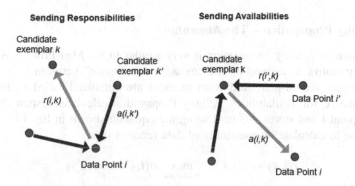

**Fig. 2.** Message passing in affinity propagation. From "Clustering By Passing Messages Between Data Points". [4]

Affinity Propagation runs iteratively through each data point passing messages and singling out exemplars until convergence is achieved. This is when the availabilities and responsibilities for each data point no longer update. At each iteration exemplars will emerge, and clusters will form. When convergence is reached a set of exemplars is selected and the clusters take their final shape. An example of convergence of Affinity Propagation is depicted in Fig. 3.

Accuracy is also very important in the clustering algorithm due to the centroids needing to represent as close as possible the data assignments of the cluster. Therefore choosing an algorithm that results in low error is preferable and Affinity Propagation will return errors that are lower compared with other clustering algorithms [5]. For these reasons, this study will focus on Affinity Propagation to create clusters from the datasets generated and use its exemplars for our prediction model.

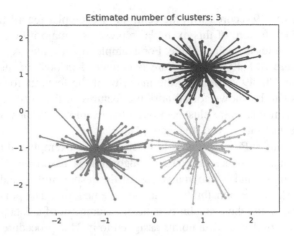

**Fig. 3.** Convergence in affinity propagation. From "Clustering By Passing Messages Between Data Points" [4].

### 2.3    Affinity Propagation – The Algorithm

The algorithm of Affinity Propagation is very similar to the Max-Product Algorithm. A similarity matrix S, defined by the "measure of distance" between two records is defined for each record-pair. Additional matrices are initialized called R, for responsibilities, and A, for availabilities. Affinity Propagation calculates responsibilities for each data point i and stores this in R using the equation shown in Eq. 1.

Equation to calculate responsibility of data record i [4].

$$r(i,k) \leftarrow s(i,k) - \max_{k' s.t. k' \neq k} \{a(i,k') + s(i,k')\} \tag{1}$$

Once all responsibilities are calculated the availability matrix must be updated. This is done using the equations shown in Eq. 2.

Equation for self-availability of data record k [4].

$$a(k,k) \leftarrow \sum_{i' s.t. \ i' \neq k} \max\{0, r(i',k)\} \tag{2}$$

Each iteration of Affinity Propagation will update the matrices until the values no longer update or update below a threshold leading to convergence.

Equation for calculating availability of data record i [4].

$$a(i,k) \leftarrow \min\left\{0, r(k,k) + \sum_{i' s.t. \ i' \notin \{i,k\}} \max\{0, r(i',k)\}\right\} \tag{3}$$

Once convergence is reached the sum of the self-availability, a(k, k) and self-responsibility, r(k, k), will determine if a data point is labeled as an exemplar.

These points represent the diagonals of the availability matrix and responsibility matrix. The responsibility matrix is used to determine which data points are assigned to these exemplars. It is due to this final procedure that Affinity Propagation returns a natural clustering result whereas other clustering algorithms initialize number of clusters to be returned a priori. Therefore, Affinity Propagation will return the number of clusters as determined by the self-availability and self-responsibility. It is possible the algorithm will not achieve convergence. This means the values for responsibility and availability are still being updated. As a result the assignments and exemplar assignments may differ if Affinity Propagation is run again. Therefore convergence is sought for any dataset as the clusters and exemplars returned will always be the same based on the damping factor and preference. Affinity Propagation has a time complexity of $O(k * n^2)$, where n is the number of records and k represents the number of iterations [6].

## 2.4    Adapting Clusters of Affinity Propagation - Preference Value

The preference value is what helps Affinity propagation know what data points to label as exemplar. If the summation of the self-responsibly and self-availability of the data point exceeds the preference value, the data point is labeled exemplar. If the preference value is too low, many data points will be labeled as exemplar thus not compressing the data enough. If the preference value is too high, too few data points will be labeled as exemplar, thus compressing the data too much and losing accuracy. Therefore it is important to determine the ideal preference value to achieve a good compression of data without losing accuracy. This can be accomplished through entropy assuming monotonic cluster formation. That is, assuming if you were to calculate the error in a series of cluster formations through Affinity Propagation using different preference values, you can use entropy to calculate the variance of change as long as this variance follows a progressive curve and does not randomly spike and dip.

## 2.5    Evaluating Preference Value Through Entropy Using the Elbow Method

The Elbow Method calculates the variance of change [8]. We can measure the variance of change by running Affinity Propagation multiple times on multiple values of a preference value. Each time Affinity Propagation is run with a new preference value, there is potential of a different clustering result. Clusters by nature will have an error which can be measured. This error is calculated by measuring the similarity between the exemplar and data points assigned of the cluster. The value from the Similarity matrix can be used for this measurement. By getting all similarity measurements between each data label of the cluster and its exemplar and computing the mean, the result is the average error of the cluster. Repeating this process for all clusters created through Affinity Propagation based on a particular preference value, then computing the mean of each cluster error, we obtain the mean error of entire cluster result for the preference value used. Entropy can be measured as a function of change between two different cluster results using different preference values resulting in a measure of information gain. The goal is to discover which preference value returns the highest information gain assuming the results for each Affinity Propagation run returns a monotonic cluster formation. This can be achieved by only including in our series of entropy measurements the preference values which return convergence. An example of this technique is shown in Fig. 4.

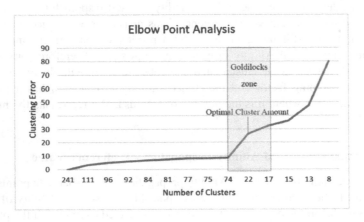

**Fig. 4.** Elbow method analysis of 3400 records of data

The x-axis represents the number of clusters formed given a specified preference value, in this case the preference values started at −8 and ended as 2. The y-axis is the measured error as a result of this returned clustered dataset. We want to discover what preference value results in high information gain, which is found by measuring the entropy at each preference value. The elbow method tells us the point where the elbow joint is greater, that is, which elbow angle is the smallest which is the preference value containing the most information gain. The Goldilocks Zone represents the preference values that will return a group of clusters with the highest information gain as this zone contains the largest entropy values.

## 2.6    Similarity of Two Records Containing Categorical Data as Features

To measure the similarity between two records is data dependent. Records containing numerical data can use Euclidian Square Distance [7] to measure the distinct distance value between the two points in N-Dimensional space. Categorical data requires a different approach to measuring similarity as the application of "distance" is not possible. For example, the distance between Male and Female cannot be mapped or visualized in N-dimensional space.

Proposed by Amir Ahmad and Lipika Dey from, "A K-Mean Clustering Algorithm for Mixed Numeric and Categorical Data" [9], the technique uses conditional probability to find a similarity measurement between two categorial values x and y in attribute i and an additional categorical pair found in another attribute. An example shown in Eq. 4 where w represents a subset of values and $\sim$ w is the complement of w of $Aj$ which maximizes the value of $P_i(\omega|x) + P_i(\sim \omega|y)$. An example shown in Eq. 4, for x = Male, and y = Female, the distance between Male and Female requires the conditional probability of all categories in attributes $j \neq i$. If the categories in attribute j are *Yes*, and *No*, then,

Distance Between Attribute values x and y for $A_i$ with respect to Attribute $A_j$ [9].

$$\delta^{ij}(x,y) = P_i(\omega|x) + P_i(\sim \omega \mid y) \tag{4}$$

Example of Eq. 4.

$$\delta ij(\text{Male}, \text{Female}) = \max(P1(\text{Yes}|\text{Male}), \ P1(\text{Yes}|\text{Female})) \\ + \max(P1(\text{No}|\text{Male}), \ P1(\text{No}|\text{Female})). \tag{5}$$

We expand the definition to include all categorical pairs found in each attribute and divide the summation of the values by the number of attributes shown by Eq. 6.

Equation to calculate all conditional probabilities for each attribute pair [9].

$$\delta(x,y) = \left(\frac{1}{m-1}\right) \sum_{j=1...m, i\neq j} \delta^{ij}(x,y) \tag{6}$$

The final step shown in Eq. 7 defines the distance between two records as the summation of the squares of the distances between each categorical pair $X_t$ and $Y_t$ for each attribute t in record D1 and D2.

Equation for the distance between record D1 and D2 [9].

$$Dist(D1, D2) = \sum_{t=1}^{m} (\delta(X_t, Y_t))^2 \tag{7}$$

## 3 Classification Model and Prediction Function

Any classification model needs three properties: a training set, a test set, and a prediction model, which is a classification algorithm like Naïve Bayes, Artificial Neural Networks, and so on. The training set for our classification model will be the collection of all returned exemplars. What exemplars are we seeking? If we are looking to classify the data we seek the features which represent each class in the data. Therefore, if we have a binary classifier, then we seek the exemplars which represent Class A and the exemplars which represent Class B. Therefore, the dataset is split on these classes to create two datasets: one with records equal to class A and the other with records equal to class B. Each dataset is then passed to Affinity Propagation and exemplars obtained. These qualities are stored in a file and labeled accordingly so that our classification model can recognize which qualities make up one class versus another. An example is shown in Fig. 5 using Breast Cancer Dataset as a core study. This represents our training set which represents a compressed dataset from the original.

Class_A = [["40-49","premeno","0-4","0-2","no","3","left","central","no","no-recurrence-events"],
["50-59","ge40","25-29","0-2","no","2","left","left_low","no","no-recurrence-events"],
["60-69","lt40","30-34","0-2","no","1","left","left_low","no","no-recurrence-events"],
["50-59","premeno","20-24","0-2","no","1","left","left_low","no","no-recurrence-events"],
["40-49","premeno","15-19","0-2","no","2","left","left_low","no","no-recurrence-events"],
["40-49","premeno","20-24","0-2","no","1","right","right_up","no","no-recurrence-events"],
["70-79","ge40","20-24","0-2","no","3","left","left_up","no","no-recurrence-events"],
["70-79","ge40","40-44","0-2","no","1","right","left_up","no","no-recurrence-events"],
["50-59","ge40","15-19","0-2","no","1","right","central","no","no-recurrence-events"],
["20-29","premeno","35-39","0-2","no","2","right","right_up","no","no-recurrence-events"],
["50-59","premeno","40-44","0-2","no","2","left","left_up","no","no-recurrence-events"],
["60-69","ge40","20-24","0-2","no","1","left","left_low","no","no-recurrence-events"],
["60-69","ge40","15-19","0-2","no","2","right","left_up","no","no-recurrence-events"],
["40-49","premeno","10-14","0-2","no","1","right","right_low","no","no-recurrence-events"],
["40-49","premeno","30-34","0-2","no","2","right","right_low","no","no-recurrence-events"],
["30-39","premeno","30-34","0-2","no","3","left","left_low","no","no-recurrence-events"],

Class_B = [["60-69","ge40","30-34","0-2","no","3","right","central","no","recurrence-events"],
["40-49","premeno","30-34","15-17","yes","3","left","left_low","no","recurrence-events"],
["60-69","ge40","40-44","3-5","yes","3","right","left_low","no","recurrence-events"],
["50-59","ge40","35-39","0-2","no","2","left","left_low","no","recurrence-events"],
["30-39","premeno","35-39","0-2","no","3","left","left_low","no","recurrence-events"],
["40-49","ge40","30-34","3-5","no","3","left","left_low","no","recurrence-events"],
["60-69","ge40","20-24","3-5","no","2","left","left_low","yes","recurrence-events"],
["40-49","premeno","20-24","3-5","yes","2","left","left_low","yes","recurrence-events"],
["50-59","premeno","25-29","3-5","yes","3","left","left_low","yes","recurrence-events"],
["40-49","premeno","30-34","0-2","no","1","left","left_low","yes","recurrence-events"],
["40-49","premeno","25-29","9-11","yes","3","right","left_up","no","recurrence-events"],
["60-69","ge40","30-34","3-5","yes","2","left","central","yes","recurrence-events"],
["40-49","ge40","25-29","10-14","yes","3","left","right_low","yes","recurrence-events"],
["40-49","premeno","30-34","0-2","yes","3","right","right_up","no","recurrence-events"],
["60-69","ge40","30-34","0-2","yes","2","right","right_up","yes","recurrence-events"]]]

**Fig. 5.**  Training set file where exemplars are split and stored in arrays.

To test the classification model, a test record is sent to compute which class the test record belongs. To classify the record this paper uses Cosine Similarity [10] shown in Eq. 8.

Cosine Similarity Equation

$$similarity = \cos(\theta) = \frac{A \cdot B}{\|A\|\|B\|} = \frac{\sum_{i=1}^{n} A_i B_i}{\sqrt{\sum_{i=1}^{n} A_i^2} \sqrt{\sum_{i=1}^{n} B_i^2}} \tag{8}$$

Cosine Similarity uses vector dot products which can be represented using binary values. Therefore we convert the records into a vector representation. This can be done by using a One Hot Encoding process. This process takes the number of features found in the compressed data set, particular to the class to represent the number of bits. For example, if attribute 2 contains 3 features, then the number of bits is 3. Each feature will toggle one of the bits. For example, attribute feature "premeno" will be represented as 100, attribute feature "lt40" is represented as 010, and attribute feature "ge40" is represented as 001. By expressing each attribute into its One Hot Encoded value and combining the encoded values together we get a binary vector representation of the data record. The binary vector representation for a test record is shown in Fig. 6 as well as a binary vector representation for a training set record, shown in Fig. 7.

001000001000100010000010001010001001

**Fig. 6.**  One hot encoding of whole test record.

010000001000100010000101000100100100

**Fig. 7.** One hot encoding of first training record from Fig. 5.

Taking the dot product of both vectors according the numerator of Cosine Similarity. Each vector pair both containing "1" will remain and the summation becomes the numerator. The denominator multiplies the magnitude of both vectors, taking each vector value squaring it and adding them together. Since the vectors only contain 1s, the summation will equal the number of 1s in the vector. Because both records will contain the same number of attributes and each attribute converted to One-Hot Encoding will have one of its bits converted to "1", the vector will always contain the same number of 1s as the number of attributes. Therefore the magnitude of a vector is the number of vectors, which becomes the denominator. Thus Cosine Similarity will return a value which calculates the number of matches over the number of attributes. We match the test record to each record in the training set for Class A, sum them together and calculate the mean. The process is done for all records in Class B with the same test record. A mean value for both classes will emerge for the test record. Whichever class returns the highest mean value is the class for the record.

## 4 Results

### 4.1 Breast Cancer Dataset

The first dataset tested is a breast cancer study done in 1992 by Dr. William H. Wolberg from the University of Wisconsin, Madison [11]. The number of records in the data set is 699 with 10 attributes and a binary classification for benign or malignant. The dataset was split on the classes, a preference value was calculated using the elbow method, a similarity matrix calculated using conditional probabilities between each categorical pair, and Affinity Propagation was run on both datasets with exemplars returned. These exemplars represent a condensed dataset of the most relevant qualities that represent 37% of the whole training dataset. The results of running cosine similarity on test records with this condensed dataset are shown in Table 1 along with comparison classification runs which use the full training dataset using Naïve Bayes and Cosine Similarity.

**Table 1.** Breast cancer & Mushroom dataset.

| | Breast cancer dataset | | | Mushroom dataset | | |
|---|---|---|---|---|---|---|
| | AP condensed | Naïve Bayes | Cosine similarity | AP condensed | Naïve Bayes | Cosine similarity |
| Precision | 86% | 83% | 88% | 85% | 56% | 83% |
| Recall | 98% | 94% | 99% | 97% | 93% | 96% |
| Accuracy | 88% | 84% | 91% | 90% | 59% | 88% |
| F-Score | 92% | 88% | 93% | 91% | 70% | 89% |

Table 1 illustrates that a condensed dataset of 37% of the total dataset still provides a comparable classification model. AP Condensed stands for "Affinity Propagation Condensed" and uses Cosine Similarity as a prediction model as well.

## 4.2  Mushroom Dataset

The second dataset comes from the same repository, created by J. Schillmmer, from the Audubon Society Field Guide to North American Mushrooms, contains over 8000 records of different types of mushrooms with 22 attributes of features of mushrooms [12]. The classification used in this test are Poisonous or Edible. In this case the training set is 7% of the total training set. The results are shown in Table 1 along with comparison classification runs which use the full training dataset using Naïve Bayes and Cosine Similarity.

A final dataset comes from a private institution, Saint Vincent College in Latrobe Pennsylvania. The records consist of around 19000 enrollment records with 15 attributes. The data was truncated to 8000 records as 17000 out of the 19000 records are related to a single class feature. This would have biased the data greatly. Therefore the data was truncated so the class distribution is 25% for "Applied" and 75% for "No Decision". Removing more records with the Attribute feature "No Decision" in my mind might corrupt the data. The idea of this prediction is to decide if a student who fills out an inquiry form will formally apply to the school. The results of this test are shown in Table 2.

**Table 2.**  Student enrollment dataset.

|           | AP condensed | Naïve Bayes | Cosine similarity |
|-----------|--------------|-------------|-------------------|
| Precision | 98%          |             | 97%               |
| Recall    | 77%          |             | 74%               |
| Accuracy  | 94%          | 77%         | 93%               |
| F-Score   | 86%          |             | 84%               |

# 5  Conclusions and Future Work

## 5.1  Conclusions

This paper illustrates a novel way to build a training set for classification purposes by using a reduced dataset represented as exemplars from the output of Affinity Propagation. It shows how to discover the relevant data points using clustering by extracting exemplars of the clustered dataset by using Affinity Propagation. It shows that by calculating entropy through clustering error, the exemplars chosen can represent the highest information gain, resulting in comparable or better accuracy at times when training a classifier. Therefore for classification purposes concerning categorical data, it is not necessary to have the entire training set, only a subset of the training set is required. However, it is important that relevant data records are chosen to represent this subset.

The training dataset is also shown to be transparent and thus visually understood and practical. It can be used as a means to describe the data through the actual features from the database itself. Naïve Bayes offers this transparency as well as classification is done through conditional probability. Other classification tools like Artificial Neural Networks and Support Vector Machines use weights or N-Dimensional representation of the classes, therefore the explanation of features making up the classes is usually represented by a black box, a set of features not readily understood without further analysis.

Weaknesses of the current technique are spare datasets where attributes are blank. This technique requires a clean, detailed dataset with all attributes with data values. Small datasets do not produce excellent accuracy. For example, an initial breast cancer dataset taken from the UCI Repository, the dataset contains 286 records, 9 attributes and all data records are categorical [13]. Due to the small size of the dataset the accuracy of this technique scored 61%. A cosine similarity classification produced an accuracy of 72%. Table 1 shows a breast cancer study of 8000 results scoring 88% accuracy therefore showing this technique produces better results with larger data.

## 5.2   Future Work

Expanding this technique to be able to include numeric attributes would lend itself well to many more datasets. The categorical algorithm used to calculate a categorical pair's distance can be expanded to include numerical values [6]. Related to the Affinity Propagation, other ways to zero in on an ideal damping factor that could help increase accuracy. The damping factor controls oscillations occurring when the responsibility and availably values are updating. Controlling the damping factor can help achieve convergence more efficiently and can be adjusted in the case where convergence does not occur with a default setting. The preferred setting to the damping factor is 0.5 but can be raised to 0.9. Finding an ideal preference value will ultimately increase the strength of the training set providing better accuracy in training during the classification process. These ideas all relate to the dynamic behavior of the technique which leads to the need for control theoretic or adaptive techniques. Because of Affinity Propagation's strong connection to the values of the preference and damping factor, it is possible to get entirely different cluster returns, and in some cases no cluster returns at all. By applying adaptive techniques to these dynamic elements which can gauge the damping factor's effect on convergence, and the preference value's effect on the quality of the returned clusters, it is possible to increase the accuracy of the exemplars chosen and thus increase the performance and accuracy of the classifier.

# References

1. Mulier, F.M., Cherkassky, V.: Learning From Data. Wiley-IEEE Press (2007)
2. Kalechofsky, H.: A simple framework for building predictive models. M Squared Consulting (2016). http://www.msquared.com/wp-content/uploads/2017/01/A-Simple-Framework-for-Building-Predictive-Models.pdf

3. Kumar, V.: Introduction to data mining. In: Cluster Analysis: Basic Concepts and Methods. Pearson (2005)
4. Dueck, D., Frey, B.J.: Clustering by passing messages between data points. Sci. Mag. **315**, 972–976 (2007)
5. Trono, J., Kronenberg, D., Redmond, P.: Affinity propagation, and other data clustering techniques, Saint Michael's College. http://academics.smcvt.edu/jtrono/Papers/SMCClustering%20Paper_PatrickRedmond.pdf
6. Refianti, R., Mutiara, A.B., Gunawan, S.: Time complexity comparison between affinity propagation algorithms. J. Theor. Appl. Inf. Technol. **95**(7), 1497–1505 (2017)
7. Barrett, P.: Euclidian distance. Technical Whitepaper (2005). https://www.pbarrett.net/techpapers/euclid.pdf
8. Kumar, A., Bholowalia, P.: EBK-means: a clustering technique based on Elbow Method and K-means in WSN. Int. J. Comput. Appl. **105**(9), 17–24 (2014)
9. Dey, L., Ahmad, A.: A k-mean clustering algorithm for mixed numeric and categorical data. Data Knowl. Eng. **63**, 503–527 (2007)
10. Garcia, E.: Cosine Similarity Tutorial, 04 October 2015. http://www.minerazzi.com/tutorials/cosine-similarity-tutorial.pdf. Accessed 15 Sept 2018
11. Wolberg, W.H.: UCI machine learning repository. University of Wisconsin Hospitals, Madison WI, 15 July 1992. https://archive.ics.uci.edu/ml/datasets/Breast+Cancer+Wisconsin+%28Original%29. Accessed 01 Mar 2019
12. Schlimmer, J.: UCI machine learning repository. The Audubon Society Field Guide to North American Mushrooms, 27 April 1987. https://archive.ics.uci.edu/ml/datasets/Mushroom. Accessed 01 Mar 2019
13. Soklic, M., Zwitter, M.: Breast cancer data set. UCI Machine Learning Repository, 11 July 1988. https://archive.ics.uci.edu/ml/datasets/BreastCancer. Accessed 15 Apr 2019

# Clustering-Based Ensemble Pruning and Multistage Organization Using Diversity

Paweł Zyblewski(✉) and Michał Woźniak

Department of Systems and Computer Networks, Faculty of Electronics,
Wrocław University of Science and Technology,
Wybrzeże Wyspiańskiego 27, 50-370 Wrocław, Poland
{pawel.zyblewski,michal.wozniak}@pwr.edu.pl

**Abstract.** The purpose of ensemble pruning is to reduce the number of predictive models in order to improve efficiency and predictive performance of the ensemble. In clustering-based approach, we are looking for groups of similar models, and then we prune each of them separately in order to increase overall diversity of the ensemble. In this paper we propose two methods for this purpose using classifier clustering on the basis of a criterion based on diversity measure. In the first method we select from each cluster the model with the best predictive performance to form the final ensemble, while the second one employs the multistage organization, where instead of removing the classifiers from the ensemble each classifier group makes the decision independently. The final answer of the proposed framework is the result of the majority voting of the decisions returned by each group. Experimentation results validated through statistical tests confirmed the usefulness of the proposed approaches.

**Keywords:** Ensemble pruning · Classifier ensemble · Clustering · Multistage organization

## 1 Introduction

Ensemble methods have been a very popular and fast-growing research topic. Their success is related to the fact that they offer solutions to learning problems, such as improving classification accuracy, learning from distributed data sources and learning from data streams [8]. In contrast to classic approach, where one classifier is trained for a given problem, ensemble methods construct multiple models from the training data and combine them. Each classifier ensemble consists of a number of the base learners, that are trained in such a way to ensure their appropriate diversity [14]. An ensemble may consist of either homogeneous or heterogeneous classifiers. Homogeneous models derive from different executions of the same learning algorithm and heterogeneous models derive, e.g., from running different learning algorithms on the same training data.

By combining classifiers we aim to achieve better classification accuracy at the expense of increased complexity. Instead of looking for the best set of features

H. Pérez García et al. (Eds.): HAIS 2019, LNAI 11734, pp. 287–298, 2019.
https://doi.org/10.1007/978-3-030-29859-3_25

and the best classifier, we look for the best set of classifiers and the best combination rule. Typically, an ensemble is constructed in two steps, i.e., generating the base learners, and then combining them (e.g., by voting or stacked generalization). To obtain a valuable ensemble, the base learners should be accurate and diverse.

This work focuses on the issue of classifier ensemble pruning (specifically clustering-based methods), in which we aim to reduce the number of the base learners in the ensemble. Thanks to this, it is possible to significantly reduce the computational overhead, as well as improve the predictive performance. An example of other reduction methods used to avoid issues like over-fitting in high dimensional space, may be the Principal Component analysis [13]. PCA is a method used to reduce number of variables in your data by extracting important one from a large pool.

In [16] the following taxonomy of ensemble pruning methods was proposed:

- Ranking-based pruning is conceptually the simplest method, which orders the individual models of the ensemble according to an evaluation function and select models in the front part.
- Optimization-based pruning formulates the ensemble pruning problem as an optimization problem which aims to find the subset of the original ensemble that optimizes a measure indicative of its generalization performance. Optimization techniques such as heuristic methods have been used.
- Clustering-based pruning identifies prototype base learners that are representative yet diverse among a given pool of classifiers, and then use only these prototypes to constitute the new ensemble. A clustering process is employed to find groups of base learners, where individual learners in the same group behave similarly while different groups have large diversity. Then, from each cluster, the prototype learner is selected, which is placed in the final ensemble.

This paper is focusing on the last approach and its main contributions are as follows:

- The proposition a novel measure based on the non-pairwise and averaged pairwise diversity, which allows to evaluate an impact of a particular model on the diversity of a given classifier ensemble.
- The formalization of an algorithm that uses the proposed measure for ensemble pruning and multistage organization of majority voting.
- An extensive experimental analysis on a large number of dataset benchmarks comparing the performance of proposed methods to the state-of-the-art ensemble methods.

The manuscript is organized as follows. Section 2 presents the papers relevant to presented work. Then the details of the proposed clustering-based ensemble pruning and multistage organization methods are discussed. The next section presents the experimental study and the last part summarizes the concluding remarks.

## 2    Related Works

Because this work focuses on employing clustering-based classifier ensemble pruning methods to improve a predictive performance of combined classifiers then let us briefly present the main works related to the problem under consideration. Basically, clustering-based pruning consists of the two steps. The first one groups base classifiers into a number of clusters on the basis of a criterion, which should take into consideration their impact on the ensemble performance. For this purpose, various clustering methods were used, such as hierarchical agglomerative clustering [3], $k$-means clustering [2,10], deterministic annealing [1] and spectral clustering [15] and most of them employ a kind of diversity based criteria. Giacinto et al. [3] estimated the probability that classifiers do not make coincident errors in a separate validation set, while Lazarevic and Obradovic [10] used the Euclidean distance in the training set. Kuncheva proposed employing matrix of pairwise diversity for hierarchical and spectral methods [9].

In the second step, prototype base learner is selected from each cluster. In [1] a new model was trained for each cluster, based on clusters centroids. Classifier, that is the most distant to the rest of clusters is selected in [3]. In [10] models were iteratively removed from the least to the most accurate. Model with the best classification accuracy was chosen in [2].

The last issue is the choice of the number of clusters. This could be determined based on the performance of the method on a validation set [2]. In case of fuzzy clustering methods, we can use indexes based on membership values and dataset or statistical indexes to automatically select the number of clusters [7].

The alternative proposal is multiple stage organization, which was briefly mentioned in [5] and described in detail by Ruta and Gabrys [12], where authors refer to such systems as a multistage organization with majority voting (MOMV) since the decision at each level is given by majority voting. Initially all outputs are allocated to different groups by permutation and majority voting is applied for each group producing single binary outputs, forming the next layer. In the next layers exactly the same way of grouping and combining is applied with the only difference being that the number of outputs in each layer is reduced to the number of groups formed previously. This repetitive process is continued until the final single decision is obtained. In this research we employ this approach but to form groups of voting classifier we use clustering methods.

## 3    Proposed Methods

In this section we propose two methods for increasing ensemble's accuracy using clustering based on diversity-based criterion.

### 3.1    Diversity

Diversity is one of a key factors for generating a valuable classifier ensemble, but the main problem is how to measure it. In this work we decided to use the

**Table 1.** A table of the relationship between a pair of classifiers.

|  | $\Psi_k$ correct (1) | $\Psi_k$ wrong (0) |
|---|---|---|
| $\Psi_i$ correct (1) | $N^{11}$ | $N^{10}$ |
| $\Psi_i$ wrong (0) | $N^{01}$ | $N^{00}$ |

diversity-based criterion for base classifier clustering. Basically, known diversity measures may be divided into two groups: pairwise and non-pairwise diversity measures. Pairwise diversity measures determine the diversity between pair of base models, ensemble consisting of $L$ classifiers will have $L(L-1)/2$ values of pairwise diversity. To get the value for the entire ensemble, we calculate the average. Non-pairwise measures take into consideration all base classifiers and give one diversity value for the entire ensemble. Let $\Psi_i$ denote the $i$th base classifier and $\Pi = \{\Psi_1, \Psi_2, ..., \Psi_L\}$ is the ensemble of base models. Let us present the selected diversity measures used in this work.

The averaged disagreement measure [4] over all pairs of classifiers

$$Dis_{av}(\Pi) = \frac{2}{L(L-1)} \sum_{i=1}^{L-1} \sum_{k=i+1}^{L} Dis(\Psi_i, \Psi_k), \tag{1}$$

where

$$Dis(\Psi_i, \Psi_k) = \frac{N^{01} + N^{10}}{N^{11} + N^{10} + N^{01} + N^{00}}, \tag{2}$$

The averaged disagreement measure is the ratio between the number of observations on which one classifier is correct and the other is incorrect to the total number of observations. $Dis$ varies between 0 and 1, where 0 indicates no difference and 1 indicates the highest possible diversity. Relationship between a pair of classifiers is denoted according to Table 1.

Kohavi-Wolpert variance [6] is defined as

$$KW(\Pi) = \frac{1}{NL^2} \sum_{j=1}^{N} l(z_j)(L - l(z_j)), \tag{3}$$

where $N$ is the number of instances, $L$ stands for the number of base models in the ensemble and $l(z_j)$ denotes the number of classifiers that correctly recognize $z_j$. The higher the value of KW, the more diverse the classifiers in the ensemble. Also, $KW$ differs from the averaged disagreement measure $Dis_{av}$ by a coefficient, i.e.,

$$KW(\Pi) = \frac{L-1}{2L} Dis_{av}(\Pi), \tag{4}$$

## 3.2    Clustering Criterion

Firstly, let us propose the measure which may be used for the clustering-based pruning. As the non-pairwise and averaged pairwise diversity measures consider

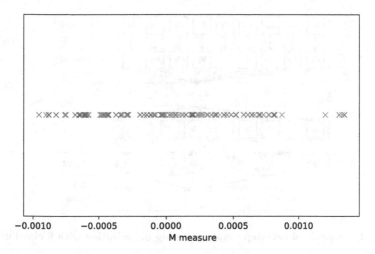

**Fig. 1.** Visualization of the proposed clustering space for Vowel dataset, where the clustering criterion is based on the averaged disagreement measure.

all the base models together and calculate one value for the entire ensemble, thus they could not be used to pruning, because they do not present an impact of a particular base classifier on the ensemble diversity. Therefore we propose a novel measure $M$ as the clustering criterion, which is the difference between the value of diversity measure for the whole ensemble $\Pi$ $(Div(\Pi))$ and the value of diversity for the ensemble without a given classifier $\Psi_i$ $(Div(\Pi - \Psi_i))$.

$$M(\Psi_i) = Div(\Pi) - Div(\Pi - \Psi_i). \tag{5}$$

Thanks to this proposition the impacts of each base learners on the ensemble diversity are presented in a one-dimensional spaces, shown in Fig. 1. Each marker represents one of the one hundred base classifiers, placed in the space according to it's value of $M$ measure based on the averaged disagreement.

### 3.3 Diversity Based One-Dimensional Clustering Space and Cluster Pruning

In this proposition, the chosen clustering algorithm is applied to the obtained clustering space. The pruned ensemble consists of the base models with the best classification accuracy in each cluster (one for each cluster).

### 3.4 Two-Step Majority Voting Organization

The second proposed method is a modification of the MOMV structure described in [12]. Instead of allocating outputs to different groups by permutation, we treat base models in each cluster as a separate ensemble combined by majority voting. Then we collect predictions from each cluster and apply the majority voting rule for the second time, to obtain the final decision (Fig. 2).

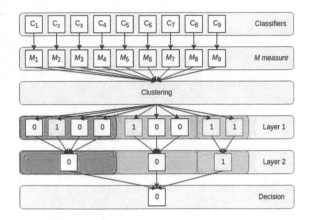

**Fig. 2.** Example of two-step majority voting organization with 9 classifiers.

# 4    Experimental Study

In this section we present the experimental study performed in order to evaluate the effectiveness of proposed clustering-based ensemble pruning and multistage organization methods. As the reference, two state-of-the-art methods: majority voting and the aggregation of probabilities were used. Experiments were designed to answer the following research questions:

- Which set of parameters (approach, diversity measure, base learner type, number of clusters) yields the best results for the given dataset?
- How the number of clusters affects the performance of methods?
- Does the proposed ensemble pruning and multistage organizations methods lead to improvements in accuracy over state-of-the-art methods?

## 4.1    Datasets Benchmarks

We have used 28 datasets from KEEL and UCI repositories to evaluate the performance of the proposed methods. We have selected a diverse set of benchmarks with varying characteristics, including different number of instances, features and classes, which are shown in Table 2.

## 4.2    Set-Up

As base learners we used two popular types of classifiers: Classification and regression trees (CART) and $k$-nearest neighbors classifier (KNN). Learners from Scikit-learn machine learning library [11] with the default parameters were used. Classifier pool always consists of 100 base models. Diversity between learners is based on the *random subspace method, Ho1998*, where classifiers are trained on pseudorandomly selected subsets of components of the feature vector. The percentage of features for training a single model has been selected depending

**Table 2.** Datasets characteristics.

| Dataset | Instances | Features | Classes | Dataset | Instances | Features | Classes |
|---|---|---|---|---|---|---|---|
| Australian | 690 | 14 | 2 | Pima | 768 | 8 | 2 |
| Bands | 365 | 19 | 2 | Saheart | 462 | 9 | 2 |
| Bupa | 345 | 6 | 2 | Segment | 2310 | 19 | 7 |
| Cleveland | 303 | 13 | 5 | Sonar | 208 | 60 | 2 |
| Contraceptive | 1473 | 9 | 3 | Spambase | 4596 | 57 | 2 |
| Dermatology | 366 | 7 | 8 | Spectfheart | 267 | 44 | 2 |
| Ecoli | 336 | 7 | 8 | Vehicle | 846 | 18 | 4 |
| Glass | 214 | 9 | 7 | Vowel | 990 | 13 | 11 |
| Heart | 270 | 13 | 2 | wdbc | 596 | 30 | 2 |
| HouseVotes | 232 | 16 | 2 | Wine | 178 | 13 | 3 |
| ILPD | 583 | 10 | 2 | WineRed | 1599 | 11 | 11 |
| Ionosphere | 351 | 33 | 2 | Winconsin | 683 | 9 | 2 |
| Libras | 360 | 90 | 15 | Yeast | 1484 | 8 | 10 |
| MuskV1 | 476 | 166 | 2 | ZOO | 101 | 16 | 7 |

on the number of features in the dataset. For majority of datasets it is 50%, the only exceptions being: Libras dataset - 20%, MuskV1 dataset - 10%, Sonar dataset - 25% and Spectfheart dataset - 35%.

Based on 3 parameters (approach, diversity measure and base learner type) we distinguish 8 different methods for improving classification score of the ensemble (4 for pruning and 4 for multistage organization). For the sake of simplicity, for each method, we take into account only the number of clusters that obtained the best classification accuracy. Name of each method is based on abbreviations of parameters values (*ApproachClassifierDiversityMeasure* format) including two state-of-the-art methods (majority voting and aggregation of probabilities) for each base learner, which gives us 12 methods overall. The following abbreviations have been used:

- *Approach*: MV – majority voting, Aggr – aggregation of probabilities, Mo - multistage voting organization and Pr - clustering based pruning,
- *Classifier*: Cart – classification and regression trees and Knn - $k$-nearest neighbors classifier,
- *DiversityMeasure*: Kw – the Kohavi-Wolpert variance and Dis – the averaged disagreement measure.

### 4.3   Experiment: Statistical Evaluation

Proposed methods were divided in 2 groups of 6 based on base learner type, and in 2 groups of 4 based on the proposed methods and diversity measure used. For each group Nemenyi post-hoc test, based on the average ranks according to classification score, was performed.

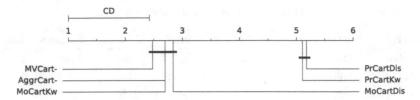

**Fig. 3.** Diagram of critical difference (CD) for Nemeyi post-hoc test at $\alpha = 0.05$ for the CART methods. $CD = 1.42$

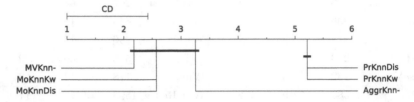

**Fig. 4.** Diagram of critical difference (CD) for Nemeyi post-hoc test at $\alpha = 0.05$ for the KNN methods. $CD = 1.42$

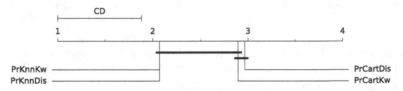

**Fig. 5.** Diagram of critical difference (CD) for Nemeyi post-hoc test at $\alpha = 0.05$ for the pruning methods. $CD = 0.89$

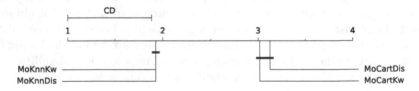

**Fig. 6.** Diagram of critical difference (CD) for Nemeyi post-hoc test at $\alpha = 0.05$ for the multistage organization methods. $CD = 0.89$

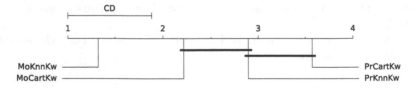

**Fig. 7.** Diagram of critical difference (CD) for Nemeyi post-hoc test at $\alpha = 0.05$ for the Kohavi-Wolpert variance methods. $CD = 0.89$

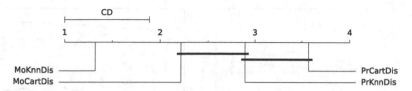

**Fig. 8.** Diagram of critical difference (CD) for Nemeyi post-hoc test at $\alpha = 0.05$ for the averaged disagreement methods. $CD = 0.89$

**Table 3.** The classification accuracy of the best performing method for each dataset, depending on the number of clusters. The highest achieved values of the classification accuracy have been marked.

| Dataset | BestMethod | 2C | 3C | 4C | 5C | 6C | 7C | 8C | 9C | 10C |
|---|---|---|---|---|---|---|---|---|---|---|
| Australian | PrCartKw | 80.87 | 85.38 | 83.04 | 87.39 | 87.26 | 89.13 | 87.69 | 89.28 | 88.27 |
| Bands | PrCartKw | 72.05 | 79.45 | 80.27 | 80.27 | 80.27 | 81.92 | 83.01 | 81.37 | 81.64 |
| Bupa | PrCartKw | 67.25 | 71.3 | 68.99 | 73.91 | 71.3 | 73.91 | 74.2 | 72.75 | 72.75 |
| Cleveland | PrCartKw | 62.28 | 62.95 | 62.58 | 62.95 | 63.62 | 64.3 | 66.34 | 63.31 | 62.32 |
| Contraceptive | PrKnnKw | 50.1 | 53.09 | 52.28 | 52.88 | 53.22 | 53.63 | 54.24 | 53.5 | 54.1 |
| Dermatology | PrKnnKw | 95.01 | 96.94 | 96.4 | 97.23 | 98.63 | 98.9 | 98.89 | 98.34 | 99.45 |
| Ecoli | PrKnnKw | 77.5 | 80.75 | 78.95 | 81.03 | 83.69 | 82.81 | 84.88 | 83.7 | 83.71 |
| Glass | PrCartKw | 76.53 | 81.68 | 82.28 | 84.05 | 82.59 | 84.08 | 84.07 | 85.4 | 85.51 |
| Heart | PrCartKw | 77.78 | 87.04 | 82.96 | 87.41 | 84.81 | 90.0 | 87.78 | 86.67 | 87.04 |
| HouseVotes | PrCartKw | 94.01 | 97.85 | 95.25 | 96.13 | 94.45 | 96.59 | 94.88 | 95.29 | 95.31 |
| ILPD | PrCartKw | 75.29 | 74.78 | 75.64 | 75.48 | 74.43 | 75.3 | 73.93 | 75.81 | 74.27 |
| Ionosphere | PrCartDis | 93.16 | 95.73 | 97.44 | 98.01 | 98.01 | 97.15 | 98.29 | 97.73 | 98.58 |
| libras | PrCartDis | 72.6 | 78.33 | 80.2 | 82.53 | 84.2 | 85.33 | 84.27 | 85.07 | 84.27 |
| MuskV1 | PrCartKw | 85.52 | 89.73 | 90.14 | 91.82 | 91.42 | 91.81 | 92.67 | 94.97 | 93.92 |
| Pima | PrKnnKw | 74.09 | 77.35 | 76.31 | 77.86 | 76.56 | 76.7 | 76.83 | 76.05 | 75.26 |
| Saheart | PrKnnKw | 70.36 | 72.3 | 72.52 | 74.03 | 72.09 | 73.17 | 73.38 | 73.38 | 72.3 |
| Segment | PrCartKw | 95.19 | 97.23 | 97.45 | 97.53 | 98.05 | 98.31 | 98.48 | 98.48 | 98.53 |
| Sonar | PrCartKw | 80.77 | 92.82 | 86.02 | 90.37 | 87.48 | 93.69 | 92.28 | 93.23 | 92.75 |
| Spambase | PrCartKw | 88.71 | 90.75 | 90.21 | 92.56 | 91.19 | 92.97 | 92.21 | 93.58 | 92.97 |
| Spectfheart | PrCartKw | 80.51 | 88.39 | 85.02 | 87.64 | 86.14 | 88.76 | 89.13 | 89.13 | 88.38 |
| Vehicle | PrCartKw | 74.35 | 78.37 | 78.96 | 79.78 | 80.86 | 80.49 | 81.69 | 81.69 | 80.51 |
| Vowel | MoCartKw | 94.34 | 94.65 | 94.34 | 95.25 | 95.56 | 95.96 | 95.96 | 96.57 | 95.35 |
| wdbc | PrCartKw | 95.78 | 97.72 | 96.13 | 97.9 | 97.02 | 97.73 | 97.19 | 97.9 | 97.55 |
| Wine | PrCartKw | 96.11 | 96.08 | 99.46 | 98.87 | 98.32 | 99.43 | 100.0 | 99.43 | 100.0 |
| WineRed | PrCartKw | 66.35 | 67.48 | 68.79 | 69.29 | 69.17 | 69.8 | 69.73 | 70.24 | 69.73 |
| Wisconsin | PrKnnKw | 97.81 | 98.1 | 97.66 | 98.39 | 98.24 | 98.24 | 97.95 | 98.24 | 97.95 |
| Yeast | PrKnnKw | 51.82 | 55.58 | 54.51 | 55.45 | 54.51 | 56.73 | 55.04 | 56.19 | 55.65 |
| ZOO | PrCartKw | 96.99 | 99.05 | 100.0 | 100.0 | 100.0 | 100.0 | 100.0 | 100.0 | 100.0 |

Figure 3 shows that, for CART classifiers, multistage organization methods are not significantly better than state-of-the-art methods. The two best-ranked methods, deemed as statistically significantly the best, are pruning methods.

In Fig. 4 we can see that for KNN methods MMOV, again, is not statistically significantly better than majority voting and aggregation of probabilities. Proposed pruning methods achieved statistically significantly better classification scores than other methods.

Figure 5 shows that for pruning, method than employs CART classifiers as base models and the averaged disagreement as clustering criterion (PrCartDis) is statistically significantly better than methods employing KNN classifiers. KNN methods and PRCartKw are deemed statistically equal.

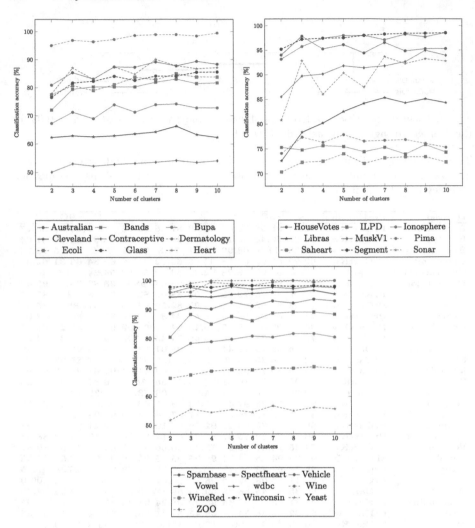

**Fig. 9.** The classification accuracy of the best performing methods for different number of clusters.

In Fig. 6 we can see that in the case of multistage organization of majority voting, methods using decision trees achieved higher ranks than methods employing KNN classifiers as base models. Both CART methods are deemed statistically equivalent as well as both KNN methods.

Figures 7 and 8 present the critical difference for methods depending on the diversity measure used. In case of both Kohavi-Wolpert variance and averaged disagreement, multistage organization method employing KNN as base classifiers is deemed as statistically the worst one. Pruning using CART classifiers achieved the highest average rank and achieved statistically the same results as pruning employing KNN classifiers. PrKnnDis and MoCartDis are deemed as statistically equivalent.

Table 3 presents the impact of the number of clusters on the best performing proposed methods, according to classification score, for each tested dataset. In the case where several methods achieved the same classification accuracy, the first one was chosen according to the order: *PrCartKw, PrKnnKw, PrCartDis, PrKnnDis, MoCartKw, MoKnnKw, MoCartDis, MoKnnDis*. Proposed multistage organization voting method achieved the highest classification score only in case of the Vowel dataset. In majority of cases we can observe a slight upward trend in the achieved classification score with an increase in the number of clusters. Figure 9 presents how the performance of methods vary depending on the number of clusters. The low classification score in the case of two clusters and any other equal number of clusters may be caused by using only two classifiers for majority voting. In that case, when there is no agreement between classifiers, the first label in the order is chosen.

# 5 Conclusions

The main aim of this work was to propose a novel, effective classifier pruning method based on clustering. We proposed the one-dimensional clustering space based on ensemble diversity measures, which is later used in order to prune the existing classifier pool or to perform a multistage majority voting. The computer experiments confirmed the usefulness of the proposed pruning method and on the basis of statistical analysis we may conclude that it is statistically significantly better than state-of-art ensemble methods. Proposed multistage organization voting scheme did not achieved statistically better results than state-of-art methods.

The results presented in this paper are quite promising therefore they encourage us to continue our work on employing clustering based methods for ensemble pruning.

**Acknowledgement.** This work was supported by the Polish National Science Centre under the grant No. 2017/27/B/ST6/01325 as well as by the statutory funds of the Department of Systems and Computer Networks, Faculty of Electronics, Wroclaw University of Science and Technology.

# References

1. Bakker, B., Heskes, T.: Clustering ensembles of neural network models. Neural Netw. **16**(2), 261–269 (2003)
2. Fu, Q., Hu, S.X., Zhao, S.: Clustering-based selective neural network ensemble. J. Zhejiang Univ. Sci. **6**(5), 387–392 (2005)
3. Giacinto, G., Roli, F., Fumera, G.: Design of effective multiple classifier systems by clustering of classifiers. In: 15th International Conference on Pattern Recognition, ICPR 2000 (2000)
4. Ho, T.K.: The random subspace method for constructing decision forests. IEEE Trans. Pattern Anal. Mach. Intell. **20**, 832–844 (1998)
5. Ho, T.K., Hull, J.J., Srihari, S.N.: Decision combination in multiple classifier systems. IEEE Trans. Pattern Anal. Mach. Intell. **16**(1), 66–75 (1994)
6. Kohavi, R., Wolpert, D.: Bias plus variance decomposition for zero-one loss functions. In: Proceedings of the Thirteenth International Conference on International Conference on Machine Learning, ICML 1996, pp. 275–283. Morgan Kaufmann Publishers Inc., San Francisco (1996)
7. Krawczyk, B., Cyganek, B.: Selecting locally specialised classifiers for one-class classification ensembles. Pattern Anal. Appl. **20**(2), 427–439 (2017)
8. Krawczyk, B., Minku, L.L., Gama, J., Stefanowski, J., Wozniak, M.: Ensemble learning for data stream analysis: a survey. Inf. Fusion **37**, 132–156 (2017)
9. Kuncheva, L.I.: Combining Pattern Classifiers: Methods and Algorithms. Wiley, Hoboken (2004)
10. Lazarevic, A., Obradovic, Z.: The effective pruning of neural network classifiers. In: 2001 IEEE/INNS International Conference on Neural Networks, IJCNN 2001 (2001)
11. Pedregosa, F., et al.: Scikit-learn: machine learning in Python. J. Mach. Learn. Res. **12**, 2825–2830 (2011)
12. Ruta, D., Gabrys, B.: A theoretical analysis of the limits of majority voting errors for multiple classifier systems. Pattern Anal. Appl. **2**(4), 333–350 (2002)
13. Topolski, M.: Algorithm of principal component analysis PCA with fuzzy observation of facility features detection of carcinoma cells multiple myeloma. In: Burduk, R., Kurzynski, M., Wozniak, M. (eds.) Progress in Computer Recognition Systems (2019)
14. Woźniak, M., Graña, M.: A survey of multiple classifier systems as hybrid systems. Inf. Fusion **16**, 3–17 (2014)
15. Zhang, H., Cao, L.: A spectral clustering based ensemble pruning approach. Neurocomputing **139**, 289–297 (2014)
16. Zhou, Z.H.: Ensemble Methods: Foundations and Algorithms. Chapman & Hall CRC, Boca Raton (2012)

# Towards a Custom Designed Mechanism for Indexing and Retrieving Video Transcripts

Gabriel Turcu[1], Stella Heras[2]($\boxtimes$), Javier Palanca[2], Vicente Julian[2], and Marian Cristian Mihaescu[1]

[1] Faculty of Automatics, Computers and Electronics, Craiova, Romania
gabriel.turcu97@gmail.com, cmihaescu@software.ucv.ro
[2] Departamento de Sistemas Informáticos y Computación,
Universitat Politècnica de València, Valencia, Spain
{sheras,jpalanca,vinglada}@dsic.upv.es

**Abstract.** Finding appropriate e-Learning resources within a repository of videos represents a critical aspect for students. Given that transcripts are available for the entire set of videos, the problem reduces to obtaining a ranked list of video transcripts for a particular query. The paper presents a custom approach for searching the 16.012 available video transcripts from https://media.upv.es/ at Universitat Politècnica de València. An inherent difficulty of the problem comes from the fact that transcripts are in the Spanish language. The proposed solution embeds all the transcripts using feed-forward Neural-Net Language Models, clusters the embedded transcripts and builds a Latent Dirichlet Allocation (LDA) model for each cluster. We can then process a new query and find the transcripts that have the LDA results closest to the LDA results for our query.

**Keywords:** Latent Dirichlet Allocation · NNLM word embeddings · Clustering

## 1 Introduction

Searching for appropriate e-Learning resources (i.e., videos, quizzes, presentations, etc.) is one of the most critical activities for students that are willing to improve their knowledge. General purpose search engines may do an excellent job, but custom designed professional search tools are more advisable for better results. Thus, within the area of e-Learning, the Video Base Learning (VBL) [19] stands a particular place which gets more and more attention due to its effectiveness in teaching and learning. The significant technological advances have given the VBL a vital role in improving learning outcomes and properly designing VBL environments.

One critical aspect in VBL is the retrieval of relevant videos given an input query. This problem has been addressed in [3] by reviewing a wide range of

© Springer Nature Switzerland AG 2019
H. Pérez García et al. (Eds.): HAIS 2019, LNAI 11734, pp. 299–309, 2019.
https://doi.org/10.1007/978-3-030-29859-3_26

machine learning algorithms that have been used for indexing and retrieving learning materials. Among the most utilised indexing parameters, there are the ones that refer to text coming from natural language, documents or images. The various formats under which text may be shaped are web document, logs, XML, structured or semi-structured data. The general picture is completed by a wide range of indexing algorithms such as clustering, Ant colony, semantic index, SemTree, text index tree, B-Tree, etc.

The most critical shortcoming of search systems is that they are general and for existing implementations, they highly depend upon underlying data. Therefore, we have developed a specific search system over a dataset of video transcripts from a Spanish public university. A particularity of our input dataset consists of the fact that their transcripts are in the Spanish language. This poses new challenges as few developments that were done for the English language were also implemented for the Spanish language.

This paper presents a custom designed mechanism for indexing and retrieving video transcripts. The task is to index available video transcripts such that for an input query provided by a user, the retrieval mechanism should return a list with the most representative videos. In our particular case, the input is represented by a broad set of educational video transcripts and the search is accomplished by learners who are seeking learning materials.

The proposed approach takes as input 16.012 available video transcripts and builds a dataset by extracting necessary features. The dataset consists of a bag-of-words (BoW) and its corresponding matrix of NNLM embedding results. Given K, representing the number of domains which span the video transcripts, we run a clustering algorithm to obtain a partitioning. Thus, available video transcripts are assigned into clusters in an attempt to group by instances (i.e., video transcripts) into domains. Once the domains are obtained, we further run an LDA [6] algorithm to get a list of topics and their score. Then, given a query, we determine the closest centroid and therefore obtain the domain of the query to which it belongs. Finally, by searching into the acquired domain's instances, we get a ranked list of video transcripts that are closest to the query and return them to the user. Preliminary validation of the proposed solution has been performed manually by comparison with the outputs provided by the currently existing search mechanism.

The main contribution of the paper consists in designing and implementing a custom data analysis pipeline whose ingredients are a Spanish transcripts corpus, the processed BoW, the matrix of NNLM embedding, a partitioning of transcripts obtained by running a clustering algorithm and LDA for ranking.

## 2    Related Work

In recent years, the rise of Massive Online Open Courses (MOOCs), and Technology Enhanced Learning (TEL) systems, in general, has highlighted even more the importance of having efficient and accurate information retrieval systems.

Usage of information retrieval [2] for indexing and retrieval textual and multi-media content from various sources represents an essential method for improving TEL systems.

In *content-based* multimedia IR, the primary objective is to identify and extract features related to image contents. Following this approach, in [7], authors compare 'traditional' engineered (hand-crafted) features (or descriptors) and learned features for content-based semantic indexing of video documents. Learned (or semantic) features are obtained by training classifiers in the context of the TRECVID[1] semantic indexing task.

In [9], authors propose a combination of content-based video indexing approaches: text-based, feature-based, and semantic-based. The text-based approach focuses on using keywords or tags to describe video content. The feature-based approach aims to extract low-level features such as colour, texture, shape and motion from the video data and use them as indexing keys.

Following an approach similar to the proposed in this paper, the work presented in [10] highlights the shortcomings of current multimedia indexing and retrieval techniques, mainly based on sparse tagging, and deals with content-based video indexing and retrieval using an LDA probabilistic framework [6].

In our same domain of lecture video retrieval, [15] presents an indexing method for recorded videos of computer science courses. This proposal uses the automatic transcriptions from a speech-recognition engine to create a chain index for detailed browsing inside a lecture video. Also, in [18], authors presented a method for content-based video indexing and retrieval in sizable German lecture video archives. This paper applies a combination of automatic video segmentation and key-frame detection with a technique to extract textual meta-data by applying video Optical Character Recognition (OCR) technology on key-frames and Automatic Speech Recognition (ASR) on lecture audio tracks. Applying a technique based on multi-modal language models for lecture video retrieval, in [8] authors demonstrated that this method outperforms LDA-based methods when speech transcripts are error-free, but the model shows similar performance for noisy text.

Other works use BoW based methods to classify and retrieve videos (with keywords [4], subjects [17] or visual features [12]). However, BOW based models cannot describe the content of an image objectively and neglect the spatial distribution of visual words and the order of words in transcripts.

To the best of our knowledge, related works refer only to English language corpus and do not work specifically on educational transcripts. Therefore, a comparative analysis with already existing language models or machine translation encoders that perform similar tasks may not be yes possible.

## 3   Proposed Approach

Searching within a large available set of videos represents a challenging task for a student. The current video search mechanism implemented in the multimedia

---

[1] https://www-nlpir.nist.gov/projects/tv2018.

platform https://media.upv.es/ performs a full-text search over the title and keyword fields of the videos. It examines the words stored in such fields and tries to match with the search query made by the user. These full-search techniques may suffer from the lack of semantics and context on the search since they only take into account the word included in the query as it is. In the approach proposed in this work, we present a procedure that is much more context-aware, being able to classify videos according to their content using their transcripts.

**1. Build the Bag-of-Words.** As the video transcripts represent the primary input for the data analysis process, the first step is to build a BoW by tokenizing the transcripts to remove stop-words and words that are too short. Once the BoW has been created the next task is to determine the domains in which the transcripts may group. As no labels are being provided, the most effective solution is to implement an unsupervised learning algorithm (i.e., clustering) for grouping the transcripts. The most significant limitation of this approach is represented by the fact that the BoW cannot be used directly as an input to a clustering algorithm and why we have to embed the transcripts using state of the art feed-forward neural net language models.

Another issue may regard the fact that we are dealing with a large number of words from BoW and with a reasonably large number of transcripts, which will end up in having a sufficiently large input dataset for clustering. From the application domain perspective, a cluster should group transcripts that belong to a particular domain. By observing the university's media website, we have identified four domains: Arts & Humanities, Engineering & Architecture, Sciences (Biological Sciences), and Social & Legal.

**2. Compute Matrix with NNLM Word Embedding Results.** We want to divide the transcripts into four different domains and to do that we have to run a clustering algorithm on the transcripts, but since we can't run K-means on string objects, we have to transform our transcripts into a data format that our clustering algorithm understands. To accomplish this, we use the NNLM word embeddings. NNLM word embedding saves much space by learning a distributed representation for words which allows each training sentence to inform the model about an exponential number of semantically neighbouring sentences. The model learns simultaneously a distributed representation for each word along with the probability function for word sequences, expressed in terms of these representations. The generalisation is obtained because a sequence of words that have never been seen before gets high probability if it is made of words that are similar (in the sense of having a nearby representation) to words forming an already seen sentence [5]. The output from the NNLM embedding is a vector of 128 float numbers for each transcript, which amounts in total to a matrix of float numbers that is 16.012 lines (one for each transcript) and 128 columns. The output vector forms the input for our clustering algorithm (Fig. 1).

**3. Build Clusters of Transcripts.** The next step consists of running a clustering method for assigning items (i.e., transcripts) into groups. From this perspective, each obtained cluster represents a domain within the entire set of

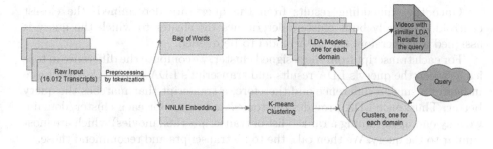

**Fig. 1.** Data analysis pipeline-cool

available videos. The primary purpose of the clustering algorithm is to bring together items (i.e., transcripts) for which the terms have similar NNLM embedding results over the entire transcripts corpus. One option is to use a standard simple k-means clustering algorithm [11] and provide value for K as a domain knowledge person provides or use other algorithms that do not require for the number of clusters, such as xMeans [14]. Finding the optimal number of clusters in a particular dataset represents itself a challenging research issue and is not covered by the current works. Another parameter that needs setup within the clustering process is the distance function (i.e., Euclidean, Jaccard, cosine, edit, etc.). Taking into account that the embedding results represent the input data, the current approach uses the Euclidean distance along with standard clustering quality metrics: SSE, homogeneity, completeness, Adjusted Rand-Index or Silhouette coefficient.

**4. Model Each Cluster and Find Its Topics and Scores.** For each cluster of transcripts (i.e., BoW corpus of that particular cluster and the dictionary of that cluster), we use LDA to determine its topics and associated scores. From an LDA perspective, the *words* are represented by the tokens obtained in the first step, the *documents* are represented by the transcripts and the *corpus* is represented by the set of transcripts for a particular cluster. As output, LDA determines the *latent topics* and their characterization in terms of *words*. Intuitively, for each cluster (i.e., domain) the LDA model creates a list of topics defined by scores whose values add up to one. More, each topic is defined by a list of words with their coefficients. Both in the case of the topic's scores and coefficient's values, the interpretation is that a more significant number represents a more critical topic or word. Finally, the model (i.e., the topic's scores and their list of coefficients and words) is serialized for later querying.

**5. Query and Retrieve a Ranked List of Results.** Once a query is obtained from a user, we need to find the cluster/domain to which it belongs. Since the query is regarded as a transcript, it is firstly preprocessed and its cluster is being determined. Determining the cluster to which the query belongs needs computing the NNLM embedding results of the words that make up the query. The embedding values for the query are being computed by considering the query as a transcript.

Once the embedding results from the query are determined, the closest centroid of already built clusters determines the cluster to which the query is assigned and therefore the LDA model to be queried.

For each transcript from the assigned cluster, we compute the difference in topics between the query's LDA results and transcript's LDA results. A lower score indicates a smaller difference and therefore, a transcript that matches the query better. Thus, once LDA provides the topics and scores for each cluster/domain, we may end up obtaining a ranked list of transcripts (i.e., movies) which are most similar to the query. We then take the top 5 transcripts and recommend those.

## 4    Experimental Results

### 4.1    Input Dataset

The input dataset consists of 16.012 video transcripts that are accessible through a *json* file. The structure and raw data for a record from the *json* file are presented in the following example.

```
1  {"video": {
2    "_id": "00054a38-5a32-4db2-ae9c-85c296015c3b",
3    "hidden": "False",
4    "title": "Programa Mathematica gratis y online",
5    "type": "polimedia",
6    "mimetype": "video/mp4",
7    "width": "640",
8    "height": "480",
9    "src": "politube.mp4",
10   "slides": {
11     "menuitem": [
12       {"mimetype":"image/jpg","url":"frame.0.jpg","
            time":"0"},
13       {"mimetype":"image/jpg","url":"frame.48.jpg","
            time":"48"},
14       {"mimetype":"image/jpg","url":"frame.431.jpg",
            "time":"431"}
15     ]
16   "duration": "564.523537",
17   "transcription": "...very long text... (or empty)
        "
18   }
19  }}
```

This dataset comes from an online e-learning platform (i.e., https://media.upv.es) from Universitat Politecnica de Valencia (UPV). The UPV's platform has recording facilities to create educational videos (most of them in Spanish) which are finally publicly available as MOOCs.

The dataset includes the information necessary to process the video transcription and also includes additional information such as the identifiers needed to download the video (such as the _id field, the src field or the type field). Similarly, the fields mimetype, width and height are the meta-information of the video file. Information such as the title of the video or its duration in seconds is also available. However, something that characterizes these videos, which are educational videos recorded in a room dedicated to this activity, is that they are accompanied by information related to the slide presentation if used during the recording of the video.

During the recording of the videos, a software that detects the change of slide in the screen of the presenter was used. Thus, each slide detected in the dataset was captured and tagged. Therefore, the slides field contains a list of identified slides with their filename to be downloaded, the mime-type of the file and the instant (in seconds) of the video in which the slide change was detected.

What makes this dataset interesting is that it not only includes a collection of videos with a specific theme (educational videos), but also includes the transcription of the audio of these videos, and even each of the slides that have been used in the presentation, along with its temporary label. This approach enables a much more in-depth analysis of this data set.

## 4.2  Numerical Results

Each transcript from the raw input dataset has been transformed into a BoW for which NNLM values were computed and saved into a matrix. The dimension of the matrix is 16.012 (i.e., number of transcripts) x 128. What the embedding algorithm does is it maps from text to 128-dimensional embedding vectors.

Then, the embedding matrix has been used as input for simple k-means algorithms from Scikit-learn [13] to obtain a distribution of items (i.e., transcripts) into four clusters. For the clustering process, the items are represented by transcripts, features are represented words, and embedding results represent the values that build up the input dataset. Table 1 presents several BoW samples along with computed embedding results and their corresponding CA (cluster assignments).

The running of simple K-means algorithm provides the following distribution of transcripts into clusters: 22.339%, 32.466%, 13.343% and 31.852%. Once we have determined the four clusters of transcripts, we further run the LDA algorithm to determine a lexical model of each cluster as a list of topics along with their scores and a list of words that make up that topic. The obtained model also consists of computed scores for each word within the topic. Table 2 presents sample numerical results for the transcripts described in Table 1.

For each transcript presented in Table 2, the computed topic scores have the sum equal to 1. This approach makes interpretation straightforward in the way that a topic score of 0.844 is a big score indicating that the topic represented is an excellent representative for that transcript. Further, each topic is represented by a list of words and their coefficients. In the same line of approach, a more significant value in coefficient is an indication of a more critical word among the words that make up the topic.

**Table 1.** Sample BoW with embedding results and cluster assignments.

| ID | Sample BoW | Sample embedding results | CA |
|----|-----------|--------------------------|----|
| 3 | hola vamos a ver la parte cinco de la documentacion del software basicamente seria otro programa que nos quedaba veamos por explicar seria el cloc... | 0.8655922, 0.77715206, −0.16605175, −1.6408461, −0.8250744, 0.02193201, 0.700683, 0.7213742, −0.3716459, 1.5170424, 0.13685325, −0.20364942, 0.5022141, ... | 1 |
| 4 | hola vamos a ver ahora la parte politica de calidad es este digamos que el objeto que se realizaria simplifica de tareas basicamente estos se utiliza mucho... | 0.65827113, 0.5848848, 0.12338758, −1.5416939, −0.59042096, −0.17713968, 0.78873384, 0.5304664, −0.37286106, 1.0820895, 0.5255295, −0.27624443, 0.50664973, 0.15184559, ... | 1 |
| 6 | bienvenidos y bienvenidas a esta unidad de formacion en la cual trataremos sobre los usos de la letra cursiva somos sepalo assange del servicio de promocion y normalizacion... | 0.33565775, 0.48877674, 0.17549452, −1.6629573, −0.91474277, 0.3601234, 0.68089503, 0.55267304, −0.40141803, 1.3872166, 0.2669923, −0.25762805, 0.7387372, 0.17457546, ... | 3 |

Once the query is being obtained from the user, it is preprocessed and embedding results are computed. At this stage, the *corpus* is represented by the available transcripts and the query. Thus, the query is reduced to an array of words (i.e. a BoW corpus) which is assigned to the closest cluster (i.e. nearest centroid). Determining the cluster (i.e. domain) of the query opens the way to further investigating its associated LDA model.

Table 3 presents the query results: the query, the assigned cluster (i.e., the domain), the LDA scores (i.e., the topic IDs and its score) and the ranked results (transcript ID and score). The results are ranked by the computed score from the fourth column as this score represents the difference between the query's LDA score and transcript's LDA score. A lower value in the transcript's LDA score represents a smaller difference; therefore, a transcript that is a better match for the query.

## 4.3   Validation

Validation is the part of the project that has proved to be the most difficult mostly because we are working with real-life data and therefore we do not possess the ground truth needed so that we may validate our work with ease. For validation, a couple of approaches have been investigated and tested.

Firstly, manual validation has been performed by looking at the results from a query which represent the top transcripts whose LDA scores are the closest to our query's LDA scores and seeing if the transcripts' content reflect the query.

**Table 2.** Sample LDA models with their topics and scores

| ID | LDA results: topics and scores |
|---|---|
| 3 | **Score:** 0.8441817164421082 **Topic:** 0.005*"filtr" + 0.004*"estad" + 0.002*"clas" + 0.002*"tension" + 0.002*"senal" |
| | **Score:** 0.0837591215968132 **Topic:** 0.002*"dat" + 0.002*"registr" + 0.001*"formulari" + 0.001*"eolic" + 0.001*"electr" |
| | **Score:** 0.06878719478845596 **Topic:** 0.001*"datagr" + 0.001*"estil" + 0.001*"motor" + 0.001*"dataset" + 0.001*"wrait" |
| 4 | **Score:** 0.6999539732933044 **Topic:** 0.005*"filtr" + 0.004*"estad" + 0.002*"clas" + 0.002*"tension" + 0.002*"señal" |
| | **Score:** 0.24506276845932007 **Topic:** 0.002*"dat" + 0.002*"registr" + 0.001*"formulari" + 0.001*"eolic" + 0.001*"electr" |
| | **Score:** 0.053080759942531586 **Topic:** 0.001*"datagr" + 0.001*"estil" + 0.001*"motor" + 0.001*"dataset" + 0.001*"wrait" |
| 6 | **Score:** 0.6111074686050415 **Topic:** 0.001*"oracion" + 0.001*"pronombr" + 0.001*"agu" + 0.001*"sistem" + 0.001*"instal" |
| | **Score:** 0.22719042003154755 **Topic:** 0.001*"plan" + 0.001*"edifici" + 0.001*"sistem" + 0.001*"element" + 0.001*"derech" |
| | **Score:** 0.049981363117694855 **Topic:** 0.001*"plan" + 0.001*"control" + 0.001*"derech" + 0.001*"punt" + 0.001*"nod" |
| | **Score:** 0.04900844022631645 **Topic:** 0.001*"fibr" + 0.001*"atom" + 0.001*"molecul" + 0.001*"temperatur" + 0.001*"sistem" |
| | **Score:** 0.04536473751068115 **Topic:** 0.001*"arrend" + 0.001*"derech" + 0.001*"pacient" + 0.001*"valencian" + 0.001*"anten" |
| | **Score:** 0.010526408441364765 **Topic:** 0.001*"sistem" + 0.001*"pec" + 0.001*"ulcer" + 0.001*"presion" + 0.001*"aliment" |

Secondly, we have used TextRazor's state-of-the-art Natural Language Processing and Artificial Intelligence API [16] to parse, analyze and extract semantic meta-data from the transcripts to see if the detected Categories and Topics are related to our query. Finally, We applied the LSTM Siamese Text Similarity [1] algorithm on the top 3 results from a couple of queries to check if the recommended transcripts for those queries are similar to each other. We had to use the English transcripts because there wasn't anything already trained for the Spanish language. We downloaded the transcripts that were translated from Spanish to English and applied the algorithm. If they would have been marked as similar by the LSTM Siamese algorithm and if the categories and topics from TextRazor would be relatable to our queries, that would tell us that our program does a good job at recommending transcripts based on the content in them. The results from the LSTM Siamese algorithm were inconclusive, however. This happened because the LSTM Siamese algorithm works best if the sentences are short and have a definite meaning. The issue with available transcripts was that they were detected imperfectly by the speech recognition engine, and then imperfectly translated into English. Also, they are very long, with an average number of words a bit over 1000 words per transcript.

**Table 3.** Query results.

| Query | CA | LDA scores (topic ID and score) | Ranked list of transcripts (transcript ID and score) |
|---|---|---|---|
| Ciencias de la Computacion | 2 | (0,0.4422385) (1,0.053041767) (2,0.04431011) (6,0.033507172) (7,0.42583835) | (41107,0.10977402180433274) (765,0.11200470328330994) (18460,0.11249668002128602) (3895,0.11415494084358216) (1236,0.11463153958320618) |
| Aprendizaje automatico | 2 | (9,0.6999294) (8,0.033338577) (7,0.03334638) | (29906,3.75695526599e-05) (16256,4.62792813777e-05) (35066,5.43296337155e-05) (3212,6.037577986724e-05) (17793,0.0001398846507487) |
| Permanente para la proteccion de los animales en cria instituido | 2 | (6,0.014290362) (7,0.8714059) (8,0.014287368) (9,0.014287801) | (44794,1.2902542948724e-05) (27277,1.7145648598676e-05) (44466,0.03214275650680065) (26721,0.04286431334912777) (41793,0.04287060722708702) |

## 5  Conclusions and Future Works

In this paper, we have implemented a custom procedure for indexing and retrieving video transcripts. The input transcripts are from educational videos available from https://media.upv.es/. The data analysis pipeline consists of stemming, computing NNLM embedding results, clustering transcripts and building one LDA model for each transcript. Once the LDA models are available, the query provided by the user is preprocessed and labelled (i.e., assigned to a cluster) and the corresponding LDA model is used for obtaining the most similar topics (i.e., with highest scores) and therefore the most important words with their coefficients. These results trace back to the original transcripts, and consequently, a ranked list of videos is obtained.

A critical limitation of the current approach is that it uses a fixed number (i.e., four) of clusters/domains. This approach is due to practical reasons since we do not have any ground truth indicating the exact number of clusters of the 16.012 transcripts. Finding the correct number of clusters from the dataset may bring significant improvements regarding the relevance of the obtained ranked list. Besides, a future goal is to improve the current search mechanism, which takes into consideration only the title of the video.

**Acknowledgements.** This work was partially supported by the project TIN2017-89156-R of the Spanish government, and by the grant program for the recruitment of doctors for the Spanish system of science and technology (PAID-10-14) of the Universitat Politècnica de València.

# References

1. Aman Srivastava: LSTM Siamese Text Similarity, April 2019. https://github.com/amansrivastava17/lstm-siamese-text-similarity
2. Baeza-Yates, R., Ribeiro, B.D.A.N., et al.: Modern information retrieval. ACM Press, New York. Addison-Wesley, Harlow (2011)
3. Bakar, Z.A., Kassim, M., Sahroni, M.N., Anuar, N.: A survey: framework to develop retrieval algorithms of indexing techniques on learning material. J. Telecommun. Electron. Comput. Eng. (JTEC) **9**(2–5), 43–46 (2017)
4. Basu, S., Yu, Y., Singh, V.K., Zimmermann, R.: Videopedia: lecture video recommendation for educational blogs using topic modeling. In: Tian, Q., Sebe, N., Qi, G.-J., Huet, B., Hong, R., Liu, X. (eds.) MMM 2016. LNCS, vol. 9516, pp. 238–250. Springer, Cham (2016). https://doi.org/10.1007/978-3-319-27671-7_20
5. Bengio, Y., Ducharme, R., Vincent, P., Jauvin, C.: A neural probabilistic language model. J. Mach. Learn. Res. **3**(Feb), 1137–1155 (2003)
6. Blei, D.M., Ng, A.Y., Jordan, M.I.: Latent Dirichlet allocation. J. Mach. Learn. Res. **3**(Jan), 993–1022 (2003)
7. Budnik, M., Gutierrez-Gomez, E.L., Safadi, B., Quénot, G.: Learned features versus engineered features for semantic video indexing. In: 2015 13th International Workshop on Content-Based Multimedia Indexing (CBMI), pp. 1–6. IEEE (2015)
8. Chen, H., Cooper, M., Joshi, D., Girod, B.: Multi-modal language models for lecture video retrieval. In: Proceedings of the 22nd ACM International Conference on Multimedia, pp. 1081–1084. ACM (2014)
9. Elleuch, N., Ammar, A.B., Alimi, A.M.: A generic framework for semantic video indexing based on visual concepts/contexts detection. Multimedia Tools Appl. **74**(4), 1397–1421 (2015)
10. Iyer, R.R., Parekh, S., Mohandoss, V., Ramsurat, A., Raj, B., Singh, R.: Content-based video indexing and retrieval using corr-lda. arXiv preprint arXiv:1602.08581 (2016)
11. MacQueen, J., et al.: Some methods for classification and analysis of multivariate observations. In: Proceedings of the Fifth Berkeley Symposium on Mathematical Statistics and Probability, Oakland, CA, USA, vol. 1, pp. 281–297 (1967)
12. Ngo, C.W., et al.: Experimenting VIREO-374: bag-of-visual-words and visual-based ontology for semantic video indexing and search. In: TRECVID (2007)
13. Pedregosa, F., et al.: Scikit-learn: machine learning in Python. J. Mach. Learn. Res. **12**(Oct), 2825–2830 (2011)
14. Pelleg, D., Moore, A.W., et al.: X-means: Extending k-means with efficient estimation of the number of clusters. In: ICML, vol. 1, pp. 727–734 (2000)
15. Repp, S., Grob, A., Meinel, C.: Browsing within lecture videos based on the chain index of speech transcription. IEEE Trans. Learn. Technol. **1**(3), 145–156 (2008)
16. Crayston, T.: The Natural Language Processing API, April 2019. https://www.textrazor.com/technology
17. Van Nguyen, N., Coustaty, M., Ogier, J.M.: Multi-modal and cross-modal for lecture videos retrieval. In: 2014 22nd International Conference on Pattern Recognition (ICPR), pp. 2667–2672. IEEE (2014)
18. Yang, H., Meinel, C.: Content based lecture video retrieval using speech and video text information. IEEE Trans. Learn. Technol. **2**, 142–154 (2014)
19. Yousef, A.M.F., Chatti, M.A., Schroeder, U.: Video-based learning: a critical analysis of the research published in 2003–2013 and future visions (2014)

# Active Image Data Augmentation

Flávio Arthur Oliveira Santos[1](✉)[ID], Cleber Zanchettin[1](✉)[ID],
Leonardo Nogueira Matos[2](✉)[ID], and Paulo Novais[3](✉)[ID]

[1] Universidade Federal de Pernambuco, Recife, Brazil
{faos,cz}@cin.ufpe.com
[2] Universidade Federal de Sergipe, São Cristóvão, Brazil
leonardo@dcomp.ufs.br
[3] University of Minho, Braga, Portugal
pjon@di.uminho.pt

**Abstract.** Deep neural networks models have achieved state-of-the-art results in a great number of different tasks in different domains (e.g., natural language processing and computer vision). However, the notions of robustness, causality, and fairness are not measured in traditional evaluated settings. In this work, we proposed an active data augmentation method to improve the model robustness to new data. We use the Vanilla Backpropagation to visualize what the trained model consider important in the input information. Based on that information, we augment the training dataset with new data to refine the model training. The objective is to make the model robust and effective for important input information. We evaluated our approach in a Spinal Cord Gray Matter Segmentation task and verified improvement in robustness while keeping the model competitive in the traditional metrics. Besides, we achieve the state-of-the-art results on that task using a U-Net based model.

**Keywords:** Data augmentation · Robustness · Interpretability

## 1 Introduction

Deep Learning (DL) [7] is a subfield of Machine Learning (ML) [2] which models achieve state-of-the-art results in many problems. For example, object recognition [5], visual question & answering [1] and super-resolution [24]. Models based on Convolutional Neural Networks (CNNs) [8] has achieved excellent performance in computer vision tasks, such as object classification and detection. However, the end-to-end way of training Deep Learning models make them a black box approach, turning it difficult to explain "why" the model produces some output. Some recent approaches such as Vanilla Backpropagation [18] and Guided Backpropagation [19] deal with that difficulty trying to visualize what is most important to network predictions using the gradient information. Zang et al. [23] proposes a method to build an explanatory graph to clarify CNNs predictions.

© Springer Nature Switzerland AG 2019
H. Pérez García et al. (Eds.): HAIS 2019, LNAI 11734, pp. 310–321, 2019.
https://doi.org/10.1007/978-3-030-29859-3_27

Despite the success, in many situations, we do not have guarantees that the excellent DL model performance is based on information we consider essential to the related task. For example, using a U-Net model [17] to gray matter segmentation [15] we achieve a Dice score [3] of 0.91 on validation the data, which is an excellent result considering this dataset. However, if we exclude some pixels of the input image (not related to gray matter pixels), even keeping all the gray matter pixels in the image, the U-Net model is no longer able to segment the input image correctly as presented in Fig. 1. This behaviour can be a problem because the model is not using the gray matter pixels to segment the image; it is using contextual pixels of the gray matter.

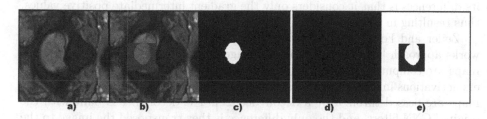

**Fig. 1.** (a) Original input image; (b) altered input image to remove contextual pixels; (c) original segmentation (considering original image); (d) new segmentation (considering altered image); (e) mask used to alter input image.

From the above observations, we assume that enriching the training dataset with images containing augmented information will improve the model robustness. So, we propose a method, called active image data augmentation (ADA) to suppress the contextual pixels and force the model to self-adapt and consider the important image pixel information. The method uses the Vanilla Gradient [18] representation to obtain the degree of importance of each input image pixel and remove some contextual pixels of this region, keeping the area with only the crucial information (e.g., gray matter region).

The proposed approach is evaluated in the dataset of the Spinal Cord Grey Matter Segmentation (SCGM) competition [15] compared with different literature approaches to the problem. As our goals are to improve model robustness, we test our method in two settings: (i) standard configuration (SCGM) to ensure model quality at original dataset and (ii) robustness setting to evaluate robustness directly. The remaining of the paper is structured as follow: Sect. 2 presents the related works. In Sect. 3, we discuss the proposed method, while Sects. 4 and 5 show the experiments and obtained results, respectively. The conclusions are presented in Sect. 6.

## 2   Related Works

This section goal is to present more relevant related works. To a better organization, we grouped the works in two main categories: (i) Interpretability models; and (ii) Data augmentation methods.

## 2.1   Interpretability Methods

There are many interpretability methods, however, we will focus on methods that identify the more relevant input information for the neural network inference.

Vanilla Backpropagation [18] is a method to create saliency maps of neural networks. The saliency maps represent what input image pixels are more important for a CNN model. Considering $c$ as the score of class activation inferred by a CNN and $i$ the input image, to obtain the saliency maps is necessary to compute the gradient of $c$ with relation to $i$. Thus, the method is similar to backpropagation, except that gradient is calculated concerning the input image instead of CNN weights. Guided Backpropagation [19] is an evolution of that method; its differences is that it considers only the gradient intermediate positive values, thus resulting in a saliency map with less noise.

Zeiler and Fergus [21] proposed a method based on Deconvolutional Networks approach [22]. Firstly, an input image is fed to a CNN and their features maps are computed. In following, to examine a specific unit activation, all others activations in the same layer are modified to zero, and the resultant features maps are passed through by a deconvnet [22]. The deconvnet is built with the original CNN filters, and the only difference is they transposed the image to the deconvnet get the same input image dimension. Thus, successively applying the following operations: (i) Unpooling, (ii) ReLU and (iii) Filtering. In operation (i) is necessary to save all local max pooling activations of CNN, while (ii) use the same ReLU function. At step (iii) it is used the convolutional layer with the transposed CNN filters.

## 2.2   Data Augmentation Methods

The small availability of data is a problem to some problems categories, for example, medical image problems such as [10,11,16,20]. A possible way to deal with that problem is to use Data Augmentation methods. In the image context problems, classical image processing transformations such as rotation, cropping, scale, channel shift, elastic deformation, and histogram based methods are used. Miko lajczyk and Grochowski [9] proposed a data augmentation approach which used style transfer to generate new images with high perceptive quality. Perez and Wang [12] evaluate the quality of traditional methods of data augmentation and use Generative Adversarial Networks [4] to generate new data.

## 3   Active Image Data Augmentation

Active Image Data Augmentation (ADA) is a method to enrich a training database based on the original input data and the trained neural network. The method produces new training data dynamically during each neural network training cycle. This method uses the computed gradient of the network output concerning the input data to obtain the degree of importance of each input vector position. In following, the method selects automatically in the input image the more important contiguous region of $N \times N$ dimension to the neural network inference and replace the pixels values this region to zero. This procedure creates new data with some important input information to network inference removed.

Since part of the most important information was removed, we use the generated data to refine the network training in the next training epoch. This training strategy forces the neural network to use other input information to adjust the weights to produce new predictions. The Eqs. 1–4 represent the ADA method step-by-step.

$$y = f(x; \theta) \tag{1}$$

Equation 1 presents a neural network with a input vector $x$ and the parameters set represented by $\theta$.

$$maps(i, x) = \left\| \frac{\partial y[i]}{\partial x} \right\| \tag{2}$$

The Eq. 2 describes how to obtain the absolute value of the gradient maps computed from the output $i$ with relation to the input vector $x$.

$$mask = build\_mask(maps(i, x), ground\_truth, n, z) \tag{3}$$

$$x\_new = x * mask \tag{4}$$

At Eq. 3, the function $build\_mask$ goals is to build a mask of $N \times N$ dimension composed of values 1 and 0. Every position has values 1, except the contiguous region of dimension $Z \times Z$ that represent the most important information to the neural network inference. The algorithm 1 present a pseudo-code describing how we implemented the $build\_mask$ function efficiently.

Considering $*$ as a point-wise multiplication between the vector $x$ and the $mask$, the vector $x\_new$ obtained in the Eq. 4 represents a new input data similar to the vector $x$, except the most critical information to neural network inference was removed.

The algorithm 1 present a function to select the most import region of $N \times N$ dimension considering the neural network gradient map. Given a gradient map and a ground truth of the input image, firstly, we compute the sum of the matrix stored in the variable $sum$ (lines 4–14). Calculate that partial sum matrix is important because it allows reusing the partial sums. Then, from line 17 to 34, we search for the region of $N \times N$ dimension inside the maps whose amount is maximum. After we find that region, we replace their positions in the mask to zero. Thus, building a mask to replace the original information in the input data $x$.

The Fig. 2 presents a step-by-step of ADA method. In that example, we use an input data $x$ of dimension 1D due to simplicity. First, we compute the gradient maps of position $i$ with relation to the input vector $x$ using the function $maps(i, x)$. So, we build a mask to exclude the region more important of dimension 2 using the function $build\_mask(maps, n = 2)$. Finally, we apply the mask to the input data $x$ and we obtain the new train data.

**Algorithm 1.** Build Mask Function

```
1: function Build_Mask(maps, ground_truth, n, z)
2:      sum ← matrix_zeros[n][n]
3:      mask ← matrix_ones[n][n]
4:      for i ← 0; i < n do
5:          for j ← 0; j < n do
6:              sum[i][j] ← maps[i][j]
7:              if i > 0 then sum[i][j] += sum[i − 1][j]
8:              end if
9:              if j > 0 then sum[i][j] += sum[i][j − 1]
10:             end if
11:             if i > 0 and j > 0 then sum[i][j] −= sum[i − 1][j − 1]
12:             end if
13:         end for
14:     end for
15:     high ← 0
16:     pos ← (0, 0)
17:     for i ← 0; i < n do
18:         for j ← 0; j < n do
19:             k, l ← i + z, j + z
20:             if k ≥ n or l ≥ n then continue
21:             end if
22:             sub = sum[k][l]
23:             if i > 0 then sum[i][j] −= sum[i − 1][j]
24:             end if
25:             if j > 0 then sum[i][j] −= sum[i][j − 1]
26:             end if
27:             if i > 0 and j > 0 then sum[i][j] += sum[i − 1][j − 1]
28:             end if
29:             if sub > high then
30:                 high ← sub
31:                 pos ← (i, j)
32:             end if
33:         end for
34:     end for
35:     for i ← pos[0]; i < pos[0] + z + 1 do
36:         for j ← pos[1]; j < pos[1] + z + 1 do
37:             if k ≥ n or l ≥ n then continue
38:             end if
39:             if ground_truth[i][j] ≠ 1 then mask[i][j] = 0
40:             end if
41:         end for
42:     end for
43:     return mask
44: end function
```

## 3.1    Active Image Data Augmentation Training

We present a training method to use ADA in a generic model in the algorithm 2. The approach consists of two steps, in the first one (lines 3–4) we train the model during *standard_epochs* using only the original data. During the second step, we train the model with *new_data*, considering renewed data at the beginning of each cycle. In general, we training the model in a total of (*standard_epochs* + *cycles* * *ada_epochs*) epochs.

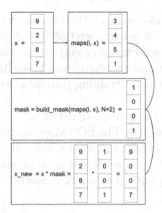

**Fig. 2.** Step-by-step of ADA method

---

**Algorithm 2.** ADA Training Method

---

1: **function** $Ada\_Training(model, data, cycles, ada\_epochs)$
2:    $conventional\_data \leftarrow data + classic\_augmentations(data)$
3:    **for** $i \leftarrow 0; i < standard\_epochs$ **do**
4:       $model.train(conventional\_data)$
5:    **end for**
6:    **for** $i \leftarrow 0; i < cycles$ **do**
7:       $new\_data \leftarrow conventional\_data + Ada(data, model)$
8:       **for** $j \leftarrow 0; j < ada\_epochs$ **do**
9:          $model.train(new\_data)$
10:      **end for**
11:    **end for**
12:    **return** $model$
13: **end function**

---

The first step (lines 3–4) in algorithm 2 is necessary because the neural network weights are initialized randomly, and in this case, the obtained saliency maps are not accurate. Thus, training the model during *standard_epochs* epochs is a good start point.

## 4 Experiments

We used the spinal cord grey matter segmentation task as the reference dataset to perform experiments with the ADA method on a U-Net [17]. In the following sections, we explain the dataset used, evaluation metrics, and U-Net architecture employed.

### 4.1 Spinal Cord Grey Matter Segmentation

We used the spinal cord grey matter segmentation (SCGM) dataset [15] to evaluate our method. This competition released magnetic resonance imaging (MRI) data of different subjects and evaluated the best model to spinal cord grey matter segmentation. The dataset is composed of 80 different subject datasets, and it is divided into 40 datasets do training and 40 to test. Different sites acquired every set of 20 datasets. Details concerning how these datasets were acquired can be found in [15]. Between the 40 training datasets, we used 20% to validation and 80% to train.

Four specialists defined the ground-truth of SCGM dataset. Each specialist segmented all data independently. The SCGM evaluation system uses the majority vote of all specialist to obtain the final result. SCGM evaluation process is executed in an online system [1], so we do not have access to test set ground truth. In the following, we will explain all evaluation metric used by SCGM challenge.

## 4.2    Metrics

SCGM challenge uses ten metrics to evaluate the models. The metrics are grouped into three different categories: (i) Overlapping, (ii) Distance and (iii) Statistical. Table 1 shows a resume of all used metrics. Details can be found in [15].

**Table 1.** Resume of evaluation metrics. Adapted from [13]

| Metric name | Abbr. | Range | Category |
| --- | --- | --- | --- |
| Dice Similarity Coefficient | DSC | 0–1 | Overlap |
| Jaccard Index | JI | 0–100 | Overlap |
| Conformity Coefficient | CC | < 100 | Overlap |
| Mean Absolute Surface Distance | MSD | > 0 | Distance |
| Hausdorff Surface Distance | HSD | > 0 | Distance |
| Skeletonized Hausdorff Distance | SHD | > 0 | Distance |
| Skeletonized Median Distance | SMD | > 0 | Distance |
| Sensitivity (TP) | TPR | 0–100 | Statistical |
| Specificity (TN) | TNR | 0–100 | Statistical |
| Precision | PPV | 0–100 | Statistical |

The lower values in the distance metrics mean better results. Besides, SHD and SMD can be interpreted, such as an indicator of maxima local error and maxima global error, respectively. Specificity metric is important to obtain the quality of the segmented background. To clarify the meaning of some results, we will explain some of the statistical indexes. Given a ground truth (GT) mask and that mask obtained by neural network model output (MO), a voxel can be considered:

---

[1] http://niftyweb.cs.ucl.ac.uk/program.php?p=CHALLENGE.

- True Positive (TP): classified as GM voxel in GT and MO;
- True Negative (TN): classified as non-GM voxel in GT and MO;
- False Positive (FP): classified as non-GM voxel in GT but was classified as GM voxel in MO;
- False Negative (FN): classified as GM voxel in GT but was classified as non-GM voxel in MO.

## 4.3 Models and Training Details

We use a U-Net based architecture to execute the experiments. The U-Net was chosen due to DeepSeg [14] be similar to it and achieve excellent results on SCGM Challenge. The U-Net used is composed of 3 downsample models followed by three upsample models and an output layer with a logistic sigmoid activation function. Besides, every convolutional layer is followed by batch normalization and dropout layer.

In addition to uses all (80%) original training data, we used the following data augmentation data: Rotation, Shift, Scale, Chanel shift, Elastic deformation. Table 2 presents all the parameters used by each method. Every parameter was chosen according to parameters present in [13].

**Table 2.** Data augmentation parameters.

| Method | Parameter |
|---|---|
| Rotation (degrees) | $[-4.6, 4.6]$ |
| Shift (%) | $[-0.03, 0.03]$ |
| Scaling | $[0.98, 1.02]$ |
| Channel Shift | $[-0.17, 0.17]$ |
| Elastic Deformation | $\alpha = 30.0, \sigma = 4.0$ |

We used a dropout rate of 0.5 in every dropout layer and a batch normalization momentum of 0.4. The optimization method used was Adam [6] with an initial learning rate of 0.001 and a batch size 16. Two versions of U-Net were trained to evaluate the impact of ADA method, and they are: (i) U-Net and (ii) U-Net-ADA, both are the same architecture. The model (i) was trained during 1000 epochs using only original training data with traditional data augmentation cited above, while that model (ii) was trained through 1,000 epochs too, but we added data generated by ADA method. To apply the ADA method on the model (ii), first, we train U-Net-ADA with the same data then (i) during 50 epochs and execute 31 cycles of 30 epochs with data from ADA using a region of $20 \times 20$. Thus, every cycle beginning, we apply ADA method using original training data and the actual state of the neural network. The $20 \times 20$ value refers to the parameter z in Eq. 3, as future work we intend proposes a method to choose the best value z. We select the epoch of each model with better results in validation data to produce the test results.

## 5   Results and Discussions

In this section, we discuss the results obtained using the ADA method and compared to other models found in the literature. Table 3 shows the results achieved by U-Net trained using (U-Net-ADA) and without (U-Net) using the ADA method. All results of DeepSeg [14] and Dilated Convolutions [13] was obtained directly from the papers.

**Table 3.** Results of SCGM challenge

| Metric | DEEPSEG | Dillated | U-Net-ADA | U-Net |
|--------|---------|----------|-----------|-------|
| DSC | 0.80 | 0.85 | 0.81 | **0.87** |
| MSD | 0.46 | 0.36 | 0.56 | **0.31** |
| HSD | 4.07 | **2.61** | 4.00 | 4.79 |
| SHD | 1.26 | **0.85** | 0.89 | 0.86 |
| SMD | 0.45 | **0.36** | 0.37 | 0.38 |
| TPR | 78.89 | 94.97 | **99.49** | 96.92 |
| TNR | **99.97** | 99.95 | 99.94 | **99.97** |
| PPV | 82.78 | 77.29 | 67.75 | **78.95** |
| JI | 0.68 | 0.74 | 0.68 | **0.77** |
| CC | 49.52 | 64.24 | 50.81 | **69.65** |

From Table 3, we can see that our U-Net model achieved state-of-art results in six of ten metrics, including essential ones such as Dice score, Precision, Jaccard Index, and Conformity. This is an important contribution to the spinal cord gray matter segmentation research. Besides, it is essential to demonstrate that our baseline model is competitive and better than other spinal cord grey matter segmentation models. As expected, U-Net-ADA did not achieve better results than the U-Net model in these settings, but it presents competitive results in all metrics and state-of-the-art results considering the true positive rate.

Robustness is a major constraint in machine learning, it is very important to achieve generalization. So, in this work, we realized that the spinal cord grey matter segmentation model did not segment the correct region because of correct information, it was using context information to segment the grey matter. Thus, our ADA method goal is to force the model to search for other information beyond the grey matter context.

The main ADA goal is to improve the model robustness. So, it is necessary to evaluate the models according to this property. To perform this evaluation, we generate two test dataset from the validation data: (i) Augmented Data 1 and (ii) Augmented Data 2. To create the data from the set (i), we use U-Net-ADA model and apply the ADA method using the validation data.

We follow the same process to generate data to set (ii), but we use U-Net instead of U-Net-ADA. We used the validation data because we do not have access to the ground-truth of test data, so it is impossible to perform this evaluation. We have been used these models to generate the data because we need trained models to visualize what is important to them and generate ADA data. Thus, we generate data (i) from U-Net-ADA because it is a data where U-Net-ADA most important information will be deleted and in data (ii) the most important information to U-Net will be deleted. So, if the U-Net-ADA present good results even in data (i) is because of it generalized to use other pieces of information, thus being most robust.

**Table 4.** Results with only augmented data

| Metric | Augmented data 1 | | Augmented data 2 | |
|--------|----------|-------|----------|-------|
|        | U-Net-ADA | U-Net | U-Net-ADA | U-Net |
| DSC  | **0.88**  | 0.45  | **0.85**  | 0.52  |
| HSD  | **1.75**  | 3.01  | 1.91      | **1.67** |
| PPV  | **0.81**  | 0.39  | **0.82**  | 0.68  |
| TNR  | **99.91** | 99.56 | 99.93     | **99.95** |

Table 4 presents all results obtained from the robustness experiment. We can see that U-Net-ADA model is better than the U-Net model in a robust evaluation approach, reinforcing our assumption. The dynamic way of ADA training method can be adjusted U-Net-ADA weights to different possible inputs, thus, making the model robust to different input image data. Since robustness is related to regularization and biased models, as future work, it is important to evaluate the ADA method with relation to those properties.

# 6   Conclusion

Robustness is a property where the model reaches certain levels of performance in the face of the parameters or input variation. In this work, we propose a method called Active Image Data Augmentation (ADA) to improve model robustness, a major constraint to applications such as health care system, security, and self-driving car. ADA method consists of generating new training data based on gradient information in a dynamic way. The data generated from ADA forces the deep neural network model to self-adapt to new input data based on the most important input information.

We evaluated the proposed approach using a U-Net network to spinal cord gray matter segmentation task. The obtained results suggest that using ADA in the U-Net output in a competitive model considering the original data and reach excellent results in a robustness evaluation scenario, significantly better than U-Net without ADA in training. That results may imply that the dynamic data generated from ADA method is forcing the model to adapt the weights to new input data.

**Acknowledgments.** This work has been supported by FCT - Fundação para a Ciência e Tecnologia within the Project Scope: UID/CEC/00319/2019. The authors also thanks CAPES and CNPq for the financial support.

# References

1. Antol, S., et al.: VQA: visual question answering. In: Proceedings of the IEEE International Conference on Computer Vision, pp. 2425–2433 (2015)
2. Bishop, C.M.: Pattern Recognition and Machine Learning. Springer, New York (2006)
3. Dice, L.R.: Measures of the amount of ecologic association between species. Ecology **26**(3), 297–302 (1945)
4. Goodfellow, I., et al.: Generative adversarial nets. In: Advances in Neural Information Processing Systems, pp. 2672–2680 (2014)
5. He, K., Zhang, X., Ren, S., Sun, J.: Deep residual learning for image recognition. CoRR abs/1512.03385 (2015). http://arxiv.org/abs/1512.03385
6. Kingma, D.P., Ba, J.: Adam: a method for stochastic optimization. arXiv preprint arXiv:1412.6980 (2014)
7. LeCun, Y., Bengio, Y., Hinton, G.: Deep learning. Nature **521**(7553), 436 (2015)
8. LeCun, Y., Bengio, Y., et al.: Convolutional networks for images, speech, and time series. In: The Handbook of Brain Theory and Neural Networks, vol. 3361, no. 10 (1995)
9. Mikołajczyk, A., Grochowski, M.: Data augmentation for improving deep learning in image classification problem. In: 2018 International Interdisciplinary PhD Workshop (IIPhDW), pp. 117–122. IEEE (2018)
10. Pereira, S., Meier, R., Alves, V., Reyes, M., Silva, C.A.: Automatic brain tumor grading from MRI data using convolutional neural networks and quality assessment. In: Stoyanov, D., et al. (eds.) MLCN/DLF/IMIMIC -2018. LNCS, vol. 11038, pp. 106–114. Springer, Cham (2018). https://doi.org/10.1007/978-3-030-02628-8_12
11. Pereira, S., Pinto, A., Alves, V., Silva, C.A.: Brain tumor segmentation using convolutional neural networks in MRI images. IEEE Trans. Med. Imaging **35**(5), 1240–1251 (2016)
12. Perez, L., Wang, J.: The effectiveness of data augmentation in image classification using deep learning. arXiv preprint arXiv:1712.04621 (2017)
13. Perone, C.S., Calabrese, E., Cohen-Adad, J.: Spinal cord gray matter segmentation using deep dilated convolutions. Sci. Rep. **8**(1), 5966 (2018)
14. Porisky, A., et al.: Grey matter segmentation in spinal cord MRIs via 3D convolutional encoder networks with shortcut connections. In: Cardoso, M.J., et al. (eds.) DLMIA/ML-CDS -2017. LNCS, vol. 10553, pp. 330–337. Springer, Cham (2017). https://doi.org/10.1007/978-3-319-67558-9_38
15. Prados, F.: Spinal cord grey matter segmentation challenge. Neuroimage **152**, 312–329 (2017)
16. Rieke, J., Eitel, F., Weygandt, M., Haynes, J., Ritter, K.: Visualizing convolutional networks for MRI-based diagnosis of Alzheimer's disease. CoRR abs/1808.02874 (2018). http://arxiv.org/abs/1808.02874
17. Ronneberger, O., Fischer, P., Brox, T.: U-net: convolutional networks for biomedical image segmentation. In: Navab, N., Hornegger, J., Wells, W.M., Frangi, A.F. (eds.) MICCAI 2015. LNCS, vol. 9351, pp. 234–241. Springer, Cham (2015). https://doi.org/10.1007/978-3-319-24574-4_28

18. Simonyan, K., Vedaldi, A., Zisserman, A.: Deep inside convolutional networks: visualising image classification models and saliency maps. arXiv preprint arXiv:1312.6034 (2013)
19. Springenberg, J.T., Dosovitskiy, A., Brox, T., Riedmiller, M.: Striving for simplicity: the all convolutional net. arXiv preprint arXiv:1412.6806 (2014)
20. Xie, X., Li, Y., Shen, L.: Active learning for breast cancer identification. CoRR abs/1804.06670 (2018). http://arxiv.org/abs/1804.06670
21. Zeiler, M.D., Fergus, R.: Visualizing and understanding convolutional networks. In: Fleet, D., Pajdla, T., Schiele, B., Tuytelaars, T. (eds.) ECCV 2014. LNCS, vol. 8689, pp. 818–833. Springer, Cham (2014). https://doi.org/10.1007/978-3-319-10590-1_53
22. Zeiler, M.D., Krishnan, D., Taylor, G.W., Fergus, R.: Deconvolutional networks (2010)
23. Zhang, Q., Cao, R., Shi, F., Wu, Y.N., Zhu, S.C.: Interpreting CNN knowledge via an explanatory graph. In: Thirty-Second AAAI Conference on Artificial Intelligence (2018)
24. Zhang, Y., Li, K., Li, K., Wang, L., Zhong, B., Fu, Y.: Image super-resolution using very deep residual channel attention networks. CoRR abs/1807.02758 (2018). http://arxiv.org/abs/1807.02758

# A Novel Density-Based Clustering Approach for Outlier Detection in High-Dimensional Data

Thouraya Aouled Messaoud[1]([✉]), Abir Smiti[2]([✉]), and Aymen Louati[1]

[1] Institut Supérieur d'Informatique du Kef, Université de Jendouba,
Jendouba, Tunisie
Thouraya.aouled.messaoud@gmail.com
[2] LARODEC, Institut Supérieur de Gestion de Tunis,
Tunis, Tunisie
Abir.smiti@gmail.com

**Abstract.** Outlier detection is a primary aspect in data-mining and machine learning applications, also known as outlier mining. The importance of outlier detection in medical data came from the fact that outliers may carry some precious information however outlier detection can show very bad performance in the presence of high dimensional data. In this paper, a new outlier detection technique is proposed based on a feature selection strategy to avoid the curse of dimensionality, named Infinite Feature Selection DBSCAN. The main purpose of our proposed method is to reduce the dimensions of a high dimensional data set in order to efficiently identify outliers using clustering techniques. Simulations on real databases proved the effectiveness of our method taking into account the accuracy, the error-rate, F-score and the retrieval time of the algorithm.

**Keywords:** Outliers · Feature selection · Clustering · DBSCAN

## 1 Introduction

Data analytics' results could be affected by misleading or wrong data, also known as outliers. In fact, outliers are data instances that extremely deviate from well defined norms of a data set or given concepts of expected behavior, see Fig. 1. Therefore, the presence of outliers has become a major challenge, as outliers could really mislead analysis results. These outliers can be caused by measurement errors, malfunctioning equipment, inherent data variability or human mistake.

Therefore, outlier detection techniques are used and can be defined as the process of identifying outlying observations. Outlier detection methods are widely adapted in different fields from which we can mention the medical, security, business and industrial fields. For instance, in the medical field unexpected behavior of a patient's health condition can be identified and treated in an early stage.

© Springer Nature Switzerland AG 2019
H. Pérez García et al. (Eds.): HAIS 2019, LNAI 11734, pp. 322–331, 2019.
https://doi.org/10.1007/978-3-030-29859-3_28

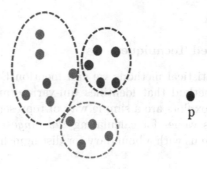

**Fig. 1.** Data object p is considered as outlier

Different techniques have been proposed and can be categorized into four main categories. Statistical-based techniques [1,2] consider outliers as observations that extremely deviates from a standard distribution. In distance-based techniques [3,4] distance metrics are used to decide whether an observation is an outlier or not. Other techniques measure the local density [5,6] to identify outliers and are known as density-based techniques. On the other hand, we can mention clustering-based techniques depend on the process of finding different clusters where the objects that does not fit into any cluster are considered as outliers [7–11]. Despite the variety of outlier detection methods, most of them suffer from bad behavior when applied to high dimensional data.

Therefore, we propose a novel density-based clustering approach for outlier detection in high-dimensional medical data called INfinite Feature Selection DBSCAN (INFS-DBSCAN). Our proposed approach is considered as an intelligent hybrid system that combines the qualities of two different techniques. A feature selection strategy is used in order to reduce dimensionality by selecting $k$ most relevant features and removing those that are less important, then the DBSCAN clustering approach is applied to identify outliers. Hence, the feature selection strategy compensates the weakness of DBSCAN in high dimensional data. In this way, better clustering accuracy is obtained and miss-classification is clearly reduced.

Our method was tested using two real world medical datasets from the UCI repository [12] compared to two other well-known techniques which are DBSCAN [7] and FS-DBSCAN [8]. Our method outperformed both techniques in terms of clustering accuracy for both datasets.

This paper is organized as follows: In Sect. 2, some strategies for outlier detection will be approached. Section 3 describes in detail our new approach INFS-DBSCAN. Section 4 presents and analyzes experimental results carried out on datasets from U.C.I. repository. Finally, Sect. 5 conclude this work and presents future work.

# 2  Related Work

## 2.1  Statistical-Based Techniques

Among parametric statistical methods we can mention Boxplots. Jorma et al. [1] have proposed a method that identifies uni-variate and multi-variate outliers using boxplots. Boxplots are a simple way of representing the five-number summary and provides values for calculating *the range* and *the inter quartile* *IQR*, which can provide us with a boundary for distinguishing normal data from outliers.

Histograms are non-parametric statistical methods that could effectively identify outliers. The Histogram Based Outlier Score (HBOS) algorithm was proposed by Markus and Andreas [2] in order to detect outlying observations by calculating an outlier score for each data instance.

## 2.2  Distance-Based Techniques

In these methods outliers are detected by calculating distances among all data objects based on various distance-related metrics. Afterward, objects that haven't enough neighbors are more likely to be outliers. Nearest-neighbor approaches are the most commonly used.

Kriegel et al. [3] proposed a novel approach called the Angle-Based Outlier Detection (ABOD) method which still uses distances but also considers the variances of angles of all data objects in order to avoid the curse of dimensionality.

Another distance-based algorithm was introduced by Ke et al. [4]. The algorithm computes LDOF factor on a data instance which indicate how much it deviates from its neighborhood. Data instances obtaining high scores are more likely considered as outliers.

## 2.3  Density-Based Techniques

In density-based techniques density is measured to detect outliers, an outlier is detected when its local density differs from its neighborhood. We can mention Local Outlier Factor (LOF) [5] and INFLuenced Outlierness degree (INFLO) as well-known approaches which are based on this technique. LOF aims to assign a degree of being an outlier to each data object in a multidimensional data set. The LOF approach needs well-separated clusters in order to perform well otherwise it suffers from wrong outlier scores, therefore Jin et al. [6] proposed INFLO based on the symmetric neighborhood relationship. In other words, neighbors and reverse neighbors are both considered when estimating its density.

## 2.4  Clustering-Based Techniques

Clustering-based techniques depend on the process of finding different clusters where the objects that does not fit into any cluster are considered as outliers. Several approaches are developed, Density Based Spatial Clustering of Applications with Noise (DBSCAN) [7] can identify noise in low density areas and can identify clusters of arbitrary shapes. Feature Selection - DBSCAN (FS-DBSCAN) [8] came with the same principle of classic DBSCAN and added a feature selection strategy in order to solve the curse of dimensionality problem. The Relative Outlier Cluster Factor (RCOF) method was introduced by Huang et al. [9] and is based on the idea that clusters which are smaller than a normal cluster are considered as outliers.

## 3    INFS-DBSCAN Approach

Our approach INFS-DBSCAN (Infinite Feature Selection DBSCAN) came from the fact that DBSCAN is an efficient outlier detection approach however it suffers from the curse of dimensionality. DBSCAN can't handle a huge amount of features but to overcome this issue a simple feature selection strategy can be applied to the dataset. Our approach combines the qualities of DBSCAN [7] for outlier detection and Inf-FS [13] for feature selection.

Thus, INFS-DBSCAN is a two-step algorithm, see Algorithm 3:

1. Feature selection: A datamining/machine learning method used to automatically select most relevant features (attributes) to a given predictive model. Simply, unneeded, irrelevant and redundant features are identified or removed from the dataset as such attributes do not contribute to the efficiency of the model. For example, when you are trying to predict patient whom suffer from a disease, person's age may be important but his phone number may not.
2. Cluster-based outlier detection: DBSCAN is implemented on the preprocessed dataset to detect outlying points.

Infinite feature selection is a recent filter-based algorithm proposed by Roffo et al. [13], see Algorithm 3, also known as Inf-FS. Inf-FS main idea is to assign a ranking score to each feature followed by a cross-validation step to select most relevant features. Unlike most filter methods, the algorithm evaluates the importance of a given feature while considering all the possible subsets of features (Fig. 2).

Suppose having a set of feature distributions $F = f(1), \ldots, f(n)$ and $x \in R$ represents a simple of the generic distribution f.

A graph can be defined as $G = (V, E)$, where V is the set of vertices corresponding to each feature distribution and E assign weights to edges. G is represented as an adjacency matrix A to define pairwise energy terms $(a_{ij})(1 < i, j < n)$ as follow :

$$a_{ij} = \alpha\sigma_{ij} + (1 - \alpha)c_{ij} \tag{1}$$

Where:

- $\alpha$ represent a loading coefficient $\in [0, 1]$
- $\sigma_{ij} = max(\sigma_i, \sigma_j)$, $\sigma_i$ and $\sigma_j$ are the standard deviation over the set $x \in f(i)$ and $x \in f(j)$.

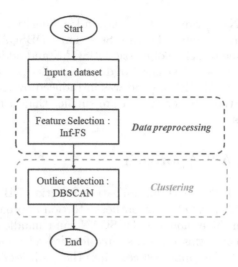

**Fig. 2.** INFS-DBSCAN algorithm

– $c_{ij} = 1 - |Spearman(f(i), f(j))|$, where Spearman indicates Spearman's rank correlation coefficient.

Then matrix $\check{S}$ is calculated which encodes all the information about the energy of our set of features and can be defined as follows:

$$\check{S} = (I - rA^l)^{-1} - I \tag{2}$$

– where $I$ represents an identity matrix and $l$ the path length.
– $r$ represents a real-valued regularization factor and must be chosen appropriately in order for the infinite sum to converge. For all their experiments G. Roffo and all fixed $r$ as follows:

$$r = 0.9/\rho(A) \tag{3}$$

– Let $\rho(A)$ be a spectral radius of the eigenvalues of matrix $A$.

Finally a final energy score can be calculated and is defined as:

$$\check{s}(i) = [\check{S}e]_i \tag{4}$$

– Let e be a 1D array of ones. A generating function is needed to assign a consistent value for the sum of infinite $A^l$ terms to avoid divergence.

The final energy scores are ranked decreasingly to select most relevant features.

The principle objective of DBSCAN algorithm is to efficiently cluster scattered data i.e. it can clearly separate clusters of arbitrary forms. Thus, DBSCAN

**Algorithm 1.** Infinite Feature Selection (Inf-FS)

---

**Input**: $F = f(1), \ldots, f(n)$ , $\alpha$

**Output**: $\check{s}$ energy scores, for each feature

Building the graph

for $i = 1 : n$ do

  for $j = 1 : n$ do

    $\sigma_{ij} = max(std(f^{(i)}), std(f^{(j)}))$

    $c_{ij} = 1 - |Spearman(f(i), f(j))|$

    $a_{ij} = \alpha\sigma_{ij} + (1 - \alpha)c_{ij}$

  end for

end for

Passing to infinite

$r = 0.9/\rho(A)$

$\check{S} = (I - rA^l)^{-1} - I$

$\check{s} = \check{S}e$

**return** $\check{s}$

---

can identify noise in low density areas, i.e., the density of noisy data must be lower than that of the clusters. DBSCAN's effectiveness relies on the three main concepts proposed by Martin et al., the directly density reachability concept (see Definition 1), the density reachability concept (see Definition 2) and finally the density connectivity concept (see Definition 3).

Eps, the maximum radius of the neighborhood, and MinPts, the minimum number of points belonging to Eps neighborhood, are the two required input parameters for DBSCAN.

---

**Algorithm 2.** Basic DBSCAN Algorithm

---

Begin

Arbitrary select a point p.

Retrieve all points that are density-reachable from p w.r.t Eps and MinPts.

If p is a core point, form a cluster.

If p is a border point, no points are density-reachable from p and DBSCAN visits the next point of the database.

Continue the process until each point of the data set has been processed.

End

---

**Definition 1 (Directly density reachability)**

A data object $p$ is considered as being directly-density reachable from a data object $q$ w.r.t Eps and MinPts when [7]:

- $p \in N_{Eps}(q)$
- Core point condition- $-N_{Eps}(q)| \geq MinPts$

**Definition 2 (Density reachability)**
"A point $p$ is density-reachable from a point w.r.t Eps and MinPts if there is a chain of points $p_1, p_2, ..., p_n$ such that $p_{i+1}$ is directly-density reachable from $p_i$." [7]

**Definition 3 (Density connectivity)**
A data instance $p$ is considered as being density-connected to a data instance $q$ w.r.t Eps and MinPts when data object $p$ and $q$ are both density-reachable from a data object $o$ w.r.t Eps and MinPts [7].

---

**Algorithm 3.** Infinite Feature Selection DBSCAN (INFS-DBSCAN)

---

**Input**: $F = f(1), ..., f(n)$, $MinPts$, $Eps$
**Output**: $n$ outliers
**Step 1: Perform Feature Selection**
(* See Inf-FS algorithm 1*)
Define $k$ features
**Step 2: Perform Clustering On Preprocessed Dataset**
Create a new Matrix $Z$ of $k$ dimensions
(* See DBSCAN algorithm 2*)

---

## 4    Experimental Results

In this section the effectiveness of our approach will be highlighted by experimenting INFS-DBSCAN on two medical datasets from U.C.I. repository, analyzing the results according to given evaluation criteria and comparing it to those of DBSCAN [7] and FS-DBSCAN [8].

To evaluate the performance of our algorithm, in comparison with DBSCAN [7] and FS-DBSCAN [8], we decided to use some basic measures derived from the confusion matrix of our model, the *error-rate*, *accuracy* and *F-score*. In addition, the *run time* of each algorithm has been measured.

### 4.1    Datasets Description

Our method was experimented on real medical datasets to provide a detailed analysis of the experimental results. To analyze our proposed method we experimented our algorithm on Thyroid Disease and Diagnostic Breast Cancer Datasets from UCI repository [12].

*Thyroid Disease Dataset.* The original thyroid disease dataset is a commonly used in machine learning for predictive classification models. The dataset contains three main classes to describe patients: **Hyperfunction, subnormal functioning** and **normal**. For outlier detection we defined **hyperfunction** patients as outliers and both other types as inliers, because **hyperfunction** is a clear minority class.

*Breast Cancer Wisconsin (Diagnostic) Dataset.* The original Breast Cancer (Diagnostics) dataset is also used to distinguish **malignant** cases from the **benign**. For outlier detection we defined the **malignant** class as outliers and the **benign** class as inliers.

## 4.2    Experimental Results

Our first experiment was done on the thyroid dataset, *MinPts* and *Eps* were fixed as 10 and 0.1 respectively for all algorithms. For FS-DBSCAN and INFS-DBSCAN, both implementing a feature selection strategy, the number of selected features was set as $k = 2$.

**Table 1.** Results on the thyroid dataset

|                  | DBSCAN   | FS-DBSCAN | INFS-DBSCAN |
|------------------|----------|-----------|-------------|
| Accuracy         | 0.95     | 0.96      | **0.97**    |
| Error-rate       | 0.05     | 0.04      | **0.03**    |
| F-score          | 0.97     | 0.97      | **0.98**    |
| Run time (Sec)   | **0.87** | 842.57    | 1.58        |

For a dataset of 6 dimensions all methods behaved in an acceptable manner, as shown in Table 1, but we can clearly notice that our method provided the best results in terms of clustering accuracy (0.97), Error-rate (0.03) and F-Score (0.98). DBSCAN used less run-time than the other two methods, but INFS-DBSCAN shows very satisfying results compared to FS-DBSCAN. Hence, we can conclude that our method's run-time depends on the number of instances of a dataset and not the number of features.

By excluding unneeded features from our data, we only select 2 features ($k = 2$) out of 6 for FS-DBSCAN and INFS-DBSCAN. By reducing the number of features INFS-DBSCAN did increase the visualization interpretability, i.e. the output of the method was simply visualized and can be clearly interpreted, as shown in Fig. 3.

In our second experiment on the Wisconsins breast cancer (Diagnotic) dataset, our method shows again top-performance with a clustering accuracy of 0.95, an error-rate of 0.05 and an F-score of 0.97. INFS-DBSCAN did obtain a better run-time than FS-DBSCAN, however it takes more time in comparison with DBSCAN. Results are represented in Table 2.

Dimensions were reduced from 30 to only 13 for FS-DBSCAN and INFS-DBSCAN and clustering results were shown in Fig. 4. For WBC dataset we fixed *MinPts* as 10 and *Eps* as 0.6 for all three techniques.

From the experimental results we can conclude that the feature selection strategy implemented in INFS-DBSCAN is more efficient than the strategy of FS-DBSCAN, since Inf-FS was able to identify better features which improved the clustering results.

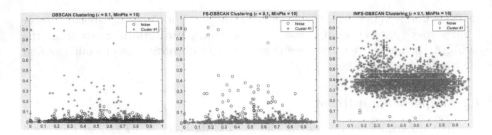

**Fig. 3.** Clustering results for $MinPts = 10$, $Eps = 0.1$ and $k = 2$

**Table 2.** Results on the breast cancer dataset

|                | DBSCAN | FS-DBSCAN | INFS-DBSCAN |
|----------------|--------|-----------|-------------|
| Accuracy       | 0.93   | 0.94      | **0.95**    |
| Error-rate     | 0.07   | 0.06      | **0.05**    |
| F-score        | 0.95   | 0.96      | **0.97**    |
| Run time (Sec) | **0.34** | 24.14   | 1.18        |

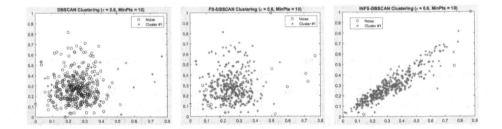

**Fig. 4.** Clustering results for $MinPts = 10$, $Eps = 0.6$ and $k = 13$

## 5    Conclusion

Our method was experimented on two real world datasets in order to compare the results with other well-known methods. In both experiments our proposed method INFS-DBSCAN behaved in a satisfying manner by providing better results than existing approaches. INFS-DBSCAN does effectively select most relevant features which did not only improve outlier detection but also the clustering accuracy and clustering interpretability.

Our outlier detection policy can still be improved by automating DBSCAN's parameter setting as this may require certain knowledge and wrong parameter settings can lead to misbehavior. Moreover, it could be useful to implement techniques to make our approach incremental, as in the medical field new data will be continuously generated.

# References

1. Laurikkala, J., Juhola, M., Kentala, E., Lavrac, N., Miksch, S., Kavsek, B.: Informal identification of outliers in medical data. In: Fifth International Workshop on Intelligent Data Analysis in Medicine and Pharmacology, vol. 1, pp. 20–24 (2000)
2. Goldstein, M., Dengel, A.: Histogram-based outlier score (HBOS): a fast unsupervised anomaly detection algorithm. In: KI-2012: Poster and Demo Track, pp. 59–63 (2012)
3. Kriegel, H.-P., Zimek, A., et al.: Angle-based outlier detection in high-dimensional data. In: Proceedings of the 14th ACM SIGKDD International Conference on Knowledge Discovery and Data Mining, pp. 444–452. ACM (2008)
4. Zhang, K., Hutter, M., Jin, H.: A new local distance-based outlier detection approach for scattered real-world data. In: Theeramunkong, T., Kijsirikul, B., Cercone, N., Ho, T.-B. (eds.) PAKDD 2009. LNCS (LNAI), vol. 5476, pp. 813–822. Springer, Heidelberg (2009). https://doi.org/10.1007/978-3-642-01307-2_84
5. Breunig, M.M., Kriegel, H.-P., Ng, R.T., Sander, J.: LOF: identifying density-based local outliers. In: ACM SIGMOD Record, vol. 29, pp. 93–104. ACM (2000)
6. Jin, W., Tung, A.K.H., Han, J., Wang, W.: Ranking outliers using symmetric neighborhood relationship. In: Ng, W.-K., Kitsuregawa, M., Li, J., Chang, K. (eds.) PAKDD 2006. LNCS (LNAI), vol. 3918, pp. 577–593. Springer, Heidelberg (2006). https://doi.org/10.1007/11731139_68
7. Ester, M., Kriegel, H.-P., Sander, J., Xiaowei, X., et al.: A density-based algorithm for discovering clusters in large spatial databases with noise. In: KDD, vol. 96, pp. 226–231 (1996)
8. Xianting, Q., Pan, W.: A density-based clustering algorithm for high-dimensional data with feature selection. In: 2016 International Conference on Industrial Informatics-Computing Technology, Intelligent Technology, Industrial Information Integration(ICIICII), pp. 114–118. IEEE (2016)
9. Huang, J., Zhu, Q., Yang, L., Cheng, D.D., Quanwang, W.: A novel outlier cluster detection algorithm without top-n parameter. Knowl. Based Syst. **121**, 32–40 (2017)
10. Smiti, A., Elouedi, Z.: COID: maintaining case method based on clustering, outliers and internal detection. In: Lee, R., Ma, J., Bacon, L., Du, W., Petridis, M. (eds.) Software Engineering, Artificial Intelligence, Networking and Parallel/Distributed Computing 2010. SCI, vol. 295, pp. 39–52. Springer, Heidelberg (2010). https://doi.org/10.1007/978-3-642-13265-0_4
11. Smiti, A., Elouedi, Z.: WCOID: maintaining case-based reasoning systems using weighting, clustering, outliers and internal cases detection. In: International Conference on Intelligent Systems Design and Applications (ISDA), pp. 356–361. IEEE Computer Society (2011)
12. UCI machine learning repository. https://archive.ics.uci.edu/ml/index.php/
13. Roffo, G., Melzi, S., Cristani, M.: Infinite feature selection. In: Proceedings of the IEEE International Conference on Computer Vision, pp. 4202–4210 (2015)

# Visual Analysis and Advanced Data Processing Techniques

# Convolutional CARMEN: Tomographic Reconstruction for Night Observation

Francisco García Riesgo[1,2], Sergio Luis Suárez Gómez[2,3],
Fernando Sánchez Lasheras[2,3(✉)], Carlos González Gutiérrez[2,4],
Carmen Peñalver San Cristóbal[1], and Francisco Javier de Cos Juez[1,2]

[1] Department of Prospecting and Exploitation of Mines,
University of Oviedo, Independencia 13, 33004 Oviedo, Spain
[2] Instituto Universitario de Ciencias y Tecnologías Espaciales de Asturias
(ICTEA), University of Oviedo, Independencia 13, 33004 Oviedo, Spain
sanchezfernando@uniovi.es
[3] Department of Mathematics, University of Oviedo,
Calvo Sotelo s/n, 33007 Oviedo, Spain
[4] Department of Computer Science, University of Oviedo,
Campus of Viesques s/n, 33024 Gijón, Spain

**Abstract.** To remove the distortion that the atmosphere causes in the observations performed with extremely large telescopes, correction techniques are required. To tackle this problem, adaptive optics systems uses wave front sensors obtain measures of the atmospheric turbulence and hence, estimate a reconstruction of the atmosphere when this calculation is applied in deformable mirrors, which compensates the aberrated wave front. In Multi Object Adaptive Optics (MOAO), several Shack-Hartmann wave front sensors along with reference guide stars are used to characterize the aberration produced by the atmosphere. Typically, this is a two-step process, where a centroiding algorithm is applied to the image provided by the sensor and the centroids from different Shack-Hartmanns wave front sensors are combined by using a Least Squares algorithm or an Artificial Neural Network, such as the Multi-Layer Perceptron. In this article a new solution based on Convolutional Neural Networks is proposed, which allows to integrate both the centroiding and the tomographic reconstruction in the same algorithm, getting a substantial improvement over the traditional Least Squares algorithm and a similar performance than the Multi-Layer Perceptron, but without the need of previously computing the centroiding algorithm.

**Keywords:** Convolutional Neural Networks (CNN) · Adaptive Optics (AO) ·
Multi Object Adaptive Optics (MOAO) · Extremely Large Telescopes (ELT) ·
Durham Adaptive Optics Simulation Platform (DASP)

## 1 Introduction

The observations of the sky with the help of telescopes, except for those built for solar observation, is performed by night. During the night hours, the atmospheric turbulence decreases if compared with daily hours when sun heats the atmosphere and turbulence increases. Despite this fact, one of the main problems of night observation with ground

© Springer Nature Switzerland AG 2019
H. Pérez García et al. (Eds.): HAIS 2019, LNAI 11734, pp. 335–345, 2019.
https://doi.org/10.1007/978-3-030-29859-3_29

telescopes is the deformation suffered for the wavefront [1] when passing through the terrestrial atmosphere layers. Nowadays, the development of techniques able to correct the wavefront deformation is still a research issue.

Adaptive Optics (AO) is a methodology which aim is the correction of the distortion produced by the atmosphere in the wavefront that comes from celestial bodies. In general, AO can be defined as a technology devoted to the improvement of optical devices accuracy by means of the correction of the wavefront distortions [2]. AO is not only employed in astronomical observation, but also it has applications in other fields such as microscopy [3], where it is employed in order to correct sample-induced aberrations, ophthalmology [4] for the measurement and correction of ocular aberrations, laser communication systems [5], etc.

There is no perfect AO system, as all of them has a certain error ratio. The reason for this, is two-fold, on the one hand, it is not going to be able to follow without delay the atmospheric turbulence which is always changing following a random pattern, and on the other hand, the AO system is not able to measure in an exact way the aberration suffered by the wavefront. Therefore, and without considering its architecture, it is not possible to measure the whole of the aberration suffered by the wave-front sensor. Considering these facts, the aim of all the AO configurations will be to reduce as much as possible the system error.

Nowadays, the Extremely Large Telescope (ELT) of the European Organization for Astronomical Research in the Southern Hemisphere (ESO) is under development. This telescope will have a diameter of 39 m, and it will be placed at Cerro Arma-zones in the Atacama Desert in Chile. One of the main challenges for this device, will be the availability of an AO system fast enough to achieve good correction results in real time.

## 2   Materials and Methods

### 2.1   Adaptive Optics in Nocturnal Observations

Adaptive Optics are techniques developed in order to correct the effect that atmospheric distortions produce on the images taken from grounded telescopes [6]. The atmosphere, working as a fluid, has movement of air masses at different scales, with different wind speeds and directions and at different altitudes and densities. The energy from the turbulences at greater scales implies the formation of other turbulences with inherited energy at lower scales. This conforms a non-linear situation where the atmosphere becomes a volume with continuously changing refraction index, and consequently, any light wavefront that passes through it suffers a distortion.

The techniques from AO search the correction of these effects to recover a high-quality image of the observed scientific object. In general, the idea is setting a configuration where the incoming wavefront is measured by a wavefront sensor, as a Shack-Hartmann WaveFront Sensor (SH-WFS) that measures the slopes of the incoming light by considering differential areas; The sensor is divided into subapertures, which locally focus the incoming light with convergent lens into a light spot, whose position is measured by a CCD sensor, allowing to determine the local slope.

The wavefront sensor gives the information to a reconstruction algorithm, which translates the received information into information of how the wavefront of the scientific object should be corrected. Nowadays, some of the most common reconstruction techniques includes Least Squares (LS) [7], Learn and Apply (L+A) [8], Linear-Quadratic Gaussian Control (LQG) [9] and Complex Atmospheric Reconstructor based on Machine lEarNing (CARMEN) [10]. The CARMEN Convolutional reconstructor has been used in other type of sensors like the TPI-WFS [11].

The information from the reconstruction algorithm is given to the deformable mirror (DM) to correct the wavefront. A DM is a mirror designed in sections, each of them with mechanical actuators that allow to modify locally the position and shape of the surface of the mirror. Giving certain voltages to each actuator allow to deform its shape in order to conform the inverse shape of the distortion to reflect a corrected wavefront.

These three components are set in different configurations to match the problem and the active optics. The most basic configuration is Single Conjugate Adaptive Optics (SCAO) [12], based on a single wavefront sensor on axis with the telescope, to measure directly the turbulences from the scientific object; the correction is given to the DM that corrects all the turbulence. It works in closed loop with feed-forward information.

Other configurations are based on correcting different regions of the turbulence, such a Ground-Layer Adaptive Optics (GLAO) [13], or using more stars as reference (either natural or laser guide stars) for gathering more information to perform the correction, such as Multi-Conjugate Adaptive Optics (MCAO) [14] or Multi-Object Adaptive Optics (MOAO) [15].

In this work, MOAO configuration is considered. MOAO uses a series of off-axis guide stars, allowing both natural and laser guide stars, and its correspondent wavefront sensors, to cover a wide field of view, gathering information from the surroundings of the turbulence in the path to the scientific object. The reconstruction algorithm is intended then, to mix all this information together and estimate a correction of the on-axis path for obtaining the correction for the scientific object.

## 2.2 Simulation Platform: DASP

For simulation, the considered software was the Durham Adaptive optics Simulation Platform (DASP) [16], which models and simulates AO systems, from the generation of the turbulence to the implementation of the AO system components and its different configurations.

DASP generates turbulences based on the hypothesis presented in the previous section, which corresponds with the Kolmogorov atmospheric model [17], and it is implemented with Monte-Carlo simulations. It allows the calculations of the physical propagation of the wavefront through the atmosphere, and how it affects the components of the system.

The turbulence of the atmosphere can be determined by the user; defining the turbulence by discrete sections of the atmosphere, with parameters as the coherence length (Fried's parameter $r_0$), a number of layers in the atmosphere, the altitude of the layers, wind direction and wind speed for each layer, and the relative strength of these layers, as well as other parameters such as scales $L_0$ and $l_0$.

This platform allows modelling systems like SCAO, GLAO, MCAO and MOAO, and the user can choose the number of SHWS, DMs, natural or laser guide stars, and their specifications, such as size, sub-apertures, resolution, wavelength, etc. This let the user to obtain the information in the preferred format, such as SHWFS images, DM actuator voltages, SHWFS slopes, centroids, etc.

The modular nature of the simulator let the user to easily implement in the AO system their own development, such as reconstruction techniques; in particular for this work, both CARMEN and the convolutional approach, Convolutional CARMEN, have been implemented to evaluate its performance compared to other reconstruction techniques. Also, there are available some work in progress options for simulation, such as solar implementations.

## 2.3    Convolutional Neural Networks

An artificial neural network (ANN) can be defined as an interconnected group of computational units, that tries to mimic the behaviour of biological neural networks [18]. An ANN is created by connecting one or more neurons to the input. Each pair of neurons may or may not have a connection between them, but in general there is a higher degree of connectivity in this kind of topology.

ANN makes use of supervised learning. In other words, they require of certain amount of information for its training. Please note that not only the value of input variable is required, but also their outputs. The training of an ANN can be performed with the help of different algorithms. In the case of the present research, backpropagation was employed [19].

The rise of Deep Learning in the last years means an important advance in the field of Machine Learning. Deep Learning can be considered as a field of Machine Learning. The hardware advances and the use of Graphical Processing Units (GPUs) in parallel computing have made possible the implementation of some theorical models well-known a few decades ago but that was impossible to apply for solving real problems.

Deep neural networks [20] make use of a large number of layers. Layers are interconnected via nodes or neurons. The input information of a layer is the output of the previous one. In the case of the present research, Convolutional Neural Networks (CNN) are employed.

CNN are a kind of neural networks [21] developed by LeCun in 1989 [22]. The aim of CNN is processing a data set in which the relative position of elements is relevant. This type of ANN has shown a good results in image recognition, language processing, etc. [23, 24].

CNN makes use of a mathematical operation called convolution. A convolution is an operation of two functions of a real-valued argument [25].

Figure 1 shows the scheme of a convolutional layer. As it can be observed, pixels in the original image are convolved resulting in the feature maps. A pooling layer reduces the dimensionality of each feature map separately.

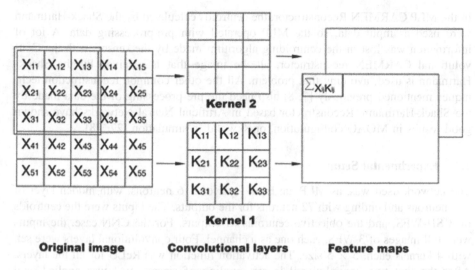

**Fig. 1.** Convolutional layer scheme.

In the case of images as in the present research, that has two dimensions, a two-dimensional kernel $k$ is employed, and the equation is written as follows:

$$s(i,j) = (I * k)(i,j) = \sum_m \sum_n I(m,n) \cdot k(i-m, j-n) \tag{1}$$

The aim of the pooling operation is the reduction of the dimensionality of the image. This operation is applied on each feature map separately, giving as output a new set of feature maps.

In a CNN, the use of the padding operator is also required. Padding consist on adding extra pixels outside the image. As the use of filters make images dimensions smaller after each operation, one way of avoiding such reduction is just adding in the border of the image one more row and column with all their pixel in 0.

The outputs of each convolution operation are passed through a nonlinear activation function. In the case of the present research, the Rectified Linear Unit Activation Function (ReLU) is employed. It can be expressed as $f(x) = \max(0, x)$ where $x$ is the input value [25].

The function to be optimized by the CNN is called error function. In the case of the present research we have made use of the mean square error (MSE) [25]. A CNN can be represented as a fully connected neural network but with infinitely strong prior over their weights. In the case of the present research, CNN are used to output a high-dimensional structural object. The implementation performed in this research, also exploits the current capabilities of parallel computing. Finally, the output obtained from the CNN is employed as input information for training a multilayer perceptron (MLP) neural network.

The main advantage that is provided with this new reconstructor compared with the previous developed by this group [26] lies in the use of this kind of neural networks.

In the MLP CARMEN Reconstructor, the centroids calculated by the Shack-Hartmann were used as input data, so the MLP operated with pre-processing data. A lot of information was lost in the centroiding algorithm made by the sensor; with the Convolutional CARMEN reconstructor, all the image that is received by the Shack-Hartmann is used, avoiding that problem. All the other common reconstruction techniques mentioned previously [7, 8] also need the pre-processing of the data made by the Shack-Hartmann. Reconstructor based in Artificial Neural Networks have shown good results in MOAO configuration, on-sky and in simulation [27, 28].

## 2.4 Experimental Setup

The network used was, as MLP, an input layer of 216 neurons, with hidden layer of 216 neurons and ending with 72 neurons for the outputs. The inputs were the centroids of 3 SH-WFS, and the objective centroids as outputs. For the CNN case, the inputs were full images of 3 WFS, each one as a channel. Four convolutional layers were set, with 4 kernels each, $5 \times 5$ size. The activation function was ReLU for all the layers. After the first two convolutional layers, pooling of sizes $2 \times 2$ was applied, and pooling of $4 \times 4$ was applied after the last two; the fully-connected layers had 3072 neurons as input, with 216 neurons in the hidden layer and 72 for the output layer.

The experiments were performed on a computer running on Ubuntu LTS 14.04.3, with an Intel Xeon CPU E5-1650 v3 @ 3.50 GHz, 128 Gb DDR4 memory, Nvidia GeForce GTX TitanX, and SSD hard drive. The computing language used was Python 3, in which DASP is implemented.

Simulations were performed to obtain the training set. Layers with heights of turbulence ranging from 0 to 15500 m with steps of 100 m formed the turbulence. Fried's coherence length was set ranging from 8 cm to 20 cm, with steps of 1 cm. This was sampled 100 times per combination. The training set was formed by 186000 images of the subapertures of the Wavefront with the simulated output for each one that needed 3 h to be created. The network was trained with different topologies looking for one of the best solves the problem. Each training process took from 24 to 48 h, depending on the number of epochs.

Two different situations were considered for the test sets. First batch of tests were performed as Table 1 shows; the atmospheric parameters for the three test cases. For this research, the outer scale was set to 30 m.

In order to assess the performance of the proposed algorithm, it is compared with different methodologies. Also, test with fixed altitudes were computed to corroborate the performance. The fixed heights turbulence layers are set at 5000, 10000 and 15000 m. The three are combined with another turbulence layer set at 0 m. The relative strength of each layer is the same. For the simulation, the $r_0$ values range from 5 cm to 20 cm. Each of the combinations was sampled 1000 times. In this case 15000 images were needed for the test, made by the simulator in approximately 15 min.

Table 1. Atmospheric parameters for the three test cases.

| Parameter | Values | | | Units |
|---|---|---|---|---|
| Test name | atm1 | atm2 | atm3 | |
| $r_0$ at (0.5 μm) | 0.16 | 0.12 | 0.085 | m |
| **Layer 1** | | | | |
| Altitude | 0 | 0 | 0 | m |
| Relative strength | 0.65 | 0.45 | 0.80 | |
| Wind speed | 7.5 | 7.5 | 10 | m/s |
| Wind direction | 0 | 0 | 0 | degrees |
| **Layer 2** | | | | |
| Altitude | 4000 | 2500 | 6500 | m |
| Relative strength | 0.15 | 0.15 | 0.05 | |
| Wind speed | 12.5 | 12.5 | 15 | m/s |
| Wind direction | 330 | 330 | 330 | degrees |
| **Layer 3** | | | | |
| Altitude | 10000 | 4000 | 10000 | m |
| Relative strength | 0.10 | 0.30 | 0.10 | |
| Wind speed | 15 | 15 | 17.5 | m/s |
| Wind direction | 135 | 135 | 135 | degrees |
| **Layer 4** | | | | |
| Altitude | 0 | 0 | 0 | m |
| Relative strength | 0.65 | 0.45 | 0.80 | |
| Wind speed | 7.5 | 7.5 | 10 | m/s |
| Wind direction | 0 | 0 | 0 | degrees |

# 3  Results and Discussion

Three different atmospheres already employed in a previous research [29] has been employed. In Table 2, results for the three atmospheres profiles are shown, regarding the reconstructors used; also the information of the uncorrected wavefront is included.

Results are presented in terms of Strehl ratio, Full Width at Half Maximum (FWHM), E50d, Wave-Front Error (WFE) [30, 31].

Table 2. PFS metrics for each tomographic reconstructor and test scenario (all metrics except WFE are defined at 1650 nm).

| Test name | Reconstructor | Metrics | | | |
|---|---|---|---|---|---|
| | | Strehl ratio | FWHM (arcsec) | E50d (arcsec) | WFE (nm) |
| atm 1 | LS | 0,0444 | 0,2677 | 0,6332 | 630,2798351 |
| | CNN | 0,0321 | 0,3053 | 0,7142 | 545,1956026 |
| | MLP | 0,0416 | 0,244 | 0,6926 | 544,2764828 |
| | Uncorrected | 0,0099 | 0,7982 | 0,9439 | 295,3788419 |

(*continued*)

**Table 2.** (*continued*)

| Test name | Reconstructor | Metrics | | | |
|---|---|---|---|---|---|
| | | Strehl ratio | FWHM (arcsec) | E50d (arcsec) | WFE (nm) |
| atm2 | LS | 0,1988 | 0,1043 | 0,4223 | 340,1268457 |
| | CNN | 0,2766 | 0,0957 | 0,3857 | 301,1824269 |
| | MLP | 0,2783 | 0,0961 | 0,3831 | 300,46288 |
| | Uncorrected | 0,0245 | 0,4519 | 0,6618 | 814,550224 |
| atm3 | LS | 0,3008 | 0,1018 | 0,2916 | 295,3788419 |
| | CNN | 0,3891 | 0,0921 | 0,293 | 259,7432544 |
| | MLP | 0,3908 | 0,0921 | 0,292 | 257,1959796 |
| | Uncorrected | 0,0388 | 0,3989 | 0,4876 | 657,6433445 |

Results in the same terms for the tomographic test at different altitudes are presented in Table 3.

**Table 3.** PFS metrics for each tomographic reconstructor and test scenario (all metrics except WFE are defined at 1650 nm).

| Test name | Reconstructor | Metrics | | | |
|---|---|---|---|---|---|
| | | Strehl ratio | FWHM (arcsec) | E50d (arcsec) | WFE (nm) |
| 5000 m | LS | 0.1993 | 0.1041 | 0.4561 | 338.6720 |
| | CNN | 0.3637 | 0.0946 | 0.3250 | 267.9234 |
| | MLP | 0.3686 | 0.0946 | 0.3210 | 268.2175 |
| | Uncorrected | 0.0225 | 0.4785 | 0.6899 | 949.7711 |
| 10000 m | LS | 0.0647 | 0.2318 | 0.5274 | 480.2900 |
| | CNN | 0.1731 | 0.1027 | 0.4887 | 361.1064 |
| | MLP | 0.1739 | 0.1025 | 0.4877 | 361.7813 |
| | Uncorrected | 0.0306 | 0.3396 | 0.6364 | 789.0039 |
| 15000 m | LS | 0.0572 | 0.2426 | 0.5320 | 506.2564 |
| | CNN | 0.0690 | 0.1480 | 0.5374 | 471.9628 |
| | MLP | 0.0653 | 0.1589 | 0.5575 | 474.1660 |
| | Uncorrected | 0.0323 | 0.3351 | 0.6279 | 726.8478 |

Results for both artificial intelligence models are quite similar, in all cases improving significantly the original phase, without correction, and also improving the correction performed by the LS reconstructor. This is particularly noticeable in some parameters as Strehl and WFE. Between both of them, the results in terms of reconstruction error are very similar, however, the principal improvement of the CNN over the MLP resides in the treatment of data.

Although both methods work in "real time", under two milliseconds of recall time, the CNN works with the full WFS image, which implies that it considers more information, and also it does not require some processes as the centroiding algorithm, needed for obtaining the input data of the MLP. Thus, the CNN improves the computational cost needed outside the reconstruction technique in order to be used.

# 4   Conclusions

In this article, a hybrid artificial system is presented, in which different devices such as a telescope, the WFS's, the DM's and the artificial intelligence-based reconstructor system work together in order to correct the atmospheric turbulence. Results of MLP models were already proven in previous works, however, the comparison with CNNs in optical terms is presented. The CNNs advantages includes the consideration of all the information from the WFS image, as well as avoiding the necessity of several pre-processing operations for the data as the centroiding algorithm.

Considering a continuous simulation for the training data could, in future developments, imply a better modeling of the problem when used in networks with a recurrent component.

# References

1. Osborn, J., et al.: Open-loop tomography with artificial neural networks on CANARY: On-sky results. Mon. Not. R. Astron. Soc. **441**, 2508–2514 (2014). https://doi.org/10.1093/mnras/stu758
2. Guzmán, D., de Cos Juez, F.J., Lasheras, F.S., Myers, R., Young, L.: Deformable mirror model for open-loop adaptive optics using multivariate adaptive regression splines. Opt. Express **18**, 6492–6505 (2010)
3. Booth, M.J.: Adaptive optical microscopy: the ongoing quest for a perfect image. Light Sci. Appl. **3**, e165 (2014)
4. Carroll, J., Dubis, A.M., Godara, P., Dubra, A., Stepien, K.E.: Clinical applications of retinal imaging with adaptive optics. Clin. Appl. Retin. Imaging Adapt. Opt. **4**(2), 78–83 (2011). https://doi.org/10.17925/USOR.2011.04.02.78. US Ophthalmic Review
5. Wang, Y., et al.: Performance analysis of an adaptive optics system for free-space optics communication through atmospheric turbulence. Sci. Rep. **8**, 1124 (2018)
6. Dipper, N., Basden, A., Bitenc, U., Myers, R.M., Richards, A., Younger, E.J.: Adaptive optics real-time control systems for the E-ELT. In: Proceedings of the Third AO4ELT Conference, vol. 1, p. 41 (2013)
7. Ellerbroek, B.L.: First-order performance evaluation of adaptive-optics systems for atmospheric-turbulence compensation in extended-field-of-view astronomical telescopes. JOSA A **11**, 783–805 (1994)
8. Vidal, F., Gendron, E., Rousset, G.: Tomography approach for multi-object adaptive optics. JOSA A **27**, A253–A264 (2010)
9. Sivo, G., et al.: First on-sky SCAO validation of full LQG control with vibration mitigation on the CANARY pathfinder. Opt. Express **22**, 23565–23591 (2014)

10. Osborn, J., et al.: First on-sky results of a neural network based tomographic reconstructor: Carmen on Canary. In: Marchetti, E., Close, L.M., Véran, J.-P. (eds.) Adaptive Optics Systems IV. International Society for Optics and Photonics, vol. 9148, p. 91484M (2014)
11. Suárez Gómez, S.L., et al.: Compensating atmospheric turbulence with convolutional neural networks for defocused pupil image wave- front sensors. In: de Cos Juez, F., et al. (eds.) HAIS 2018. LNCS, vol. 10870, pp. 411–421. Springer, Cham (2018). https://doi.org/10.1007/978-3-319-92639-1_34
12. Hippler, S., et al.: Single conjugate adaptive optics for the ELT instrument METIS. Exp. Astron. (2018). https://doi.org/10.1007/s10686-018-9609-y
13. Bendek, E.A., Hart, M., Powell, K.B., Vaitheeswaran, V., McCarthy, D., Kulesa, C.: Latest GLAO results and advancements in laser tomography implementation at the 6.5 m MMT telescope. In: Astronomical Adaptive Optics Systems and Applications IV, vol. 8149, p. 814907 (2011)
14. Beckers, J.M. Increasing the size of the isoplanatic patch with multiconjugate adaptive optics. In: European Southern Observatory Conference and Workshop Proceedings, vol. 30, p. 693 (1988)
15. Gendron, E., et al.: MOAO first on-sky demonstration with CANARY. Astron. Astrophys. **529**, L2 (2011). https://doi.org/10.1051/0004-6361/201116658
16. Basden, A.: DASP the Durham Adaptive optics Simulation Platform: Modelling and simulation of adaptive optics systems
17. Zilberman, A., Golbraikh, E., Kopeika, N.S.: Propagation of electromagnetic waves in Kolmogorov and non-Kolmogorov atmospheric turbulence: three-layer altitude model. Appl. Opt. **47**, 6385–6391 (2008)
18. Aghdam, H.H., Heravi, E.J.: Guide to Convolutional Neural Networks. Springer, New York (2017). vol. 10, pp. 973–978
19. Lasheras, J.E.S., Donquiles, C.G., Nieto, P.J., et al.: A methodology for detecting relevant single nucleotide polymorphism in prostate cancer with multivariate adaptive regression splines and backpropagation artificial neural networks. Neural Comput. Appl., 1–8 (2018). https://doi.org/10.1007/s00521-018-3503-4
20. Suárez-Gómez, S.L., et al.: An approach using deep learning for tomographic reconstruction in solar observation. In: Proceedings of the Adaptive Optics for Extremely Large Telescopes 5; Instituto de Astrofísica de Canarias (IAC) (2017)
21. Suárez Gómez, S.L.: Técnicas estadísticas multivariantes de series temporales para la validación de un sistema reconstructor basado en redes neuronales (2016)
22. LeCun, Y., et al.: Backpropagation applied to handwritten zip code recognition. Neural Comput. **1**, 541–551 (1989)
23. Mirowski, P.W., LeCun, Y., Madhavan, D., Kuzniecky, R.: Comparing SVM and convolutional networks for epileptic seizure prediction from intracranial EEG. In: 2008 IEEE Workshop on Machine Learning for Signal Processing, pp. 244–249 (2008)
24. Nagi, J., et al.: Max-pooling convolutional neural networks for vision-based hand gesture recognition. In: 2011 IEEE International Conference on Signal and Image Processing Applications (ICSIPA), pp. 342–347 (2011)
25. Goodfellow, I., Bengio, Y., Courville, A.: Deep Learning. MIT Press, Cambridge (2016)
26. Suárez Gómez, S.L., et al.: Improving adaptive optics reconstructions with a deep learning approach. In: de Cos Juez, F.J., et al. (eds.) Hybrid Artificial Intelligent Systems, pp. 74–83. Springer International Publishing, Cham (2018)
27. Osborn, J., et al.: Open-loop tomography using artificial neural networks. Proc. Adapt. Opt. Extrem. Large Telesc. **2**, 2420–2434 (2011)

28. Suárez Gómez, S.L., et al.: Analysing the performance of a tomographic reconstructor with different neural networks frameworks. In: International Conference on Intelligent Systems Design and Applications, pp. 1051–1060 (2016)

29. Osborn, J., et al.: Using artificial neural networks for open-loop tomography. Opt. Express **20**, 2420–2434 (2012). https://doi.org/10.1364/oe.20.002420

30. Hardy, J.W.: Adaptive optics for astronomical telescopes. Oxford University Press on Demand, vol. 16 (1998)

31. Alonso Fernández, J.R., Díaz Muñiz, C., Garcia Nieto, P.J., de Cos Juez, F.J., Sánchez Lasheras, F., Roqueñí, M.N.: Forecasting the cyanotoxins presence in fresh waters: A new model based on genetic algorithms combined with the MARS technique. Ecol. Eng., 68–78 (2013). https://doi.org/10.1016/j.ecoleng.2012.12.015

# A Proof of Concept in Multivariate Time Series Clustering Using Recurrent Neural Networks and SP-Lines

Iago Vázquez[1], José R. Villar[2(✉)], Javier Sedano[1], Svetlana Simić[3], and Enrique de la Cal[2]

[1] Instituto Tecnológico de Castilla y León,
Pol. Ind. Villalonquejar, 09001 Burgos, Spain
{iago.vazquez,javier.sedano}@itcl.es
[2] Computer Science Department, EIMEM, University of Oviedo, Oviedo, Spain
{villarjose,delacal}@uniovi.es
[3] Department of Neurology, Clinical Centre of Vojvodina Novi Sad,
University of Novi Sad, Novi Sad, Republic of Serbia
svetlana.simic@mf.uns.ac.rs

**Abstract.** Big Data and the IoT explosion has made clustering multivariate Time Series (TS) one of the most effervescent research fields. From Bio-informatics to Business and Management, multivariate TS are becoming more and more interesting as they allow to match events the co-occur in time but that is hardly noticeable. This study represents a step forward in our research. We firstly made use of Recurrent Neural Networks and transfer learning to analyze each example, measuring similarities between variables. All the results are finally aggregated to create an adjacency matrix that allows extracting the groups. In this second approach, splines are introduced to smooth the TS before modeling; also, this step avoid to learn from data with high variation or with noise. In the experiments, the two solutions are compared suing the same proof-of-concept experimentation.

## 1 Introduction

Time Series (TS) clustering is one of the most effervescent research fields due to the Big Data and the IoT explosion. Until recently, the problem was focused on univariate TS clustering. For instance, [10] proposed use dynamic time warping and k-means to cluster the performance of a photovoltaic power plant, so to predict the meteorological conditions. Similarly, k-means was used to cluster TS and then predict the weather conditions [8]. Interested readers would read the review in [1] for a good review on this topic.

However, TS clustering has been moving from univariate to multivariate TS problems. In these problems, a TS includes more than one variable; i.e., the pollution measurements in a medium or big city includes several physical and chemical variables registered in several stations placed all around of a city. Clustering multivariate TS has been found interesting in order to perform complex

© Springer Nature Switzerland AG 2019
H. Pérez García et al. (Eds.): HAIS 2019, LNAI 11734, pp. 346–357, 2019.
https://doi.org/10.1007/978-3-030-29859-3_30

event detection or to classify the current scenario. For instance, [4] proposed a Partitioning around Meroids and Fuzzy C-Meroids clustering for the problem of detecting high-value pollution records or alarms in the city of Rome.

The similarity among the variables within the TS is one of the most studied topics. PCA similarity factor was combined with the average based Euclidean distance together with a fuzzy clustering scheme in [6]. Discords have been used in multivariate TS to identify anomalies and introduce more efficient search processes [7]. Hash functions have also been used to index and to measure the similarities in multivariate TS searches [16].

Interestingly, models have been also used in measuring the similarity between multivariate TS, i.e., Gaussian Mixture Models [11]. A different approach is based on extracting features and then using these features to group the multivariate TS [5]. Feature extraction together with Self-Organized Maps [14], Hidden Markov Models [9] or Fuzzy Linear [3] are techniques that have been also proposed in solving multivariate TS. Still, this problem cannot be considered solved and a recent study found out that the combination of feature extraction and a classification stage performs better than the current approaches [2].

In this study, a similar idea of that proposed in [11] is revisited for multivariate TS. Recurrent Neural Networks (RNN) are learned to predict a variable from an example and then used to measure the similarity between the different variables. Afterward, the adjacency matrix is found for each example, then aggregated for all the examples and finally binarized to generate the final adjacency matrix. The groups are proposed based on the variables that mutually dependent. To reduce the complexity of the solution transfer learning is proposed.

The organization of this manuscript is as follows. The next section describes both the previous research and the proposal for this study. Section 3 details the dataset and some method's parameters, while Sect. 4 includes the figures and the discussion on the results. The study ends with the conclusions.

## 2    Clustering Multivariate TS

Basically, we use SP-Lines before modelling each of the features in an instance of a multivariate TS. Afterwards, we use the same procedure proposed in [15]. This sections explains the whole process by, first, introducing the previous study in Subsect. 2.1 and then describing how the SP-Lines are used in Subsect. 2.2.

Let's define multivariate TS dataset as the dataset containing examples, each example is a multivariate TS. A multivariate TS is an arrangement of several TS, each one belonging to a different variable. We assume all the examples having the same variables and, without loss of generality, the same sampling frequency and the same number of samples. Therefore, a multivariate TS example is a matrix of $m$ rows of $n$ variables, where each column represents a univariate TS. However, each example has its own number of samples.

## 2.1    RNN Applied in Multivariate TS Clustering

A two stages solution was proposed in [15]: (i) the first stage is devoted to find the similarities between variables in a single example, that is, in a single multivariate TS, and (ii) the second stage aims to aggregate the results among the examples and extract the relationships. To find the similarities between features a RNN predicts the test subsequence of a variable, the prediction error over the remaining variables is a measurement of similarity among variables. For this preliminary study, the aggregation of the results was performed with simple thresholding followed by a graph representation. To make the process feasible, we propose to use transfer learning [12].

**Finding Similarities Between Variables from an Example.** The procedure is depicted in Algorithm 1. Let's $TS^i$ be the current example, $TS^i = \{X_1^i, \ldots, X_n^i\}$ $\forall i : 1, \ldots, N$, where $N$ is the number of examples, $n$ is the number of variables. Moreover, each variable $X_j^i$ can be written as $X_j^i = (x_{j1}^i, \ldots, x_{jm_i}^i)$ $\forall j : 1, \ldots, n$, with $m_i$ being the number of samples of the TS for each variable in the example i.

---

**Algorithm 1.** Computing similarities between features in an example

---

1: **procedure** IN-EXAMPLE-SIMILARITY($TS^i$, LoRNN)  ▷ LoRNN list of pre-learnt RNNs, if available
2:    $sim \leftarrow$ zeroes matrix of size $n \times n$
3:    **for** each variable $j$ in $TS^i$ **do**
4:        $X_j^i \leftarrow$ normalize($X_j^i$)
5:        $RNN_j^i \leftarrow$ **Train-RNN**($X_j^i$, LoRNN[j])
6:        $LoRNN[j] \leftarrow RNN_j^i$
7:        $e_j^i \leftarrow$ RMSE($RNN_j^i$, test($X_j^i$) )
8:        **for** each variable $k$ in $TS^i$, $k \neq j$ **do**
9:            $X_k^i \leftarrow$ normalize($X_k^i$)
10:           $e_{jk}^i \leftarrow$ RMSE($RNN_j^i$, test($X_k^i$) )
11:           $sim[j, k] \leftarrow abs(\frac{e_{jk}^i - e_j^i}{e_j^i})$
12:       **end for**
13:   **end for**
14:   **return** $sim$
15: **end procedure**
16:
17: **procedure** TRAIN-RNN($X_j^i$, $RNN$)  ▷ $RNN$ is a RNN, if available
18:    **if** is.NULL($RNN$) **then**
19:        $RNN \leftarrow$ *full train* RNN for the train part of $X_j^i$
20:    **else**
21:        $RNN \leftarrow$ *tune RNN* for the train part of $X_j^i$
22:    **end if**
23:    **return** $RNN$
24: **end procedure**

---

Let's also assume that a given percentage (%TRN) of the samples of a TS is kept for training and the remaining for testing. In other words, for any variable $X_j^i$ in example i, $(x_{j1}^i, \ldots, x_{j(\%TRN \times m_i)}^i)$ are kept for training and $(x_{j(\%TRN \times m_i)}^i, x_{jm_i}^i)$ are kept for testing.

It is possible to learn an RNN using the training part of $X_j^i$ to predict its behavior in the testing part, let's call this $RNN_j^i$. Let us suppose we obtain a good model, and that the aggregation of the prediction error along the test subset for variable $X_j^i$ is $e_j^i$. This prediction error can be any well-known measurement, as the Root Mean Square Error (RMSE) or similar.

The $RNN_j^i$ is applied to predict each of the remaining variables $X_k^i$ with $k : 1 \cdots n$ and $k \neq j$. The error obtained with $RNN_j^i$ when predicting the test subsequence of the variable $X_k^i$ is denoted as $e_{jk}^i$. This error is scaled wrt the $e_j^i$ in order to obtain a similarity value: $E_{jk}^i = |(e_{jk}^i - e_j^i)/e_j^i|$. Values close to 0.0 means the TS can be successfully predicted by $RNN_j^i$. This prediction is also repeated for $-X_k^i$, that is, the normalized test sequence $X_k^i$ is swapped wrt the time axis to consider the case the two TS $X_j^i$ and $X_k^i$ have a negative correlation. Consequently, the minimum of both errors is kept.

Therefore, the similarity between variable j and the remaining variables in the example i is obtained as the vector $sim_j^i = (E_{j1}^i, \cdots, E_{jn}^i)$, with $E_{jj}^i = 0$. Finally, repeating this procedure for each of the variables in the example i, a distance $n \times n$ matrix is obtained, which represents the outcome of this stage.

**RNN and Transfer Learning.** As seen in Algorithm 1 and in the previous subsection, an RNN is trained using the train part of variable $X_j^i$ from example $TS^i$. As for our previous research [15]; this study makes use of the *rnn* R-package [13]. For each training process a simple grid of 12 different learning rate values (from 1/12 to 1.0), 1 to 12 as the number of epochs and 1 to 12 hidden neurons.

However, training a complete RNN from scratch for each variable and for each example makes this approach unfeasible for even small multivariate TS datasets. A simplification is clearly needed.

To do so, a simple transfer learning scheme [12] is used. For the first time, the Train-RNN is call, a NULL value is given as current RNN; thus, full learning of the RNN is performed. However, when it is not NULL, then it is the $RNN_j^i$ trained in the first iteration of the process for variable j and example $i == 1$. We reuse this RNN model, fitting it to the current $X_j^i$. This adaptation is just a simple weight tuning during a reduced number of iterations (20 in this study).

**Computing the Similarities Within a Multivariate TS Dataset.** Once the similarity matrix between the variables from an example is obtained, computing the similarity between the variables for the whole multivariate TS dataset is a matter of choosing the method. In the preliminary research, each matrix was converted to an adjacency matrix and then to aggregate the adjacency matrices.

The similarity matrix for example $TS^i$ is converted into the SIM adjacency matrix $SIM_i$ as follows. For each pair of variables j and k from the example

$i$, if similarity between $j$ and $k$ is smaller or equal than $th1$ ($sim_i[j,k] \leq th1$), then variable $j$ can predict variable $k$ (denoted as $k \lesssim j$). Thus, $SIM_i[j,k] = 1$; otherwise, $SIM_i[j,k] = 0$.

The adjacency matrices are then aggregated as just the sum of all of them. Therefore, the final aggregated adjacency matrix is $SIM_{ag} = \sum_{\forall i} SIM_i$, each cell contains an integer from 0 to $N$. Finally, the outcome adjacency matrix $SIM_{final}$ is obtained by thresholding $SIM_{ag}$ such that whenever $SIM_{ag}[j,k] \geq th2$ then $SIM_{final} = 1$, otherwise $SIM_{final} = 0$.

This binarization produces an adjacency matrix that can be represented in a graph. This visualization can help in deciding what to do with those variables that have not been grouped yet.

## 2.2   Using SP-Lines in Multivariate TS Clustering

The main difference in this new study is that the variables in each instance are modeled using SP-lines. The idea was developed after the fluctuation of the signal and also because of the effects in a focused problem related with photovoltaic solar power plants. In this problem, the power of a photovoltaic panel has a strong dependence on the irradiance; the clouds that might partially cover the panel clearly decrease the generated power. But as long as the clouds can move rather fast, the consequence is that the generated power behaves with many peaks and valleys and with more fluctuation than with steady weather conditions. Thus, using SP-lines part of this extra fluctuation is absorbed but the peaks and valleys are kept.

The main part of the previous approach is used, but small variations are introduced to include the SP-lines. Previous to the RNN modeling, each TS of an instance is processed as follows:

- Filter outlier values and smooth the TS with sliding window of 5 time samples.
- Normalize the TS with the mean and standard deviation of the TS.
- If needed, segment the TS according to the behaviour of the signal. The main point is to do not substitute stationary segments with a SP-line.
- For each non-stationary segment, determine the SP-line that best fit the TS with 100 degrees of freedom.
- Join all the segments (stationary and converted to SP-lines) in a single TS; this TS is used as the variable's TS within the instance.

An example of a part of a TS and the calculated SP-line is shown in Fig. 1. The instances with the new computed TS for each variable are then used as the instance to evaluate the similarity among the variables.

## 3   Experiment and Methods

To evaluate this preliminary study a real-world multivariate TS dataset has been used; it is the same data set used in [15], so comparisons can be presented. This dataset includes up to $n = 11$ variables in each TS example. These multivariate

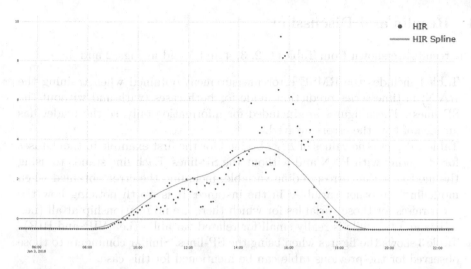

**Fig. 1.** Example to show the behaviour of the SP-Lines (continuous lines) compared to the original TS (dotted lines). The original TS have two stationary segments at the beginning and the end of the TS. The final TS would contain the stationary segments and the inner part of the SP-line curve.

TS have been extracted from a photovoltaic solar power plant, including the following variables:

- Indoor and outdoor temperatures in the weather station (TIN, TOUT)
- Horizontal and Vertical Irradiance reference measurement (HIR and VIR)
- The voltage at the weather station's battery (BV)
- The temperature of 4 photovoltaic panels linked to an inverter (T1 to T4)
- An In-panel Horizontal and Vertical Irradiance measurement (PHI, VHI)

This a toy problem used as a proof of concept. For this problem, the relationships among variables are known: the temperatures ($T_x$, $x \in \{1, 2, 3, 4\}$) are interrelated and so does PVI and VIR. Besides, PVI influences the behaviour of HRI and PHI (these two are interrelated). TIN and TOUT are also mutually dependent, while BV is totally independent of the others.

Each example includes data from the evolution of the magnitude of these variables for a period of four days. This period has been chosen to include long enough TS, however, larger periods could have been chosen as well. Although data are available for more than three months, in this preliminary study only the $N = 5$ examples of these data are considered. Clearly, this is a proof of concept study. A more in-depth research is needed to extract more solid conclusions. However, using this toy problem allows us to evaluate how the method behaves, its weak points and the enhancements needed to be valid for general problems.

The values of the thresholds have been set before any further analysis to $th1 = 0.07$ and $th2 = 3$ (equivalent to require that the 60% of the examples must include that relationship in order to accept a dependence).

# 4    Results and Discussion

The results are shown from Tables 1, 2, 3, 4 and 5 and in Figs. 2 and 3:

- Table 1 includes the RMSE error measurement obtained when training the $RNN_j^i$ in time series prediction mode for both cases (with and without the SP-lines). These figures are included for information only, so the reader has an idea of how the errors evolved.
- Table 2 depicts the values of $E_{jk}^i$ obtained for the first example in the dataset for clustering with RNN and without the SP-lines. Each line stands to using the model for the corresponding variable, including the error obtained when modelling the other variables in the instance. It is worth noticing how the error raises for those variables for which there are no relationship at all (i.e., VIR and T1), while is really small for related variables (i.e., VIR and HIR).
- Table 3 shows the figures when using the SP-lines. Similar comments to those observed for the previous table can be mentioned for this case.
- Tables 4 and 5 show the adjacency matrix obtained after the aggregation of the different examples and pruning with $th2 = 3$, without and with the SP-lines, respectively. The two methods produce the same adjacency matrix.
- Figs. 2 and 3 show the graph obtained from that adjacency matrix for each case; the two graphs are the same. More importantly, the graphs show the existent relationships among the features.

Aside the type of toy problem that we are using in this proof-of-concept and comparison, with a very short multivariate TS without much complexity, the behaviour of both the approach and the results seem valid. In both cases the found relationships are the expected (with variations from one method to the other). Perhaps the most suited grouping is the original one (using only the RNN with the multivariate TS raw data), but the used threshold was higher than that of the clustering with the sp-lines pre-processing.

Nevertheless, it is clear that the proposal still needs plenty of amendments as well as the structure of the method. However, the obtained results seem to be promising. Items such as different type of TS prediction techniques that might be applied provided transfer learning can be deployed, automatic setting the thresholds to adapt to the problem faced, or the definition of similarity measurements that might be more promising than the scaled RMSE are included among others in the next research to be performed. Furthermore, we do believe that this method could be directly applied in medicine and biology, especially in problems where the experts need support in the analysis of big volumes of multivariate TS.

**Table 1.** The RMSE error measurement for each of the fully trained RNN. Upper part: the previous research in [15]. Lower part: results using the SP-lines.

| RNN clustering | | | |
|---|---|---|---|
| Variable | Train error | Variable | Train error |
| VIR | 0.1194 | PHI | 0.1471 |
| HIR | 0.1481 | PVI | 0.1346 |
| T1 | 0.0449 | TIN | 0.04212 |
| T2 | 0.0536 | TOUT | 0.0505 |
| T3 | 0.0461 | BV | 0.1669 |
| T4 | 0.0461 | | |
| SP-lines + RNN clustering | | | |
| Variable | Train error | Variable | Train error |
| VIR | 2.5075 | PHI | 1.3457 |
| HIR | 1.3251 | PVI | 2.5258 |
| T1 | 0.3993 | TIN | 0.1931 |
| T2 | 0.4336 | TOUT | 0.2268 |
| T3 | 0.4353 | BV | 0.1669 |
| T4 | 0.4265 | | |

**Table 2.** RNN clustering. The similarity matrix obtained with the first example from the multivariate TS dataset.

| | T1 | T2 | T3 | T4 | TIN | BV | TOUT | VIR | VHI | HIR | PHI |
|---|---|---|---|---|---|---|---|---|---|---|---|
| T1 | 0 | 0.037 | 0.044 | 0.056 | 0.038 | $> 10^{14}$ | 0.084 | $> 10^{14}$ | $> 10^{14}$ | $> 10^{13}$ | $> 10^{14}$ |
| T2 | 0.038 | 0 | 0.005 | 0.006 | 0.199 | $> 10^{14}$ | 0.065 | $> 10^{14}$ | $> 10^{14}$ | $> 10^{14}$ | $> 10^{14}$ |
| T3 | 0.046 | 0.007 | 0 | 0.009 | 0.122 | $> 10^{14}$ | 0.003 | $> 10^{14}$ | $> 10^{14}$ | $> 10^{14}$ | $> 10^{13}$ |
| T4 | 0.058 | 0.019 | 0.012 | 0 | 0.149 | $> 10^{14}$ | 0.023 | $> 10^{14}$ | $> 10^{14}$ | $> 10^{14}$ | $> 10^{14}$ |
| TIN | 0.142 | 0.184 | 0.191 | 0.196 | 0 | $> 10^{14}$ | 0.151 | $> 10^{14}$ | $> 10^{14}$ | $> 10^{14}$ | $> 10^{14}$ |
| BV | 0.224 | 0.194 | 0.195 | 0.210 | 0.374 | 0 | 0.284 | 0.154 | 0.176 | 0.183 | 0.151 |
| TOUT | 0.036 | 0.074 | 0.080 | 0.078 | 0.129 | $> 10^{14}$ | 0 | $> 10^{14}$ | $> 10^{14}$ | $> 10^{14}$ | $> 10^{14}$ |
| VIR | 0.503 | 0.489 | 0.487 | 0.486 | 0.498 | 0.343 | 0.438 | 0 | 0.008 | 0.011 | 0.006 |
| VHI | 0.207 | 0.163 | 0.172 | 0.197 | 0.075 | 0.648 | 0.233 | 0.065 | 0 | 0.007 | 0.070 |
| HIR | 0.189 | 0.146 | 0.155 | 0.180 | 0.112 | 0.685 | 0.271 | 0.075 | 0.007 | 0 | 0.080 |
| PHI | 0.231 | 0.188 | 0.196 | 0.219 | 0.005 | 0.601 | 0.148 | 0.003 | 0.059 | 0.065 | 0 |

**Table 3.** Clustering using the SP-lines. The similarity matrix obtained with the first example from the multivariate TS dataset.

|      | T1    | T2    | T3    | T4    | TIN   | BV    | TOUT  | VIR    | VHI    | HIR   | PHI   |
|------|-------|-------|-------|-------|-------|-------|-------|--------|--------|-------|-------|
| T1   | 0     | 0.071 | 0.073 | 0.048 | 0.521 | 0.08  | 0.432 | 5.281  | 5.327  | 2.321 | 2.373 |
| T2   | 0.097 | 0     | 0.013 | 0.036 | 0.562 | 0.159 | 0.48  | 4.784  | 4.826  | 2.058 | 2.106 |
| T3   | 0.099 | 0.018 | 0     | 0.039 | 0.561 | 0.156 | 0.479 | 4.763  | 4.805  | 2.047 | 2.094 |
| T4   | 0.081 | 0.002 | 0.004 | 0     | 0.553 | 0.139 | 0.47  | 4.881  | 4.924  | 2.11  | 2.158 |
| TIN  | 1.042 | 1.224 | 1.229 | 1.178 | 0     | 0.904 | 0.177 | 11.987 | 12.081 | 5.865 | 5.973 |
| BV   | 0.102 | 0.199 | 0.202 | 0.175 | 0.447 | 0     | 0.35  | 5.937  | 5.987  | 2.67  | 2.727 |
| TOUT | 0.731 | 0.887 | 0.891 | 0.847 | 0.163 | 0.629 | 0     | 10.062 | 10.142 | 4.85  | 4.942 |
| VIR  | 0.841 | 0.827 | 0.827 | 0.831 | 0.919 | 0.854 | 0.908 | 0      | 0.007  | 0.472 | 0.463 |
| VHI  | 0.839 | 0.826 | 0.825 | 0.828 | 0.914 | 0.854 | 0.904 | 0.007  | 0      | 0.475 | 0.467 |
| HIR  | 0.691 | 0.665 | 0.664 | 0.671 | 0.833 | 0.721 | 0.809 | 0.892  | 0.906  | 0     | 0.016 |
| PHI  | 0.7   | 0.674 | 0.673 | 0.68  | 0.843 | 0.726 | 0.823 | 0.863  | 0.877  | 0.015 | 0     |

**Table 4.** RNN clustering. Final adjacency matrix obtained using $th2 = 3$.

|      | T1 | T2 | T3 | T4 | TIN | BV | TOUT | VIR | VHI | HIR | PHI |
|------|----|----|----|----|-----|----|------|-----|-----|-----|-----|
| T1   | 0  | 1  | 1  | 1  | 0   | 0  | 0    | 0   | 0   | 0   | 0   |
| T2   | 1  | 0  | 1  | 1  | 0   | 0  | 0    | 0   | 0   | 0   | 0   |
| T3   | 1  | 1  | 0  | 1  | 0   | 0  | 0    | 0   | 0   | 0   | 0   |
| T4   | 1  | 1  | 1  | 0  | 0   | 0  | 0    | 0   | 0   | 0   | 0   |
| TIN  | 0  | 0  | 0  | 0  | 0   | 0  | 1    | 0   | 0   | 0   | 0   |
| BV   | 0  | 0  | 0  | 0  | 0   | 0  | 0    | 0   | 0   | 0   | 0   |
| TOUT | 0  | 0  | 0  | 0  | 1   | 0  | 0    | 0   | 0   | 0   | 0   |
| VIR  | 0  | 0  | 0  | 0  | 0   | 0  | 0    | 0   | 1   | 0   | 0   |
| VHI  | 0  | 0  | 0  | 0  | 0   | 0  | 0    | 1   | 0   | 1   | 1   |
| HIR  | 0  | 0  | 0  | 0  | 0   | 0  | 0    | 0   | 0   | 0   | 1   |
| PHI  | 0  | 0  | 0  | 0  | 0   | 0  | 0    | 0   | 0   | 1   | 0   |

**Table 5.** Sp-line clustering. Final adjacency matrix obtained using $th2 = 2$.

|      | T1 | T2 | T3 | T4 | TIN | BV | TOUT | VIR | VHI | HIR | PHI |
|------|----|----|----|----|-----|----|------|-----|-----|-----|-----|
| T1   | 0  | 1  | 1  | 1  | 0   | 0  | 0    | 0   | 0   | 0   | 0   |
| T2   | 1  | 0  | 1  | 1  | 0   | 0  | 0    | 0   | 0   | 0   | 0   |
| T3   | 1  | 1  | 0  | 1  | 0   | 0  | 0    | 0   | 0   | 0   | 0   |
| T4   | 1  | 1  | 1  | 0  | 0   | 0  | 0    | 0   | 0   | 0   | 0   |
| TIN  | 0  | 0  | 0  | 0  | 0   | 1  | 0    | 0   | 0   | 0   | 0   |
| BV   | 0  | 0  | 0  | 0  | 0   | 0  | 1    | 0   | 0   | 0   | 0   |
| TOUT | 0  | 0  | 0  | 0  | 0   | 0  | 0    | 0   | 0   | 0   | 0   |
| VIR  | 0  | 0  | 0  | 0  | 0   | 0  | 0    | 0   | 1   | 0   | 0   |
| VHI  | 0  | 0  | 0  | 0  | 0   | 0  | 0    | 1   | 0   | 0   | 0   |
| HIR  | 0  | 0  | 0  | 0  | 0   | 0  | 0    | 0   | 0   | 0   | 0   |
| PHI  | 0  | 0  | 0  | 0  | 0   | 0  | 0    | 0   | 0   | 0   | 0   |

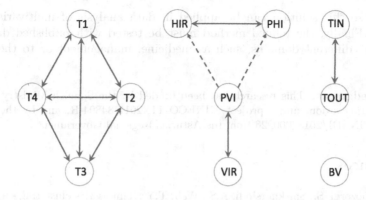

**Fig. 2.** RNN clustering. Final graph: the groups are clearly remarked.

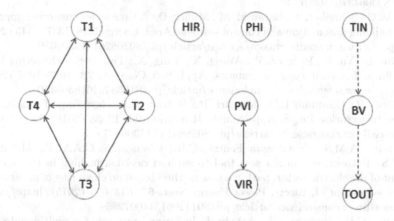

**Fig. 3.** SP-line clustering. Final graph: the groups are clearly remarked.

## 5    Conclusions

In this study two solutions for multivariate TS clustering has been compared, both of them using RNN. In the first solution, the prediction error of RNN learned for a variable within an example is used to define a similarity measurement. The aggregation of the obtained similarities among all the examples in the dataset allows developing an adjacency matrix that, finally, is used to group the variables. The second solution introduced an intermediate stage of sp-lines modeling to smooth the evolution of the TS.

A simple proof of concept has been presented, showing that the performance of the method perfectly groups the different variables. Interestingly, the first approach only has two parameters (two thresholds) that were easily tuned, while the second has the extra parameter of the degrees of freedom of the sp-lines.

We do expect to perform improvements in the algorithm, avoiding the use of thresholds, enhancing and improving the modeling method and in the similarity function, so this solution can be applied in data analysis of multivariate TS datasets. Finally, the refined method must be tested with published datasets concerning different domains, such as medicine, management or to the stock market.

**Acknowledgment.** This research has been funded by the Spanish Ministry of Science and Innovation, under project MINECO-TIN2017-84804-R, and by the Grant FC-GRUPIN-IDI/2018/000226 from the Asturias Regional Government.

# References

1. Aghabozorgi, S., Shirkhorshidi, A.S., Wah, T.Y.: Time-series clustering - a decade review. Inf. Syst. **53**, 16–38 (2015). http://www.sciencedirect.com/science/article/pii/S0306437915000733
2. Bode, G., Schreiber, T., Baranski, M., Müller, D.: A time series clustering approach for building automation and control systems. Appl. Energy **238**, 1337–1345 (2019). http://www.sciencedirect.com/science/article/pii/S0306261919302089
3. Duan, L., Yu, F., Pedrycz, W., Wang, X., Yang, X.: Time-series clustering based on linear fuzzy information granules. Appl. Soft Comput. **73**, 1053–1067 (2018). http://www.sciencedirect.com/science/article/pii/S1568494618305490
4. D'Urso, P., Giovanni, L.D., Massari, R.: Robust fuzzy clustering of multivariate time trajectories. Int. J. Approximate Reasoning **99**, 12–38 (2018). http://www.sciencedirect.com/science/article/pii/S0888613X17306977
5. Ferreira, A.M.S., de Oliveira Fontes, C.H., Cavalcante, C.A.M.T., Marambio, J.E.S.: Pattern recognition as a tool to support decision making in the management of the electric sector. part ii: a new method based on clustering of multivariate time series. Int. J. Electr. Power Energy Syst. **67**, 613–626 (2015). http://www.sciencedirect.com/science/article/pii/S0142061514007285
6. Fontes, C.H., Budman, H.: A hybrid clustering approach for multivariate time series - a case study applied to failure analysis in a gas turbine. ISA Trans. **71**, 513–529 (2017). http://www.sciencedirect.com/science/article/pii/S0019057817305530
7. Hu, M., Feng, X., Ji, Z., Yan, K., Zhou, S.: A novel computational approach for discord search with local recurrence rates in multivariate time series. Inf. Sci. **477**, 220–233 (2019). http://www.sciencedirect.com/science/article/pii/S0020025516320849
8. Lee, Y., Na, J., Lee, W.B.: Robust design of ambient-air vaporizer based on time-series clustering. Comput. Chem. Eng. **118**, 236–247 (2018). http://www.sciencedirect.com/science/article/pii/S0098135418308822
9. Li, J., Pedrycz, W., Jamal, I.: Multivariate time series anomaly detection: a framework of hidden Markov models. Appl. Soft Comput. **60**, 229–240 (2017). http://www.sciencedirect.com/science/article/pii/S1568494617303782
10. Liu, G., Zhu, L., Wu, X., Wang, J.: Time series clustering and physical implication for photovoltaic array systems with unknown working conditions. Solar Energy **180**, 401–411 (2019). http://www.sciencedirect.com/science/article/pii/S0038092X19300532
11. Mikalsen, K.Ø., Bianchi, F.M., Soguero-Ruiz, C., Jenssen, R.: Time series cluster kernel for learning similarities between multivariate time series with missing data. Pattern Recogn. **76**, 569–581 (2018). http://www.sciencedirect.com/science/article/pii/S0031320317304843

12. Pan, S.J., Yang, Q.: A survey on transfer learning. IEEE Trans. Knowl. Data Eng. **22**(10), 1345–1359 (2010). https://doi.org/10.1109/TKDE.2009.191
13. Quast, B.: Recurrent neural networks in R, February 2019. https://github.com/bquast/rnn
14. Salvo, R.D., Montalto, P., Nunnari, G., Neri, M., Puglisi, G.: Multivariate time series clustering on geophysical data recorded at Mt. Etna from 1996 to 2003. J. Volcanol. Geoth. Res. **251**, 65–74 (2013). Flank instability at Mt. Etna. http://www.sciencedirect.com/science/article/pii/S0377027312000443
15. Váquez, I., Villar, J.R., Sedano, J., Simić, S.: A preliminary study on multivariate time series clustering. In: Martínez Álvarez, F., Troncoso Lora, A., Sáez Muñoz, J.A., Quintián, H., Corchado, E. (eds.) SOCO 2019. AISC, vol. 950, pp. 473–480. Springer, Cham (2020). https://doi.org/10.1007/978-3-030-20055-8_45
16. Yu, C., Luo, L., Chan, L.L.H., Rakthanmanon, T., Nutanong, S.: A fast LSH-based similarity search method for multivariate time series. Inf. Sci. **476**, 337–356 (2019). http://www.sciencedirect.com/science/article/pii/S0020025518308430

# On the Influence of Admissible Orders in IVOVO

Mikel Uriz[1,2](✉), Daniel Paternain[1,2], Humberto Bustince[1,2], and Mikel Galar[1,2]

[1] Department of Statistics, Computer Science and Mathematics,
Public University of Navarre, Campus Arrosadia s/n, 31006 Pamplona, Spain
{mikelxabier.uriz,daniel.paternain,bustince,mikel.galar}@unavarra.es
[2] Institute of Smart Cities, Public University of Navarre,
Campus Arrosadia s/n, 31006 Pamplona, Spain

**Abstract.** It is known that when dealing with interval-valued data, there exist problems associated with the non-existence of a total order. In this work we investigate a reformulation of an interval-valued decomposition strategy for multi-class problems called IVOVO, and we analyze the effectiveness of considering different admissible orders in the aggregation phase of IVOVO. We demonstrate that the choice of an appropriate admissible order allows the method to obtain significant differences in terms of accuracy.

**Keywords:** Multi-class classification problems · One-vs-one strategy · Interval-valued fuzzy sets · Admissible order

## 1 Introduction

The objective of a classification problem consists in learning a mapping (classifier) from a set of labeled data (examples) into a set of labels (classes), being able of correctly predicting the class of new unseen instances. Depending on the number of classes (labels) the classifier must deal with, the classification problem can either be a two-class (binary) or multi-class problem. From the point of view of complexity, it is much more complex to deal with multi-class problems due to the overlapping between decision boundaries of the classes [16]. Under this context, decomposition strategies, which divide multi-class problems into binary ones, are usually applied to reduce the inherent complexity. The two main decomposition strategies are One-Vs-All (OVA), that generates as many binary problems as number of classes, and One-Vs-One (OVO), that generates as many binary problems as pair of classes, being the latter the most widely used one [10]. Then, each binary problem is solved by an independent base classifier. Finally, each new example is submitted to each classifier and their outputs are fused in order to predict the final class of the example.

This work has been partially supported by the Spanish Ministry of Science and Technology under the project TIN2016-77356-P and the Public University of Navarre under the project PJUPNA13.

H. Pérez García et al. (Eds.): HAIS 2019, LNAI 11734, pp. 358–369, 2019.
https://doi.org/10.1007/978-3-030-29859-3_31

In this work we focus on Fuzzy Rule-Based Classification Systems (FRBCSs) [13], which are known for their interpretability due to the use of linguistic labels in the antecedent of the produced rules. In order to better deal with the uncertainty associated with the choice of fuzzy membership functions in the reasoning method of FRBCSs, in [19] IVTURS, a new method based on the use of interval-valued fuzzy sets, was presented. This model must deal with interval data and therefore, many difficulties arises. One of the most important issue comes from the fact that the usual order between intervals is a partial order.

FRBCSs have been proven to successfully deal with multi-class problems applying an OVO decomposition strategy (see, for example [7–9]). Moreover, the first solution to handle multi-class problems with IVTURS was given in [8] with the name of IVOVO. It is worth noting that the aggregation phase of IVOVO manages interval-valued confidences and accordingly, the problems related to the order needs to be addressed. The arrangement of interval-valued confidences in IVOVO was solved by the usage of a total order given by Xu and Yager [20], but no further analysis about its suitability has been carried out yet. Moreover, the order given by Xu and Yager has been proven to be a particular case of a larger set of total orders between intervals called *admissible orders* [4]. Considering all these facts, the purpose of this work is to study the influence of different admissible orders (apart from the one given by Xu and Yager) in the performance of IVOVO. To do this, we will consider several admissible orders under two aggregation strategies for OVO, Voting and Win Weighted Voting (WinWV). We will not only evaluate the final performance of each admissible order, but also how influential is on each specific dataset. The experimental study will consider 22 datasets from the KEEL dataset repository [1] and the analysis will be supported by non-parametric statistical tests [11].

The structure of the paper is as follows. In Sect. 2 we recall the main mathematical concepts related with intervals. In Sect. 3 we describe FRBCSs, OVO and IVOVO. In Sect. 4 we explain the concept of admissible order and we provide several construction methods. We finish this paper with an experimental study on the influence of admissible orders in IVOVO in Sect. 5 and the conclusions and future research lines in Sect. 6.

## 2  Preliminaries

In this section we recall the theoretical basis of this work. Since we will deal with interval data, we will denote by $L([0,1])$ the set of all closed subintervals of the unit interval $[0,1]$, that is,

$$L([0,1]) = \{\mathbf{x} = [\underline{x}, \overline{x}] \mid (\underline{x}, \overline{x}) \in [0,1]^2 \text{ and } \underline{x} \le \overline{x}\}.$$

*Remark 1.* Although in this work we focus on $L([0,1])$, we can consider also $L$ as the set of all positive closed intervals, i.e., $L = \{\mathbf{x} = [\underline{x}, \overline{x}] \mid 0 \le \underline{x} \le \overline{x}\}$.

Notice that $L([0,1])$ is a partially ordered set with respect to the order relation $\le_L$ defined in the following way: given $\mathbf{x}, \mathbf{y} \in L([0,1])^2$

$$\mathbf{x} \le_L \mathbf{y} \text{ if and only if } \underline{x} \le \underline{y} \text{ and } \overline{x} \le \overline{y}.$$

Considering this order relation, we have that $(L([0,1]), \leq_L)$ is a complete lattice whose smallest element is $\mathbf{0} = [0,0]$ and the greatest is $\mathbf{1} = [1,1]$ [6,12].

The fact that $L([0,1])$ is a partially ordered set means that it is not always possible to establish an order relation between two arbitrary intervals $\mathbf{x}, \mathbf{y} \in L([0,1])$. We will say that $\mathbf{x}$ and $\mathbf{y}$ are incomparable, denoted by $\mathbf{x} \parallel \mathbf{y}$, whenever $\mathbf{x} \not\leq_L \mathbf{y}$ and $\mathbf{y} \not\leq_L \mathbf{y}$ hold simultaneously.

*Example 1.* The intervals $\mathbf{x} = [0.1, 0.8]$ and $\mathbf{y} = [0.2, 0.3]$ are incomparable. In fact, $\mathbf{x} \parallel \mathbf{y}$ whenever $(\underline{x} < \underline{y}$ and $\overline{y} < \overline{x})$ or $(\underline{y} < \underline{x}$ and $\overline{x} < \overline{y})$.

## 3   IVOVO: Interval-Valued One-Vs-One

IVOVO stands for Interval-Valued One-Vs-One, and is based on the application of the OVO strategy to IVTURS fuzzy classifier, which outputs interval-valued confidence degrees instead of real-valued ones. For this reason, in this section we recall IVOVO and its main components: IVTURS and OVO.

### 3.1   Fuzzy Rule-Based Classification Systems: IVTURS

Among classification algorithms, Fuzzy Rule-Based Classification Systems (FRBCSs) try to create models whose rules are interpretable by humans due to the use of linguistic labels [13]. These linguistic rules are extracted by a learning algorithm from a training dataset $\mathcal{D}_T$ having $P$ labeled examples $x_p = (x_{p1}, \dots, x_{pn}), p = \{1, \dots, P\}$, where $x_{pi}$ is the value of the $i$-th attribute $(i = \{1, 2, \dots, n\})$ of the $p$-th training example. Each example is associated with a class $y_p \in \mathbb{C} = \{C_1, C_2, \dots, C_m\}$, being $m$ is the number of classes of the problem.

IVTURS algorithm [19] is based on FARC-HD (Fuzzy Association Rule-based Classification model for High-Dimensional problems) [1]. Both use rules with the following structure:

$$\text{Rule } R_j : \text{ If } x_1 \text{ is } A_{j1} \text{ and } \dots \text{ and } x_{n_j} \text{ is } A_{jn_j} \text{ then Class} = C_j \text{ with } RW_j \tag{1}$$

where $R_j$ is the label of the $j$-th rule, $x = (x_1, \dots, x_n)$ is a vector representing the example, $A_{ji} \in \mathbb{X}_i$ is a linguistic label modeled by a triangular membership function (where $\mathbb{X}_i = \{X_{i1}, \dots, X_{il}\}$ is the set of linguistic labels for the $i$-th antecedent, being $l$ the number of linguistic labels in this set), $C_j$ is the class label and $RW_j$ is the rule weight computed using the certainty factor defined in [14].

The main difference in the rule representation between FARC-HD and IVTURS is that the latter take advantage of Interval-Valued Fuzzy Sets (IVFSs) to model the uncertainty under the definition of the linguistic labels, and hence its membership functions are defined by IVFSs instead of FSs. Accordingly, the whole Fuzzy Reasoning Method (FRM) needs to be adapted to work with interval along all its steps. As a consequence, the confidence (association) degree for each class obtained in the final step is also an interval. Therefore, the final class is taken as the one with the largest confidence degree (according to an admissible order, see Sect. 4).

With respect to the rule learning algorithm, FARC-HD was composed of three steps (see [1] for more details): a fuzzy association rule extraction, a candidate rule pre-screening, and a genetic rule selection and lateral tuning. IVTURS makes use of FARC-HD for carrying out the rule extraction, but without performing the last step. Then, it introduces IVFSs and finally uses a genetic algorithm to tune the interval FRM and carry out a rule selection.

### 3.2 One-Versus-One (OVO)

In OVO the original $m$ class problem is transformed into a $m(m-1)/2$ subproblems (all possible pair of classes). Therefore, each base classifier will learn to distinguish a pair of classes $\{C_i, C_j\}$. To predict the class of a new examples, each classifier is expected to provide a pair confidence degrees $r_{ij}, r_{ji} \in [0,1]$ in favor of classes $C_i$ and $C_j$, respectively. For simplicity, these outputs are stored in a *score-matrix* $R$. In the case of fuzzy classifiers, these pairs are rarely normalized [7,9]. This fact requires a normalization step so that the outputs of all the base classifiers are in the same scale. Normalization with real-valued confidence degrees is direct, but it is not so straightforward with intervals.

### 3.3 IVOVO: Interval-Valued One-Vs-One

IVOVO [8] refers to the combination of IVTURS and OVO to enhance the performance of the former in multi-class problems. Nevertheless, there are three main issues when using OVO with IVTURS because the score-matrix is filled by interval confidence scores: (1) there is no consensus on which normalization strategy should be applied; (2) the aggregations needs to be adapted to work with intervals; (3) there is no a total order defined to compare intervals.

Hereafter we recall how these issues were addressed in [8]. Recall that the score-matrix is formed of intervals ($\mathbf{R}$):

$$\mathbf{R} = \begin{pmatrix} - & \mathbf{r}_{12} & \cdots & \mathbf{r}_{1m} \\ \mathbf{r}_{21} & - & \cdots & \mathbf{r}_{2m} \\ \vdots & & & \vdots \\ \mathbf{r}_{m1} & \mathbf{r}_{m2} & \cdots & - \end{pmatrix} \tag{2}$$

$\mathbf{r}_{ij}, \mathbf{r}_{ji} \in L$ corresponding to the confidence degrees for classes $C_i, C_j$, respectively.

In IVOVO, the score-matrix $\mathbf{R}$ was normalized to a new score-matrix $\mathbf{R}^u$ in such a way that all the elements are closed sub-intervals in $[0,1]$, that is, $\mathbf{r}_{ij}^u \in L([0,1])$ for every $i,j$, $i \neq j$ (according to the theory described in [19]). This was done by normalizing them according to the upper bounds:

$$\mathbf{r}_{ij}^u = \begin{cases} \left[ \dfrac{\underline{r}_{ij}}{\overline{r}_{ij} + \overline{r}_{ji}}, \dfrac{\overline{r}_{ij}}{\overline{r}_{ij} + \overline{r}_{ji}} \right] & \text{if } \overline{r}_{ij} \neq 0 \quad \text{or} \quad \overline{r}_{ji} \neq 0 \\ [0.5, 0.5] & \text{otherwise} \end{cases} \tag{3}$$

This normalization allows one to maintain the proportion of ignorance and satisfies the property $\overline{r}_{ij}^u + \overline{r}_{ji}^u = 1$.

Regarding the adaptation of the aggregations methods for OVO, they mainly consisted in using the interval arithmetic. We recall the voting strategy and the WinWV strategy as they will be the ones considered in the experimental study (notice that WV was shown to perform worse than WinWV when considering fuzzy classifiers and IVOVO).

- *Voting strategy (Vote): Class* $= \arg \max\limits_{i=1,\ldots,m} \sum\limits_{1 \leq j \neq i \leq m} s_{ij}$, where $s_{ij}$ is 1 if $\mathbf{r}_{ij}^u > \mathbf{r}_{ji}^u$ and 0 otherwise.
- *WinWV: Class* $= \arg \max\limits_{i=1,\ldots,m} \sum\limits_{1 \leq j \neq i \leq m} \mathbf{s}_{ij}$, where $\mathbf{s}_{ij}$ is $\mathbf{r}_{ij}^u$ if $\mathbf{r}_{ij}^u > \mathbf{r}_{ji}^u$ and 0 otherwise.

*Remark 2.* Observe that in both voting strategies the need to compare intervals appears, thus the necessity of using total orders between intervals.

## 4   Admissible Orders for Comparing Interval Data

In the preliminaries of this paper we have seen that the set $L([0,1])$ is usually equipped with a partial order $\leq_L$. This order, which is not a total order, does not allow neither to compare any arbitrary pair of intervals, nor to calculate the maximum from an arbitrary set of intervals (see, for example [4,5,15,18]). These two aspects are a crucial step of the IVOVO algorithm, specially in the aggregation phase, where intervals have to be compared in both the Voting and the WinWV strategies.

If we focus on the Voting strategy, we observe that each vote is given to the class whose interval-valued confidence is greater. In the original IVOVO, this was done using Xu and Yager's total order. With respect to the WinWV strategy, apart from comparing which interval-valued confidence is greater (in the same way as in the Voting strategy), we must obtain the most voted class, where the vote for each class is again given by an interval. Again, this last calculation was originally performed with Xu and Yager's total order.

In fact, the total order given by Xu and Yager has been widely used in many decision-making procedures both based on interval-valued and intuitionistic fuzzy sets. However, in [4], a wider concept that encompasses Xu and Yager's order, among others, was given establishing therefore the theoretical framework of *admissible orders*. An admissible order is as a linear (total) order that refines the partial order $\leq_L$.

**Definition 1.** *Consider the partially ordered set* $(L([0,1]), \leq_L)$. *The order* $\preceq$ *on* $L([0,1])$ *is called an admissible order if*

*(i)* $\preceq$ *is a linear order on* $L([0,1])$;
*(ii) for all* $\mathbf{x}, \mathbf{y} \in L([0,1])$, $\mathbf{x} \preceq \mathbf{y}$ *whenever* $\mathbf{x} \leq_L \mathbf{y}$.

Consider the following order: we say that $\mathbf{x} \preceq_{XY} \mathbf{y}$ if and only if $\underline{x}+\overline{x} \leq \underline{y}+\overline{y}$ or $(\underline{x} + \overline{x} = \underline{y} + \overline{y}$ and $\overline{y} - \underline{y} < \overline{x} - \underline{x}$. We have that $\preceq_{XY}$ is an admissible order and it is actually the order given by Xu and Yager.

*Example 2.* Let $\mathbf{x} = [0, 0.6], \mathbf{y} = [0.2, 0.4], \mathbf{z} = [0.3, 0.3]$. Observe that any two intervals can be compared by means of $\leq_L$. However, according to $\preceq_{XY}$, we have that $\mathbf{x} \preceq_{XY} \mathbf{y} \preceq \mathbf{z}$, and so $\min_{\preceq_{XY}}\{\mathbf{x}, \mathbf{y}, \mathbf{z}\} = \mathbf{x}$ and $\max_{\preceq_{XY}}\{\mathbf{x}, \mathbf{y}, \mathbf{z}\} = \mathbf{z}$.

Apart form the well-known admissible order given by Xu and Yager, there exist two examples of admissible orders which are derived from the usual lexicographic rules used in $\mathbb{R}^2$.

*Example 3.* The following are examples of admissible orders:

(i) $\mathbf{x} \preceq_{Lex1} \mathbf{y}$ if and only if $\underline{x} < \underline{y}$ or $(\underline{x} = \underline{y}$ and $\overline{x} \leq \overline{y})$.
(ii) $\mathbf{x} \preceq_{Lex2} \mathbf{y}$ if and only if $\overline{x} < \overline{y}$ or $(\overline{x} = \overline{y}$ and $\underline{x} \leq \underline{y})$.

*Example 4.* Having again $\mathbf{x} = [0, 0.6], \mathbf{y} = [0.2, 0.4], \mathbf{z} = [0.3, 0.3]$, we have that $\mathbf{x} \preceq_{Lex1} \mathbf{y} \preceq_{Lex1} \mathbf{z}$, while $\mathbf{z} \preceq_{Lex2} \mathbf{y} \preceq_{Lex2} \mathbf{x}$.

Notice that the admissible orders given by Lex1 and Lex2 are, in some sense, extreme cases. This means that, for any $\mathbf{x}, \mathbf{y} \in L([0, 1])$ with $\mathbf{x} \parallel \mathbf{y}$, we always have that if $\mathbf{x} \preceq_{Lex1} \mathbf{y}$, then $\mathbf{y} \preceq_{Lex2} \mathbf{y}$, and viceversa.

Besides these well-known examples of admissible orders, in [4], a construction method of admissible orders considering two continuous functions defined on $K([0, 1]) = \{(x, y) \in [0, 1]^2 \mid x \leq y\}$ was given.

**Definition 2.** *Let $\preceq$ be an admissible order on $L([0, 1])$. The order $\preceq$ is called a generated admissible order if there exist two continuous functions $f, g$ : $K([0, 1]) \to R$ such that for all $[a, b], [c, d] \in L([0, 1])$,*

$$[a, b] \preceq [c, d] \text{ if and only if } [f(a, b), g(a, b)] \preceq_{Lex1} [f(c, d), g(c, d)].$$

*We will denote the admissible order generated by the pair $(f, g)$ as $\preceq_{f,g}$.*

*Example 5.* Let $f(x, y) = \frac{x^2 + y^2}{2}$ and $g(x, y) = y$. If we consider $\mathbf{x} = [0, 0.6], \mathbf{y} = [0.2, 0.4], \mathbf{z} = [0.3, 0.3]$, we have that $f(\underline{x}, \overline{x}) = 0.18$, $f(\underline{y}, \overline{y}) = 0.1$ and $f(\underline{z}, \overline{z}) = 0.09$ and, therefore, $\mathbf{z} \preceq_{f,g} \mathbf{y} \preceq_{f,g} \mathbf{z}$.

Finally, a much simpler and parametrizable family of admissible orders by considering only two real numbers in the unit interval was also presented in [4]. The admissible order applies the so called $K_\alpha$ operator, which is a mapping $K_\alpha : [0, 1]^2 \to [0, 1]$ given by $K_\alpha(a, b) = a + \alpha(b - a)$ [3].

**Definition 3.** *Let $\alpha, \beta \in [0, 1]$ with $\alpha \neq \beta$. For any $\mathbf{x}, \mathbf{y} \in L([0, 1])$, we say that $\mathbf{x} \preceq_{\alpha,\beta} \mathbf{y}$ if and only if $K_\alpha(\underline{x}, \overline{x}) < K_\alpha(\underline{y}, \overline{y})$ or $(K_\alpha(\underline{x}, \overline{x}) = K_\alpha(\underline{y}, \overline{y})$ and $K_\beta(\underline{x}, \overline{x}) < K_\beta(\underline{y}, \overline{y}))$.*

*Example 6.* Let $\alpha = 1/3$ and $\beta = 2/3$. Then, if we have $\mathbf{x} = [0, 0.6], \mathbf{y} = [0.2, 0.4], \mathbf{z} = [0.3, 0.3]$, then $\mathbf{x} \preceq_{1/3,2/3} \mathbf{y} \preceq_{1/3,2/3} \mathbf{z}$, while $\mathbf{z} \preceq_{2/3,1/3} \mathbf{y} \preceq_{2/3,1/3} \mathbf{x}$.

## 5    Experimental Study

The goal of this experimental study is to analyze the influence of admissible orders in the arrangement of interval-valued confidences in both Vote and WinWV strategies. To do so, we will make use of the same experimental framework as in [8] considering a set of 6 different admissible orders, the six admissible orders presented in the examples of Sect. 4, namely $\preceq_{Lex1}$, $\preceq_{Lex2}$, $\preceq_{1/3,2/3}$, $\preceq_{f,g}$, $\preceq_{2/3,1/3}$ and $\preceq_{XY}$. We recall that any admissible order is a refinement of the partial order $\leq_L$, which means that if any pair of intervals can be arranged by means of $\leq_L$, then any admissible order $\preceq$ will produce exactly the same arrangement, i.e., the admissible orders have no influence in the experiment. On the contrary, as the frequency of appearance of incomparable interval-valued confidences increases, the different admissible orders may yield a larger variety of results, thus justifying this analysis.

### 5.1    Experimental Framework

The usage of different admissible orders is evaluated using twenty-two datasets from the KEEL dataset repository [2]. Notice that the same datasets were considered in previous studies [7–9]. Table 1 presents a summary description of the datasets, including the number of examples (#Ex.), the number of attributes (#Atts.), the number of numerical (#Num.) and nominal (#Nom.) attributes, and the number of classes (#Class.).

**Table 1.** Summary description of the datasets.

| Id. | Dataset | #Ex. | #Atts. | #Num. | #Nom. | #Class. | Id. | Dataset | #Ex. | #Atts. | #Num. | #Nom. | #Class. |
|-----|---------|------|--------|-------|-------|---------|-----|---------|------|--------|-------|-------|---------|
| aut | autos | 159 | 25 | 15 | 10 | 6 | bal | balance | 625 | 4 | 4 | 0 | 3 |
| cle | cleveland | 297 | 13 | 13 | 0 | 5 | con | contraceptive | 1473 | 9 | 6 | 3 | 3 |
| der | dermatology | 358 | 34 | 1 | 33 | 6 | eco | ecoli | 336 | 7 | 7 | 0 | 8 |
| gla | glass | 214 | 9 | 9 | 0 | 7 | hay | hayes-roth | 132 | 4 | 4 | 0 | 3 |
| iri | iris | 150 | 4 | 4 | 0 | 3 | lym | lymphography | 148 | 18 | 3 | 15 | 4 |
| new | newthyroid | 215 | 5 | 5 | 0 | 3 | pag | pageblocks | 548 | 10 | 10 | 0 | 5 |
| pen | penbased | 1100 | 16 | 16 | 0 | 10 | sat | satimage | 643 | 36 | 36 | 0 | 7 |
| seg | segment | 2310 | 19 | 19 | 0 | 7 | shu | shuttle | 2175 | 9 | 9 | 0 | 7 |
| tae | tae | 151 | 5 | 3 | 2 | 3 | thy | thyroid | 720 | 21 | 21 | 0 | 3 |
| veh | vehicle | 846 | 18 | 18 | 0 | 4 | vow | vowel | 990 | 13 | 13 | 0 | 11 |
| win | wine | 178 | 13 | 13 | 0 | 3 | yea | yeast | 1484 | 8 | 8 | 0 | 10 |

The results are evaluated using accuracy performance measure obtained by a *stratified 5-fold cross-validation model* following the *Distribution Optimally Balanced Cross Validation* procedure [17]. As recommended in the literature [11], our conclusions are supported by the proper statistical analysis using nonparametric statistical tests. Wilcoxon rank test is used for pairwise comparisons, whereas Aligned Friedman test is considered for multiple method comparisons.

With respect to the configuration of IVTURS for generating the base classifiers of IVOVO, the configuration recommended by the authors is used: 5 fuzzy labels for each variable, 3 as maximum depth of the tree, a minimum support of 0.05, a minimum confidence of 0.8, 50 individuals as population size, 30 bits per gene for the Gray codification and a maximum of 20000 evaluations.

## 5.2  Influence of Normalization Strategies in IVOVO

Tables 2 and 3 present the classification accuracy obtained by each admissible order tested using both Vote and WinWV aggregations methods, respectively. Additionaly, in each table we include two columns (loss and ad%). Loss refers to the percentage of accuracy loss between the best admissible order and the worst (computed as $\frac{best-worst}{best}$), whereas ad% refers to the percentage of times that the usage of admissible orders can change the interval arrangement (that is, incomparable intervals are present). This information is useful for understanding the possible effect of the admissible order in the loss. Obviously, if ad% is 0, it means that all admissible orders will return the same results and hence, the loss will also be 0.

**Table 2.** Results VOTE

| Dataset | $\preceq_{\text{Lex1}}$ | $\preceq_{\text{Lex2}}$ | $\preceq_{1/3,2/3}$ | $\preceq_{f,g}$ | $\preceq_{2/3,1/3}$ | $\preceq_{XY}$ | loss | ad% |
|---------|------|------|------|------|------|------|------|------|
| aut | 75.78 | 72.17 | 76.52 | 75.92 | 75.23 | **77.13** | 6.42 | 1.70 |
| bal | 81.10 | 84.15 | 84.31 | 85.27 | **85.44** | 85.12 | 5.08 | 6.68 |
| cle | **56.93** | 53.54 | 55.26 | 55.25 | 54.58 | 54.57 | 5.97 | 7.07 |
| con | **54.65** | 52.62 | 53.84 | 53.23 | 53.16 | 53.64 | 3.72 | 11.33 |
| der | **95.29** | **95.29** | **95.29** | **95.29** | **95.29** | **95.29** | 0.00 | 0.01 |
| eco | 80.18 | **83.47** | 82.57 | 83.17 | 82.86 | 81.67 | 3.94 | 3.81 |
| gla | 70.50 | 63.40 | **71.93** | 69.11 | 70.55 | 70.98 | 11.87 | 7.78 |
| hay | **74.45** | **74.45** | **74.45** | **74.45** | **74.45** | **74.45** | 0.00 | 0.00 |
| iri | **97.33** | 94.00 | 96.00 | 95.33 | 95.33 | 95.33 | 3.42 | 1.39 |
| lym | 80.47 | 79.83 | **80.52** | **80.52** | 79.83 | **80.52** | 0.86 | 0.14 |
| new | 91.63 | 91.63 | **95.35** | 91.16 | 93.02 | 94.88 | 4.39 | 7.21 |
| pag | 89.56 | 93.81 | 94.17 | **94.35** | **94.35** | **94.35** | 5.08 | 3.65 |
| pen | 93.92 | 90.93 | 95.10 | 93.92 | 94.19 | **95.19** | 4.48 | 6.23 |
| sat | 74.52 | 69.55 | 79.02 | 71.69 | 71.69 | **81.98** | 15.16 | 12.55 |
| seg | 90.69 | 86.71 | **92.25** | 90.95 | 91.34 | 92.16 | 6.01 | 5.69 |
| shu | 89.30 | 82.46 | 91.87 | 86.34 | 86.80 | **94.35** | 12.60 | 6.11 |
| tae | 55.53 | 56.73 | 56.13 | 56.13 | **56.77** | **56.77** | 2.19 | 12.58 |
| thy | 93.48 | 94.03 | 94.03 | 94.03 | 94.03 | **94.17** | 0.74 | 2.92 |
| veh | 70.32 | 61.80 | **72.21** | 65.58 | 68.07 | 70.91 | 14.41 | 17.04 |
| vow | 84.04 | 84.85 | 89.09 | 88.18 | 89.29 | **89.90** | 6.52 | 7.21 |
| win | 94.85 | 95.54 | **96.63** | 96.09 | 96.09 | **96.63** | 1.84 | 3.51 |
| yea | 56.53 | 55.07 | **59.57** | 58.43 | 58.90 | **59.57** | 7.55 | 10.20 |
| AVG | 79.59 | 78.00 | 81.19 | 79.75 | 80.06 | **81.34** | 5.56 | 6.13 |

Attending at these tables, several points can be highlighted:

**Table 3.** Results WINWV

| Dataset | $\preceq_{Lex1}$ | $\preceq_{Lex2}$ | $\preceq_{1/3,2/3}$ | $\preceq_{f,g}$ | $\preceq_{2/3,1/3}$ | $\preceq_{XY}$ | loss | ad% |
|---------|------|------|------|------|------|------|------|------|
| aut | 75.03 | 70.90 | 75.17 | 75.83 | **76.44** | 75.85 | 7.25 | 1.70 |
| bal | 71.95 | 83.34 | 79.96 | 83.99 | **84.31** | 82.54 | 14.66 | 6.68 |
| cle | 49.47 | 53.53 | 55.24 | **55.93** | 55.26 | 54.58 | 11.54 | 7.07 |
| con | 53.70 | 52.42 | 53.70 | 53.57 | 53.37 | **53.71** | 2.40 | 11.33 |
| der | **96.96** | 96.69 | 96.69 | 96.69 | 96.69 | 96.69 | 0.28 | 0.01 |
| eco | 76.29 | **84.36** | 81.35 | 82.25 | 83.13 | 81.37 | 9.57 | 3.81 |
| gla | 59.90 | 58.74 | 62.21 | 65.01 | **67.32** | 64.53 | 12.75 | 7.78 |
| hay | **75.22** | **75.22** | **75.22** | **75.22** | **75.22** | **75.22** | 0.00 | 0.00 |
| iri | **96.67** | 94.00 | 96.00 | 95.33 | 95.33 | 95.33 | 2.76 | 1.39 |
| lym | 80.49 | 80.54 | **81.23** | **81.23** | 80.54 | **81.23** | 0.91 | 0.14 |
| new | 87.91 | 90.70 | 93.02 | 91.63 | 93.02 | **93.95** | 6.44 | 7.21 |
| pag | 74.11 | **94.17** | 83.37 | 92.10 | 92.29 | 88.01 | 21.30 | 3.65 |
| pen | 88.92 | 93.28 | 92.47 | 94.01 | **94.56** | 93.93 | 5.96 | 6.23 |
| sat | 71.41 | 66.28 | 78.71 | 72.95 | 73.73 | **81.83** | 19.00 | 12.55 |
| seg | 85.54 | 84.81 | 88.66 | 88.57 | 88.96 | **89.31** | 5.04 | 5.69 |
| shu | 67.94 | 82.37 | 81.58 | 89.35 | **89.99** | 89.46 | 24.51 | 6.11 |
| tae | 52.35 | 56.77 | 56.19 | **56.82** | **56.82** | 56.79 | 7.87 | 12.58 |
| thy | 92.51 | 93.89 | 93.47 | **93.90** | **93.90** | 93.48 | 1.48 | 2.92 |
| veh | 65.96 | 63.22 | **71.03** | 67.23 | 68.89 | **71.03** | 10.99 | 17.04 |
| vow | 75.45 | 82.83 | 81.31 | 84.44 | **85.56** | 84.24 | 11.81 | 7.21 |
| win | 93.17 | 95.54 | 96.63 | 96.09 | 96.09 | **97.17** | 4.12 | 3.51 |
| yea | 52.23 | 55.89 | 56.74 | 57.15 | 57.69 | **58.03** | 9.99 | 10.20 |
| AVG | 74.69 | 77.70 | 78.64 | 79.51 | **79.96** | 79.92 | 8.66 | 6.13 |

- In both Vote and WinWV, results are affected by the admissible order selected. Interestingly, WinWV is more affected by this decision, leading to worse results than in Vote (e.g., $\preceq_{Lex1}$ drops to 74.69% in average with WinWV vs. 79.59% in Vote).
- $\preceq_{XY}$ obtains the highest accuracy in Vote, whereas $\preceq_{2/3,1/3}$ leads in WinWV. $\preceq_{2/3,1/3}$ and $\preceq_{1/3,2/3}$ behave alternatively between Vote and WinWv. Hence, the order seems to be dependent on the OVO aggregation method.
- Observing the maximum loss values we note that deciding the appropriate admissible order is definitely a key factor. In Vote, a 15% of accuracy loss can be obtained if $\preceq_{Lex2}$ is used instead of $\preceq_{XY}$. Similarly, in WinWV this value increases to 24% in shuttle dataset when considering $\preceq_{Lex1}$ instead of $\preceq_{2/3,1/3}$.

- Overall, for IVOVO using either $\preceq_{Lex1}$ or $\preceq_{Lex2}$ is harmful (both for Vote and WinWV). This fact is interesting because it tell us that the interval has more information than the one encoded in its bounds. All the other orders are considering values insider the interval and not letting only one bound decide the final order.

Anyway, the previous statements needs to be supported by the appropriate statistical analysis. Table 4 shows the results of the Aligned Friedman Ranks tests, one for each OVO aggregation to focus on the differences among admissible orders.

**Table 4.** Aligned Friedman test

| Method | Vote | WinWV |
|---|---|---|
| | Rank (p-value) | Rank (p-value) |
| $\preceq_{XY}$ | 41.61 (0.8422) | 43.80 (−) |
| $\preceq_{1/3,2/3}$ | 39.32 (−) | 68.98 (0.0870) |
| $\preceq_{2/3,1/3}$ | 67.57 (0.0286+) | 44.11 (0.9780) |
| $\preceq_{f,g}$ | 73.59 (0.0089+) | 51.95 (0.9585) |
| $\preceq_{Lex1}$ | 76.02 (0.0058+) | 106.70 (0.0000+) |
| $\preceq_{Lex2}$ | 100.89 (0.0000+) | 83.45 (0.0023+) |

+near the p-value means that statistical differences are found at 95% confidence.

According to the tests, results are rather different in each OVO aggregation. In Vote, $\preceq_{1/3,2/3}$ gets the lowest ranks (better), but it does not show significant differences with $\preceq_{XY}$. The performance of the other orders is significantly worse. On the contrary, in WinWV $\preceq_{XY}$ shows its robustness in this case being at the top accompanied by $\preceq_{2/3,1/3}$ (no significant differences). The rest of the methods

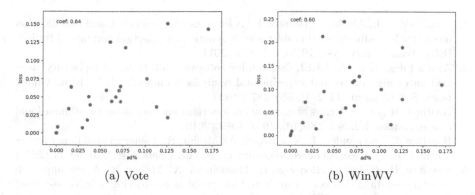

(a) Vote                    (b) WinWV

**Fig. 1.** Scatter plot of ad% vs. loss of all datasets.

are again significantly worse than $\preceq_{XY}$. Notice that in both cases, $\preceq_{Lex1}$ and $\preceq_{Lex2}$ does not perform well, whereas $\preceq_{2/3,1/3}$ and $\preceq_{1/3,2/3}$ have a different behavior depending on the OVO aggregation considered.

Finally, to better understand the results obtained, we tried to find a correlation between the percentage of times the selection of the admissible order may change the decision (ad%) and the percentage of loss obtained. Figure 1 shows these plots for both Vote and WinWV. As we can observe, there is a tendency for greater losses as the ad% increases, which could be expected but should be checked. Pearson correlation coefficients for Vote and WinWV between ad% and loss were 0.6569 and 0.4011, respectively.

# 6   Conclusions

In this work we have analyzed the influence of admissible orders for arranging interval-valued data in IVOVO. To do so, we have considered six different admissible orders, that comes from the most used construction methods, and we have carried out an experimental study that has proven the large differences in terms of accuracy that can be obtained. This result justifies itself the importance of the performed study and points out that deeper studies can be carried out to determine the most appropriate admissible order. From the results obtained, we have seen that the usual order of Xu and Yager can be outperformed by other admissible orders, such as $\preceq_{2/3,1/3}$ (even $\preceq_{1/3,2/3}$ obtains very good results). Moreover, we have observed that extreme orders, such as $\preceq_{Lex1}$ and $\preceq_{Lex2}$, are not suitable for IVOVO.

For future work, we aim to carry out a deeper study including more admissible orders and a further analysis of the most suitable order for each aggregation strategy. Moreover, we plan to continue introducing new theoretical developments of interval theory into IVOVO, as interval aggregation functions or interval normalization methods.

# References

1. Alcalá-Fdez, J., Alcalá, R., Herrera, F.: A fuzzy association rule-based classification model for high-dimensional problems with genetic rule selection and lateral tuning. IEEE Trans. Fuzzy Syst. **19**(5), 857–872 (2011)
2. Alcalá-Fdez, J., et al.: KEEL data-mining software tool: Data set repository, integration of algorithms and experimental analysis framework. J. Multiple-Valued Logic Soft Comput. **17**(2–3), 255–287 (2011)
3. Bustince, H., et al.: A class of aggregation functions encompassing two-dimensional owa operators. Inf. Sci. **180**(10), 1977–1989 (2010)
4. Bustince, H., Fernandez, J., Kolesárová, A., Mesiar, R.: Generation of linear orders for intervals by means of aggregation functions. Fuzzy Sets Syst. **220**, 69–77 (2013)
5. Bustince, H., Galar, M., Bedregal, B., Kolesárová, A., Mesiar, R.: A new approach to interval-valued choquet integrals and the problem of ordering in interval-valued fuzzy set applications. IEEE Trans. Fuzzy Syst. **21**(6), 1150–1162 (2013)

6. Cornelis, C., Deschrijver, G., Kerre, E.E.: Advances and challenges in interval-valued fuzzy logic. Fuzzy Sets Syst. **157**, 622–627 (2016)
7. Elkano, M., Galar, M., Sanz, J., Bustince, H.: Fuzzy rule-based classification systems for multi-class problems using binary decomposition strategies: on the influence of n-dimensional overlap functions in the fuzzy reasoning method. Inf. Sci. **332**, 94–114 (2016)
8. Elkano, M., Galar, M., Sanz, J., Lucca, G., Bustince, H.: IVOVO: a new interval-valued one-vs-one approach for multi-class classification problems. In: 17th International Fuzzy Systems Association (IFSA), pp. 1–6 (2017)
9. Elkano, M., et al.: Enhancing multiclass classification in FARC-HD fuzzy classifier: on the synergy between $n$-dimensional overlap functions and decomposition strategies. IEEE Trans. Fuzzy Syst. **23**(5), 1562–1580 (2015)
10. Galar, M., Fernández, A., Barrenechea, E., Bustince, H., Herrera, F.: An overview of ensemble methods for binary classifiers in multi-class problems: experimental study on one-vs-one and one-vs-all schemes. Pattern Recogn. **44**(8), 1761–1776 (2011)
11. García, S., Fernández, A., Luengo, J., Herrera, F.: A study of statistical techniques and performance measures for genetics-based machine learning: accuracy and interpretability. Soft. Comput. **13**(10), 959–977 (2009)
12. Bustince, H., Montero, J., Pagola, M., Barrenechea, E., Gomez, D.: A survey on interval-valued fuzzy sets. In: Pedrycz, W. (ed.) Handbook of Granular Computing, vol. 1. Wiley, New Jersey (2008)
13. Ishibuchi, H., Nakashima, T., Nii, M.: Classification and Modeling with Linguistic Information Granules: Advanced Approaches to Linguistic Data Mining. Springer, Heidelberg (2004). https://doi.org/10.1007/b138232
14. Ishibuchi, H., Yamamoto, T., Nakashima, T.: Hybridization of fuzzy GBML approaches for pattern classification problems. IEEE Trans. Syst. Man Cybern. B **35**(2), 359–365 (2005)
15. Lizasoain, I., Moreno, C.: Owa operators defined on complete lattices. Fuzzy Sets Syst. **224**, 36–52 (2013)
16. Lorena, A., Carvalho, A., Gama, J.: A review on the combination of binary classifiers in multiclass problems. Artif. Intell. Rev. **30**(1–4), 19–37 (2008)
17. Moreno-Torres, J., Saez, J., Herrera, F.: Study on the impact of partition-induced dataset shift on k-fold cross-validation. IEEE Trans. Neural Netw. Learn. Syst. **23**(8), 1304–1312 (2012)
18. Paternain, D., Miguel, L.D., Ochoa, G., Lizasoain, I., Mesiar, R., Bustince, H.: The interval-valued choquet integral based on admissible permutations. IEEE Trans. Fuzzy Syst. (in press)
19. Sanz, J., Fernández, A., Bustince, H., Herrera, F.: IVTURS: a linguistic fuzzy rule-based classification system based on a new interval-valued fuzzy reasoning method with tuning and rule selection. IEEE Trans. Fuzzy Syst. **21**(3), 399–411 (2013)
20. Xu, Z.S., Yager, R.R.: Some geometric aggregation operators based on intuitionistic fuzzy sets. Int. J. General Syst. **35**(4), 417–433 (2006)

# Does the Order of Attributes Play an Important Role in Classification?

Antonio J. Tallón-Ballesteros[1]([⊠]), Simon Fong[2], and Rocío Leal-Díaz[3]

[1] Department of Electronic, Computer Systems and Automation Engineering,
University of Huelva, Huelva, Spain
antonio.tallon.diesia@zimbra.uhu.es
[2] Department of Computer and Information Science, University of Macau,
Taipa, Macao, Special Administrative Region of China
[3] Higher School of Computer Science, University of Seville, Seville, Spain

**Abstract.** This paper proposes a methodology to feature sorting in the context of supervised machine learning algorithms. Feature sorting is defined as a procedure to order the initial arrangement of the attributes according to any sorting algorithm to assign an ordinal number to every feature, depending on its importance; later the initial features are sorted following the ordinal numbers from the first to the last, which are provided by the sorting method. Feature ranking has been chosen as the representative technique to fulfill the sorting purpose inside the feature selection area. This contribution aims at introducing a new methodology where all attributes are included in the data mining task, following different sortings by means of different feature ranking methods. The approach has been assessed in ten binary and multiple class problems with a number of features lower than 37 and a number of instances below than 106 up to 28056; the test-bed includes one challenging data set with 21 labels and 23 attributes where previous works were not able to achieve an accuracy of at least a fifty percent. ReliefF is a strong candidate to be applied in order to re-sort the initial characteristic space and C4.5 algorithm achieved a promising global performance; additionally, PART -a rule-based classifier- and Support Vector Machines obtained acceptable results.

**Keywords:** Data mining · Feature sorting · Feature ranking ·
Feature selection · Data preparation ·
Knowledge Discovery in Databases

## 1 Introduction

Hybrid artificial intelligent systems comprise the development and application of symbolic and sub-symbolic techniques aimed at the building of highly and reliable problem-solving techniques [8]. Hybridisation of intelligent approaches, coming from different computational intelligent areas [12], has become popular due to the increasing awareness that such combinations perform frequently

© Springer Nature Switzerland AG 2019
H. Pérez García et al. (Eds.): HAIS 2019, LNAI 11734, pp. 370–380, 2019.
https://doi.org/10.1007/978-3-030-29859-3_32

better than the individual techniques such as neurocomputing [1], fuzzy systems [17], rough sets [23], evolutionary algorithms [3], agents and multiagent systems [40], etc. In a hybrid intelligent system, a synergistic combination of multiple approaches is used to create an efficient solution to confront a particular problem [5].

Data mining [37] goals to discover relationships inside the data in order to predict effectively with unseen data. There are different phases in a Knowledge Discovery in Databases (KDD) task [2] although data preparation represents a significant computational burden [42]; there are newer methodologies than KDD such as CRoss Industry Standard Process for Data Mining (CRISP-DM) [41] and its extension Data Mining Methodology for Engineering applications (DMME) which also include the data pre-processing stage [13]. Feature engineering has an outstanding function in data analytics; it encompasses many fields such as feature transformation, feature generation, feature selection, feature analysis and many others [36]. By its part, feature construction has been successfully combined with ontologies to improve the classification performance in the context of daily living activities [27].

This paper applies feature selection procedures based on feature ranking to obtain an initial arrangement of the characteristics space which is entered into the data mining stage to build a decision-making model. Particularly, the performance of the original and proposed sortings are analysed in order to shed light on the recommendation to conduct or not a prior feature ranking method with certain classifiers which differ in the way to represent the knowledge. The remaining of this paper is organised as follows. Section 2 reviews different concepts about feature selection and presents the motivation of this work. Section 3 describes the proposed approach. Then, Sect. 4 depicts the experimental results. Lastly, Sect. 5 draws some conclusions.

## 2 Attribute Selection and Motivation

Feature selection aims at getting a subset of attributes which are able to keep the same performance or even to obtain an improved one, compared to the accuracy using the original feature space [21]. There are many criteria to create a taxonomy within the feature selection field. According to the usage or not of a learning algorithm in the data preparation step of feature selection, three methods may be distinguish such as filters [28], semi-wrappers [35] and wrappers [15]. Filters score the features, individually, or a subset, according to an inner measure in the data such as the correlation or any type of statistical measure; semi-wrappers compute the performance of the feature set with a supervised machine learning method which is different from the target learner while in the wrappers the supervised strategy to evaluate the potential solution is exactly the same as the target machine learning algorithm.

Depending on the generation procedure of the solution, attribute selection methods may be divided into feature ranking, feature subset selection and extended feature subset selection [29]. Feature ranking creates an ordered list where the first features in the list have a higher value meaning that are more suitable and the potentially profitable for the data mining stage; feature subset selection gives as output a subset of attributes with a good global performance following any type of fitness measure like the correlation, the consistency and so on; extended feature subset selection where the feature subset selection is applied more than once and hence the characteristic spaces are partially overlapped, in the sense that some attributes may be the input for the feature selector some times.

Most of machine learning algorithms [14] iterate over the feature space following an ascendant or descendant loop which may start from the first or the last attribute of the characteristic space; it means that features which are closer to the starting point are more likely to be explored and in case of ties, according to any measure to guide the model learning, would be first kept in the final decision-making model. This work has some connections with the paper written by Liu et al. [18] in the sense that they apply a feature selection method to sort out the top features inside a clustering task. Nonetheless, the current paper is sorting the attribute space of the training set via a feature ranking procedure without dropping features. Ortega and Fisher [22] perform a sorting with the training instances. Typically, the reduced feature set tends to get better results than the full feature set.

From previous empirical analysis, we have extracted as motivating example the Breast problem [20] since we have been working with it from a long time ago and the data preparation techniques at the feature level yielded not very promising results [31]. Additionally, we analysed deeply some ad-hoc scenarios in order to get a better understanding of the problem two years ago in order to retain as much features as possible following a feature ranking method as an alternative to feature subset selection procedures [33]; results without feature selection are superior to those with data pre-processing. This paper goals to introduce a new approach where all attributes are included in the data mining task although following different sortings by means of different feature ranking methods, in order to catch up some scenarios where the behaviour of the application of specific machine learning algorithms is favourable and to propose some general guidelines. Feature sorting has been applied in the context of Visual Data Mining although with a different aim [19].

# 3    The Proposed Methodology

This paper presents a new approach in the data pre-processing context at the feature level. More concretely, it introduces a feature sorting stage which may be defined as a procedure to order the initial arrangement of the attributes according to any sorting algorithm to assign an ordinal number to every feature, depending on its importance; later the initial features are sorted following the ordinal numbers from the first to the last, which are provided by the sorting method. The initial feature space is sorted according to a feature ranking method where no threshold is required since no attributes are discarded after the

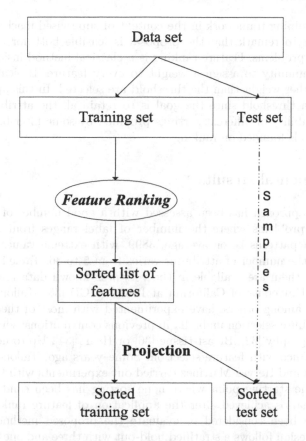

**Fig. 1.** Proposed methodology: Feature Ranking to sort the Feature Space (FR2sortFS).

feature selection step. Very often, there is a trade-off between the threshold, the different methods and the classification algorithms -just to cite a type of prediction technique where feature selection is usually applied nowadays-, since small changes may unbalance the race in favour of a particular element. The intention of this paper is to compare fairly some feature ranking approaches to assess the convenience or not of applying one of such methods as the first step inside a KDD task just to sort the characteristic space. By its part, feature ranking methods have been chosen since they are very efficient and not very demanding in terms of computing time.

Figure 1 depicts the proposed methodology which has been named Feature Ranking to sort the Feature Space (FR2sortFS). The flowchart comprises the following steps: (a) first, the data set is splitted into training and test sets, (b) second, the training set is submitted to a feature ranking and the output is a sorted list of attributes and (c) third, the aforementioned list is projected onto the initial training and test sets to obtain the sorted training and test sets, respectively. Now both the new training and test data are ready to be entered into any data mining framework in the context of supervised machine learning. It is important to remark that the proposal is feasible both for classification and regression problems. Feature ranking is a classical method inside data pre-processing community to assign a weight to every feature. Basically, features that have a higher weight than the threshold are selected. In this approach, we do not define a threshold since the goal is to retain all the attributes and to assess the suitability of different sortings provided by some filter-based feature selection methods founded on ranking.

## 4    Experimental Results

The proposed approach has been assessed with a good number of binary and multiple class problems where the number of label ranges from 2 up to 18. The number of patterns is, on average, 3899, with extreme values of 106 and 28056. Lastly, the number of attributes varies from 5 to 36. Table 1 depicts the test-bed; all of them are available in the very well-known data mining repository from the University of California at Irvine (UCI) [38]. Tallón-Ballesteros and Riquelme, among others, have experimented with most of these data sets under the attribute selection umbrella in previous contributions; some examples are Cardiotocography [32], Breast-tissue [36] or Heart [34]. Up to now, we have not tested Balance with feature selection. Some years ago, Tallón-Ballesteros, Gutiérrez-Peña and Hervás-Martínez carried out experiments with the raw data [30]. Besides the starting point where none sorting has been conducted, three additional scenarios achieved after the application of feature ranking without threshold have been considered to evaluate the proposed methodology. The experimental design follows a stratified hold-out with three and one quarters for the training and test sets, respectively; it was initially suggested by L. Prechelt in the context of neural networks [24] although is also valid for other kinds of classifiers [9]. Waikato Environment for Knowledge Analysis (Weka) [11] has been the visual exploitation framework [4] to run the learners. As supervised machine learning algorithms related to the classification task some approaches following different ways to represent the discovered knowledge such as C4.5 [26], $k$NN ($k$ Neareast Neighbour; the experiments have been conducted with $k = 1$) [6], PART (PARTial Decision Tree) [10] or Support Vector Machines (SVM) [39], which belong to decision trees, distance-based classifiers, rule-based learners or classifiers that build hyper-planes, respectively have been tested.

Table 2 summarises the sorting methods and shows some interesting properties as well as the associated reference. ReliefF works by randomly sampling an instance from the data and then locating its nearest neighbours from the same and the remaining classes. The attributes' values and then compared to the sample instance and are used to update the weight for each feature [16]. Gain-Ratio is supported by the concepts of expected information needed to classify a given sample and the entropy based on the partitioning into subsets by every attribute; the latter value is subtracted to the former one to get the gain of the branching on all attributes. The feature with the highest gain ratio is selected as the splitting attribute [25]. InformationGain, also named as InfoGain, is a very common attribute evaluation method which computes the entropy of the class before and after observing the attribute [7].

The sorting methods and the classifiers have been run in Weka with the default setting as this is the setup recommended by the own authors. The running environment has been a computer equipped with an Intel i7 processor with 8 GB main memory and a Solid State Drive (SSD) with 120 GB. The results report the accuracy in the test set measured as the percentage of the number of hits. The computational time is negligible both in the data preparation phase, given that the feature ranking methods are very fast, as well as the data mining stage and we do not conduct any type of comparison as the characteristic space has the same number of features.

Table 3 shows the performance of the proposed methodology. There are some data sets where the sorting algorithm does not affect the performance like Balance or Newthyroid. In other cases, like Primary-tumor, Heart and Breast-tissue some changes take place compared to the control method where none sorting is done. The last row includes a brief outline of the global performance of every classifier with the different feature ranking approaches; as it can be seen, ReliefF is a good option for C4.5 classifier, followed by GainRatio or InfoGain which have a similar performance, generally speaking, though there are individual differences. PART and SVM learners may take advantage of the new proposal with ReliefF and some ties may be switched to wins. Finally, $k$NN does not alter the behaviour with or without any sorting method which means that is a very robust classifier.

**Table 1.** Classification test-bed

| Data set | Patterns | Attributes | Labels |
|---|---|---|---|
| Balance | 625 | 4 | 3 |
| Breast | 286 | 15 | 2 |
| Breast-tissue | 106 | 9 | 6 |
| Cardiotocography | 2126 | 22 | 3 |
| Chess | 3196 | 36 | 2 |
| Heart | 270 | 13 | 2 |
| Hypothyroid | 3772 | 29 | 4 |
| Krkopt | 28056 | 6 | 18 |
| Newthyroid | 215 | 5 | 3 |
| Primary-tumor | 339 | 23 | 21 |
| Average | 3899.1 | 16.2 | 6.4 |

**Table 2.** List of sorting methods to arrange the feature space before data mining phase

| Name | Type of data selection method | Kind of procedure | Reference |
|------|-------------------------------|-------------------|-----------|
| ReliefF | Feature level | Feature ranking | [16] |
| GainRatio | Feature level | Feature ranking | [25] |
| InfoGain | Feature level | Feature ranking | [7] |

**Table 3.** Test results for the classification test-bed with different algorithms and data pre-processing methods, if any

| Data set | Sorting method | Classifier | | | |
|----------|----------------|------|-----|------|-----|
| | | C4.5 | kNN | PART | SVM |
| Balance | None | 83.33 | 87.82 | 85.26 | 88.46 |
| | ReliefF | 83.97 | 87.82 | 85.26 | 88.46 |
| | GainRatio | 83.33 | 87.82 | 85.26 | 88.46 |
| | InfoGain | 83.33 | 87.82 | 85.26 | 88.46 |
| Breast | None | 70.42 | 64.79 | 69.01 | 64.79 |
| | ReliefF | 70.42 | 64.79 | 70.42 | 64.79 |
| | GainRatio | 70.42 | 64.79 | 67.61 | 64.79 |
| | InfoGain | 70.42 | 64.79 | 67.61 | 64.79 |
| Breast-tissue | None | 52.00 | 60.00 | 44.00 | 52.00 |
| | ReliefF | 56.00 | 60.00 | 44.00 | 52.00 |
| | GainRatio | 56.00 | 60.00 | 44.00 | 56.00 |
| | InfoGain | 56.00 | 60.00 | 44.00 | 60.00 |
| Cardiotocography | None | 82.71 | 76.32 | 82.52 | 83.65 |
| | ReliefF | 83.08 | 76.32 | 82.52 | 83.65 |
| | GainRatio | 83.08 | 76.32 | 82.52 | 83.65 |
| | InfoGain | 83.08 | 76.32 | 82.52 | 83.65 |
| Chess | None | 99.00 | 96.62 | 99.12 | 95.74 |
| | ReliefF | 99.00 | 96.62 | 99.12 | 95.74 |
| | GainRatio | 99.00 | 96.62 | 99.12 | 95.74 |
| | InfoGain | 99.00 | 96.62 | 99.12 | 95.87 |
| Heart | None | 70.59 | 73.53 | 73.53 | 76.47 |
| | ReliefF | 69.12 | 73.53 | 73.53 | 76.47 |
| | GainRatio | 69.12 | 73.53 | 75.00 | 76.47 |
| | InfoGain | 69.12 | 73.53 | 75.00 | 76.47 |
| Hypothyroid | None | 99.15 | 90.99 | 98.83 | 93.85 |
| | ReliefF | 99.26 | 90.99 | 98.83 | 93.85 |
| | GainRatio | 99.15 | 90.99 | 98.83 | 93.85 |
| | InfoGain | 99.15 | 90.99 | 98.83 | 93.85 |

**Table 3.** (*continued*)

| Data set | Sorting method | Classifier | | | |
|---|---|---|---|---|---|
| | | C4.5 | kNN | PART | SVM |
| Krkopt | None | 78.27 | 60.37 | 74.96 | 28.25 |
| | ReliefF | 78.42 | 60.37 | 75.23 | 28.35 |
| | GainRatio | 78.38 | 60.37 | 74.96 | 28.07 |
| | InfoGain | 78.38 | 60.37 | 75.62 | 27.85 |
| Newthyroid | None | 96.30 | 94.44 | 92.59 | 88.89 |
| | ReliefF | 96.30 | 94.44 | 92.59 | 88.89 |
| | GainRatio | 96.30 | 94.44 | 92.59 | 88.89 |
| | InfoGain | 96.30 | 94.44 | 92.59 | 88.89 |
| Primary-tumor | None | 45.45 | 40.91 | 43.18 | 47.73 |
| | ReliefF | 44.32 | 40.91 | 45.45 | 48.86 |
| | GainRatio | 45.45 | 40.91 | 43.18 | 48.86 |
| | InfoGain | 45.45 | 40.91 | 43.18 | 47.73 |
| W/T/L | ReliefF | 5/3/2 | 0/10/0 | 3/7/0 | 2/8/0 |
| Control method | *GainRatio* | 3/6/1 | 0/10/0 | 1/8/1 | 2/7/1 |
| (None) | *InfoGain* | 3/6/1 | 0/10/0 | 2/7/1 | 2/7/1 |

# 5  Conclusions

This paper presented a new approach to feature sorting under the umbrella of supervised machine learning problems in order to re-arrange the initial feature space according to the ranking achieved by a good number of feature selection procedures. The empirical study comprised a test-bed of ten binary and multiple class data sets with problems of special difficulty like Primary tumor where the classifiers are not able to achieve a fifty percent hit rate. ReliefF is a good method to sort the characteristic space along with the classifiers C4.5, PART or SVM. InfoGain and GainRatio are also competitive options although with a lower performance.

**Acknowledgments.** This work has been partially subsidised by TIN2014-55894-C2-R and TIN2017-88209-C2-2-R projects of the Spanish Inter-Ministerial Commission of Science and Technology (MICYT), FEDER funds and the P11-TIC-7528 project of the "Junta de Andalucía" (Spain).

# References

1. Amari, S.-I.: Mathematical foundations of neurocomputing. Proc. IEEE **78**(9), 1443–1463 (1990)
2. Azevedo, A.: Data mining and knowledge discovery in databases. In: Advanced Methodologies and Technologies in Network Architecture, Mobile Computing, and Data Analytics, pp. 502–514. IGI Global (2019)

3. Bäck, T.: Evolutionary Algorithms in Theory and Practice: Evolution Strategies, Evolutionary Programming, Genetic Algorithms. Oxford University Press, New York (1996)
4. Cho, S.-B., Tallón-Ballesteros, A.J.: Visual tools to lecture data analytics and engineering. In: Ferrández Vicente, J.M., Álvarez-Sánchez, J.R., de la Paz López, F., Toledo Moreo, J., Adeli, H. (eds.) IWINAC 2017. LNCS, vol. 10338, pp. 551–558. Springer, Cham (2017). https://doi.org/10.1007/978-3-319-59773-7_56
5. Corchado, E., Corchado Rodrguez, J.M., Abraham, A.: Innovations in Hybrid Intelligent Systems, vol. 44. Springer Science & Business Media, Berlin (2007)
6. Cover, T., Hart, P.: Nearest neighbor pattern classification. IEEE Trans. Inf. Theory **13**(1), 21–27 (1967)
7. Cover, T.M., Thomas, J.A.: Elements of information theory. New York **68**, 69–73 (1991)
8. Corchado, E., Kurzyński, M., Woźniak, M. (eds.): HAIS 2011. LNCS (LNAI), vol. 6678. Springer, Heidelberg (2011). https://doi.org/10.1007/978-3-642-21219-2
9. Di Ruberto, C., Putzu, L., Arabnia, H.R., Quoc-Nam, T.: A feature learning framework for histology images classification. In: Emerging Trends in Applications and Infrastructures for Computational Biology, Bioinformatics, and Systems Biology: Systems and Applications, pp. 37–48. Elsevier Press (2016)
10. Frank, E., Witten, I.H.: Generating accurate rule sets without global optimization. In: Shavlik, J. (ed.) Fifteenth International Conference on Machine Learning, pp. 144–151. Morgan Kaufmann (1998)
11. Hall, M., Frank, E., Holmes, G., Pfahringer, B., Reutemann, P., Witten, I.H.: The weka data mining software: an update. ACM SIGKDD Exp. Newsl. **11**(1), 10–18 (2009)
12. He, J., Yang, Z., Yao, X.: Hybridisation of evolutionary programming and machine learning with k-nearest neighbor estimation. In: 2007 IEEE Congress on Evolutionary Computation, pp. 1693–1700. IEEE (2007)
13. Huber, S., Wiemer, H., Schneider, D., Ihlenfeldt, S.: DMME: data mining methodology for engineering applications-a holistic extension to the CRISP-DM model. Procedia CIRP **79**, 403–408 (2019)
14. Jordan, M.I., Mitchell, T.M.: Machine learning: trends, perspectives, and prospects. Science **349**(6245), 255–260 (2015)
15. Kohavi, R., John, G.H.: Wrappers for feature subset selection. Artif. Intell. **97**(1–2), 273–324 (1997)
16. Kononenko, I.: Estimating attributes: analysis and extensions of RELIEF. In: Bergadano, F., De Raedt, L. (eds.) ECML 1994. LNCS, vol. 784, pp. 171–182. Springer, Heidelberg (1994). https://doi.org/10.1007/3-540-57868-4_57
17. Kruse, R., Gebhardt, J.E., Klawon, F.: Foundations of Fuzzy Systems. John Wiley & Sons Inc., New York (1994)
18. Liu, W., Liu, S., Gu, Q., Chen, X., Chen, D.: FECS: a cluster based feature selection method for software fault prediction with noises. In: 2015 IEEE 39th Annual Computer Software and Applications Conference, vol. 2, pp. 276–281. IEEE (2015)
19. May, T., Bannach, A., Davey, J., Ruppert, T., Kohlhammer, J.: Guiding feature subset selection with an interactive visualization. In: 2011 IEEE Conference on Visual Analytics Science and Technology (VAST), pp. 111–120. IEEE (2011)
20. Michalski, R.S., Mozetic, I., Hong, J., Lavrac, N.: The multi-purpose incremental learning system AQ15 and its testing application to three medical domains. In: Proceedings of the AAAI 1986, pp. 1–041 (1986)
21. Narendra, P.M., Fukunaga, K.: A branch and bound algorithm for feature subset selection. IEEE Trans. Comput. **9**, 917–922 (1977)

22. Ortega, J., Fisher, D.: Flexibly exploiting prior knowledge in empirical learning. In: IJCAI, pp. 1041–1049 (1995)
23. Pawlak, Z.: Rough sets. Int. J. Comput. Inf. Sci. **11**(5), 341–356 (1982)
24. Prechelt, L.: Proben 1-a set of benchmarks and benchmarking rules for neural network training algorithms (1994)
25. Quinlan, J.R.: Induction of decision trees. Mach. Learn. **1**(1), 81–106 (1986)
26. Quinlan, J.R.: C4.5: Programs for Machine Learning, vol. 1. Morgan Kaufmann, San Mateo (1993)
27. Salguero, A.G., Medina, J., Delatorre, P., Espinilla, M.: Methodology for improving classification accuracy using ontologies: application in the recognition of activities of daily living. J. Ambient Intell. Humaniz. Comput. **10**(6), 2125–2142 (2019)
28. Sánchez-Maroño, N., Alonso-Betanzos, A., Tombilla-Sanromán, M.: Filter methods for feature selection – a comparative study. In: Yin, H., Tino, P., Corchado, E., Byrne, W., Yao, X. (eds.) IDEAL 2007. LNCS, vol. 4881, pp. 178–187. Springer, Heidelberg (2007). https://doi.org/10.1007/978-3-540-77226-2_19
29. Tallón-Ballesteros, A.J., Cavique, L., Fong, S.: Addressing low dimensionality feature subset selection: ReliefF(-k) or Extended Correlation-Based Feature Selection(eCFS)? In: Martínez Álvarez, F., Troncoso Lora, A., Sáez Muñoz, J.A., Quintián, H., Corchado, E. (eds.) SOCO 2019. AISC, vol. 950, pp. 251–260. Springer, Cham (2020). https://doi.org/10.1007/978-3-030-20055-8_24
30. Tallón-Ballesteros, A.J., Gutiérrez-Peña, P.A., Hervás-Martínez, R.: Distribution of the search of evolutionary product unit neural networks for classification. arXiv preprint arXiv:1205.3336 (2012)
31. Tallón-Ballesteros, A.J., Hervás-Martínez, C., Riquelme, J.C., Ruiz, R.: Improving the accuracy of a two-stage algorithm in evolutionary product unit neural networks for classification by means of feature selection. In: Ferrández, J.M., Álvarez Sánchez, J.R., de la Paz, F., Toledo, F.J. (eds.) IWINAC 2011. LNCS, vol. 6687, pp. 381–390. Springer, Heidelberg (2011). https://doi.org/10.1007/978-3-642-21326-7_41
32. Tallón-Ballesteros, A.J., Riquelme. J.C.: Deleting or keeping outliers for classifier training? In: 2014 Sixth World Congress on Nature and Biologically Inspired Computing (NaBIC 2014), pp. 281–286. IEEE (2014)
33. Tallón-Ballesteros, A.J., Riquelme, J.C.: Low dimensionality or same subsets as a result of feature selection: an in-depth roadmap. In: Ferrández Vicente, J.M., Álvarez-Sánchez, J.R., de la Paz López, F., Toledo Moreo, J., Adeli, H. (eds.) IWINAC 2017. LNCS, vol. 10338, pp. 531–539. Springer, Cham (2017). https://doi.org/10.1007/978-3-319-59773-7_54
34. Tallón-Ballesteros, A.J., Riquelme, J.C., Ruiz, R.: Accuracy increase on evolving product unit neural networks via feature subset selection. In: Martínez-Álvarez, F., Troncoso, A., Quintián, H., Corchado, E. (eds.) HAIS 2016. LNCS (LNAI), vol. 9648, pp. 136–148. Springer, Cham (2016). https://doi.org/10.1007/978-3-319-32034-2_12
35. Tallón-Ballesteros, A.J., Riquelme, J.C., Ruiz, R.: Semi-wrapper feature subset selector for feed-forward neural networks: applications to binary and multi-class classification problems. Neurocomputing **353**, 28–44 (2019)
36. Tallón-Ballesteros, A.J., Tuba, M., Xue, B., Hashimoto, T.: Feature selection and interpretable feature transformation: a preliminary study on feature engineering for classification algorithms. In: Yin, H., Camacho, D., Novais, P., Tallón-Ballesteros, A.J. (eds.) IDEAL 2018. LNCS, vol. 11315, pp. 280–287. Springer, Cham (2018). https://doi.org/10.1007/978-3-030-03496-2_31

37. Tan, P.-N.: Introduction to Data Mining. Pearson Education India, India (2018)
38. ML UCI. Repository, the uc irvine machine learning repository (2017). http://archive.ics.uci.edu/ml/
39. Vapnik, V.N.: The Nature of Statistical Learning Theory. Springer, New York (1995). https://doi.org/10.1007/978-1-4757-2440-0
40. Weiss, G.: Multiagents systems (1999)
41. Wirth, R., Hipp, J.: CRISP-DM: towards a standard process model for data mining. In: Proceedings of the 4th International Conference on the Practical Applications of Knowledge Discovery and Data Mining, pp. 29–39. Citeseer (2000)
42. Xu, G., Zong, Y., Yang, Z.: Applied Data Mining. CRC Press, Boca Raton (2013)

# The Contract Random Interval Spectral Ensemble (c-RISE): The Effect of Contracting a Classifier on Accuracy

Michael Flynn[✉], James Large, and Tony Bagnall

University of East Anglia, Norwich, UK
Michael.Flynn@uea.ac.uk

**Abstract.** The Random Interval Spectral Ensemble (RISE) is a recently introduced tree based time series classification algorithm, in which each tree is built on a distinct set of Fourier, autocorrelation and partial autocorrelation features. It is a component in the meta ensemble HIVE-COTE [9]. RISE has run time complexity of $O(nm^2)$, where $m$ is the series length and $n$ the number of train cases. This is prohibitively slow when considering long series, which are common in problems such as audio classification, where spectral approaches are likely to perform better than classifiers built in the time domain. We propose an enhancement of RISE that allows the user to specify how long the algorithm can have to run. The contract RISE (c-RISE) allows for check-pointing and adaptively estimates the time taken to build each tree in the ensemble through learning the constant terms in the run time complexity function. We show how the dynamic approach to contracting is more effective than the static approach of estimating the complexity before executing, and investigate the effect of contracting on accuracy for a range of large problems.

**Keywords:** Time Series Classification · Spectral features · Contract classifier

## 1   Introduction

Data sets of increasing size are now common within machine learning. Big data undeniably has its benefits. However, as advancements in processing capabilities begin to slow and the complexity of algorithms increase, we are often faced with more data than we are capable of processing. Even with the rise in popularity of cloud computing platforms and high performance computer facilities, it often becomes infeasible to construct a full learning model on all of the available data. A particularly common area in which the problem arises is the spectral/audio domain. This is typically down to the sample rate used to record data. Consider that the standard audio sample rate is 44.1 kHz. Creating models from audio data requires either extensive bespoke preprocessing or adaptations of the learning algorithms to compensate for the volume. We do not look to challenge or reaffirm

© Springer Nature Switzerland AG 2019
H. Pérez García et al. (Eds.): HAIS 2019, LNAI 11734, pp. 381–392, 2019.
https://doi.org/10.1007/978-3-030-29859-3_33

the traditional, volume of data/increase in accuracy paradigm. Instead, we aim to investigate the relationship between reduced train time and accuracy, assuming a fixed volume of data.

All experimental processes strive to make complete use of the training set and in ideal conditions this will always be preferable. However, in our work with large datasets we have experienced first hand the problems of extreme train and test times of approaches. In working through these problems it has become apparent that very little research has been undertaken in understanding how reduced training time affects accuracy. Homogeneous ensembles typically require a large number of trees to be effective. The most basic way of managing train times is simply to build base models until the time has expired. However, for very large problems, this may result in very small ensembles. The Random Interval Spectral Ensemble (RISE) is a Time Series Classification (TSC) algorithm that uses spectral features. It selects a different random interval of the series for each base classifier, then calculates spectral coefficients to be used as features. For large problems, if we happen to select intervals close to the full series length, we can use all available computation on very few models. To compensate for this, we investigate whether we can predict the run time, then use this prediction to guide the interval sampling, ensuring a minimum size ensemble.

Our aims are twofold; firstly, we aim to make RISE more useful by making it a contract classifier, i.e. a classifier where you can specify approximately the amount of computational time allowed to build the model. Secondly, we aim to compare the basic approach and our adaptive, dynamic approach with respect to their effect on accuracy and ability to adhere to the time contract.

In Sect. 2 we describe RISE in detail and further motivate the need for contracting. In Sect. 3 we introduce our contract version, c-RISE, and describe both timing models. In Sect. 4 we outline the experimental procedure and present the results and in Sect. 6 we conclude.

## 2   The RISE Algorithm

RISE draws on ideas from tree-based ensembles such as random forest [4] and the TSC interval feature classifier time series forest (TSF) [6]. Like TSF, we build trees on random intervals from the data to construct a random forest classifier. A key difference is that TSF uses time domain features by calculating the mean, variance, and slope of each interval, but RISE extracts spectral features over each random interval instead. Once we have derived the spectral features, we build a decision tree using the random tree algorithm used by random forest. The base RISE algorithm is described in Algorithm 1. The first tree in RISE is a special case that uses the whole series. We include this step for continuity with the previous spectral classifiers used in The Collective of Transformation-Based Ensembles [3] (COTE) classifier which only used the whole series. The procedure for building RISE is outlined in Algorithm 1.

RISE uses several forms of spectral features: the power spectrum, the autocorrelation function, the partial autocorrelation and the autoregressive model.

---

**Algorithm 1.** BuildRISE(Training data *train*, number of classifiers *r*, minimum interval length *minLen*)

---

1:  Let $\mathbf{F} \leftarrow\ <F_1 \ldots F_r>$ be the trees in the forest.
2:  Let $m$ be the length of series in *train*
3:  *wholeSeriesFeatures* $\leftarrow$ getSpectralFeatures(*train*)
4:  buildRandomTreeClassifier($\mathbf{F}_1$,*wholeSeriesFeatures*)
5:  **for** $i \leftarrow 2$ to $r$ **do**
6:      *startPos* $\leftarrow$ randBetween($1, m - minLen$)
7:      *endPos* $\leftarrow$ randBetween($startPos + minLen, m$)
8:      *train* $\leftarrow$ removeAttributesOutsideOfRange(*train*,*startPos*,*endPos*)
9:      *intervalFeatures* $\leftarrow$ getSpectralFeatures(*train*)
10:     buildRandomTreeClassifier($\mathbf{F}_i$,*intervalFeatures*)

---

New classes are classified using a simple majority vote. Further details can be found in [9]. The run time for transforming a series is quadratic in the interval length.

# 3   The Contracted RISE Algorithm

In many areas it may be advantageous or even necessary to constrict the run time of a classification algorithm. Generally, it is not well understood how long classification algorithms take to run for a given problem. It is of practical importance when considering which algorithm to use or how much preprocessing to perform. This is of particular difficulty when using cloud services where the computation is charged for per time period, or situations in which there is a hard deadline or where there is a limit on how long a process is allowed to run. Two solutions to these problems are check-pointing, periodically saving a partial version of the classification model to disk, and contracting, limiting the amount of computational time an algorithm is allowed. Used together, they make a classifier more flexible and useful to the practitioner. Check-pointing RISE is simple, especially with a Java implementation; we can simply serialise the constructed trees at certain points and adapt RISE to allow the loading from file. Contracting is also simple: we can keep building trees until we run out of time or reach the maximum number. However, this simple contracting approach can result in very small ensembles if the series are very long. We propose an adaptive scheme that avoids this problem by dynamically estimating the build time for each particular tree.

## 3.1   Algorithmic Improvements

A number of small but influential changes are implemented in c-RISE, with the goal of not significantly decreasing accuracy whilst drastically improving runtime. These changes are outlined below and a more thorough description provided in Algorithm 3.

RISE uses power spectrum (PS), autocorellation function (ACF), partial autocorellation function (PACF) and autoregressive model (AR) features over each interval. These features are clearly related, but it was found that combining them created a more accurate classifier than just using one set [9]. However, the drawback is that although the PS can be found in $O(nlog(n))$ time if the series length is a power of 2, there is no way to do this for the PACF and AR terms. Hence c-RISE does not derive PACF or AR features, and only selects intervals that are a power of 2. An interval is still selected randomly, but now it is rounded to the nearest power of two. To correct for intervals exceeding the series length, the interval is then divided by 2, ensuring a valid interval and favouring shorter intervals.

## 3.2    Timing Models

**Non-adaptive Model.** The simplest way to limit the train time of tree based ensemble is to simply set a timer and keep adding trees until the contract is met, or a maximum number of trees have been built. This is described in Algorithm 2.

---

**Algorithm 2.** Build c-RISE_Non-adaptive(Training data *train*, number of classifiers *r*, minimum interval length *min*)

---

1: Let $\mathbf{F} \leftarrow < F_1 \ldots F_{500} >$ be the trees in the forest.
2: Let $m$ be the length of series in *train*
3: startForestTimer()
4: **for** $i \leftarrow 1$ to 500 AND queryForestTimer() **do**
5:     $validLengths \leftarrow$ getValidPowersOf2($min, instanceLength$)
6:     $randomLength \leftarrow$ randBetween(maxValue($validLengths$)/2)
7:     $r \leftarrow$ findClosest($validLengths, randomLength$)
8:     $startPos \leftarrow$ randBetween($1, m - r$)
9:     $interval \leftarrow$ removeAttributesOutsideOfRange($train, startPos, r$)
10:     $intervalFeatures \leftarrow$ getSpectralFeatures($interval$)
11:     buildRandomTreeClassifier($\mathbf{F}_i, intervalFeatures$)
12:     updateTimer()

---

**Adaptive Model.** c-RISE performs two transformations: A discrete Fourier transform (DFT) to find the power spectrum and construction of the autocorrelation function (ACF). With the simplest implementation, each of these is $O(r^2)$, where $r$ is the interval width. We can improve the efficiency of the Fourier transform to $O(rlog(r))$ by using the fast Fourier Transform (FFT). To gain the full benefit we restrict c-RISE to intervals of length the power of 2. However, the best average case complexity for the ACF is $O(r^2)$. Hence, the transformations will dominate the runtime in relation to the decision tree, so we can reasonably model the runtime $t$ for a single member of the ensemble of interval length $r$ as

$$t = a \cdot r^2 + b \cdot r + c.$$

If we fix the contract time $t$ and knew the constant factors $a$, $b$ and $c$ we could find the positive root of the quadratic and use that as the maximum allowable interval for the tree. The quadratic terms will of course be both problem and hardware dependent. Hence, we use an adaptive algorithm to learn these parameters. For each tree we build, we record the selected interval and the observed run time. Using this data, we refit a least squares linear regression model. For clarity, we let $x_1 = r_1^2$ and $x_2 = r_1$, our dependent variable matrix is then,

$$X = \begin{bmatrix} 1, x_{11}, x_{12} \\ 1, x_{21}, x_{22} \\ \dots \\ 1, x_{k1}, x_{k2} \end{bmatrix}$$

the estimates of the parameters are $B = (\hat{a}, \hat{b}, \hat{c})^T$ and our response variable is $Y = (y_1, y_2, \dots, y_k)^T$. The least squares estimates are then,

$$B = (X^T X)^{-1} X^T Y.$$

Since $(X^T X)$ is based on sums of squares, we do not have to recalculate if from scratch each time. It is also possible to update $(X^T X)^{-1}$ online with the Sherman-Morrison formula, but we leave that for future work. After the construction of each tree, we update the remaining contracted time $t$, re-estimate the coefficients $t = \hat{a} \cdot r^2 + \hat{b} \cdot r + \hat{c}$, and calculate a new maximum allowable interval $r$. This is used as the maximum for the next iteration.

---

**Algorithm 3.** Build c-RISE_Adaptive(Training data *train*, number of classifiers $r$, minimum interval length *min*)

---

1: Let $\mathbf{F} \leftarrow\ <F_1 \dots F_{500}>$ be the trees in the forest.
2: Let $m$ be the length of series in *train*
3: startForestTimer()
4: **for** $i \leftarrow 1$ to 500 AND queryForestTimer() **do**
5:     startTreeTimer()
6:     buildAdaptiveModel()
7:     $max \leftarrow$ findMaxIntervalLength()
8:     $validLengths \leftarrow$ getValidPowersOf2($min$, $max$)
9:     $randomLength \leftarrow$ randBetween(maxValue($validLengths$)/2)
10:    $r \leftarrow$ findClosest($validLengths$, $randomLength$)
11:    $startPos \leftarrow$ randBetween($1, m - r$)
12:    $interval \leftarrow$ removeAttributesOutsideOfRange($train$, $startPos$, $r$)
13:    $intervalFeatures \leftarrow$ getSpectralFeatures($interval$)
14:    buildRandomTreeClassifier($\mathbf{F}_i$, $intervalFeatures$)
15:    $y \leftarrow$ queryTreeTimer()
16:    updateAdaptiveModel($r$, $y$)

---

# 4   Results

In order to ensure the following results are comparable, all approaches were evaluated on 10 stratified random re-samples, of the same 85 datasets from the UCR archive. All code is available from the UEA TSC repository[1], whereas, raw results and analysis can be downloaded from this repository[2]. The experimental pipeline and algorithms used were all implemented in Java and the experiments were carried out on the HPC system at the University of East Anglia.

## 4.1   RISE vs c-RISE

RISE has been shown to be significantly better than other spectral based approaches on the TSC archive data and on simulated data [2] and therefore was selected as the spectral component for HIVE-COTE. However, RISE is computationally expensive, since each transformed series is based on an $O(r^2)$ operation (finding the PACF), where $r$ is the series length. This is further impacted by the derivation of auto regressive and spectral features. In a summary of experiments over 82 datasets from the UCR archive it was concluded that these features do not significantly effect accuracy, as shown in Fig. 1. However computation of these features represent significant complexity in the algorithm and as such they have a detrimental effect on runtime. Table 1 presents both the mean, median train time for RISE and cRISE over 85 datasets from the UCR archive.

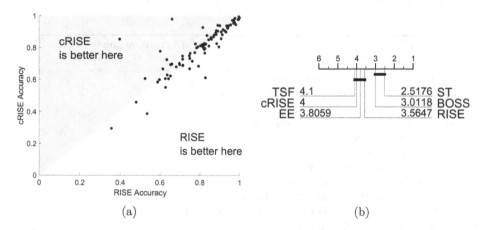

(a)                                        (b)

**Fig. 1.** (a) A pairwise scatter diagram showing average train time over 10 random re-samples of 85 UCR datasets for both RISE and cRISE. (b) A pairwise critical difference diagram showing the ranks of TSF [6], cRISE, EE [7], RISE [9], BOSS [10] and ST [8] over the same 10 random resamples of 85 UCR datasets.

---

[1] https://github.com/TonyBagnall/uea-tsc.
[2] https://tinyurl.com/y3robhbk.

However, on investigation of the impact each of these features has on accuracy and runtime, it became apparent that they did not contribute equally. Fundamentally cRISE behaves in the same manner as RISE when not under contract. This allows us to attribute any changes in runtime or accuracy to the removed transformations. The impact on runtime when deriving the PACF and AR features is unsurprising. Figure 1(b) is a critical difference diagram displaying various state of the art classifiers, including both RISE and cRISE. The impact of removing PACF and AR features has on accuracy was unexpected. The outcome of these experiments was the removal of PACF and AR derivations.

**Table 1.** A table showing mean and median train times of RISE and cRISE over all UCR datasets, as well as average speed up.

|  | cRISE | RISE | Difference |
|---|---|---|---|
| Mean (seconds) | 144 | 3554 | 3410 |
| Median (seconds) | 145 | 3794 | 3649 |

Moving forward all experimental results are achieved with the updated architecture of cRISE.

## 4.2   Naive vs Adaptive Timing Models

In this section we evaluate how the accuracy of cRISE changes as a function of total training time for both approaches. We also asses how well each approach adheres to the contract itself.

In order to achieve this, nine pairs of experiments were carried out. For both approaches a contract is set representing 10%–90% total training time per dataset in 10% increments. This allowed us to examine changes in accuracy at 10 evenly spaced points in time as well test each approaches ability to stay within the contract.

Figure 2(a) shows how the actual train time changes over different contracts. Each point represents the mean train time over 85 datasets over 10 folds for each contract. The contracts themselves are defined a percentage of full train time per dataset. This represents 8,500 experiments per approach over 10 contract percentages.

Figure 2(a) also shows that the Adaptive approach displays much more predictable behaviour in the context of adhering to contract time. Adhering to relatively small contract times presents more of a challenge to both approaches. This can be explained by the limit imposed on the minimum interval size. Small contracts are better fulfilled by small intervals as each iteration represents a smaller proportion of the contract, allowing for finer control over the total time taken.

Initially this appears a major flaw in both approaches. However, this problem is largely exacerbated by the existence of many small problems in the UCR

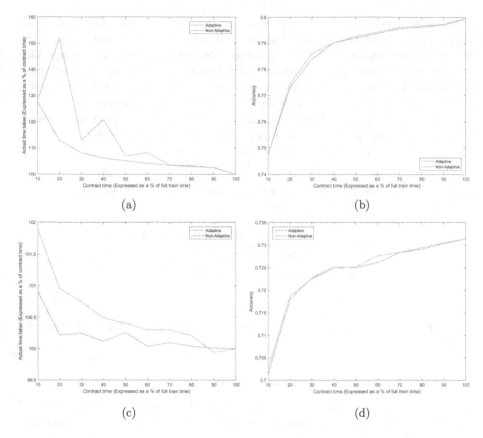

**Fig. 2.** The graphs on the left show the performance of the Adaptive and Non-adaptive approaches in respect to their ability to adhere to contract time. The graphs on the right show the predictive accuracy of the Adaptive and Non-adaptive approaches over 10 contracts. The top row displays results averaged over all datasets from the UCR database. The bottom row displays results averaged over all problems in the UCR datasets with at least 700 attributes.

85 database. Problems that without intervention take between 1 and 4 h to complete.

In order to remove the bias introduced by smaller datasets the same experiments were repeated with all datasets containing data over 700 attributes.

Figure 2(c) shows how the actual time taken changes over different contract times for problems from the UCR archive with 700 or more attributes. Figure 2(c) shows that both approaches were confounded by smaller problems. It also illustrates the Adaptive approaches superior ability to adhere more closely to the contract than the Non-adaptive approach.

Interestingly, these changes in contract accuracy have very little to no effect on accuracy. Figure 2(b) and (d) show how accuracy changes as a function of contract time for all UCR datasets and datasets over 700 attributes respectively.

This is important as it confirms that the superior ability of adhering to contract time comes at no cost to accuracy for the Adaptive approach.

## 5    Eigenworms Case Study

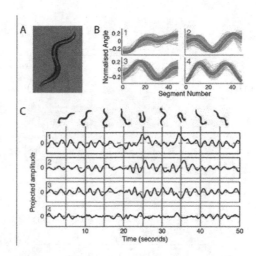

**Fig. 3.** An image showing an example of Caenorhabditis elegans, the 4 Eigenworm shapes and an illustration of a corresponding 4 dimensional time series with positional images [5]

### 5.1    The Dataset

Caenorhabditis elegans is commonly used in genetic studies as a model organism. Their movements have proven to be a robust indicator in determining behavioural traits. [5] outline a number of human-defined features [11] as well a procedure to derive them. It has been shown that all positional variants that the organism adopts can be expressed as combinations of 4 base shapes, known as eigenworms, shown in Fig. 3(B). The raw worm outline is captured and similarity value assigned to each of the 6 base shapes at each frame of motion.

Using the data collected in [5] worms can be classified as either wild type, or 1 of 4 mutant types: goa-1, unc-1, unc-38, unc-63. The dataset is split into a 131 instance train set and a 128 instance test set. The original classification dataset, available at [1], consists of 6 dimensions of 17,984 attributes. However, these were concatenated to create instance lengths of 107,903 for this problem.

### 5.2    Results

Figure 4(a) shows how the actual train time changes over different contract times. Each point represents the mean train time over the Eigenworms dataset over

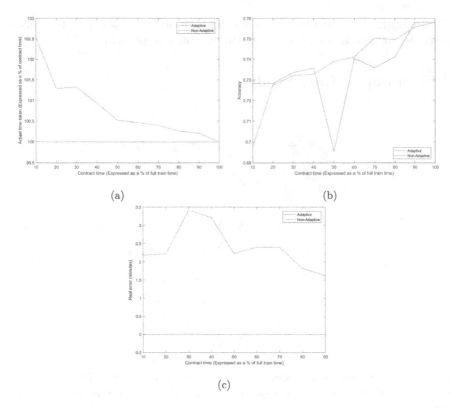

**Fig. 4.** (a) shows the performance of the Adaptive and Non-adaptive approaches in respect to their ability to adhere to contract time. (b) shows displays the predictive accuracy of the Adaptive and Non-adaptive approaches over 10 contracts. (c) shows the real error, in minutes, of the Adaptive and Non-adaptive models over the same contracts.

10 folds for each contract. Interestingly, the Non-adaptive approach appears to follow a similar downward trend to that displayed in Fig. 2(c), this is in contrast to the Adaptive model which has shown improvement as series length increases.

Figure 4(c) displays the real error in minutes of the 2 approaches. Although the Non-adaptive approach's error does vary in respect to contract time, it does not display a convincing downward trend. This suggests that the error may be bound to either the series length, number of cases, or both. Conversely, the Adaptive method produces results that indicate that it is robust to increases in series length. Figure 4(b) shows how accuracy changes as a function of contract time. The Non-adaptive approach displays a more consistent assent in accuracy as the contract time increases. The performance of the Adaptive approach contradicts the results shown in Fig. 2(b) and (d) which suggest the accuracy of both models over all contracts do not deviate from one another.

# 6 Conclusions

In conclusion, we present changes to RISE that consistently result in a significantly faster train time. This result was achieved using an established experimental design on an established Time Series database. These changes remove two costly derivations making cRISE at least twice as fast whilst the cost to test accuracy is shown to be insignificant.

We also present and compare the Adaptive and Non-Adaptive timing models in the context of cRISE. On an established Time Series Database of 85 problems the experimental design does not show any significant difference in accuracy between models. Although, one approach did show a superior ability to adhere to contract.

This superior ability to adhere to a contract time was made further evident when considering a problem consisting of significantly longer series. It was shown that the Adaptive model is robust to scaling of series length, whereas the Non-adaptive approach is bound by series length, number of cases or both.

Although relatively incremental, these changes collectively represent a significant improvement in the usability of cRISE and consequently HIVE-COTE. They also serve to address the commonly unanswered question of, how best to contract?

**Acknowledgements.** This work is supported by the Biotechnology and Biological Sciences Research Council [grant number BB/M011216/1], and the UK Engineering and Physical Sciences Research Council (EPSRC) [grant number EP/M015807/1]. The experiments were carried out on the High Performance Computing Cluster supported by the Research and Specialist Computing Support service at the University of East Anglia and using a Titan X Pascal donated by the NVIDIA Corporation.

# References

1. Bagnall, A., et al.: The UEA multivariate time series classification archive, 2018. ArXiv e-prints arXiv:1811.00075 (2018). http://arxiv.org/abs/1809.06705
2. Bagnall, A., Lines, J., Bostrom, A., Large, J., Keogh, E.: The great time series classification bake off: a review and experimental evaluation of recent algorithmic advances. Data Min. Knowl. Disc. **31**(3), 606–660 (2017)
3. Bagnall, A., Lines, J., Hills, J., Bostrom, A.: Time-series classification with COTE: The collective of transformation-based ensembles. IEEE Trans. Knowl. Data Eng. **27**, 2522–2535 (2015)
4. Breiman, L.: Random forests. Mach. Learn. **45**(1), 5–32 (2001)
5. Brown, A.E., Yemini, E.I., Grundy, L.J., Jucikas, T., Schafer, W.R.: A dictionary of behavioral motifs reveals clusters of genes affecting caenorhabditis elegans locomotion. Proc. Natl. Acad. Sci. **110**(2), 791–796 (2013)
6. Deng, H., Runger, G., Tuv, E., Vladimir, M.: A time series forest for classification and feature extraction. Inf. Sci. **239**, 142–153 (2013)
7. Lines, J., Bagnall, A.: Time series classification with ensembles of elastic distance measures. Data Min. Knowl. Disc. **29**, 565–592 (2015)

8. Lines, J., Davis, L., Hills, J., Bagnall, A.: A shapelet transform for time series classification. In: Proceedings of the 18th ACM SIGKDD International Conference on Knowledge Discovery and Data Mining (2012)
9. Lines, J., Taylor, S., Bagnall, A.: Time series classification with HIVE-COTE: the hierarchical vote collective of transformation-based ensembles. ACM Trans. Knowl. Discov. Data **12**(5) (2018)
10. Schäfer, P.: The BOSS is concerned with time series classification in the presence of noise. Data Min. Knowl. Disc. **29**(6), 1505–1530 (2015)
11. Yemini, E., Jucikas, T., Grundy, L., Brown, A., Schafer, W.: A database of caenorhabditis elegans behavioral phenotypes. Nat. Methods **10**, 877–879 (2013)

# Graph-Based Knowledge Inference
# for Style-Directed Architectural Design

Agnieszka Mars, Ewa Grabska, Grażyna Ślusarczyk, and Barbara Strug[✉]

Jagiellonian University, Lojasiewicza 11, Krakow, Poland
{agnieszka.mars,ewa.grabska,grazyna.slusarczyk,barbara.strug}@uj.edu.pl

**Abstract.** This paper deals with style-oriented approach to computer aided architectural design. The generated 3D-models of architectural forms are composed of perceptual primitives determined on the basis of Biederman's geons theory. The knowledge about the designed models is represented by the labelled graphs. In the provided CAD environment the designer selects and marks the model parts characterizing the considered style. Then the system infers subgraphs corresponding to these parts from graph representations of the models and encodes them into graph grammar rules. The additional graph grammar rules are constructed on the basis of creative design actions taken by the designer during the design process. The system supports the designer in his creative process by automatically generating graphs corresponding to new architectural forms in the desired style. The approach is illustrated by examples of designing objects in the Neoclassical style.

## 1 Introduction

CAD software is widely used to support the process of generating design drawings and product models. However CAD systems still lack generative methods to easily generate classes of design variants. Moreover, the process of architectural design requires a lot of imagination and is usually directed not only by the desire to obtain an optimal artefact but also such a one that meets certain requirements. Design requirements are often related to styles [8]. The representation of style in design is described in [3].

Thus the problem discussed in this paper lies on the border of computational science and humanities. On one hand we want to obtain an optimal design in the sense of cost, construction time and other computable factors and on the other hand we also need to have the design be in a required style or at least be similar in terms of style to other designs. Therefore the hybrid approach which combines methods of knowledge management, intelligent human interaction, syntactic pattern analysis and grammar inference is proposed.

In this paper 3D-models of architectural forms are composed of perceptual primitives determined on the basis of Biederman's Recognition-By-Components theory. During the design process the architect can interactively select parts of a created design and designate them as important features for expected/required style for current task solutions.

© Springer Nature Switzerland AG 2019
H. Pérez García et al. (Eds.): HAIS 2019, LNAI 11734, pp. 393–403, 2019.
https://doi.org/10.1007/978-3-030-29859-3_34

To make the manipulation and transformation of design objects possible, a graph based representation of knowledge about designed buildings is proposed. Each designed object has its representation in the form of an attributed composition graph, where nodes denote components, bonds assigned to nodes represent component surfaces, while edges describe spatial relations between these surfaces. Such a representation enables us not only to express geometrical properties of an object but also its attributes (like size, material etc.). In composition graphs representing architectural forms some subgraphs representing features characteristic for building styles can be distinguished. These subgraphs correspond to the parts of designs selected by the designer as important features. Composition graphs representing designed objects can be generated and modified by composition graph transformation rules. Design requirements related to styles, which are inferred from building graph representations on the basis of the features selected by the designer, are encoded into composition graph grammar rules. Grammar rules with their right-hand sides being graphs representing style requirements are created. The other grammar rules are inferred on the basis of design actions taken by the designer in the building design process. The obtained graph rules are used to automatically generate graphs corresponding to new architectural forms with geometry and material properties specified by graph attributes. The possibility of relating attributes of right-hand sides of CP-graph rules to attributes of their left-hand sides enables the system to capture parametric modeling knowledge. The proposed method of controlling the order of applying grammar rules ensures the conformity of the buildings corresponding to generated graphs to the desired style. The proposed generative method allows the system supporting building design to automatically model alternative forms preserving the required characteristics.

Graph grammars are typically difficult to construct for people without a considerable background knowledge, especially when the analysis of characteristic features and designing production rules for buildings of a certain style is taken into account. This paper contributes to the field of computational artificial intelligent methods by proposing the method, which supports the grammar design process.

## 2    Architectural Form Representation

The architectural building designs are currently created with the use of CAD tools. The majority of buildings are presented in the form both of drawings understandable to users and of internal representations useful for automated processing [7]. In our approach to support maintaining styles of building designs the building models are generated by the designer on the basis of an alphabet of elementary shapes related to elements from the universal architectural dictionary (UAD) and transformed into their internal representations in the form of graphs. Analysis of these graphs enables us to construct a tool for automatic generation of architectural forms which preserve the required features. This tool is composed of graph rules describing features of the considered style.

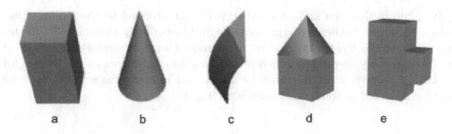

**Fig. 1.** Examples of geons (a), (b), (c); relations between geons: (d) end-to-end, (e) end-to-side

In the proposed method the alphabet of primitives used for visual representation of architectural forms consists of geons. According to Biederman's Recognition-By-Components (RBC) model of visual perception [2], geons are idealized elementary shapes lacking sharp concavities that we divide perceived objects into. Object identification is performed by investigation of geon types and relations between geons. A geon type is defined by non-accidental properties: edge (straight S or curved C), symmetry (rotational ++, vertical + or none-), cross section size (constant ++, contracted - or expanded and contracted −) and axis (straight + or curved -). These four properties can be quickly and accurately recognized independently on the observer's viewpoint. Each geon can be also described by a set of metric properties, like height or width. Visual perception of metric properties is relatively slow and prone to errors.

Figures 1a–c show three geons of different types with the following combinations of non-accidental properties values: [S ++ ++ +] in Fig. 1a, [C ++ - +] in Fig. 1b and [S - ++ -] in Fig. 1c. There are two non-accidental relations between geons: *end-to-end*, in which one geon is prolonged by another one as in Fig. 1d, and *end-to side*, in which one geon is attached to another one as in Fig. 1e. RBC model assumes existence of 36 3D geon types and two types of spatial relations. Such an alphabet can be used to generate infinite number of different configurations and is quite an effective tool for visual representation of architecture.

Using geons as basic primitives offers at least two advantages. As geons are based on object properties that are viewpoint independent, a single geon description describes an object from all possible viewpoints. Moreover, a relatively small set of geons forms an alphabet of shapes that can be combined to form complex objects, so the representation is efficient.

## 3   Style Representation

Our approach to design is style-oriented. In this paper it is assumed that in most cases style features are based on non-accidental properties, i.e., only geon types are taken into account, disregarding metric information. Therefore, the most important part of the building description constitute non-accidental attributes and relation types, although metric parameters are of course necessary to visualize a designed object.

It is known that "an architectural style is characterized by the features that make a building or other structure notable and historically identifiable". In this paper we consider style in the context of a building form. It means that we aim at the automatic design process which preserves a set of elements characteristic for the form in a given style. It should be noted that in many cases these elements must be arranged in some predefined way [1, 9, 10].

**Fig. 2.** An example of a building in the Neoclassical style

In this paper we formalize rules related to composition of elements determining the Neoclassical style. Neoclassical architecture was an architectural style that began in the mid-18th century. It was a reaction against the Rococo style of naturalistic ornament. Neoclassicism draw inspiration from the architecture of Classical antiquity, the Vitruvian principles and the architecture of the Italian architect Andrea Palladio. This style is characterized by removal of all unnecessary ornamentation, enhancement of geometric forms, and repeating series of arches and/or columns. There exist three types of neoclassical architecture: temple style based on an ancient temple, Palladian style, and classical block style [4]. We consider Neoclassicism, as objects in this style can be represented by compositions of distinct forms, which are arranged using an explicit set of rules. It is assumed here that a building is in the Neoclassical style if it has a facade with a porch having at least two columns and exactly one pediment [5].

An example building shown in Fig. 2 has all characteristic features of the classical block style.

## 4    Graph Representation of Buildings and Style Features

The proposed approach uses composition graphs (CP-graphs) as the representation of design architectural forms [6]. Graph nodes represent components (parts) of the object being designed, bonds assigned to nodes represent component surfaces, while directed edges connecting bonds express relations between these surfaces. Nodes are labelled by names of components (or types of components) and edges are labelled by names of relations between them. In our approach the graph nodes are labelled by the names of the building elements represented by geons. To each node two groups of attributes are assigned. The attributes of the first group describe non-accidental properties of geons, while attributes of the second group describe their metric parameters. Node bonds represent types of geon surfaces and their number varies depending on the cross section shape. The edges labelled *e-to-e* represent the end-to-end relation, while the edges labelled *e-to-s* represent the end-to-side relation.

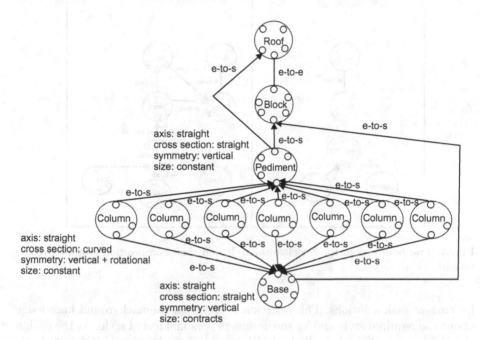

**Fig. 3.** A composition graph representation of the building from Fig. 2

Figure 3 presents a composition graph representing the building shown in Fig. 2. The nodes labelled Block, Roof and Base represent cuboids, the node labelled Pediment represents a prism, while nodes labelled Column represent cylinders. The edge labelled e-to-e connects bonds representing the basis and the top basis of cuboids. All other edges are labelled e-to-s. They connect bonds representing the slanted side of the cuboid with the back side of the prism, the back side of the small cuboid and the front side of the bigger cuboid, as well as top bases of all cylinders and the base of the prism, and bottom bases of all cylinders and the top basis of the small cuboid. To each node a set of attributes is assigned. The non-accidental attributes for the cuboid, the prism, and the cylinder are shown. Two other buildings in the Neoclassical style and their composition graph representations are shown in Fig. 4.

During the design process the architect interactively selects parts of a created design and designate them as important features for expected/required style

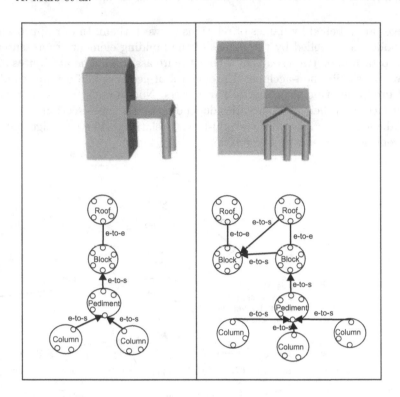

**Fig. 4.** Two buildings in the Neoclassical style and their composition graph representations

for current task solutions. This selection is based on the background knowledge about the required style and/or the designers personal feel of style. As the design object is represented internally by a CP-graph, the nodes corresponding to basic solids (geons) belonging to the selected parts are marked with a special attribute feature. Therefore the subgraphs corresponding to the features characteristic for a required style can be easily identified. Analysis of composition graphs representing several designed buildings in a given style (with designated style features) allows the system to infer subgraphs representing style requirements by finding all consistent subgraphs composed of nodes with attribute feature set to on.

As in the Neoclassical style the main building block must be connected with a porch composed of a pediment and at least two columns, it can be deduced that the style features which occur in buildings from Figs. 2 and 4 are represented by composition graphs shown in Fig. 5, which are subgraphs of CP-graphs presented in Figs. 3 and 4, respectively.

In Fig. 5 three examples of composition graphs representing the Neoclassical style requirements are shown. The first one represents a porch with seven columns standing on a base, the second one corresponds to a porch with three columns, while the third one represents a porch with two columns.

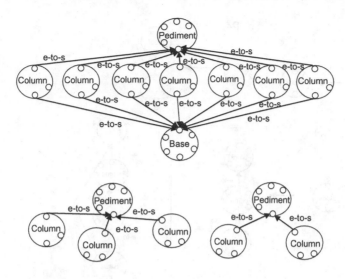

**Fig. 5.** Examples of composition graphs representing style requirements

# 5   Graph Transformation Rules

When structures of building designs are described in terms of CP-graphs, CP-graph grammars can serve as efficient tools for generating these structures. A CP-graph grammar is composed of a set of productions and an axiom being an initial CP-graph. Applying rules of a CP-graph grammar generates graph structures representing many architectural forms, some of which can be unexpected and innovative.

In order to apply grammar rules some bonds of CP-graphs occurring in these rules are specified as external ones. A CP-graph transformation rule is of the form $p = (l, r, sr)$, where $l$ and $r$ are attributed CP-graphs with the same number of different values of external bonds and $sr$ is a set of functions specifying the way in which attributes assigned to nodes of $l$ are transferred to the attributes assigned to nodes of $r$. Each attribute has a set of possible values, which are established during the rule application. The application of the rule $p$ to a CP-graph $G$ consists in substituting $r$ for a CP-graph being an isomorphic image of $l$ in $G$, replacing external bonds of the CP-graph being removed with the external bonds of $r$ with the same numbers, and specifying values of attributes assigned to elements of $r$ according to functions of $sr$. After inserting $r$ into a host CP-graph all edges which were coming into (or out of) a bond with a given number in the CP-graph $l$, are coming into (or out of) bonds of $r$ with the same number.

Taking into account the inferred composition graphs representing style requirements, a set of CP-graph grammar rules generating representations of buildings, which contain subgraphs corresponding to desired features, can be constructed. If in the set of found subgraphs there are more than two CP-graphs with different numbers of nodes with the same label a set of rules which allows

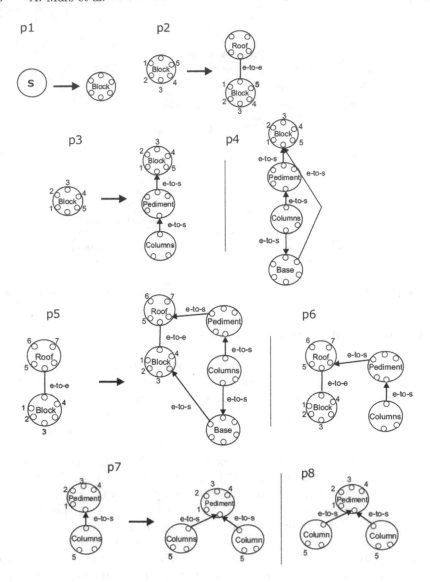

**Fig. 6.** CP-graph rules preserving features of the Neoclassical style

to generate subgraphs containing an arbitrary number of nodes with that label is created. For example rules *p9* and *p10* of the CP-graph grammar presented in Fig. 7 allow to generate an arbitrary number (but at least two) of nodes representing columns and connected to the node representing a pediment. In other cases each of the found subgraphs is directly used as a right side of a production. As successive design actions performed by the designer in the building design process can be seen as steps corresponding to application of grammar rules which develop the generated solution, the additional CP-graph grammar rules can be specified on the basis of the designer activity.

A set of CP-graph rules preserving features of the Neoclassical style is shown in Fig. 6. Production $p2$ adds a roof to a block, productions $p3$ and $p4$ (combined with $p7$ and $p8$) add a pediment with at least two columns and with or without a base, respectively, where a pediment and a base are adjacent to a block. Productions $p5$ and $p6$ add a pediment with at least two columns and with or without a base, respectively, where a pediment is adjacent to a roof and a base is adjacent to a block. Two additional rules are specified on the basis of the successive design actions. The first of these rules adds the next floor to a building, while the second one adds a new wing.

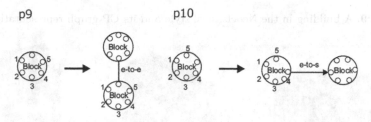

**Fig. 7.** Additional CP-graph rules for the grammar generating structures of architectural forms

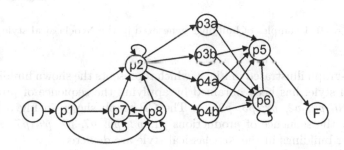

**Fig. 8.** A control diagram for the grammar rules shown in Figs. 6 and 7

The order of applying grammar rules can be specified by a control diagram. To ensure the conformity of the buildings corresponding to generated graphs with the desired style each path in the control diagram should contain one of the productions $p3$, $p4$, $p5$ or $p6$.

An example of a control diagram for the CP-graph grammar presented in Figs. 6 and 7 is shown in Fig. 8. With exception of the initial and the final node, all other nodes are labelled with numbers of rules. Applying a derivation process according to the order stated in the control diagram, we start with a rule specified by the label of a direct successor of the initial node. The derivation process stops when the final node is reached. As it is assumed that for each axiom CP-graph all attributes have specified values, each path from the initial node to the final node allows us to generate the CP-graph corresponding to one building model.

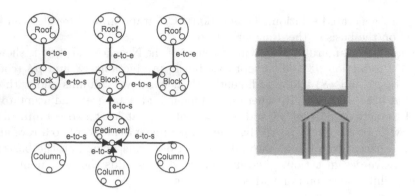

**Fig. 9.** A building in the Neoclassical style and its CP-graph representation

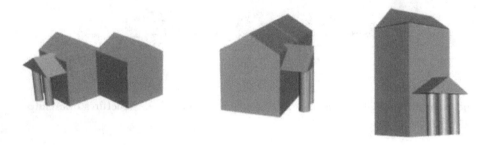

**Fig. 10.** Examples of buildings generated in the Neoclassical style

The CP-graph illustrated in Fig. 9, which represents the shown building in the Neoclassical style, has been obtained by applying the sequence of productions *p1, p10, p10, p2, p2, p2, p3, p7, p8*. The CP-graph shown in Fig. 3 has been obtained by the sequence of productions *p1, p2, p5, p7, p7, p7, p7, p7, p8*. In Fig. 10 other buildings in the Neoclassical style are depicted.

## 6     Conclusions

This paper proposes a formal computational approach to describing and analyzing styles in architecture. Style is often perceived as a human perceptible and not computational factor/feature so it is relatively difficult to analyse in a computational model. In this paper the Recognition-by-Components perception model, developed for explaining human ability to identify structured objects, is used in combination with a graph based representation of building forms. Style is here proposed to be defined as a set of patterns that are required to be present in any design in a given style. It has to be added that the term style does not have to represent an actual style as known from history of architecture but can, for example, represent a style typical for a region, architect or other specific aspects.

In our approach the patterns defining a style are represented in the form of subgraphs of a graph representing a design. Moreover, a generative method based on an inferred grammar is proposed in this paper. Such a grammar is able to generate objects conforming to a given style. After each design session, a CP-graph grammar which allows the system supporting building design to automatically model buildings with the required characteristics is created. Therefore, we plan to build a database of style grammars. Using such a database the designer will be able to select a group of grammars to generate eclectic designs combining all rules of these grammars into one grammar.

# References

1. Alexander, C., Ishikawa, S., Silverstein, M., Jacobson, M., Fiksdahl-King, I., Angel, S.: A Pattern Language: Towns, Buildings. Construction. Oxford University Press, USA (1977)
2. Biederman, J.: Recognition-by-components: a theory of human image understanding. Psychol. Rev. **94**, 115–147 (1987)
3. Ding, L., Gero, J.: The emergence of the representation of style in design. Environ. Planning B: Urban Anal. City Sci. **28**, 707–731 (2001)
4. Dowling, E.: New Classicism: The Rebirth of Traditional Architecture. Rizzoli International Publications, New York (2004)
5. Carson Dunlop Associates: Architectural Styles. Dearborn Real Estate (2003)
6. Grabska, E., Borkowski, A.: Assisting creativity by composite representation. In: Gero, J.S., Sudweeks, F. (eds.) Artificial Intelligence in Design '96, pp. 743–759. Kluwer Academic Publishers, The Netherlands (1996)
7. Grabska, E., Lachwa, A., Ślusarczyk, G.: New visual languages supporting design of multi-storey buildings. Adv. Eng. Inf. **26**, 681–690 (2012)
8. Jupp, J., Gero, J.: Let's look at style: visual and spatial representation and reasoning in design. In: Argamon, S., Burns, K., Dubnov, S. (eds.) The Structure of Style. Springer, Heidelberg (2010). https://doi.org/10.1007/978-3-642-12337-5_8
9. Simon, H.: Style in design. In: Eastman, C.M. (ed.) Spatial Synthesis in Computer-Aided Building Design, pp. 287–309. Wiley, New York (1975)
10. Ślusarczyk, G., Strug, B., Stasiak, K.: An ontology-based graph approach to support buildings design conformity with a given style. Appl. Ontol. **11**(4), 279–300 (2016)

# A Convex Formulation of SVM-Based Multi-task Learning

Carlos Ruiz[1]([✉]), Carlos M. Alaíz[1], and José R. Dorronsoro[1,2]

[1] Department of Computer Engineering,
Universidad Autónoma de Madrid, Madrid, Spain
`carlos.ruizp@estudiante.uam.es`, {`carlos.alaiz,jose.dorronsoro`}`@uam.es`
[2] Inst. Ing. Conocimiento, Universidad Autónoma de Madrid, Madrid, Spain

**Abstract.** Multi-task learning (MTL) is a powerful framework that allows to take advantage of the similarities between several machine learning tasks to improve on their solution by independent task specific models. Support Vector Machines (SVMs) are well suited for this and Cai *et al.* have proposed additive MTL SVMs, where the final model corresponds to the sum of a common one shared between all tasks, and each task specific model. In this work we will propose a different formulation of this additive approach, in which the final model is a convex combination of common and task specific ones. The convex mixing hyper-parameter $\lambda$ takes values between 0 and 1, where a value of 1 is mathematically equivalent to a common model for all the tasks, whereas a value of 0 corresponds to independent task-specific models. We will show that for $\lambda$ values between 0 and 1, this convex approach is equivalent to the additive one of Cai *et al.* when the other SVM parameters are properly selected. On the other hand, the predictions of the proposed convex model are also convex combinations of the common and specific predictions, making this formulation easier to interpret. Finally, this convex formulation is easier to hyper-parametrize since the hyper-parameter $\lambda$ is constrained to the $[0, 1]$ region, in contrast with the unbounded range in the additive MTL SVMs.

## 1 Introduction

The goal of Multi-task Learning (MTL) is to simultaneously solve several related problems, the **tasks**, sharing information among them in such a way that it enhances the overall learning process. MTL has grown enormously since its proposal by Caruana [3] and MTL approaches have been proposed for the main Machine Learning (ML) paradigms; recent surveys can be found in [11,12]. More recently, MTL has also been proposed as a tool to achieve algorithmic fairness [9] by building single procedures that, nevertheless, are able to take into account sensitive information across groups. This is in contrast with the use of different models for different groups, something that for some applications may simply not be allowed.

© Springer Nature Switzerland AG 2019
H. Pérez García et al. (Eds.): HAIS 2019, LNAI 11734, pp. 404–415, 2019.
https://doi.org/10.1007/978-3-030-29859-3_35

Among the ML methods better suited to MTL are Support Vector Machines (SVMs). Starting with the work of Evgeniou and Pontil [6], various approaches have been proposed for MTL on SVMs; see [2,5,7]. Of particular interest here is [1], which uses models, which we will call *additive* MTL-SVRs, that act combining a common part $w$ with task specific components $v_t$, and where the global cost function has an error term upon which a penalty $C$ acts, plus a regularization term given by the squared weight norms which in the case of the common part also include a penalty term $\mu$ (see Sect. 2 for more details). As we will see, the practical use of MTL-SVR models is done in a kernel setting. This allows to consider them from a hybrid perspective. In fact, different kernels can be used for the common and independent tasks, which implies that each single model works with a different set of extended kernel features. In other words, each component in a kernel MTL-SVR model implicitly builds its own extended features, and the final model is a hybrid combination of individual models acting on different feature spaces.

Clearly, a good understanding of the interplay between the common and independent models is an important issue when working with MTL-SVRs. In principle, a high $\mu$ value would force the common vector norm to be close to 0 and, hence, result in an MTL model dominated by the $T$ independent tasks. On the other hand, a small $\mu$ gives more space for the common weight $w$ to explore and, therefore, for it to have an impact greater than those of the independent task models. In any case, this discussion is more of a description than of a quantitative explanation. Our main contribution here is to make more precise this common vs. independent task interplay. To do so we will consider models of the form $\lambda u + (1-\lambda)u_t, 0 \le \lambda \le 1$, that is, a convex combination of a common $u$ and $T$ independent $u_t$ weights. This formulation relies on the tunable parameter $\lambda$ and makes more precise the extreme situations of just a single common model ($\lambda = 1$) or just $T$ independent models ($\lambda = 0$).

As we will show, this approach, which we will call *convex* MTL-SVR, is equivalent to the additive one if we take

$$\mu = \frac{(1-\lambda)^2}{\lambda^2}.$$

This results in two clear advantages. The first one is the easier interpretation of the common vs. independent task interplay which, as mentioned, can be measured through the optimal $\lambda^*$ value. The second one is that it allows a much simpler exploration of the $\lambda$ hyper-parameter space. In fact, while in the additive case it may be difficult to explore the entire $(0, \infty)$ potential range of the $\mu$ parameter, the convex formulation naturally constrains $\lambda$ to the $[0, 1]$ interval, which can be simply explored using, for instance, a uniform grid where we can explicitly test the common $\lambda = 1$ and independent $\lambda = 0$ task extremes.

The rest of the paper is organized as follows. We will briefly discuss the additive and convex approaches to MTL-SVR in Sect. 2 while in Sect. 3 we will show the equivalence between them. In Sect. 4 we will compare both over several regression problems showing numerically that the additive and convex approaches give essentially the same results and, also, that they outperform a single common model and a set of independent, task-specific models. The paper ends with a brief discussion as well as pointers to further work.

## 2    Additive and Convex MTL-SVMs

In this section we will introduce the additive MTL-SVM and the convex MTL-SVM problems. We recall that in the multi-task (MTL) framework we assume to have $T$ tasks $t = 1, \ldots, T$, that are different but share some relation among them; we also assume that each task can have a different number of examples $m_t$. The goal in MTL is to learn all the tasks at the same time so that we can improve on either a single, global task or on a set of task-specific models; we will call these Common-Task Learning (CTL) and Independent-Task Learning (ITL) respectively.

Following [8], we first introduce some notation that allows to write Support Vector Classification (SVC) and Support Vector Regression (SVR) problems in an unified way. Consider a sample $S = \{(x_i, y_i, p_i), 1 \leq i \leq N$, where $y_i = \pm 1$, and a primal problem of the form

$$
\begin{aligned}
\operatorname*{arg\,min}_{w,b,\xi} \quad & J(w, b, \xi) = C \sum_{n=1}^{N} \xi_i + \frac{1}{2} \|w\|^2 \\
\text{s.t.} \quad & y_i(w \cdot x_i + b) \geq p_i - \xi_i, \ i = 1, \ldots, N, \\
& \xi_i \geq 0, \ i = 1, \ldots, N.
\end{aligned}
\tag{1}
$$

Then, it can be easily checked [8] that for a classification sample $(x_i, y_i)$,, $i = 1, \ldots, M$, problem (1) is equivalent to the SVC primal problem when choosing $N = M$ and $p_i = 1$ for all $i$. Similarly, for a regression sample $(x_i, t_i)$, $i = 1, \ldots, M$, problem (1) is equivalent to the $\epsilon$-insensitive SVR primal problem when we choose $N = 2M$ and we select

$$
y_i = 1, \ p_i = t_i - \epsilon, \ i = 1, \ldots, M,
$$
$$
y_{M+i} = -1, \ p_{M+i} = -t_i - \epsilon, \ i = 1, \ldots, M.
$$

With this notation any result obtained for (1) will be valid for both SVC and SVR.

### 2.1    Additive MTL-SVMs

The Multi-Task Learning SVM (MTL-SVM), first presented in [6] and extended by Cai *et al.* in [2], considers a multi-task framework for SVMs in which a model is constructed for each task $t$, $t = 1, \ldots, T$. These task models, in its primal version, are formed by a weight vector $w_t$ and a task specific bias $b_t$, which are divided into a common part $w$ and $b$, and task specific additions $v_t$ and $d_t$. When coupled with a kernel setting, this formulation permits the (implicit) transformation of the original data $x$ into two different spaces: the space corresponding to the common part of the models, whose elements are represented by the transformation $\phi(x)$, and the space corresponding to the task specific deviations with the corresponding representation $\phi_t(x)$. Of course, as in standard SVMs, we do not explicitly work with the $\phi$ or $\phi_t$ transformations but instead with their corresponding kernels. With this formulation the MTL-SVM primal problem is the following:

$$\underset{w, v_t, \xi}{\arg\min} \quad J(w, v_t, \xi) = C \sum_{t=1}^{T} \sum_{i=1}^{m_t} \xi_i^t + \frac{1}{2} \sum_{t=1}^{T} \|v_t\|^2 + \frac{\mu}{2} \|w\|^2$$

$$\text{s.t.} \quad y_i^t (w \cdot \phi(x_i^t) + b + v_t \cdot \phi_t(x_i^t) + d_t) \geq p_i^t - \xi_i^t,$$

$$\xi_i^t \geq 0; \ i = 1, \dots, m_t, \ t = 1, \dots, T. \tag{2}$$

We can notice that in the regularization term we penalize the common part and the task specific deviations separately, using the parameter $\mu$ to tune the importance given to each part. Assigning large values to $\mu$ leads to a smaller common part $w$ and, thus, to strongly task-independent models; conversely, small $\mu$ values should result in a large common part $w$ and, therefore, to a stronger common model. We observe that the MTL-SVM model is quite flexible, because of the multiple biases $b_t = b + d_t$ and the possibility of working with different common and task-specific extended spaces, which can be adapted to the characteristics of each task. Using standard Lagrangian theory we can obtain the dual problem of (2) as:

$$\underset{\alpha}{\arg\min} \quad \Theta(\alpha) = \frac{1}{2} \alpha^{\mathsf{T}} \hat{Q} \alpha - p\alpha$$

$$\text{s.t.} \quad 0 \leq \alpha_i^t \leq C, \ i = 1, \dots, m_t, \ t = 1, \dots, T \ \text{(box constraints)}, \tag{3}$$

$$\sum_{i=1}^{n_t} \alpha_i^t y_i^t = 0, \ t = 1, \dots, T \ \text{(equality constraints)}.$$

Here $\alpha$ is the vector with all values $\alpha_i^t$ and we have $\hat{Q} = (1/\mu)Q + K$, where $Q$ is the matrix in which $Q_{ij} = k(x_i, x_j) = \phi(x_i) \cdot \phi(x_j)$ and $K$ is the following block-diagonal matrix

$$K = \begin{pmatrix} K_1 & \mathbf{0}^{m_1 \times m_2} & \cdots & \mathbf{0}^{m_1 \times m_T} \\ \mathbf{0}^{m_2 \times m_1} & K_2 & \cdots & \mathbf{0}^{m_2 \times m_T} \\ \vdots & \vdots & \ddots & \vdots \\ \mathbf{0}^{m_T \times m_1} & \mathbf{0}^{m_T \times m_2} & \cdots & K_T \end{pmatrix}, \ K_t = \begin{pmatrix} k_t(x_1^t, x_1^t) & \cdots & k_t(x_1^t, x_{m_t}^t) \\ k_t(x_2^t, x_1^t) & \cdots & k_t(x_2^t, x_{m_t}^t) \\ \vdots & \ddots & \vdots \\ k_t(x_{m_t}^t, x_1^t) & \cdots & k_t(x_{m_t}^t, x_{m_t}^t) \end{pmatrix}$$

here $k_t(x_i^t, x_j^t) = \phi_t(x_i^t) \cdot \phi_t(x_j^t)$. Notice the two main differences with the standard SVM dual problem: the use of a different kernel $\hat{Q}$ and the existence of $T$ task-related equality constraints. These multiple constraints do not allow the direct application of the standard SMO algorithm and, to solve this problem, Cai *et al.* developed instead the Generalized SMO algorithm in [2]. In what follows we will call this approach *additive* MTL-SVM.

## 2.2   Convex MTL-SVMs

In this work we propose an alternative approach for Multi-Task SVM, which will be named *convex* MTL-SVM. As in the previous additive approach, we have one model for each task, which is divided into a common part $(u, \hat{b})$ and a task specific part $(u_t, \hat{d}_t)$. However, our approach combines these two parts in a convex way, as done in the following primal problem:

$$
\begin{aligned}
\underset{u, u_t, \xi}{\arg\min} \quad & J(u, u_t, \xi) = C \sum_{t=1}^{T} \sum_{i=1}^{m_t} \xi_i^t + \frac{1}{2} \sum_{t=1}^{T} \|u_t\|^2 + \frac{1}{2} \|u\|^2 \\
\text{s.t.} \quad & y_i^t (\lambda \{ u \cdot \phi(x_i^t) + \hat{b} \} + (1 - \lambda) \{ u_t \cdot \phi_t(x_i^t) + \hat{d}_t \}) \geq p_i^t - \xi_i^t, \\
& \xi_i^t \geq 0; \; i = 1, \dots, m_t, \; t = 1, \dots, T.
\end{aligned}
\tag{4}
$$

Notice that we use the convex mixing parameter $\lambda \in [0, 1]$ to regulate the interplay between the common and independent components; moreover, there are no regularization hyper-parameters. The use of $\lambda$ makes clearer the influence of each model, in contrast with the rather indirect influence of $\mu$ in additive MTL-SVM. Moreover, while $\mu$ can range in the unbounded interval $(0, \infty)$, $\lambda$ varies in the bounded $[0, 1]$. The dual problem corresponding to (4) is now

$$
\begin{aligned}
\underset{\alpha}{\arg\min} \quad & \Theta(\alpha) = \frac{1}{2} \alpha^{\mathsf{T}} \tilde{Q} \alpha - p\alpha \\
\text{s.t.} \quad & 0 \leq \alpha_i^t \leq C, \; i = 1, \dots, m_t, \; t = 1, \dots, T \text{ (box constraints)}, \\
& \sum_{i=1}^{n_t} \alpha_i^t y_i^t = 0, \; t = 1, \dots, T \text{ (equality constraints)},
\end{aligned}
\tag{5}
$$

where $\tilde{Q} = \lambda^2 Q + (1 - \lambda)^2 K$, with $Q$ and $K$ being the same full and block diagonal matrices defined in the previous additive approach. Observe that the dual problems (5) and (3) are formally the same, the only difference being the kernel matrices $\tilde{Q}$ and $Q$ involved.

# 3   Equivalence of Additive and Convex MTL-SVMs

In this section we will present two results; the first one, Proposition 1, proves the equivalence between the additive and the convex approaches; while the second one proves the equivalence of the convex MTL-SVM with a common standard SVM and an independent standard SVM for each task when setting $\lambda = 1$ and $\lambda = 0$, respectively.

**Proposition 1.** *The additive MTL-SVM primal problem with parameters $C_{add}$ and $\mu$ (and possibly $\epsilon$) and the convex MTL-SVM primal problem with parameters $C_{conv}$ and $\lambda$ (and possibly $\epsilon$), with $\lambda \in (0, 1)$, are equivalent when $C_{add} = (1 - \lambda)^2 C_{conv}$ and $\mu = (1 - \lambda)^2 / \lambda^2$.*

*Proof.* Making the change of variables $w = \lambda u$, $v_t = (1 - \lambda)u_t$, $b = \lambda\hat{b}$ and $d_t = (1 - \lambda)\hat{d}_t$ in the convex primal problem, we can write it as

$$\underset{w,v_t,\xi}{\arg\min} \quad J(w, v_t, \xi) = C_{\text{conv}} \sum_{t=1}^{T} \sum_{i=1}^{m_t} \xi_i^t + \frac{1}{2(1-\lambda)^2} \sum_{t=1}^{T} \|v_t\|^2 + \frac{1}{2\lambda^2} \|w\|^2$$

$$\text{s.t.} \quad y_i^t(w \cdot \phi(x_i^t) + b + v_t \cdot \phi_t(x_i^t) + d_t) \geq p_i^t - \xi_i^t,$$

$$\xi_i^t \geq 0; \ i = 1, \ldots, m_t, \ t = 1, \ldots, T.$$

Multiplying now the objective function by $(1-\lambda)^2$ we obtain the additive MTL-SVM primal problem with $\mu = (1-\lambda)^2/\lambda^2$ and $C_{\text{add}} = (1-\lambda)^2 C_{\text{conv}}$. Conversely, we can start at the primal additive problem and make the inverse changes to arrive now to the primal convex problem. □

Although this result focuses on the equivalence of the primal problems, there is also a corresponding equivalence between the dual problems. For instance, we can set $\hat{\alpha}_i^t = (1 - \lambda)^2 \alpha_i^t$ in the dual convex problem to arrive to the following dual problem:

$$\underset{\hat{\alpha}}{\arg\min} \quad \Theta(\hat{\alpha}) = \left( \frac{1}{2} \frac{1}{(1-\lambda)^2} \hat{\alpha}^{\mathsf{T}} \left( \frac{\lambda^2}{(1-\lambda)^2} Q + \frac{(1-\lambda)^2}{(1-\lambda)^2} K \right) \hat{\alpha} - \frac{1}{(1-\lambda)^2} p\hat{\alpha} \right)$$

$$\text{s.t.} \quad 0 \leq \hat{\alpha}_i^t \leq C_{\text{add}}, \ i = 1, \ldots, m_t, \ t = 1, \ldots, T \text{ (box constraints)},$$

$$\sum_{i=1}^{n_t} \hat{\alpha}_i^t y_i^t = 0, \ t = 1, \ldots, T \text{ (equality constraints)}.$$

Conversely, multiplying the dual convex objective function by $(1-\lambda)^2$ we recover the dual additive problem with $\mu = (1 - \lambda)^2/\lambda^2$.

The previous proposition shows an equivalence between the additive and convex approaches when $\lambda \in (0, 1)$; we now consider the situations arising when setting $\lambda = 0, 1$ in the convex approach. First, solving the convex MTL-SVM problem with $\lambda = 1$ is equivalent to solving a standard SVM primal problem: in fact the primal problem would become

$$\underset{u,u_t,\xi}{\arg\min} \quad J(u, u_t, \xi) = C \sum_{t=1}^{T} \sum_{i=1}^{m_t} \xi_i^t + \frac{1}{2} \sum_{t=1}^{T} \|u_t\|^2 + \frac{1}{2} \|u\|^2 \tag{6}$$

$$\text{s.t.} \quad y_i^t(u \cdot \phi(x_i^t) + \hat{b}) \geq p_i^t - \xi_i^t; \ \xi_i^t \geq 0; \ i = 1, \ldots, m_t, \ t = 1, \ldots, T.$$

Note that we obtain a standard SVM primal problem with an additional regularization on the $u_t$ terms that disappears when solving the optimization problem since the variables $u_t$ do not have any influence on the error of the model.

The convex MTL-SVM is also equivalent to solving one independent SVM primal problem for each task when setting $\lambda = 0$, since the resultant problem becomes:

$$\underset{u,u_t,\xi}{\arg\min} \quad J(u, u_t, \xi) = C \sum_{t=1}^{T} \sum_{i=1}^{m_t} \xi_i^t + \frac{1}{2} \sum_{t=1}^{T} \|u_t\|^2 + \frac{1}{2} \|u\|^2$$

$$\text{s.t.} \quad y_i^t(u_t \cdot \phi_t(x_i^t) + \hat{d}_t) \geq p_i^t - \xi_i^t; \ \xi_i^t \geq 0; \ i = 1, \ldots, m_t, \ t = 1, \ldots, T.$$

Since the variables $u_t$ are independent and $u$ disappears because of the regularization, solving this problem is equivalent to solving separately a standard SVM primal problem for each task.

# 4    Experiments

Because of space constrains, all the examples considered here will correspond to regression problems, with Support Vector Regression (SVR) being therefore the underlying SVM model. This section will be divided in two parts. In the first one we will illustrate empirically the equivalence between the convex and additive approaches. The second one will compare the convex multi-task SVR, which we will denote as cvxMTL, against a single SVR model over all tasks, which we will call the ct1SVR model, and also against the application of an independent SVR on each one, which we will call the it1SVR model.

Before doing so, we first comment on some computational simplifications which we will use when building our experimental MTL-SVR models. In the original MTL-SVR formulations in (2) and (4), we have $T$ different task specific tolerances $\epsilon_t$ and biases $b_t$, as well as a common bias $b$. These task specific parameters have two consequences in the dual problems shown in (3) and (5). The use of multiple tolerances implies that they should be properly hyper-parametrized, with the corresponding rather high computational cost. To alleviate this we will consider a common tolerance $\epsilon$ for all the tasks. Moreover, the multiple biases $b_t$ are present in the dual problem through the $T$ different equality constraints. That is, instead of solving a dual problem with a single equality constraint, as in the SMO algorithm, we would need to ensure that the $T$ task specific equality constraints are met, as proposed in the GSMO algorithm of [2]. Since there seems to be no efficient GSMO code available, we will use instead the scikit-learn wrapper [10] of the well-known LIBSVM [4] library. For this we will just use a common tolerance $\epsilon$ and, also, a common bias $b$ for all tasks so that a single equality constraint has to be met.

## 4.1    Equivalence Experiments

In this subsection we will illustrate empirically the equivalence between the convex and additive approaches over a synthetic problem shown in Fig. 1, left. It is a four task problem in which we derive four different functions $f_0, f_1, f_2, f_3$ from a linear function $f(x) = 2x - 1$ by adding zero mean, unit variance Gaussian noise to the slope and bias of $f(x)$. Task $t$, $t = 0, 1, 2, 3$, consists then in estimating the slope $\hat{a}_t$ of each $\hat{f}_t(x) = a_t x + b_t$ and a common intercept $\hat{b}$. To do so, we randomly generate 30 points $x_i \in [1, 50)$ for each task with targets $y_i = f_t(x_i)$. We thus have a 120 point sample divided in four 30-point tasks, as shown in Fig. 1, left. The sample will be randomly split in a train set with two thirds of the points and the remaining ones as a test set; the task size proportions are kept in both splits.

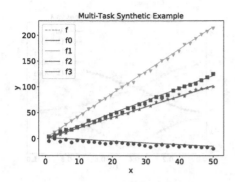 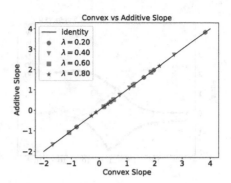

**Fig. 1.** Left: Synthetic example dataset, where the data of each task (corresponding to a different function $f_i$) are represented with a different color. Right: Comparison of the weights obtained by the convex and additive approaches.

We will build four convex models, one for each $\lambda \in \{0.2, 0.4, 0.6, 0.8\}$, and the corresponding additive models for which we set $\mu = (1 - \lambda)^2/\lambda^2$ and $C_{\text{add}} = (1 - \lambda)^2 C_{\text{conv}}$. In order for the regularization to influence on the final models and, hence, highlight the interplay between the common and independent tasks, we will consider a relatively small convex penalty $C_{\text{conv}} = 10^{-2}$.

We use linear kernels so that we can easily visualize the joint common and task dependent weights; here they correspond to the task slopes estimated by the additive and convex models. The estimated slopes are shown in Fig. 1, right, with the convex ones on the $x$ axis and the additive ones on the $y$ axis. Thus, for each $\lambda$ there are four points, all in the same color, one for each task slope. As it can be seen, these points are perfectly aligned with the $y = x$ diagonal, that is, as expected the additive and convex slopes coincide.

Figure 2 offers an alternative visualization of the slope estimates on the $y$-axis, now for a finer $\lambda$ grid (right) and their corresponding $\mu$ values (left) on the $x$-axis. In the right picture we can see how the common coefficient increases from 0 with $\lambda = 0$ to a value of 0.86 while the independent ones decrease to 0 at $\lambda = 1$, as the MTL models evolve from four essentially independent task models at $\lambda = 0$ to just the common one at $\lambda = 1$.

## 4.2 Convex MTL Experiments

We now test the convex MTLSVR (`cvxMTL`) models on the problems `abalone` and `crime` taken from the UCI repository, `boston_housing` taken from the Kaggle repository and two other problems, which we call `majorca` and `tenerife`, where we must forecast hourly photovoltaic (PV) energy production in the islands of Majorca and Tenerife in Spain. Table 1 gives problem dimensions and sample sizes.

The table also gives the number of tasks and their minimum, average and maximum sizes. These tasks are defined as follows. In `majorca` and `tenerife` there is one task for each UTC hour with solar radiation and it consists in

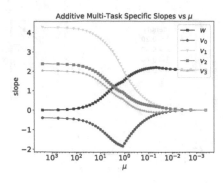
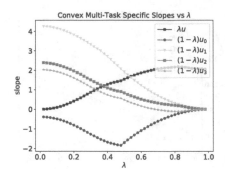

**Fig. 2.** Convex (right) and additive (left) MTLSVR slope estimates weights as a function of $\lambda$. We represent the common part of the models, $w$ for the additive and $\lambda u$ for the convex, as well as each task specific deviation $v_t$ and $(1 - \lambda)u_t$.

**Table 1.** Sample sizes, dimensions and number of tasks of the datasets used.

| Name | Size | No. feat. | No. tasks | Avg. task size | Min. t. s. | Max. t. s. |
|------|------|-----------|-----------|----------------|------------|------------|
| majorca | 15330 | 765 | 14 | 1095 | 1095 | 1095 |
| tenerife | 15330 | 765 | 14 | 1095 | 1095 | 1095 |
| boston | 506 | 14 | 2 | 253 | 35 | 471 |
| abalone | 4177 | 8 | 3 | 1392 | 1307 | 1527 |
| crime | 1195 | 127 | 9 | 132 | 60 | 278 |

forecasting the PV energy for that hour. In `majorca` we select the UTC hours from 06:00 to 19:00, while in `tenerife` we use UTC hours from 07:00 to 20:00; in both cases we have 14 tasks. The target in the `crime` dataset is the number of violent crimes per capita in different cities of the USA. We group these cities by their states and the individual tasks are precisely to predict the crime rate for all the cities in a given state. We will only work with states that have at least 50 cities in the database.

The target in the `abalone` problem is the number of rings in the shells of a particular kind of shellfish. Here we will divide the overall problem in three tasks, the prediction of the number of rings for the male, female and infant specimens. Finally, in the `boston` problem we have to predict the median price value of the houses in a number of Boston neighbors. We consider two prediction tasks, one for the neighbors bound by the Charles river and the other for the remaining neighbors. The first task is rather small, with just 35 samples.

We will compare the `cvxMTL` model performance with those of the `ctlSVR` model, which builds a common Gaussian SVR for all the tasks, and the `itlSVR` model, which trains a specific Gaussian SVR for each task. The `ctlSVR` model requires to estimate the optimal $C$, $\gamma$ and $\epsilon$ hyper-parameters, and for the `itlSVR` models we have to do the same for the $C_t$, $\gamma_t$ and $\epsilon_t$ of each task. Here $\gamma$ and $\gamma_t$ refer to the common and independent Gaussian kernel widths (see [4] for more

**Table 2.** Hyper-parameters, grids used to find them (when appropriate) and hyper-parameter selection method for each model.

| Par. | Grid | ctlSVR | itlSVR | cvxMTL |
|------|------|--------|--------|--------|
| $C_{\mathrm{conv}}$ | $\left\{10^k : -5 \leq k \leq 6\right\}$ | CV | CV | CV |
| $\epsilon_{\mathrm{conv}}$ | $\left\{\frac{\sigma}{2^k} : 1 \leq k \leq 6\right\}$ | CV | CV | CV |
| $\gamma$ | $\left\{\frac{4^k}{d} : -2 \leq k \leq 3\right\}$ | CV | - | ctlSVR |
| $\gamma_t$ | $\left\{\frac{4^k}{d} : -2 \leq k \leq 3\right\}$ | - | CV | itlSVR |
| $\lambda$ | $\left\{10^{-1}k : 1 \leq k \leq 9\right\}$ | - | - | CV |

details). To hyper-parametrize the cvxMTL models, we will retain the optimal common $\gamma$ and independent $\gamma_t$ hyper-parameters and optimize their $C_{\mathrm{conv}}$ and $\epsilon_{\mathrm{conv}}$ values and the convex mixing parameter $\lambda$. The grids considered for this are given in Table 2.

In majorca and tenerife, we will use data from the years 2013, 2014 and 2015 as training, hyper-parameter validation and test subsets, respectively. The other datasets do not have predefined train-test splits and we will evaluate all models by three-fold nested cross-validation. More precisely, we first define three outer folds and cyclically use one of them as the test set and split the other two in another three fold inner subsets, of which two are used for training and the remaining one for hyper-parameter validation. Once the best hyper-parameters are found, they are used to define a model which is now trained on the two outer folds and evaluated on the remaining test outer fold. In all cases the folds have been built in a stratified fashion to maintain task size proportions.

We will compare the model results in terms of the mean absolute error and of the R2 score, whose means are given in Table 3; this table also gives the standard deviations where folds have been used. As it can be seen, the cvxMTL models give the best MAE (in bold) for all problems but tenerife, where it essentially ties with itlSVR, Similarly, they have the best R2 scores for all datasets but abalone, where it again basically ties now with ctlSVR. Notice also that, in general, the itlSVR models tend to give better results than ctlSVR in those problems where task sizes are more or less the same (but not for abalone) while ctlSVR outperforms itlSVR when task sizes are unbalanced. On the other hand, the multi-task approach can handle both situations well. To show statistical significance we give in Table 4 the Wilcoxon $p$-values of absolute errors of a model and the one following it in the MAE ranking (top), and of quadratic errors and the R2 score ranking (bottom); $p$-value rankings shown in parenthesis.

**Table 3.** Test MAE's (top) and R2 scores (bottom) of the models considered.

|        | majorca | tenerife | boston | abalone | crime |
|--------|---------|----------|--------|---------|-------|
| MAE    |         |          |        |         |       |
| ctlSVR | 5.264   | 5.785    | $2.254 \pm 0.055$ | $1.483 \pm 0.036$ | $0.078 \pm 0.001$ |
| itlSVR | 5.102   | **5.351** | $2.574 \pm 0.083$ | $1.499 \pm 0.041$ | $0.081 \pm 0.006$ |
| cvxMTL | **4.966** | 5.352  | $\mathbf{2.244 \pm 0.041}$ | $\mathbf{1.466 \pm 0.027}$ | $\mathbf{0.074 \pm 0.003}$ |
| R2     |         |          |        |         |       |
| ctlSVR | 0.831   | 0.901    | $0.856 \pm 0.030$ | $\mathbf{0.563 \pm 0.016}$ | $0.743 \pm 0.022$ |
| itlSVR | 0.843   | 0.903    | $0.777 \pm 0.049$ | $0.506 \pm 0.060$ | $0.714 \pm 0.008$ |
| cvxMTL | **0.849** | **0.906** | $\mathbf{0.860 \pm 0.047}$ | $0.561 \pm 0.017$ | $\mathbf{0.755 \pm 0.015}$ |

**Table 4.** Wilcoxon $p$-values of absolute errors of a model and the one following it in the MAE ranking (top), and of quadratic errors and the R2 score ranking (bottom); $p$-value rankings shown in parenthesis.

|        | majorca | | tenerife | | boston | | abalone | | crime | |
|--------|---------|-----|----------|-----|--------|-----|---------|-----|--------|-----|
| ctlSVR | 0.0000 | (3) | 0.0000 | (3) | 0.8165 | (1) | 0.0116 | (2) | 0.0000 | (2) |
| itlSVR | 0.8740 | (1) | -      | **(1)** | 0.1151 | (1) | 0.2401 | (2) | 0.4811 | (2) |
| cvxMTL | -      | **(1)** | 0.0003 | (2) | -   | **(1)** | -    | **(1)** | -    | **(1)** |
| ctlSVR | 0.0011 | (3) | 0.0000 | (3) | 0.8355 | (1) | -      | **(1)** | 0.0004 | (2) |
| itlSVR | 0.0000 | (2) | 0.2411 | (1) | 0.0522 | (1) | 0.0479 | (3) | 0.4944 | (2) |
| cvxMTL | -      | **(1)** | -   | **(1)** | -   | **(1)** | 0.0119 | (2) | -    | **(1)** |

## 5 Conclusions

In this work we have proposed convex Multi-task Learning for SVMs, an alternative formulation of Cai and Cherkassky's additive MTL for SVMs. We have proved that both result in the same final models but convex MTL has the advantage of a much simpler interpretation of the interplay between the common and independent components through the optimal value of the convex hyper-parameter $\lambda^*$. Moreover, the exploration of the $\lambda$ hyper-parameter is much easier that of the $\mu$ hyper-parameter of additive MTL, as the former is bound to the $[0, 1]$ interval while $\mu$ ranges in principle in the unbounded $(0, \infty)$.

In any case, and while our formulation of convex MTL deals simultaneously with classification and regression problems, we have considered only the latter in our experiments; thus, a first obvious point of further work is to study convex MTL on classification settings. Also, a more precise study should be made of the influence of the parameter $\lambda$ on the performance of convex MTL models, both by themselves and in their relation with those of the standard common and task specific SVM models. In parallel to this, our convex formulation of MTL also suggests to compare the performance of either one of the equivalent additive or convex MTL models with a straight convex combination of the outputs of the single and task specific SVM models. Finally, while the additive MTL formulation works in terms of a common model to which task-dependent differences are added, the convex formulation makes possible to decompose the final MTL model into an overall full common SVM model and its single task counterparts. This decomposition should lead to a better understanding of their separate contributions to the global MTL model. We are currently considering these issues.

**Acknowledgments.** With partial support from Spain's grant TIN2016-76406-P. Work supported also by the UAM–ADIC Chair for Data Science and Machine Learning. We thank Red Elctrica de Espaa for making available solar energy data and the Agencia Estatal de Meteorologa, AEMET, and the ECMWF for access to the MARS repository. We also gratefully acknowledge the use of the facilities of Centro de Computación Científica (CCC) at UAM.

# References

1. Cai, F., Cherkassky, V.: SVM+ regression and multi-task learning. In: International Joint Conference on Neural Networks, IJCNN 2009, Atlanta, Georgia, USA, 14–19 June 2009, pp. 418–424 (2009)
2. Cai, F., Cherkassky, V.: Generalized SMO algorithm for SVM-based multitask learning. IEEE Trans. Neural Netw. Learn. Syst. **23**(6), 997–1003 (2012)
3. Caruana, R.: Multitask learning. Mach. Learn. **28**(1), 41–75 (1997)
4. Chang, C.C., Lin, C.J.: LIBSVM: a library for support vector machines. ACM Trans. Intell. Syst. Technol. (TIST) **2**(3), 27 (2011)
5. Donini, M., Martinez-Rego, D., Goodson, M., Shawe-Taylor, J., Pontil, M.: Distributed variance regularized multitask learning. In: 2016 International Joint Conference on Neural Networks (IJCNN), pp. 3101–3109. IEEE (2016)
6. Evgeniou, T., Pontil, M.: Regularized multi-task learning. In: Proceedings of the Tenth ACM SIGKDD International Conference on Knowledge Discovery and Data Mining, pp. 109–117. ACM (2004)
7. Liang, L., Cai, F., Cherkassky, V.: Predictive learning with structured (grouped) data. Neural Netw. **22**(5–6), 766–773 (2009)
8. Lin, C.J.: On the convergence of the decomposition method for support vector machines. IEEE Trans. Neural Networks **12**(6), 1288–1298 (2001)
9. Oneto, L., Donini, M., Elders, A., Pontil, M.: Taking advantage of multitask learning for fair classification. CoRR abs/1810.08683 (2018). http://arxiv.org/abs/1810.08683
10. Pedregosa, F., et al.: Scikit-learn: machine learning in python. J. Mach. Learn. Res. **12**, 2825–2830 (2011)
11. Ruder, S.: An overview of multi-task learning in deep neural networks. CoRR abs/1706.05098 (2017). http://arxiv.org/abs/1706.05098
12. Zhang, Y., Yang, Q.: A survey on multi-task learning. CoRR abs/1707.08114 (2017). http://arxiv.org/abs/1707.08114

# Influence Maximization and Extremal Optimization

Tamás Képes, Noémi Gaskó$^{(\boxtimes)}$, Rodica Ioana Lung,
and Mihai-Alexandru Suciu

Babeş-Bolyai University, Cluj-Napoca, Romania
kepestamas@gmail.com, rodica.lung@econ.ubbcluj.ro,
{gaskonomi,mihai-suciu}@cs.ubbcluj.ro

**Abstract.** Influence Extremal Optimization (InfEO) is an algorithm based on Extremal Optimization, adapted for the influence maximization problem for the independent cascade model. InfEO maximizes the marginal contribution of a node to the influence set of the model. Numerical experiments are used to compare InfEO with other influence maximization methods, indicating the potential of this approach. Practical results are discussed on a network constructed from publication data in the field of computer science.

**Keywords:** Influence maximization · Extremal Optimization · Independent cascade model

## 1 Introduction

In recent years network analysis [16] earned an important role due to its multiple applications not only in science but also in the every day life. The study of various networks (e.g. social, biological, and ecological networks) from different perspectives: concerning on degree-properties, maximum clique detection, on community structure detection, etc. reveals important information that has become essential in any decision making process.

In social network analysis a crucial task is the computation of the most influential nodes in an information diffusion process. The influence maximization problem is the following: given a graph $G = (V, E)$, a propagation model (e.g. Independent Cascade, Weighted Cascade, Linear Threshold), and a number $k$ ($k \geq 1$), which is the number of initial seeder nodes, identify the set of the $k$ nodes which have the maximum influence on the graph $G$. In this form the problem was first proposed in [6], and proven to be NP-hard in [10] for the independent cascade model and the linear threshold model.

In this paper we propose the use of the independent cascade model (ICM) and a novel optimization method based on extremal optimization to analyze journal citation data and to identify influential journals in a field by constructing a citation journal network. Thus the main goal of this article is two-sided: on

© Springer Nature Switzerland AG 2019
H. Pérez García et al. (Eds.): HAIS 2019, LNAI 11734, pp. 416–427, 2019.
https://doi.org/10.1007/978-3-030-29859-3_36

**Algorithm 1.** Independent Cascade Model

---

   **Input:** $G, A_0, p$

   **Output:** $\sigma(A_0)$ - **the size of set of active nodes** $A$

1:  $finished := false$
2:  $A := A_0$
3:  $B := A$
4:  **while** not finished **do**
5:    $nextB := \emptyset$
6:    **for each** $v \in B$ **do**
7:      **for** each direct neighbor $w$ of $v$, where $w \notin A$ **do**
8:        with probability $p$ add $w$ to $nextB$
9:      **end for**
10:   **end for**
11:   $B := nextB$
12:   $A := A \cup B$
13:   **if** $B = \emptyset$ **then**
14:     $finished := true$
15:   **end if**
16: **end while**
17: **return** $\sigma(A_0)$ equal to the size of A

---

one side to propose a new algorithm, called Influence Extremal Optimization, designed to detect the promising seeder nodes, and on the other side to use this algorithm to analyze a real-world journal citation network.

The article is organized as follows: in the next section we present the independent cascade model and in the third section the Influence Extremal Optimization algorithm is described. The fourth section describes the experimental results and finally we conclude the paper.

## 2   Independent Cascade Model

The Independent Cascade Model (ICM) is one of the most popular propagation models in the study of information diffusion in social networks due to its versatility and basic intuitive approach. It has been shown to reliably model information diffusion in social networks such as Twitter [1,14,19,24] or Facebook [12,13]. Examples of the use of this model in marketing can be found in [7], where the authors study the product information process on a network of tweets regarding a particular phone product, finding that the ICM is suitable in modeling the spread of information.

ICM has also been employed in the study and analysis of collaboration networks. In [8] the authors propose a general threshold for which the independent cascade model is a particular case and use the DPLB network to illustrate the behavior of the model. In [7] a model based on probabilistic maximum coverage and clustering to maximize information propagation integrated in the ICM and linear threshold model is proposed and tested on large collaboration networks.

The independent cascade model (Algorithm 1) is a simple, intuitive cascade diffusion model that constructs a set of active nodes starting from an initial seed set by using a probability of diffusion from a node to another. An initial set of active nodes $A_0$ is used to start an activation iterative process: an active node $v$ at iteration/step $t$ has one chance to activate each of its inactive neighbors with a propagation probability $p$ - parameter of the method. If the activation succeeds then the activated neighbors are added to the active set of nodes $A$ in step $t + 1$; whether or not $v$ has activated any neighbors it will not attempt to activate them again in this process. The ICM ends when no more activations can be performed. The size of the resulting set of active nodes $A$ is denoted by $\sigma(A_0)$ and it depends on the probability $p$ of 'activation' and on the structural properties of the network. The influence maximization problem consists in identifying the set $A_0$ that maximizes $\sigma(A_0)$. There are many methods that attempt the maximization of influence based on ICM. Heuristics based on genetic algorithms can be found in [3,4,9,20,21,23].

# 3   Influence Extremal Optimization - InfEO

Extremal optimization (EO) [2] is a powerful optimization tool for combinatorial optimization problems for which a solution $s$ can be expressed as a set of components $s_i, i = 1, \ldots, n$ with individual fitnesses $f_i(s)$ assigned to each one of them. The standard EO maximizes each component of a potential solution by randomly reassigning the one having the worst fitness. An individual $s_{best}$ is used to preserve the best solution found so far based on an overall fitness $f()$. An outline of the standard EO is presented in Algorithm 2.

The fact that for the ICM it is shown in [11] that $\sigma(A)$ (Algorithm 1, line 17) is a submodular function, makes it particularly suitable to be optimized by an extremal optimzation approach in which the seeder set corresponding to $s$ is evolved by maximizing $\sigma(s)$ while the fitness of each component - node in the set - is computed as the marginal contribution of the node to $\sigma(s)$, i.e. the difference of the number of activated nodes when node $i$ is the part of the seeder nodes and the number of activated nodes when node $i$ is not a part of the seeder nodes.

In this context we propose the Influence Extremal Optimization algorithm (InfEO), to identify the maximum influence set of a network for the independent cascade model. In what follows InfEO is described in detail, and outlined in Algorithm 3.

*Encoding.* Binary representation is used for $s$, having length $n$ equal to the number of the nodes of the graph. A 0 value on the position $i$ means that node $i$ is not a part of the seeders set, the value 1 at the position $i$ means that node $i$ is the part of the seeder set.

---

**Algorithm 2.** Standard EO

---

   **Input:** $s$
   **Output:** $s_{best}$
1: Initialize $s$ random;
2: $s_{best} := s$; $//s_{best}$ preserves the best solution found so far
3: **while** a termination condition is not met **do**
4:    find $i_{min}$ component with the smallest fitness value
5:    randomly reassign $s_{i_{min}}$ in $s$;
6:    **if** $f(s) > f(s_{best})$ **then**
7:       $s_{best} := s$
8:    **end if**
9: **end while**

---

*Fitness Function.* Within EO two fitness functions are used: one for individual $s$ and one for evaluating components of $s$. InfEO aims to maximize $\sigma(s)$, where we consider individual $s$ equivalent to its set of seeder nodes. Because $\sigma(s)$ depends on probability $p$, an approximation $\bar{\sigma}(s)$ computed as the average of $\sigma(s)$ on 30 independent runs of ICM is used to evaluate $s$.

The fitness value of an active node $i$ in $s$ is computed as its marginal contribution to $\bar{\sigma}$:

$$f_i(s) = \bar{\sigma}(s) - \bar{\sigma}(s \backslash i) - 1, \tag{1}$$

where $s \backslash i$ denotes the set of active nodes in $s$ without node $i$.

*Variation Operator.* Each iteration the seeder node having the lowest fitness value $f_{i_{min}}$ is removed from the set and another, random node $j$ is added to the seeder set. If the new solution will replace $s$ or $s_{best}$ if it has a better $\bar{\sigma}$ value. If no replacement is made, another node $j$ is randomly chosen to replace $i_{min}$.

*Outline.* The outline of InfEO is presented in Algorithm 3. InfEO uses only one parameter, the maximum number of iterations $MaxGen$. The independent cascade model uses two parameters: the size $k$ of the seed set $A_0$ and the probability of activation $p$.

# 4   Experimental Results

Experimental results illustrate the behaviour of InfEO by using the following set of networks: Netscience [17], UCsocial [18], Polbooks[1], Karate [22], and Dolphins [15]. These are networks that are frequently used in evaluating various algorithms. Particularly for the influence maximization problem, while they can be considered as benchmarks, there is no 'ground truth' information available. To compare the results of InfEO with those provided by other methods we use as performance measure the average size of the ICM cascade $\bar{\sigma}()$ for each

---

[1] http://www.orgnet.com/divided2.html, accessed April, 2019.

---

**Algorithm 3. InfEO**

---

   **Input:** $s$
   **Output:** $s_{best}, \bar{\sigma}(s_{best})$
1: Initialize $s$ random;
2: $s_{best} := s$
3: changed := true; $gen := 0$;
4: **while** $gen < MaxGen$ **do**
5:    **if** changed **then**
6:       find $i_{min}$ index with the smallest fitness value $f_{i_{min}}$ from the active nodes of
       $s$
7:    **end if**
8:    $s' := s$
9:    $s'_{i_{min}} := 0$
10:   Choose $j$ random such that $s'_j = 0$ and set $s'_j := 1$
11:   **if** $\bar{\sigma}(s') > \bar{\sigma}(s)$ **then**
12:     $s := s'$
13:     changed:=true
14:     **if** $\bar{\sigma}(s) > \bar{\sigma}(s_{best})$ **then**
15:       $s_{best} := s$
16:     **end if**
17:   **else**
18:     changed:=false
19:   **end if**
20:   $gen := gen + 1$
21: **end while**

---

method, for 10 independent runs. A Wilcoxon sign-rank test is used to test the statistical significance of differences between medians with $\alpha = 0.05$.

To further analyze the performance of InfEO we constructed a set of real-world data-sets from highly cited journal publication data in computer science according to Web of Science[2].

## 4.1 Benchmarks Comparisons with Other Methods

*Datasets.* Table 1 presents the main characteristics of five networks used to illustrate the behavior of InfEO compared with other methods from the literature.

*Parameters.* InfEO uses only one specific parameter - $MaxGen$ the maximum number of generations. In all experiments $MaxGen = 10000$. The parameters of ICM tested were the size of the seeder set $k = 3, 5, 8, 10$ and propagation probability $p = 1\%, 3\%$. For each network and each ICM parameter the average number of activated nodes in the last iteration $\bar{\sigma}$ is computed for 10 independent runs.

---

[2] www.webofknowledge.com, accessed February, 2019.

**Table 1.** The networks used in experiments and some basic properties

| Name | No. nodes | No. edges | Avg. degree | Density | Diameter |
|---|---|---|---|---|---|
| Netscience [17] | 1461 | 2742 | 1.877 | 0.001 | 7 |
| Ucsocial [18] | 1899 | 20296 | 10.688 | 0.006 | 8 |
| Polbooks | 105 | 441 | 8.4 | 0.081 | 8.4 |
| Dolphins [15] | 62 | 159 | 5.129 | 0.084 | 8 |
| Karate [22] | 34 | 78 | 4.588 | 0.139 | 5 |

**Table 2.** Wilcoxon sign-rank test results. A • indicates that the corresponding method provided the best results. If there are more methods with results that are not statistically different from the best one, they are also marked with a •.

| Network | $p = 1\%$ | | | | | $p = 3\%$ | | | | |
|---|---|---|---|---|---|---|---|---|---|---|
| | InfEO | EO | Highdeg | Sdisc | Random | InfEO | EO | Highdeg | Sdisc | Random |
| **k = 3** | | | | | | | | | | |
| Netscience | • | • | • | - | - | • | - | - | • | - |
| Ucsocial | • | - | - | - | - | • | - | • | - | - |
| Polbooks | • | • | - | - | - | • | • | - | - | - |
| Dolphins | • | • | - | - | - | • | • | - | - | - |
| Karate | • | • | - | - | - | • | - | - | • | - |
| **k = 5** | | | | | | | | | | |
| Netscience | • | • | • | • | - | • | - | • | • | - |
| Ucsocial | • | • | - | • | - | • | - | - | - | - |
| Polbooks | • | • | - | - | - | • | • | - | - | - |
| Dolphins | • | • | - | - | - | • | • | - | - | - |
| Karate | • | • | • | - | - | • | • | - | - | - |
| **k = 8** | | | | | | | | | | |
| Netscience | • | • | • | - | - | • | - | • | • | - |
| Ucsocial | - | - | • | • | - | - | - | • | • | - |
| Polbooks | • | • | • | - | - | • | • | - | - | - |
| Dolphins | • | • | - | - | - | • | • | - | - | - |
| Karate | • | • | - | - | - | • | • | - | - | - |
| **k = 10** | | | | | | | | | | |
| Netscience | • | • | • | • | - | • | - | - | - | - |
| Ucsocial | - | - | • | • | - | - | - | • | • | - |
| Polbooks | • | • | - | • | - | • | • | - | • | - |
| Dolphins | • | - | - | - | - | • | • | - | - | - |
| Karate | • | • | - | - | - | • | • | - | - | - |

*Comparisons with Other Methods.* Results provided by InfEO are compared with those reported by four other methods:

- EO - a standard variant of EO as presented in Algorithm 2 that uses the same encoding as InfEO but the classic EO replacement scheme;
- HighDeg - a standard heuristic that chooses nodes in the seed set in descending order of their degree;
- SDisc - the SingleDiscount algorithm of ICM in [5] that uses a discount heuristic - discounts nodes already in the seed set when considering the degree of new nodes to be added to the seed set;
- a Random ICM starting with a random seed set.

*Results and Discussion.* Figures 1 and 2 present boxplots of the number of activated nodes reported by each method on 10 independent runs. HighDeg and SDisc are deterministic methods, the difference in results among runs results from the 30 ICMs used to compute the size of the active node set. Table 2 complements the figures with Wilcoxon sign-rank non-parametric statistical tests comparing the medians for pairs of results with a 0.05 significance level.

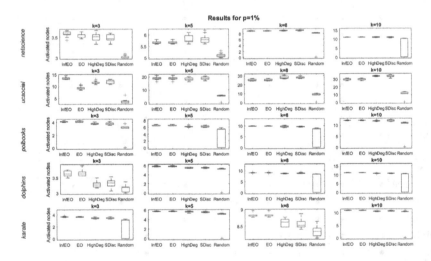

**Fig. 1.** Number of active nodes reported by the five methods for different values of initial set size $k$ and propagation probability $p = 1\%$, 10 independent runs.

We find that InfEO reports results better or as good as the other methods in 36 out of the 40 cases tested. In 8 cases InfEO results are significantly better than those reported by the standard EO. Compared with other methods, InfEO performs best for $k = 3$, indicating its capability to identify influential nodes that are different from those having the highest degrees. However, in terms of computational time, a run for the smaller networks takes approximately 60 min while for the 'Ucsocial' network about 9 h on a Desktop computer, i7 CPU, 3.30 GHz, 32 GB RAM, but using an un-optimized code.

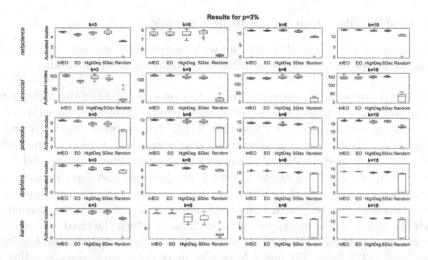

**Fig. 2.** Number of active nodes reported by the five methods for different values of initial set size $k$ and propagation probability $p = 3\%$, 10 independent runs.

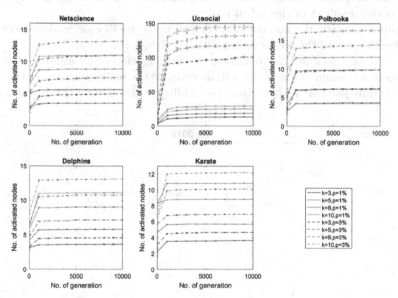

**Fig. 3.** InfEO - Number of activated nodes in 10000 generations, for $k = 3, 5, 8, 10$ and $p = 1\%, 3\%$.

Figure 3 presents the evolution of average number of activated nodes for the five benchmarks, 10 independent runs. The number of activated nodes increases fast at the beginning of the search and stagnates afterwards. This indicates that few iterations are enough to get a good seed set and also that results are consistent with a small standard error in 10 independent runs.

## 4.2  Highly Cited Journal Citation Network

In this section we construct a new network and analyze it from the influence maximization point of view using InfEO and HighDeg/SDisc methods.

*Data.* The network was constructed using the reference data of the *highly cited* articles from the domain of Computer Science listed on Web of Science (https:// apps.webofknowledge.com/) for the year 2018. Articles were selected using a search query specifying the category (computer science), and the year; the *highly cited* label is assigned by Web of Science to most cited articles in each field[3].

In our approach nodes of the networks are journals and links indicate the relation between the journals in which the citing and cited articles have appeared. The network is directed, with links from nodes that represent the journal in which the citing article has appeared to journals in which the cited (referenced) articles have appeared. The network was constructed from 606 articles, has 7482 nodes, and 14479 links. 131 nodes have a positive out-degree, corresponding to 131 distinct journals in the article list. This particularity makes the network interesting to study as any influence maximization method should report as seeder nodes a subset of this set of nodes.

Figure 4 presents the number of activated nodes in 10 independent runs and for different sizes of the seeder nodes set: $k = 3, 5, 8, 10$.

An ICM in this case may illustrate the propagation of information through citations in a domain: a researcher will investigate the references of an article and find a certain journal; with a given probability will continue the investigation

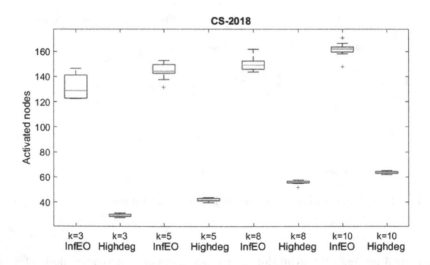

**Fig. 4.** Average number of activated nodes for the CS network, for $k = 3, 5, 8, 10$ as the size of the initial activated nodes

---

[3] https://images.webofknowledge.com/images/help/WOS/hs_citation_applications. html.

**Table 3.** Journals more frequent reported by InfEO for each $k$ value. An * indicates that the journal has appeared more than 5 times in the 10 runs. A ^ indicates that the journal has also appeared in corresponding HighDeg/SDisc results.

| No | Journal/venue name | $k = 3$ | $k = 5$ | $k = 8$ | $k = 10$ |
|----|--------------------|---------|---------|---------|----------|
| 1  | INT J of Inf and Coding Theory | * | - | - | - |
| 2  | P 31 ANN ACM S APPL | *^ | *^ | *^ | *^ |
| 3  | INT J IMAG SYST TECH | * | - | * | * |
| 4  | P 8 INT C MOB SYST A | - | * | - | *^ |

from this journal/cited article to reach/discover a new journal. The list of most influential nodes in this network represents a list of journals that are found to have a larger set of journals that are reached through such a process. For each tested $k$ value a list of the journals/venues reported by InfEO more frequent is presented in Table 3. We find that one node, line 2 of Table 3, appears in all results, ranking the third in descending order of degree in the network. Journals on lines 3 and 4 have a high degree, but not the highest, which, considering results in Fig. 4, indicates that InfEO is capable of identifying interesting influential nodes of a network.

## 5 Conclusions

This paper presents an algorithm based on Extremal Optimization - InfEO - adapted for the influence maximization problem for the independent cascade model, which is a NP-hard problem. InfEO maximizes the marginal contribution of each node in the seeder set to the total size of the cascade. The method is compared with two high-degree based methods (SDisc and HighDeg), a standard version of the EO and a simple random ICM with results showing the effectiveness of the proposed method.

As an application we constructed a highly cited journal citation network. InfEO algorithm provided in several runs stable solutions, offering a list of influential nodes that can be further used in decision making processes. Further work will address the improvement, as well as the adaptation of InfEO for other types of networks and diffusion models.

**Acknowledgement.** This work was supported by a grant of the Romanian National Authority for Scientific Research and Innovation, CNCS - UEFISCDI, project number PN-III-P1-1.1-TE-2016-1933.

## References

1. Berry, G., Cameron, C.J., Park, P., Macy, M.: The opacity problem in social contagion. Soc. Netw. **56**, 93–101 (2019). https://doi.org/10.1016/j.socnet.2018.09.001

2. Boettcher, S., Percus, A.G.: Optimization with extremal dynamics. Complexity **8**(2), 57–62 (2002)
3. Bucur, D.: Influence Maximization in Social Networks with Genetic Algorithms Influence Maximization in Social Networks with Genetic Algorithms, March 2016. https://doi.org/10.1007/978-3-319-31204-0
4. Bucur, D., Iacca, G., Marcelli, A., Squillero, G.: Multi-objective Evolutionary Algorithms for Influence Maximization in Social Networks Multi-Objective Evolutionary Algorithms for Influence Maximization in Social Networks, August 2017. https://doi.org/10.1007/978-3-319-55849-3
5. Chen, W., Wang, Y., Yang, S.: Efficient influence maximization in social networks. In: Proceedings of the 15th ACM SIGKDD International Conference on Knowledge Discovery and Data Mining, KDD 2009, pp. 199–208. ACM, New York (2009). https://doi.org/10.1145/1557019.1557047. http://doi.acm.org/10.1145/1557019.1557047
6. Domingos, P., Richardson, M.: Mining the network value of customers. In: Proceedings of the Seventh ACM SIGKDD International Conference on Knowledge Discovery and Data Mining, pp. 57–66. ACM (2001)
7. Fan, X., Li, V.O.: The probabilistic maximum coverage problem in social networks. In: GLOBECOM - IEEE Global Telecommunications Conference (2011). https://doi.org/10.1109/GLOCOM.2011.6133985
8. Gao, J., Ghasemiesfeh, G., Schoenebeck, G., Yu, F.Y.: General threshold model for social cascades: analysis and simulations. In: EC 2016 - Proceedings of the 2016 ACM Conference on Economics and Computation, pp. 617–634 (2016). https://doi.org/10.1145/2940716.2940778
9. Gong, M., Yan, J., Shen, B., Ma, L., Cai, Q.: Influence maximization in social networks based on discrete particle swarm optimization. Inf. Sci. **367–368**, 600–614 (2016). https://doi.org/10.1016/j.ins.2016.07.012
10. Kempe, D., Kleinberg, J., Tardos, É.: Maximizing the spread of influence through a social network. In: Proceedings of the Ninth ACM SIGKDD International Conference on Knowledge Discovery and Data Mining, pp. 137–146. ACM (2003)
11. Kempe, D., Kleinberg, J., Tardos, É.: Maximizing the spread of influence through a social network, p. 137 (2004). https://doi.org/10.1145/956750.956769
12. Kuo, T.T., Hung, S.C., Lin, W.S., Lin, S.D., Peng, T.C., Shih, C.C.: Assessing the quality of diffusion models using real-world social network data. In: Proceedings - 2011 Conference on Technologies and Applications of Artificial Intelligence, TAAI 2011, pp. 200–205 (2011). https://doi.org/10.1109/TAAI.2011.42
13. Leung, T., Chung, F.L.: Persuasion driven influence propagation in social networks, pp. 548–554 (2014). https://doi.org/10.1109/ASONAM.2014.6921640. cited By 5
14. Liu, W., Chen, X., Jeon, B., Chen, L., Chen, B.: Influence maximization on signed networks under independent cascade model. Appl. Intell. **49**(3), 912–928 (2019). https://doi.org/10.1007/s10489-018-1303-2
15. Lusseau, D., Schneider, K., Boisseau, O.J., Haase, P., Slooten, E., Dawson, S.M.: The bottlenose dolphin community of doubtful sound features a large proportion of long-lasting associations. Behav. Ecol. Sociobiol. **54**(4), 396–405 (2003)
16. Newman, M.: Networks. Oxford University Press, New York (2018)
17. Newman, M.E.: Finding community structure in networks using the eigenvectors of matrices. Phys. Rev. E **74**(3), 036104 (2006)
18. Panzarasa, P., Opsahl, T., Carley, K.M.: Patterns and dynamics of users' behavior and interaction: network analysis of an online community. J. Am. Soc. Inform. Sci. Technol. **60**(5), 911–932 (2009)

19. Sumith, N., Annappa, B., Bhattacharya, S.: A holistic approach to influence maximization in social networks: STORIE. Appl. Soft Comput. J. **66**, 533–547 (2018). https://doi.org/10.1016/j.asoc.2017.12.025
20. Tang, J., et al.: Maximizing the spread of influence via the collective intelligence of discrete bat algorithm. Knowl.-Based Syst. **160**, 88–103 (2018). https://doi.org/10.1016/j.knosys.2018.06.013
21. Tong, G., et al.: Adaptive influence maximization in dynamic social networks. IEEE/ACM Trans. Networking **25**(1), 112–125 (2017). https://doi.org/10.1109/TNET.2016.2563397
22. Zachary, W.W.: An information flow model for conflict and fission in small groups. J. Anthropol. Res. **33**(4), 452–473 (1977)
23. Zhang, K., Du, H., Feldman, M.W.: Maximizing influence in a social network: improved results using a genetic algorithm. Physica A (2017). https://doi.org/10.1016/j.physa.2017.02.067
24. Zhang, L., Luo, M., Boncella, R.J.: Product information diffusion in a social network. Electron. Commer. Res. (2018). https://doi.org/10.1007/s10660-018-9316-9

# Data Mining Applications

Data Mining Applications

# Forecast Daily Air-Pollution Time Series with Deep Learning

Miguel Cárdenas-Montes[✉] [ID]

Centro de Investigaciones Energéticas Medioambientales y Tecnológicas,
Madrid, Spain
miguel.cardenas@ciemat.es

**Abstract.** Air-quality in urban areas is one of the most critical concern for governs. Wide spectrum measures are implemented in relation to this issue, from laws and promotion of renewal of heating and transport systems, to stablish monitoring and prediction systems. When air-pollutant levels excess from healthy thresholds, traffic limitations are activated with non-negligible nuisances, and social and economic impacts. For this reason, high-pollution episodes must be appropriately anticipated. In this work, deep learning-based implementations are evaluated for forecasting daily values of three pollutants: $CO$, $NO_2$, and $O_3$, at three types of monitoring station from the air-quality time series provided by Madrid City Council. In this analysis, the influence of working-non-working days and the use of multivariant input, composed of multiple-pollutants time series, is also evaluated. As a consequence, a rank of the most suitable algorithms for forecasting air-quality time series is stated.

**Keywords:** Time series analysis · Deep learning · Forecasting ·
Air quality · Recurrent Neural Networks

## 1 Introduction

Nowadays all the activities related with the Big Data ecosystem are generating a growing interest in industry and science. Larger data volumes must be faster analysed with larger efficiencies for extracting valuable information. Time Series modelling and forecasting, as part of the Big Data ecosystem, are forced to timely process larger data volumes offering accurate predictions for the future values of the observables.

Air pollution is one of the most critical health issue in urban areas. The scientific literature shows its relation with the population health [2,7,13]. For this reason monitoring systems have been deployed in the cities and outskirts, at the same time that numerous laws and protocols have been implemented for reducing the level of certain chemical components: $NO_2$, $O_3$, $CO$, benzene and others.

Depending on the level threshold of the these components, grading protocol scenarios are proposed for mitigating the adverse level. These scenarios range

© Springer Nature Switzerland AG 2019
H. Pérez García et al. (Eds.): HAIS 2019, LNAI 11734, pp. 431–443, 2019.
https://doi.org/10.1007/978-3-030-29859-3_37

from the speed limit reduction in the accesses to the city centre and promotion of the public transport, to the prohibition of use of public parking in the lowest levels; to partial or total access restriction depending on the plate numbers or the year of matriculation in the most adverse scenarios.

It has been demonstrated that the efficiency of these mitigation measurements largely varies [4]. Only the most restrictive ones have shown a relevant impact on the improvement of the air quality. Therefore, the prediction of the values of pollutants of these adverse episodes could help in the decision-making of the most efficient protocols and the handling of the mitigating actions. Taking into account that those actions, which include traffic access restriction to the city centres or the suspension of motor-vehicle traffic, have strong economic implication, the social and economic impact of the research proposed in the current paper is underlined.

In this work, mean daily values of $NO_2$, $O_3$, $CO$ for the period from 1-1-2010 to 31-5-2018 for three monitoring stations in Madrid are used [1]. The monitoring stations are *Arturo Soria*, *Fernández Ladreda*, and *Casa de Campo*. In relation to the placement of these monitoring stations, they are labelled in different categories: *Casa de Campo* is a suburban station (term for stations in parks in urban areas), *Fernández Ladreda* is a traffic station (term for stations affected by traffic and close to a principal street or road), and finally *Arturo Soria* is an urban background station affected by both traffic and background pollution, mainly due to heating. The type of station (suburban, traffic or background) points out also the pollutant with more impact on the area: $O_3$ for suburban, $CO$ for background stations, and $NO_2$ traffic ones.

Multilayer Perceptron [3], and Deep Learning algorithms: Convolutional Neural Networks (CNN) [10,12] and Bidirectional Recurrent Neural Networks (BRNN) [10,15] are the architectures used in this work for forecasting pollutant concentrations. Keras [5] has been used for implementing the MLP, CNN, and BRNN architectures. MLP is configured with two hidden layer—dense ones— with 32 and 16 neurons, and the output layer with a single neuron. BRNN is composed of one single layer—bidirectional Long Short-Term Memory LSTM (LSTM)—with 50 neurons, plus a dense layer of 100 neurons and the same previous output layer [11]. These networks are trained with 50 epochs. Finally, CNN is configured with two convolutional blocks of 32 kernels of size 5, and 64 kernels of size 3, each one with a pooling layer of size 2, plus a dense layer of 100 neurons, and finally the output layer. It is trained with 25 epochs. In all cases the test size is composed 1000 examples, created using a loopback of 30 days. For reducing the processing time and the risk of overfitting, and a batch size of 100 has been used.

The rest of the paper is organised as follows: The models comparison and the results obtained are presented and analysed in Sect. 2. Finally, Sect. 3 contains the conclusions of this work.

## 2    Experimental Results and Models Comparison

### 2.1    Univariant Implementations

In Figs. 1, 2 and 3, the Mean Squared Error (MSE) for 25 independent runs
of univariant implementations of MLP, CNN and BRNN are shown. The mean
value and standard deviation of the MSE are presented in Table 1. As can be
appreciated MLP and BRNN perform similarly, whereas both clearly outperform
CNN. Except for *Casa de Campo* station, BRNN achieves a lower mean MSE
than MLP (Table 1).

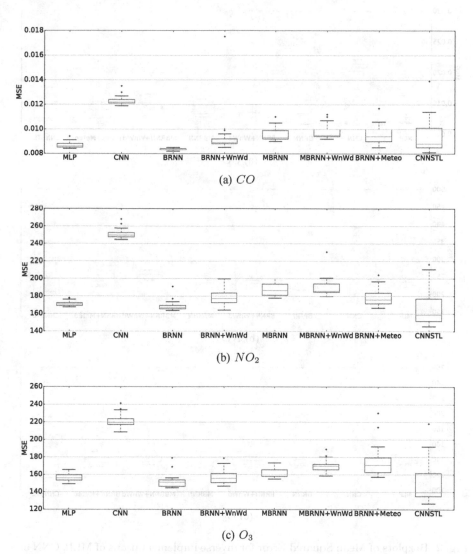

(a) $CO$

(b) $NO_2$

(c) $O_3$

**Fig. 1.** Boxplots of Mean Squared Error for diverse implementations of MLP, CNN and
BRNN when forecasting the pollutants $CO$, $NO_2$ and $O_3$ for the monitoring station
*Arturo Soria*.

The application of the Wilcoxon signed-rank test [8,9] to the values of the MSE of MLP and BRNN implementations points that the differences between the median of the MSE are significant for a confidence level of 95% (p-value under 0.05), which means that the differences are unlikely to have occurred by chance with a probability of 95%, for the three pollutants in the *Arturo Soria* station; is not significant for $NO_2$ nor $O_3$ in the *Casa de Campo* station, whereas it is significant for $CO$; and finally for *Fernández Ladreda* station are significant only for $NO_2$ and $O_3$ pollutant, whereas is not significant for $CO$ pollutant.

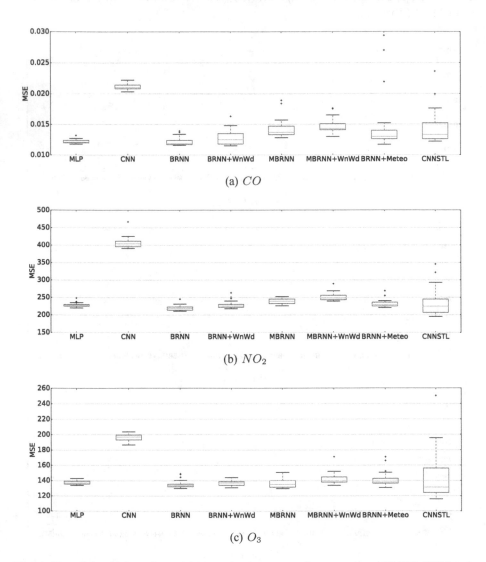

(a) $CO$

(b) $NO_2$

(c) $O_3$

**Fig. 2.** Boxplots of Mean Squared Error for diverse implementations of MLP, CNN and BRNN when forecasting the pollutants $CO$, $NO_2$ and $O_3$ for the monitoring station *Fernandez Ladreda*.

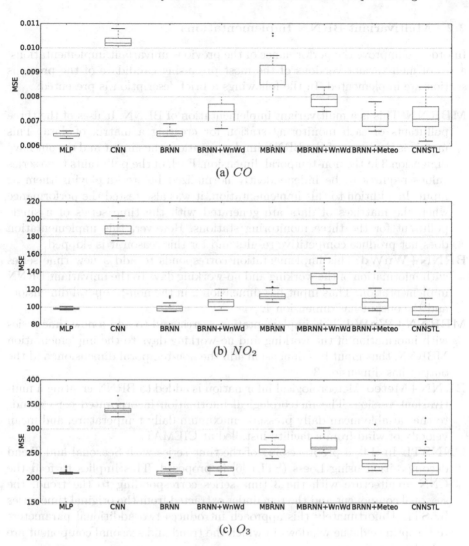

(a) $CO$

(b) $NO_2$

(c) $O_3$

**Fig. 3.** Boxplots of Mean Squared Error for diverse implementations of MLP, CNN and BRNN when forecasting the pollutants $CO$, $NO_2$ and $O_3$ for the monitoring station *Casa de Campo*.

In Figs. 4, 5 and 6, the original data and the prediction of BRNN architecture for the last 1000 days (test set) of the time series are visualized. As observed, predictions reproduce acceptable well the values of the time series. They overlap the original data in the test set. In most of the cases the general behaviour of the time series is acceptably reproduced.

## 2.2   Multivariant BRNN Implementations

In order to improve the performance of the previous univariant implementations, a set of multivariant versions of the most promising candidate of the previous section are implemented. In the followings a brief description is presented:

**MBRNN** This is a multivariant implementation of BRNN. It uses of the three pollutants of each monitoring station for creating a matrix of data. This matrix is the input of the BRNN implementation, thus input and output have dimension 3 in the non-temporal dimension. Each of the pollutants time series values requires to be independently normalized before employing them as input. In addition to this implementation, it was also tested the performance when the matrixes of data are generated with the time series of a single pollutant for the three monitoring stations. However, this implementation does not produce competitive results and for this reason it is skipped.

**BRNN+WnWd** This implementation corresponds to add a new time series with information of the working and no-working days to the univariant BRNN implementation. Thus input has dimension 2 in the non-temporal dimension, and the output has dimension 1.

**MBRNN+WnWd** This implementation correspond to add a new time series with information of the working and no-working days to the implementation MBRNN, thus input has dimension 4 in the non-temporal dimension, and the output has dimension 3.

**BRNN+Meteo** Meteorological information is added to BRNN creating a multivariant version. The meteorological information incorporated corresponds to the variable mean daily pressure, maximum daily temperature, and mean velocity of wind from a facility installed at CIEMAT.

**CNNSTL** In [14], a preprocessing of the time series with Seasonal and Trend decomposition using Loess (STL) [6] is proposed. This implies to feed the CNN architecture with the 3 time series corresponding to the trend, the seasonal component and the remainder extracted from the original time series by STL. Unfortunately this approach introduces two additional parameters to be optimized: the windows in which the trend and seasonal component are calculated.

In Figs. 1, 2 and 3, the boxplots of the MSE of BRNN and the previously described multivariant implementations are shown. As can be observed, MBRNN, UBRNN+WnWd and MBRNN+WnWd are clearly outperformed by BRNN. Thus, additional information introduced as input, multiple pollutant and working-no-working days time series, does not seem improve the forecasting capacity of BRNN.

With regard to CNNSTL, this approach produces some competitive results in the *Arturo Soria* time series for $NO_2$ and $O_3$, and $NO_2$ in *Casa de Campo*, but with the penalty of large increment in the standard deviation of the MSE achieved (Table 2). Although promising, the performance of CNNSTL is not competitive.

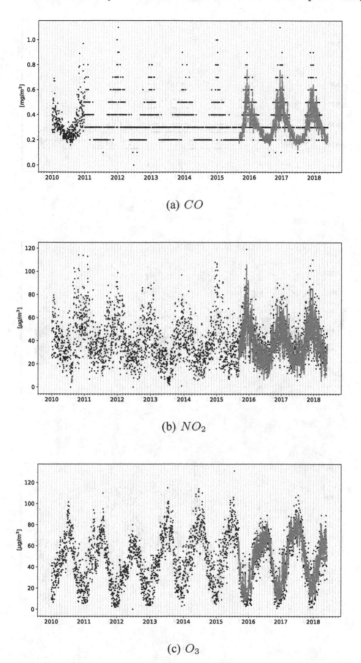

(a) $CO$

(b) $NO_2$

(c) $O_3$

**Fig. 4.** Daily air-quality data sets, expanding from 01/01/2010 to 31/05/2018, and the forecasting for the 1000 last days based on BRNN implementation. Datasets correspond to three pollutants ($CO$, $NO_2$ and $O_3$) and monitoring station *Arturo Soria*.

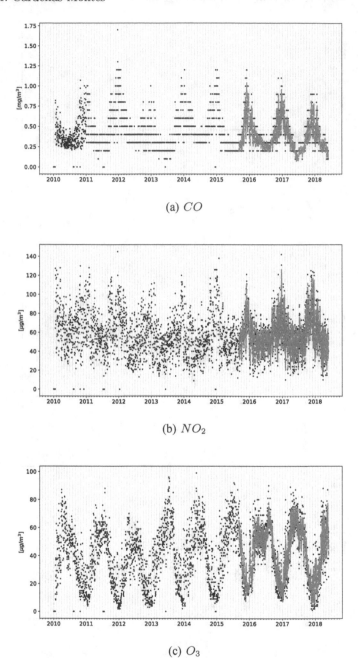

(a) $CO$

(b) $NO_2$

(c) $O_3$

**Fig. 5.** Daily air-quality data sets, expanding from 01/01/2010 to 31/05/2018, and the forecasting for the 1000 last days based on BRNN implementation. Datasets correspond to three pollutants ($CO$, $NO_2$ and $O_3$) and monitoring station *Fernández Ladreda*.

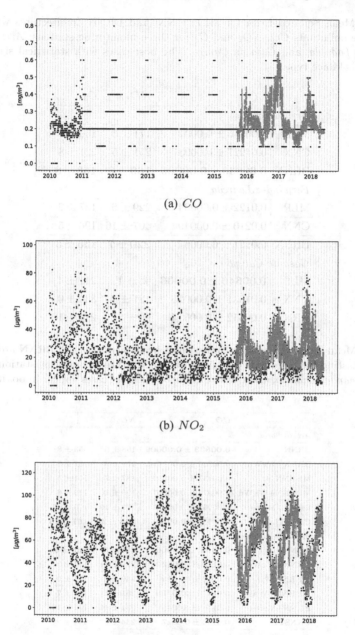

(a) $CO$

(b) $NO_2$

(c) $O_3$

**Fig. 6.** Daily air-quality data sets, expanding from 01/01/2010 to 31/05/2018, and the forecasting for the 1000 last days based on BRNN implementation. Datasets correspond to three pollutants ($CO$, $NO_2$ and $O_3$) and monitoring station *Casa de Campo*.

**Table 1.** Mean Squared Error for MLP, CNN, and BRNN architectures when forecasting the pollutants $CO$, $NO_2$ and $O_3$ for three monitoring stations: *Arturo Soria*, *Fernández Ladreda*, and *Casa de Campo*. The best cases with statistical significance appear in boldface type.

|  | $CO$ | $NO_2$ | $O_3$ |
|---|---|---|---|
| *Arturo Soria* | | | |
| MLP | $0.00866 \pm 0.00024$ | $171 \pm 3$ | $156 \pm 4$ |
| CNN | $0.01227 \pm 0.00037$ | $251 \pm 5$ | $221 \pm 8$ |
| BRNN | $\mathbf{0.00833 \pm 0.00009}$ | $\mathbf{169 \pm 6}$ | $\mathbf{153 \pm 8}$ |
| *Fernández Ladreda* | | | |
| MLP | $0.0122 \pm 0.00034$ | $229 \pm 6$ | $137 \pm 2$ |
| CNN | $0.0210 \pm 0.00049$ | $407 \pm 16$ | $196 \pm 5$ |
| BRNN | $0.0121 \pm 0.00064$ | $\mathbf{220 \pm 7}$ | $\mathbf{135 \pm 5}$ |
| *Casa de Campo* | | | |
| MLP | $\mathbf{0.00646 \pm 0.00006}$ | $98 \pm 1$ | $211 \pm 4$ |
| CNN | $0.01031 \pm 0.00025$ | $187 \pm 7$ | $339 \pm 9$ |
| BRNN | $0.00652 \pm 0.00014$ | $99 \pm 5$ | $213 \pm 10$ |

**Table 2.** Mean Squared Error for mutivariant implementations of BRNN architecture when forecasting the pollutants $CO$, $NO_2$ and $O_3$ for three monitoring stations: *Arturo Soria*, *Fernández Ladreda*, and *Casa de Campo*. The best cases appear in boldface type.

|  | $CO$ | $NO_2$ | $O_3$ |
|---|---|---|---|
| *Arturo Soria* | | | |
| BRNN | $\mathbf{0.00833 \pm 0.00009}$ | $168 \pm 6$ | $153 \pm 8$ |
| BRNN+WnWd | $0.00935 \pm 0.00171$ | $179 \pm 9$ | $157 \pm 8$ |
| MBRNN | $0.00957 \pm 0.00050$ | $187 \pm 7$ | $161 \pm 5$ |
| MBRNN+WnWd | $0.00977 \pm 0.00057$ | $190 \pm 10$ | $170 \pm 6$ |
| BRNN+Meteo | $0.00958 \pm 0.00073$ | $179 \pm 10$ | $175 \pm 17$ |
| CNNSTL | $0.00935 \pm 0.00129$ | $\mathbf{166 \pm 20}$ | $\mathbf{151 \pm 23}$ |
| *Fernández Ladreda* | | | |
| BRNN | $\mathbf{0.0122 \pm 0.0006}$ | $\mathbf{220 \pm 7}$ | $\mathbf{135 \pm 5}$ |
| BRNN+WnWd | $0.0128 \pm 0.0012$ | $229 \pm 11$ | $137 \pm 3$ |
| MBRNN | $0.0142 \pm 0.0015$ | $240 \pm 8$ | $136 \pm 6$ |
| MBRNN+WnWd | $0.0147 \pm 0.0011$ | $251 \pm 11$ | $142 \pm 7$ |
| BRNN+Meteo | $0.0146 \pm 0.0045$ | $232 \pm 10$ | $141 \pm 10$ |
| CNNSTL | $0.0144 \pm 0.0027$ | $234 \pm 37$ | $144 \pm 29$ |
| *Casa de Campo* | | | |
| BRNN | $\mathbf{0.00652 \pm 0.00014}$ | $99 \pm 5$ | $\mathbf{213 \pm 10}$ |
| BRNN+WnWd | $0.00734 \pm 0.00035$ | $105 \pm 6$ | $220 \pm 9$ |
| MBRNN | $0.00897 \pm 0.00064$ | $114 \pm 5$ | $263 \pm 15$ |
| MBRNN+WnWd | $0.00787 \pm 0.00031$ | $134 \pm 8$ | $271 \pm 15$ |
| BRNN+Meteo | $0.00705 \pm 0.00030$ | $107 \pm 10$ | $235 \pm 18$ |
| CNNSTL | $0.00725 \pm 0.00062$ | $\mathbf{97 \pm 7}$ | $224 \pm 29$ |

Considering the execution time of the most promising implementations, MLP, CNN and CNNSTL architectures take around 8 s, whereas BRNN implementations takes around 330 s. For real scenarios, where accuracy in prediction and speed would be equally valuable, the processing times will become relevant.

### 2.3 Forecasting

In order to evaluate the forecasting capacity of the best previous implementation, they are evaluated by forecasting and comparing with the concentration of $CO$, $NO_2$, and $O_3$ for the following three, five and seven days. For the sake of the brevity, only the evaluation corresponding to the next three days is shown (Table 3). In this table the mean value and standard deviation of the selected implementation are compared with the real concentration achieved these days, from firth to third of June 2018.

**Table 3.** Mean values and standard deviation of 25 independent runs for forecasting 3 ahead days. Due to precision reasons, $CO$ concentration forecasting are shown without standard deviation.

|  |  | BRNN | ANN |
|---|---|---|---|
| *Arturo Soria* | | | |
| $CO$ | (**0.3, 0.2, 0.2**) | (0.3, 0.2, 0.2) | (0.3, 0.2, 0.2) |
| $NO_2$ | (**43, 16, 23**) | $(29 \pm 2, 21 \pm 3, 23 \pm 3)$ | $(31 \pm 1, 27 \pm 1, 24 \pm 2)$ |
| $O_3$ | (**43, 73, 56**) | $(55 \pm 1, 57.1 \pm 2, 61 \pm 2)$ | $(54 \pm 2, 57 \pm 2, 59 \pm 2)$ |
| *Fernandez Ladreda* | | | |
| $CO$ | (**0.2, 0.1, 0.1**) | (0.2, 0.2, 0.2) | (0.2, 0.2, 0.2) |
| $NO_2$ | (**52, 34, 36**) | $(38 \pm 3, 31 \pm 6, 41 \pm 7)$ | $(42 \pm 1, 38 \pm 2, 39 \pm 3)$ |
| $O_3$ | (**39, 73, 47**) | $(55 \pm 2, 53 \pm 3, 53 \pm 4)$ | $(51 \pm 2, 53 \pm 3, 56 \pm 3)$ |
| *Casa de Campo* | | | |
| $CO$ | (**0.2, 0.2, 0.2**) | (0.2, 0.2, 0.2) | (0.2, 0.2, 0.2) |
| $NO_2$ | (**19, 8, 10**) | $(7 \pm 2, 7 \pm 3, 11 \pm 4)$ | $(9 \pm 1, 11 \pm 1, 12 \pm 1)$ |
| $O_3$ | (**59, 87, 61**) | $(78 \pm 2, 75 \pm 4, 73 \pm 4)$ | $(73 \pm 2, 74 \pm 3, 72 \pm 4)$ |

As can be appreciated, ANN and BRNN show a remarkable forecasting capacity, specially taking into account the range of the pollutant concentrations. Rawly $CO$ is the easiest to forecast, whereas $O_3$ and $NO_2$ are more difficult.

## 3  Conclusions

In this work, an evaluation of the most suitable deep learning algorithm for forecasting the concentration of the main pollutant in three monitoring stations from the Madrid air-quality monitoring network has been undertaken. The evaluation expands for the time series of three pollutants: $CO$, $NO_2$, and $O_3$; and more than 8 years; whereas from the algorithm side, MLP, CNN and BRNN have been included in the evaluation. They have been tested with univariant time series (a single pollutant) as input, as well as with multivariant inputs: multiple pollutant and pollutants plus working-no-working days information and meteorological information.

Finally, the evaluation has shown the excellent performance of Bidirectional Recurrent Neural Networks with univariant inputs. In some tests, additional improvements are achieved when using as input for CNN the components of STL decomposition.

**Acknowledgment.** The research leading to these results has received funding by the Spanish Ministry of Economy and Competitiveness (MINECO) for funding support through the grant FPA2016-80994-C2-1-R, and "Unidad de Excelencia María de Maeztu": CIEMAT - FÍSICA DE PARTÍCULAS through the grant MDM-2015-0509. Author would like to thank Dra. Begoña Artíñano Rodríguez de Torres Head of the Unit of Atmospheric Pollution and POP Characterization and Elías Díaz Ramiro of the Unit for Characterization of Atmospheric Pollution of CIEMAT for useful comments and the meteorological data provided for this work.

# References

1. Open data Madrid, August 2018. https://datos.madrid.es/portal/site/egob
2. Alberdi Odriozola, J.C., Díaz Jiménez, J., Montero Rubio, J.C., Mirón Pérez, I.J., Pajares Ortíz, M.S., Ribera Rodrigues, P.: Air pollution and mortality in Madrid, Spain: a time-series analysis. Int. Arch. Occup. Environ. Health **71**(8), 543–549 (1998). https://doi.org/10.1007/s004200050321
3. Bishop, C.M.: Neural Networks for Pattern Recognition. Oxford University Press Inc., New York (1995)
4. Borge, R., et al.: Application of a short term air quality action plan in Madrid (Spain) under a high-pollution episode - part I: Diagnostic and analysis from observations. Sci. Total Environ. **635**, 1561–1573 (2018). https://doi.org/10.1016/j.scitotenv.2018.03.149
5. Chollet, F., et al.: Keras (2015). https://github.com/fchollet/keras
6. Cleveland, R.B., Cleveland, W.S., McRae, J., Terpenning, I.: STL: a seasonal-trend decomposition procedure based on loess. J. Off. Statist. **6**, 3–73 (1990)
7. Díaz, J., et al.: Modeling of air pollution and its relationship with mortality and morbidity in Madrid, Spain. Int. Arch. Occup. Environ. Health **72**(6), 366–376 (1999). https://doi.org/10.1007/s004200050388
8. García, S., Fernández, A., Luengo, J., Herrera, F.: A study of statistical techniques and performance measures for genetics-based machine learning: accuracy and interpretability. Soft. Comput. **13**(10), 959–977 (2009)
9. García, S., Molina, D., Lozano, M., Herrera, F.: A study on the use of non-parametric tests for analyzing the evolutionary algorithms' behaviour: a case study on the CEC'2005 special session on real parameter optimization. J. Heuristics **15**(6), 617–644 (2009)
10. Goodfellow, I., Bengio, Y., Courville, A.: Deep Learning. MIT Press (2016). http://www.deeplearningbook.org

11. Hochreiter, S., Schmidhuber, J.: Long short-term memory. Neural Comput. **9**(8), 1735–1780 (1997). https://doi.org/10.1162/neco.1997.9.8.1735
12. LeCun, Y.: Generalization and network design strategies. Technical report. University of Toronto (1989)
13. Linares, C., Díaz, J., Tobías, A., Miguel, J.M.D., Otero, A.: Impact of urban air pollutants and noise levels over daily hospital admissions in children in Madrid: a time series analysis. Int. Arch. Occup. Environ. Health **79**(2), 143–152 (2006). https://doi.org/10.1007/s00420-005-0032-0
14. Méndez-Jiménez, I., Cárdenas-Montes, M.: Time series decomposition for improving the forecasting performance of convolutional neural networks. In: Herrera, F., et al. (eds.) CAEPIA 2018. LNCS (LNAI), vol. 11160, pp. 87–97. Springer, Cham (2018). https://doi.org/10.1007/978-3-030-00374-6_9
15. Schuster, M., Paliwal, K.: Bidirectional recurrent neural networks. Trans. Sig. Proc. **45**(11), 2673–2681 (1997). https://doi.org/10.1109/78.650093

# Botnet Detection on TCP Traffic Using Supervised Machine Learning

Javier Velasco-Mata[1,2]([✉]), Eduardo Fidalgo[1,2], Víctor González-Castro[1,2],
Enrique Alegre[1,2], and Pablo Blanco-Medina[1,2]

[1] Department of Electrical, Systems and Automation Engineering,
Universidad de León, León, Spain
{jvelm,eduardo.fidalgo,victor.gonzalez,enrique.alegre,pblanm}@unileon.es
[2] Researcher at INCIBE (Spanish National Cybersecurity Institute), León, Spain

**Abstract.** The increase of botnet presence on the Internet has made it necessary to detect their activity in order to prevent them to attack and spread over the Internet. The main methods to detect botnets are traffic classifiers and sinkhole servers, which are special servers designed as a trap for botnets. However, sinkholes also receive non-malicious automatic online traffic and therefore they also need to use traffic classifiers. For these reasons, we have created two new datasets to evaluate classifiers: the TCP-Int dataset, built from publicly available TCP Internet traces of normal traffic and of three botnets, Kelihos, Miuref and Sality; and the TCP-Sink dataset based on traffic from a private sinkhole server with traces of the Conficker botnet and of automatic normal traffic. We used the two datasets to test four well-known Machine Learning classifiers: Decision Tree, k-Nearest Neighbours, Support Vector Machine and Naïve Bayes. On the TCP-Int dataset, we used the F1 score to measure the capability to identify the type of traffic, i.e., if the trace is normal or from one of the three considered botnets, while on the TCP-Sink we used ROC curves and the corresponding AUC score since it only presents two classes: non-malicious or botnet traffic. The best performance was achieved by Decision Tree, with a 0.99 F1 score and a 0.99 AUC score on the TCP-Int and the TCP-Sink datasets respectively.

## 1 Introduction

The development of the Internet as a common resource has caused a constant grow of the number of cyberthreats. Therefore, the need of automatic solutions to supply the lack of human resources to detect these threats has fomented the use of Machine Learning (ML) algorithms in cybersecurity [1]. One of the most hazardous menaces are *botnets*.

A botnet is a network of malware-infected computer-like devices called *bots*, which are controlled by a remote user, known as *botmaster*. Botnets are used to perform different types of cyber-attacks such as DDoS (Distributed Denial of Service) to targeted servers or massive credential collection from online users. For this reason, botnets have become one of the most remarkable security concerns

© Springer Nature Switzerland AG 2019
H. Pérez García et al. (Eds.): HAIS 2019, LNAI 11734, pp. 444–455, 2019.
https://doi.org/10.1007/978-3-030-29859-3_38

in the last years [2,3]. Depending on the structure of the network [2], botnets can be classified as:

- **Centralized:** In this structure all the infected devices communicate directly with the botmaster.
- **Distributed:** In this case, the botnet use peer-to-peer protocols for intra-communication and the botmaster only connects to a small subset of botnets to transmit the orders.
- **Hybrid:** This type of network is divided into subsets of bots. Each subset follows a centralized hierarchy where all the *client* bots communicate with a *server* bot. On a higher level, the server bots communicate with each other and with the botmaster using a distributed architecture.

Botnets need to receive orders from online connections, and the generated traffic can be exploited to detect them. Early solutions used a signature-based approach [4] and thus analysed plain network communications in search of known malicious patterns, but they can be bypassed using obfuscation or encrypted protocols. The supervised ML based detectors can overcome this drawback and provide a more efficient solution. A trained ML model is capable of classifying the network traffic according to general traffic characteristics instead of looking for specific signatures [5].

Beside of traffic classifiers, botnets also can be detected using *sinkhole servers*. A sinkhole is a server designed to deceive botnets by disguising as a real botnet controller server, and thus it collects the attempts of bots to communicate with the botmaster [6]. However, sinkholes also receive non-malicious connections from automatic programs like web crawlers or *spiders*, i.e., programs that systematically explore the Internet usually for web indexing [7]. Because of this, it is also necessary to apply network classifiers to the traffic received by sinkholes servers to separate non-malicious traces from the botnet ones.

This work explores the automatic classification of TCP network traffic, both in regular Internet communications and in sinkhole traffic. Besides, we have not found another work on botnet detection in sinkhole traffic focused on the TCP protocol.

We present two main contributions. First, we constructed two new datasets from TCP network traffic: one from public Internet traffic captures and another from private sinkhole data. And second, we tested on the two datasets four well-known ML classifiers, namely Decision Tree (DT), k-Nearest Neighbours (k-NN), Support Vector Machine (SVM) and Naïve Bayes (NB), using a scheme applicable to the three types of botnet architectures we previously described. The general scheme of this work is illustrated on Fig. 1.

The rest of the paper is organized as follows. Section 2 presents the literature review about botnet detection and classification. In Sect. 3 we present the datasets we made for this work, and we describe the features used to model the traffic samples, as well as the tested algorithms and the performance metrics we have used. The experimental settings are described on Sect. 4, and the results are discussed on Sect. 5. Finally, our conclusions and future work are on Sect. 6.

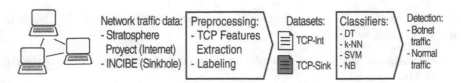

**Fig. 1.** Methodology followed in this work

## 2   Related Works

Over the last years, different works have proposed various ML algorithms to detect botnets with varying results. These algorithms have been compared in a variety of network traffic classification problems, such as detection of anomalies, network intrusions or botnet detection.

For example, the experiments of Kirubavathi and Anitha [5] used a mixture of the ISOT botnet dataset[1] and a dataset of private laboratory traces, obtaining an accuracy of 99.14% for NB while the accuracies for DT and SVM were 95.86% and 92.02% respectively.

However, Sangkatsanee et al. reported an accuracy on their work about intrusion detection [8] of 99.00% using DT, outperforming NB, Bayesian Networks (BN) and shallow Neural Networks (NN), which got an accuracy of 78.70%, 89.30% and 93.00% respectively. These results were obtained with a private dataset with 7200 data records. Based on these results, they implemented a DT in a real-time Intrusion Detection System (IDS) that showed a good performance, with a Total Detection Rate (TDR) of 99.33%.

Moreover, Kim et al. [9] concluded that SVM surpasses k-NN, NB, BN, DT and shallow NN in network anomaly detection with an accuracy comprised between 94.2% and 97.8% on their datasets of collected network traces from different entities in the USA, Japan and Korea.

Recently, the appearance of the Internet of Things (IoT) has offered another environment for botnet propagation. The research of Doshi et al. [10] on the Mirai botnet propagation over IoT devices demonstrated the effectiveness of five ML classifiers, namely k-NN, SVM, DT, Random Forest (RF) and NN, on self generated IoT traffic with Mirai traces. In particular, RF achieved the highest mean F1 score (0.99).

## 3   Methodology

### 3.1   Datasets and Features

The Internet-alike network traffic data used in this work is publicly available by courtesy of the of the Stratosphere Project, formerly Malware Capture Facility Project [11]. The traffic captures that we have selected for this work are listed on Table 1. Our dataset is composed of traces from these data where each sample

---

[1] https://www.uvic.ca/engineering/ece/isot/datasets/.

corresponds to a TCP connection or *TCP flow* and is characterized with the features described in Table 2. The dataset, which we called *TCP-Int dataset*, contemplates four classes of data: the Normal class represents non-malicious traffic, while the Kelihos, Miuref and Sality classes represent traffic traces from the botnets with the same name. We built a balanced dataset where each class contains 44231 samples since this is the number of TCP connections of the least represented class, *Miuref*.

**Table 1.** List of the selected traffic captures including the type of their traffic (i.e., from Normal communications or from activity of the malware Kelihos, Miuref or Sality) and the number of TCP flows present. The source of the captures is the Stratosphere Project [11]

| Traffic capture | Type | TCP flows |
|---|---|---|
| ctu-normal-21.pcap | Normal | 5886 |
| ctu-normal-22.pcap | Normal | 45461 |
| ctu-normal-23.pcap | Normal | 4059 |
| ctu-normal-24.pcap | Normal | 1721 |
| ctu-normal-25.pcap | Normal | 1092 |
| ctu-normal-26.pcap | Normal | 1866 |
| ctu-normal-27.pcap | Normal | 9200 |
| ctu-normal-28.pcap | Normal | 4034 |
| ctu-normal-30.pcap | Normal | 13965 |
| ctu-normal-31.pcap | Normal | 14982 |
| ctu-normal-32.pcap | Normal | 22000 |
| 2015-12-09_capture-win4-1.pcap | Kelihos | 95575 |
| 2015-12-09_capture-win4-2.pcap | Kelihos | 94219 |
| 2015-06-07_capture-win12.pcap | Miuref | 1285 |
| 2015-06-07_capture-win8.pcap | Miuref | 4899 |
| 2015-06-19_capture-win12.pcap | Miuref | 4807 |
| 2015-07-08_capture-win8.pcap | Miuref | 33240 |
| 2014-04-07_capture-win13.pcap | Sality | 57808 |

Besides, we have constructed a dataset using traffic data from a sinkhole server which we called *TCP-Sink dataset*. These data was supplied by INCIBE (Spanish National Cybersecurity Institute)[2], and have both non-malicious traces from automatic online programs and traces from the Conficker botnet. The TCP-Sink dataset contains 4027 samples of TCP flows from non-malicious connections (Normal class) and the same amount of samples from Conficker connections (Botnet class). The samples are described according to the TCP features shown in Table 2.

---

[2] https://incibe.es/.

**Table 2.** List and descriptions of the TCP flow features used in this work

| Feature(s) | Description |
|---|---|
| sPort, dPort [12,13] | Source and destination ports in the TCP connection |
| mLen, vLen [12,13] | Mean and variance of the payload lengths of all packets in the TCP connection |
| mTime, vTime [13] | Mean and variance of the interval times between sent packets in the TCP connection |
| mResp, vResp [5] | Mean and variance of the interval times between a packet reaches the machine that started the connection and it responds to that packet |
| nBytes [12] | Total number of bytes exchanged in the TCP connection |
| nSYN [13] | Total number of SYN flags exchanged during the TCP connection |
| nPackets [12] | Total number of packets exchanged in the TCP connection |
| mVel, vVel [13] | Mean and variance of the number of packets exchanged per second in the TCP connection |

## 3.2 Classifiers

We have chosen four ML algorithms that achieve a high detection rate in other works [5,8–10] on botnet detection to study their performance on our two datasets, the TCP-Int and the TCP-Sink ones. In spite of finding more algorithms in the network classification literature, for a fair comparison we have avoided comparing assembled methods with their non-assembled versions, for example Random Forest with Decision Tree.

*Decision Tree (DT):* This algorithm can be represented as a tree of logical decisions [14], where each decision splits the path in two or more [15]. A new query starts its route from the root, and depending on the result in each decision, it will follow a certain path. The class of the data is determined by the end of its path.

*K-Nearest Neighbours (k-NN):* Let each sample of the labelled training data has $m$ features and thus, it can be represented as a labelled vector in a $m$-dimensional space. Given an input data sample, also represented as a $m$-dimensional vector, k-NN [16] searches for the $k$ nearest vectors from the training data. The input sample is classified depending on the labels of these vectors, and it is regular to used an odd $k$ value to avoid a tie between neighbours.

*Support Vector Machine (SVM):* Given that each sample from the training dataset has $m$ features, it can be viewed as a point in a $m$-dimensional space. The SVM model [17] looks for the hyperplane or set of hyperplanes that best separates points from the same class from points of different classes. If necessary, SVM uses a function called *kernel* or *kernel function* that projects the points into a higher dimensional space where they can be separated more easily by a hyperplane. Depending on the problem characteristics, SVM can use either a generic kernel such as the linear kernel [18] or a custom kernel [19–21].

*Naïve Bayes (NB):* Given a value $x$ for a certain feature $f$, the probability that this value corresponds to a certain class $C_j$ is $P(C_j|f = x)$. This probability is calculated by fitting the training data into a probability distribution defined by the event model of NB [22,23]. To calculate the probability that an input data corresponds to a certain class, NB supposes that each feature contributes independently [24].

### 3.3   Evaluation

In this work we have used three metrics to measure the performance of the classification algorithms, which are described as follows:

*F1 Score:* This score is defined as the harmonic mean of the precision and recall scores [25] as shown in (1). The precision is defined in (2) from the number of True Positives (TP) and the number of False Positives (FP), while the recall is defined in (3) from the number of TP and the number of False Negatives (FN). The F1 score can be applied with multiclass datasets to measure the performance of a classifier over each class, where the ideal score is 1 and the worst is 0.

$$F_1 = 2 \times \frac{\text{precision} \times \text{recall}}{\text{precision} + \text{recall}} \tag{1}$$

$$\text{precision} = \frac{\text{TP}}{\text{TP} + \text{FP}} \tag{2}$$

$$\text{recall} = \frac{\text{TP}}{\text{TP} + \text{FN}} \tag{3}$$

*Receiver Operating Characteristic (ROC) Curves:* ROC curves [26] usually plot the True Positive Rate (TPR) on the Y axis against the False Positive Rate (FPR) on the X axis using the results of cross-validation. This method is used to evaluate the performance of a classifier on datasets with two types of data such as the TCP-Sink dataset. The ideal measure maximizes the TPR while minimizing the FPR.

*Area Under the Curve (AUC):* This measurement corresponds with the total area under the ROC curve [26]. The ideal measure is an area of 1, which means a perfect TPR and FPR scores, while a score under 0.5 means that the performance is worse than a random classifier.

# 4   Experiments

All the experiments were performed on a machine with 128 GB of RAM and two Intel(R) Xeon(R) E5-2630v3 CPUs running an Ubuntu 14.04.5 LTS.

The software was developed using Python 3. To extract traffic characteristics from traffic captures in .pcap format, the script used the module PyShark[3] which is a python wrapper for TShark[4], and the module PyTables[5] to manage intermediary HDF5 files. The Machine Learning Python module used was sklearn[6], and the most relevant optimization settings for the four used algorithms in the sklearn libraries are described as follows:

1. For DT, the main setting is the maximum depth which refers to the maximum number of splits in a path from the root to a leaf. While reducing this depth allows decreasing the computational cost, the downside is a decrease on the detection capability. In this work we use the default option of unlimited depth, i.e., it is not established a restriction on the maximum number of branches in the tree.
2. In k-NN the main hyper-parameters are the number $k$ of neighbours considered in a search and the distance metric used to determine the similarity with the neighbours. Before comparing k-NN with the rest of the classifiers, we experiment with different values of $k$ to determine the optimal one for each of the two datasets used in this work. Besides, to determine the similarity with the neighbours, we use the Euclidean distance in the 11-dimensional space of the features from Table 2.
3. The main settings for the SVM algorithm are the employed kernel and the error penalty parameter or cost $C$. In this work we have chosen an optimised implementation of the linear kernel that uses the liblinear library [27] due to its better escalation on a large number of samples, which is beneficial on network traffic classification problems. With respect to the cost variable, we have tried several values ($C = \{0.1, 0.5, 0.9, 1.0, 5.0, 10.0, 100.0\}$) without observing any significant difference, and thus we have selected the default value of $C = 1.0$.
4. The only relevant setting for NB is the employed event model, whose selection is based on the type of the used data. In this work, we use features represented as continuous variables that can not be converted into discrete ones, such as all the features on Table 2 that represent a mean or a variance, and thus we use a Gaussian event model.

We tested the four algorithms using stratified 10-folds cross-validation on the TCP-Int and the TCP-Sink datasets introduced on Sect. 3.1.

---

[3] https://pypi.org/project/pyshark/.
[4] https://www.wireshark.org/.
[5] https://www.pytables.org/.
[6] http://scikit-learn.org/stable/index.html.

# 5   Results and Discussion

We have tested the performance of the four classifiers on Internet-alike traffic using the TCP-Int dataset. This dataset presents four classes: Normal, Kelihos, Miuref and Sality, and we used the F1 score to measure the class-identification ability of each algorithm.

First, we need to find the optimal $k$ parameter of the k-NN classifier. Figure 2 shows that the highest mean F1 score of k-NN on the TCP-Int dataset is achieved with $k = 1$.

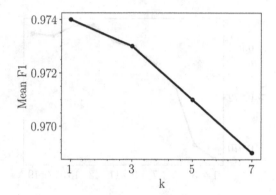

**Fig. 2.** Mean F1 score achieved by k-NN with different $k$ values over the TCP-Int dataset

Afterwards, we tested the four classification algorithms (with $k = 1$ in k-NN) on the TCP-Int dataset and the results are displayed on Table 3. DT achieves the best detection rate with a mean F1 score of 0.99 with a stable detecting performance among the four classes, meaning that the current used features are easy to categorize. On the one hand, k-NN also presents a relatively high and stable performance, implying that the data is well separated into clusters on the feature space. On the other hand, SVM with a linear kernel shows a relatively poor detection rate over the Normal and Sality classes. This suggest that the Normal and Sality classes are not linearly separable in the feature space. Besides, NB only achieves a relatively good result on the Kelihos class, while it shows a poor performance on the other three classes with an approximate 50% F1 score. Since the NB classifier is based on the idea that the features are independent, we infer that the Normal, Miuref and Sality classes contain features that present correlations.

To test the performance of the four classifiers on sinkhole traffic we used the TCP-Sink dataset. Since this dataset only contains two classes, Normal and Botnet traffic, it is preferred to measure the performance using ROC curves and AUC scores because they are more informative in binary classification than the F1 score. First, we need to optimize the $k$ parameter of k-NN. As depicted in Fig. 3, we observe that the AUC score first increases with $k$ and then stabilises after $k = 13$.

**Table 3.** F1 scores by class achieved by the four models tested over the TCP-Int dataset. [†]Appears as perfect score due to two decimal rounding

| Class | Samples | DT | k-NN | SVM-linear | NB-gauss |
|-------|---------|------|------|------------|----------|
| Normal | 44231 | **0.99** | 0.96 | 0.66 | 0.49 |
| Kelihos | 44231 | **1.00**[†] | 0.99 | 0.86 | 0.75 |
| Miuref | 44231 | **1.00**[†] | 0.98 | 0.83 | 0.51 |
| Sality | 44231 | **0.99** | 0.97 | 0.64 | 0.51 |
| *Mean* | | **0.99** | 0.97 | 0.74 | 0.56 |

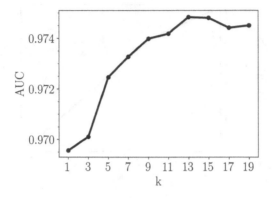

**Fig. 3.** AUC scores achieved by k-NN with different $k$ values over the TCP-Sink dataset

After optimizing k-NN with $k = 13$, we compared the accuracy of the four classifiers on the TCP-Sink dataset to discern between non-malicious traces from Conficker traces. The achieved AUC scores are tabulated on Table 4 and the ROC curves are shown on Fig. 4. We notice that all the four classifiers achieve a high accuracy, where the best result is obtained by DT with an AUC of 0.99. Taking into account the poor performance of SVM and NB on the TCP-Int dataset and their good results on the TCP-Sink dataset, we deduce that the non-malicious traces from automatic connections of online services are easily identifiable from the Conficker botnet ones.

**Table 4.** AUC scores for binary classifications over the TCP-Sink dataset

| Model | AUC |
|-------|------|
| DT | **0.99** |
| k-NN | 0.97 |
| SVM-linear | 0.89 |
| NB-gauss | 0.88 |

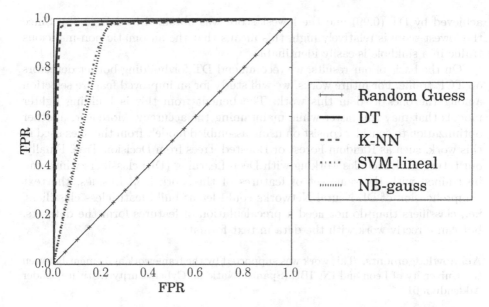

**Fig. 4.** ROC curves for Conficker botnet detection over the TCP-Sink dataset

# 6    Conclusions and Future Work

In this paper we explored the problem of botnet detection on TCP traffic in two different environments: regular Internet communications and traffic captures on sinkhole servers. For this purpose, we constructed two datasets: the TCP-Int dataset with public Internet traffic captures from the Stratosphere Project [11], and the TCP-Sink dataset with traces from a private sinkhole server gathered by INCIBE.

We have tested four supervised ML algorithms, namely DT, k-NN, SVM and NB, on the two datasets using an approach independent from the three possible network architectures used by botnets.

The TCP-Int dataset contained traces from three botnets: Kelihos, Miuref and Sality, aside from non-malicious data. We used the F1 score to evaluate the capability of the algorithms to identify the four classes. Both DT and k-NN showed a stable performance among the classes, and DT achieved the best mean F1 score (0.99). However, the performance of SVM and NB varied significantly between classes: the worst F1 score of SVM was 0.64 for the Sality class while the best score was 0.86 for the Kelihos class, and the best F1 score of NB was 0.49 for the Normal class while the best score was 0.75 for the Kelihos class. It is noticeable that the Kelihos class was the best identified by the four classifiers, revealing that its characteristics are more distinguishable from the other classes.

The TCP-Sink dataset contained Conficker botnet traces and normal traces from automatic Internet programs such as web crawlers. The four algorithms were evaluated on this dataset using the AUC score, where the highest score was achieved by DT (0.99) and the lowest score was obtained by NB (0.88). Since the lowest score is relatively high, this means that the automatic non-malicious traffic in a sinkhole is easily identifiable.

On the basis of our results, we recommend DT for building botnet detectors on TCP traffic. For future works, we will study for an improved feature selection among the ones used in this work. The benefit from this is building lighter models that may run faster while maintaining the accuracy. Moreover, another optimization to consider consist on using assembled models from the ones used in this work, such as Random Forest or Boosted Trees from Decision Tree. Finally, our future work includes working with Deep Learning (DL) classifiers which can be trained with the same set of features of this work [28]. Besides, the text comprehension of DL Neural Networks could let us build featureless classifiers, i.e., classifiers than do not need a precalculation of features form the datasets, but can directly work with the data in text format.

**Acknowledgements.** This work was supported by the framework agreement between the University of León and INCIBE (Spanish National Cybersecurity Institute) under Addendum 01.

# References

1. Martínez, J., Iglesias, C., García-Nieto, P.: Machine learning techniques applied to cybersecurity. Int. J. Mach. Learn. Cybern. **1**–14 (2019)
2. Silva, S.S., Silva, R.M., Pinto, R.C., Salles, R.M.: Botnets: a survey. Comput. Netw. **57**(2), 378–403 (2013)
3. Boshmaf, Y., Muslukhov, I., Beznosov, K., Ripeanu, M.: Design and analysis of a social botnet. Comput. Netw. **57**(2), 556–578 (2013)
4. Bujlow, T., Carela-Español, V., Barlet-Ros, P.: Independent comparison of popular DPI tools for traffic classification. Comput. Netw. **76**, 75–89 (2015)
5. Kirubavathi, G., Anitha, R.: Botnet detection via mining of traffic flow characteristics. Comput. Electr. Eng. **50**, 91–101 (2016)
6. Kim, H., Choi, S.S., Song, J.: A methodology for multipurpose DNS Sinkhole analyzing double bounce emails. In: International Conference on Neural Information Processing, pp. 609–616 (2013)
7. Fetzer, C., Felber, P., Rivière, É., Schiavoni, V., Sutra, P.: UniCrawl: a practical geographically distributed web crawler. In: International Conference on Cloud Computing, pp. 389–396 (2015)
8. Sangkatsanee, P., Wattanapongsakorn, N., Charnsripinyo, C.: Practical real-time intrusion detection using machine learning approaches. Comput. Commun. **34**(18), 2227–2235 (2011)
9. Kim, H., Claffy, K.C., Fomenkov, M., Barman, D., Faloutsos, M., Lee, K.: Internet traffic classification demystified: myths, caveats, and the best practices. In: Proceedings of the 2008 ACM CoNEXT Conference, pp. 11:1–11:12 (2008)
10. Doshi, R., Apthorpe, N., Feamster, N.: Machine learning DDoS detection for consumer internet of things devices. In: IEEE Security and Privacy Workshops, pp. 29–35 (2018)
11. García, S., Grill, M., Stiborek, J., Zunino, A.: An empirical comparison of botnet detection methods. Comput. Secur. **45**, 100–123 (2014)

12. Saad, S., et al.: Detecting P2P botnets through network behavior analysis and machine learning. In: 2011 Ninth Annual International Conference on Privacy, Security and Trust, pp. 174–180 (2011)
13. Zhao, D., et al.: Botnet detection based on traffic behavior analysis and flow intervals. Comput. Secur. **39**, 2–16 (2013)
14. Buntine, W., Niblett, T.: A further comparison of splitting rules for decision-tree induction. Mach. Learn. **8**, 75–85 (1992)
15. Friedman, J.H.: Lazy decision trees. In: Proceedings of the Thirteenth National Conference on Artificial Intelligence, vol. 1, pp. 717–724 (1996)
16. Dong, W., Moses, C., Li, K.: Efficient K-nearest neighbor graph construction for generic similarity measures. In: Proceedings of the 20th International Conference on World Wide Web, pp. 577–586 (2011)
17. Cherkassky, V., Ma, Y.: Practical selection of SVM parameters and noise estimation for SVM regression. Neural Netw. **17**(1), 113–126 (2004)
18. Al Nabki, M.W., Fidalgo, E., Alegre, E., de Paz, I.: Classifying illegal activities on TOR network based on web textual contents. In: Proceedings of the 15th Conference of the European Chapter of the Association for Computational Linguistics, vol. 1, pp. 35–43 (2017)
19. Fidalgo, E., Alegre, E., González-Castro, V., Fernández-Robles, L.: Compass radius estimation for improved image classification using Edge-SIFT. Neurocomputing **197**, 119–135 (2016)
20. Fidalgo, E., Alegre, E., González-Castro, V., Fernández-Robles, L.: Illegal activity categorisation in darknet based on image classification using CREIC method. In: Pérez García, H., Alfonso-Cendón, J., Sánchez González, L., Quintián, H., Corchado, E. (eds.) SOCO/CISIS/ICEUTE -2017. AISC, vol. 649, pp. 600–609. Springer, Cham (2018). https://doi.org/10.1007/978-3-319-67180-2_58
21. Fidalgo, E., Alegre, E., González-Castro, V., Fernández-Robles, L.: Boosting image classification through semantic attention filtering strategies. Pattern Recogn. Lett. **112**, 176–183 (2018)
22. Schneider, K.: A comparison of event models for Naive Bayes Anti-spam e-Mail Filtering. In: Proceedings of the Tenth Conference on European Chapter of the Association for Computational Linguistics, vol. 1, pp. 307–314 (2003)
23. Xu, S.: Bayesian Naïve Bayes classifiers to text classification. J. Inf. Sci. **44**(1), 48–59 (2018)
24. Ren, J., Lee, S.D., Chen, X., Kao B., Cheng, R., Cheung, D.: Naive Bayes classification of uncertain data. In: 2009 Ninth IEEE International Conference on Data Mining, pp. 944–949 (2009)
25. Sasaki, Y.: The truth of the F-measure. Teach Tutor mater **1**(5), 1–5 (2007)
26. Fawcett, T.: An introduction to ROC analysis. Pattern Recogn. Lett. **27**(8), 861–874 (2006)
27. Fan, R.E., Chang, K.W., Hsieh, C.J., Wang, X.R., Lin, C.J.: LIBLINEAR: a library for large linear classification. J. Mach. Learn. Res. **9**, 1871–1874 (2008)
28. van Roosmalen, J., Vranken, H., van Eekelen, M.: Applying deep learning on packet flows for botnet detection. In: Proceedings of the 33rd Annual ACM Symposium on Applied Computing, pp. 1629–1636 (2018)

# Classifying Pastebin Content Through the Generation of PasteCC Labeled Dataset

Adrián Riesco[1](✉), Eduardo Fidalgo[2,3](✉), Mhd Wesam Al-Nabki[2,3](✉),
Francisco Jáñez-Martino[3](✉), and Enrique Alegre[2,3](✉)

[1] Summer Internship at the Universidad de León with the VARP Research Group,
León, Spain
ariesv00@estudiantes.unileon.es
[2] Department of Electrical, Systems and Automation, Universidad de León,
León, Spain
{efidf,mnab,enrique.alegre}@unileon.es
[3] Researcher at INCIBE (Spanish National Cybersecurity Institute), León, Spain
fjanem00@estudiantes.unileon.es

**Abstract.** Online notepad services allow users to upload and share free text anonymously. Reviewing Pastebin, one of the most popular online notepad services websites, it is possible to find textual content that could be related to illegal activities, such as leaks of personal information or hyperlinks to multimedia files containing child sexual abuse images or videos. An automatic approach to monitor and to detect these activities in such an active and a dynamic environment could be useful for Law Enforcement Agencies to fight against cybercrime. In this work, we present Pastes Content Classification 17K (PasteCC_17K), a dataset of 17640 textual samples crawled from Pastebin, which are classified in 15 categories, being 6 of them suspicious to be related to illegal ones. We used PasteCC_17K to evaluated two well-known text representation techniques, ensembled with three different supervised approaches to classify the pastes of the Pastebin website. We found that the best performance is achieved ensembling TF-IDF encoding with Logistic Regression obtaining an accuracy of 98.63%. The proposed model could assist the authorities in the detection of suspicious content shared in Pastebin.

**Keywords:** Natural Language Processing · Machine learning ·
Pastebin · Text classification · Bag of Words ·
Term Frequency-Inverse Document Frequency

## 1 Introduction

In the digital age where all information is shared the online notepad services are an essential meeting point. These web sites allow users to upload and to share anonymously text with a limit of extension far superior to other online platforms. The most common use of such services is to share code snippets or personal experiences. However, although the majority of the shared text is legitimate,

H. Pérez García et al. (Eds.): HAIS 2019, LNAI 11734, pp. 456–467, 2019.
https://doi.org/10.1007/978-3-030-29859-3_39

these services have been abused by sharing suspicious activities including, but not limited to, child pornography multimedia hyperlinks, drug sales and leaked information. The three most known web applications that offer online notepad services are PiratePad[1], codepad[2], and Pastebin[3]. The majority of these services offer a free membership for users which allows them to publish pastes and to access pastes uploaded by other users. The *paste* is known in Pastebin community as any piece of a text published by users. Pastebin was the first one to provide online notepad services since the year 2002, and in the year 2016, Pastebin registered 18 million visitors, more than 95 million of pastes and a monthly mean of 14.430.000 pastes published. According to Herath [9], pastes from Pastebin could contain suspicious activities such as leaked personal data, credit cards, login credentials or confidential financial documents. Figure 1 shows an example of an anonymous paste leaking Netflix accounts to the public.

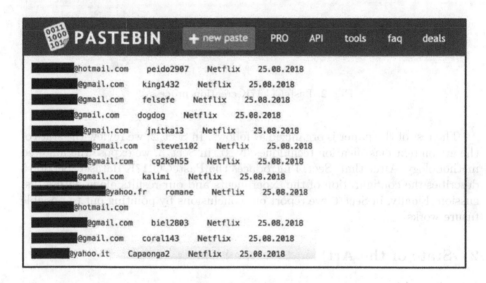

**Fig. 1.** Paste from Pastebin that leaks Netflix accounts, passwords and verification dates

Motivated by the huge amount of public text data which Pastebin handles, and the possibility of finding suspicious text inside motived us to propose a fully-automatic approach to classify and detect this possible malicious content. To support this task, we first built **P**astes **C**ontent **C**lassification 17K (PasteCC_17K), a new publicly available dataset with 17640 pastes extracted from Pastebin. Then, we used PasteCC_17K data to train a classifier using six different classification pipelines. We ensembled two encoding methods, Term Frequency–Inverse

---

Document Frequency (TF-IDF) and Bag of Words (BoW), with three different classifiers, Logistic Regression (LR), Support Vector Machine (SVM), and Näive Bayes(NB). Figure 2 illustrates the described process.

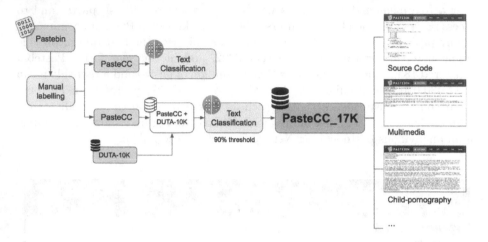

**Fig. 2.** PasteCC_17K creation process

The rest of the paper is organized as follows. In Sect. 2, we review the state of the art on text classification techniques. Next, in Sect. 3, we expose the followed methodology. After that, Sect. 4 introduces the PasteCC_17K dataset. Section 5 describes the configuration of the experiments and our results along with a discussion. Finally, in Sect. 6, we report our conclusions by pointing out to possible future works.

## 2   State of the Art

### 2.1   Pastebin

Pastebin was created in 2002 by Paul Dixon as a web application for handling pastes. Pastebin does not need authentication, the interface allows users to post pastes without much effort and create a unique URL for each paste, without any limitation on the number of characters, as happens in Twitter, Matic et al. [14].

In recent years, it is increasingly common to find illicit content within Pastebin; hackers tend to publish everything and often use Pastebin for sharing their information. A group of hackers published on Pastebin thousands of names, usernames, passwords, addresses, and phone numbers of students, faculty, and staff members from 53 universities, including Harvard, Stanford, Cornell, Princeton, Johns Hopkins, the University of Zurich and other universities around the world, Perlroth [18].

To the best of our knowledge, the conventional classification and monitoring methods used with Pastebin content depend on a predefined list of keywords, which creation needs a significant amount of time, its maintenance is very costly and become outdated in the short term. For example, *PasteHunter* tool[4] and *AIL-framework*[5] depend heavily on a sequence of rules and regular expressions to parse particular tokens such as email addresses, IBAN account numbers, passwords, or even phone numbers. Furthermore, Herath [9] proposed a framework called *LeakHawk*[6] that employs a number of filters and classifiers whereas the first two components screen out the pastes according to a predefined list of keywords, which makes these approaches susceptible to be hacked.

## 2.2 Machine Learning and Text Classification

Text classification is an important task of Natural Language Processing (NLP), which aim is to separate text documents in different classes or categories based on their content, in a similar way as a human been could do that. Sebastiani [19], presented a survey where discusses the main approaches to text classification using machine learning, discussing in detail the namely document representation, classifier construction, and classifier evaluation. Joachims [11] analyzed the particular properties of SVM and text classification, and recommended using SVM for the classification of text because most of the problems of text classification are linearly separable.

Text classification has been studied extensively during the last few years. Mironczuk and Protasiewicz [16] reviewed every element of text classification and its state-of-art; they also presented a quantitative analysis of studies including the research topics more used, articles per year or articles per journal. Diab and Hindi [7] used three different meta-heuristic methods to find better estimations for the probability terms of NB, these methods were differential evolution, genetic algorithms, and simulated annealing. Silva et al. [20] proposed a novel multinomial text classification technique based on the MDL principle due to the limitations of the traditional classifiers such as SVM or NB with high dimensionality corpus and computational costs for training. The statistical analysis of their results indicated that the proposed technique called MDLText outperformed all of the methods that they evaluated. Zhu and Wong [25] evaluated the five commonly used text classifiers, which are SVM, NB, KNN, AdaBoost and Neural Network. They used an automatically generated text document collection which is labeled by a group of experts. The results verify that different algorithms performed differently depend on document datasets and the need for feature selection method to reduce the high dimensionality of features.

Text Classification is widely used in detection, categorization, and translation domains. Lochter et al. [13] presented an ensemble of text expansion method and the well-known and established classification methods available in literature such as LR, NB, SVM, C4.5 and KNN to detect opinions inside social network. The overall results showed that their ensemble enhanced the performance of

---

[4] https://github.com/kevthehermit/PasteHunter.
[5] https://github.com/CIRCL/AIL-framework.
[6] https://github.com/isuru-c/LeakHawk.

the classifiers used individually. Bui et al. [4] presented a different approach to recognize the structure of a PDF to extract raw text in order to be able to classify it later. Panchenko et al. [17] took English text from Common Crawl and constructed a large web-scale corpus using text classification.

### 2.3   Text Classification Using Deep Learning Techniques

During the last years appeared some works using Deep Learning models for the text classification process. Since word representation is an essential step for success, Meng et al. [15] proposed an RNN-based generative model for predicting key-phrases in scientific text. Also, Zhang et al. [23] presented a Deep Learning method based on recurrent neural networks to perform automatic key-phrase extraction ensembling keywords and context information to perform the key-phrase extraction task, especially useful in Twitter domains. In terms of text classification, Zhang et al. [24] empirically evaluated the advantages of using ConvNets for the detection of characteristics at character-level. They built several large-scale datasets to demonstrate that the ConvNets model offers better results.

Additionally, Facebook AI Research created two methods based on Deep Learning oriented to text classification. Joulin et al. [12] presented and compared the text classifier called FastText with the recent text classification models based on Deep Learning or Wu et al. [22], who presented StarSpace, a general-purpose neural embedding model that can solve a wide variety of problems. Stein et al. [21] evaluated and compared FastText with different algorithms, including SVM, in the hierarchical text classification. FastText achieved very satisfactory results that showed that the use of word inlays is a very promising approach in hierarchical text classification. These works show that it is necessary to use large datasets to use these Deep Learning methods to obtain good results. Despite the superiority of the broad learning approach in solving various NLP-related problems, there is a need for an enormous number of labeled samples for training, which is not available in the context of our problem.

## 3   Methodology

We proposed the following methodology to (i) build PasteCC_17K dataset and (ii) to assess the performance of the six pipelines evaluated. First, we applied a preprocessing stage to the text crawled from the Pastebin website. Then, we encoded the previous text, obtaining the features that represent each paste. Finally, we applied supervised learning to train a model with Pastebin samples and classify the online notepad content. We used two well-known features extractors ensembling with three different supervised classifiers, i.e. six different text classification pipelines. To build the classifier, we only worked with one pipeline, i.e. one encoding and classifier, filtering the classified pastes if they obtained an accuracy higher than a predefined threshold. To define what is the best approach to classifying the Pastebin content, we evaluated the six pipelines.

## 3.1  Text Preprocessing

Text preprocessing was used to delete no important characters before starting the text classification process. The text preprocessing *cleans* the text of words known as stop-words. We built a text preprocessing that removed all the format files, URL links, currency units, a single letter, long words, repeated text, numbers, words with numbers and special characters.

## 3.2  Feature Extraction

In this Section, we have reviewed two text coding techniques, Bag of Words (BOW) and Term Frequency-Inverse Document Frequency (TF-IDF). **Bag of Words** was proposed by Harris [8], is a form of simplified text representation that creates a list of vocabulary that is found within the corpus or body of the message. Once created, the frequency of each word in the text to be processed is measured and a vector with said data is obtained, using each word of the text as a characteristic to train the classifiers. Aizawa [1] presented **Term Frequency-Inverse Document Frequency** (TF-IDF), a method to assign weights for the dataset vocabulary statistically. It assigns a high weight value for words which occur frequently in a given text and with small frequency in all the other documents.

## 3.3  Classifiers

Our work uses three well-known supervised classifiers: Logistic Regression (LR), Support Vector Machines (SVM) and Naive Bayes (NB). **Logistic Regression**. LR is a supervised classification technique, which function was created by Cox [6]. It tries to obtain an accurate prediction for one or a set of variables by obtaining a logistic function. This function is the one that finds the probability that a certain variable presents the predicted characteristic. **Support Vector Machines**. SVM is a form of supervised classification firstly developed by Cortes and Vapnik [5], and it is based on maximizing the distance between the decision boundary and the data. SVM machines have a kernel function that defines how data will be split. **Naïve Bayes**. NB is a classifier based on the Bayes theorem, which was developed in the 18th century, and it applies the principle of independence assumption. It can be explained as a calculation of the probability of a document of belonging to a certain category, and then it assigns to this document the class with the best probability.

# 4  Pastes Content Classification

## 4.1  PasteCC_17K Building Procedure

To best of our knowledge, there is no labeled dataset related the content on the Pastebin web service. Hence, we created the first public Pastebin dataset. To build the dataset we followed the procedure used to generate a dataset with

Darknet content by Wesam et al. [3]. We crawled pastes from Pastebin site via the website official API along with premium (PRO) license. In particular, we created a customized scraper to download pastes along with their meta-data. Our crawler downloaded $500K$ pastes from Pastebin, but we only labeled manually 804 samples into 14 categories as shown in Table 1. We refer to this dataset by Pastes Content Classification (PasteCC). PasteCC is publicly available on our web page[7].

Among PasteCC samples, we found that the major category is **Source Code** which is expected, given that Pastebin is basically oriented for developers to share code snippets and logs. The category **Multimedia** contains pastes which share links to access videos, images or documents, and television hosting. **Encoded Text** category groups pastes whose content is related to sets of letters and numbers without semantic meaning. **Forum** category includes pastes in relation to any topic coming from a community as videogames or recommendations. **Personal** category has conversations or personal experiences. The pastes outside the previous categories belong to the category **Others**.

Interestingly, we noticed that suspicious activities are the minority. However, despite there being a minority, they might be dangerous and holding critical information that the authorities are interested in monitoring [9]. This fact makes the problem of building a supervised approach to detect and to classify suspicious activities even more challenging. Because of the lack of training examples hiders the capability of training a classifier to recognize them.

**Table 1.** Classification of the manually labelled pastes. Initial version of PasteCC dataset

| Category | # Samples |
|---|---|
| Source Code | 298 |
| Multimedia | 230 |
| Forum | 90 |
| Logs | 60 |
| Others | 33 |
| Leaked-Data | 16 |
| Cryptocurrency | 12 |
| Encoded Text | 12 |
| Drugs | 11 |
| Child-pornography | 11 |
| Counterfeit Credit-Cards | 10 |
| Hacking | 10 |
| General-pornography | 6 |
| Personal | 5 |
| **Total** | **804** |

---

[7] http://gvis.unileon.es/dataset/paste-bin.

To overcome this drawback, we incorporated an external dataset that contains activities similar to the ones we are looking for in Pastebin because they are samples of activities suspicious to be illegal. We made use of the Darknet Usage Text Addresses (DUTA-10K) dataset[8], which is the state-of-the-art dataset in terms of suspicious activities in the Tor Darknet and holds 26 classes with regular and suspicious categories found in 10367 unique onion domains in the Tor Darknet. We hypothesize that if we joined DUTA-10K dataset with PasteCC dataset, we would have more samples per category and more classes as well. The resulting dataset is a union of *PasteCC* and *DUTA-10K*, and in total it holds 11171 samples distributed in a total of 28 categories.

After joining both dataset DUTA-10K and PasteCC, we trained models based on an ensemble of two encoding techniques, TF-IDF and BoW, together with three classifiers, namely SVM, LR, and Naive Bayes. We chose TF-IDF with LR since their overall results overcame the rest of classifiers. Therefore, we applied trained TF-IDF and LR classifier with PasteCC + DUTA-10K samples on 100000 unlabelled pastes extracted by our search engine feature from Pastebin. The classifier predicted the classes of pastes above three probability-of-success thresholds, which are 85%, 90% and, 95%. We checked 100 pastes per threshold in order to determine the achievement of the classifier. The pastes predicted by the threshold based on the 85% probability of success had many classification mistakes and the pastes obtained by the thresholds with success probability of 95% despite their correct prediction, the number of contributed pastes was scant. We discarded the pastes classified by both thresholds. We selected the pastes predicted with the 90% probability of success due to the correct pastes prediction was similar to 95% threshold and contributed with many more number of pastes. Finally, we removed DUTA dataset examples in order to keep only the text content provided by Pastebin, the resulting dataset was called PasteCC_17K. Table 2 shows the number of samples per category.

## 4.2 Statistics

PasteCC_17K comprises 15 classes, including 17640 samples taken from Pastebin, the most of the samples belongs to "Source Code" category, 70.27%, the next most influential class is "Multimedia", 26.75%. "Counterfeit Personal-Identification" is a class imported from DUTA and represents the categories with fewer examples. About suspicious content, there are 217 suspect pastes i.e. 1.23% of pastes. To the best of our knowledge and the extensions of our current evaluation, we can summarize that 98.7% of the Pastebin content could be considered safe. However, although unlawful content represents a small percentage, since 481, 000 of pastes are uploaded per day, it results in a significant amount of suspicious content that should be considered, since it is publicly available and visible for the whole Web. A text classifier capable of finding these suspicious content would represent a useful tool for monitoring Pastebin.

---

[8] http://gvis.unileon.es/dataset/duta-darknet-usage-text-addresses/.

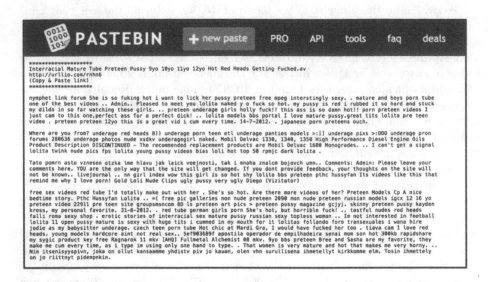

**Fig. 3.** Example of illicit categories examples provided to PasteCC_17K, Child-pornography category

**Table 2.** Influence of the categories in PasteCC_17K. The categories of Counterfeit Credit-Cards and Counterfeit Personal-Identification are named respectively as Counterfeit C-C and Counterfeit P-I.

| Category type | Category | # Samples | Percent(%) |
|---|---|---|---|
| Suspicious 1.23%; 217 Samples | Hacking | 86 | 0,49 |
| | Counterfeit C-C | 67 | 0,38 |
| | Child-pornography | 29 | 0,16 |
| | Leaked-Data | 19 | 0,11 |
| | Drugs | 14 | 0,08 |
| | Counterfeit P-I | 2 | 0,01 |
| Normal 98.70%; 17423 Samples | Source Code | 12396 | 70,27 |
| | Multimedia | 4719 | 26,75 |
| | Forum | 90 | 0,51 |
| | Others | 81 | 0,46 |
| | Logs | 65 | 0,37 |
| | Encoded Text | 25 | 0,14 |
| | Crytocurrency | 20 | 0,11 |
| | General-pornography | 17 | 0,10 |
| | Personal | 10 | 0,06 |
| Total | | **17640** | |

# 5   Experimentation

## 5.1   Experimental Settings

We used Python 3 with PyCharm IDE. To build the text vectorizers and the classifiers, we used Scikit-learn library[9], Numpy and Pandas. For classification, we used 70% of the samples as a training set and the remaining 30% as a set of testing, following the work of Al Nabki et al. [2,3].

## 5.2   Results

Table 3 shows the results of the different models. Given the low number of samples and the unbalanced classes, the learning curve of the models created presented overfitting.

**Table 3.** Results of training classifier with PasteCC_17K (%)

|  |  | TF-IDF LR | TF-IDF SVM | TF-IDF NB | BOW LR | BOW SVM | BOW NB |
|---|---|---|---|---|---|---|---|
| Accuracy |  | **98.63** | 97.80 | 97.52 | 98.37 | 97.62 | 98.05 |
| Precision | Macro | 71.60 | 72.50 | 21.10 | 58.80 | 62.90 | 41.70 |
|  | Micro | 98.60 | 97.80 | 97.50 | 98.40 | 97.60 | 98.10 |
|  | Weighted | 98.70 | 98.80 | 95.40 | 98.70 | 98.70 | 97.50 |
| Recall | Macro | 72.10 | 63.20 | 21.00 | 63.20 | 59.90 | 38.10 |
|  | Micro | 98.60 | 97.80 | 97.50 | 98.40 | 97.60 | 98.10 |
|  | Weighted | 98.60 | 97.80 | 97.50 | 98.40 | 97.60 | 98.10 |
| F1 Score | Macro | **72.10** | 62.20 | 21.10 | 58.70 | 58.30 | 35.64 |
|  | Micro | **98.70** | 97.80 | 97.50 | 98.40 | 97.60 | 98.10 |
|  | Weighted | **98.70** | 98.10 | 97.50 | 98.50 | 98.00 | 97.70 |

## 5.3   Discussion

Due to we are working with an unbalanced multiclass dataset where two categories, Source Code and Multimedia, represent 97.02% of the total, we present the precision, recall, and F1 score for each ensemble. The model with the best performance is the LR together with TF-IDF as feature extractor, yielding a cross-validation accuracy of 98.63% and micro F1 Score of 98.60%. The ensemble of TF-IDF with LR overcomes the rest of ensembles in overall results. The results show that NB obtains higher accuracy with BOW than TF-IDF, which only occurs with NB classifier. The SVM classifier was less influenced by feature extractors due to the linear kernel function of SVM we used that mapped

---

[9] Machine Learning library for Python. (Source: http://scikit-learn.org/stable).

the data in order to be able to separate linearly. All classifier achieved low performance when the evaluation does not account imbalanced dataset as *Macro* measures show with both feature extraction, NB classifier presents the lowest performance.

## 6    Conclusions and Future Works

In this work, we proposed a solution to solve the problem of classifying Pastebin the content, with the aim of identifying suspicious content. We first crawled and created labeled dataset in a semi-automatic way, Pastes Content Classification 17K (PasteCC_17K), which comprises 17640 pastes grouped in 15 categories, being six of them suspicious to contain illicit content that represents the 1.23% of the total classified pastes. The category that shows the habitual use of Pastebin, sharing source code, constitutes the 70, 27% of our dataset. The small percentage of suspicious content over the total can be misleading since applied over the daily Pastebin pastes would imply a significant amount of illegal and publicly available content that ought to be considered. We have found pastes with leaked data, hacking offers, and even child pornography (see Fig. 3) links which should be monitored. We evaluated the ensemble of two encoding techniques, TF-IDF and BoW, together with three well-known classifiers, namely SVM, LR and Naive Bayes. We found that the ensemble of TF-IDF together with LR achieves the best performance in PasteCC_17K, with an accuracy of 98.90%.

The obtained results encourage us to extend the PasteCC_17K dataset using Active Learning techniques considering the work of Hu et al. [10], to find new patterns in Pastebin. Moreover, given the variety of the categories in Pastebin, we are planning to use different text encoding techniques for each category. In future works, we will explore the use of Active Learning techniques to monitor Pastebin, trying to decrease the labelling cost of unknown samples while the classifier accuracy increases.

**Acknowledgements.** This research is supported by the INCIBE grant "INCIBEI-2015-27359", corresponding to the "Ayudas para la Excelencia de los Equipos de Investigación avanzada en ciberseguridad" and also by the framework agreement between the University of León and INCIBE (Spanish National Cybersecurity Institute) under Addendum 22 and 01.

## References

1. Aizawa, A.: An information-theoretic perspective of tf-idf measures. Inf. Process. Manage. **39**(1), 45–65 (2003)
2. Al-Nabki, M.W., Fidalgo, E., Alegre, E., Fernández-Robles, L.: Torank: identifying the most influential suspicious domains in the tor network. Expert Syst. Appl. **123**, 212–226 (2019)
3. Al Nabki, M.W., Fidalgo, E., Alegre, E., de Paz Centeno, I.: Classifying illegal activities on tor network based on web textual contents. In: Proceedings of the 15th Conference of the European Chapter of the Association for Computational Linguistics. Association for Computational Linguistics, Valencia, Spain, April 2017

4. Bui, D.D.A., Fiol, G.D., Jonnalagadda, S.: Pdf text classification to leverage information extraction from publication reports. J. Biomed. Inform. **61**, 141–148 (2016)
5. Cortes, C., Vapnik, V.: Support-vector networks. Mach. Learn. **20**(3), 273–297 (1995)
6. Cox, D.R.: The regression analysis of binary sequences. J. Roy. Stat. Soc. B **20**, 215–242 (1958)
7. Diab, D.M., Hindi, K.: Using differential evolution for fine tuning naïve bayesian classifiers and its application for text classification. Appl. Soft Comput. **54**, 183–199 (2016)
8. Harris, Z.S.: Distributional structure. Word **10**(2–3), 146–162 (1954)
9. Herath, H.: Web information extraction system to sense information leakage. Master's thesis, University of Moratuwa, Sri Lanka (2003)
10. Hu, R., Jane Delany, S., Mac Namee, B.: EGAL: exploration guided active learning for TCBR. In: Bichindaritz, I., Montani, S. (eds.) ICCBR 2010. LNCS (LNAI), vol. 6176, pp. 156–170. Springer, Heidelberg (2010). https://doi.org/10.1007/978-3-642-14274-1_13
11. Joachims, T.: Text categorization with support vector machines: learning with many relevant features. In: Nédellec, C., Rouveirol, C. (eds.) ECML 1998. LNCS, vol. 1398, pp. 137–142. Springer, Heidelberg (1998). https://doi.org/10.1007/BFb0026683
12. Joulin, A., Grave, E., Bojanowski, P., Mikolov, T.: Bag of tricks for efficient text classification. CoRR abs/1607.01759 (2016)
13. Lochter, J.V., Zanetti, R.F., Reller, D., Almeida, T.A.: Short text opinion detection using ensemble of classifiers and semantic indexing. Expert Syst. Appl. **62**, 243–249 (2016)
14. Matic, S., Fattori, A., Bruschi, D., Cavallaro, L.: Peering into the muddy waters of pastebin. ERCIM News **90**, 16 (2012)
15. Meng, R., Zhao, S., Han, S., He, D., Brusilovsky, P., Chi, Y.: Deep keyphrase generation. CoRR abs/1704.06879 (2017)
16. Mironczuk, M., Protasiewicz, J.: A recent overview of the state-of-the-art elements of text classification. Expert Syst. Appl. **106**, 36–54 (2018)
17. Panchenko, A., Ruppert, E., Faralli, S., Ponzetto, S.P., Biemann, C.: Building a web-scale dependency-parsed corpus from commoncrawl. CoRR abs/1710.01779 (2017)
18. Perlroth, N.: Hackers breach 53 universities and dump thousands of personal records online. New York Times, New York (2012)
19. Sebastiani, F.: Machine learning in automated text categorization. ACM Comput. Surv. **34**(1), 1–47 (2002)
20. Silva, R.M., Almeida, T.A., Yamakami, A.: Mdltext: an efficient and lightweight text classifier. Knowl.-Based Syst. **118**, 152–164 (2017)
21. Stein, R.A., Jaques, P.A., Valiati, J.F.: An analysis of hierarchical text classification using word embeddings. CoRR abs/1809.01771 (2018)
22. Wu, L., Fisch, A., Chopra, S., Adams, K., Bordes, A., Weston, J.: Starspace: Embed all the things! CoRR abs/1709.03856 (2017)
23. Zhang, Q., Wang, Y., Gong, Y., Huang, X.: Keyphrase extraction using deep recurrent neural networks on twitter. In: EMNLP (2016)
24. Zhang, X., Zhao, J., LeCun, Y.: Character-level convolutional networks for text classification. In: Advances in Neural Information Processing Systems, pp. 649–657. Neural Information Processing Systems Foundation, January 2015
25. Zhu, D., Wong, K.W.: An evaluation study on text categorization using automatically generated labeled dataset. Neurocomputing **249**, 321–336 (2017)

# Network Traffic Analysis for Android Malware Detection

José Gaviria de la Puerta^(✉), Iker Pastor-López, Borja Sanz,
and Pablo G. Bringas

University of Deusto, Avenida de las Universidades 24, 48007 Bilbao, Spain
{jgaviria,iker.pastor,borja.sanz,pablo.garcia.bringas}@deusto.es

**Abstract.** The possibilities offered by the management of huge quantities of equipment and/or networks is attracting a growing number of developers of *malware*. In this paper, we propose a working methodology for the detection of malicious traffic, based on the analysis of the flow of packets circulating on the network. This objective is achieved through the parameterization of the characteristics of these packages to be analyzed later with supervised learning techniques focused on traffic labeling, so as to enable a proactive response to the large volume of information handled by current filters.

**Keywords:** Malware Detection · Machine Learning ·
Network Traffic Analysis

Since mobile devices are terminals continuously connected to the network of networks, it can be thought that the analysis of the packets coming from the applications could be modelled to obtain a valid classification in terms of *goodware* or *malware.*

As proven in previous publications [7], the use of traffic from applications on conventional computers is a good approach to detecting malicious applications. Throughout this paper we will again use the methodology shown in the past to see if a change in technology continues to behave efficiently.

The structure of this paper is organized as follows. In the Sect. 1 we will detail the work done by other authors for traffic modelling. In the Sect. 2.1 we will specify the scope of this research describing the data capture processes. In the Sect. 2 we will define the techniques and methods used to define the captures.

In the Sect. 3 we will explain the automatic learning methods used and we will bring together the set of experiments carried out and the results obtained after their execution. Finally, in the Sect. 4 we will collect the conclusions drawn during the execution of this work.

## 1 State of the Art

Real-time traffic classification has the potential to solve the problems associated with network management that *Internet* providers and system administrators

© Springer Nature Switzerland AG 2019
H. Pérez García et al. (Eds.): HAIS 2019, LNAI 11734, pp. 468–479, 2019.
https://doi.org/10.1007/978-3-030-29859-3_40

have to manage. For this reason, network administrators need to know what is flowing through their networks.

Methods of classifying network traffic can be a fundamental part of intrusion detection systems or *IDS* [32,37]. These systems are used for: (i) detecting patterns indicative of denial of service attacks, (ii) reallocating network resources to priority customers, and (iii) identifying network customers who make malicious use of the network.

Many applications use increasingly unpredictable port numbers [19]. As a result, the most sophisticated classification techniques check the type of application by searching for application-specific data within the *payload* of *TCP* or *UDP* [39].

When we do a traditional classification of traffic, we rely on: (i) the inspection of packets coming from the *TCP* or *UDP* protocol, or (ii) the reconstruction of the signature of the *payload*.

- *Port based inspection. TCP* and *UDP* allow multiplexing of flows between different IP terminals using the port number. Therefore, many applications use a known port on the *host* as an entry point for other terminals to initiate a communication. From this point, it is possible to obtain the port number of the connection and then search for it within a list of registered ports (*IANA* or *Internet Assigned Numbers Authority*) [43].
  This type of approach has its limitations, for example, some applications may not have their ports registered in *IANA* as may be the case of applications *P2P* [38].
  On the other hand, Moore et al. [28] noted that the use of the *IANA* for this type of classification was not a good approach as it only obtained 70% accuracy in the classification. In turn, Madhukar et al. [24] showed that port-based analysis cannot identify between 30 and 70% of the traffic that exists in *Internet*.
- *The reconstruction of the signature of the payload.* Currently, many applications to avoid having to make use of ports, use a system of reconstructing session state from the information contained in the package. Sen et al. [39] showed that the classification of *payload* of traffic on P2P networks could reduce the number of false positives and false negatives.
  In turn Moore et al. [28] did a hybrid check between the ports and the *payload* of the package. In a first approximation they checked if the port belonged to the list *IANA*, if it does not belong it would pass to the next phase. In the second phase, they would examine the connection packets. In them they would analyze whether the signature of the *payload* is known or unknown.
  Although this approach gives us a dependency on the need to know port numbers, it incorporates a complexity in processing the *payload* on the device where we want to analyse the apps. This device will have to be continuously updated, in addition to the need for sufficient power to be able to analyze a large number of network packets. However, this approach may be completely useless if for example the traffic is encrypted or uses a private protocol.

The above approaches are limited by the semantics of the information obtained, due to the necessary inspection of the obtained packets, as well as the *payload* and port numbers. On the other hand, more recent studies are based on obtaining the statistical characteristics of traffic to identify an application. In other words, network traffic has a number of statistical properties, such as the distribution of the duration of the flow, the flow of downtime or time between packet arrivals and packet lengths, that make them unique for certain types of applications, thus allowing them to differentiate from each other.

Paxson et al. [31] analyzed and built empirical models on the characteristics of the connection. Within this approach, they made use of the bytes of the connection, the duration of the connection, as well as the periodicity of the arrival, extracted from a specific number of applications that used text. Also, Dewes et al. [12] conducted a study on the *chat* of *Internet*. In it, they focused on the characteristics of the traffic, more specifically on the duration of the packet flow, the time between packet arrivals and the size of each packet. Later studies would make evolutions of this type of approaches, but adding in turn the lengths of the packages, as well as the difference between the arrivals of the same [9,21,22]. The results of this work have stimulated new classification techniques based on traffic flow statistical properties.

In the past a number of studies have been carried out in which they analysed the use of spectrum and the characteristics of the application in mobile data networks [26,33,36]. Most of these previous studies attempted to understand the use of the wireless spectrum and in turn characterize the performance of the network.

In order to create a pattern of network usage with its behavior, it is necessary to make a comprehensive analysis of all traffic generated on that network. Although some recent studies [20,42] have been analysing the samples collected on the Internet, focusing specifically on voice-generated traffic or traffic generated by the user's browsing, detailed studies of network traffic are needed.

On the other hand, McGregor et al. [27] published one of the first works related to the use of Machine Learning for the classification of network traffic. The approach groups traffic with similar properties for different types of applications.

Later, Moore et al. [29] proposed the use of the classifier *Naive Bayes* to categorize the traffic generated by the applications. The traffic used for the dataset is manually tagged to allow accurate evaluation by the classifier.

To carry out the classification, they made use of 248 extracted characteristics to train the classifier. The traffic selected from different applications was grouped into different categories to get a good ranking.

Also Bernaille et al. [2] proposed an approach using the classifier *K-Means* or *K-neighbors* to classify different types of applications based on *TCP* using only the first communication packets of the same. In contrast to other techniques carried out by other authors, Bernaille et al. could detect and classify applications earlier thanks to the use of the first network packets. In short, the focus is on getting the first packets that are usually predefined sequences of messages between applications.

Nguyen et al. [30] described a method for addressing the classification of network traffic across a finite number of network packets. As a direct consequence of the use of a small number of packets, it is the decrease in the time used for classification, thus reducing the space needed for intermediate storage of information related to the packets needed for classification. The use of this approach allows the classification process to begin at any time, not just at the beginning of the application as in previous approaches.

The work of Erman et al. [14] was aimed at the challenge of classifying traffic at the core of the network, where available information about packets and their connections might be limited. Therefore, in the work they proposed the use only of the outgoing information from the device.

Also, Crotti et al. [11] proposed a flow classification mechanism based on three properties of captured *IP* packets: (i) packet length, (ii) time between arrivals and (iii) order of arrival of packets. They defined a structure called protocol footprints that express the three traffic properties in a compact form and use an algorithm based on normalized thresholds for flow classification.

Haffner et al. [18] proposed a method for the automatic generation of signatures from applications using automatic learning techniques. Unlike other approaches, this work makes use of the first *N-Bytes* of a packet stream as features for signature generation.

Erman et al. [13] demonstrated that accuracy in text is crucial when evaluating classification algorithms. In this hypothesis, they pointed out that most *Internet* flows are reduced, representing only a small portion of the total *bytes* and packets in the network. On the other hand, they also observed that most of the *bytes* of traffic are generated by a small number of large network packet flows.

Also, Brezo [6] showed that the use of network packets for the detection of *Command&Control* channels was a good approach. In this research, the author made use of *N-Packets* to classify them using supervised learning algorithms.

## 2   Methodology

The methodology will focus on collecting information on two clearly differentiated types of traffic: (i) on the one hand, packets belonging to legitimate connections carried out by applications obtained from the *Google* application market and validated as benign, and (ii) on the other hand, packets belonging to malicious applications obtained from a dataset of the scientific community.

Following the installation of the *goodware* and *malware* applications, we proceeded to run them within a controlled test environment, avoiding the inclusion of unwanted biases by other applications.

## 2.1    Environment Preparation

Once we have defined the obtaining of samples necessary for the execution of the experimentation, we have to do the preparation of the environment in which these samples will be executed to analyse them.

The work philosophy proposed in previous works [7] is based on *Intrusion Detection System*, based on host. That is to say, the network traffic listening point is located in the infected nodes of the network.

Since the applications that are going to run are based on the operating system *Android*, you need to run a virtual machine in order to execute its code. To this end, we have made use of the virtual machine system *Genymotion*[1]. This is based on the virtualization platform *VirtualBox*[2] of the company *Oracle*. *Genymotion*provides a number of operating system virtual machines *Android* modified to have access *root*.

The virtual machine used for this experiment has a processor, with a text memory of 1024 MB, a screen resolution of $720 \times 1280$ and a text density of 320. The system we used was *Android* 4.1.1 *Jelly Bean*.

Traffic monitoring will be carried out using one of the best known tools: *Wireshark*. *Wireshark* is a multi-platform package analysis tool. It is commonly used for monitoring problems related to network connectivity, traffic analysis and the development of *software* or new communication protocols.

## 2.2    Automation Process

Getting legitimate and malicious traffic was done by automating the virtual machine. To create this process, we have made use of the programming language C#. Figure 1 shows the automation process carried out to obtain the traffic.

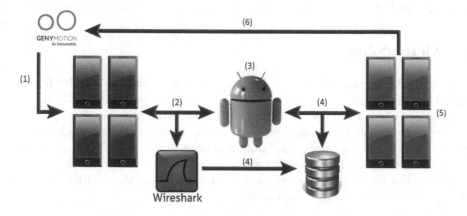

**Fig. 1.** Virtual machine automation process

---

[1] https://www.genymotion.com/.
[2] https://www.virtualbox.org/.

In the first section of the system, we have made use of the application *Geny-motion* to create the initial virtual machine with the features discussed in the Subsect. 2.1. To the initial virtual machine we had to install the *Google Apps*, a series of native applications of *Android*, through which, the applications can access the *APIs* that *Google* offers.

Next, we make a copy of this virtual machine (1), which will be where we install the necessary applications. After creating and starting the machine, we will associate an instance of *Wireshark* to the machine for the purpose of getting all the network packets running (2). Subsequently, we will make an installation of one of the applications with its consequent execution inside the system *Android*, leaving it in execution during a time of one minute (3). Once this time has elapsed, we have proceeded to store all the network files, one for each application executed (4). Finally, the virtual machine used to run an application is removed (5) to restart the process by making a copy of the original virtual machine (6).

## 3 Empirical Validation

In this section we will proceed to collect data from the experiments carried out for experiment. As explained above, the process consists of the analysis of the results for the representation of individual packages from the applications. To do this we will represent the results as malicious or benign traffic within the applications.

As a learning model, we have used different supervised learning algorithms with cross validation. Specifically:

- *Bayesian networks (BN):* With regards to Bayesian networks, we utilize different structural learning algorithms: K2 [10] and Tree Augmented Naïve (TAN) [17]. Moreover, we also performed experiments with a Naïve Bayes Classifier [3].
- *Support Vector Machines (SVM):* We performed experiments with a polynomial kernel [1], a normalized polynomial Kernel [25], a Pearson VII function-based universal kernel [41] and a radial basis function (RBF) based kernel [8].
- *K-nearest neighbor (KNN):* We performed experiments with $k = 1$, $k = 2$, $k = 3$, $k = 4$, $k = 5$, $k = 6$, $k = 7$, $k = 8$, $k = 9$, and $k = 10$.
- *Decision Trees (DT):* We performed experiments with J48 (the WEKA [16] implementation of the *C4.5* algorithm [35]) and Random Forest [5], an ensemble of randomly constructed decision trees. In particular, we tested random forest with a 100 trees.
- *Meta models:* We use in this case *Bagging* method that is a meta-algorithm designed to improve the stability and accuracy of other automatic learning algorithms in statistical classification and regression processes [4].

We have made use of the following algorithms: (i) *Random Forest*, (ii) *J48*, (iii) *Naïve Bayes*, (iv) *BayesNet*, (v) *IBK* with values of $K = 1, 2, 3, 4, 5, 10$.

We have extracted the individual atomic characteristics that define each of the packets of a given connection. In total, the total features extracted are 23. Because the amount of features is not numerous it is not necessary to make use of *Information Gain* techniques [15] to make sorting easier. In our case, the use of this technique would imply a computational cost giving a marginal improvement to the results.

The evaluation of each of the methods will be carried out on the basis of different parameters usually used in the evaluation of the performance of the classification methods [34].

- *True Positive Ratio (TPR)*. Conceptually, in *text mining*, is identified as the probability that a randomly chosen document is truly relevant in the search. In our case, it can be defined as the rate of correctly detected malicious packages among all those analyzed by the classifier. It is calculated by dividing the number of correctly classified malicious packages ($TP$) by the total number of extracted malicious samples, either *true positives* or *false negatives* ($TP + FN$) (Eq. 1):

$$TPR = \frac{TP}{TP + FN} \tag{1}$$

- *Accuracy*. It is calculated by dividing, in this case, the total number of hits by the number of instances that make up the entire data set (Eq. 2).

$$Acc = \frac{TP + TN}{TP + TN + FP + FN} \tag{2}$$

- *Precision or Accuracy*. Precision is a value that presents the possibility that a positive result reflects the condition that has been tested, i.e. that a sample labelled as positive is indeed a true positive. It is defined as (equation):

$$PPV = \frac{TP}{TP + FP} \tag{3}$$

- *F-measure (F)*. F-measure is a typical *Information Retrieval* metric. (*IR*) to evaluate the result of a search that determines to what extent the groups obtained resemble those that would have been achieved with manual categorization. This measure can be defined as 4:

$$F = 2 \cdot \frac{precision \cdot recall}{precision + recall} = 2 \cdot \frac{PPV \cdot TPR}{PPV + TPR} \tag{4}$$

This way if we substitute in $F$ the values of $PPV$ and $TPR$ we will be able to simplify the formula as follows:

$$F = 2 \cdot \frac{TP}{2 \cdot TP + FN + FP} \tag{5}$$

- The meaning of the area below the curve $ROC$ is the establishment of the relationship between false negatives and false positives [40] and is obtained by representing, for each possible choice of cut-off values the $TPR$ on the ordinate axis and the *FPR* on the abscissa axis. Originally used by the US military in investigations following the 1941 Pearl Harbor attack to correctly detect radar signatures on Japanese aircraft, it has been used since the late 20th century as a comparison and evaluation metric for different classification algorithms. It is often used to generate statistics representing the performance of a classifier in a binary system as a tool for selecting possibly optimal models and discarding suboptimal models. Although it does not provide information on the good behaviour of the model, the text AUC helps to determine the validity of a data distribution over a given set of predictive conditions [23].

## 3.1 Capacity of the Classifier for Classifying Traffic

The Table 1 shows the results obtained with the methodology discussed in the Sect. 2 for the network traffic generated by the applications for the system *Android*.

The best result in terms of *Accuracy* is obtained by the classifier *Bagging* with a value greater than 80%. But this result is an exception, since most classifiers range between 55% and 65%. At the same time, it is observed that the worst classifier of all is *Voted Perceptron*, which slightly exceeds 50% of well classified samples.

As far as the true positive rates are concerned, the results are very high, with rates as high as 0.99 in various classifiers, such as Bayesian nets or decision trees, and around values close to and above 0.90 for the rest of the classifiers, assuming a classification reliability higher than 80% in all classifiers. On the other hand, the worst classifier is still *Voted Perceptron*, which with a true positive rate of 0.69 and a reliability of 76% lets us see that it is not a good classifier for this type of methodology.

The *F-Measure* presents fairly homogeneous results. There is a variation between 0.81 and 0.93 in all classifiers except, once again, *Voted Perceptron*, which gets a value of 0.71, ten points less than the worst classifier of the rest. On the other hand, the best classifiers are found again in Bayesian networks, using the search algorithms *K2* and *Hill Climbing*, and the decision trees, adding also the classifier *KN* with the value of $k = 1$, all of them with a value of 0,93.

The value of the area under the curve *ROC* higher obtained was 0,71 for the classifiers *Bagging* and *KNN* with values of $k = 1, 2, 3, 4$. Even so, the values we find are much lower, ranging from 0.53 obtained by vector support machines and between 0.60 and 0.70 from Bayesian networks and the rest of the values of $k = 5, 6, 7, 8, 9, 10$ from *KNN*. Again, the worst classifier is *Voted Perceptron* with a value for the area of 0.48, which only confirms what was said earlier about the ineffectiveness of the classifier in this type of problem.

## 4   Summary

In this paper, we have shown the results obtained with the representation models proposed above. First, we have carried out a study of the most relevant research techniques in the use of communication protocols and the detection of *malware*.

**Table 1.** Results obtained by classifiers for the analysis of network traffic in *Android* applications

| Classifier | Accuracy (%) | TRP | Precision | F-Measure | AUC |
|---|---|---|---|---|---|
| SMO Polik. | 60,40 | 0,85 | 0,83 | 0,81 | 0,53 |
| SMO Polik. N. | 60,56 | 0,88 | 0,83 | 0,82 | 0,53 |
| SMO RBF | 58,85 | 0,88 | 0,80 | 0,81 | 0,52 |
| SMO Pear-VII | 63,63 | 0,80 | 0,90 | 0,82 | 0,56 |
| Bagging REPT | 81,69 | 0,96 | 0,95 | 0,92 | 0,71 |
| Naive Bayes | 59,58 | 0,93 | 0,87 | 0,90 | 0,60 |
| BayesNet K2 | 65,79 | 0,99 | 0,87 | 0,93 | 0,66 |
| Bayes Net Hill | 65,89 | 0,99 | 0,87 | 0,93 | 0,66 |
| Bayes Net TAN | 66,80 | 0,97 | 0,88 | 0,89 | 0,67 |
| J48 | 64,70 | 0,99 | 0,87 | 0,93 | 0,65 |
| Random Forest | 67,47 | 0,99 | 0,87 | 0,93 | 0,67 |
| Voted Per. | 53,36 | 0,69 | 0,76 | 0,71 | 0,48 |
| KNN $k = 1$ | 65,79 | 0,96 | 0,95 | 0,93 | 0,71 |
| KNN $k = 2$ | 65,79 | 0,97 | 0,96 | 0,92 | 0,71 |
| KNN $k = 3$ | 65,79 | 0,96 | 0,95 | 0,92 | 0,71 |
| KNN $k = 4$ | 65,79 | 0,96 | 0,94 | 0,92 | 0,71 |
| KNN $k = 5$ | 65,79 | 0,95 | 0,94 | 0,91 | 0,70 |
| KNN $k = 6$ | 65,79 | 0,95 | 0,92 | 0,91 | 0,69 |
| KNN $k = 7$ | 65,79 | 0,94 | 0,94 | 0,91 | 0,69 |
| KNN $k = 8$ | 65,79 | 0,94 | 0,92 | 0,90 | 0,69 |
| KNN $k = 9$ | 65,79 | 0,94 | 0,92 | 0,90 | 0,69 |
| KNN $k = 10$ | 65,79 | 0,93 | 0,91 | 0,89 | 0,68 |

Next, we have explained the methodology created in previous publications for the detection of *malware* on conventional computers. Since this approach is intended for mobile devices, the methodology has been modified to focus on this type of terminal. In addition, the two sets of data used for experimentation have been explained at this point. The automation process used to obtain the information from the experiments was then explained.

We then proceeded to define the problem as a problem of supervised classification. To this end, the training packages used have been adapted by tagging traffic samples knowingly from malicious or benign applications.

Then, we defined the 13 automatic learning algorithms used as classifiers: from vector support machines, to decision trees, through perceptrons, Bayesian networks or the analysis of the nearest $k$ in the n-dimensional space analyzed. Next, we have analyzed in advance which have been the metrics to use to compare the results obtained in the different experimental processes choosing for it some of the most used in the field of *machine learning*. Finally, we have proceeded to represent the results obtained through the use of the selected classifiers.

# References

1. Amari, S.I., Wu, S.: Improving support vector machine classifiers by modifying kernel functions. Neural Netw. **12**(6), 783–789 (1999)
2. Bernaille, L., Teixeira, R., Akodkenou, I., Soule, A., Salamatian, K.: Traffic classification on the fly. ACM SIGCOMM Comput. Commun. Rev. **36**(2), 23–26 (2006)
3. Bishop, C.M.: Pattern Recognition and Machine Learning. Springer, New York (2006)
4. Breiman, L.: Bagging predictors. Mach. Learn. **24**(2), 123–140 (1996)
5. Breiman, L.: Random forests. Mach. Learn. **45**(1), 5–32 (2001)
6. Brezo, F.: Detección de tráfico de control de botnets modelizando el flujo de los paquetes de red. Ph.D. thesis, University de Deusto, Febrero 2014
7. Brezo, F., de la Puerta, J.G., Barroso, D.: BRIANA: Botnet detection Relying on an Intelligent Analysis of Network Architecture. Master's thesis, University de Deusto, España (2012)
8. Cho, B., Yu, H., Lee, J., Chee, Y., Kim, I., Kim, S.: Nonlinear support vector machine visualization for risk factor analysis using nomograms and localized radial basis function kernels. IEEE Trans. Inf Technol. Biomed. **12**(2), 247–256 (2008)
9. Claffy, K.C.: Internet traffic characterization. Ph.D. thesis, University of California, San Diego (1994)
10. Cooper, G.F., Herskovits, E.: A bayesian method for constructing bayesian belief networks from databases. In: Proceedings of the Seventh conference on Uncertainty in Artificial Intelligence, pp. 86–94. Morgan Kaufmann Publishers Inc. (1991)
11. Crotti, M., Dusi, M., Gringoli, F., Salgarelli, L.: Traffic classification through simple statistical fingerprinting. ACM SIGCOMM Comput. Commun. Rev. **37**(1), 5–16 (2007)
12. Dewes, C., Wichmann, A., Feldmann, A.: An analysis of internet chat systems. In: Proceedings of the 3rd ACM SIGCOMM Conference on Internet Measurement, pp. 51–64. ACM (2003)
13. Erman, J., Mahanti, A., Arlitt, M.: Byte me: a case for byte accuracy in traffic classification. In: Proceedings of the 3rd Annual ACM Workshop on Mining Network Data, pp. 35–38. ACM (2007)
14. Erman, J., Mahanti, A., Arlitt, M., Williamson, C.: Identifying and discriminating between web and peer-to-peer traffic in the network core. In: Proceedings of the 16th International Conference on World Wide Web, pp. 883–892. ACM (2007)
15. Föllmer, H.: On entropy and information gain in random fields. Probab. Theory Relat. Fields **26**(3), 207–217 (1973)
16. Garner, S.: Weka: the waikato environment for knowledge analysis. In: Proceedings of the 1995 New Zealand Computer Science Research Students Conference, pp. 57–64 (1995)
17. Geiger, D., Goldszmidt, M., Provan, G., Langley, P., Smyth, P.: Bayesian network classifiers. Mach. Learn. **29**, 131–163 (1997)
18. Haffner, P., Sen, S., Spatscheck, O., Wang, D.: ACAS: automated construction of application signatures. In: Proceedings of the 2005 ACM SIGCOMM Workshop on Mining Network Data, pp. 197–202. ACM (2005)

19. Karagiannis, T., Broido, A., Brownlee, N., Claffy, K., Faloutsos, M.: Is p2p dying or just hiding? [p2p traffic measurement]. In: Global Telecommunications Conference 2004. GLOBECOM 2004, vol. 3, pp. 1532–1538. IEEE. IEEE (2004)
20. Keralapura, R., Nucci, A., Zhang, Z.L., Gao, L.: Profiling users in a 3G network using hourglass co-clustering. In: Proceedings of the Sixteenth Annual International Conference on Mobile Computing and Networking, pp. 341–352. ACM (2010)
21. Lang, T., Armitage, G., Branch, P., Choo, H.Y.: A synthetic traffic model for half-life. In: Australian Telecommunications Networks & Applications Conference, vol. 2003 (2003)
22. Lang, T., Branch, P., Armitage, G.: A synthetic traffic model for quake3. In: Proceedings of the 2004 ACM SIGCHI International Conference on Advances in Computer Entertainment Technology, pp. 233–238. ACM (2004)
23. Lobo, J.M., Jiménez-Valverde, A., Real, R.: AUC: a misleading measure of the performance of predictive distribution models. Glob. Ecol. Biogeogr. 17(2), 145–151 (2008)
24. Madhukar, A., Williamson, C.: A longitudinal study of p2p traffic classification. In: 14th IEEE International Symposium on Modeling, Analysis, and Simulation of Computer and Telecommunication Systems 2006. MASCOTS 2006, pp. 179–188. IEEE (2006)
25. Maji, S., Berg, A., Malik, J.: Classification using intersection kernel support vector machines is efficient. In: IEEE Conference on Computer Vision and Pattern Recognition (CVPR), pp. 1–8. IEEE (2008)
26. Mattar, K., Sridharan, A., Zang, H., Matta, I., Bestavros, A.: TCP over CDMA2000 networks: a cross-layer measurement study. In: Uhlig, S., Papagiannaki, K., Bonaventure, O. (eds.) PAM 2007. LNCS, vol. 4427, pp. 94–104. Springer, Heidelberg (2007). https://doi.org/10.1007/978-3-540-71617-4_10
27. McGregor, A., Hall, M., Lorier, P., Brunskill, J.: Flow clustering using machine learning techniques. In: Barakat, C., Pratt, I. (eds.) PAM 2004. LNCS, vol. 3015, pp. 205–214. Springer, Heidelberg (2004). https://doi.org/10.1007/978-3-540-24668-8_21
28. Moore, A.W., Papagiannaki, K.: Toward the accurate identification of network applications. In: Dovrolis, C. (ed.) PAM 2005. LNCS, vol. 3431, pp. 41–54. Springer, Heidelberg (2005). https://doi.org/10.1007/978-3-540-31966-5_4
29. Moore, K.: 71% of online adults now use video-sharing sites. Pew Internet and American Life Project (2011)
30. Nguyen, T.T., Armitage, G.: Training on multiple sub-flows to optimise the use of machine learning classifiers in real-world IP networks. In: Proceedings 2006 31st IEEE Conference on Local Computer Networks, pp. 369–376. IEEE (2006)
31. Paxson, V.: Empirically derived analytic models of wide-area TCP connections. IEEE/ACM Trans. Networking (TON) 2(4), 316–336 (1994)
32. Paxson, V.: Bro: a system for detecting network intruders in real-time. Comput. Netw. 31(23), 2435–2463 (1999)
33. Pentikousis, K., Palola, M., Jurvansuu, M., Perala, P.: Active goodput measurements from a public 3G/UMTS network. IEEE Commun. Lett. 9(9), 802–804 (2005)
34. Powers, D.: Evaluation: From precision, recall and f-factor to ROC, informedness, markedness & correlation (Technical report). Adelaide, Australia (2007)
35. Quinlan, J.: C4.5: Programs for Machine Learning. Morgan kaufmann, San Mateo (1993)

36. Reichl, P., Umlauft, M.: Project WISQY: a measurement-based end-to-end application-level performance comparison of 2.5G and 3G networks. In: Wireless Telecommunications Symposium 2005, pp. 9–14. IEEE (2005)
37. Roesch, M., et al.: Snort: lightweight intrusion detection for networks. In: LISA, vol. 99, pp. 229–238 (1999)
38. Roughan, M., Sen, S., Spatscheck, O., Duffield, N.: Class-of-service mapping for QOS: a statistical signature-based approach to IP traffic classification. In: Proceedings of the 4th ACM SIGCOMM Conference on Internet Measurement, pp. 135–148. ACM (2004)
39. Sen, S., Spatscheck, O., Wang, D.: Accurate, scalable in-network identification of p2p traffic using application signatures. In: Proceedings of the 13th International Conference on World Wide Web, pp. 512–521. ACM (2004)
40. Singh, Y., Kaur, A., Malhotra, R.: Comparative analysis of regression and machine learning methods for predicting fault proneness models. Int. J. Comput. Appl. Technol. **35**(2), 183–193 (2009)
41. Üstün, B., Melssen, W.J., Buydens, L.M.: Facilitating the application of support vector regression by using a universal pearson vii function based kernel. Chemometr. Intell. Lab. Syst. **81**(1), 29–40 (2006)
42. Willkomm, D., Machiraju, S., Bolot, J., Wolisz, A.: Primary users in cellular networks: a large-scale measurement study. In: 3rd IEEE Symposium on New Frontiers in Dynamic Spectrum Access Networks 2008. DySPAN 2008, pp. 1–11. IEEE (2008)
43. Zeilenga, K.D.: Internet assigned numbers authority (IANA) considerations for the lightweight directory access protocol (LDAP) (2002)

# Inferring Knowledge from Clinical Data for Anesthesia Automation

Jose M. Gonzalez-Cava[1(✉)], Iván Castilla-Rodríguez[1],
José Antonio Reboso[2], Ana León[2], María Martín[2],
Esteban Jove-Pérez[1,3], José Luis Calvo-Rolle[3],
and Juan Albino Méndez-Pérez[1]

[1] Departamento de Ingeniería Informática y de Sistemas,
Universidad de La Laguna (ULL), 38200 La Laguna, Tenerife, Spain
jgonzalc@ull.edu.es
[2] Hospital Universitario de Canarias,
San Cristóbal de La Laguna, Tenerife, Spain
[3] Department of Industrial Engineering, Universidade da Coruña, Coruña, Spain

**Abstract.** The use of Hybrid Artificial Intelligent techniques in medicine has increased in recent years. Specifically, one of the main challenges in anesthesia is achieving new controllers capable of automating the drug titration during surgeries. This work deals with the development of a Takagi-Sugeno fuzzy controller to automate the drug infusion for the control of hypnosis in patients undergoing anesthesia. To do that, a combination of Neural Networks and optimization techniques were applied to tune the internal parameters of the fuzzy controller. For the training process, data from 20 patients undergoing surgery were used. Finally, the controller proposed was tested over 16 virtual surgeries. It was concluded that the fuzzy controller was able to meet both clinical and control objectives.

**Keywords:** Artificial Intelligence · Control of anesthesia · Fuzzy controller · ANFIS

## 1 Introduction

The use of automatic control techniques based on Artificial Intelligence (AI) in engineering has been gradually grown in recent years [1, 2]. Regarding the biomedical engineering field, the use of these tools has grown considerably within the scope of anesthesiology [3, 4]. Anesthesia is aimed at protecting patients from the aggression that a surgical intervention involves. The main objective of automatic control in anesthesia is to develop a controller that allows the dosage of drug automatically according to the real needs of patients [5].

One of the drawbacks in the design of a controller for the anesthetic process is the difficulty in having a reliable model able to simulate the response of patients [6]. In this sense, recent studies have been based on AI techniques to propose new models from real data [7]. Fuzzy logic together with genetic algorithms has been applied to deal with patient uncertainty in models [8]. Different machine learning algorithms were also used

© Springer Nature Switzerland AG 2019
H. Pérez García et al. (Eds.): HAIS 2019, LNAI 11734, pp. 480–491, 2019.
https://doi.org/10.1007/978-3-030-29859-3_41

in [9] to model perioperative hypoxia. Stochastic population model for neuromuscular blockade based on artificial intelligence algorithm has been also proposed [10]. Clustering as well as regression techniques have been used for identifying intrinsic features of patient in order to improve the prediction of models [11].

The main objective of this research is the use of Artificial Intelligence to design an automatic controller able to decide the dose of drug to be supplied. Specifically, a controller to compute the accurate dose of propofol (hypnotic drug) that results in the appropriate level of hypnosis is aimed. Among the different types of controllers available in the control theory, a Takagi-Sugeno fuzzy logic controller is proposed. The main reason for the use of fuzzy controllers lies in the versatility to study complex systems in which physical-chemical knowledge is limited. Furthermore, it is easy to incorporate medical experience in an intuitive way.

Previous proposals have been based on traditional control methods [12]. Several adaptive and robust controllers have been developed to improve previous results [13, 14]. The main problem found is the difficulty to tune the parameters of the controllers intuitively due to the absence of a straightforward relationship with clinical factors that affect the process. Regarding fuzzy controllers, main research has been focused on the design of Mamdani fuzzy controllers based on the theoretical criteria of the clinicians [15, 16]. The main novelty of this research lies in the capability of including practical knowledge from real clinical data to design a fuzzy controller. To define the structure of the fuzzy controller, Adaptive Neuro Fuzzy Inference System (ANFIS) is proposed. This type of algorithm let combine the learning capacity of Neural Networks with Fuzzy Logic from a dataset of training samples. In addition, optimization techniques are also included to tune some internal parameters. Dataset for training the algorithm has been captured from real surgical interventions.

This paper is structured as follows. First, the Case Study is presented. Then, the Methodology used in this study is described. After that, the design process of the fuzzy logic controller is explained. Finally, the results obtained in this study are analysed.

# 2  Case Study

Anesthesia can be regarded as a multivariable control scenario in which a combination of drugs must be supplied to induce the desired level of hypnosis (loss of consciousness), analgesia (absence of pain) and muscular blockade [17]. The standard work scheme in anesthesiology consists of administering a drug, analyzing the drug-patient interaction and, finally, adapting the pharmacological requirements to the characteristics of the patient. From the anesthetic point of view, three phases are generally identified during an intervention. First, the induction involves the transition from the waking state to the hypnosis state. Then, during the maintenance, the anesthetic situation achieved after induction must be guaranteed throughout the intervention. Finally, at the recovery, the administration of drug ceases and patient returns to the waking state.

This research is focused on the automatic control of hypnosis. Specifically, the description of the main variables involved in this study is shown in Fig. 1. The feedback variable to measure the hypnotic state of the patient was the Bispectral Index (BIS). By means of the analysis of the electroencephalographic activity, this monitor is capable of computing an index that ranges from 0 (absence of brain activity) to 100 (awake patient). In General Anesthesia, BIS values within the range from 40 to 60 are recommended. In this research, a target value of BIS = 50 was proposed for the design of the controller. Furthermore, the manipulated variable to control the level of hypnosis was the propofol infusion rate. It was possible to modify this variable through an infusion pump. From the clinical perspective, maximum dose of opioids was limited to minimize possible interactions with BIS signal [18]. Despite of this fact, an adequate level of analgesia was ensured for patients enrolled in this study.

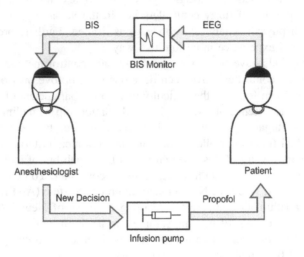

**Fig. 1.** Scheme of the anesthetic process for the control of hypnosis.

## 3   Methodology

### 3.1   Basis of Fuzzy Logic Controllers

Fuzzy controllers are tools specifically designed for situations in which is difficult to have the complete analytical description of a process, but in which empirical knowledge, expertise or patterns are available to control it. The basis of fuzzy logic lies in fuzzy sets. A fuzzy set is defined as a class that represents a gradual progression from "belonging" to "non-belonging" to a certain set [19]. Therefore, an element is part of a set in a certain degree of belonging. A fuzzy system with $n$ inputs $u_i \in U_i$ where $i = 1$, $2, \ldots, n$ and $m$ outputs $y_i \in Y_i$ with $i = 1, 2, \ldots, m$ is assumed. All the possible values that an input $u_i$ and output $y_i$ can reach is called the Universe of Discourse, expressed as $U_i$ and $Y_i$ respectively. Then, to be able to represent the knowledge through a set of rules, it is necessary to use linguistic variables. A linguistic variable is used to express features of the inputs and outputs by means of words or sentences [20]. Linguistic variables are denoted by $\tilde{u}_i$ to describe the input $u_i$ and by $\tilde{y}_i$ for the output $y_i$. In addition, linguistic variables are described by means of linguistic values, so that $\tilde{A}_i^j$ denotes the j-th linguistic value of the input linguistic variable $\tilde{u}_i$, defined on the universe of discourse $U_i$. Finally, according to fuzzy set theory, a membership function is able to describe the degree of membership of an element $u_i$ within the Universe of Discourse in terms of linguistic variables as follows:

$$\mu_{A_i^j}(u_i) = X \rightarrow [0, 1] \tag{1}$$

Different shapes can be used to express membership functions such as triangular, Gaussian or trapezoidal depending on the problem. Then, fuzzy inputs and outputs are related through a set of rules. Considering a Takagi-Sugeno structure for a MISO system, rules are expressed by means of if-else statements as follows:

$$\textit{If } u_1 \textit{ is } A_1^j \textit{ and}, \ldots, \textit{ and } u_n \textit{ is } A_n^l \textit{ then } b_i = g_i(\cdot) \tag{2}$$

where $b_i$ represents the output expressed as a mathematical function. Finally, a crisp value is obtained at the output by means of the defuzzification process according to the following expression:

$$y = \frac{\sum_{i=1}^{R} b_i \mu_i}{\sum_{i=1}^{R} \mu_i} \tag{3}$$

The general scheme of a fuzzy controller structure is depicted in Fig. 2

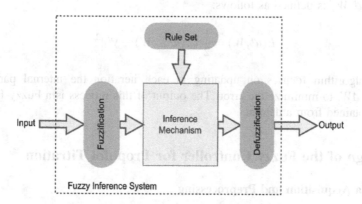

**Fig. 2.** General structure of a Fuzzy Inference System

### 3.2 Basis of ANFIS

Tuning a fuzzy controller consist of specifying some parameters such as type and number of membership functions or the definition of the set of rules. The Adaptive Neuro-Fuzzy Inference System (ANFIS) is a kind of artificial Neural Network that is based on Takagi-Sugeno Fuzzy Inference System (FIS) capable of tuning the fuzzy system from a dataset [21]. Specifically, the structure of Neural Networks is applied to a dataset in terms of linguistic representation and fuzzy logic rules. The architecture of ANFIS is subdivided into five layers:

- Layer 1: It is formed by adaptive nodes so that the parameters of the input membership functions are modified to obtain the best adjustment with the output samples.
- Layer 2: It is formed by fixed nodes that apply an AND function to all incoming signals from layer 1.
- Layer 3: Formed by fixed nodes, this layer applies a normalization step to the degree of belonging obtained in layer 2.
- Layer 4: This layer is formed by a series of adaptive nodes that calculates the output Takagi-Sugeno inference function.
- Layer 5: Formed by a single fixed node, it is responsible for adding all incoming signals from the previous layer and computing the outcome of the system.

For the minimization of the error between the dataset and the output generated by the ANFIS scheme, a hybrid algorithm based on backpropagation and least squares is used. This algorithm is capable of adjusting the adaptive parameters $(W)$ of the network. Considering a training set of known input-output data series $(u, y)$ formed by $p$ data $(q = 1, ..., p)$, the main objective of the algorithm is to minimize the total error obtained from the following expression:

$$E(W) = \sum_{q=1}^{p} E(u^q; W) \tag{4}$$

where $E(u^q, W)$ is defined as follows:

$$E(u^q, W) = \frac{1}{2} \|g(u^q - W) - y^q\|^2 \tag{5}$$

This algorithm focuses on updating in each iteration the internal parameters, $W = W + \Delta W$, to minimize the error. The output of this process is a Fuzzy Inference System obtained from a dataset.

## 4    Design of the Fuzzy Controller for Propofol Titration

### 4.1    Data Acquisition and Preprocessing

For this study, the dataset was obtained from 20 patients undergoing general surgery at the Hospital Universitario de Canarias. Bispectral Index and propofol dose were automatically registered every 5 s. A laptop ran an application developed in Matlab in order to communicate it through RS232 interfaces with the infusion pump as well as with the monitor.

After the analysis of the data obtained from the surgical interventions, it was observed that the BIS signal presented two main problems. On the one hand, BIS index rarely displayed an erroneous value (BIS = 0) because of a poor signal quality during the measurement process. To recompute these values, a linear interpolation between the points immediately before and after the appearance of the artefact was applied.

On the other hand, BIS signal was affected by a noise component. To reduce the effect of this random variation, the Savitzky-Golay filter was used [22]. This filter was based on the use of polynomial approximations through least squares adjustment. This type of filter is capable of maintaining the general characteristics of the original signal, such as maximum, minimum or width of the peaks. It is possible to approximate the signal at a point $t$ by a linear combination $g_t$ by means of the following expression [23]:

$$g_t = \sum_{n=nl}^{nr} c_n f_{t+n} \tag{6}$$

where $c_n$ refers to the outcome polynomials obtained by means of the least squares adjustment. In addition, $f_{t+n}$ refers to the original signal at time $t+n$. Therefore, this method adjusts a polynomial to the information within the data window ($n_l + n_r + 1$). Then, the value at time $t$ is calculated as $g_t$. For BIS signal, after testing different parameters, a Savitzky-Golay filter based on $2^{nd}$ order polynomials and a window of 31 samples (150 s.) was proposed.

## 4.2 Selection of Features

One of the most relevant aspects when applying AI techniques lies in the definition of the features involved in the decision-making process. Particularly a 2 inputs – 1 output controller was proposed in this research according to the criteria of the anesthesiologist:

- Input 1. *BIS error, e(t):* This input considered the difference between real BIS of the patient at time $t$, *BIS(t)*, and the target, *BIS$_r$*.
- Input 2. *BIS increment, ΔBIS(t).* This input regarded the variations of BIS during the anesthetic process. *ΔBIS(t)* was computed as the slope of the linear regression line obtained through the last 30 s of BIS measurements.
- Output. *Infusion rate increment, Δv.* The output of the Fuzzy controller proposed was the variation of the propofol infusion rate with respect to the previous instant $t$.

The general scheme of the closed loop system based on the inputs and outputs described, is depicted in Fig. 3.

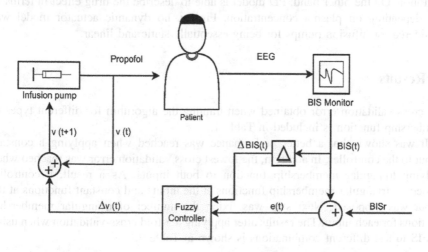

**Fig. 3.** Closed-loop system considering the inputs and outputs proposed in this study.

**4.3   Tuning the Fuzzy Controller**

For the training step, the toolbox ANFIS Editor GUI of Matlab was used according to the algorithm described in Sect. 3.2. For a Takagi-Sugeno fuzzy system, the objective was to determine the type, number and distribution along the Universe of Discourse of the membership functions, as well as the set of rules and the output functions. First, three types of membership functions were tested: triangular, trapezoidal and Gaussian. Analogously, the performance of the algorithm was studied by trying both constant values and linear functions at the output of the system. Once the best input-output combination was defined, a variation in the amount of membership functions for each input was tried. Finally, the performance of the system was determined by studying the Mean Square Error. A training based on 5-fold cross validation was proposed.

In addition, the importance or weight of each rule in the rule set obtained from the ANFIS was studied. Each rule is characterized by a weight that ranges from 0 to 1, specifying a relative importance among rules. By default, the ANFIS toolbox set the weight of the rules to 1. In order to adjust the weight of the rules, the application of an optimization technique to minimize the error between the reference and the hypnotic level of the patient was proposed.

$$J = \sum_{i=t_0}^{i=t_{end}} (50 - BIS_i)^2 \tag{7}$$

A constrained optimization based on *fmincon* function available in Matlab was performed. In order to minimize the cost function in (7) by means of adjusting the weight of the fuzzy rules, it was necessary to compare the evolution of the Bispectral Index ($BIS_i$) with the target BIS = 50. To obtain $BIS_i$, a virtual closed-loop based on the scheme depicted in Fig. 3 was simulated. First, it was necessary to model the patient response to propofol infusion rate. This model was based on pharmacokinetic and pharmacodynamics structures (PK-PD models) and Schnider parameters [24, 25]. On the one hand, the PK model correlates propofol infusion rate with plasma concentration. On the other hand, PD model is able to describe the drug effect in terms of BIS depending on plasma concentration. Finally, no dynamic actuator model was considered for infusion pumps for being essentially static and linear.

## 5   Results

The cross-validation error obtained when training the algorithm for different types of membership function is included in Table 1.

It was shown that a better performance was reached when applying a constant output to the controller. In addition, the lowest cross validation error was reached when applying triangular membership function to both inputs. As a result, a controller defined by triangular membership functions at the inputs and constant functions at the output was proposed. Next step was fixing the number of triangular membership functions for each input. The results after applying a 5-fold cross-validation when using ANFIS to test different combinations is shown in Table 2.

**Table 1.** Cross-validation error obtained when applying different combinations of membership functions for inputs and output.

| Membership functions | | Cross validation error [Mean ± SD] |
|---|---|---|
| Inputs | Output | |
| Triangular | Constant | 0.153 ± 0.000504 |
| Triangular | Linear | 0.611 ± 0.761 |
| Trapezoidal | Constant | 0.230 ± 0.0298 |
| Trapezoidal | Linear | 0.418 ± 0.340 |
| Gaussian | Constant | 0.167 ± 0.00244 |
| Gaussian | Linear | 0.446 ± 0.421 |

**Table 2.** Cross-validation error obtained when testing different number of membership functions for both inputs.

| Number of membership functions | | Cross validation error [Mean ± SD] |
|---|---|---|
| $e(t)$ | $\Delta BIS(t)$ | |
| 3 | 3 | 0.155 ± 0.000179 |
| 4 | 4 | 0.153 ± 0.000394 |
| 5 | 5 | 0.153 ± 0.000504 |
| 6 | 6 | 0.149 ± 0.000224 |
| 4 | 5 | 0.206 ± 0.0191 |

It was shown that the best results were obtained when applying 6 triangular membership functions to both inputs. Finally, an optimization process was applied to the Fuzzy Controller in order to fix the weight of each rule according to Sect. 4.3. The response surface of the resulting Fuzzy Controller is depicted in Fig. 4.

It was observed that, generally, for $e(t) < 0$, propofol infusion rate increased. Conversely, when $e(t) > 0$, it was observed that the dose of drug decreased. Finally, when $e(t)$ is close to 0, propofol infusion should keep at the same value ($\Delta v = 0$). Despite of the acceptable general behaviour in normal situations, there were some combinations that did not correspond to the expected behaviour of the controller. This problem was mainly due to the absence of these abnormal situations in the training data.

After the training process, the Takagi-Sugeno controller was tested over 16 virtual patients according to PK-PD models based on Schnider parameters. For the induction, a manual bolus of 2 mg/kg was supplied. The induction phase finished after one minute unless a BIS < 50 was reached before. Then, the automatic control started during the maintenance phase. A normally distributed random noise was added to simulated BIS signal to mimic the presence of noise. Recovery phase was not taken into account in the simulation. The evolution of BIS signal as well as the variations of propofol infusion rate according to the fuzzy controller is shown in Fig. 5.

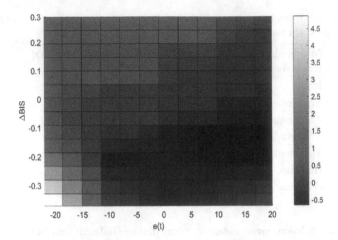

**Fig. 4.** Surface of response of the Fuzzy Controller after the training step. Grey bar represents the increments of propofol infusion rate in mg/kg/h. $\Delta BIS(t)$ and e(t) corresponds to the inputs of the controller.

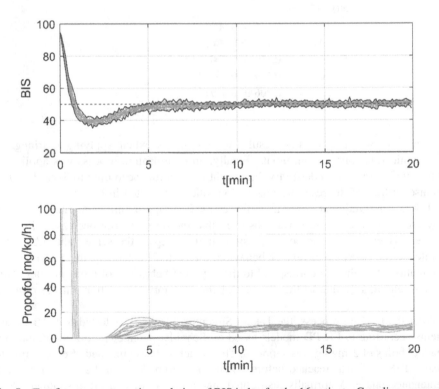

**Fig. 5.** Top figure represents the evolution of BIS index for the 16 patients. Grey lines represent the individual responses. Black lines are the highest and lowest values (bounds) reached at every sample during simulation. Dotted line shows the reference BIS = 50. Bottom figure represents the propofol infusion rates [mg/kg/h] provided by the controller to the individual patients.

It was shown that all the 16 patients reached a BIS $\cong$ 50 during the maintenance phase. Small divergences are mainly due to the presence of noise. In addition, it was checked that the infusion rate profiles met the recommended values for propofol infusion rate during maintenance phase for adult patients (3 to 12 mg/kg/h) [26]. Specifically, mean propofol infusion rate during maintenance was 7.45(1.57) mg/kg/h (expressed as Mean(SD)). As a result, the fuzzy controller obtained by means of the ANFIS technique was validated in simulation.

# 6    Conclusions

This study presents the development of a Hybrid Intelligent System based on a fuzzy structure capable of controlling the hypnotic process during anesthesia. Specifically, Neural Networks were used to design the structure of the Takagi-Sugeno controller. Then, optimization techniques were used to tune internal parameters of the fuzzy structure. Data for the training phase were obtained from 20 patients undergoing general surgery. Finally, a 2-inputs 1-output fuzzy controller was obtained. The Fuzzy Inference System was tested in virtual surgeries over 16 patients. It was concluded that the solution proposed was able to meet both clinical and control objectives.

This study opens new possible future research lines. On the one hand, the results could be outperformed testing new combinations of membership functions, including data from new patients or including abnormal situations in the training by means of virtual simulations. As a result, this controller could be finally tested in real practice.

On the other hand, one of the main current trends in the automation of anesthesia involves multivariable control. Specifically, main efforts focus on the control of both hypnosis and analgesia simultaneously. However, the main problem lies in the absence of a reliable index capable of quantifying analgesia objectively. Nevertheless, the methodology explained in this article could be applied once a reliable index is finally developed for the supply of analgesic. In addition, fuzzy methodology could be also helpful to try to correlate this hypothetical new indexes with traditional practice criteria as a tool for the analgesia monitors validation.

**Acknowledgements.** José Manuel Gonzalez-Cava's research was supported by the Spanish Ministry of Science, Innovation and Universities (http://www.ciencia.gob.es/) under the "Formación de Profesorado Universitario" grant FPU15/03347.

# References

1. González Gutiérrez, C., Sánchez Rodríguez, M.L., Fernández Diáz, R.Á., Calvo Rolle, J.L., Roqueñí Gutiérrez, N., Javier De Cos Juez, F.: Rapid tomographic reconstruction through GPU-based adaptive optics. Log. J. IGPL **27**, 214–226 (2019)
2. Vega Vega, R., Quintián, H., Calvo-Rolle, J.L., Herrero, Á., Corchado, E.: Gaining deep knowledge of Android malware families through dimensionality reduction techniques. Log. J. IGPL **27**, 160–176 (2019)

3. Méndez Pérez, J.A., Torres, S., Reboso, J.A., Reboso, H.: Estrategias de Control en la Práctica de Anestesia. Rev. Iberoam. Automática e Informática Ind. RIAI **8**(3), 241–249 (2011)
4. Reboso, J.A., Gonzalez-Cava, J.M., Leon, A., Mendez-Perez, J.A.: Closed loop administration of propofol based on a Smith predictor: a randomized controlled trial. Minerva Anestesiol. **85**, 585–593 (2018)
5. Le Guen, M., et al.: Automated sedation outperforms manual administration of propofol and remifentanil in critically ill patients with deep sedation: a randomized phase II trial. Intensive Care Med. **39**, 454–462 (2013)
6. Engbers, F.H.M., Dahan, A.: Anomalies in target-controlled infusion: an analysis after 20 years of clinical use. Anaesthesia **73**, 619–630 (2018)
7. Casteleiro-Roca, J.-L., Jove, E., Gonzalez-Cava, J.M., Méndez Pérez, J.A., Calvo-Rolle, J.L., Blanco Alvarez, F.: Hybrid model for the ANI index prediction using Remifentanil drug and EMG signal. Neural Comput. Appl. **2018**, 1–10 (2018)
8. Marrero, A., Méndez, J.A., Reboso, J.A., Martín, I., Calvo, J.L.: Adaptive fuzzy modeling of the hypnotic process in anesthesia. J. Clin. Monit. Comput. **31**(2), 319–330 (2017)
9. Sippl, P., Ganslandt, T., Prokosch, H.U., Muenster, T., Toddenroth, D.: Machine learning models of post-intubation hypoxia during general anesthesia. Stud. Health Technol. Inform. **243**, 212–216 (2017)
10. Rocha, C., Mendonça, T., Eduarda Silva, M.: Modelling neuromuscular blockade: a stochastic approach based on clinical data. Math. Comput. Model. Dyn. Syst. **19**, 540–546 (2013)
11. Esteban Jove, J.L.C.-R., et al.: Modelling the hypnotic patient response in general anaesthesia using intelligent models. Log. J. IGPL **27**, 189–201 (2018)
12. Ilyas, M., Butt, M.F.U., Bilal, M., Mahmood, K., Khaqan, A., Ali Riaz, R.: A review of modern control strategies for clinical evaluation of propofol anesthesia administration employing hypnosis level regulation. Biomed. Res. Int. **2017**, 12 (2017)
13. van Heusden, K., et al.: Optimizing robust PID control of propofol anesthesia for children; design and clinical evaluation. IEEE Trans. Biomed. Eng. **PP**, 1 (2019)
14. Dineva, A., Tar, J.K., Várkonyi-Kóczy, A., Piuri, V.: Adaptive controller using fixed point transformation for regulating propofol administration through wavelet-based anesthetic value. In: Proceedings of the 2016 IEEE International Symposium on Medical Measurements and Applications, MeMeA 2016 (2016)
15. Mendez, J.A., et al.: Improving the anesthetic process by a fuzzy rule based medical decision system. Artif. Intell. Med. **84**, 159–170 (2018)
16. Mendez, J.A., Marrero, A., Reboso, J.A., Leon, A.: Adaptive fuzzy predictive controller for anesthesia delivery. Control Eng. Pract. **46**, 1–9 (2016)
17. Chang, J.J., Syafiie, S., Kamil, R., Lim, T.A.: Automation of anaesthesia: a review on multivariable control. J. Clin. Monit. Comput. **29**(2), 231–239 (2015)
18. Bouillon, T.W., et al.: Pharmacodynamic interaction between propofol and remifentanil regarding hypnosis, tolerance of laryngoscopy, bispectral index, and electroencephalographic approximate entropy. Anesthesiology **100**(6), 1353–1372 (2004)
19. Passino, K.M., Yurkovich, S.: Fuzzy control. In: Levine, W.S. (ed.) The Control Systems Handbook: Control System Advanced Methods, 2nd edn. Addison-Wesley, Boston (2010)
20. Zadeh, L.A.: The concept of a linguistic variable and its application to approximate reasoning-I. Inf. Sci. (Ny). **8**, 199–249 (1975)
21. Jang, J.S.R.: ANFIS: adaptive-network-based fuzzy inference system. IEEE Trans. Syst. Man Cybern. **23**(3), 665–685 (1993)
22. Savitzky, A., Golay, M.J.E.: Smoothing and differentiation of data by simplified least squares procedures. Anal. Chem. **36**, 1627–1639 (1964)

23. Press, W.H., Teukolsky, S.A., Vetterling, W.T., Flannery, B.P.: Numerical Recipes 3rd Edition: The Art of Scientific Computing. Cambridge University Press, Cambridge (2007)
24. Gambús, P.L., Trocõniz, I.F.: Pharmacokinetic-pharmacodynamic modelling in anaesthesia. Br. J. Clin. Pharmacol. **79**, 72–84 (2015)
25. Minto, C.F., Schnider, T.W.: Contributions of PK/PD modeling to intravenous anesthesia. Clin. Pharmacol. Ther. **84**, 27–38 (2008)
26. DiLorenzo, A.N., Schell, R.M.: Morgan & Mikhail's Clinical Anesthesiology. Anesth. Analg. **119**, 495–496 (2014)

# Anomaly Detection Over an Ultrasonic Sensor in an Industrial Plant

Esteban Jove[1,4(✉)], José-Luis Casteleiro-Roca[1], Jose Manuel González-Cava[4],
Héctor Quintián[1], Héctor Alaiz-Moretón[2], Bruno Baruque[3],
Juan Albino Méndez-Pérez[4], and José Luis Calvo-Rolle[1]

[1] Department of Industrial Engineering, University of A Coruña,
Avda. 19 de febrero s/n, 15405 Ferrol, A Coruña, Spain
esteban.jove@udc.es, jose.luis.casteleiro@udc.es

[2] Departamento de Ingeniería Eléctrica y de Sistemas y Automática,
Universidad de León, Edificio Tecnológico - Campus de Vegazana s/n,
24071 León, Spain

[3] Departamento de Ingeniería Civil, Universidad de Burgos,
Calle Francisco de Vitoria, s/n, 09006 Burgos, Spain

[4] Department of Computer Science and System Engineering, Universidad de La
Laguna, Avda. Astrof. Francisco Sánchez s/n, 38200 S/C de Tenerife, Spain

**Abstract.** The significant industrial developments in terms of digitalization and optimization, have focused the attention on anomaly detection techniques. This work presents a detailed study about the performance of different one-class intelligent techniques, used for detecting anomalies in the performance of an ultrasonic sensor. The initial dataset is obtained from a control level plant, and different percentage variations in the sensor measurements are generated. For each variation, the performance of three one-class classifiers are assessed, obtaining very good results.

**Keywords:** Anomaly detection · Control system · Ultrasonic sensor

## 1 Introduction

In recent decades, most companies present industrial processes whose operation can be optimized [1,18,28]. Technological advances in fields like instrumentation or digitalization are important tools to achieve optimization tasks [16]. In a context where the energy optimization or the economic competitiveness are promoted, the use of intelligent techniques to detect anomalies in actuators, sensors, and so on, plays an important role [12,21,35].

Theoretically, anomalies are data patterns that does not represent an expected behaviour in a given application [8]. The anomaly detection must face some problems, like: choosing a boundary between anomalous and normal data, the appearance of noise or the availability of data from fault situations [8].

Anomaly detection techniques are used in many different fields and applications, such as fault detection in industrial processes, intrusion detection in

© Springer Nature Switzerland AG 2019
H. Pérez García et al. (Eds.): HAIS 2019, LNAI 11734, pp. 492–503, 2019.
https://doi.org/10.1007/978-3-030-29859-3_42

surveillance systems or fraud detection of credit cards [13, 26, 38]. The anomaly detection problem can be solved using intelligent models [14, 24] or projection methods [10, 22, 39].

This work deals the anomaly detection over an ultrasonic sensor used to measure the liquid level in a tank. In this case, the initial dataset is recorded from a laboratory plant used to control the liquid level in a tank. As the data is registered from normal operation, it is proposed the use of one-class techniques to obtain anomaly classifiers. The anomalies are artificially generated, by modifying the ultrasonic sensor measurements a certain percentage. The performance of the classifiers were assessed varying the percentage deviation from 5% to 50%. The used techniques are: Autoencoder, Approximate Convex Hull (ACH), and Support Vector Machine (SVM).

This work is organized as follows. After the present section, the case of study is detailed. Then, Sect. 3 describes the used techniques to obtain the classifiers. Section 4 presents the experiments and results and then, conclusions and future works are exposed in Sect. 5.

## 2   Case of Study

The description of the plant under study and the dataset used to obtain the classifiers are detailed in this section.

### 2.1   Laboratory Plant

This works deals the anomaly detection in a laboratory plant used to control the liquid level in a tank. The used plant is shown in Fig. 1. The liquid is pumped to the objective tank (1) using a three phase pump (2), driven by a variable frequency driver (3). An ultrasonic sensor (4) measures the real state of the liquid level. The plant has two built-in output valves (5), in charge of emptying the objective tank to the feed tank (6).

The system is implemented using a virtual controller programmed with Matlab software. The control signal represents the pump speed, and the process value is the level percentage of the objective tank. The chosen controller is an adaptive PID, whose parameters are determined according to the transfer function coefficients, obtained with Recursive Least Square method [20]. A National Instruments data acquisition card (model USB-6008 12-bit 10 KS /s Multifunction I/O) was used to connect the plant and the computer.

### 2.2   Ultrasonic Sensor Performance

The sensor used in this work is a ultrasonic sensor of Banner $^{TM}$, model S18UUA, whose running principle is the following: an ultrasonic signal is sent to the liquid and, once it reaches the surface, it is reflected. The sensor captures this signal after a certain time and then, it calculates the distance to the surface.

**Fig. 1.** Control level plant

In this specific application, due to different reasons like the liquid flow rate, it would be possible the non desirable ripple or curling formation at the liquid surface. This fact could lead to wrong measurements, since the ultrasonic sensor does not receive the right signal. This situation is shown in Fig. 2.

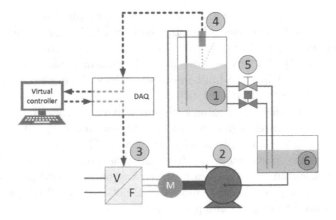

**Fig. 2.** Sensor misreading situation

## 2.3   Dataset Descritpion

To achieve the one-class classification, it is considered an initial dataset corresponding to three different operating points:

- Tank level at 30%: 5400 samples.
- Tank level at 50%: 5400 samples.
- Tank level at 70%: 5400 samples.

For each operating point, the control signal, the tank level measured by the sensor and the three coefficients of the plant transfer function, are registered with a sample rate of 2 Hz. All this variables are considered as inputs to the classifier.

# 3    Techniques Applied to Validate the Proposed Model

The three different intelligent techniques used to achieve the one-class classifiers are explained in this section. Furthermore, the method followed to generate the sensor wrong measurements is detailed.

## 3.1    One-Class Techniques

**Artificial Neural Networks Autoencoder.** The Autoencoder configuration using Artificial Neural Networks (ANN) for one-class has led to succesful results in different applications [15, 34, 40].

One of the most commonly used ANN is the Multilayer Perceptron (MLP) [6, 7], whose structure has an input layer, an output layer and a hidden layer [17]. Each layer is composed by neurons connected with weighted links between adjacent layers, which have nonlinear activation functions [2, 5, 19].

The Autoencoder is a MLP that reconstructs the input patterns in the output with a intermediate nonlinear dimensional reduction in the hidden layer. Hence, the input and output neurons are the same as the number of variables, and the hidden layer have at most one less neuron.

The hidden layer reduction aims to eliminate the data that is not consistent with the dataset. The difference between the input and the reconstructed output, represents the reconstruction error. Therefore, anomalous data should lead to high reconstruction error.

**Approximate Convex Hull.** The Approximate Convex Hull (ACH) is a one-class classification technique, that has been applied with successful results in previous works [4, 10]. The main basis of this method is to approximate the boundaries of a dataset $S \in \mathbb{R}^n$ using the convex hull. Given the fact that the convex limits of a dataset with $k$ samples and $p$ variables has a computational cost of $O(k^{(p/2)+1})$ [4], an approximation can be done using $t$ random projections on $2D$ planes and then, determine the convex limits on that plane. Hence, once the convex hull is modelled by $t$ projections, the criteria to determine if a new test data is an anomaly, is the following: if the data is out of at least one of the $t$ projections, the data is considered anomalous. In this study, the computational cost problem is solved using this approximation, since the computing time is lower than system sample rate.

Also, the possibility to expand or contract the limits of the projection from its centroid, can be considered using a parameter $\lambda$. With values of $\lambda$ lower than 1, the limits are reduced, and values higher than 1 lead to wider limits.

The appearance of an anomaly in $\mathbb{R}^3$ space is shown in Fig. 3.

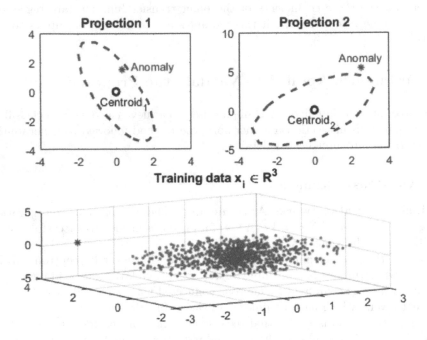

**Fig. 3.** Anomaly point in $\mathbb{R}^3$

**Support Vector Machine.** The Support Vector Machine (SVM) is implemented in classification and regression applications [9,23,27]. The use of SVM for one class classification aims to map the dataset into a high dimensional space using a kernel function. In this high dimensional space, a hyper-plane that maximises the distance between the origin and the data is implemented [11,33].

Once the SVM traning process is finished, the criteria to determine if a test data is anomalous, is based on the distance between the point and the hyper-plane. A negative distance means that, the data does not belong to the target class.

## 3.2   Artificial Outlier Generation

As the main objective of this study is to detect anomalies in the ultrasonic sensor, the variations on its measurements are artificially generated [25]. From a initial dataset with $n$ samples and $p$ attributes, whose content corresponds to correct measurements, the anomaly generation is achieved according to the following steps:

1. Select the attribute $k$ to be modified.
2. Select the percentage $pct$ of samples that will be converted to anomalies.
3. The samples subjected to the anomaly generation are randomly selected.
4. The anomaly generation consist of deviate those samples a certain percentage $\pm var\%$.

The use of this technique can be very useful to evaluate the deviation that a one-class can identify as anomaly in a specific variable.

## 4    Experiments and Results

From the initial dataset, corresponding to the operating points described in Sect. 2, 10% of the samples were converted to anomalies, according to the process described in Sect. 3.2. To ensure the correct assessment, each anomaly generation was implemented 10 times, and the classifier was trained using a $k - fold$ with $k = 10$.

The anomaly generation, train and test processes are presented in Fig. 4. The percentage variation of the ultrasonic sensor is swept from 5% to 50%, with a 5% step. Then, the process followed in Fig. 4 is repeated for each deviation.

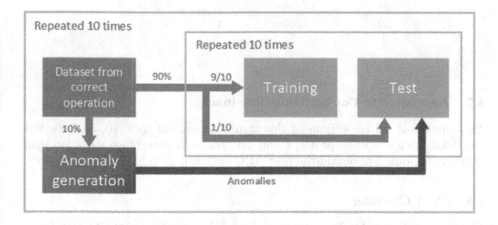

**Fig. 4.** Anomaly generation, train and test process

The classifiers performance are evaluated with the Area Under Curve (AUC) parameter, that establishes a relationship between true positives and false positives [3].

## 4.1    Artificial Neural Network Autoencoder Classifier

The Autoencoder ANN was implemented with the using the Matlab function *trainAutoencoder* [29]. In this case, the number of neurons in the hidden layer was set to 1, 2, 3 and 4. The data was normalised in a 0 to 1 interval and also using the z-score [37]. The possibility of not normalizing the data is also considered. The criteria to select if a test data is an anomaly, is based on the reconstruction error. When the reconstruction error is higher than the one obtained with 99% of the training set, it is considered anomalous. The best configuration for each percentage variation is Table 1.

**Table 1.** Results obtained with Autoencoder Classifier

| Variation (%) | Normalization | Neurons | AUC (%) |
|---|---|---|---|
| 5 | NoNorm | 2 | 78,041 |
| 10 | Zscore | 4 | 93,916 |
| 15 | NoNorm | 4 | 95,551 |
| 20 | Zscore | 4 | 95,976 |
| 25 | Zscore | 4 | 95,989 |
| 30 | NoNorm | 3 | 96,009 |
| 35 | NoNorm | 3 | 96,001 |
| 40 | NoNorm | 2 | 96,008 |
| 45 | Zscore | 3 | 95,999 |
| 50 | NoNorm | 3 | 96,001 |

## 4.2    Approximate Convex Hull Classifier

The number of $2D$ projections of this technique was set to 5, 10, 50, 100, 500 and 1000, with a $\lambda$ value of 0.9, 1 and 1.1. For each percentage variation, the ACH configuration to obtain the best AUC is shown in Table 2.

## 4.3    SVM Classifier

The Support Vector Machine classifier was implemented using the Matlab function $fitcsvm$ [30]. In this case, the kernel function was configured as Gaussian, the outlier fraction percentage of was set from 0 to 5. The normalization configuration was the same as for the Autoencoder technique. The criteria to the determine if a data is anomalous, is based in the distance to the hyperplane. This is evaluated with the Matlab function *predict* [31]. The results are presented in Table 3.

**Table 2.** Results obtained with ACH Classifier

| Variation (%) | Parameter $\lambda$ | Projections | AUC (%) |
|---|---|---|---|
| 5 | 1 | 1000 | 79,152 |
| 10 | 1 | 1000 | 97,584 |
| 15 | 1,1 | 1000 | 99,911 |
| 20 | 1,1 | 50 | 99,977 |
| 25 | 1,1 | 50 | 99,975 |
| 30 | 1,1 | 50 | 99,978 |
| 35 | 1,1 | 50 | 99,976 |
| 40 | 1,1 | 10 | 99,986 |
| 45 | 1,1 | 50 | 99,978 |
| 50 | 1,1 | 10 | 99,991 |

**Table 3.** Results obtained with SVM

| Variation (%) | Normalization | Outlier fr (%) | AUC (%) |
|---|---|---|---|
| 5 | Zscore | 3 | 50,072 |
| 10 | NoNorm | 4 | 50,256 |
| 15 | Zscore | 4 | 51,002 |
| 20 | Norm | 4 | 52,734 |
| 25 | Zscore | 4 | 57,492 |
| 30 | Zscore | 4 | 62,437 |
| 35 | Norm | 3 | 65,378 |
| 40 | Norm | 4 | 72,933 |
| 45 | NoNorm | 4 | 76,897 |
| 50 | NoNorm | 4 | 81,722 |

## 5    Conclusions and Future Works

This paper evaluates the capability of three one-class intelligent techniques to detect deviations in an ultrasonic sensor. Different percentage deviations were artificially generated to identify the threshold of each technique. Regardless the sensor measurement deviation, the classifiers obtained with ACH presented better results than Autoencoder and SVM. For ACH technique, the lowest AUC (79,152%) corresponds to a 5% variation. As this value increases, the number of projections are significantly lower, and the AUC presents higher values. The Autoencoder results are slightly lower than ACH for all percentage variations. The SVM does not perform as good as the rest of techniques, obtaining a highest AUC of 81,722% with a 50% variation.

Then, it is concluded that ACH is able to identify slight deviations in the ultrasonic sensor. If a single sample is classified as anomalous, this method can be presented as a very important tool, as previous step to apply imputation techniques and hence, recover the wrong data. This could lead to better system performance, increasing the energy efficiency, and control optimization. In addition to sensor misreadings, it is also possible to detect sensor or DAQ failures.

The use of local classifiers for each operating point can be considered as future works. It can also taken into account the use of Dimensional Reduction Techniques (DRT) [32,36] to reduce the computational cost of the techniques proposed.

Given the fact that system evolves with its use, the initial dataset may not be representative after a certain time. Then, the possibility of retrain the classifiers with new data can be taken into consideration.

# References

1. Alaiz Moretón, H., Calvo Rolle, J., García, I., Alonso Alvarez, A.: Formalization and practical implementation of a conceptual model for PID controller tuning. Asian J. Control **13**(6), 773–784 (2011)
2. Baruque, B., Porras, S., Jove, E., Calvo-Rolle, J.L.: Geothermal heat exchanger energy prediction based on time series and monitoring sensors optimization. Energy **171**, 49–60 (2019). http://www.sciencedirect.com/science/article/pii/ S0360544218325817
3. Bradley, A.P.: The use of the area under the roc curve in the evaluation of machine learning algorithms. Pattern Recogn. **30**(7), 1145–1159 (1997)
4. Casale, P., Pujol, O., Radeva, P.: Approximate convex hulls family for one-class classification. In: Sansone, C., Kittler, J., Roli, F. (eds.) MCS 2011. LNCS, vol. 6713, pp. 106–115. Springer, Heidelberg (2011). https://doi.org/10.1007/978-3-642-21557-5_13
5. Casteleiro-Roca, J.L., Barragán, A.J., Segura, F., Calvo-Rolle, J.L., Andújar, J.M.: Fuel cell output current prediction with a hybrid intelligent system. Complexity **2019**, 10 (2019)
6. Casteleiro-Roca, J.L., et al.: Short-term energy demand forecast in hotels using hybrid intelligent modeling. Sensors **19**(11), 2485 (2019)
7. Casteleiro-Roca, J.L., Pérez, J.A.M., Piñón-Pazos, A.J., Calvo-Rolle, J.L., Corchado, E.: Modeling the electromyogram (EMG) of patients undergoing anesthesia during surgery. In: Herrero, Á., Sedano, J., Baruque, B., Quintián, H., Corchado, E. (eds) 10th International Conference on Soft Computing Models in Industrial and Environmental Applications. Advances in Intelligent Systems and Computing, vol. 368, pp. 273–283. Springer, Cham (2015). https://doi.org/10. 1007/978-3-319-19719-7_24
8. Chandola, V., Banerjee, A., Kumar, V.: Anomaly detection: a survey. ACM Comput. Surv. (CSUR) **41**(3), 15 (2009)
9. Chen, Y., Zhou, X.S., Huang, T.S.: One-class SVM for learning in image retrieval. In: Proceedings 2001 International Conference on Image Processing 2001, vol. 1, pp. 34–37. IEEE (2001)
10. Fernández-Francos, D., Fontenla-Romero, Ó., Alonso-Betanzos, A.: One-class convex hull-based algorithm for classification in distributed environments. IEEE Trans. Syst. Man. Cybern. Syst. 1–11 (2018)

11. Fernández-Serantes, L.A., Estrada Vázquez, R., Casteleiro-Roca, J.L., Calvo-Rolle, J.L., Corchado, E.: Hybrid intelligent model to predict the SOC of a LFP power cell type. In: Polycarpou, M., de Carvalho, A.C.P.L.F., Pan, J.-S., Woźniak, M., Quintian, H., Corchado, E. (eds.) HAIS 2014. LNCS (LNAI), vol. 8480, pp. 561–572. Springer, Cham (2014). https://doi.org/10.1007/978-3-319-07617-1_49

12. Garcia, R.F., Rolle, J.L.C., Castelo, J.P., Gomez, M.R.: On the monitoring task of solar thermal fluid transfer systems using nn based models and rule based techniques. Eng. Appl. Artif. Intell. **27**, 129–136 (2014)

13. González, G., Angelo, C.D., Forchetti, D., Aligia, D.: Diagnóstico de fallas en el convertidor del rotor en generadores de inducción con rotor bobinado. Revista Iberoamericana de Automática e Informática Industrial **15**(3), 297–308 (2018). https://polipapers.upv.es/index.php/RIAI/article/view/9042

14. Gonzalez-Cava, J.M., Reboso, J.A., Casteleiro-Roca, J.L., Calvo-Rolle, J.L., Méndez Pérez, J.A.: A novel fuzzy algorithm to introduce new variables in the drug supply decision-making process in medicine. Complexity **2018**, 15 (2018)

15. Goodfellow, I., Bengio, Y., Courville, A., Bengio, Y.: Deep Learning, vol. 1. MIT press, Cambridge (2016)

16. Hobday, M.: Product complexity, innovation and industrial organisation. Res. Policy **26**(6), 689–710 (1998)

17. Jove, E., Aláiz-Moretón, H., Casteleiro-Roca, J.L., Corchado, E., Calvo-Rolle, J.L.: Modeling of bicomponent mixing system used in the manufacture of wind generator blades. In: Corchado, E., Lozano, J.A., Quintián, H., Yin, H. (eds.) IDEAL 2014. LNCS, vol. 8669, pp. 275–285. Springer, Cham (2014). https://doi.org/10.1007/978-3-319-10840-7_34

18. Jove, E., Alaiz-Moretón, H., García-Rodríguez, I., Benavides-Cuellar, C., Casteleiro-Roca, J.L., Calvo-Rolle, J.L.: PID-ITS: an intelligent tutoring system for PID tuning learning process. In: Pérez García, H., Alfonso-Cendón, J., Sánchez González, L., Quintián, H., Corchado, E. (eds.) SOCO/CISIS/ICEUTE -2017. AISC, vol. 649, pp. 726–735. Springer, Cham (2018). https://doi.org/10.1007/978-3-319-67180-2_71

19. Jove, E., Antonio Lopez-Vazquez, J., Isabel Fernandez-Ibanez, M., Casteleiro-Roca, J.L., Luis Calvo-Rolle, J.: Hybrid intelligent system to predict the individual academic performance of engineering students. Int. J. Eng. Educ. **34**(3), 895–904 (2018)

20. Jove, E., Casteleiro-Roca, J.L., Quintián, H., Méndez-Pérez, J.A., Calvo-Rolle, J.L.: A fault detection system based on unsupervised techniques for industrial control loops. Expert Syst. **2019**, e12395 (2019)

21. Jove, E., Casteleiro-Roca, J.-L., Quintián, H., Méndez-Pérez, J.A., Calvo-Rolle, J.L.: A new approach for system malfunctioning over an industrial system control loop based on unsupervised techniques. In: Graña, M., López-Guede, J.M., Etxaniz, O., Herrero, Á., Sáez, J.A., Quintián, H., Corchado, E. (eds.) SOCO'18-CISIS'18-ICEUTE'18 2018. AISC, vol. 771, pp. 415–425. Springer, Cham (2019). https://doi.org/10.1007/978-3-319-94120-2_40

22. Jove, E., Casteleiro-Roca, J.-L., Quintián, H., Méndez-Pérez, J.A., Calvo-Rolle, J.L.: Outlier generation and anomaly detection based on intelligent one-class techniques over a bicomponent mixing system. In: Martínez Álvarez, F., Troncoso Lora, A., Sáez Muñoz, J.A., Quintián, H., Corchado, E. (eds.) SOCO 2019. AISC, vol. 950, pp. 399–410. Springer, Cham (2020). https://doi.org/10.1007/978-3-030-20055-8_38

23. Jove, E., et al.: Modelling the hypnotic patient response in general anaesthesia using intelligent models. Logic J. IGPL **27**, 189–201 (2018)

24. Jove, E., Gonzalez-Cava, J.M., Casteleiro-Roca, J.L., Pérez, J.A.M., Calvo-Rolle, J.L., de Cos Juez, F.J.: An intelligent model to predict ANI in patients undergoing general anesthesia. In: Pérez García, H., Alfonso-Cendón, J., Sánchez González, L., Quintián, H., Corchado, E. (eds.) SOCO/CISIS/ICEUTE -2017. AISC, vol. 649, pp. 492–501. Springer, Cham (2018). https://doi.org/10.1007/978-3-319-67180-2_48

25. Jove, E., Gonzalez-Cava, J.M., Casteleiro-Roca, J.-L., Quintián, H., Méndez-Pérez, J.A., Calvo-Rolle, J.L.: Anomaly detection on patients undergoing general anesthesia. In: Martínez Álvarez, F., Troncoso Lora, A., Sáez Muñoz, J.A., Quintián, H., Corchado, E. (eds.) CISIS/ICEUTE -2019. AISC, vol. 951, pp. 141–152. Springer, Cham (2020). https://doi.org/10.1007/978-3-030-20005-3_15

26. Moreno-Fernandez-de Leceta, A., Lopez-Guede, J.M., Ezquerro Insagurbe, L., Ruiz de Arbulo, N., Graa, M.: A novel methodology for clinical semantic annotations assessment. Logic J. IGPL **26**(6), 569–580 (2018). http://dx.doi.org/10.1093/jigpal/jzy021

27. Li, K.L., Huang, H.K., Tian, S.F., Xu, W.: Improving one-class SVM for anomaly detection. In: 2003 International Conference on Machine Learning and Cybernetics, vol. 5, pp. 3077–3081. IEEE (2003)

28. Manuel Vilar-Martinez, X., Aurelio Montero-Sousa, J., Luis Calvo-Rolle, J., Luis Casteleiro-Roca, J.: Expert system development to assist on the verification of "tacan" system performance. Dyna **89**(1), 112–121 (2014)

29. MathWorks: Autoencoder, 29 January 2019. https://es.mathworks.com/help/deeplearning/ref/trainautoencoder.html

30. MathWorks: fitcsvm, 29 January 2019. https://es.mathworks.com/help/stats/fitcsvm.html

31. MathWorks: predict, 29 January 2019. https://es.mathworks.com/help/stats/classreg.learning.classif.compactclassificationsvm.predict.html

32. Quintián, H., Corchado, E.: Beta scale invariant map. Eng. Appl. Artif. Intell. **59**, 218–235 (2017)

33. Rebentrost, P., Mohseni, M., Lloyd, S.: Quantum support vector machine for big data classification. Phys. Rev. Lett. **113**, 130503 (2014). https://link.aps.org/doi/10.1103/PhysRevLett.113.130503

34. Sakurada, M., Yairi, T.: Anomaly detection using autoencoders with nonlinear dimensionality reduction. In: Proceedings of the MLSDA 2014 2nd Workshop on Machine Learning for Sensory Data Analysis, p. 4. ACM (2014)

35. Sánchez-González, L., et al.: Use of classifiers and recursive feature elimination to assess boar sperm viability. Logic J. IGPL **26**(6), 629–637 (2018)

36. Segovia, F., Górriz, J.M., Ramírez, J., Martinez-Murcia, F.J., García-Pérez, M.: Using deep neural networks along with dimensionality reduction techniques to assist the diagnosis of neurodegenerative disorders. Logic J. IGPL **26**(6), 618–628 (2018). http://dx.doi.org/10.1093/jigpal/jzy026

37. Shalabi, L.A., Shaaban, Z.: Normalization as a preprocessing engine for data mining and the approach of preference matrix. In: 2006 International Conference on Dependability of Computer Systems, pp. 207–214, May 2006

38. Vega Vega, R., Quintián, H., Calvo-Rolle, J.L., Herrero, A., Corchado, E.: Gaining deep knowledge of Android malware families through dimensionality reduction techniques. Logic J. IGPL **27**(2), 160–176 (2018). https://doi.org/10.1093/jigpal/jzy030

39. Vega Vega, R., Quintián, H., Cambra, C., Basurto, N., Herrero, Á., Calvo-Rolle, J.L.: Delving into android malware families with a novel neural projection method. Complexity **2019**, 10 (2019)
40. Vincent, P., Larochelle, H., Lajoie, I., Bengio, Y., Manzagol, P.A.: Stacked denoising autoencoders: learning useful representations in a deep network with a local denoising criterion. J. Mach. Learn. Res. **11**, 3371–3408 (2010)

# A Machine Learning Approach
# to Determine Abundance of Inclusions
# in Stainless Steel

Héctor Mesa[1], Daniel Urda[1(✉)], Juan J. Ruiz-Aguilar[1],
José A. Moscoso-López[1], Juan Almagro[2], Patricia Acosta[2],
and Ignacio J. Turias[1]

[1] Dpto. de Ingeniería Informática, EPS de Algeciras, Universidad de Cádiz,
Cádiz, Spain
{hector.mesa,daniel.urda}@uca.es
[2] Dpto. Técnico, Polígono Industrial Los Barrios, ACERINOX Europa, S.A.U.,
Los Barrios, Spain

**Abstract.** Steel-making process is a complex procedure involving the
presence of exogenous materials which could potentially lead to non-
metallic inclusions. Determining the abundance of inclusions in the ear-
liest stage possible may help to reduce costs and avoid further post-
processing manufacturing steps to alleviate undesired effects. This paper
presents a data analysis and machine learning approach to analyze data
related to austenitic stainless steel (Type 304L) in order to develop a
decision-support tool helping to minimize the inclusion content present
in the final product. Several machine learning models (generalized linear models with regularization, random forest, artificial neural networks
and support vector machines) were tested in this analysis. Moreover,
two different outcomes were analyzed (average and maximum abundance
of inclusions per steel cast) and two different settings were considered
within the analysis based on the input features used to train the models
(full set of features and more relevant ones). The results showed that
the average abundance of inclusions can be predicted more accurately
than the maximum abundance of inclusions using linear models and the
reduced set of features. A list of the more relevant features linked to the
abundance of inclusions based on the data and models used in this study
is additionally provided.

**Keywords:** Machine learning · Stainless steel · Inclusions ·
Data mining

## 1 Introduction

Austenitic stainless steel is the largest family of stainless steels which is formed
by sufficient nickel and/or manganese and nitrogen in order to maintain an
austenitic micro-structure [13]. In particular, chromium-nickel alloys are labelled

© Springer Nature Switzerland AG 2019
H. Pérez García et al. (Eds.): HAIS 2019, LNAI 11734, pp. 504–513, 2019.
https://doi.org/10.1007/978-3-030-29859-3_43

under the *300-series* tag, being the *Type 304* the best known grade with an 18% of chromium and between 8%–10% of nickel. This austenitic stainless steel labelling may add the "*L*" designator which means that the carbon content of the alloy is below 0.03%. In general, steel-making process is a complex procedure involving the presence of exogenous materials which may potentially lead to non-metallic inclusions. These inclusions could be mainly classified according to (i) their origin (exogenous and endogenous inclusions), and (ii) their size (micro-inclusions when they are below 20 mm or macro-inclusions in any other case, where both could again be exogenous and endogenous). In this sense, the demand of cleaner steels requires to analyze and take into account the impact of these inclusions on physical properties with the aim of minimizing the presence of inclusions [10]. The use of accurate evaluation methods which are applied at all stages of the steel-making process becomes relevant in order to analyze and control steel cleanliness.

Nowadays, data availability and/or exchange and automation of processes in manufacturing is a reality in the Industry 4.0 [8] supported by the Internet of Things [1], which mainly allows every machine, tool, robot or equipment present in the manufacturing process to be equipped with sensors in order to measure and provide data. In this sense, Computer Science (CS) and Artificial Intelligence (AI) domains arise as powerful tools to transform the manufacturing domain improving manufacturing processes (minimizing costs and time) and the quality of final products [17]. Particularly, Machine Learning (ML) techniques are a data-driven approach, part of the AI domain, which allows to find highly complex and non-linear patterns in data sources of different types for different purposes such as prediction, detection, classification, regression, or forecasting [15]. Furthermore, ML applications have continuously increased during the last two decades linked to manufacturing processes [5,11,16] and, in fact, several initiatives have already been proposed by the European Commission to revamp the manufacturing sector [4].

This paper aims at developing a decision-support system to determine the abundance of inclusions in austenitic stainless steel using features linked to the first stage of the manufacturing process. In particular, several ML models and techniques are tested in order to predict the abundance of inclusions (continuous outcome) based on input features available at the melting shop stage (independent variables). Although direct measurement techniques such as Metallographic Microscope Analysis (MMA) are more accurate than indirect methods, the latter may potentially act as relative indicators techniques [18] which, on one hand are cheaper than direct measurement techniques and, on the other hand, may allow to perform any required action during the steel-making process rather than at the end of it once the final product is finished. This approach is tested using data provided by ACERINOX Europa S.A.U., a stainless steel factory located in Los Barrios (Spain). In spite of developing this kind of decision-support system based on data provided by this factory, in principle the approach described in this paper (or a similar one adapted to the specific manufacturing environment) may also be useful to any other stainless steel factory.

The rest of the paper is organized as follows: Sect. 2 describes the data used and the ML methods tested in the analysis. Then, Sect. 3 shows the results obtained in this paper after fitting several ML methods to the industrial data and, finally, some concluding remarks are provided in Sect. 4.

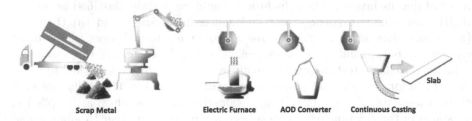

**Fig. 1.** Schematic overview of the different parts involved in the melting shop stage of the manufacturing process.

## 2    Materials and Methods

The data used in this analysis was provided by ACERINOX Europa S.A.U. consisting of a set of variables measured during the first stage of the steel-making process (see Fig. 1) for each steel coil made, which is the final product obtained at the end of the process. In concrete, the initial dataset $\mathcal{D} = \{x^{(i)}, y^{(i)}\}_{i=1}^{N}$ is composed by $N = 454$ steel casting samples, where a given steel cast $x^{(i)}$ is described by $P = 342$ variables linked to the melting shop stage. At the beginning of this stage, scrap metal is first transported and melted in the electric furnace. Next, some chemical elements are added to the molten steel in the Argon Oxygen Decarburization (AOD) converter to provide some required properties of austenitic stainless steel. Finally, the molten steel gets to a continuous casting machine where the steel is continuously poured through a mold of adjustable width in order to obtain a steel slab. In this sense, $K$ different steel slabs are obtained from a single steel cast. Moreover, each steel slab is further transformed to produce the final steel coil in other stages of the steel-making process, although these parts are not included in this work as they are not considered to be relevant in the problem under study. Additionally, each steel cast in the dataset also includes the abundance of inclusions, $y^{(i)}$, which is determined off-line by staff members in ACERINOX Europa S.A.U. through MMA. In particular, for each steel cast two different outcomes were defined by considering the average and the worst case scenario of the abundance of inclusions in steel slabs that come from the same steel cast. Equations 1 and 2 show how these outcomes were defined for each $i$-th sample of $\mathcal{D}$:

$$y_{avg}^{(i)} = \frac{1}{k}\sum_{j=1}^{k} abundance_{j}^{(i)} \ , \ \ \forall k \in steel\ cast\ i \tag{1}$$

$$y_{max}^{(i)} = max\left(abundance_{1}^{(i)}, \ ... \ , abundance_{k}^{(i)}\right) \ , \ \ \forall k \in steel\ cast\ i \tag{2}$$

The methodology used in this ML approach was the Standard Cross-Industry for Process Data Mining (CRISP-DM) which was developed in year 2000 by a consortium of European companies [3]. The first stage of this methodology named *Business Understanding* consisted of knowledge acquisition of the stainless steel manufacturing process. Subsequently, a second stage known as *Data Understanding* was performed to identify different data sources available (e.g. electric furnace, AOD converter, continuous casting, etc.), collect as much as data as possible from them and understand each single data point measured in the entire melting shop stage. In this sense, the definition of the most important variables was done by staff members of ACERINOX Europa S.A.U. with high expertise trying to cover most of the possible sources which are known to be connected to endogenous and exogenous inclusions [14]. The main block of features with potential relevance in the problem under study includes chemical composition of the austenitic stainless steel, tapping conditions, steel-making parameters (temperatures, time of treatment, amount and type of additions, etc.), casting speed, tundish variables, special events occurred during casting (alumina detection, etc.), slab dimensions and defects classification in laboratory tests. All this information easily allows to monitor the status of a steel cast or steel slab throughout the process. The third stage in the CRISP-DM methodology is known as *Data Preparation* which basically aimed at pre-processing and curating the raw data collected to make it ready for the analysis phase through the application of well-know techniques such as outliers exclusion, missing data imputation, or categorical variables manipulation, among others.

Once the final curated dataset was obtained, several well-known ML models were estimated in order to, on one hand, develop a decision-support system which may potentially help the staff members to perform required actions during the steel-making process and, on the other hand, identify relevant features which may have some relationship to the problem under study. Next, a brief description of the ML models used is provided:

- Lasso (GLM-$l_1$): it is a generalized linear regression model with regularization which applies a $l_1$ penalty to perform feature selection during the model estimation procedure.
- Elastic net (GLM-$l_{1,2}$): it is a generalized linear regression model with regularization which applies a mixed $l_1$-$l_2$ penalties to perform both feature selection and control parameters' magnitude.
- Random forest (RF): it is a tree-based bagging ensemble in which multiple decision trees are fitted to different views of the observed data and predictions are made by averaging the individual predictions provided by the multiple decision trees [2].
- Artificial Neural Networks (ANN): it is a neural network-based model where network architectures are "bio-inspired" in the human brain [6] consisting of several fully-connected layers of artificial neurons which allows to address complex non-linear problems.

– Support Vector Machines (SVM): it is a kernel-based method that uses a kernel function (radial basis, linear, polynomial, or any other) to map the original input space into a new space where predictions can be made more accurately [12].

Three different well-known performance metrics for regression tasks are used to measure the goodness of the models: the Mean Squared Error ($MSE$), the Mean Absolute Error ($MAE$) and the Pearson's correlation coefficient ($\sigma$). Equations 3–5 show how each performance measure is calculated given the observed ($\boldsymbol{y}$) and predicted ($\hat{\boldsymbol{y}}$) vector values. Low values for either the MSE or MAE ($MSE \approx 0$, $MAE \approx 0$) and high values of the Pearson's correlation coefficient ($\sigma \approx 1$) indicate a better performance than low values of $\sigma$ and high values of $MSE$, $MAE$.

$$MSE(\boldsymbol{y}, \hat{\boldsymbol{y}}) = \frac{1}{N} \sum_{i=1}^{N} (\hat{y}_i - y_i)^2 \tag{3}$$

$$MAE(\boldsymbol{y}, \hat{\boldsymbol{y}}) = \frac{1}{N} \sum_{i=1}^{N} |\hat{y}_i - y_i| \tag{4}$$

$$\sigma(\boldsymbol{y}, \hat{\boldsymbol{y}}) = \frac{\sum_{i=1}^{N} (y_i - \bar{y})(\hat{y}_i - \bar{\hat{y}})}{\sqrt{\sum_{i=1}^{N} (y_i - \bar{y})^2 \sum_{i=1}^{N} (\hat{y}_i - \bar{\hat{y}})^2}} \tag{5}$$

## 3  Results

The analysis was carried out performing three repetitions of 10-fold cross-validation [7], thus partitioning the entire datasets in 10 folds of equal sizes in order to estimate the performance of each model. In this sense, models were fitted 3 times in 9 folds (train set) and tested in the unseen test fold left apart (test set) within an iterative procedure that rotates the train and test folds used. Furthermore, independent variables were normalized with zero mean and unit variance to avoid side effects linked to possible different variable scales. Models' hyper-parameters were chosen through an exhaustive grid search performed within the train set.

On one hand, the analysis was carried out using the complete set of features available ($P = 342$) for both outcomes considered in this work ($y_{max}$ and $y_{avg}$). In this setting, only models such as GLM-$l_1$ or GLM-$l_{1,2}$ which internally perform feature selection, or RF which is less sensitive to over-fitting issues were estimated. For both outcomes studied, the average performance of the trained models is shown in Table 1, where the low Pearson's correlation coefficient and the poor $MSE$ clearly pointed out the need to remove non-relevant features in order to first reduce the dimensionality of the dataset and, second, aim at achieving higher performance rates. Therefore, Generalized Linear Models with regularization and RF were used as feature selection methods in a pre-processing step. For the former, the absolute value of the magnitude of the coefficients were

**Table 1.** Average test performance metrics obtained by the different ML models tested to address both problems analyzed ($y_{avg}$, $y_{max}$) using the complete set of features available (left part of the table) and a reduced subset of selected features (right part of the table).

$y_{avg}$

| Model | All features ($P = 342$) | | | Selected features ($P = 37$) | | |
|---|---|---|---|---|---|---|
| | MSE | MAE | $\sigma$ | MSE | MAE | $\sigma$ |
| GLM-$l_{1,2}$ | 0.680 | 0.622 | 0.566 | 0.549 | 0.570 | 0.674 |
| GLM-$l_1$ | 0.750 | 0.665 | 0.540 | **0.542** | 0.563 | **0.679** |
| RF | 0.676 | 0.622 | 0.572 | 0.624 | 0.595 | 0.614 |
| ANN | - | - | - | 0.560 | **0.561** | 0.666 |
| SVM | - | - | - | 0.565 | 0.579 | 0.665 |

$y_{max}$

| Model | All features ($P = 342$) | | | Selected features ($P = 29$) | | |
|---|---|---|---|---|---|---|
| | MSE | MAE | $\sigma$ | MSE | MAE | $\sigma$ |
| GLM-$l_{1,2}$ | 0.710 | 0.650 | 0.542 | 0.637 | 0.613 | 0.603 |
| GLM-$l_1$ | 0.759 | 0.672 | 0.504 | 0.647 | 0.619 | 0.595 |
| RF | 0.727 | 0.661 | 0.523 | 0.682 | 0.643 | 0.564 |
| ANN | - | - | - | 0.651 | 0.620 | 0.593 |
| SVM | - | - | - | **0.626** | **0.610** | **0.612** |

used to measure variable importance, while the mean decrease impurity criterion was used for the latter [9]. Furthermore, each of these models was fitted 50 times to average the obtained features scores and increase robustness of the feature selection step. Features with scores below a specific threshold were discarded, thus reducing the feature space to $P = 37$ and $P = 29$ input features for the studied outcomes $y_{avg}$ and $y_{max}$, respectively. Hence, a second analysis was carried out complementing the models used in the first setting with ANN and SVM as more complex models which may capture better the non-linear relationships present in the data.

Table 1 shows both the performance obtained in the first step when all features available in the dataset were used to train the models, and the one obtained when only features obtained through the feature selection step were used to estimate them. The best result was obtained by Generalized Linear Models when addressing the average abundance of inclusions in the steel cast ($y_{avg}$). As shown in Figs. 2a and b, although the predicted and observed values are moderately correlated ($\sigma = 0.679$), the largest residuals are associated mostly with high values of abundance of inclusions. With respect to the outcomes analyzed, the overall performance obtained showed that predictions of the average abundance of inclusions are more accurate ($MSE = 0.542$) than predictions of the maximum abundance of inclusions ($MSE = 0.626$), either for the complete or reduced data set settings. Furthermore, dimensionality reduction allowed getting lower generalization error with all models tested: $y_{avg}$ and $y_{max}$ errors were reduced in 0.13 and 0.08 units in the worst of the cases, respectively. In general, the results obtained for the $y_{avg}$ outcome showed lower error using linear models and the reduced set of features, which may suggests more linearity between the selected features and $y_{avg}$ than $y_{max}$, where all models behaved similarly.

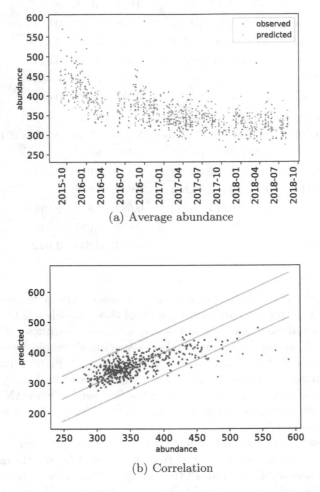

(a) Average abundance

(b) Correlation

**Fig. 2.** (a) Average abundance of inclusions per steel cast (blue) and the predicted values for the best setting tested (yellow), and (b) scatter plot representation of the observed and predicted values of the average abundance of inclusions, where red data points correspond to predicted values out of the 95% confidence interval. (Color figure online)

According to the best performance result obtained with the models and settings tested, the most important features linked to the average abundance of inclusions $y_{avg}$ are listed in Table 2. Although a few hundred variables are initially involved in the melting shop stage, the analysis performed suggests that a subset of features related to the steel chemical composition, additions, fatigue, stirring and some others may be more relevant when trying to reduce the abundance of inclusions in the final steel slab manufactured.

**Table 2.** List of the most relevant features linked to the average abundance of inclusions ($y_{avg}$) outcome.

| Category | Feature |
| --- | --- |
| Composition | Nickel content (%) |
| | Phosphorus content (%) |
| Additions | Total weight ($Kg$) |
| | Manual/automatized additions weight ($Kg$) |
| Fatigue | Relative mold lifespan (%) |
| | Presence of molybdenum in the ladle previous casting (yes/no) |
| Stirring | Flow variance (($Nm^3/h)^2$) |
| | Molten steel temperature at start (°C) |
| | Molten steel temperature at end (°C) |
| Others | Casting speed ($m/min$) |
| | Tundish heating duration ($min$) |

# 4  Conclusions

This paper has presented a data analysis and machine learning approach in order to help obtaining austenitic stainless steel (Type 304L), used for deep drawing application, with minimum inclusion content. In concrete, data provided by ACERINOX Europa S.A.U. linked to the melting shop stage of the entire steel-making process was used to test several ML models and to develop a decision-support tool which may help staff members to take required actions during the process in this or any other steel factory. In this sense, the average and maximum abundance of inclusions per steel slab were defined as two different outcomes in the analysis. Moreover, two different settings were considered according to the input feature space used to estimate the ML models (full set of features or most relevant ones). Models' performance was evaluated by performing three repetitions of 10-fold cross-validation.

In general, the results showed that the average abundance of inclusions can be predicted more accurately than the maximum abundance of inclusions using linear models and the reduced set of features, which may suggests more linearity between the selected features and this outcome. In particular, predicted and observed values of the average abundance of inclusions presented a Pearson's correlation coefficient of $\sigma = 0.679$ and a mean squared error of $MSE = 0.542$

in the best of the settings tested. Furthermore, dimensionality reduction helped to reduce the error obtained by all models tested in 0.13 and 0.08 units in the worst of the cases and for the two outcomes analyzed (average and maximum abundance of inclusions, respectively). Additionally, a list of features related to the steel chemical composition, additions, fatigue, stirring and some others were found to be more relevant to the abundance of inclusions in stainless steel.

In future works, other ways of measuring the inclusion content could be considered as it may complement the results obtained in this analysis. Moreover, a similar approach could be address in such a way that each sample is directly mapped to a steel slab instead of a steel cast, thus eliminating the need of summarizing (by averaging or computing the maximum) the inclusion content present in steel slabs which come from the same steel cast.

**Acknowledgments.** This work is part of the ACERINOX EUROPA S.A.U research project AUSINOX IDI-20170081 - "Obtaining austenitic stainless steels with minimum inclusion content from the development of new advanced simulation models in melting shop processes", supported by CDTI (Centro para el Desarrollo Tecnológico Industrial), Spain. This project has been co-financed by the European Regional Development Fund (FEDER), within the Intelligent Growth Operational Program 2014–2020, with the aim of promoting research, technological development and innovation. Authors acknowledge support through grant RTI2018-098160-B-I00 from MINECO-SPAIN which include FEDER funds.

# References

1. Atzori, L., Iera, A., Morabito, G.: The internet of things: a survey. Comput. Netw. **54**(15), 2787–2805 (2010). https://dx.doi.org/10.1016/j.comnet.2010.05.010
2. Breiman, L.: Random forests. Mach. Learn. **45**(1), 5–32 (2001). https://doi.org/10.1023/A:1010933404324
3. Chapman, P., Clinton, J., Khabaza, T., Reinartz, T., Rüdiger, W.: The CRISP-DM process model. CRISP-DM discussion paper (1999)
4. European Commission: Factories for the future (2016). http://ec.europa.eu/research/industrial_technologies/factories-of-the-future_en.html
5. Hansson, K., Yella, S., Dougherty, M., Fleyeh, H.: Machine learning algorithms in heavy process manufacturing. Am. J. Intell. Syst. **6**(1), 1–13 (2016)
6. Hassabis, D., Kumaran, D., Summerfield, C., Botvinick, M.: Neuroscience-inspired artificial intelligence. Neuron **95**(2), 245–258 (2017). http://www.sciencedirect.com/science/article/pii/S0896627317305093
7. Kohavi, R.: A study of cross-validation and bootstrap for accuracy estimation and model selection. In: Proceedings of the 14th International Joint Conference on Artificial Intelligence IJCAI 1995, vol. 2, pp. 1137–1143 (1995)
8. Liao, Y., Deschamps, F., de Freitas Rocha Loures, E., Ramos, L.F.P.: Past, present and future of industry 4.0 - a systematic literature review and research agenda proposal. Int. J. Prod. Res. **55**(12), 3609–3629 (2017)
9. Louppe, G., Wehenkel, L., Sutera, A., Geurts, P.: Understanding variable importances in forests of randomized trees. In: Advances in Neural Information Processing Systems, vol. 26, pp. 431–439 (2013)

10. Park, J.H., Kang, Y.: Inclusions in stainless steels - a review. Steel Res. Int. **88**(12), 1700130 (2017)
11. Pham, D.T., Afify, A.A.: Machine-learning techniques and their applications in manufacturing. Proc. Inst. Mech. Eng. Part B: J. Eng. Manufact. **219**(5), 395–412 (2005)
12. Rossi, F., Villa, N.: Support vector machine for functional data classification. Neurocomputing **69**(7), 730–742 (2006)
13. Saravanan, M., Devaraju, A., Venkateshwaran, N., Krishnakumari, A., Saarvesh, J.: A review on recent progress in coatings on AISI austenitic stainless steel. Mater. Today: Proc. **5**(6), 14392–14396 (2018). http://www.sciencedirect.com/science/article/pii/S221478531830600X. International Conference on Advanced Functional Materials 2017 (ICAFM 2017), Part 2
14. da Costa e Silva, A.L.V.: Non-metallic inclusions in steels - origin and control. J. Mater. Res. Technol. **7**(3), 283–299 (2018). http://www.sciencedirect.com/science/article/pii/S2238785418300280
15. Susto, G.A., Schirru, A., Pampuri, S., McLoone, S., Beghi, A.: Machine learning for predictive maintenance: a multiple classifier approach. IEEE Trans. Ind. Inform. **11**(3), 812–820 (2015)
16. Wuest, T., Irgens, C., Thoben, K.D.: An approach to monitoring quality in manufacturing using supervised machine learning on product state data. J. Intell. Manufact. **25**(5), 1167–1180 (2014)
17. Wuest, T., Weimer, D., Irgens, C., Thoben, K.D.: Machine learning in manufacturing: advantages, challenges, and applications. Prod. Manufact. Res. **4**(1), 23–45 (2016)
18. Zhang, L., Thomas, B., Wang, X., Cai, K.: A comparison of forecasting methods for RO-RO traffic: a case study in the strait of gibraltar. In: 85th Steelmaking Conference. Steelmaking Conference, Warrendale, PA (2002)

# Measuring Lower Limb Alignment and Joint Orientation Using Deep Learning Based Segmentation of Bones

Kamil Kwolek[1]([✉]), Adrian Brychcy[1], Bogdan Kwolek[2],
and Wojciech Marczyński[3]

[1] Professor Adam Gruca Teaching Hospital, Otwock, Poland
kwolekamil@gmail.com
[2] Department of Computer Science, AGH University of Science and Technology,
30-059 Kraków, Poland
[3] Centre of Postgraduate Medical Education, Professor Adam Gruca Teaching
Hospital, Otwock, Poland
kl.ortopedii@spskgruca.pl

**Abstract.** Deformities of the lower limbs are a common clinical problem encountered in orthopedic practices. Several methods have been proposed for measuring lower limb alignment and joint orientation clinically or using computer-assisted methods. In this work we introduce a new approach for measuring lower limb alignment and joint orientation on the basis of bones segmented by deep neural networks. The bones are segmented on X-ray images using an U-net convolutional neural network. It has been trained on forty manually segmented images. Afterwards, the segmented bones are post-processed using fully connected CRFs. Finally, lines are fitted to pruned skeletons representing the bones. We discuss algorithms for measuring lower limb alignment and joint orientation. We present both qualitative and quantitative segmentation results on ten test images. We compare the results that were obtained manually using a computer-assisted program and by the proposed algorithm.

## 1 Introduction

Deformities of the lower limbs are a common clinical problem encountered in orthopedic [1]. Several methods were proposed for measuring lower limb alignment and joint orientation clinically or using computer-assisted methods.

In this work we propose a new algorithm for measuring lower limb alignment on the basis of deep learning based segmentation of bones. We propose an automatic algorithm for lower limb bone segmentation on X-ray images. The bones are segmented using an U-Net convolutional network. It has been trained on forty manually segmented images. Fully connected Conditional Random Fields (CRFs) were used to enhance detailed local structure of the segmented bones. We show both qualitative and quantitative segmentation results on ten test images. We present algorithm for aLDFA, MPTA and FTA angle estimation. We compare results that were obtained manually and by the proposed algorithm. The

© Springer Nature Switzerland AG 2019
H. Pérez García et al. (Eds.): HAIS 2019, LNAI 11734, pp. 514–525, 2019.
https://doi.org/10.1007/978-3-030-29859-3_44

contribution of this work is an algorithm for measuring lower limb alignment. To the best of our knowledge, this is seminal paper on application of such techniques in orthopedic. In [2] fine-tuned deep networks were applied to classify X-ray images. The networks achieved at least 90% accuracy in identifying laterality, body part, and exam view, whereas accuracy for fractures was about 83% for the best performing neural network. Our work significantly goes beyond the standard classification of X-ray images and introduces deep-learning based X-ray image analysis for orthopedics and radiologists.

## 2 Lower Limb Alignment

During human life, the axes of the lower limbs usually undergo various changes and deformations as part of skeletal development as well as aging. The term alignment is utilized to quantify how far an axis deviates from a straight line joining two points in the coronal, sagittal or axial planes of the lower limb. This straight line can refer to the mechanical axis or an anatomical axis. The whole limb has a mechanical axis, whereas each bone has an anatomic and a mechanical axis. The mechanical axis coincides with the line from the center of the proximal joint to the center of the distal joint [1]. It is determined by drawing a line from the center of the femoral head to the center of the ankle joint [3]. Deformity of lower limb affects the mechanical axis of the limb, i.e. how the load is transferred by joints and bones. Since the mechanical axis of the femur in the sagittal plane usually goes beyond due to femoral neck anteversion, the anatomical axis is utilized more frequently. The anatomic axis of a bone is a mid-diaphyseal path, i.e. equidistant from the cortices within the diaphysis. The anatomical axis is straight in coronal plane, but curved in the sagittal plane. In the tibia bone, anatomical axis is straight in both frontal and sagittal planes. The discussed axis passes through the middle of the bone rather than joining only end points. Deviation from the normal axis can be due to either translation or angulation.

There are several methods of estimating bone alignment, including clinical examinations, magnetic resonance imaging (MRI), computed tomography (CT), conventional standing knee radiographs (SKR) and intraoperative navigation. The use of SKR is a well-established method for measuring bone alignment. The lower limb alignment is assessed two-dimensionally on the basis of gray scale radiographic images of the whole lower limb in stance. Traditional methods require a clinician to draw lines on the radiographs representing the femoral and tibial mechanical or anatomic axes and to manually define the resulting angles. Recent research demonstrated that when comparing measurements of axial alignment between conventional and digital images, digital-based approaches are as good and in some cases can be better than conventional ones [4].

In this work, anatomical lateral-distal femoral angle (aLDFA), medial-proximal tibial angle (MPTA), and anatomic femorotibial angle (FTA) are estimated to evaluate the degree of alignment deformity of the knee. The anatomic lateral–distal femoral angle is defined as the angle between the anatomical femoral axis and the distal femoral articular axis [1]. The anatomic medial proximal tibial angle is defined as the medial angle between the tibial anatomic axis and the joint line of the proximal tibia. The femorotibial angle in the frontal plane is formed by the intersection of the anatomic axes of the femur and tibia [3]. This angle is considered as one of the most representative angles of lower limb geometry. In regular knee alignment, there is an roughly 5°–7°valgus femorotibial angle [5]. It is significant measurement to assess the varus or valgus deformity of the knee [6]. It is widely known that knee alignment disorders (malalignment) are significant biomechanical factors in the progression of knee osteoarthritis [7]. Thus, proper estimation of this angle is important problem in orthopedics.

Although medical software, like PreOPlan or mediCAD can measure the lower limb alignment, there is no precise method to define the beginning and end points of the axis. Existing approaches are based on assumptions that the points may vary from person to person. For instance, in [8] a reference point is assigned to the center of the knee. Moreland et al. [9] identified five points, which may be considered as the centers of the knee: the center of the femoral notch (trochlea), center of tibial spines, center of femoral condyles, center of soft tissue, and center of the tibia. They found that all points were within 5 mm of one another.

## 3  Methods

At the beginning of this Section we discuss bone segmentation on X-ray images. Next, we present our dataset. Afterwards, we outline our neural network for bone segmentation. Next, we explain how dense CRFs were used in the postprocessing. Finally, we present our approach to anatomic axis estimation.

### 3.1  Bone Segmentation on X-Ray Images

Computer delineation of bones from X-ray radiographs is not an easy task [10]. Usually X-ray images are noisy and have insufficient contrast, or intensities. Moreover, images have different sizes and differ considerably by intensity ranges, see Fig. 1. The images can contain some text, e.g. labels or remarks. Otherwise, objects like surgical nails, rods, pins that are routinely used in internal fixation of fractures can also appear on images, and even in bones. Marking out of bone contours on X-ray radiographs needs distinguishing soft tissue from bones. However, the boundaries between soft tissue and bones are not easily distinguishable on X-ray images. Typical approaches to bone segmentation are based on active shape model (ASM). Usually such algorithms are not fully automatic and due to unsatisfactory results, refinement of results by human is needed. In recent work [10], Wu et al. employed ASM to extract the distal femur and proximal tibia in knee radiographs with various quality. The X-ray images were denoised using a spectral clustering algorithm that has been based on the eigensolution of an affinity matrix. Finally, ASM-based segmentation has been employed on denoised X-ray images. In a recently published article in Scientific Reports [11],

Aganj et al. introduced a new approach to unsupervised image segmentation on the basis of on the computation of the local center of mass. The algorithm has been validated on a 2D X-ray image and a 3D abdominal magnetic resonance (MR) image. To the best of our knowledge, convolutional neural networks and deep learning were not used for automatic bone segmentation on X-ray images.

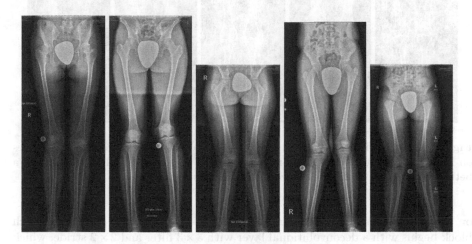

**Fig. 1.** Example samples of lower limb images on X-ray radiographs.

## 3.2 Dataset

The dataset includes fifty annotated X-ray images. Forty images are in training subset and remaining images are in the test subset. The dataset contains X-ray images of the legs of children aged 8 to 15 years. On every X-ray image, femoral and tibial bones were manually delineated by the orthopedist. Figure 2 illustrates sample examples of the annotated images, see also corresponding images on Fig. 1. For learning and evaluation of neural networks for bone segmentation both X-ray and annotated images were rescaled to size $512 \times 256$ pixels. In order to make possible determining angles describing deformation of legs as well as to compare results, the dataset contains also X-ray images of width equal to 256 pixels and the image height resulting from the height/width ratio for the original X-ray images. The images depicted on Fig. 4 have different number of rows since they were rescaled according the height/width ratios from original X-ray images.

## 3.3 Our Neural Network for Bone Segmentation

The architecture of neural network has been based on the U-Net in which we can distinguish a down-sampling (encoding) path and an up-sampling (decoding) path, see Fig. 3. In the down-sampling path there are five convolutional blocks. Each block has two convolutional layers with $3 \times 3$ filters and stride equal to 1. Down-sampling is realized by max pooling with stride $2 \times 2$ that is applied

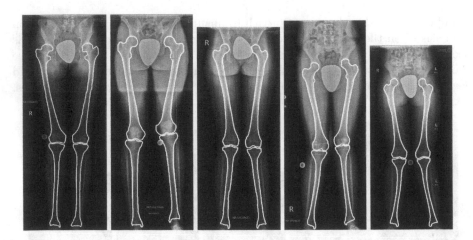

**Fig. 2.** Segmented images from our Deformed-legs dataset, see corresponding images on Fig. 1. The illustrated annotations are stored in vectorized graphics format and a better viewing can be obtained by zooming this figure.

on the end of every blocks except the last one. In the up-sampling path, each block begins with a deconvolutional layer with $3 \times 3$ filter and $2 \times 2$ stride, which doubles the dimension of feature maps in both directions and decreases the number of feature maps by two. In each up-sampling block, two convolutional layers decrease the number of feature maps, which arise as a result of concatenation of deconvolutional feature maps and the feature maps from corresponding block in the encoding path. In contrast to the original U-Net architecture [12], we utilize zero padding to preserve the output dimension for all the convolutional layers of both down-sampling and up-sampling path. Finally, a $1 \times 1$ convolutional layer is used to diminish feature number to two. The neural network was trained on images of size $512 \times 256$. In order to reduce training time, prevent overfitting and increase performance of the U-Net we added Batch Normalization (BN) [13] after each Conv2D. BN is a kind of supplemental layer that adaptively normalizes the input values of the following layer, mitigating the risk of overfitting. Since it improves gradient flow through the network, it reduces dependence on initialization and higher learning rates are achieved. Data augmentation is useful for the reduction of overfitting and it has also been applied during the training.

The pixel-wise cross-entropy has been used as the loss function for bone structure segmentation:

$$\mathcal{L}_{\text{CE}} = -\frac{1}{N} \sum_{i=1}^{N} [y_i \log(\hat{y}_i) + (1 - y_i) \log(1 - \hat{y}_i)] \tag{1}$$

where $N$ stands for the number of training samples, $y$ is true value and $\hat{y}$ denotes predicted value. In X-ray images, it is not uncommon that the bones of interest occupy smaller region of the images in comparison to image backgrounds. Thus, training of the network may yield a network whose predictions be biased towards

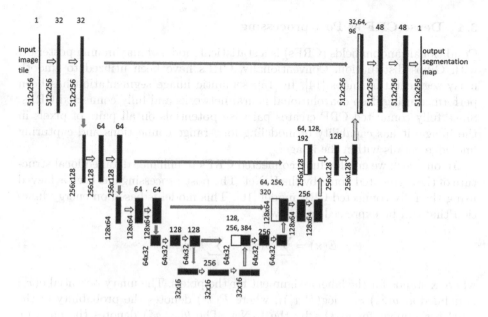

**Fig. 3.** Architecture of U-Net used for bone segmentation.

background. As a result, the background can be over-segmented, i.e. can be incompletely delineated. To copy with such undesirable outcomes we utilized a Dice loss, which can be expressed in the following manner:

$$\mathcal{L}_{\text{Dice}} = -\frac{\sum_{i=1}^{N} p_i g_i}{\sum_{i=1}^{N} p_i^2 + \sum_{i=1}^{N} g_i^2} \quad \text{s.t.} \quad p_i \in \{0,1\} \wedge g_i \in \{0,1\} \tag{2}$$

where the sums run over the $N$ pixels of the predicted segmentation image $p_i \in P$ and ground truth one $g_i \in G$. The derivative with respect to predicted pixel $p_j$ can be expressed in the following manner:

$$\frac{\partial \mathcal{L}_{\text{Dice}}}{\partial p_j} = \frac{\partial}{\partial p_j} \left( -\frac{\sum_{i=1}^{N} p_i g_i}{\sum_{i=1}^{N} p_i^2 + \sum_{i=1}^{N} g_i^2} \right)$$

$$= -\left[ \frac{g_j \left( \sum_{i=1}^{N} p_i^2 + \sum_{i=1}^{N} g_i^2 \right) - 2p_j \sum_{i}^{N} p_i g_i}{\left( \sum_{i=1}^{N} p_i^2 + \sum_{i=1}^{N} g_i^2 \right)^2} \right] \tag{3}$$

Using such a form of Dice we do not need to account for class imbalance between image regions, i.e. there is no need to employ loss weights for pixels of different classes. One of the benefits of cross-entropy is that it is easier to optimize using the backpropagation algorithm, whereas the Dice can give better results for class imbalanced data. In our approach the loss function takes the following form:
$\mathcal{L} = \mathcal{L}_{\text{CE}} + \mathcal{L}_{\text{Dice}}$.

## 3.4    Dense CRF as Post-processing

Conditional random fields (CRFs) is a statistical model of maximum a posteriori with Gibbs distribution. Conventionally, CRFs have been utilized to smooth noisy segmentation maps [14]. In [15], semantic image segmentation has been performed using deep convolutional neural networks and fully connected CRFs. Since fully connected CRF creates pairwise potentials on all pairs of pixels in the image, it has capability of modeling long-range connections and capturing fine edge details within the image.

In our work we employ fully connected CRFs to enhance detailed local structure of the segmented bones by the U-Net. The post-processing has been achieved using the fully connected CRF model [16]. This model is based on energy function that can be expressed as follows:

$$E(\mathbf{x}) = \sum_i \theta(x_i) + \sum_{ij} \theta(x_i, x_j) \tag{4}$$

where $\mathbf{x}$ stands for the label assignment for the pixels. The unary potential $\theta(x_i)$ is defined as $\theta(x_i) = -\log(P(x_i))$, where $P(x_i)$ denotes the probability of the label assignment for pixel $i$ by the U-Net. The $\theta(x_i, x_j)$ denotes the pairwise potential $\theta(x_i, x_j) = \mu(x_i, x_j) \sum_{m=1}^{K} w_m \cdot k^m(\mathbf{f}_i, \mathbf{f}_j)$, where $\mu(x_i, x_j) = 1$ if $x_i \neq x_j$, and zero otherwise. In such fully connected CRFs there is one pairwise term of every pixels pair $i, j$, no matter at what distance they are apart. Each $k^m$ is a Gaussian kernel that hinges on the features $\mathbf{f}$ that are determined for pixels $i$ and $j$ and is weighted by the $w_m$. The Gaussian kernels have the following form:

$$w_1 \exp\left(-\frac{|p_i - p_j|^2}{2\sigma_\alpha^2} - \frac{|I_i - I_j|^2}{2\sigma_\beta^2}\right) + w_2 \exp\left(-\frac{|p_i - p_j|^2}{2\sigma_\gamma^2}\right) \tag{5}$$

The first (appearance) kernel depends both on pixels positions $p$ and pixel intensities $I$, whereas the second (smoothness) kernel depends only on position of pixels. The $\sigma_\alpha, \sigma_\beta, \sigma_\gamma$ parameters determine the scale of the Gaussian kernels. Following [16] we utilize a mean field method for approximate Maximum Posterior Marginal (MPM) inference.

## 3.5    Anatomic Axis Estimation

Given the segmented bones we executed the image skeletonization. The skeletons were determined using morphological thinning, i.e. by shrinking the binary image until the area of interest was 1 pixel wide. The major drawback of skeleton representation is that it is sensitive to noise and artefacts in shape boundaries. Small perturbations of the shape boundary may lead to redundant skeleton branches, which change the topology of the skeleton graph. We coped with such unnecessary branches by executing a skeleton pruning. We utilized the geodesic distance transform [17] to calculate the longest continuous path in the thinned image. Through calculating the longest path we robustly eliminated artifacts and spurs from the skeleton. It is worth noting that such skeleton artifacts are close to hip joint, knee, ankle, and not in femur's and tibia's parts that were employed in angle measurements.

After skipping begin as well the end part of the skeletons representing femurs and tibia bones we fit a line by linear regression (by minimizing the sum of squared errors between the best fit line and data points belonging to the limb segment). The femoral anatomic axis tibial anatomic axis were represented by lines estimated in such a way. The line parameters were determined as follows:

$$
\begin{bmatrix} m \\ b \end{bmatrix} = \frac{1}{(\sum_{i=1}^{n} x_i)^2 - n \sum_{i=1}^{n} x_i^2} \begin{bmatrix} \sum_{i=1}^{n} x_i & -n \\ -\sum_{i=1}^{n} x_i^2 & \sum_{i=1}^{n} x_i \end{bmatrix} \begin{bmatrix} \sum_{i=1}^{n} y_i \\ \sum_{i=1}^{n} x_i y_i \end{bmatrix} \quad (6)
$$

where $n$ stands for the number of pixels. They have been then utilized to determine the values of aLFDA, MPTA and FTA angles. The knee joint axes have been determined by finding least-energy pathes across the joint spaces and then fitting lines to the paths. The least-energy path on the joint space has been determined by the Dynamic Programming (DP) [18] algorithm operating on images with weighted sums of pixel intensities and the binary mask values. The seed location for DP algorithm has been determined on the basis of the segmented images, see also Fig. 4 that illustrates how bones are separated by the segmentation algorithm.

## 4  Experimental Results

In order to obtain good delineation of the bones on X-ray we investigated numerous U-Net architectures and assessed them in terms of mean IoU (Intersection over Union) and mean Dice metrics. At this stage of experiments we evaluated also various loss functions, including weighted cross-entropy, Dice Loss, sum of the cross-entropy and the Dice-loss, Generalized Dice Loss, pixel-wise Wasserstein distance. We trained models and compared the results for various optimizers and learning parameters. In this way we designed the architecture of the neural network, c.f. Subsect. 3.3, which is capable of delineating the bones of the lower limbs on X-ray images with sufficient accuracy.

Figure 4 depicts segmentation results for images selected from the test subset of our Deformed-legs dataset. The bones were segmented using U-Net and then processed by the fully connected CRFs. The CRFs-based post-processing has been executed on images with width equal to 512 and the number of rows resulting from the height/width ration in x-ray images of original resolution. The mean IoU on all images from test subset of Deformed-legs dataset was equal to 0.954 and mean Dice was equal to 0.968. The mean IoU and mean Dice on images segmented by U-Net were equal to 0.945 and 0.962, respectively. As we can observe on depicted images, promising results were achieved using the proposed framework for bone segmentation on X-ray images.

The neural network yields images of size $512 \times 256$. The images delivered by the neural network were scaled so that the number of pixels in rows is equal to

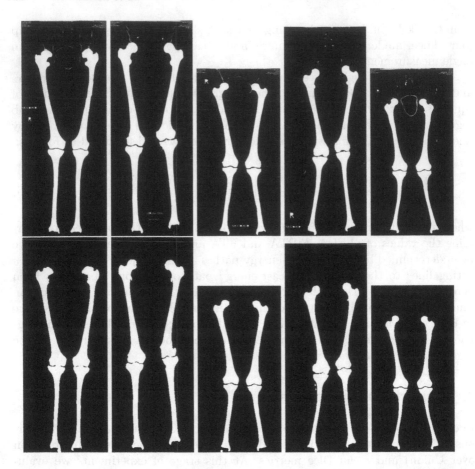

**Fig. 4.** Manual bone segmentation by orthopedist (1st row), bone segmentation by our network (2nd row), see X-ray images on Fig. 1 and manually delineated bones on Fig. 2. From left to right the images #2, 3, 6, 7 and 9 from test subset of our Deformed-legs dataset are shown.

256 whereas the image heights are calculated using the height/widths ratios of the input X-ray images. Hereby, the images with segmented bones preserve the joint angles.

As we can observe on Fig. 4, on manually segmented images there are some artifacts. They appear since the well known flood-fill algorithm has been used to segment the bones from the background on the basis of bone borders determined manually, see also Fig. 2. This means that white text or pixels with values 255 were treated as borders of the bones on the images. On the other hand, our neural network is able to properly delineate the bones despite noisy X-ray images, as well as the text and remarks that were included by radiologists.

Figure 5 depicts skeletons fitted to the segmented bones. The segmentation masks generated by our algorithm are overlaid transparently on the X-ray images

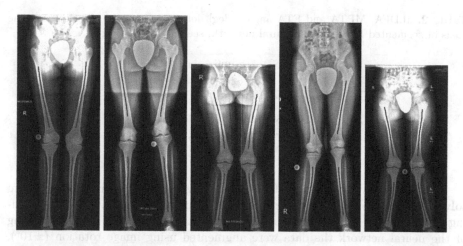

**Fig. 5.** Skeletons of the bones. The segmentation masks generated by our algorithm were overlaid transparently on the X-ray images (yellowish color). Skeletons representing the segmented bones were marked with blue. From left to right the images #2, 3, 6, 7 and 9 are shown. The illustrated annotations are stored in vectorized graphics format and a better viewing can be obtained by zooming this figure. (Color figure online)

**Table 1.** aLDFA, MPTA and FTA angles [deg] determined on X-ray images by orthopedist on basis of X-ray images, see Fig. 1.

|  | im2 | | im3 | | im6 | | im7 | | im9 | |
|---|---|---|---|---|---|---|---|---|---|---|
|  | R | L | R | L | R | L | R | L | R | L |
| aLDFA | 76 | 82 | 78 | 68 | 75 | 79 | 72 | 75 | 77 | 79 |
| MPTA | 89 | 91 | 85 | 85 | 96 | 96 | 96 | 92 | 96 | 95 |
| FTA | 167 | 173 | 173 | 165 | 161 | 163 | 160 | 164 | 160 | 163 |

(yellowish color). As we can observe, the automatically extracted skeleton represents well the extracted bones. It is worth noting that radiologists and orthopedics pointed a considerable usefulness of the skeleton determined in this way for the manual determination of the angles describing deformed legs.

Table 1 illustrates the aLFDA, MPTA and FTA angles determined by the orthopedist. An open-source DICOM software (OsiriX, PIXMEO SARL, Switzerland) has been employed for manual angle measurements.

Table 2 illustrates the aLFDA, MPTA and FTA angles determined by our algorithm on the basis of segmented bones on images with 256 columns and row number determined on the basis of height/with ratios in X-ray images, see also images in 2nd row on Fig. 4. The axes were determined on the basis of bone segments starting at 15% length of the bone and ending at 85% bone length. The images indexed in tables are from test subset of our Deformed-legs dataset and were not used to train the neural network.

**Table 2.** aLDFA, MPTA and FTA angles [deg] determined by our algorithm on the basis of segmented bones by our neural network, see 2nd row on Fig. 4.

|        | im2 |     | im3 |     | im6 |     | im7 |     | im9 |     |
|--------|-----|-----|-----|-----|-----|-----|-----|-----|-----|-----|
|        | R   | L   | R   | L   | R   | L   | R   | L   | R   | L   |
| aLDFA  | 77  | 85  | 77  | 71  | 79  | 81  | 70  | 76  | 77  | 77  |
| MPTA   | 89  | 92  | 86  | 85  | 97  | 97  | 91  | 90  | 94  | 97  |
| FTA    | 168 | 173 | 172 | 167 | 165 | 165 | 161 | 165 | 163 | 166 |

The network was trained in 200 epochs using 8 images per batch, the Adam solver and the proposed loss function. Momentum was set to 0.9, while a learning rate of $1 \times 10^{-4}$ was used and a weight decay of $5 \times 10^{-4}$. While learning of the neural network the data were augmented using image rotation ($\pm 10°$), horizontal and vertical shift (shift range set to 0.1), zoom (range set to 0.2), rescale and horizontal flip. The training of neural networks has been performed on TitanX GPU. The neural network has been implemented in Python using Keras framework with TensorFlow backend. The algorithms for aLDFA, MPTA and FTA angle estimation have been implemented in Python.

## 5    Conclusions

In this work we presented a new approach to measuring lower limb alignment and joint orientation. The originality of the introduced approach consists in innovative application of a deep U-Net convolutional network for the extraction of lower limb bones on X-ray images. Automatic bone extraction on X-ray images with lower limbs is a very difficult problem, and to best of our knowledge, until now deep learning techniques have not been applied to bone extraction on X-ray images. In order to obtain satisfactory results we designed an U-Net based neural architecture to segment the bones, fully connected CRFs to post-process the segmented bones as well as we prepared a Deformed-legs dataset. Considerable effort has been devoted for proper delineation of bones on X-ray images. It contains fifty annotated X-ray images. The dataset has a training subset of images, which can be used for training new neural architectures for bone segmentation, and test subset that can be utilized for evaluation such new network architectures as well as evaluating new algorithms for lower limb alignment and joint orientation estimation. We showed both qualitative and quantitative segmentation results on ten test images. Our experiments demonstrated that the experimental results are promising and the proposed approach has a potential.

## References

1. Paley, D.: Normal lower limb alignment and joint orientation In: Principles of Deformity Correction, pp. 1–18. Springer, Heidelberg (2002). https://doi.org/10.1007/978-3-642-59373-4_1

2. Olczak, J., et al.: Artificial intelligence for analyzing orthopedic trauma radiographs. Acta Orthop. **88**, 581–586 (2017)
3. Cherian, J.J., Kapadia, B.H., Banerjee, S., Jauregui, J.J., Issa, K., Mont, M.A.: Mechanical, anatomical, and kinematic axis in TKA: concepts and practical applications. Curr. Rev. Musculoskelet. Med. **7**, 89–95 (2014)
4. Lohman, M., Tallroth, K., Kettunen, J.A., Remes, V.: Changing from analog to digital images: does it affect the accuracy of alignment measurements of the lower extremity? Acta Orthop. **82**, 351–355 (2011)
5. Yang, N.H., Nayeb-Hashemi, H., Canavan, P.K., Vaziri, A.: Effect of frontal plane tibiofemoral angle on the stress and strain at the knee cartilage during the stance phase of gait. J. Orthop. Res. **28**, 1539–47 (2010)
6. Ariumi, A.: Three-dimensional lower extremity alignment in the weight-bearing standing position in healthy elderly subjects. J. Orthop. Sci. **15**, 64–70 (2010)
7. Sharma, L., Song, J., Felson, D.T., Cahue, S., Shamiyeh, E., Dunlop, D.D.: The role of knee alignment in disease progression and functional decline in knee osteoarthritis. JAMA **286**, 188–195 (2001)
8. Pape, D., Rupp, S.: Preoperative planning for high tibial osteotomies. Operative Techn. Orthop. **17**, 2–11 (2007)
9. Moreland, J., Bassett, L., Hanker, G.: Radiographic analysis of the axial alignment of the lower extremity. J. Bone Joint Surg. Am. **69**, 745–749 (1987)
10. Wu, J., Mahfouz, M.R.: Robust x-ray image segmentation by spectral clustering and active shape model. J. Med Imaging **3**, 034005 (2016)
11. Aganj, I., Harisinghani, M., Weissleder, R., Fischl, B.: Unsupervised medical image segmentation based on the local center of mass. Sci. Rep. **8**, 13012 (2018)
12. Ronneberger, O., Fischer, P., Brox, T.: U-Net: convolutional networks for biomedical image segmentation. In: Navab, N., Hornegger, J., Wells, W.M., Frangi, A.F. (eds.) MICCAI 2015. LNCS, vol. 9351, pp. 234–241. Springer, Cham (2015). https://doi.org/10.1007/978-3-319-24574-4_28
13. Ioffe, S., Szegedy, C.: Batch normalization: accelerating deep network training by reducing internal covariate shift. In: ICML, vol. 37, pp. 448–456 (2015)
14. Liu, C., Szeliski, R., Bing Kang, S., Zitnick, C.L., Freeman, W.T.: Automatic estimation and removal of noise from a single image. IEEE Trans. Pattern Anal. Mach. Intell. **30**, 299–314 (2008)
15. Chen, L., Papandreou, G., Kokkinos, I., Murphy, K., Yuille, A.L.: Semantic image segmentation with deep convolutional nets and fully connected CRFs. In: 3rd International Conference on Learning Representations, ICLR 2015 (2015)
16. Krähenbühl, P., Koltun, V.: Efficient inference in fully connected CRFs with Gaussian edge potentials. In: Proceedings of the 24th International Conference on Neural Information Processing Systems, USA, Curran Associates Inc., pp. 109–117 (2011)
17. Soille, P.: Morphological Image Analysis: Principles and Applications, 2nd edn. Springer, Heidelberg (2003). https://doi.org/10.1007/978-3-662-05088-0
18. Felzenszwalb, P.F., Zabih, R.: Dynamic programming and graph algorithms in computer vision. IEEE Trans. Pattern Anal. Mach. Intell. **33**, 721–740 (2011)

# Constraint Programming Based Algorithm for Solving Large-Scale Vehicle Routing Problems

Bochra Rabbouch[1,2](✉), Foued Saâdaoui[2,3], and Rafaa Mraihi[4]

[1] Institut Supérieur de Gestion de Tunis, Université de Tunis,
Cité Bouchoucha, 2000 Tunis, Tunisia
bochra.rabbouch@gmail.com
[2] Laboratoire d'Algèbre, Théorie de Nombres et Analyse Non-linéaire,
Faculté des Sciences, University of Monastir, 5019 Monastir, Tunisia
[3] Department of Statistics, Faculty of Sciences, King Abdulaziz University,
P.O. BOX 80203, Jeddah 21589, Saudi Arabia
[4] Ecole Supérieure de Commerce de Tunis, Université de Manouba,
Campus Universitaire de La Manouba, 2010 Tunis, Tunisia

**Abstract.** Smart cities management has become currently an interesting topic where recent decision aid making algorithms are essential to solve and optimize their related problems. A popular transportation optimization problem is the Vehicle Routing Problem (VRP) which is high complicated in such a way that it is categorized as a NP-hard problem. VRPs are famous and appear as influential problems that are widely present in many real-world industrial applications. They have become an elemental part of economy, the enhancement of which arises in a significant reduction in costs.

The basic version of VRPs, the Capacitated VRP (CVRP) occupies a central position for historical and practical considerations since there are important real-world systems can be satisfactorily modeled as a CVRP. A Constraint Programming (CP) paradigm is used to model and solve the CVRP by applying interval and sequence variables in addition to the use of a transition distance matrix to attain the objective. An empirical study over 52 CVRP classical instances, with a number of nodes that varies from 16 to 200, and 20 CVRP large-scale instances, with a number of nodes that varies from 106 to 459, shows the relative merits of our proposed approach. It shows also that the CP paradigm tackles successfully large-scale problems with a percentage deviation varying from 2% to 10% where several exact and heuristic algorithms fail to tackle them and only a few meta-heuristics can probably solve instances with a such big number of customers.

## 1 Introduction

Combinatorial optimization problems can be expressed on a declarative or on a procedural formulation. The declarative formulation directly expresses

© Springer Nature Switzerland AG 2019
H. Pérez García et al. (Eds.): HAIS 2019, LNAI 11734, pp. 526–539, 2019.
https://doi.org/10.1007/978-3-030-29859-3_45

constraints and the objective function then attempts to find the solution without the distraction of algorithmic features. Whereas, the procedural formulation defines the manner to solve the problem by providing an algorithm. The artificial intelligence community invented ways to twist declarative and procedural formulations to work together. This idea had brought about the introduction of logic programming, constraint logic programming, constraint handling rules and CP. This last is a new paradigm evolved within operations research, computer sciences, algorithms and artificial intelligence. As its name reveals, CP is a two level architecture including a constraint and a programming component. It is considered as an emerging programming language for describing feasibility problems. It consists of parameters, sets and deals with discrete decision variables (binary or integer). Each problem is expressed using three entities: variables, their corresponding domains and constraints. The problems can then be solved by applying complete techniques such as depth-first search for satisfaction and branch and bound for optimization. Ham and Cakici [12] affirm in their recent study: "[We] also provide the reasons why CP would be soon welcomed by the practitioners. Firstly, CP natural formulation is closer to the problem description than the restricted linear programming formulation. Secondly, a concise CP code provides a flexibility and scalability to practitioners. Thirdly, unlike meta-heuristics which are tailor-made requiring a fine tuning of parameters to reach at the best performance, CP is off-the-rack. Namely, practitioners provide a high level description of the problem only. All settings of search algorithms and detailed tunings are automatically done by CP engine. Finally, CP outperforms MIP in the scheduling problems as we demonstrate in this study". In addition, CP has proven very competent in different applications and is now used in a wide range of science applications, businesses and industries including manufacturing, transportation, health care, financial services, telecommunications, energy, utilities, marketing and sales. It is used also to fix industrial problems and is on the basis of many industrial applications (e.g., gate allocation at the Hong Kong airport, container scheduling at Port of Singapore, timetabling of Dutch Railways (INFORMS Edelman-award), etc). In addition, several international companies like BMW, eBay, France Telecom, HP, orange, Porsche, Shell, Royal Bank of Scotland, Xerox, Yves Rocher, etc use the CP to optimize their business systems. Moreover, the need to use the CP models appears especially since the appearance of complex logical and arithmetic relationships between decision variables, activities and resources. CP models have proven very useful when dealing with the complexity of many real-world sequencing and scheduling problems such as the VRP which is a famous combinatorial optimization problem that can successfully resolved through the CP paradigm. The VRP is a widely studied NP-hard problem and has become, since its appearance by Dantzig and Ramser [7] in 1959, one of the most challenging research topics where a great number of researchers investigate thousands of models, approaches, algorithms and papers to solve it as it can directly applied in the real world scenarios for distribution and transportation systems that are considered as important parts of economy, the improvement of which results in an appreciable reduction in costs. The VRP can reflect different aspects present in realistic scenarios since it considers different variants such as time windows, different depots, loading constraints, etc. The CVRP is the generalization of other VRP variants and contains basic constraints that must be satisfied to obtain feasible solutions. It occupies a central position for historical [7] and practical considerations since there are important real-world systems can be satisfactorily modeled as a CVRP. In addition, it is a common wind tunnel for new several good ideas and concepts that can be next applied on other variants. In addition, several solution methods have been raised to solve the CVRP and are basically exact, heuristic and meta-heuristic approaches. An interesting recent survey for the CVRP and its solution methods is well explored on the paper of Rabbouch et al. [20]. Since this paper, several approaches have presented to fix the CVRP problem [1,3,4,6,10,13–15,18,21–23].

The actual paper is organized as follows. Section 2 presents a mathematical formulation for the CVRP. Section 3 is reserved to a constraint programming model for the VRP. Section 4 describes benchmark instances and different results obtained through the mathematical and the constraint programming models. Finally, Sect. 5 concludes the paper.

## 2    Mathematical Programming Model for the CVRP

In this section, we present a Mixed Integer Linear Problem (MILP) for the CVRP.

Let us consider a vehicle routing network defined as a directed graph $G = (V, A)$. $K$ designs an homogenous fleet of vehicles with a capacity $c$ where a used vehicle is indexed by $k \in K$.

The graph $G$ includes a set of vertices $V = \{v_1, \ldots, v_n\}$ and $A = \{(v_i, v_j): v_i, v_j \in V, i \neq j\}$ is the arc set.

The customer set $V_c = \{v_2, \ldots, v_n\}$ represents customers, while $V_1$ corresponds to the central depot where are parked the $K$ vehicles.

Each vertex $V_i \in V_c$ has a non-negative fixed loads such as demand $d_i$ with $(d_i = 0 \ \forall \ i = 1)$.

The distance between points $i$ and $j$ for all $i, \ j \in V$ is computed in a variable $D_{ij}$.

The following CVRP mathematical model is proposed,

- **decision variables**
    - The assignment variable $X_{ijk}$
      $X_{ijk} = 1$ if a vehicle $k(k = 1, .., K)$ is traveling along an arc (i, j) ($i \in V$; $j \in V$) and $X_{ijk} = 0$ otherwise.
    - The load variable $Y_{ij}$ which is a non-negative continuous variable representing the total load remaining in the vehicle before reaching the node $j$ while traveling the arc $(i, j)$ with ($i \in V$; $j \in V$).
- **The objective function:**

$$Minimize \sum_{i=1}^{n} \sum_{j=1}^{n} \sum_{k=1}^{K} D_{ij} X_{ijk} \qquad (1)$$

The objective function (1) aims to minimize the total distance which is the sum of all consecutive points serviced by $K$ vehicles starting and ending at a central depot.

- **Constraints:**

$$\sum_{i=1}^{n}\sum_{k=1}^{K}X_{ijk} = 1, \quad \forall j = 2,\ldots,n; \quad i \neq j; \tag{2}$$

$$\sum_{j=1}^{n}\sum_{k=1}^{K}X_{ijk} = 1, \quad \forall i = 2,\ldots,n; \quad i \neq j; \tag{3}$$

- The constraints (2) and (3) ensure that each customer is visited and serviced only once by a given vehicle

$$\sum_{i=1}^{n}X_{ijk} = \sum_{i=1}^{n}X_{jik}, \quad \forall k = 1,\ldots,K; \quad \forall j = 1,\ldots,n; \quad i \neq j; \tag{4}$$

- The constraint (4) states that servicing an arc $(ij)$ implies servicing the arc $(ji)$ (a symmetric VRP).

$$\sum_{j=2}^{n}Y_{1j} \geq \sum_{j=2}^{n}d_j \tag{5}$$

- The constraint (5) refers that the total vehicle's load when leaving the depot is superior or equal to the total customer demand.

$$\sum_{i=1}^{n}Y_{ij} - \sum_{i=1}^{n}Y_{ji} = d_j, \quad \forall j = 2,\ldots,n \tag{6}$$

- The constraint (6) indicates that the quantity remaining after visiting a customer $j$ is exactly the load before visiting this customer minus its demand.

$$Y_{ij} \leq \sum_{k=1}^{K}cX_{ijk}, \quad \forall i = 1,\ldots,n; \quad \forall j = 2,\ldots,n \tag{7}$$

- The constraint (7) guarantees that the vehicles capacity may not be violated.

$$X_{ijk} \in \{0,1\}, \quad \forall i,j = 1,\ldots,n; \quad \forall k = 1,\ldots,K; \tag{8}$$

- The constraint (8) refers to the binary of the assignment decision variable.

$$Y_{ij} \geq 0; \quad \forall i,j = 1,\ldots,n \tag{9}$$

- The constraint (9) refers to the non negativity of the load decision variable.

# 3   A Constraint-Based Model for the CVRP

A constraint is a logical relation among several variables, each taking a value in a given domain. Its purpose is to restrict the possible values that variables can take. Thus, CP is a logic-based method to model and solve large combinatorial optimization problems based on constraints. The idea of CP is to solve problems by stating constraints that must be satisfied by the solution. Considering its roots in computer science, CP basic concept is to state the problem constraints as invoking procedures then constraints are viewed as relations or procedures that operate on the solution space. Constraints are stored in the constraint store and each constraint contributes some mechanisms (enumeration, labeling, domain filtering) to reduce the search space by discarding infeasible solutions using constraint satisfiability arguments.

## 3.1   The CP Solution Process

The CP is an expressive language for formulating constraints where they interact naturally and each constraint is encapsulated. The CP model is usually expressed through: variables, corresponding domains for those variables, constraints between those variables and objectives that can be either for minimization or maximization (it focuses on the constraints and variables rather than the objective function). Our actual paper tackles a constraint optimization problem which aims to minimize an objective function by comparing the value of the objective for all feasible solutions search procedures and then display the best solution it can found.

The solution process of CP includes both a search mechanism and a constraint propagation technique. The problem is tackled firstly by applying propagation algorithms that aim to decrease the search space by filtering out the values in variable domains that can generate infeasible solutions and then a set of possible values of decision variables that satisfy all constraints is obtained. Because a single variable can model more than one variable, propagation algorithms are triggered every time a change occurs in the domain of a variable. Precisely, when a constraint has already propagated and at least one value has been eliminated from the domain of the variable, propagation algorithms of all the other constraints involving the deleted value are triggered. Once propagation is ended, some failing values can still occur in the variable domains, hence, a systematic search phase is run. Both propagation algorithms and search procedures are combined to reach all feasible solutions. When a feasible solution is already better than the best current one, a new constraint is added to the search tree to state that remaining feasible solutions must have better values and excluding then worst remaining values and the search proceeds to improve the value of the objective function.

## 3.2   The CP_CVRP Model

Given $m$ the set of available vehicles with a fixed capacity $K$. The model aims to serve $n$ customers where each customer has an index in $N = \{1 \ldots n\}$.

A solution is made up of routes and we assume that we have a route per vehicle. Each route has an index in $M = \{1 \ldots m\}$ and an unassigned route has an index 0. The subset of routes including unassigned routes is expressed by $M \cup \{0\}$.

In addition, the following variables are proposed:

- $V = V_1 \ldots V_n$ with an integer domain $[1 \ldots m]$
- $Q = Q_1 \ldots Q_m$ with a real domain $[0 \ldots q]$

Each route is made up of sequence of visits: $V$ is a record of $n$ variables that corresponds to the route decision variable to serve the customers where each $V_i$ value precise which vehicle serves the $i$th customer. This set includes indices corresponding to all visits needed. Each route has a start visit and an end visit and when a visit is serving a given customer, it is called a customer visit. A start visit for a route $k$ has the index $N + k$ and an end visit for a route $k$ has the index $N + m + k$. $D_i$ is the amount of goods to pick up in visit $i$ and $Q_i$ is the quantity of goods in the vehicle serving the visit. Hence, $Q$ corresponds to the accumulation variable and is the record of $m$ variables needed to determine the cumulative capacity at every route from the $m$ routes.

A finite-domain variables $next_i$ associated to each position $i$ where $i.next_i = j$ if $j$ is a direct successor to $i$. In order to represent visit to depot, a constraint based model introduces two visits where $S = \{n+1 \ldots n+m\}$ denotes indices illustrated the start visits (departure from depot) and $E = \{n+m+1 \ldots n+2m\}$ denotes the indices indicated the end visits. Note that $V = N \cup S \cup E$ is the set of all visits. Hence, $next_i = j$ is meaningful only if $i \in V^s = S \cup N$ and $j \in V^e = E \cup N$. In addition, it is important to mention that $j$ is called a *successor* of another visit $i$ if $j$ comes somewhere after $i$ in the path and is called *next* if it is visited exactly after $i$.

**Decision Variables.** The IBM ILOG CP Optimizer introduces different mathematical concepts as interval and sequence variables to capture the temporal aspects of scheduling:

- **Interval variables:** Define an interval of time during which a particular activity executes. The value of an interval variable is an integer variable that has a start and end points. Its domain $dom(a)$ is a subset of $\{\perp\} \cup \{[s, e) | s, e \in \mathbb{Z}, s \leq e\}$. A such interval variable can be absent ($a = \perp$) and that means that it is not considered by any expression or constraint on the interval variable it is included in. However, by default, the interval variable is supposed present and $a = [s, e)$ where $s$ denotes the start and $e$ denotes the end of the interval and the length's interval $e - s$ can be computed. Moreover, an interval variable has the ability to be optional, thus it has an additional possible value in its domain known as the absence value $\perp$. It is an important feature that allows the problem to decide whether the variable will be present or absent in the solution. This concept is crucial in scheduling applications to model optional activities or alternative execution modes for activities. In the CP_CVRP model, the interval variables are exploit to express the route variables aiming to assign customers to routes to be served. Route decision variable is denoted $V$ in the proposed model.

- **Sequence variable** is defined on a set of interval variables and its value is generally a permutation of the present set of intervals to model the sequence of elementary operations and activities. It aims to model constraints that enforce a tempering ordering of a set of interval variables. In the CP_CVRP proposed model, the sequence variable is employed to describe the accumulation variable described the sequence of visited customers by a given vehicle allowing to accumulate the quantity of goods picked up to customers to not violate the capacity constraints.

**Constraints.** The CP model must obey to some constraints to guarantee the feasibility of the generated solutions.

- Each route in a feasible solution for the CVRP must start and end at a central depot. A depot is designed by a start visit and an end visit then, we need to illustrate two constraints through the two functions $first$ and $last$ on interval variables to constrain the first and the last components of an interval variable.
- The second constraint in our model is that enforcing to a vehicle to finish servicing a given customer before starting the service of the next customer. Hence, a feasible solution is composed of a set of no overlapping fixed distances between every two positions. Therefore, the $noOverlap$ function is applied to constrain to each component of the sequence to end before the beginning of the next interval in the permutation. As the sequence is composed of customers visits, it presents then a query of non overlapping distances where each visit is enforced to be handled after the end of the previous visit in the sequence.
- The next constraint is that enforcing the non violation of the vehicles capacity. The sum of demands of customers assigned to routes is computed and a comparison between the vehicles capacity is made. This constraint ensure that accumulated commodities loaded are already inferior or equal to the vehicles capacity. Formally, in addition to the $sum$ function and the arithmetic and logical concepts, we applied the $presenceOf(x)$ constraint which states that an interval $x$ is present $(x \neq \perp)$ and it is used to confirm the customers visits to accumulate then goods loaded and compare them with vehicles capacity.
- The last constraint is to allocate vehicles to visit customers where each customer is visited once by one vehicle. This option is ensured using the $alternative$ constraint $alternative\ (x, \{x_1, \ldots, x_n\})$ which presents an exclusive alternative interval variable between $\{x_1, \ldots, x_n\}$ to state that if the interval $x$ is present, then exactly one of the intervals $\{x_1, \ldots, x_n\}$ is present

and the interval $x$ and the chosen interval begins and ends together in the same time and when the interval $x$ is absent then all the $x_i$ are absent.

**Objective Function.** The proposed objective to the problem is to minimize the total traveled distance by all vehicles. We assume that the distance between two cities $i$ and $j$ is an Euclidean distance. Classical CP problems for traveling salesman problems and some VRPs build objective functions based on interval variables by exploiting the integer expression $endOf$ used to access the end time of the interval variable designing the visits by indexing the end visit. The objective function minimizes then the sum of the end time of the routes and is formulated as below, where the $tvisit[< 1 >][k]$ designs the end customer visit for each route as in the example illustrated in [17]. In our approach, we presented a transition cost (distance) matrix that computes the traveled distance by each vehicle. It accumulates transition distances by switching from a customer visit to its successor in the same route until it finishes. This method is one of the main novelties of our approach that computes successfully the traveled distance. The total cost expression of all routes traveled is minimized to attain best objective values.

The transition distance matrix is used by the expression $typeOfNext$ on both the sequence and interval variable to constrain the length of the setup activity and extract the transition distance between immediate successors while accumulate them in a variable designing the distance traveled by a single vehicle. Then, by the mean of the $min$ function, the sum of distances traveled by all vehicles is then minimized.

### 3.3 Influencing the Search

As for mathematical formulations, there are various methods to model the same problem using the CP optimizer and that can lead to different results. For complex problems, in addition to efficient formulation exploiting the power of scheduling concepts to reason on time-lines (used above in our formulation e.g. sequence variables, transition cost matrix, optionally..), there are few concepts that can be used to influence the search. In our approach, we select to order the decision variables using search phases then the CP optimizer can fix the key decision variables early in the process and so on it may be easy to extend the partial solution to the remaining variables. In our approach, search phases are applied on the sequence variable used to define the sequence of different cities visited by each vehicle described above.

$$cp.setSearchPhases(cp.factory.searchPhase(route));$$

Search phases specify important decision variables for the scheduling step and hierarchy between them then the constructive search is performed and the search mechanism for the CP process is enhanced. Meanwhile, another strategy is used simultaneously with the search phase to perform the constraint propagation technique for the CP process in order to achieve full domain reduction.

This strategy aims to adjust the constraint propagation inference levels to the extended level then the search space is decreased and not-promising variables are filtered out. In addition, the model is run using parallel search with 4 workers to exploit parallelism by starting from different random seeds for the random generator. The 4 workers communicate their solutions so that if a particular worker finds an improving solution, the other workers can benefit from it then are synchronized. Obtained results show that the strategies precisely above enhance the problem solving and provide especially a rapid convergence toward optimal and good solutions.

```
CP_CVRP Model : CP model for the Capacitated VRP
Begin
    1 : using CP;
    2 : tuple Position {key int id;  float x;   float y;};
    3 : tuple Demand { key int id; int q; };
    4 : int n= ...;
    5 : int m=...;
    6 : int cap=...;
    7 : {Position} Pos= ...;
    8 : {Demand} Demands= ...;
    9 : {int} Customers= { p.id | p in Pos };
   10: tuple triplet { int c1; int c2; int d; };
   11: {triplet} Dist= { < p1.id,p2.id,ftoi (round (sqrt(pow (p2.x-p1.x,2)+pow (p2.y-p1.y,2)))) >
        |p1, p2 in Pos};
   12: execute { writeln(Dist); };
   13: dvar interval visit [d in Demands] size 1;
   14: dvar interval tvisit[d in Demands][k in 1..m] optional(d.id>1) size 1;
   15: dvar sequence route[k in 1..m] in all(d in Demands) tvisit[d][k] types all(d in Demands) d.id;
   16: int distMatrix[p1 in Pos][p2 in Pos] = ftoi (round (sqrt (pow(p2.x-p1.x,2) + pow(p2.y-p1.y,2))));
   17: int distMatrixById[p1 in Customers][p2 in Customers] = distMatrix[<p1>][<p2>];
   18: dexpr float routeDistance[k in 1..m] =
        sum(d in Demands) distMatrixById[d.id][typeOfNext(route[k], tvisit[d][k], d.id, d.id)];
   19: dexpr float totalDistance = sum(k in 1..m) routeDistance[k];
   20: minimize totalDistance;
   21: constraints {
   22:    forall(k in 1..m) {
   23:       noOverlap(route[k], Dist);
   24:       first(route[k],tvisit[<0>][k]);
   25:       last(route[k],tvisit[<1>][k]);
   26:       sum(d in Demands) presenceOf(tvisit[d][k])*d.q <= cap;
   27:    }
   28:    forall(d in Demands: d.id>1) {
   29:       alternative(visit[d], all(k in 1..m) tvisit[d][k]);
   30:    }
end
```

# 4   Computational Results

## 4.1   Instances

To validate the proposed approach, we select firstly the use of the well-known *ABEFMP* classical instances[1] and the characteristics of the selected instances are listed in Table 1.

---

[1] http://neo.lcc.uma.es/radi-aeb/WebVRP/Problem_Instances/CVRPInstances. html.

**Table 1.** Characteristics of a set of selected instances from the ABEMFP data sets

| No | Instances | Number of nodes | Number of vehicles | Capacity | Optimal solution | No | Instances | Number of nodes | Number of vehicles | Capacity | Optimal solution |
|---|---|---|---|---|---|---|---|---|---|---|---|
| 1. | A-n32-k5 | 32 | 5 | 100 | 784 | 27. | B-n38-k6 | 38 | 6 | 100 | 805 |
| 2. | A-n33-k5 | 33 | 5 | 100 | 661 | 28. | B-n41-k6 | 41 | 6 | 100 | 829 |
| 3. | A-n33-k6 | 33 | 6 | 100 | 742 | 29. | B-n44-k7 | 44 | 7 | 100 | 909 |
| 4. | A-n37-k5 | 37 | 5 | 100 | 669 | 30. | B-n45-k5 | 45 | 5 | 100 | 751 |
| 5. | B-n31-k5 | 31 | 5 | 100 | 672 | 31. | B-n45-k6 | 45 | 6 | 100 | 678 |
| 6. | B-n39-k5 | 39 | 5 | 100 | 549 | 32. | B-n50-k7 | 50 | 7 | 100 | 741 |
| 7. | E-n22-k4 | 22 | 4 | 6000 | 375 | 33. | B-n52-k7 | 52 | 7 | 100 | 747 |
| 8. | E-n23-k3 | 23 | 3 | 4500 | 569 | 34. | B-n56-k7 | 56 | 7 | 100 | 707 |
| 9. | E-n30-k3 | 30 | 3 | 4500 | 534 | 35. | B-n66-k9 | 66 | 9 | 100 | 1316 |
| 10. | F-n45-k4 | 45 | 4 | 2010 | 724 | 36. | B-n68-k9 | 68 | 9 | 100 | 1272 |
| 11. | P-n16-k8 | 16 | 8 | 35 | 450 | 37. | B-n78-k10 | 78 | 10 | 100 | 1221 |
| 12. | P-n19-k2 | 19 | 2 | 160 | 212 | 38. | E-n33-k4 | 33 | 4 | 8000 | 835 |
| 13. | P-n20-k2 | 20 | 2 | 160 | 216 | 39. | E-n51-k5 | 51 | 5 | 160 | 521 |
| 14. | P-n21-k2 | 21 | 2 | 160 | 211 | 40. | E-n76-k7 | 76 | 7 | 220 | 682 |
| 15. | P-n22-k2 | 22 | 2 | 160 | 216 | 41. | E-n76-k10 | 76 | 10 | 140 | 830 |
| 16. | A-n34-k5 | 34 | 5 | 100 | 778 | 42. | E-n101-k8 | 101 | 8 | 200 | 815 |
| 17. | A-n36-k5 | 36 | 5 | 100 | 799 | 43. | E-n101-k14 | 101 | 14 | 112 | 1067 |
| 18. | A-n44-k6 | 44 | 6 | 100 | 937 | 44. | P-n22-k8 | 22 | 8 | 3000 | 603 |
| 19. | A-n45-k6 | 45 | 6 | 100 | 944 | 45. | P-n40-k5 | 40 | 5 | 140 | 458 |
| 20. | A-n45-k7 | 45 | 7 | 100 | 1146 | 46. | P-n45-k5 | 45 | 5 | 150 | 510 |
| 21. | A-n60-k9 | 60 | 9 | 100 | 1354 | 47. | P-n70-k10 | 70 | 10 | 135 | 827 |
| 22. | A-n63-k9 | 63 | 9 | 100 | 1616 | 48. | P-n76-k4 | 76 | 4 | 350 | 593 |
| 23. | A-n64-k9 | 64 | 9 | 100 | 1401 | 49. | P-n76-k5 | 76 | 5 | 280 | 627 |
| 24. | A-n69-k9 | 69 | 9 | 100 | 1159 | 50. | P-n101-k4 | 101 | 4 | 400 | 681 |
| 25. | A-n80-k10 | 80 | 10 | 100 | 1763 | 51. | M-n200-k16 | 200 | 16 | 200 | 1274 |
| 26. | B-n34-k5 | 34 | 5 | 100 | 788 | 52. | M-n200-k17 | 200 | 17 | 200 | 1275 |

## 4.2 Results

All the experiments have been realized on 2.3 GHz Intel Core 2, i5 processor (2.3 GHz) 8 GB RAM PC. The $CP\_CVRP$ model has been designed with the IBM CP Optimizer and the optimization is performed with four workers. The $CVRP\_MILP$ model (proposed in the second subsection of the actual paper) has been formulated with the IBM ILOG OPL optimizer. (The IBM CP Optimizer and the IBM ILOG OPL optimizer are two parts of IBM ILOG CPLEX Optimization Studio version 12.8).

**Table 2.** The algorithms used to compare with the proposed approaches

| N | Abbreviation | Authors | Year | Approach |
|---|---|---|---|---|
| 1. | CWS | Clarke & Wrights [5] | 1964 | The Clarke and Wrights Savings heuristic |
| 2. | E_CWS | Altinel & Öncan[2] | 2005 | An enhancement of the Clarke and Wright savings heuristic |
| 3. | ELECT | Faulin & García del Valle [8] | 2008 | An electrostatic algorithm |
| 4. | E_Sweep | Na et al.[19] | 2011 | Extensions to the sweep algorithm |
| 5. | P_VNS | Guimarans et al. [11] | 2011 | A probabilistic variable neighborhood search algorithm |
| 6. | H_VNS | Tilli et al. [24] | 2014 | A hybrid variable neighborhood search & swarm optimization |
| 7. | TS&ALNS | Kir et al. [15] | 2017 | A heuristic algorithm based on the tabu search & adaptive large neighborhood |
| 8. | VTPSO | Akhand et al. [1] | 2017 | Sweep clustering and velocity tentative Particle Swarm Optimization |
| 9. | IWFA | Kerwad et al. [14] | 2018 | Improved water flow-like algorithm |
| 10. | $\gamma$-EMSA | Rabbouch et al. [21] | 2019 | A $\gamma$ empirical mode simulated annealing |

To evaluate the quality of our proposed approach, the obtained results displayed on Tables 3 and 4 are compared with results obtained using methods summarized in Table 2. The Table 3 presents the CP_CVRP solutions for 15 small instances which are compared to solutions obtained by algorithms shown in Table 2 in addition to results obtained by the proposed mathematical model proposed in the second section of this paper. The mathematical model fails to resolve the instances presented in Tables 4 and 5. Note that the results with **bold** characters indicate that the obtained result is optimal whereas results into () indicates that the number of routes is inadequate and are obtained by using a number of vehicles greater than the number required in the instances.

**Table 3.** Obtained results on the first set from the ABEFMP CVRP instances

| No | Instances | BKS | CWS | E_CWS | ELECT | E_Sweep | P_VNS | | H_VNS | TS&ALNS | VTPSO | IWFA | γ-EMSA | MILP | | CP_CVRP | |
|---|---|---|---|---|---|---|---|---|---|---|---|---|---|---|---|---|---|
| | | | | | | | Cost | Time | | | | | | Cost | Time | Cost | Time |
| 1 | A-n32-k5 | 784 | 843 | 827 | 1032 | 810 | 784 | 01:37 | 807 | 784 | 882 | 787 | - | 786 | 00:08:38 | 784 | 00:16 |
| 2 | A-n33-k5 | 661 | 712 | 700 | 789 | 686 | 661 | 00:47 | 685 | 661 | 698 | 662 | 661 | 661 | 02:46:23 | 661 | 00:35 |
| 3 | A-n33-k6 | 742 | (775) | 743 | 834 | 743 | 742 | 00:36 | 762 | 742 | 751 | 742 | - | 744 | 02:54:15 | 742 | 02:31 |
| 4 | A-n37-k5 | 669 | 707 | 708 | 783 | 670 | 669 | 01:07 | 691 | 669 | 754 | 672 | - | 669 | 01:15:25 | 669 | 12:12 |
| 5 | B-n31-k5 | 672 | 681 | 673 | - | 677 | - | - | - | - | - | - | - | 686 | 00:56:38 | 672 | 03:37 |
| 6 | B-n39-k5 | 549 | 564 | 552 | - | 575 | 553 | 01:57 | - | - | - | - | - | 600 | 01:16:53 | 549 | 17:44 |
| 7 | E-n22-k4 | 375 | 388 | 375 | - | 375 | - | - | - | - | - | 375 | - | 375 | 00:02:35 | 375 | 03:17 |
| 8 | E-n23-k3 | 569 | 621 | 574 | - | 569 | - | - | - | - | - | 569 | - | 569 | 00:00:04 | 569 | 06:02 |
| 9 | E-n30-k3 | 534 | 534 | - | - | 543 | - | - | - | - | - | 535 | 534 | 539 | 01:03:33 | 534 | 17:01 |
| 10 | F-n45-k4 | 724 | 737 | - | - | 750 | - | - | - | - | - | - | - | 729 | 02:18:15 | 724 | 00:28 |
| 11 | P-n16-k8 | 450 | (478) | (472) | - | 513 | 453 | 01:02 | - | - | - | 549 | - | 450 | 00:01:57 | 450 | 00:02 |
| 12 | P-n19-k2 | 212 | 237 | 219 | - | 219 | 219 | 00:57 | - | - | - | 246 | - | 212 | 00:00:05 | 212 | 02:47 |
| 13 | P-n20-k2 | 216 | 234 | 247 | - | 217 | 216 | 01:35 | - | - | - | 249 | - | 216 | 00:00:08 | 216 | 01:31 |
| 14 | P-n21-k2 | 211 | 236 | 233 | - | 211 | 211 | 03:15 | - | - | - | 211 | - | 211 | 00:00:03 | 211 | 03:13 |
| 15 | P-n22-k2 | 216 | 239 | 234 | - | 216 | 216 | 05:08 | - | - | - | 216 | - | 217 | 00:00:06 | 216 | 00:10 |

The CP_CVRP model resolves successfully the first set of instances in Table 3 and attains the optimal solutions for all the 15 instances where the number of customers varies from 22 to 45 customers. The CP is more efficient than the mathematical model results which has the ability to obtain optimal results for 7 instances but it requires a large computational time when comparing with the CP paradigm.

**Table 4.** Obtained results on the second set from the ABEFMP CVRP instances

| No | Instances | BKS | CWS | E_CWS | ELECT | E_Sweep | P_VNS | H_VNS | TS&ALNS | VTPSO | IWFA | γ-EMSA | CP_CVRP |
|---|---|---|---|---|---|---|---|---|---|---|---|---|---|
| 16. | A-n34-k5 | 778 | (810) | 793 | 835 | 785 | - | - | 778 | 785 | - | - | 785 |
| 17. | A-n36-k5 | 799 | 826 | 806 | 908 | 826 | - | 811 | 799 | 881 | 802 | - | 824 |
| 18. | A-n39-k5 | 822 | 898 | 894 | 990 | - | - | 846 | 822 | 877 | - | - | 822 |
| 19. | A-n44-k6 | 937 | (974) | (985) | - | 957 | - | - | 939 | 1056 | - | - | 960 |
| 20. | A-n45-k6 | 944 | (1005) | (985) | 1040 | 991 | 944 | 977 | 955 | 1073 | 944 | - | 977 |
| 21. | A-n45-k7 | 1146 | 1200 | 1178 | 1258 | 1173 | - | 1197 | 1153 | 1305 | - | - | 1146 |
| 22. | A-n60-k9 | 1354 | 1416 | 1372 | 1503 | 1420 | - | - | 1366 | 1503 | 1355 | 1354 | 1381 |
| 23. | A-n63-k9 | 1616 | (1684) | 1648 | 1745 | 1712 | 1622 | - | 1644 | 1823 | 1633 | - | 1652 |
| 24. | A-n64-k9 | 1401 | 1489 | 1441 | 1521 | 1499 | - | - | 1442 | 1598 | 1426 | - | 1446 |
| 25. | A-n80-k10 | 1763 | 1859 | 1810 | 1901 | 1866 | - | - | 1790 | 2136 | 1786 | - | 1840 |
| 26. | B-n34-k5 | 788 | 794 | 788 | - | 802 | - | - | - | - | - | - | 788 |
| 27. | B-n38-k6 | 805 | 837 | 820 | - | 817 | 809 | - | - | - | 808 | - | 820 |
| 28. | B-n41-k6 | 829 | (896) | 869 | - | 843 | - | - | - | - | 834 | - | 880 |
| 29. | B-n44-k7 | 909 | 936 | 932 | - | 942 | - | - | - | - | - | - | 930 |
| 30. | B-n45-k5 | 751 | 754 | 751 | - | 797 | 753 | - | - | - | 756 | 751 | 754 |
| 31. | B-n45-k6 | 678 | (723) | (742) | - | 732 | - | - | - | - | - | - | 732 |
| 32. | B-n50-k7 | 741 | 745 | 746 | - | 779 | 744 | - | - | - | - | - | 741 |
| 33. | B-n52-k7 | 747 | 761 | 754 | - | 758 | - | - | - | - | - | - | 761 |
| 34. | B-n56-k7 | 707 | 727 | 718 | - | 726 | - | - | - | - | - | - | 715 |
| 35. | B-n66-k9 | 1316 | (1419) | 1354 | - | 1363 | - | - | - | - | 1329 | - | 1363 |
| 36. | B-n68-k9 | 1272 | 1311 | 1309 | - | 1308 | - | - | - | - | 1286 | 1272 | 1301 |
| 37. | B-n78-k10 | 1221 | 1259 | 1255 | - | 1268 | - | - | - | - | 1230 | 1236 | 1254 |
| 38. | E-n33-k4 | 835 | 841 | 841 | - | 852 | 837 | - | - | - | 837 | - | 841 |
| 39. | E-n51-k5 | 521 | (582) | - | - | 532 | 527 | - | - | - | 527 | 523 | 560 |
| 40. | E-n76-k7 | 682 | 733 | 703 | - | 703 | - | - | - | - | 694 | 696 | 729 |
| 41. | E-n76-k10 | 830 | 903 | (854) | - | 907 | 843 | - | - | - | 854 | - | 903 |
| 42. | E-n101-k8 | 817 | 879 | 845 | - | 850 | 841 | - | - | - | 838 | - | 871 |
| 43. | E-n101-k14 | 1071 | 1130 | 1120 | - | 1152 | - | - | - | - | 1112 | - | 1130 |
| 44. | P-n22-k8 | 603 | (591) | (590) | - | (560) | - | - | - | - | 633 | - | 603 |
| 45. | P-n40-k5 | 458 | 516 | 484 | - | 467 | 461 | - | - | - | 483 | - | 490 |
| 46. | P-n45-k5 | 510 | 569 | 519 | - | - | 512 | - | - | - | 524 | - | 521 |
| 47. | P-n70-k10 | 827 | (892) | - | - | - | - | - | - | - | 911 | 841 | 850 |
| 48. | P-n76-k4 | 593 | 684 | (638) | - | 612 | - | - | - | - | 612 | 610 | 617 |
| 49. | P-n76-k5 | 627 | 705 | 678 | - | - | 633 | - | - | - | 647 | 650 | 701 |
| 50. | P-n101-k4 | 681 | 754 | 708 | - | 715 | 693 | - | - | - | 699 | - | 699 | 725 |
| 51. | M-n200-k16 | 1274 | - | - | - | - | - | - | - | - | - | - | 1496 |
| 52. | M-n200-k17 | 1275 | - | - | - | - | 1324 | - | 1331 | - | 1390 | - | 1413 |

In addition, the Table 4 contains solutions for 36 instances where the number of customers varies from 22 to 200 customers. The CP_CVRP model was run for 5 h to obtain displayed results whereas some solutions need just few minutes to converge to final solutions. For this subset of instances, the CP_CVRP model extracts the optimal solutions for four instances, obtained general competitive and performing solutions and has designed better solutions than all those generated by the ALGELECT algorithm [8], by the CWS heuristic [5] and by the VTPSO [1] (expect 4 instances from the proposed 52 instances).

However, exact methods cannot efficiently solve VRP instances with more than 50–100 customers in a reasonable time, which remains generally a small number in real-life applications. The CP Optimizer search algorithm was designed from the beginning with the main objective to efficiently solve complex industrial scheduling problems [17]. The proposed approach was applied to some large-scale problems which are quite different from academic benchmarks. The instance from the $G$ class ($G - n262 - k25$) proposed by Gillet and Johnson [9] which has no optimal solution yet, is tackled in addition to a set from the class of instances recently proposed by Uchoa et al. [25]. Results on 20 large-scale CVRP instances are displayed in Table 5 where 3 of them have no optimal solutions yet (instances with *). For each instance, we presented the number of nodes (N), the number of vehicles (V), the vehicles capacity (Q), the best obtained result (Best) and solutions obtained by the CP_CVRP model (Cost and GAP). The GAP is expressed as the percentage between the cost and the best solution presented, using the formulation:

$$GAP = \frac{(cost - best)}{best} \times 100$$

**Table 5.** Obtained results on large-scale CVRP instances

| No. | Instances | N | V | Q | Best | CP_CVRP Cost | GAP | No. | Instances | N | V | Q | Best | CP_CVRP Cost | GAP |
|---|---|---|---|---|---|---|---|---|---|---|---|---|---|---|---|
| 1. | X-n106-k14 | 106 | 14 | 600 | 26362 | 27101 | 2.80 | 11. | X-n209-k16 | 209 | 16 | 101 | 30656 | 32985 | 7.59 |
| 2. | X-n110-k13 | 110 | 13 | 66 | 14971 | 16431 | 9.75 | 12. | X-n237-k14 | 237 | 14 | 18 | 27042 | 29015 | 7.29 |
| 3 | X-n120-k6 | 120 | 6 | 21 | 13332 | 14296 | 7.23 | 13. | X-n251-k28 | 251 | 28 | 69 | 38684 | 40602 | 5.11 |
| 4. | X-n129-k18 | 129 | 18 | 39 | 28940 | 30530 | 5.49 | 14. | G-n262-k25* | 262 | 25 | 500 | 6119 | 6252 | 2.17 |
| 5. | X-n139-k10 | 139 | 10 | 106 | 13590 | 14837 | 9. | 15. | X-n284-k15 | 284 | 15 | 109 | 20215 | 22403 | 10.82 |
| 6. | X-n143-k7 | 143 | 7 | 1190 | 15700 | 17082 | 8.80 | 16. | X-n327-k20* | 327 | 20 | 128 | 27532 | 30556 | 10.98 |
| 7. | X-n162-k11 | 162 | 11 | 1174 | 14138 | 15323 | 8. | 17. | X-n331-k15 | 331 | 15 | 23 | 31102 | 34140 | 9.71 |
| 8. | X-n167-k10 | 167 | 10 | 133 | 20557 | 22633 | 10.09 | 18. | X-n393-k38 | 393 | 38 | 216 | 38260 | 42209 | 10.32 |
| 9. | X-n190-k8 | 190 | 8 | 138 | 16980 | 18073 | 6.43 | 19. | X-n439-k37 | 439 | 37 | 12 | 36391 | 39111 | 7.47 |
| 10. | X-n204-k19 | 204 | 19 | 836 | 19565 | 21313 | 8.93 | 20. | X-n459-k26* | 459 | 26 | 1106 | 24145 | 26620 | 10.25 |

The large-scale problems are usually more structured, less pure, larger than classical benchmark instances. In addition, they are generally considered as close simulation to real-world instances. Experiments show that the CP_CVRP model solves successfully this kind of instances where the number of nodes varies from 106 to 459 which is a very big number that exact algorithms, heuristics and meta-heuristics fail mostly to solve such instances with a such big number of customers.

# 5    Conclusion

The VRP is one of the most widely studied families of combinatorial optimization problems as it is vital in modern economies. It can directly applied on real world systems transporting goods and peoples. Even the CVRP is not a rich VRP as it includes only basic variants, it remains a particulary important variant to study the simplest and the most representative case of the problem then to generalize the findings for more complex cases. In this paper, we modeled the CVRP as MILP and CP models which require prior knowledge on the structure of the problem to express it with the constraints and decision/ finite-domain variables in order to resolve it. The MILP model solves only some short instances whereas the CP_CVRP model solves both benchmark and large-scale instances with more than 400 nodes. It performs to attain optimal and near-optimal solutions for benchmark instances and performs on solving large-scale problems with a GAP varying between 2% and 10 %. Obtained results are competitive and that demonstrates the clear advantages of using the constraint programming to solve routing problems and to draw the spectrum of problem characteristics. In future, we propose to combine constraint programming paradigm with local search methods to decrease the required CPU time when solving large-scale instances.

# References

1. Akhand, M.A.H., Paya, Z.J., Murase, K.: Capacitated vehicle routing problem solving using adaptive sweep and velocity tentative PSO. Int. J. Adv. Comput. Sci. Appl. **8**(12), 288–295 (2017)
2. Altinel, I.K., Öncan, T.: A new enhancement of the Clarke and Wright savings heuristic for the capacitated vehicle routing problem. J. Oper. Res. Soc. **56**(8), 954–961 (2005)
3. Baran, E.: Route determination for capacitated vehicle routing problem with two different hybrid heuristic algorithm. Int. J. Eng. Sci. Appl. **2**(2), 55–64 (2018)
4. Borčinová, Z.: Two models of the capacitated vehicle routing problem. Croatian Oper. Res. Rev. **8**, 463–469 (2017)
5. Clarke, G., Wright, J.W.: Scheduling of vehicles from a central depot to a number of delivery points. Oper. Res. **12**, 568–581 (1964)
6. Comert, S.E., Yazgan, H.R., Kir, S., Yener, F.: A cluster first-route second approach for a capacitated vehicle routing problem: a case study. Int. J. Procurement Manage. **11**(4), 399–419 (2018)
7. Dantzig, G., Ramser, J.: The truck dispatching problem. Manage. Sci. **6**(1), 80–91 (1959)
8. Faulin, J., García del Valle, A.: Solving the capacitated vehicle routing problem using the ALGELECT electrostatic algorithm. J. Oper. Res. Soc. **59**(12), 1685–1695 (2008)
9. Gillet, B.E., Johnson, J.G.: Multi-terminal vehicle-dispatch algorithm. Omega **4**, 711–718 (1976)
10. Goli, A., Aazami, A., Jabbarzadeh, A.: Accelerated cuckoo optimization algorithm for capacitated vehicle routing problem in competitive conditions. Int. J. Artif. Intell. **16**(1), 88–112 (2018)
11. Guimarans, D., Herrero, R., Riera, D., JuanJuan, A.A., Ramos, J.: Combining probabilistic algorithms, constraint programming and lagrangian relaxation to solve the vehicle routing problem. Ann. Math. Artif. Intell. **62**(3), 299–315 (2011)
12. Ham, A., Cakici, E.: Flexible job shop scheduling problem with parallel batch processing machines: MIP and CP approaches. Comput. Ind. Eng. **102**, 160–165 (2016)
13. Hannan, M.A., Akhtar, M., Begum, R.A., Basri, H., Hussain, A., Scavino, E.: Capacitated vehicle-routing problem model for scheduled solid waste collection and route optimization using PSO algorithm. Waste Manage. **71**, 31–41 (2018)

14. Kerwad, M.M., Othman, Z.A., Zainudin, S.: Improved water flow-like algorithm for capacitated vehicle routing problem. J. Theor. Appl. Inf. Technol. **96**(15), 4836–4853 (2018)
15. Kir, S., Yazgan, H.R., Tüncel, E.: A novel heuristic algorithm for capacitated vehicle routing problem. J. Ind. Eng. Int. **13**(3), 323–330 (2017)
16. Laborie, P., Rogerie, J.: Temporal linear relaxation in IBM ILOG CP optimizer. J. Sched. **19**(4), 391–400 (2014)
17. Laborie, P., Rogerie, J., Shaw, P., Vilm, P.: IBM ILOG CP optimizer for scheduling. Constraints **23**(2), 210–250 (2018)
18. Noorizadegan, M., Chen, B.: Vehicle routing with probabilistic capacity constraints. Eur. J. Oper. Res. **270**(2), 544–555 (2018)
19. Na, B., Jun, Y., Kim, B.I.: Some extensions to the sweep algorithm. Int. J. Adv. Manufact. Technol. **56**, 1057–1067 (2011)
20. Rabbouch, B., Mraihi, R., Saâdaoui, F.: A recent brief survey for the multi depot heterogenous vehicle routing problem with time windows. In: Abraham, A., Muhuri, P.K., Muda, A.K., Gandhi, N. (eds.) HIS 2017. AISC, vol. 734, pp. 147–157. Springer, Cham (2018). https://doi.org/10.1007/978-3-319-76351-4_15
21. Rabbouch, B., Saadaoui, F., Mraihi, R.: Empirical mode simulated annealing for solving the capacitated vehicle routing problem. J. Exp. Theor. Artif. Intell. (2019, forthcoming)
22. Rojas-Cuevas, I.-D., Caballero-Morales, S.-O., Martinez-Flores, J.-L., Mendoza-Vazquez, J.-R.: Capacitated vehicle routing problem model for carriers. J. Transp. Supply Chain Manage. (2018). https://doi.org/10.4102/jtscm.v12i0.345
23. Sahraeian, R., Esmaeili, M.: A multi-objective two-echelon capacitated vehicle routing problem for perishable products. J. Ind. Syst. Eng. **11**(2), 62–84 (2018)
24. Tlili, T., Faiz, S., Krichen, S.: A hybrid metaheuristic for the distance-constrained capacitated vehicle routing problem. Procedia Soc. Behav. Sci. **109**, 779–783 (2014)
25. Uchoa, E., Pecin, D., Pessoa, A., Poggi, M., Vidal, T., Subramanian, A.: New benchmark instances for the capacitated vehicle routing problem. Eur. J. Oper. Res. **257**(3), 845–858 (2017)

# Application of Extractive Text Summarization Algorithms to Speech-to-Text Media

Domínguez M. Victor[4], Fidalgo F. Eduardo[1,3], Rubel Biswas[1,3]([✉]),
Enrique Alegre[1,3], and Laura Fernández-Robles[2,3]

[1] Department of Electrical, Systems and Automatics Engineering,
Universidad de León, Leon, Spain
{eduardo.fidalgo, rbis, enrique.alegre}@unileon.es
[2] Department of Mechanical, IT and Aerospatial Engineering,
Universidad de León, Leon, Spain
l.fernandez@unileon.es
[3] Researcher at INCIBE (Spanish National Institute of Cybersecurity),
León, Spain
[4] Summer Research Stay at GVIS Research Group, León, Spain
vdomim00@estudiantes.unileon.es

**Abstract.** This paper presents how speech-to-text summarization can be performed using extractive text summarization algorithms. Our objective is to make a recommendation about which of the six text summary algorithms evaluated in the study is the most suitable for the task of audio summarization. First, we have selected six text summarization algorithms: Luhn, TextRank, LexRank, LSA, SumBasic, and KLSum. Then, we have evaluated them on two datasets, DUC2001 and OWIDSum, with six ROUGE metrics. After that, we have selected five speech documents from ISCI Corpus dataset, and we have transcribed using the Automatic Speech Recognition (ASR) from Google Cloud Speech API. Finally, we applied the studied extractive summarization algorithms to these five text samples to obtain a text summary from the original audio file. Experimental results showed that Luhn and TextRank obtained the best performance for the task of extractive speech-to-text summarization on the samples evaluated.

**Keywords:** Audio signal summarization · Speech-to-text summarization · Extractive text summarization · Natural Language Processing

## 1 Introduction

Every day new platforms, services and applications emerge and manage a large amount of information, the vast majority in multimedia format, such as images, audio or video. Speech is one of the most effective methods of communication. However, it is not very easy to reuse, review or retrieve speech documents if they are contained on an audio signal. Extracting useful insights from an audio file is a hard task, especially if the number of audio files or their length is high. Besides, conversations from audio might include redundant information, e.g. word fragments, fillers or repetitions, together with

© Springer Nature Switzerland AG 2019
H. Pérez García et al. (Eds.): HAIS 2019, LNAI 11734, pp. 540–550, 2019.
https://doi.org/10.1007/978-3-030-29859-3_46

irrelevant information not related to the topic of interest or the objective followed. For these reasons, automatic summarization of audio files could help a user to attain the essential information from an audio file without listening to the complete content.

Apart from supporting people with disabilities, speech-to-text applications serve also to transfer the content of an audio file into a readable text document. Converting audio files to text documents could allow managing the information contained in those audios for later processing. A text document can be quickly reviewed, its interesting parts can be easily extracted, and Natural Language Processing (NLP) techniques could be easily applied to extract some knowledge from the text document.

The aim of this work is the automatic summarization of speech documents recorded as audio signals with extractive text summarization techniques. With this objective, we aim at easing and reducing the time required for processing audio information. We have reviewed six text summarization methods, Luhn [1], TextRank [2], LexRank [3], LSA [4], KLSum [5] and SumBasic [6]. We evaluated their performance against OWIDSum [8] and DUC2001 [10] datasets through the following ROUGE metrics; ROUGE-N, ROUGE-L, ROUGE-W, ROUGE-S and ROUGE-SU. Next, we have applied the six algorithms to the transcriptions of five audio files from the ICSI-Corpus dataset, computing the ROUGE metrics and making a recommendation about which extractive text summarization technique would be recommended for the task of speech document summarization.

Figure 1 gives an overview of the work that will be presented in this paper.

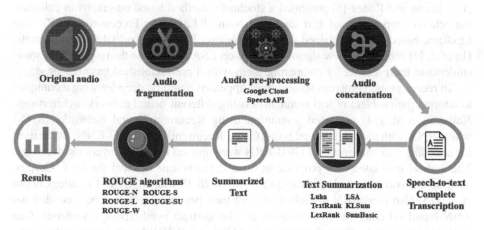

**Fig. 1.** Speech-to-text summarization through extractive text techniques

The organization of the paper is as follows. In Sect. 2, we briefly review the state-of-the-art about extractive text summarization techniques and the methods used to transcript audio signals to text. Next, in Sect. 3, we describe the six extractive text summarization methods employed in this work. Experimental setup, datasets, experimental results and discussion are presented in Sect. 4. Finally, in Sect. 5, the conclusion and future work are discussed.

## 2    Related Work

### 2.1    Extractive Text Summarization

Automatic text summarization reduces the size of the text by preserving the information it contains. Text summarization techniques are usually divided into extractive and abstractive [7], depending on how the resulting text is produced. In this work, we are going to work with the extractive approach, i.e. a summary will be generated using a selection of complete sentences from the original text.

Traditionally, text summarization algorithms are based on frequency appearance of words in the document, such as Luhn [1], or based on graphs, such as TextRank [2] and LexRank [3]. Latent Semantic Analysis (LSA) [4] method represents each document as a matrix to which it applies decomposition values. Some schemas are presented based on the probability distribution of words in documents like KLSum [5] and SumBasic [6]. Akanksha et al. evaluated the five already mentioned algorithms in the task of summarizing the content of Tor darknet. With this objective, they proposed OWIDSum [8], a dataset comprising 60 text documents from Tor domains, and grouped into six categories. Working also with DUC2002 [10] dataset, they recommended TextRank [3], since it obtained the better ROUGE metrics on the two datasets evaluated.

Haghighi and Vanderwende [5] explored probabilistic models for the synthesis of multiple documents where different algorithms like KLSum [6], SumBasic [7], TopicSum [11] and HieSum [12], working on DUC2006 dataset with ROUGE metrics [13]. Erkan and Radev [3] presented a stochastic method based on graphs to calculate the relative importance of text units for Natural Language Processing (NLP). The LexRank-based system obtained the best scores on DUC2003 and DUC2004 datasets. Hu et al. [4] proposed a new algorithm based on LSA to evaluate the responses of some students to their teacher by comparing them with a pre-established base response.

In recent years, numerous studies have appeared based on deep learning techniques to address the problem of text summaries using different neural network architectures. Nallapati et al. [14] presented a solution using Recurrent Neural Networks (RNN) where RNN with closed recurrent units, Gated Recurrent Unit (GRU-RNN), were used for encoding and unidirectional GRU-RRN were applied in hidden layers for decoding. The authors evaluated their proposal in three datasets, one of them the DUC Corpus, used in our work, where they obtained a score of 28.35 with ROUGE-1 metric. Chopra et al. [15] also used RRN in their work, but they presented it with a new conditional RNN based on convolutional attention for the abstract synthesis of a sentence. Cao et al. [16] proposed a Convolutional Neural Network (CNN) based on a novel system, named PriorSum, to capture the foregoing summary and concatenate it with the dependent characteristic of the document under a regression framework. This algorithm was applied on three datasets DUC2001, DUC2002 and DUC2004, obtaining 35.98, 36.63 and 38.91 for the ROUGE-1 metric, respectively. In another work, Cao et al. [17] presented a CNN to generate inlays of sentences to form inlays of documents in the same latent space. A hierarchical framework proposed by Cheng and Lapata [18] where CNN generated the representation and RNN represented the document. Lil et al. [19] proposed a solution based on Variable Autocoder (VAE) and Deep Recurrent Generative Decoder (DRGN) to improve the performance and quality of the abstract summary. The idea is based on the sequence-to-sequence oriented encoder-decoder framework equipped with a latent structure modelling component of variable auto-encoders. As a result, they obtained a metric of 36.25 with ROUGE-1 on the GIGA dataset and a metric of 36.71 with ROUGE-1 on the LCSTS dataset.

Akanksha et al. [9] extended their work on Tor darknet, and they proposed SummCoder, an unsupervised approach for extractive text summarization based on the use of autoencoders and sentence ranking through three different criteria of content, novelty and position. They extended the dataset proposed in [8] up to the 100 samples with two gold summaries, and they obtained state-of-the-art results on their dataset and DUC2002 benchmark.

## 2.2  Speech to Text

Voice recognition is the interdisciplinary subfield of computational linguistic that develops methodologies and technologies that allow the recognition and translation of the spoken language to text. It is also known as automatic voice recognition or Automatic Speech Recognition (ASR), computer voice recognition, voice to text or Speech-to-Text (STT). Traditionally, the ASR Mel-frequency Cepstral Coefficient (MFCC) [20] or Relative Spectral Transform-Perceptual Linear Prediction (RASTA-PLP) [21] were used as the characteristic vectors and Gaussian Mixture Model-Hidden Markov Model (GMM-HMM) [22] as an acoustic model. Currently, Deep Neural Networks (DNN) models have become very significant in this field like great computational advance and increase of data available for training. Combining these recent techniques with some of the techniques above allows us to obtain better results. Villalba et al. [23] presented a method for detecting the wrong test in a speaker verification system. In some situations, the quality of the signals involved in the verification test is not as good as it would be needed to make a reliable decision, which is based on modelling speech quality measures using Bayesian Networks. Reviewing previous works and using their system to eliminate the untrustworthy test have achieved a dramatic improvement of the actual detection cost function.

## 3  Method

A diagram representing the steps of the proposed speech-to-text summarization pipeline is depicted in Fig. 1. First, we preprocess the audio signal by segmenting it into small pieces that could be processed by the ASR method. Then, the ASR method is applied to convert the speech documents into text. Next, the resulting text is preprocessed by adding punctuation marks to separate text sequences if it is not automatically provided by the ASR method. After that, in the next step, we apply the six extractive text summarization methods to obtain a summary of the transcribed speech. Finally, we compute the ROUGE metrics using the automatically generated and the gold summary.

### 3.1    ASR

The speech processing engine Google Cloud Speech API, included in the python library SpeechRecognition3.8.1, is used in this work. This API allows both to transcribe audios directly recorded with a microphone and to import a series of files as in this work, applying neural network models to audio for speech recognition. The API recognizes 120 languages and variants, and in our work, we used the Synchronous Recognition.

### 3.2    Text Summarization Techniques

**Luhn.** One of the most popular text summarization methods is Luhn which is proposed by Luhn [1], and it is based on the frequency of appearance of the words in the text plus the distance that exists between the relevant words which depend on the amount of non-relevant words among relevant ones.

Once the meaningful words have been obtained, the algorithm assigns punctuation to each phrase of the text through a significance factor, which reflects the occurrences of the significant words and the linear distance between them due to non-significant words in the middle.

**TextRank.** Mihalcea and Tarau proposed a text summarization algorithm based on graph theory, named TextRank [2], where the sentences are represented through the vertices of the graph. Each of the vertices has a critical value which is determined by taking into account the global information calculated recursively from the entire graph. It is performed by the PageRank algorithm, which is used to calculate the rank, grade or quality of a web page through the number of existing links. In order to use PageRank in the text summarization, each node is represented by a text entity, or sentence, instead of a web page.

Since in the cases of study in our paper no links among pages exist, we defined a similarity between two sentences depending on the number of words that overlap in both.

After obtaining the similarity among all the sentences, a graph is presented where each vertex may not be connected to any other vertex because no similarity is found within the sentences. The edges that connect two vertices will have an associated weight which represents the force with which they are connected.

**LexRank.** Another graph-based text summarization technique presented by Erkan, G. and Radev is called LexRank [3]. It is based on TextRank, but edges between the vertices are obtained with the cosine similarity scores of sentences that are represented as TF-IDF measurement vectors.

Then a similarity graph is obtained, in which each sentence is represented with a vertex and the cosine similarity between sentences with an edge that holds weight information. A threshold-based approach is used to determine which sentences are considered to be similar to each other.

**LSA.** Latent Semantic Analysis (LSA) [4] is an automatic technique that obtains the statistical relationship of words in a sentence. The text is parsed into words defined as unique strings of characters and is separated into meaningful sentences.

Subsequently, the text is represented as a matrix where each cell contains the frequency of appearance of the word of such row in the passage indicated by its column. Next, each cellular frequency is weighted by a function that expresses both the importance of the word in the particular passage and information about the type of word in the domain of discourse in general. Finally, Single Value Decomposition (SVD) [24] is applied to the matrix by eliminating the coefficients in the diagonal matrix.

**SumBasic.** SumBasic [5] is an algorithm that uses a sentence selection component based on the frequency with a component to re-weight the word probabilities in order to minimize redundancy. Firstly, it calculates the probability distribution of the words that appear in the input and, then it computes its average and assigns it a weight equal to this value in each sentence. After that, the best scoring phrase that contains in the most likely word is considered. Finally, the probability of each word in the sentence that has been chosen in the previous step is updated. The algorithm can be applied iteratively to reduce further the length of the summary.

**KLSum.** The KLSum [6] algorithm for text summarization generates an output summary based on Sum = minKL(Pdoc‖Psumm) where Psumm is the empirical distribution of the unigram of the candidate summary, and minKL(P‖Q) represents the divergence of Kullback-Lieber. P represents the distribution of the document unigrams and Q is the distribution of the document summary. The criterion tries to generate the summary that closely matches the source document.

### 3.3 ROUGE Metrics

ROUGE [13] is a set of metrics used to determine the quality of text summarization methods. This metric compares the summary result obtained with an algorithm with the one created by human experts, named Gold Summaries. ROUGE metrics determine the number of units that overlap; this is, the number of units that appear in both summaries. The ROUGE metrics used in this work are introduced below.

**ROUGE-N.** This metric is based on the number of matching "n-grams" between the summary obtained and the reference summary. "n-grams" is a group of "n" units written consecutively, the units can be letters, syllables or words.

**ROUGE-L.** This method is based on the largest sub-chain (strictly increasing) measurement that is coincident between the candidate summary and the reference summary. It uses the union of the proportion of chain words in the reference summary contained in the candidate summary, and the union of the proportion of chain words in the candidate summary present in the reference summary.

**ROUGE-W.** This method is similar to ROUGE-L, but it allows differentiating chains of subsequences of different spatial relationships within the original sequence.

**ROUGE-S.** This method uses bigrams with all the possible jumps between the two bigrams.

**ROUGE-SU.** This method is an extension of the ROUGE-S that solves the problem by which no value is given to the candidate statement if it does not have any matching pairs of words, even if it has individual words that match. Thus, sentences that do not contain matching pairs of words and do not have matching individual words are given less importance in the metric calculation. To do this, ROUGE-S is extended using the unigram instead of the bigram as the counting unit.

## 4    Experimentation and Results

### 4.1    Experimental Settings

We made our experimentation with an Intel Core i7 with 16 GB of DDR4 RAM. To improve the performance of Google Cloud Speech API, the audio signals were segmented into fragments of 5–10 s, which facilitates the transcription of the audio. Once each audio is translated into text, the resulting outputs will be concatenated to obtain the final text. Besides, punctuation marks were manually added in the transcription output to separate the different sequences since ASR does not automatically provide it.

### 4.2    Datasets

In this work, we have used three different datasets: two text datasets for evaluating the extractive text summarization algorithms, and a speech dataset for the evaluation of the complete method.

**DUC2001.** Document Understanding Conference 2001 (**DUC2001**) dataset consists of 303 documents that contain news from newspapers grouped into 30 categories, such as Hurricane Andrew, mad cow disease and a plane crash in the city of Sioux.

**OWIDSum.** The Onion Web Illegal Document Summarization (OWIDSum) is the initial version of the newer Tor Illegal Domain Summarization dataset (TIDSumm) [8]. This dataset consists of texts extracted from Tor darknet domains of the Darknet Usage Text Addressed (DUTA) dataset [25]. DUTA contains 6831 domains of the dark Tor Network and is grouped into 26 categories related to legal and illegal activities. OWIDSum comprises 60 documents gathered in six categories, i.e. credit cards counterfeit, hacking, drugs selling, money counterfeit, market and crypto-currency, together with two gold summaries for each document.

**ICSI Corpus.** The International Computer Science Institute Meeting Corpus (ICSI Meeting Corpus) dataset comprises 75 audio conversations between several people with durations of approximately 30 to 70 min. It also contains 20 extractive summaries coming from 17 of the audios. Thus, it can be used to calculate ROUGE metrics. We have considered five out of 17 audios from ICSI Corpus in the experiments.

## 4.3    Results and Discussion

Table 1 presents the evaluation of the six algorithms described over DUC2001 dataset using six ROUGE metrics. It can be noticed that Luhn attained the highest performance for four ROUGE metrics, i.e. ROUGE-2, ROUGE-L, ROUGE-W and ROUGE-S, and LexRank did for the other two, i.e. ROUGE-1 and ROUGE-SU with 0.42293 and 0.17252 respectively. It can be recommended that for DUC2001, the algorithms that perform better, i.e. obtain higher quality extractive summaries are Luhn followed by LexRank and TextRank.

**Table 1.** Results obtained in dataset DUC2001. **R-1** and **R-2** stand for the ROUGE-1 and ROUGE-2 respectively. **R-L**, **R-W**, **R-S** and **R-SU** stands for the ROUGE-L, ROUGE-W, ROUGE-S and ROUGE-SU metrics respectively.

| DUC2001 | Luhn | TextRank | LexRank | LSA | SumBasic | KLSum |
|---------|---------|----------|-----------|---------|----------|---------|
| R-1 | 0.42071 | 0.40418 | **0.42293** | 0.35848 | 0.36031 | 0.35847 |
| R-2 | **0.16814** | 0.15399 | 0.15796 | 0.11992 | 0.11243 | 0.11695 |
| R-L | **0.25817** | 0.24616 | 0.25291 | 0.21103 | 0.2132 | 0.21582 |
| R-W | **0.11247** | 0.10645 | 0.10556 | 0.09747 | 0.08175 | 0.08582 |
| R-S | **0.16818** | 0.15427 | 0.16781 | 0.11911 | 0.11605 | 0.11968 |
| R-SU | 0.17218 | 0.15823 | **0.17252** | 0.12382 | 0.12123 | 0.12447 |

In the case of OWIDSum dataset, Table 2, TextRank algorithm yielded the best scores for almost all the ROUGE metrics, except ROUGE-1, in comparison to the rest of the methods. Moreover, Luhn and LexRank achieved similar scores for every ROUGE metric.

**Table 2.** Results obtained in dataset OWIDSum.

| OWIDSum | Luhn | TextRank | LexRank | LSA | SumBasic | KLSum |
|---------|---------|----------|---------|---------|----------|---------|
| R-1 | **0.39176** | 0.39015 | 0.38062 | 0.38515 | 0.21643 | 0.34228 |
| R-2 | 0.24468 | **0.26119** | 0.21066 | 0.23293 | 0.09417 | 0.19686 |
| R-L | 0.34888 | **0.35759** | 0.32921 | 0.33783 | 0.1851 | 0.30422 |
| R-W | 0.17938 | **0.18348** | 0.15871 | 0.16776 | 0.08257 | 0.14758 |
| R-S | 0.17758 | **0.18787** | 0.14754 | 0.15428 | 0.05424 | 0.12889 |
| R-SU | 0.18116 | **0.19145** | 0.15229 | 0.15903 | 0.05781 | 0.13343 |

Finally, Table 3 shows the results obtained after applying the extractive text summarization to the transcription of the five audio files from ICSI Corpus dataset. The best results were obtained by Luhn with 0.57215, 0.35778, 0.3735 0.32403 and 0.32413 for ROUGE-1, ROUGE2, ROUGE-W, ROUGE-S and ROUGE-SU, respectively. TextRank yielded the best score of 0.3368 for ROUGE-L. Luhn and TextRank achieved similar results for all ROUGE metrics while the rest of methods perform worse, especially KLSum and SumBasic. We can conclude that Luhn is the most suitable algorithm for the analyzed speech documents.

**Table 3.** Results obtained in ICSI Corpus dataset after transcribing the audio to text files

| ICSI Corpus | | | | | | |
|-------|---------|---------|---------|---------|---------|---------|
|       | Luhn    | TextRank | LexRank | LSA     | SumBasic | KLSum   |
| R-1   | **0.57215** | 0.56974 | 0.5441  | 0.52978 | 0.28808 | 0.40229 |
| R-2   | **0.35778** | 0.35741 | 0.3382  | 0.32095 | 0.16248 | 0.24832 |
| R-L   | 0.33572 | **0.33684** | 0.31893 | 0.30217 | 0.16124 | 0.23187 |
| R-W   | **0.03735** | 0.03732 | 0.03463 | 0.03111 | 0.01484 | 0.02155 |
| R-S   | **0.32403** | 0.32154 | 0.29613 | 0.26965 | 0.06006 | 0.13555 |
| R-SU  | **0.32413** | 0.32163 | 0.26923 | 0.26976 | 0.06016 | 0.13567 |

## 5    Conclusions and Future Work

In this work, we presented an automatic pipeline to summarize audio content by applying extractive text summarization algorithms to a previously transcription from audio to text using ASR.

To this end, traditional text summarization techniques have been reviewed, and from them, six methods have been selected and evaluated on two text summarization datasets, DUC2001 and OWIDSum employing six ROUGE metrics. Experimental results proved that Luhn and the TextRank algorithms mainly obtained the best results in DUC2001 and OWIDSum, respectively.

Then, five speech documents of ICSI Corpus were transcribed with Google Cloud Speech API ASR. The resulting text documents were evaluated using the same six extractive text summarization algorithms. Again, Luhn and TextRank algorithms yielded the best results, so they are the initial recommendation to solve the automatic speech-to-text summarization task using extractive text summarization techniques.

Possible future work from this study is the extension of the experimentation performed on ICSI Corpus dataset to all of the speech documents that come with gold summaries, or even the manual elaboration of gold summaries for the rest of speech documents in the dataset for further assessment.

The main issue to test this proposal in other public datasets is the lack of extractive gold summaries. As future work, other speech datasets could also be tested after creating appropriate gold extractive summaries. Moreover, predicting the punctuation marks from word sequences would be done automatically.

**Acknowledgement.** This research is supported by the INCIBE grant "INCIBEC-2015-27359" corresponding to the "Ayudas para la Excelencia de los Equipos de Investigación avanzada en ciberseguridad" and by the framework agreement between the University of León and INCIBE (Spanish National Cybersecurity Institute) under Addendum 22 and 01.

# References

1. Luhn, H.P.: The automatic creation of literature abstracts. IBM J. Res. Dev. **2**(2), 159–165 (1958)
2. Mihalcea, R., Tarau, P.: Text Rank: bringing order into texts. In: Proceedings of EMNLP-04 and Conference on Empirical Methods in Natural Language Processing (EMNLP) (2004)
3. Erkan, G., Radev, D.R.: LexRank: graph-based lexical centrality as salience in text summarization. J. Artif. Intell. Res. **22**, 457–479 (2004)
4. Xiangen, H., Zhiqiang, C., Max, L., Andrew, O., Phanni, P., Art, G.: A revised algorithm for latent semantic analysis. In: Proceedings of the 18th International Joint Conference on Artificial Intelligence (IJCAI 2003), pp. 1489–1491. Morgan Kaufmann Publishers Inc., San Francisco (2003)
5. Aria, H., Lucy, V.: Exploring content models for multi-document summarization. In: Proceedings of Human Language Technologies, Annual Conference of the North American Chapter of the Association for Computational Linguistics (NAACL 2009), pp. 362–370. Association for Computational Linguistics, Stroudsburg (2009)
6. Nenkova, A., Wandervende, L.: The impact of frequency on summarization. Technical report, Microsoft Research (2005)
7. Murthy, V., Vishnu Vardhan, M.B., Vijaypal Sreenivas, P., Reddy, V.: Text classification using text summarization–a case study on Telugu text. Int. J. Adv. Res. Comput. Sci. Softw. Eng. **3**, 1399–1403 (2013)
8. Joshi, A., Fidalgo, E., Alegre, E.: Summarization of text from illegal documents in Tor domains using extractive algorithms. In: International Conference on Applications of Intelligent Systems (APPIS), Las Palmas de Gran Canaria (2018)
9. Joshi, A., Fidalgo, E., Alegre, E., Fernández-Robles, L.: SummCoder: an unsupervised framework for extractive text summarization based on deep auto-encoders. Expert Syst. Appl. **129**, 200–215 (2019)
10. DUC 2002: Document Understanding Conference (2002). https://duc.nist.gov/. Accessed 07 Apr 2019
11. David, M.B., Andrew, Y.N., Michael, I.J.: Latent Dirichlet allocation. JMLR **3**, 993–1022 (2003)
12. Blei, D.M., Griffiths, T.L., Jordan, M.I., Tenenbaum, J.B.: Hierarchical topic models and the nested Chinese restaurant process. In: NIPS (2004)
13. Lin, C.-Y.: Rouge: a package for automatic evaluation of summaries. In: Proceedings of the ACL Workshop, pp. 74–81 (2004)
14. Nallapati, R., Zhou, B., Santos, C.D., Gulcehre, C., Xiang, B.: Abstractive text summarization using sequence-to-sequence RNNs and beyond. In: Proceedings of the 20th SIGNAL Conference on Computational Natural Language Learning, pp. 282–290 (2016)
15. Chopra, S., Auli, M., Rush, A.M.: Abstractive sentence summarization with attentive recurrent neural networks. In: Proceedings of the 2016 Conference of the North American Chapter of the Association for Computational Linguistics: Human Language Technologies, pp. 93–98 (2016)
16. Cao, Z., Wei, F., Li, S., Li, W., Zhou, M., Wang, H.: Learning summary prior representation of extractive summarization. In: Proceedings of the 53rd Annual Meeting of the Association for Computational Linguistics and the 7th International Joint Conference on Natural Language Processing, pp. 829–833 (2005)
17. Cao, Z., Wenjie, L., Sujian, L., Furu, W., Yanran, L.: AttSum: joint learning of focusing and summarization with neuronal attention. In: COLING, pp. 547–556 (2016)

18. Cheng, J., Lapata, M.: Neural summarization by extracting sentences and words. In: Proceedings of the 54th Annual Meeting of the Association for Computational Linguistics, pp. 484–494 (2016)
19. Lil, P., Lam, W., Bing, L., Wang, Z.: Deep recurrent generative decoder for abstractive text summarization. In: Proceedings of the 2017 Conference on Empirical Methods in Natural Language Proceedings, pp. 2091–2100 (2017)
20. Martinez, J., Perez-Meana, H., Escamilla-Hernandez, E., Suzuki, M.M.: Mel-frequency cepstral coefficients for speaker recognition: a review. Int. J. Adv. Eng. Res. Dev. **2**, 248–251 (2015)
21. Zulkifly, M.A., Yahya, N.: Relative spectral-perceptual linear prediction (RASTA-PLP) speech signal analysis using singular value decomposition (SVD). In: IEEE 3rd International Symposium on Robotics and Manufacturing Automation (ROMA) (2017)
22. Xuan, G., Zhang, W., Chai, P.: EM algorithms of Gaussian mixture model and hidden Markov model. In: Proceedings of the 2001 International Conference on Image Processing, pp. 145–148 (2001)
23. Villalba, J., Lleida, E., Ortega, A., Miguel, A.: Reliability estimation of the speaker verification decisions using Bayesian networks to combine information from multiple speech quality measures. In: Torre Toledano, D., et al. (eds.) IberSPEECH 2012. CCIS, vol. 328, pp. 1–10. Springer, Heidelberg (2012). https://doi.org/10.1007/978-3-642-35292-8_1
24. Gliozzo, A.M., Giuliano, C., Strapparava, C.: Domain kernels for word sense disambiguation. In: 43rd Annual Meeting of the Association for Computational Linguistics, pp. 403–410 (2005)
25. Al-Nabki, M., Fidalgo, E., Alegre, E., Fernández-Robles, L.: ToRank: identifying the most influential suspicious domains in the Tor network. Expert Syst. Appl. **123**, 212–226 (2019)

# User Profiles Matching for Different Social Networks Based on Faces Identification

Timur Sokhin(✉), Nikolay Butakov, and Denis Nasonov

ITMO University, 49 Kronverksky Pr., St. Petersburg 197101, Russia
245591@niuitmo.ru, alipoov.nb@gmail.com, denis.nasonov@gmail.com

**Abstract.** It is common practice nowadays to use multiple social networks for different social roles. Although this, these networks assume differences in content type, communications and style of speech. If we intend to understand human behaviour as a key-feature for recommender systems, banking risk assessments or sociological researches, this is better to achieve using a combination of the data from different social media. In this paper, we propose a new approach for user profiles matching across social media based on publicly available users' face photos and conduct an experimental study of its efficiency. Our approach is stable to changes in content and style for certain social media.

**Keywords:** Face detection · Profiles · Matching · Social networks · Face embedding · Clustering · Computer vision

## 1 Introduction

Nowadays, social media may differ in their capabilities to share information and express, that reflects in types of published content, conversation style, etc. For instance, About.me or LinkedIn may be used as the main page for self-presentation purposes, while Twitter or Instagram are used for informal communication of random people, publishing selfies.

Understanding how a user behaves is an important task for many applications such as the generation of recommendations, candidate assessment by HR departments or even for the analysis of further developing for social media itself. We suppose that a person comprehensively may be described with a set of profiles joined from different social networks. While some users link all their profiles together explicitly or mention in their posts somehow, mostly, people don't want to associate them.

Previous attempts to solve this problem has been directed to matching by features such as names, friend-graphs, published textual contents (e.g. topics of posts) and so on. These methods often may lack precision or recall because of differences between social networks in published content and style, absence of required interlinks for friends and so on, and lead to mismatching of expected and real person [12].

© Springer Nature Switzerland AG 2019
H. Pérez García et al. (Eds.): HAIS 2019, LNAI 11734, pp. 551–562, 2019.
https://doi.org/10.1007/978-3-030-29859-3_47

In this work, we propose and study a new approach of profiles matching based on publicly available users' images and faces identification. The face is a unique attribute for humans, that should keep almost unchanged from network to network. The existing methods of face detection and embedding allow us to detect faces on photos and compare them. But a single face image may suffer from positions, perspective, quality problem. We need more than that to reliably match profiles: we have to identify the owner's faces among others, even if there is only one person presence on a photo.

The contributions of this paper are the following: (1) we propose a novel approach to user profiles matching using face detection and comparison of face embeddings from different social media; (2) we conduct a set of experiments for two popular in Russia social networks VKontakte and Instagram and investigate limitations of our approach in terms of quality and quantity of a data. The latter includes answering the following questions:

1. How many data (photos) does effective matching require?
2. How does efficiency (precision and recall) depend on the quality and the quantity of the data?

## 2    Related Work

Mostly, previous work in this field is focused on the easily accessible information about the user: self-description, biography, name, nickname [2,6]; or on the dynamic of users behaviour: dates of posts, profiles updates [3,10]. As it noticed in [8] and [5], this kind of information (username, location, followers/followings, meta paths) are very noisy, easily faked, not required, they provide huge research of existing methods to profiles matching. The last suppose that methods of behaviour dynamic analysis show potential for further work, but they have some major disadvantages: they require collecting of information during some period of user activities and require an unusual method of data representation in different social media, which can vary in their features.

Also, it should be noticed, that there is almost no works with images, which provide a lot of additional information about the user itself and are useful for profiles matching. [3,10] and similar approaches require features, which can not be extracted from all social media - Instagram and VKontakte are different in the type and the context of the content, friends-system, etc. Our approach reveals new possibilities for comparing profiles based on photographic materials, which are more suitable in this case.

## 3    The Approach

The main idea of our approach is to form a single defining vector - representation of a user's profile based on the embeddings of his faces.

*Data Collecting.* Our approach consists of several stages. At first, we must collect data from two social media using a crawling framework (profiles, photos from albums and posts) [1]. For the purposes of validation of our results, we collect a set of profiles from VKontakte, which have an explicit link to their secondary profile in Instagram - the only possible way to build the labelled dataset.

*Face Detection and Embedding.* We process photos using two algorithms:

1. face detection - we apply MTCNN - Multi-task Cascaded Convolutional Networks [11], which achieved efficiency superior to the closest competitors and is not affected by scaling of the faces;
2. face embedding - to construct embeddings of extracted faces FaceNet neural network is applied [7].

We apply MTCNN pre-trained on the WIDER FACE dataset and FaceNet pre-trained on the VGGFace2[1]. Then this data is filtered.

*Filtering.* The extracted face embeddings are further filtered by their parameters according to several heuristics:

1. filtering by number of pixels (hereinafter, we will use the term quality of the image);
2. filtering by anchors (child faces removing).

FaceNet has limitations on the minimum required quality of images and we filter images of faces by the number of pixels of these faces. The accurate control of the above parameters allows to achieve an improved precision and recall of matching, this is partly due to the behaviour of the selected method for embedding construction. In the experimental study in Sect. 4 we found an effect of the quality of facial images on the final matching efficiency - it improves the F1-score by 4%.

The other heuristics probably can be related to the dataset limitation of VGGFace2 with which FaceNet was trained. VGGFace2 contains young and mature faces of people but does not contain the faces of babies and small children. This leads to a problem that embeddings of child's faces have a very small margin between each other. That is why we should remove their faces from the user's collection of photos to avoid mismatching of profiles. Figure 1 reveals that the distribution of distances between embeddings of children's faces has a bias from the distribution of distances between embeddings of random people's faces.

Additional filtering of data is accomplished using so-called anchors. An anchor is a vector that represents some space of embedded faces. In our study, we use the anchor to represent the faces of children. We create it by following way. A set of children faces was collected semi-automatically: we find kindergarten and photographers accounts using tags and specific usernames. For instance, tags under the photos with words "children", "kindergarten", etc. Then we build an anchor - element-wise mean of all vectors of children's faces. All face embeddings which are close to this anchor are removed from the dataset.

---

[1] Code repository used - https://github.com/davidsandberg/facenet.

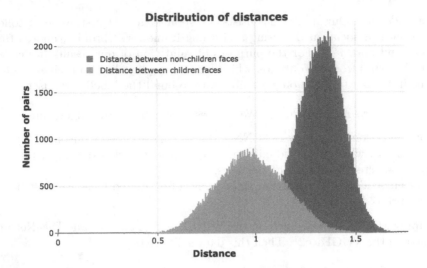

**Fig. 1.** Distribution of distances between random people faces and between children faces

*Owner Identification.* This is the main part of our approach that is performed separately for each profile in each social network. Embeddings of faces are formed in Euclidean space. We apply hierarchical clustering for each profile separately with the single linkage algorithm and distance threshold 0.8. This algorithm allows us to generate a non-fixed number of clusters based on the Euclidean distance between face embeddings.

Each cluster of the profile should belong either to a single person in the real world, whose faces have slightly different but close embeddings or to persons who look very similar due to distortions introduced by hairstyle, put on glasses, beards and other things which make them look similar.

We assume that most users publish photos with different people, but the number of their face occurrences is greater than others. Following this hypothesis, in order to find the owners' faces, we must choose the largest cluster and combine them into one vector - the defining vector (DV) of profile using faces from a chosen cluster. The DV is an element-wise mean of all generated embeddings with the same dimension (1, where V - face embedding, n - number of embeddings of the user).

$$DV = \frac{1}{n} \sum_{i=1}^{n} V_i \tag{1}$$

However, due to the possible sharpness of the DV, it is worth to take into account the other largest clusters. Sometimes people publish many similar photos, even the same photos. In case that is shown in Fig. 2(a) the first cluster only consists of two unique images. We are not able to match this profile using this cluster. But we can add the others (for instance, the second largest, that is shown in Fig. 2(b)) and form a new DV using more than two unique face embeddings. Our experiments in Sect. 4 show that this assumption and the proposed solution allow us to achieve results that exceed the use of one cluster. Experimental results give us the optimal value - 2 clusters. If after clustering there is only one cluster, we use all photos of the user, if there are all clusters with the same size (e.g. 1 element), we set this profile as "unable to set the owner" and mark as profiles without a pair.

**Fig. 2.** Examples of cluster: (a) the first largest; (b) the second largest

After that, the DV of each profile in both social media represents the user and will be used to matching. If the size of the largest cluster is less than a given threshold, this user marked as profiles without pair, because it is not possible to detect the owner's face correctly. The tuning of the threshold is also provided in the experimental study.

*Profiles Matching.* The process of profiles matching is simple: defining vectors of users from two social media are compared with each other. We calculate the L2 norm between profiles in two social media, for each profile in one social media we find the profile from the other with the smallest distance and mark as a candidate for matching (2).

$$
argmin\, L2\left(DV_i^{VK}, DV_j^{Inst}\right) = \left\{ DV_j^{Inst} | DV_k^{Inst} \in DV^{Inst} : \\
L2\left(DV_i^{VK}, DV_k^{Inst}\right) > L2\left(DV_i^{VK}, DV_j^{Inst}\right) \right\}
\tag{2}
$$

If the smallest distance is higher than the given threshold (threshold distance, hereinafter), this means there is no pair in the other social media or we could not find it.

## 4 Experimental Study

### 4.1 Details of the Experimental Part

**Our Experimental Plan.** Consists of three main steps: baseline evaluation using real names-based matching; evaluation for full profiles without any limitations; evaluation with alignment rate reduction and photos number reduction.

**Dataset Description.** We use our own dataset - Dataset4675, which consists of 4675 profiles from VKontakte and 3100 profiles from Instagram, which simulates working with partially aligned networks - only 3100 VKontakte users have a pair in other social media. These VKontakte users have explicit link to their Instagram profile, this allows us to make an accurate assessment of the precision. They have from 50 to 500 publicly available photos.

**Metrics.** We clarify definitions of precision, recall and F1-score, that we use for this classification problem, which is not fully classical. Since we are working with VKontakte as our main social media and want to saturate its profiles with additional information, all metrics are calculated with respect to the number of VKontakte users.

With V as a number of all real pairs in our dataset (3193), $K^p$ as a number of the correct predictions of the algorithm (correctly matched pairs of VKontakte and Instagram profiles) and K as a number of all predictions of the algorithm, the precision is defined as follows (3):

$$P = \frac{K^p}{K} \tag{3}$$

And the recall is defined as follows (4):

$$R = \frac{K^p}{V} \tag{4}$$

We need both the recall and precision in order to evaluate our approach, F1-score shows the balance between them and is used to choose the best parameters.

## 4.2 Baseline Evaluation. Real Names Matching

The real names of users from Dataset4675 are compared with Levenshtein distance metric and sensitivity is analyzed according to its threshold distance. For each user we are looking for the closest user from other social networks, if the closest distance exceeds the threshold value, we remain this user without a pair. The real names are processed in the following sequence: lower case translation; non-alphabetic characters removing; transliteration.

The precision and recall are shown in Table 1. The highest F1 of 0.295 is achieved with P = 0.765 and R = 0.183 and the distance threshold of 4 permutations. With a small dataset in relation to the real number of users, this approach achieves a good precision, but it should be noticed that the precision decreases with the increasing number of users. This can be explained from the fact of a large number of homonyms in the real world. Also, we have a very low recall rate.

**Table 1.** Real name based matching results

| Threshold | Precision | Recall | F1-score |
|---|---|---|---|
| 1 | 0.976 | 0.106 | 0.191 |
| 2 | 0.972 | 0.148 | 0.257 |
| 3 | 0.922 | 0.169 | 0.286 |
| 4 | **0.765** | **0.183** | **0.295** |
| 5 | 0.511 | 0.192 | 0.279 |
| 6 | 0.352 | 0.198 | 0.253 |
| 7 | 0.269 | 0.203 | 0.231 |
| 8 | 0.235 | 0.205 | 0.219 |

**Table 2.** Cluster dependence analysis

| Number of largest clusters used | Precision | Recall | F1-score |
|---|---|---|---|
| 1 | 0.9617 | 0.7885 | 0.8665 |
| 2 | **0.9782** | **0.7875** | **0.8726** |
| 3 | 0.9797 | 0.7839 | 0.8709 |
| 4 | 0.9793 | 0.7845 | 0.8712 |
| 5 | 0.9801 | 0.7842 | 0.8713 |

### 4.3 Evaluation for Full Profiles

**Cluster Analysis.** At first, we analyze the dependency on the clusters number in Table 2 with fixed parameter of threshold distance - 0.65 and image quality - 6400.

It can be seen as proof of the requirements of more than 1 clusters mentioned in Sect. 3 - the F1-score in this case is 0.855. The optimal value of the number of the cluster is 2.

**Face-Based Matching.** We also provide a sensitivity analysis of our approach in Fig. 3. We use Dataset4675 for this part of the experimental study. One can see a strong dependence between the threshold distance and efficiency. While a high precision is achieved with a smallest threshold distance value, the recall remains lower than 0.7, that can be seen in Table 3. The higher F1 is 0.0868 with image quality 80 and threshold distance 0.65.

### 4.4 Evaluation with the Reduced Alignment Rate and the Reduced Number of Photos

Here we experiment with limited data and rate of alignment of users. If our approach requires as much data as possible, it is only applicable for government and law enforcement with social media cooperation.

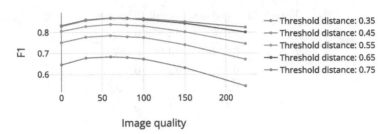

**Fig. 3.** F1-score of face-based matching depending on the image quality and the threshold distance

**Table 3.** Face-based matching results

| Image quality | Threshold distance | | | | |
|---|---|---|---|---|---|
| | 0.35 | 0.45 | 0.55 | 0.65 | 0.75 |
| Precision | | | | | |
| 0 | 0.997 | 0.989 | 0.976 | 0.951 | 0.898 |
| 30 | 1.0 | 0.999 | 0.997 | 0.984 | 0.933 |
| 60 | 1.0 | 1.0 | 1.0 | 0.995 | 0.947 |
| 80 | 1.0 | 1.0 | 1.0 | **0.994** | 0.946 |
| 100 | 1.0 | 1.0 | 1.0 | 0.992 | 0.948 |
| 150 | 1.0 | 1.0 | 1.0 | 0.992 | 0.948 |
| Recall | | | | | |
| 0 | 0.478 | 0.606 | 0.687 | 0.739 | 0.77 |
| 30 | 0.513 | 0.637 | 0.709 | 0.763 | 0.793 |
| 60 | 0.519 | 0.645 | 0.721 | 0.77 | 0.8 |
| 80 | 0.515 | 0.638 | 0.715 | **0.77** | 0.798 |
| 100 | 0.507 | 0.634 | 0.71 | 0.761 | 0.797 |
| 150 | 0.461 | 0.588 | 0.671 | 0.734 | 0.772 |

**Avatars Only Matching.** When working with facial images, using avatars can be the easiest way. This removes the need for the owner detection stage because the idea of an avatar is to present the owner. Here we use only users' avatars from Dataset4675 to evaluate this assumption in Fig. 4.

We faced the recall decrease in general and almost zero F1-score with a high value of the quality filter. We achieve 0.539 F1-score with the following parameters: threshold distance - 0.75, quality - 30.

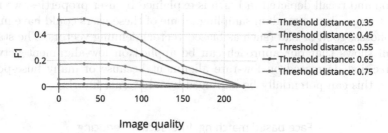

**Fig. 4.** F1-score of face-based matching depending on the image quality and the threshold distance. Avatars only

**Reducing the Number of Images for Each User.** We reduce the number of available photos of each user from Dataset4675 in order to estimate our approach in the condition of greater uncertainty in Fig. 5.

The procedure of sampling is as follows: for each user, we select X% of his/her photos for 10 times. It is interesting that the precision rate remains almost the same even with 10% of data from each user profile of both social media. The reason for the low recall rate is the owner detection part: a small amount of randomly sampled data does not allow to find the owner's face and to form a good defining vector.

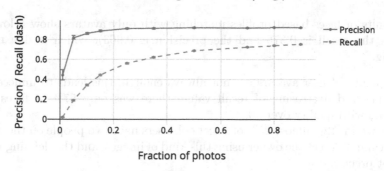

**Fig. 5.** The dependence of the efficiency of the algorithm on the proportion of user photos

**Reducing the Rate of Intersections. Partial Alignment.** In the final part of the experiments, we examine the partial alignment of social networks. As noted by [10] authors real social media are partially alignment - not all users from one social media have accounts in another one. It is impossible to investigate the real rate of this intersection, but we can consider a number of rate values and create a synthetically reduced intersection. The high variance of precision and recall depicted in Fig. 6 is explained by user properties: we match different users, due to random sampling. Some of these users could have more or fewer photos, good or bad (such as biased vector) defining vectors. The stability of recall shows that our approach can be applied on low-alignment networks. The precision decreased on low-rate alignment because of many false-positive samples, this can potentially be improved by additional filtering.

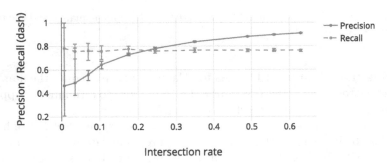

**Fig. 6.** The dependence of the efficiency of the algorithm on the proportion of user photos

## 5   Discussion

The results of faces-based profiles matching with only avatars show a low efficiency - the recall is 0.375 and the precision is 0.963), which is due to the following:

- the quality of user avatars are not always enough, this leads to unnecessary filtering and decreasing of recall value, there was only 57% of faces from avatars with quality over 80;
- as shown in [12], almost 25% of Facebook users have two people on the avatar - we cannot detect the owner using this kind of images, and the defining vector is not precise.

There is also one indirect reason why avatars are not enough even if we were able to detect the owner: as it is seen in Table 2, one cluster gives us less F1-score - 0.8665. This aspect and the analysis of results show that very homogeneous clusters lead to mistakes in matching. Using only one image would be a degenerate case of one cluster from one face.

The results of our study indicate that our approach works less efficiently without all available user's data. This is expected behaviour, because of the essence of our approach: we work with a content of profiles. The recall decreases very quickly, but the precision remains almost the same until the 5–10% of available data (P = 0.80, R = 0.18 with 5% of available photos and P = 0.84, R = 0.33). But even 18% of users still allow you to match many profiles in absolute values.

The last thing to discuss the experiments is user sampling. It should be noticed: we do not know the real intersections of people in different social media. [9] and [4] reports that there are 30 millions of Instagram users in Russia and 80 millions of VKontakte users. Also, we know that 3.3 millions of VKontakte users link their Instagram profile. So, the rough estimate of profiles alignment is 3–4%. This value allows us to achieve P = 0.49 and the average R = 0.758. The alignment rate is probably greater due to historical features: VKontakte is one of the first social media in Russia and it is very popular among active users of the Internet who can be Instagram users. In this case, the alignment is about 30% and the expected precision is 0.8 and the recall is 0.76.

## 6 Conclusion

In this paper, we propose a method to profiles matching across different social media using users' photos. Our approach use photos from the profiles to form a single feature-vector using embedding techniques and use only this vector for further profiles matching. The proposed approach achieved a high precision up to 0.994 in case of 70% of users have profiles in both social media and recall up to 0.76. According to the result of partial alignment, in the real world condition, we can achieve precision up to 0.8 and recall up to 0.78. We found the best parameters for the image quality filtering and threshold distance, which allows you to evaluate whether two profiles are a pair.

Our approach provides a large number of applications. We can match a set of criminals faces from street or security cameras with their profiles in social media. Moreover, it is very useful for scientific purposes: additional information could help to find new features of the user behaviour and open new opportunities in the research of social media impact on the person.

**Acknowledgments.** This work financially supported by Ministry of Education and Science of the Russian Federation, Agreement #14.575.21.0165 (26/09/2017). Unique Identification RFMEFI57517X0165.

## References

1. Butakov, N., Petrov, M., Mukhina, K., Nasonov, D., Kovalchuk, S.: Unified domain-specific language for collecting and processing data of social media. J. Intell. Inf. Syst. **51**, 1–26 (2018)
2. Goga, O.: Matching user accounts across online social networks methods and applications (2014)
3. Hazimeh, H., Mugellini, E., Abou Khaled, O., Cudre-Mauroux, P.: SocialMatching++: A Novel Approach for Interlinking User Profiles on Social Networks (2017)

4. Leading countries based on number of Instagram users as of October 2018 (2018). https://www.statista.com/statistics/578364/countries-with-most-instagram-users/
5. Khaled, O.A., Hazimeh, H., Mugellini, E., Cudré-Mauroux, P.: Linking user profiles in social networks: a comparative review. Int. J. Soc. Netw. Min. **2**(4), 333–361 (2017)
6. Malhotra, A., Totti, L., Meira, W., Kumaraguru, P., Almeida, V.: Studying user footprints in different online social networks. In: Proceedings of the 2012 IEEE/ACM International Conference on Advances in Social Networks Analysis and Mining, ASONAM 2012 (2012)
7. Schroff, F., Kalenichenko, D., Philbin, J.: Facenet: a unified embedding for face recognition and clustering. In: Proceedings of the IEEE Conference on Computer Vision and Pattern Recognition, pp. 815–823 (2015)
8. Shu, K., Wang, S., Tang, J., Zafarani, R., Liu, H.: User Identity Linkage across Online Social Networks: A Review (2015)
9. Number of monthly active mobile VKontakte users from March 2014 to December 2017 (2017). https://www.statista.com/statistics/425429/vkontakte-mobile-mau/
10. Zhang, J., Shao, W., Wang, S., Kong, X., Yu, P.S.: PNA: partial network alignment with generic stable matching. In: Proceedings - 2015 IEEE 16th International Conference on Information Reuse and Integration, IRI 2015 (2015)
11. Zhang, K., Zhang, Z., Li, Z., Qiao, Y.: Joint face detection and alignment using multitask cascaded convolutional networks. IEEE Sig. Process. Lett. **23**(10), 1499–1503 (2016)
12. Zhong, C., Chang, H., Karamshuk, D., Lee, D., Sastry, N.R.: Wearing many (social) hats: How different are your different social network personae? CoRR abs/1703.04791 (2017). http://arxiv.org/abs/1703.04791

# Ro-Ro Freight Prediction Using a Hybrid Approach Based on Empirical Mode Decomposition, Permutation Entropy and Artificial Neural Networks

Jose Antonio Moscoso-Lopez, Juan Jesus Ruiz-Aguilar,
Javier Gonzalez-Enrique, Daniel Urda[(✉)], Hector Mesa, and Ignacio J. Turias

Intelligent Modelling of Systems Research Group,
Polytechnic School of Engineering (Algeciras), University of Cadiz,
Avda. Ramon Puyol s/n, 11202 Algeciras, Spain
`joseantonio.moscoso@uca.es, daniel.urda@uca.es`

**Abstract.** This study attempts to create an optimal forecasting model of daily Ro-Ro freight traffic at ports by using Empirical Mode Decomposition (EMD) and Permutation Entropy (PE) together with an Artificial Neural Networks (ANNs) as a learner method.

EMD method decomposes the time series into several simpler subseries easier to predict. However, the number of subseries may be high. Thus, the PE method allows identifying the complexity degree of the decomposed components in order to aggregate the least complex, significantly reducing the computational cost. Finally, an ANNs model is applied to forecast the resulting subseries and then an ensemble of the predicted results provides the final prediction.

The proposed hybrid EMD-PE-ANN method is more robust than the individual ANN model and can generate a high-accuracy prediction. This methodology may be useful as an input of a Decision Support System (DSS) at ports as well it provides relevant information to plan in advance in the port community.

**Keywords:** Ro-Ro freight · Empirical Mode Decomposition ·
Permutation Entropy · Artificial Neural Networks · Hybrid models

## 1 Introduction

Nowadays, ports need a technological transformation, due they manage a lot of information of every process. Internet of Things (IoT) can be considered an important revolution in new generation ports 4.0 (Smart Ports) [19]. The main challenges of such a new generation port are a seamless process and intelligent decision support to deal the variability in port operation. These require, inter alia, to have reliable information to improve infrastructure, workers schedule and process management. Hence, knowledge in operational and logistic planning of freight flows is one of the basic elements in transportation systems [4].

© Springer Nature Switzerland AG 2019
H. Pérez García et al. (Eds.): HAIS 2019, LNAI 11734, pp. 563–574, 2019.
https://doi.org/10.1007/978-3-030-29859-3_48

The accurate prediction of freight traffic flow plays an important role in practical applications such as port planning, maritime management and maritime security assurance, etc. [1,10]. Such forecasts can allow port operators to formulate appropriate strategies in order to maintain competitiveness.

This work addresses in a freight Ro-Ro forecasting system which may have an important impact on the existing port activity and logistics [16]. Of interest in this work is the development of a forecasting model to guide decisions on logistics to facilitate transnational freight flows. In particular, the effect of a forecasting system could improve on the routing of truck movements passing through international borders. Here, ro-ro truck-based freight shipments between two continents are considered.

Empirical mode decomposition (EMD) is an effective technology with adaptive data processing. EMD was especially devised for nonlinear and complicated signal sequences. The original time series dataset is disassembled into a series of independent Intrinsic Mode Functions (IMFs) and after that a method such as ANNs can be applied to predict the values for IMFs in different frequencies. Different authors [8] also proposed hybrid forecasting models based on ensemble EMDs. PE is one of the most relevant and novel approach used together with EMD method in order to obtain more effective predictions in time series.

Some authors proved the ability of machine learning and hybrid methods to produce adequate predictions for traffic or freight flows in ports [10,11]. Authors have previously applied machine learning for flow forecasting at ports [13–15].

For Ro-Ro traffic volume prediction, we propose to apply a hybrid methodology that combines both EMD and PE together with Artificial Neural Network (ANN) models, and show the comparison of simple models and the hybrid methodology. Empirical results indicate that the hybrid methodology is efficient and superior for Ro-Ro traffic volume prediction.

Nowadays, ports need a technological transformation where the management of many information is considered an important challenge.

The article is organized as follows. Section 2 details the proposed methodology. We review and discuss different existing approaches. Section 3 illustrates the experimental procedure and the real data set taken from the Port of Algeciras (Spain). Section 4 addresses model approach and explains how the predictive performance of the proposed approach can be assessed. Section 5 concludes with some additional remarks and a discussion.

## 2    Methodology

### 2.1    Empirical Mode Decomposition

Empirical Mode Decomposition (EMD) is a technique designed to decompose a signal into its intrinsic modes [7]. EMD is attractive in time-series analysis because it is an empirically based technique that is a posteriori and adaptive. It is a data-driven technique that allows the data to speak for itself. No prior assumptions are required, as is the case with traditional time-frequency techniques such as Fourier or wavelet analyses. The time-frequency components obtained from EMD can simplify the problem by allowing us to investigate the series for one intrinsic mode function (IMF) at a time and over time that is optimal for the respective IMFs. Huang et al. [7] described the EMD algorithm as follows:

1. Identify the local extrema of the time series $X(t)$ and then generate the maximum $(e_{max}(t))$ and the minimum envelopes $(e_{min}(t))$ by a cubic spline interpolation.
2. Calculate the mean value $(e_{mean,i}(t))$ in a iterative proccess as:

$$e_{mean,i}(t) = (e_{max,i}(t) + e_{min,i}(t))/2 \qquad (1)$$

3. Obtained the difference between $e_{mean,i}(t)$ and $X(t)$ denominated first difference as $d_1(t) = X(t) - e_{mean,i}(t)$, where $h_1(t)$ is the pre-intrinsic mode function.
4. Check whether the pre-intrensic mode function satisfies the conditions of an IMF. If true, the $d_1(t)$ is denoted as the $i_{th}$ IMF (later called $c_i(t)$ component) and a residue $r_i(t)$ of the original time-series is obtained.
5. Repeat points 1–4 until the residue $r_i(t)$ satisfies one condition of the termination criteria (see Huang et al. [7] for better understanding). The original time-series is finally descomposed into $N$ components or IMFs and a residue as represented in Eq. (2):

$$x(t) = \sum_{i=1}^{N} c_i(t) + r_N(t) \qquad (2)$$

where $N$ is the number of IMFs, $c_i(t)$ represents IMFs mutually orthogonal and periodic, and $r_N(t)$ is the final residue.

## 2.2  Permutation Entropy

The entropy [18] is one way to quantity by a non-parametric measure the information which is contained in the data. The concept of permutation entropy (PE) was introduced by Bandt and Pompe [3]. They used a comparison of neighbour values in order to evaluate the complexity degree of a time series data. Since them, the PE method has been successfully used in different research areas [9]. The PE method can perform well with high-level noise data [20]. PE transforms the initial time series $X(t)$ into embedding vector as Eq. (3):

$$X(t) = \{X_{t+(j_1-1)\dot{\tau}} + X_{t+(j_2-1)\cdot\tau}+, ..., X_{t+(j_m-1)\cdot\tau}\} \qquad (3)$$

where $m$ is the embedding dimension and $\tau$ is the time delay. For a given $m$, there can exist $m!$ permutations. The entropy is calculated for each permutation obtaining each $PE$ value and, then, is calculated the normalized PE values as Eq. (4):

$$PE = \frac{-\sum_1^m h_i \cdot ln(h_i)}{ln(m!)} \qquad (4)$$

where $h_i$ is the relative frequency (ratio between the absolute frequency and the total number of sample observations) in each permutation $m$. The value of $PE$ ranges from 0 to 1. A threshold value $\theta$ is firstly defined in order to check whether the $PE$ exceeds or not this value. The time series has a high complexity if the PE value is greater than the threshold. In this work, the threshold value $\theta$ is used to identify the IMF components with higher complexity. These components are isolated and may contain the noise of the time series.

## 2.3   Artificial Neural Networks

Artificial Neural Networks (ANNs) are a widely used machine learning technique that simulates the structure and learning of a biological neural network. ANNs learning is based on back-propagation procedure [17] in feedforward networks. ANNs are universal approximators [6].

Feedforward ANNs consists of three (or more) layers (input, hidden and output). The hidden layer (or more) receives data from input layer and process them to the output layer. Typically, layers are fully interconnected between them with networks weights. The network is trained using the Levenberg-Marquardt (LM) algorithm to obtain the optimal weights for learning a mapping between inputs-outputs database. This training algorithm seeks to optimize performance based on mean square error. The Levenberg-Marquart algorithm used the Gauss-Newton approximation in order to provide robustness and velocity [5]. Early stopping has been used in order to avoid overfitting. In this Neural approach all nodes or neurons are connected between layers by weighted links and the outputs are expressed as Eq. (5),

$$Y = g \cdot \left( \sum_{f=0}^{M} W_{kj} \cdot f \left( \sum_{T=0}^{D} W_{tf} \cdot X_t \right) \right) \tag{5}$$

where $W_{ij}$ is the matrix of the connection weight input layer with $D$ is the total number of inputs and $W_{kj}$ is the matrix of the connection weight of the hidden layer with $M$ number of units.

## 3   Experimental Procedure

In order to evaluate the forecasting performance of the proposed approach, this study uses daily sampled Ro-Ro fresh vegetable volume (Kg) at the Port of Algeciras from 1 January 2000 to 31 December 2007 (Fig. 1). Ro-Ro vegetable freight is the largest goods in the case of the Port of Algeciras. The main features of this daily vegetable time series is the seasonality and the variation of goods weight unload in port between consecutive days. These features make it complex to develop an accuracy forecasting model. The EMD-PE-ANN model aims to improve the forecasting in comparison with simpler forecasting methods as ANNs. The experimental procedure includes the following steps:

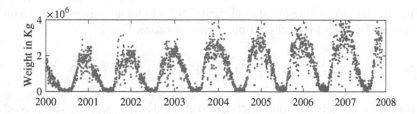

**Fig. 1.** Original time series used in this work.

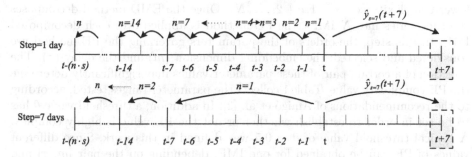

**Fig. 2.** Operational framework of the autoregressive window.

- Step 1: The EMD method decomposes the whole time series into $N$ components, named IMFs.
- Step 2: Identifying the complexity degree of the $N$ components (each IMF) by the PE method and generating the new series. Depending on this level of complexity, the $N$ IMFs are reconstructed into $c+1$ components as follows: the first $c$ components correspond to the first $c$ IMFs with higher complexity; and an aggregated component which is constructed as the sum of the last $N - c$ IMFs with lower complexity.
- Step 3: An ANNs model is used for the prediction of each component.
- Step 4: The forecasting results of each component are combined as an aggregated output.

This algorithm is repeated from step 2 depending on the aggregated component defined. In this study, the EMD method is first applied to decompose the original time series into several simpler subseries, which are assumed easier to predict. The high number of subseries generated leads to in a high computational cost. In order to overcome this problem, the PE method is applied in each IMF, identifying thereby the complexity of each IMF.

Knowing the complexity allows identifying the subseries with higher noise. The assumption in this study is that subseries with higher complexity contain most of the noise of the whole time series. Thus, the PE method acts as a denoising method. The rest of IMFs, with low complexity and thereby without noise, are aggregated in a new component. At this point, the number of subseries to predict have been reduced significantly.

Finally, the forecasting outputs of each ANN model are aggregated, resulting the final prediction of the whole time series as shows Eq. 6:

$$\hat{Y}_t = \hat{Y}_{IMF_{1,t}} + \hat{Y}_{IMF_{2,t}} + ... + \hat{Y}_{IMF_{c,t}} + \sum_{i=c+1}^{N} \left( \hat{Y}_{IMF_{i,t}} \right) \qquad (6)$$

where at time $t$, $\hat{Y}_t$ is the final aggregated prediction, $\hat{Y}_{IMF_{i,t}}$ is the prediction value of each IMF, $c$ indicates the number of IMFs with higher complexity, $N$ is the total number of IMFs obtained and $i$ depicts the number of IMFs with lower complexity $(i = c+1, c+2, ..., N)$. Once the EMD method decomposes the time series into $N$ IMFs, the PE method is applied over each decomposed IMF. In this step, the orders of the parameters governing the PE method are considered and selected: the embedding dimension $(m)$ and the delay $(\tau)$. The selection of a certain pair of these parameter values may significantly determine the PE complexity value. Table 1 collects the parameter ranges tested, according to the recommendations of Amigó et al. [2]. In addition, a threshold value $\theta$ has to be set in order to establish whether or not the PE value is highly complex. A typical threshold value of $\theta = 0.5$ was defined in this work. Since different values of PE can be obtained for one IMF, depending on the pair $[m, \tau]$ and $\theta$, the aggregated IMF component $(\sum IMFs)$ may be constructed by the sum of a larger or smaller number of individual IMFs. Therefore, different solutions have been explored and tested depending on the possible configurations of the aggregated IMF component, hereinafter called ensembles. Table 2 provides a better understanding of the composition of the ensembles and its settings.

Once the original time series was transformed into several IMFs and the aggregated component is obtained, an ANNs forecasting model is used to predict both the $c$ IMFs with higher complexity and the aggregated component. In this step, the inputs of each ANN model are configured as an auto-regressive approach using different lags of the time series in the past (the size of the auto-regressive window, $aw$). In addition, these lags were computed considering two different temporal leap sizes $(s)$ in the past (as a step size): every one-day $(s = 1)$ and every seven-day $(s = 7)$ in the past. Moreover, a seven-day prediction horizon $(ph)$ was tested in this work for each scenario. This medium-term prediction horizon may provide higher quality information to improve the port operation management. Figure 2 shows the two different configuration of the auto-regressive window and the prediction horizon used in this study. The values of the ANNs parameters tested are collected in Table 1.

In order to select the optimal lags, a two-fold cross-validation (2-CV) technique was applied using a random resampling procedure. This procedure was repeated 20 times in each subseries generated, both the individual IMFs and the aggregated component, in order to ensure the randomness of the process and to obtain the most accurate configuration of the ANNs model.

In order to asses and compare the fit of the proposed models five performance indexes are considered: the Correlation Coefficient $(R)$, the Index of Agreement $(d)$, the Root Mean Square Error $(RMSE)$ and the Mean Absolute Error $(MAE)$. The Eqs. 7–10 describe how the performances indexes are

**Table 1.** Parameters values tested.

| | Parameters | Values of the parameters |
|---|---|---|
| PE | Embedding dimension ($m$) | 2, 3, 4, 5 |
| | Delay ($\tau$) | 1, 2, 3, 4 |
| ANNs | Prediction horizon ($ph$) | 7 days |
| | nhiddens ($nh$) | 1, 2, 3, 4, 5, 6, 7, 8, 9, 10, 12, 14, 16, 18, 20 |
| | Size of the auto-regressive windows ($aw$) | 1, 2, 3, 4, 7, 14, 21, 28, 52 |
| | Steps in the auto-regressive windows ($s$) | 1, 7 days |

calculated given the observed ($O$) and forecasted ($F$) outcome and $n$ is the times compared.

$$d = 1 - \frac{\sum_{i=1}^{n} (F_i - O_i)^2}{\sum_{i=1}^{n} \left(|F_i - \overline{O}| - |O_i - \overline{O}|\right)^2} \tag{7}$$

$$R = \frac{\sum_{i=1}^{n} (O_i - \overline{O})(F_i - \overline{F})}{\sqrt{\sum_{i=1}^{n} (O_i - \overline{O})^2 \cdot \sum_{i=1}^{n} (F_i - \overline{F})^2}} \tag{8}$$

$$RMSE = \sqrt{\frac{\sum_{i=1}^{n} (F_i - O_i)^2}{n}} \tag{9}$$

$$MAE = \frac{\sum_{i=1}^{n} |F_i - O_i|}{n} \tag{10}$$

## 4    Results

In this work a database composed for full eight years have been used. This database contains the daily import records from January 2000 to December 2007 in the Port of Algeciras, both included. The results represent the values (in Kg) of daily fresh vegetables in Ro-Ro freight in the port of Algeciras Bay. Fresh vegetables is one of the most important types of goods that is handled in the port of Algeciras and has a significant impact in the supply chain port management. Hence, a prediction of this freight volume could improve the process management. To this end, a hybrid approach based on a combination of EMD-PE-ANN methods is proposed to obtain accurate predictions. The original time series was firstly pre-processed by the EMD method in order to decompose the original series into several subseries called IMFs. A total of $N = 10$ individual IMFs were obtained. Then, the PE method was applied in each IMF to detect the subseries with higher complexity, depending on the $PE$ value. The results of the $PE$ value are plotted in Fig. 4. This figure shows the $PE$ value of each IMF, considering the recommended range of parameters. It can be seen that the higher the value of the parameters $m$ and $\tau$, the lower the PE value. For an established threshold of $\theta = 0.5$, the first four IMF can be practically considered as high complexity ($PE$ value higher than 0.5), overcoming this threshold in most of its combinations of parameters. However, the fitht IMF only overcomes the threshold in some cases. It can be concluded that, in principle, the first four IMFs have high complexity. In this case, the aggregated component is formed by the sum of the rest of IMFs. In order to prove this assumption, other possible configurations are tested and compared, hereinafter referred to as ensembles. Including the above-mentioned ensemble, five ensemble-based approaches are developed. The description of these ensembles and their components (the IMFs with higher complexity and the aggregated component), are shown in Table 2.

In the final step, for each ensemble, an ANNs forecasting model is used individually in each of its components. The most accurate parameters setting obtained of the ANNs model, both the size of the auto-regressive window $aw$ and the number of hidden neurons $nh$, are collected in Table 3 for each component and temporal leap ($s$ equal to 1 and 7). For any given time $t$, the predicted result of each component is then aggregated, obtaining the final prediction result of the whole time series. The final predicted results of the proposed ensembles are collected in Table 4.

**Table 2.** Proposed Ensambles of IMFs forecasted.

| Ensemble | Combinations of IMFs forecasting |
|---|---|
| Ensemble 1 | $\sum_{i=1}^{5}\left(\hat{Y}_{IMF_i}\right) + \hat{Y}\left(\sum_{i=6}^{10} IMF_i\right)$ |
| Ensemble 2 | $\sum_{i=1}^{4}\left(\hat{Y}_{IMF_i}\right) + \hat{Y}\left(\sum_{i=5}^{10} IMF_i\right)$ |
| Ensemble 3 | $\sum_{i=1}^{3}(\hat{Y}_{IMF_i}) + \hat{Y}\left(\sum_{i=4}^{10} IMF_i\right)$ |
| Ensemble 4 | $\sum_{i=1}^{2}\left(\hat{Y}_{IMF_i}\right) + \hat{Y}\left(\sum_{i=3}^{10} IMF_i\right)$ |
| Ensemble 5 | $\hat{Y}_{IMF_1} + \hat{Y}\left(\sum_{i=2}^{10} IMF_i\right)$ |

**Table 3.** Most accurate parameter setting of the ANN models for each IMF considering the temporal leap $s$.

| | $IMF_1$ | | $IMF_2$ | | $IMF_3$ | | $IMF_4$ | | $IMF_5$ | | $\sum_{i=6}^{10} IMF_i$ | | $\sum_{i=5}^{10} IMF_i$ | | $\sum_{i=4}^{10} IMF_i$ | | $\sum_{i=3}^{10} IMF_i$ | | $\sum_{i=2}^{10} IMF_i$ | |
|---|---|---|---|---|---|---|---|---|---|---|---|---|---|---|---|---|---|---|---|---|
| | $aw$ | $nh$ | $aw$ | $nh$ | $aw$ | $nh$ | $aw$ | $nh$ | $aw$ | $nh$ | $aw$ | $nh$ | $aw$ | $nh$ | $aw$ | $nh$ | $aw$ | $nh$ | $aw$ | $nh$ |
| $s=1$ | 7 | 1 | 3 | 1 | 21 | 1 | 21 | 1 | 14 | 1 | 14 | 1 | 7 | 1 | 14 | 1 | 52 | 1 | 52 | 2 |
| $s=7$ | 7 | 1 | 4 | 1 | 21 | 1 | 7 | 12 | 3 | 2 | 14 | 1 | 7 | 1 | 52 | 1 | 21 | 10 | 52 | 1 |

Table 4 shows the values of the performance indexes of the different ensembles for each temporal leap considered. In both cases of $s$, a better prediction performance is achieved by the ensemble approach comparing with the individual ANN model. All the proposed models significantly overcome the individual ANN model. Moreover, even though the $s = 7$ day overcomes the s=1 days for the individual ANN model, there are no significant differences between the temporal leaps tested for the proposed approaches. Indeed, the $s = 7$ temporal leap performs slightly worse than the $s = 1$ day for these proposed ensemble models.

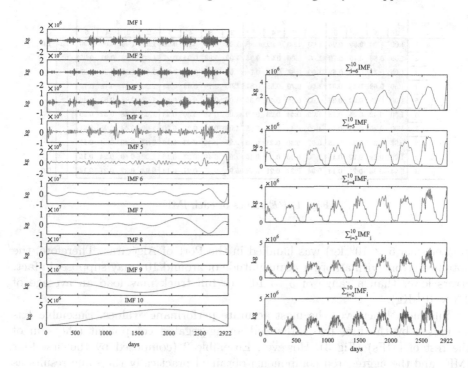

**Fig. 3.** The individual IMFs (left) and the possible aggregated components used in the ensembles (right)

**Table 4.** Performance index values for 7 days of prediction and temporal leaps of 1 and 7 days. Best values in bold.

| Step | P. Index | ANN | Ens. 1 | Ens. 2 | Ens. 3 | Ens. 4 | Ens. 5 |
|------|----------|-----|--------|--------|--------|--------|--------|
| $s = 1$ | $R$ | 0.8750 | **0.9369** | 0.9367 | 0.9331 | 0.8952 | 0.8790 |
| | $d$ | 0.9275 | **0.9670** | 0.9669 | 0.9647 | 0.9398 | 0.9294 |
| | $RMSE$ | 4.94E+5 | **3.57E+5** | 3.57E+5 | 3.67E+5 | 4.57E+5 | 4.88E+5 |
| | $MAE$ | 3.30E+5 | **2.20E+5** | 2.20E+5 | 2.27E+5 | 2.91E+5 | 3.34E+5 |
| $s = 7$ | $R$ | 0.8898 | **0.9275** | 0.9273 | 0.9160 | 0.8906 | 0.8993 |
| | $d$ | 0.9380 | **0.9618** | 0.9614 | 0.9543 | 0.9410 | 0.9427 |
| | $RMSE$ | 4.72E+5 | **3.85E+5** | 3.86E+5 | 4.14E+5 | 4.69E+5 | 4.53E+5 |
| | $MAE$ | 3.15E+5 | 2.50E+5 | **2.48E+5** | 2.67E+5 | 3.10E+5 | 3.01E+5 |

This is since the decomposition of the original time series, that may result in the loos of the weekly stationarity.

It is worth mentioning that the values of performance indices obtained in each model seems to be high in terms of RMSE and MAE. Note that the time series was processed in Kg instead of another unit of measurement such as Tons. Considering the study period, a daily average of $3 \times 10^6$ Kg (with daily peaks

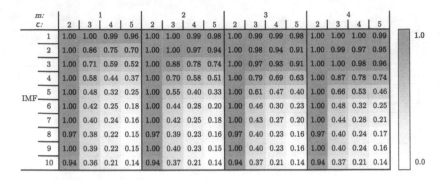

| IMF | m: 1 | | | | m: 2 | | | | m: 3 | | | | m: 4 | | | |
|---|---|---|---|---|---|---|---|---|---|---|---|---|---|---|---|---|
| c: | 2 | 3 | 4 | 5 | 2 | 3 | 4 | 5 | 2 | 3 | 4 | 5 | 2 | 3 | 4 | 5 |
| 1 | 1.00 | 1.00 | 0.99 | 0.96 | 1.00 | 1.00 | 0.99 | 0.98 | 1.00 | 0.99 | 0.99 | 0.98 | 1.00 | 1.00 | 1.00 | 0.99 |
| 2 | 1.00 | 0.86 | 0.75 | 0.70 | 1.00 | 1.00 | 0.97 | 0.94 | 1.00 | 0.98 | 0.94 | 0.91 | 1.00 | 0.99 | 0.97 | 0.95 |
| 3 | 1.00 | 0.71 | 0.59 | 0.52 | 1.00 | 0.88 | 0.78 | 0.74 | 1.00 | 0.97 | 0.93 | 0.91 | 1.00 | 1.00 | 0.98 | 0.96 |
| 4 | 1.00 | 0.58 | 0.44 | 0.37 | 1.00 | 0.70 | 0.58 | 0.51 | 1.00 | 0.79 | 0.69 | 0.63 | 1.00 | 0.87 | 0.78 | 0.74 |
| 5 | 1.00 | 0.48 | 0.32 | 0.25 | 1.00 | 0.55 | 0.40 | 0.33 | 1.00 | 0.61 | 0.47 | 0.40 | 1.00 | 0.66 | 0.53 | 0.46 |
| 6 | 1.00 | 0.42 | 0.25 | 0.18 | 1.00 | 0.44 | 0.28 | 0.20 | 1.00 | 0.46 | 0.30 | 0.23 | 1.00 | 0.48 | 0.32 | 0.25 |
| 7 | 1.00 | 0.40 | 0.24 | 0.16 | 1.00 | 0.42 | 0.25 | 0.18 | 1.00 | 0.43 | 0.27 | 0.20 | 1.00 | 0.44 | 0.28 | 0.21 |
| 8 | 0.97 | 0.38 | 0.22 | 0.15 | 0.97 | 0.39 | 0.23 | 0.16 | 0.97 | 0.40 | 0.23 | 0.16 | 0.97 | 0.40 | 0.24 | 0.17 |
| 9 | 1.00 | 0.39 | 0.22 | 0.15 | 1.00 | 0.40 | 0.23 | 0.15 | 1.00 | 0.40 | 0.23 | 0.16 | 1.00 | 0.40 | 0.24 | 0.16 |
| 10 | 0.94 | 0.36 | 0.21 | 0.14 | 0.94 | 0.37 | 0.21 | 0.14 | 0.94 | 0.37 | 0.21 | 0.14 | 0.94 | 0.37 | 0.21 | 0.14 |

**Fig. 4.** $PE$ values of each IMF.

greater than $5 \times 10^6$ kg) was handled in the Port of Algeciras. Therefore, the obtained errors with an order of magnitude of around $10^5$ may suppose, in fact, errors lower than a range of 5 to 10% (a full truck may load an average of $2.3 \times 10^4$ Kg).

Ensemble 1 achieves the most accurate performance values. Ensemble 1 is composed by the first five IMFs and an aggregated component (the sum of the rest of IMFs) (Fig. 3). However, Ensemble 2 (composed by the first four IMFs and the aggregated component) obtained practically the same results as Ensemble 1 (only slightly worse than Ensemble 1). In addition, these results are in accordance with those obtained with the study of the $PE$ value, that led to considering the first four IMFs as the higher complexity ones. It is confirmed that, in this case, no more than four individual IMFs are necessary to achieve accurate predictions.

It can be concluded that the PE method allows detecting the highest complexity IMFs with a high degree of reliability. Therefore, the PE method significantly reduces the computational cost of the simple EMD-ANN methodology, in which all the IMFs must be individually considered. The EMD-PE-ANN definitely improves the performance of the daily fresh vegetable freight volume passing through the Algeciras Port in comparison with other studies [12]. Thus, it can accurately predict the Ro-Ro traffic of the port and it also can provide the basic need for operation scheduling and resource management in advance.

## 5   Conclusions

The changes increasingly produced in recent years in the context of globalization have created an environment in which the port transport sector must adapt continuously. Knowing traffic flows that pass through the port in advance is crucial for better planning and resource allocation and to avoid congestion.

In this work, a combined procedure is proposed to forecast the Ro-Ro traffic flow at the Algeciras Port in terms of volume of fresh vegetables. The methodology assembles Artificial Neural Network, Permutation Entropy and Empirical

Mode Decomposition in a hybrid framework. As a result, the EMD-PE-ANN model is proposed and tested. The experimental results indicate that the application of PE allows computational costs to be reduced detecting the IMFs with higher complexity and aggregating the rest IMFs. Moreover, this hybrid approach leads to greater predictive accuracy in forecasting Ro-Ro traffic volumes. The EMD-PE-ANN ensemble model obtained a great performance in fresh vegetables Ro-Ro freight forecasting with high values of fitting and lower error than the individual ANN model.

The contribution of this study is to detect freight traffic peaks at ports, specifically in Ro-Ro traffic, in order to avoid delays in port operations and congestion. It can be beneficial to both the public and business sectors due to it can support better planning human and material resource allocation

**Acknowledgments.** This work is part of the ACERINOX EUROPA S.A.U research project AUSINOX IDI-20170081 - "Obtaining austenitic stainless steels with minimum inclusion content from the development of new advanced simulation models in melting shop processes", supported by CDTI (Centro para el Desarrollo Tecnológico Industrial), Spain. This project has been co-financed by the European Regional Development Fund (FEDER), within the Intelligent Growth Operational Program 2014–2020, with the aim of promoting research, technological development and innovation. Authors acknowledge support through grant RTI2018-098160-B-I00 from MINECO-SPAIN which include FEDER funds. The database has been kindly provided by the Port Authority of Algeciras Bay.

# References

1. Al-Deek, H.M.: Use of vessel freight data to forecast heavy truck movements at seaports. Transp. Res. Rec. **1804**(1), 217–224 (2002)
2. Amigó, J., Keller, K.: Permutation entropy: one concept, two approaches. Eur. Phys. J. Spec. Topics **222**(2), 263–273 (2013)
3. Bandt, C., Pompe, B.: Permutation entropy: a natural complexity measure for time series. Phys. Rev. Lett. **88**(17), 174102 (2002)
4. Blackburn, R., Lurz, K., Priese, B., Göb, R., Darkow, I.L.: A predictive analytics approach for demand forecasting in the process industry. Int. Trans. Oper. Res. **22**(3), 407–428 (2015)
5. Hagan, M.T., Menhaj, M.B.: Training feedforward networks with the Marquardt algorithm. IEEE Trans. Neural Netw. **5**(6), 989–993 (1994)
6. Hornik, K., Stinchcombe, M., White, H.: Multilayer feedforward networks are universal approximators. Neural Netw. **2**(5), 359–366 (1989)
7. Huang, N.E., et al.: The empirical mode decomposition and the hilbert spectrum for nonlinear and non-stationary time series analysis. Proc. R. Soc. Lond. A: Math. Phys. Eng. Sci. **454**(1971), 903–995 (1998)
8. Jiang, X., Zhang, L., Chen, X.M.: Short-term forecasting of high-speed rail demand: a hybrid approach combining ensemble empirical mode decomposition and gray support vector machine with real-world applications in china. Transp. Res. Part C: Emerg. Technol. **44**, 110–127 (2014)
9. Leite, G.D.N.P., Araújo, A.M., Rosas, P.A.C., Stosic, T., Stosic, B.: Entropy measures for early detection of bearing faults. Phys. A **514**, 458–472 (2019)

10. Liu, R.W., Chen, J., Liu, Z., Li, Y., Liu, Y., Liu, J.: Vessel traffic flow separation-prediction using low-rank and sparse decomposition, p. 6, October 2017

11. Mangan, J., Lalwani, C., Gardner, B.: Modelling port/ferry choice in RoRo freight transportation. Int. J. Transp. Manage. 1(1), 15–28 (2002)

12. Moscoso-López, J.A., Turias, I.J.T., Come, M.J., Ruiz-Aguilar, J.J., Cerbán, M.: Short-term forecasting of intermodal freight using ANNs and SVR: case of the port of algeciras bay. Transp. Res. Procedia 18, 108–114 (2016)

13. Moscoso-Lopez, J.A., Turias, I., Jimenez-Come, M.J., Ruiz-Aguilar, J.J., Cerban, M.D.M.: A two-stage forecasting approach for short-term intermodal freight prediction. Int. Trans. Oper. Res. 26(2), 642–666 (2016)

14. Ruiz-Aguilar, J.J., Turias, I.J., Jiménez-Come, M.J.: Hybrid approaches based on SARIMA and artificial neural networks for inspection time series forecasting. Transp. Res. Part E: Logist. Transp. Rev. 67, 1–13 (2014)

15. Ruiz-Aguilar, J.J., Turias, I.J., Moscoso-López, J.A., Come, M.J.J., Cerbán, M.M.: Forecasting of short-term flow freight congestion: a study case of Algeciras Bay Port (Spain). Dyna 83(195), 163–172 (2016)

16. Ruiz-Aguilar, J.J., Turias, I.J., Jiménez-Come, M.J.: A novel three-step procedure to forecast the inspection volume. Transp. Res. Part C: Emerg. Technol. 56, 393–414 (2015)

17. Rumelhart, D.E., Hinton, G.E., Williams, R.J.: Learning Internal Representations by Error Propagation, vol. 2. MIT press, Cambridge (1986)

18. Shannon, C.E.: A mathematical theory of communication. Bell Syst. Tech. J. 27(3), 379–423 (1948)

19. Yang, Y., Zhong, M., Yao, H., Yu, F., Fu, X., Postolache, O.: Internet of things for smart ports: Technologies and challenges. IEEE Instrum. Meas. Mag. 21(1), 34–43 (2018)

20. Yu, L., Wang, Z., Tang, L.: A decomposition-ensemble model with data-characteristic-driven reconstruction for crude oil price forecasting. Appl. Energy 156, 251–267 (2015)

# Hybrid Intelligent Applications

# Modeling a Mobile Group Recommender System for Tourism with Intelligent Agents and Gamification

Patrícia Alves[1](✉) ⬤, João Carneiro[1] ⬤, Goreti Marreiros[1] ⬤,
and Paulo Novais[2] ⬤

[1] GECAD – Research Group on Intelligent Engineering and Computing
for Advanced Innovation and Development, Institute of Engineering,
Polytechnic of Porto, 4200-072 Porto, Portugal
{prjaa,jrc,mgt}@isep.ipp.pt
[2] ALGORITMI Centre, University of Minho, 4800-058 Guimarães, Portugal
pjon@di.uminho.pt

**Abstract.** To provide recommendations to groups of people is a complex task, especially due to the group's heterogeneity and conflicting preferences and personalities. This heterogeneity is even deeper in occasional groups formed for predefined tour packages in tourism. Group Recommender Systems (GRS) are being designed for helping in situations like those. However, many limitations can still be found, either on their time-consuming configurations and excessive intrusiveness to build the tourists' profile, or in their lack of concern for the tourists' interests during the planning and tours, like feeling a greater liberty, diminish the sense of fear/being lost, increase their sense of companionship, and promote the social interaction among them without losing a personalized experience. In this paper, we propose a conceptual model that intends to enhance GRS for tourism by using gamification techniques, intelligent agents modeled with the tourists' context and profile, such as psychological and socio-cultural aspects, and dialogue games between the agents for the post-recommendation process. Some important aspects of a GRS for tourism are also discussed, opening the way for the proposed conceptual model, which we believe will help to solve the identified limitations.

**Keywords:** Group Recommender Systems · Mobile tourism · Context-awareness · Gamification · Multi-agent systems

## 1 Introduction

Since 1992 [1] that Recommender Systems (RS) have been studied to help individual users make better choices [2, 3] thus recommending items that intend to better satisfy the users tastes in various domains, each one with its specific challenges, like recommending a movie to watch, a music to listen, a place to visit, a restaurant to lunch, etc. But if to generate accurate individual recommendations is complex, to provide accurate recommendations to groups is even more. The tourism domain has many particularities and is an interesting challenge. To support groups of tourists plan and get

© Springer Nature Switzerland AG 2019
H. Pérez García et al. (Eds.): HAIS 2019, LNAI 11734, pp. 577–588, 2019.
https://doi.org/10.1007/978-3-030-29859-3_49

accompanied in their excursions can be a very complex task, especially due to the group's heterogeneity and conflicting preferences [4]. Millions of tourists participate in planned tours every day, some travel alone, others in groups, but are their needs, interests and curiosity satisfied? Do they enjoy the tours they engaged in? Boratto and Carta [5] state how a group is formed influences its modeling and the predicted recommendations. Groups formed occasionally for a common aim, like travelling together to a specific destination, and that may or may not be acquainted to each other [5] causes this heterogeneity to go deeper. Group Recommender Systems (GRS) are being designed for helping in situations like those, and if they use the capabilities of a mobile device, they can brutally improve the users' experience, bringing new possibilities to explore, like the users' context [6], i.e., the information that surrounds him [7].

In this paper, we introduce a conceptual model that intends to improve the tourists experience in a GRS for tourism by showing concern for their interests, facilitate the post-recommendation process, by proposing the use of an argumentation-based dialogue model between intelligent agents, agents that will accompany the tourists during the tour. Gamification techniques are also proposed to acquire the tourists' profile and motivate them during the tour.

In the next section we present a brief state-of-the-art in GRS for tourism and discuss some current issues. Section 3 introduces dialogue games between intelligent agents and gamification as ways of enhancing the choice process and the tourists' involvement in GRS for tourism, respectively. This section also explains the connection between choice and decision, and how important explanations are in a recommendation. The conceptual model of the GRS for tourism is presented and shortly explained. Section 4 summarizes the contents addressed in the paper and describes what will be done as future work.

## 2    Group Recommender Systems for Tourism

GRS have become an important and challenging theme in the field of RS [8–12] since the group members' preferences can vary, and therefore, to reach a solution that satisfy all the members can be hard to accomplish. It is of extreme importance to guarantee that none of the group members gets too dissatisfied, dissatisfaction that can spread within the group due to the emotional contagion phenomenon [11]. For instance, suppose a travel agency in China that has vacation packages for groups of tourists, with a set of different types of Points of Interest (POI) to visit in a certain country. It is known that Chinese tourists usually travel in groups, either by option or because of impositions [13]. Families, individuals, friends can subscribe a package. But does the package has POI that satisfy all the subscribed members? Although they share the same culture, not all members have the same personality and preferences, but they had no other choice than to choose a predefined package. A vacation that seemed exciting can easily become toilsome. A GRS capable of providing personal and contextual recommendations can be the perfect solution.

Many interesting prototypes of GRS for tourism have and are being proposed to help groups of tourists in the planning of vacations or excursions, usually presenting a list of POI to visit. For instance, looking at some of the first GRS for tourism,

*INTRIGUE* (INteractive TouRist Information GUidE) was proposed in 2003 by Ardissono et al. [14] to help (heterogenous) groups of tourists find sightseeing destinations and itineraries in Italy. It is a GRS for mobile and desktop devices where a group member configures the group size, their preferences and characteristics. The group is then divided into subgroups according to those configurations, and recommendations are given to each subgroup grounded by explanations that address potential conflicting requirements.

*CATS* (Collaborative Advisory Travel System) aims to help a group of friends in planning a ski-holiday [12, 15] using a face-to-face collaborative platform (the DiamondTouch interactive tabletop) that uses critiques as a way of giving feedback to recommended POI and iteratively find a final choice.

Garcia et al. [16] developed a GRS for tourist activities, based on the group's tastes, demographic data and places visited in former trips, by extending the *e-Tourism* tool they previously developed for individual tourists. This tool is composed by the *Generalist Recommender System Kernel* (GRSK), which is a domain-independent taxonomy-driven search engine that manages the group recommendation. It is responsible for aggregating, intersection and incrementally intersection the users' preferences and present a final list of items to recommend.

Travel Decision Forum is a GRS that uses animated characters to represent the group members [17]. The authors state that mutual-awareness and communication are important in order to reach a consensus in the post-recommendation process. For that, the group members configure their preferences incrementally and collaboratively, being able to see the other members' preferences. Since the choice of preferences can be influenced by a person's motivations, the authors implemented a simple way for the members to configure their motivational orientation regarding the other members. This is a very important factor in social interactions that other GRS do not consider, and that we will further discuss later in this paper.

It is perceptible that due to technological limitations at the time, the first GRS were totally dependent of the users' interactions and configurations. Indeed, since the mobile technology was still emerging, the users felt offended for having a "too intelligent" application and argued they could think and decide for themselves, not accepting a too much automatization of the system [18]. However, fifteen years later, the minds "evolved", the users' requirements changed, and many would like to have a more automated system that could think and decide for them, at least regarding recommendations.

In the early 2000's, wireless internet access was very limited and very expensive, but now, that is no longer a problem. The rapid evolution of the wireless internet connections, its throughput, stability, price and massification, also shifted the way (G)RS were being designed and many ideas/approaches found in literature were discontinued. This is a positive reinforcement for creating new and better (G)RS.

For example, the very recent work by Nguyen and Ricci [4] consists on a chat-based GRS for mobile devices that also allows the group members to become part of the choice process. It is similar to WhatsApp in the way users in a group can exchange messages between them, with the additional features of allowing the users to rate previously visited POI and define their mood, so a higher importance is attributed to the user in the preferences aggregation in case he is in a bad mood, tired, etc. The users can classify the recommended POI by liking/disliking them or by classifying one as the best, or comment on them with text and emoticons. This evaluation allows the system to infer users' constraints based on the attributes of the classified POI, and incrementally update the information on a recommended POI with additional explanations, based on those restrictions. Although the system provided higher perceived recommendation quality than the standard benchmark, this approach may not be practical for large and/or occasional groups, since the tested groups were very small, composed of 2 or 3 members. We think it can be very confusing for a group of 20 or more people to chat and exchange opinions in an efficient way. Something else is needed.

## 2.1    Important Aspects to Consider in a GRS for Tourism

To support groups in travel planning is not a simple process and to generate a list of recommendations based on the users' context and preferences is not enough. Other factors need to be considered for a GRS to effectively serve its purposes. For instance, in 2003, Jameson, Baldes and Kleinbauer [17] made the intelligent observation that the recommendation process does not end when a list of recommendations is presented to the user. The users need to decide what to choose from the list, so all the group members get (minimally) satisfied. The authors went even further by stating that it would be short-sighted not to include post-recommendation processes in the design of a (G)RS, like ways of persuading the other group members to follow a certain recommendation a user finds better. If the process of reaching the final choice has not been delegated to one of the group members, communication and possibly negotiation will be needed between the group members [17]. This falls into the same line of thought that the users need to be somehow involved in the recommendation process, and as mentioned before, a full automatization may not be the perfect solution.

It is evidenced that many people like to know the preferences of other group members, leaning to choose similar preferences [17], either because they would like to please other member(s) or because they tend to avoid conflicts if they previously know what the other users think, like in a real face-to-face scenario. This awareness leads to a sort of collaboration that can help reach a faster consensus. However, this type of behavior is not so linear. Like in a decision-making process, the group members in a choice process can have different intentions, which influence their behaviors and choices. Jameson, Baldes and Kleinbauer [17] address motivation as a way of influencing the choice process. However, motivation is what compels us to fulfil or not our intentions. So, a person's intentions are in the core of a choice, powered by her motivations, and we believe both need to be accounted for. For instance, Phoebe can have an intention to visit a country, but because she cannot go with her boyfriend, she doesn't feel motivated to go, and therefore she won't go unless he does.

As RS can be seen as "tools for helping people to make better choices" [2], how choices are made (the psychology of choice) and how the process of making choices can be supported is of extreme importance [2]. Some GRS are already considering group decision-making (GDM) as an indispensable factor for their success. McCarthy et al. [12] developed a face-to-face collaborative GRS for planning skiing vacations. The users reach a consensus by critiquing the items in a list of recommendations during the choice process. Castro, Quesada, Palomares and Martinez [9] proposed a consensus driven GRS, which implements a consensus reaching process used for group decision-making, to iteratively piece together individual recommendations before delivering the group recommendations. The authors concluded that applying a consensus reaching process to group recommendations undoubtedly improved the results and that GRS could benefit from the use of GDM approaches. Marques, Respício and Afonso [19] developed a mobile GRS that uses group collaborative decision-making by using votes. The users model their preferences into the system and give weights to existing restaurants recommendation' platforms. The users have then to democratically elect a restaurant from the generated list of recommended restaurants.

Another extremely important aspect for a RS are the explanations it provides. For instance, Tintarev and Masthoff [20] dedicated a paper to the explanation of recommendations in RS. Explanations can be used with many purposes like: to expose the reasoning behind a recommendation, to gain the users trust and loyalty, to persuade users to buy a recommended item, to increase satisfaction, to help users make better and faster decisions, etc. [20]. The users like to feel the system is not a black box or a computerized oracle that gives advices [21] and that they understand the system. This is even more true when decisions with some impact are involved, like when choosing a honeymoon destination: "Why is the system suggesting I should go to Galápagos in my honeymoon?". Explanations are also very helpful to detect errors in recommendations [20, 21], like suggesting Galápagos as a vacation destination because the user visited many websites related to Galápagos since he is researching on Galápagos penguins.

The GRS found in literature are also intrusive in the ways they present the recommendations and are not focused in the tourists' personal interests allied to their context. This causes the tourists to ignore recommendations or ignore the remaining group members. For instance, suppose a group of tourists is visiting a monument with tall towers at some point, and that a member is afraid of heights. The GRS should be capable of warning her that she should not climb those towers because of her fear, avoiding the tourist's discomfort. Or, suppose a tourist is constantly ignoring notifications presented by the GRS in the morning. The GRS should be capable of detecting that the tourist does not like notifications and stop showing them at that time of day.

## 3 Intelligent Agents, Dialogue Games and Gamification to Enhance a GRS

We believe the post-recommendation process can be improved by using intelligent agents and techniques from group decision-making and consensus reaching. So, we propose to solve some of the issues presented before by applying formal dialogue games [22], for agent communication and interaction using argumentation, between intelligent agents modeled to represent the group members. We intend to model each agent with the respective tourist's profile and context, acting on his behalf. So, each agent will consider the respective tourist's preferences, personality, socio-cultural aspects, mood, intentions, etc., to choose the POI to visit from the list, engaging in a real time conversation with the other agents by using argumentation. The agents argumentation will also be based on the dynamic argumentation model developed in our previous work [23, 24], and will use dialogues of different types, such as negotiation and deliberation [25], to propose solutions and reach a final consensus on the list of POI to visit that better suits the group's interests and intentions. We believe this strategy can be helpful for large groups, since the agents automatic dialogues will minimize the time the tourists will need to spend in the system to reach a consensus, and will avoid the confusion inherent to chats of large groups of people, simplifying and making the choice process more organized. For example, suppose a group of 30 members where 5 of them are from the same family. The agents from the same family can deliberate together on the POI to visit before dialoguing with the other agents, and then negotiate the POI with the other agents.

The proposed argumentation-based dialogue model will be capable of proposing recommendations and at the same time, due to its self-nature, be capable of explaining the reasons behind those recommendations. We believe this will allow the tourists to feel part of the process and understand it. The dialogue model will also have a high level of expressiveness, meaning the agents will be capable of acting according to different intentions and motivations in the same dialogue, mirroring their tourist, as mentioned in our previous works [26, 27].

Since the tourists will exchange messages in real-time with other tourists in the group, the content of those messages will be studied, content that will influence their agents' dialogues. For that, we will rely on machine-learning techniques such as text-mining and natural language processing, in order to study the human dialogues and produce important information in terms of their meaning and the sentiment existent in them.

### 3.1    The Conceptual Model

Figure 1 shows the architecture for the conceptual mobile GRS. We chose microservices because they allow a better modularity, scalability and the services can be deployed independently, each one with its own database. This means the most suitable programming language(s) can be used for each service, a better faults isolation, continuous delivery and components spread across multiple servers, among others [28]. The communication between the microservices will be asynchronous and through the REST protocol. The API Gateway will be the single-entry point into the system, simplifying the mobile clients' requests and serving as a load balancer for the microservices. The microservices will include the:

**Multi-agent Service**
This service will be responsible for modeling the intelligent agents according to the tourists' information (profile and context), and other agents necessary to the process, by using the JADE Framework. Here is where the dialogue games between the agents, to choose the POI to visit, will also be processed. The agents are also intended to learn the tourists' behavior and context, automatically improving their profile, so better and more proactive recommendations/notifications can be made to the group and/or the individual tourist.

Machine-learning classification algorithms will be applied to form (if possible) subgroups of agents/tourists with similar profiles and interests. This can minimize the group's heterogeneity and conflicts of interest, facilitating the consensus reaching in the

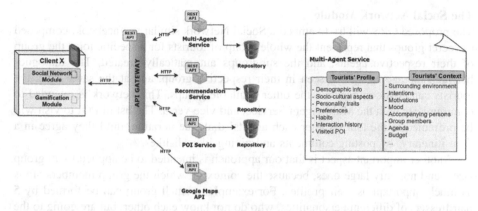

**Fig. 1.** Left: Architecture of the conceptual mobile Group Recommender System. Right: Information about the tourists, available in the Multi-Agent Service.

post-recommendation process and the generation of more precise recommendations to the (sub)groups. This aggregation also intends to promote more socialization and the creation of bounds between the group members. For instance, suppose a group of 50 tourists where 6 of them play Pokémon Go. By comparing the tourists' personality, the agents find out 5 of them have a high openness to experience, agreeableness and low neuroticism. The agents can suggest those members to meet after lunch, at the hotel entrance, to search for Pokémon in the surroundings. Also, suppose the whole group went to visit a monument. If the agents know the personality traits and mood of the tourists, they can suggest a quest for some group members to complete, where they will receive instructions that will make them perform joint tasks to better know the monument, like taking a picture from some important window or collect a certain object that represents the monument's history, promoting their socialization and opening ways of creating bounds between them.

### Recommendation Service
This service is intended to iteratively run the recommendation algorithm(s) based on the tourists' profile, context and the results obtained from the agents dialogues, presenting the processed recommendations in each iteration, until a final recommendation is accepted by the tourist, in the case of individual recommendations/notifications, or the group/subgroup, in the case of group recommendations/notifications. Since our focus is to work on the post-recommendation process that will lead to a consensus on the places to visit and on making more interesting the tourists' experience from the planning to the tour itself, we won't detail on the recommendation algorithm(s).

### POI Service
This service will be fed by the Google Maps API, or similar, and will be responsible for retrieving all the available POI that match the tour requirements. The list of POI will then be fed into the Recommendation Service so recommendation lists can be generated for the group or subgroups of tourists.

**The Social Network Module**

The proposed GRS will try to embed a Social Network similar to Facebook, composed of secret groups that represent the whole group of tourists for a specific tour, the group of their respective agents, and the subgroups automatically created. The dialogues between the agents will be seen in their respective group, and at the same time, the tourists can post comments in the other available groups. This network is intended to communicate with the Multi-Agent Service, and vice-versa. The list of POI to visit will be presented to the tourists after each agents' dialogue iteration, until they agree in a final itinerary, by posting comments and giving likes/dislikes.

Another important aspect is that our approach is intended to be applied to all group sizes, and not only large ones, because the context in which the group members are is as much important as their profiles. For example, a small group can be formed by 5 hairdressers of different personalities, who do not know each other, but are going to the same congress, i.e., are in the same context, and one of them decides to use the GRS to find other hairdressers to visit the cultural heritage in the congress's country.

**The Gamification Module**

Personalization is a key factor for the success of RS in tourism [29–31]. The more information about the tourist is known better recommendations can be made. Information like the tourists' demographics, personality traits, socio-cultural aspects, habits and preferences can be critical factors for the system's effectiveness. Personality has been evidenced to improve the recommendations made to groups and can even help in the cold-start problem [32–34], since it is demonstrated that personality is strongly related to the users preferences and therefore, correlating the users' personalities and their preferences can help find the preferences of users with similar personalities. For instance, tourists with a high Openness to Experience tend to be more appreciative of the significance of intellectual and artistic pursuits [35] and will probably be more interested in visiting an art exhibit than tourists with low Openness.

To model the tourists' profile will help form groups with similar interests, minimizing the groups' heterogeneity and conflicts of interest. However, the existing GRS are still intrusive and time-consuming in the ways they gather the tourists' profile. The challenge here will be to gather all that information in a non-intrusive and less time-consuming way, and at the same time, motivating and challenging the tourists. Gamification can be the leverage we are looking for. It is demonstrated that gamification improves the users' involvement and motivation while learning, working, among other tasks [34, 36–38]. For instance, it has been showed that challenging games motivate students to be more concentrated and committed to the studies, learning significantly better [38]. The use of achievement badges proved to affect the students' behavior motivating them to study [39]. In their work, Mortara et al. [40] present the state-of-the-art of serious games for cultural heritage and state that this approach can be of a tremendous value to learn about the history of a location, its inhabitants and their behaviors. Hence, a GRS for tourism could become more challenging and exciting if we add gaming components to it, like badges for accomplishing certain tasks or mini games to gather the tourists' profile.

Gamification techniques can also be used to personify the agent that represents the tourist in the Multi-Agent Service, transforming it into an Augmented Reality

(AR) avatar, visible through the mobile device screen[1]. The avatar would be like the tourist's companion and can play an important role in the system by accompanying the tourist throughout the whole process, helping to decide the itinerary for the group he belongs to, and motivating the tourist during the tour by presenting intelligent information (push-notifications) and proposing personalized challenges according to the tourist's intentions and interests. Why an avatar? It is evidenced that representing the tourist with an avatar can help him feel empathy towards the system [40].

Location-based AR games can have a tremendous potential, and they can be a smarter way of catching the tourists' attention to visit a country's heritage. We propose to transform the whole trip process into a sort of a location-based AR game, where the tourists will have to complete certain personalized "quests" in the POI they visit, using AR features. We hope this will also increase their interest in knowing and learning about a country's heritage, and in a more exciting way.

## 4  Summary and Future Work

In this work, we discuss on a novel approach for a Group Recommender System for tourism using agents and gamification. The aim is not to focus on a better algorithm for generating a list of recommendations, but to facilitate the consensus in the post-recommendation process so higher quality and more satisfactory choices can be made, and to enhance the tourists' experience during the whole process, from the planning to the tour itself. We intend to accomplish this by taking advantage of dialogue games using argumentation for the post-recommendation process, between intelligent agents modeled with the tourists' profile and context, and by introducing gaming components in the system that will encourage the tourists' interaction in a more appealing way. The tourists' profile and context will be used to provide more intelligent and personalized recommendations and notifications during the whole tour, to groups of any size. We believe the dialogue games between the agents will be a smarter way of explaining the recommendations to the tourists.

Travelling is an emotional experience [41] and therefore, personalization and gamification are becoming a crucial factor for the success of GRS in tourism. In fact, gamification techniques and personalized services will be a major trend for the future of tourism [42]. To motivate the tourists in planning the group tour and configure their profile and context, either implicitly or explicitly, we propose the use of gamification techniques like mini games, badges, trophies, and rankings of the best achievements. An AR avatar is also proposed to represent the tourist's agent and accompany him through the whole process, including during the tour, being responsible for providing personalized and contextual recommendations and push-notifications for the tourist's well-being.

The proposed approach will be thoroughly explained in our future work, and will include, among other tasks, the realization of questionnaires to different cultures in

---

[1] Or possibly another device, like Google Glasses®, but that is another chapter, not to be addressed in this work.

order to develop the model to correlate personality traits with (culture related) touristic preferences, and the development of mini games to implicitly acquire the tourists' personality, preferences and context. The gathered information will be used to model the agents representing the tourists and their avatar. The Social Network prototype will be developed for the post-recommendation choice process and to enable the tourists' online interaction. Intelligent push-notifications, recommendations, other mini games and tasks during the tour will be designed based on the tourists' profile and context. Experiments with real users will be conducted to test the viability of the proposed work and the users' satisfaction.

**Acknowledgements.** This work was supported by the GrouPlanner Project (POCI-01-0145-FEDER-29178) and by National Funds through the FCT – Fundação para a Ciência e a Tecnologia (Portuguese Foundation for Science and Technology) within the Projects UID/CEC/00 319/2019 and UID/EEA/00760/2019.

# References

1. Goldberg, D., Nichols, D., Oki, B.M., Terry, D.: Using collaborative filtering to weave an information tapestry. Commun. ACM **35**, 61–70 (1992)
2. Jameson, A., et al.: Human decision making and recommender systems. In: Ricci, F., Rokach, L., Shapira, B. (eds.) Recommender Systems Handbook, pp. 611–648. Springer, Boston, MA (2015). https://doi.org/10.1007/978-1-4899-7637-6_18
3. Resnick, P., Varian, H.R.: Recommender systems. Commun. ACM **40**, 56–58 (1997)
4. Nguyen, T.N., Ricci, F.: A chat-based group recommender system for tourism. Inf. Technol. Tourism **18**, 5–28 (2018)
5. Boratto, L., Carta, S.: State-of-the-art in group recommendation and new approaches for automatic identification of groups. In: Soro, A., Vargiu, E., Armano, G., Paddeu, G. (eds) Information Retrieval and Mining in Distributed Environments. Studies in Computational Intelligence, vol. 324, pp. 1–20. Springer, Heidelberg. https://doi.org/10.1007/978-3-642-16089-9_1
6. del Carmen Rodríguez-Hernández, M., Ilarri, S., Hermoso, R., Trillo-Lado, R.: Towards trajectory-based recommendations in museums: evaluation of strategies using mixed synthetic and real data. Procedia Comput. Sci. **113**, 234–239 (2017)
7. Lamsfus, C., Wang, D., Alzua-Sorzabal, A., Xiang, Z.: Going mobile: defining context for on-the-go travelers. J. Travel Res. **54**, 691–701 (2015)
8. Masthoff, J.: Group recommender systems: combining individual models. In: Ricci, F., Rokach, L., Shapira, B., Kantor, Paul B. (eds.) Recommender Systems Handbook, pp. 677–702. Springer, Boston, MA (2011). https://doi.org/10.1007/978-0-387-85820-3_21
9. Castro, J., Quesada, F.J., Palomares, I., Martinez, L.: A consensus-driven group recommender system. Int. J. Intell. Syst. **30**, 887–906 (2015)
10. Masthoff, J.: Group recommender systems: aggregation, satisfaction and group attributes. In: Ricci, F., Rokach, L., Shapira, B. (eds.) Recommender Systems Handbook, pp. 743–776. Springer, Boston, MA (2015). https://doi.org/10.1007/978-1-4899-7637-6_22
11. Delic, A., Masthoff, J.: Group recommender systems. In: Proceedings of the 26th Conference on User Modeling, Adaptation and Personalization, pp. 377–378. ACM (2018)

12. McCarthy, K., Salamó, M., Coyle, L., McGinty, L., Smyth, B., Nixon, P.: Group recommender systems: a critiquing based approach. In: Proceedings of the 11th International Conference on Intelligent User Interfaces, pp. 267–269. ACM (2006)

13. Nasolomampionona, R.F.: Profile of Chinese outbound tourists: characteristics and expenditures. Am. J. Tourism Manage. **3**, 17–31 (2014)

14. Ardissono, L., Goy, A., Petrone, G., Segnan, M., Torasso, P.: Intrigue: personalized recommendation of tourist attractions for desktop and hand held devices. Appl. Artif. Intell. **17**, 687–714 (2003)

15. McCarthy, K., McGinty, L., Smyth, B., Salamó, M.: Social interaction in the cats group recommender. In: Workshop on the Social Navigation and Community Based Adaptation Technologies (2006)

16. Garcia, I., Sebastia, L., Onaindia, E., Guzman, C.: A group recommender system for tourist activities. In: Di Noia, T., Buccafurri, F. (eds.) EC-Web 2009. LNCS, vol. 5692, pp. 26–37. Springer, Heidelberg (2009). https://doi.org/10.1007/978-3-642-03964-5_4

17. Jameson, A., Baldes, S., Kleinbauer, T.: Enhancing mutual awareness in group recommender systems. In: Proceedings of the IJCAI (2003)

18. van Setten, M., Pokraev, S., Koolwaaij, J.: Context-aware recommendations in the mobile tourist application COMPASS. In: De Bra, P.M.E., Nejdl, W. (eds.) AH 2004. LNCS, vol. 3137, pp. 235–244. Springer, Heidelberg (2004). https://doi.org/10.1007/978-3-540-27780-4_27

19. Marques, G., Respício, A., Afonso, A.P.: A mobile recommendation system supporting group collaborative decision making. Procedia Comput. Sci. **96**, 560–567 (2016)

20. Tintarev, N., Masthoff, J.: Explaining recommendations: design and evaluation. In: Ricci, F., Rokach, L., Shapira, B. (eds.) Recommender Systems Handbook, pp. 353–382. Springer, Boston, MA (2015). https://doi.org/10.1007/978-1-4899-7637-6_10

21. Herlocker, J.L., Konstan, J.A., Riedl, J.: Explaining collaborative filtering recommendations. In: Proceedings of the 2000 ACM Conference on Computer Supported Cooperative Work, pp. 241–250. ACM (2000)

22. McBurney, P., Parsons, S.: Dialogue games for agent argumentation. In: Simari, G., Rahwan, I. (eds) Argumentation in Artificial Intelligence, pp. 261–280 (2009). Springer, Boston. https://doi.org/10.1007/978-0-387-98197-0_13

23. Carneiro, J., Martinho, D., Marreiros, G., Jimenez, A., Novais, P.: Dynamic argumentation in UbiGDSS. Knowl. Inf. Syst. **55**, 633–669 (2018)

24. Carneiro, J., Alves, P., Marreiros, G., Novais, P.: A multi-agent system framework for dialogue games in the group decision-making context. In: Rocha, Á., Adeli, H., Reis, L.P., Costanzo, S. (eds.) WorldCIST'19 2019. AISC, vol. 930, pp. 437–447. Springer, Cham (2019). https://doi.org/10.1007/978-3-030-16181-1_41

25. Walton, D., Krabbe, E.C.: Commitment in Dialogue: Basic Concepts of Interpersonal Reasoning. SUNY press, New York (1995)

26. Carneiro, J., Martinho, D., Marreiros, G., Novais, P.: Arguing with behavior influence: a model for web-based group decision support systems. Int. J. Inf. Technol. Decis. Making 1–37 (2018)

27. Carneiro, J., Saraiva, P., Martinho, D., Marreiros, G., Novais, P.: Representing decision-makers using styles of behavior: an approach designed for group decision support systems. Cognit. Syst. Res. **47**, 109–132 (2018)

28. Villamizar, M., et al.: Evaluating the monolithic and the microservice architecture pattern to deploy web applications in the cloud. In: 2015 10th Computing Colombian Conference (10CCC), pp. 583–590. IEEE (2015)

29. Ricci, F.: Travel recommender systems. IEEE Intell. Syst. **17**, 55–57 (2002)

30. Schmidt-Belz, B., Nick, A., Poslad, S., Zipf, A.: Personalized and location-based mobile tourism services. In: Workshop on "Mobile Tourism Support Systems" in conjunction with Mobile HCI (2002)
31. Gavalas, D., Kenteris, M.: A web-based pervasive recommendation system for mobile tourist guides. Pers. Ubiquit. Comput. **15**, 759–770 (2011)
32. Tkalcic, M., Chen, L.: Personality and recommender systems. In: Ricci, F., Rokach, L., Shapira, B. (eds.) Recommender Systems Handbook, pp. 715–739. Springer, Boston, MA (2015). https://doi.org/10.1007/978-1-4899-7637-6_21
33. Feil, S., Kretzer, M., Werder, K., Maedche, A.: Using gamification to tackle the cold-start problem in recommender systems. In: Proceedings of the 19th ACM Conference on Computer Supported Cooperative Work and Social Computing Companion, pp. 253–256. ACM (2016)
34. de C.A. Ziesemer, A., Müller, L., Silveira, M.S.: Just rate it! gamification as part of recommendation. In: Kurosu, M. (ed.) HCI 2014. LNCS, vol. 8512, pp. 786–796. Springer, Cham (2014). https://doi.org/10.1007/978-3-319-07227-2_75
35. Friedman, H.S., Schustack, M.W.: Personality: Classic Theories and Modern Research. Allyn and Bacon, Boston (1999)
36. Hamari, J.: Transforming homo economics into homo ludens: a field experiment on gamification in a utilitarian peer-to-peer trading service. Electron. Commer. Res. Appl. **12**, 236–245 (2013)
37. Hamari, J., Koivisto, J., Sarsa, H.: Does gamification work?–a literature review of empirical studies on gamification. In: 2014 47th Hawaii International Conference on System Sciences (HICSS), pp. 3025–3034. IEEE (2014)
38. Hamari, J., Shernoff, D.J., Rowe, E., Coller, B., Asbell-Clarke, J., Edwards, T.: Challenging games help students learn: an empirical study on engagement, flow and immersion in game-based learning. Comput. Hum. Behav. **54**, 170–179 (2016)
39. Hakulinen, L., Auvinen, T., Korhonen, A.: The effect of achievement badges on students' behavior: an empirical study in a university-level computer science course. Int. J. Emerg. Technol. Learn. (iJET) **10**, 18–29 (2015)
40. Mortara, M., Catalano, C.E., Bellotti, F., Fiucci, G., Houry-Panchetti, M., Petridis, P.: Learning cultural heritage by serious games. J. Cult. Heritage **15**, 318–325 (2014)
41. Delic, A., Neidhardt, J., Nguyen, N., Ricci, F.: Research Methods for Group Recommender System. CEUR-WS (2016)
42. Xu, F., Tian, F., Buhalis, D., Weber, J., Zhang, H.: Tourists as mobile gamers: Gamification for tourism marketing. J. Travel Tourism Mark. **33**, 1124–1142 (2016)

# Orthogonal Properties of Asymmetric Neural Networks with Gabor Filters

Naohiro Ishii[1]([⊠]), Toshinori Deguchi[2], Masashi Kawaguchi[3],
Hiroshi Sasaki[4], and Tokuro Matsuo[1]

[1] Advanced Institute of Industrial Technology, Tokyo, Japan
nishii@acm.org, matsuo@aiit.ac.jp
[2] Gifu National College of Technology, Gifu, Japan
deguchi@gifu-nct.ac.jp
[3] Suzuka National College of Technology, Suzuka, Mie, Japan
masashi@elec.suzukact.ac.jp
[4] Fukui University of Technology, Fukui, Japan
hsasaki@fukui-ut.ac.jp

**Abstract.** Neural networks researches are developed for the recent machine learnings. To improve the performance of the neural networks, the biological inspired neural networks are often studied. Models for motion processing in the biological systems have been used, which consist of the symmetric networks with quadrature functions of Gabor filters. This paper proposes a model of the bio-inspired asymmetric neural networks, which shows excellent ability of the movement detection. The prominent features are the nonlinear characteristics as the squaring and rectification functions, which are observed in the retinal and visual cortex networks. In this paper, the proposed asymmetric network with Gabor filters and the conventional energy model are analyzed from the orthogonality characteristics. It is shown that the biological asymmetric network is effective for generating the orthogonality function using the network correlation computations. Further, the asymmetric networks with nonlinear characteristics are able to generate independent subspaces, which will be useful for the creation of features spaces and efficient computations in the learning.

**Keywords:** Asymmetric neural network · Gabor filter ·
Orthogonality analysis · Energy model · Linear and nonlinear pathways

## 1 Introduction

Neural networks currently plays an important role in the processing complex tasks for the visual perception and the learning. To estimate the visual motion, sensory biological information models have been studied [1, 2]. For the learning efficiently, an orthogonal projection [3] are studied in the convolutional neural networks. To improve faster and stable learning, an orthogonality regulation is introduced for deep neural networks [4]. Widrow et al. [5] showed that neurons nonlinear characteristics generate independent outputs for network learning. For the efficient deep learning, the orthogonalization in the weight matrix of neural networks are developed using optimization methods [6]. Further, the feature vectors from different classes are expected to be as

© Springer Nature Switzerland AG 2019
H. Pérez García et al. (Eds.): HAIS 2019, LNAI 11734, pp. 589–601, 2019.
https://doi.org/10.1007/978-3-030-29859-3_50

orthogonal as possible [7]. It is important to make clear the network structures how to generate the independence and orthogonality relations, which will make feature spaces, effectively. By using Gabor filters, a symmetric network model was developed for the motion detection, which is called energy model [2]. This paper develops a bio-inspired asymmetrical networks with the squaring and rectification functions [11–13], which shows excellent ability of the movement detection. This is an extended version of an earlier article [13]. It is shown that the asymmetric network with Gabor filters has orthogonal properties strongly under stimulus conditions, while the conventional energy symmetric network is weak in the orthogonality. Since the visual cortex is derived as the extended asymmetric networks with nonlinear functions, it is shown that their nonlinearities generate independent subspaces under stimulus conditions.

## 2    Asymmetric Neural Networks

### 2.1    Background of Asymmetric Neural Networks

In the biological neural networks, the structure of the network, is closely related to the functions of the network. Naka et al. [11] presented a simplified, but essential networks of catfish inner retina as shown in Fig. 1.

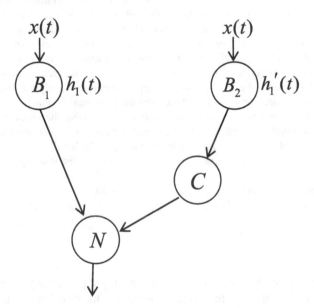

**Fig. 1.** Asymmetric network with linear and squaring nonlinear pathways

Visual perception is carried out firstly in the retinal neural network as the special processing between neurons. The asymmetric structure network with a quadratic nonlinearity in the catfish retinal network [11] is shown in Fig. 1, which composes of the pathway from the bipolar cell B to the amacrine cell N and that from the bipolar cell B, via the amacrine cell C to the N [10, 11]. Figure 1 shows a network which plays an important role in the movement perception as the fundamental network. It is shown that N cell response is realized by a linear filter, which is composed of a differentiation filter followed by a low-pass filter. Thus, the asymmetric network in Fig. 1 is composed of a linear pathway and a nonlinear pathway with the cell C, which works as a squaring function.

## 3   Orthogonal Properties of Asymmetric Networks

To make clear the asymmetric network with Gabor functions, orthogonality is computed, which shows independence characteristics of the asymmetric structure network without conventional maximizing independence in the quadratic model [2].

### 3.1   Orthogonality of Asymmetric Network Under the Stimulus Condition

The inner orthogonality under the constant value stimulus is computed in the asymmetric networks as shown in Fig. 2.

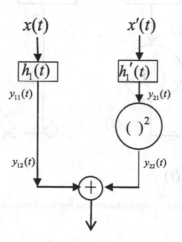

**Fig. 2.** Asymmetric network unit with Gabor filters

The vatiable $t$ in the Gabor filters is changed to $t'$, where by setting $\xi \triangleq 2\pi\omega$ in the Eq. (1), $t' = 2\pi\omega t = \xi t$ and $dt = dt/\xi$ hold. Then, Gabor filters become to the following equation.

$$G_s(t') = \frac{1}{\sqrt{2\pi}\sigma}e^{-\frac{t'^2}{2\sigma^2\xi^2}}sin(t') \text{ and } G_c(t') = \frac{1}{\sqrt{2\pi}\sigma}e^{-\frac{t'^2}{2\sigma^2\xi^2}}cos(t'). \tag{1}$$

The impulse response functions $h_1(t)$ and $h'_1(t)$ are replaced by $G_s(t')$ and $G_c(t')$ or vice versa. The outputs of these linear filters are given as follows,

$$y_{11}(t) = \int_0^\infty h_1(t')x(t - t')dt' \tag{2}$$

$$y_{21}(t) = \int_0^\infty h_1'(t')x(t - t')dt' \tag{3}$$

## 3.2    Orthogonality Between the Asymmetric Networks Units

We can compute orthogonality properties between the asymmetric networks units in Fig. 3. Orthogonal or non-orthogonal properties depend on the input stimulus to the networks.

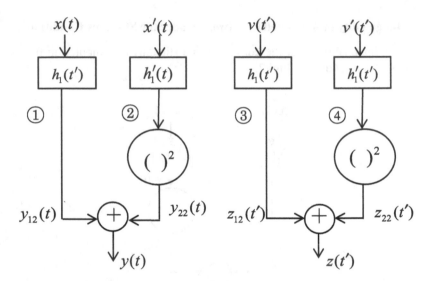

**Fig. 3.**  Orthogonality computations between asymmetric networks units

To verify the orthogonal properties between the asymmetric networks units, the white noise stimuli [9, 10] are schematically shown in Fig. 4. In the first row in Fig. 5, one white noise is a low pass filtered one with zero mean and its power $p$, which is shown in the circles only under the input variables $x(t)$, $x'(t)$, $v(t)$ and $v'(t')$. Similarly, in the second row in Fig. 4, the other white noise is a high pass filtered one with zero mean and its power $p'$, which is shown in the grayed circles under input variable $v'(t')$. The impulse response functions $h_1(t)$ and $h_1'(t)$ are replaced by the Gabor filters, $G_s(t')$ and $G_c(t')$ as shown in the Eq. (1). The stimulus with the high pass filtered noise is moved from the right to the left according to $(a)$, $(b)$, $(c)$, $(d)$ and $(e)$ in front of the visual space. Under the stimulus condition $(a)$ in Fig. 4, the correlation between outputs $y(t)$ and $z(t)$.

$$\int_{-\infty}^{\infty} y(t)z(t)\,dt$$

$$= \int_0^\infty \int_0^\infty h_1(\tau)h_1'(\sigma)\,d\tau d\sigma E[x(t-\tau)v(t-\sigma)]$$

$$+ \int_0^\infty \int_0^\infty \int_0^\infty h_1(\tau)h_1'(\sigma_1)h_1'(\sigma_2)E[x(t-\tau)v'(t-\sigma_1)v'(t-\sigma_2)]\,d\tau d\sigma_1 d\sigma_2$$

$$+ \int_0^\infty \int_0^\infty \int_0^\infty h_1(\sigma)h_1'(\tau_1)h_1'(\tau_2)E[v(t-\sigma)x'(t-\tau_1)x'(t-\tau_2)]\,d\sigma d\tau_1 d\tau_2 \qquad (4)$$

$$+ \int_0^\infty \int_0^\infty \int_0^\infty \int_0^\infty h_1'(\tau_1)h_1'(\tau_2)h_1'(\sigma_1)h_1'(\sigma_2)E[x'(t-\tau_1)x'(t-\tau_2)v'(t-\sigma_1)v'(t-\sigma_2)]\,d\tau_1 d\tau_2 d\sigma_1 d\sigma_2$$

$$= \{\int_0^\infty h_1(\tau)d\tau \int_0^\infty h_1'(\sigma)d\sigma\} \cdot p + 0 + 0 + 3p^2\{\int_0^\infty h_1'(\tau)d\tau\}^4$$

$$\quad ①③ \qquad\quad ①④ \quad ②③ \qquad ②④$$

where the first term of (4) of the path ways ① and ③, shown in ①, and the second and third terms are to be 0 by ①④ and ② ③. The fourth term is by ②④.

The terms ①③ and ②④ are not zero, because the following equations hold,

$$\int_0^\infty h_1(\tau)d\tau = \frac{1}{\sqrt{2\pi}\sigma} \int_0^\infty e^{-\frac{\tau^2}{2\sigma^2\xi^2}} \sin(\tau)d\tau = \frac{\xi}{\sqrt{\pi}} e^{-\frac{1}{2}\sigma^2\xi^2} \int_0^{\frac{1}{\sqrt{2}}\sigma\xi} e^{\tau^2}d\tau > 0 \qquad (5)$$

and

$$\int_0^\infty h_1'(\tau)d\tau = \frac{1}{\sqrt{2\pi}\sigma} \int_0^\infty e^{-\frac{\tau^2}{2\sigma^2\xi^2}} \cos(\tau)d\tau = \frac{\xi}{2} e^{-\frac{1}{2}\sigma^2\xi^2} > 0, \qquad (6)$$

where $\xi$ is the center frequency of the Gabor filter.

Thus, since two pathways are zero in the correlation, while other two pathways (25% and 25%) are non-zero, the orthogonality becomes 50% for the stimuli $(a)$ in Fig. 4.

Under the stimulus condition $(b)$ in Fig. 4, the correlation between outputs $y(t)$ and $z(t)$ between the asymmetrical networks in Fig. 3, is computed as the Eq. (7),

$$\int_{-\infty}^{\infty} y(t)z(t)dt = p\{\int_0^\infty h_1(\tau)d\tau \int_0^\infty h_1'(\sigma)d\sigma\} + 0 + 0 + 0$$

$$\quad ①③ \quad ①④ \quad ②③ \quad ②④ \qquad\qquad (7)$$

Since three pathways are zero in the correlation, while the first pathway is non-zero, (25%), the orthogonality becomes 75% for the stimuli $(b)$ in Fig. 4.

Under the stimulus condition $(c)$ in Fig. 4, the correlation between outputs $y(t)$ and $z(t)$ between the networks in Fig. 3, is computed as the Eq. (8),

$$\int_{-\infty}^{\infty} y(t)z(t)dt = 0 + 0 + 0 + pp' \cdot \{\int_0^\infty h_1'(\tau)d\tau\}^4$$

$$\quad ①③ \quad ①④ \quad ②③ \qquad ②④ \qquad\qquad (8)$$

$$x(t) \qquad x'(t) \qquad v(t') \qquad v'(t')$$

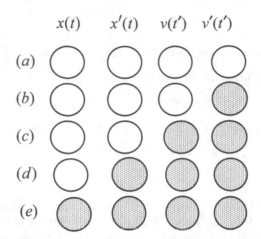

**Fig. 4.** White noise stimuli for checking orthogonality and non-orthogonality

Since three pathways are zero in the correlation (8), and the fourth pathway is non-Zero (8%), the orthogonality becomes 92% for the stimuli $(b)$ in Fig. 4.

Under the stimulus condition $(d)$ in Fig. 4, the correlation between outputs $y(t)$ and $z(t)$ between the networks in Fig. 3, is computed as the Eq. (9)

$$\int_{-\infty}^{\infty} y(t)z(t)dt = 0 + 0 + 0 + 3p^2 \cdot \{\int_0^\infty h_1'(\tau)d\tau\}^4 \tag{9}$$
$$①③ \quad ①④ \quad ②③ \qquad ②④$$

Since three pathways are zero in the correlation, while the fourth pathway is non-zero, (25%), the orthogonality becomes 75% for the stimuli $(d)$ in Fig. 4.

Under the stimulus condition $(e)$ in Fig. 4, the correlation between outputs $y(t)$ and $z(t)$ between the networks in Fig. 3, is computed as the Eq. (10)

$$\int_{-\infty}^{\infty} y(t)z(t)dt = = \{\int_0^\infty h_1(\tau)d\tau \int_0^\infty h_1'(\sigma)d\sigma\} \cdot p' + 0 + 0 + 3p'^2\{\int_0^\infty h_1'(\tau)d\tau\}^4 \tag{10}$$
$$①③ \qquad\qquad ①④ \quad ②③ \qquad ②④$$

Since two pathways are zero in the correlation, while the first and the fourth pathways are non-zero, (25% and 25%), the orthogonality becomes 50% for the stimuli $(e)$ in Fig. 4.

## 4   Orthogonal Properties of Conventional Energy Model

A symmetric network with Gabor filter were proposed by Adelson and Bergen [2] as the energy model of the perception of the visual motion. Their model is extensively applied to the motion detection and physiological models as a fundamental model of networks. Further, their models are developed for feature and learning networks.

## 4.1 Orthogonality Under the Stimulus Condition

Symmetric network is shown under the conditions of the stimulus, which is called energy model [2].

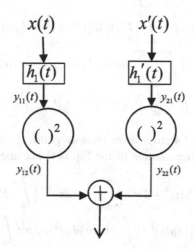

**Fig. 5.** Symmetric network unit, which is called energy model

## 4.2 Orthogonality Between the Energy Model Units

To verify the orthogonal properties between the conventional symmetric network units, the white noise stimuli are schematically shown in Fig. 6.

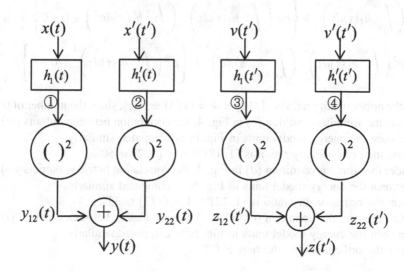

**Fig. 6.** Orthogonality computations between energy model units

Under the stimulus condition $(a)$ in Fig. 4, the correlation between outputs $y(t)$ and $z(t)$ between the energy model units in Fig. 6, is computed as follows,

$$
\int_{-\infty}^{\infty} y(t)z(t)dt = \int_{-\infty}^{\infty} dt\{\int_0^{\infty}\int_0^{\infty} h_1(\tau_1)h_1(\tau_2)x(t-\tau_1)x(t-\tau_2)d\tau_1 d\tau_2
$$
$$
+ \int_0^{\infty}\int_0^{\infty} h_1'(\tau_1')h_1'(\tau_2')x'(t-\tau_1')x'(t-\tau_2')d\tau_1' d\tau_2'\}
$$
$$
\cdot\{\int_0^{\infty}\int_0^{\infty} h_1(\sigma_1)h_1(\sigma_2)v(t-\sigma_1)v(t-\sigma_2)d\sigma_1 d\sigma_2
$$
$$
+ \int_0^{\infty}\int_0^{\infty} h_1'(\sigma_1')h_1'(\sigma_2')v'(t-\sigma_1')v'(t-\sigma_2')d\sigma_1' d\sigma_2'\}
\tag{11}
$$

The Eq. (11) consists of 4 cross-terms (4cross- pathways, ①③, ①④, ②③, ②④ in Fig. 6) computations. The solution of the Eq. (11) becomes

$$
3p^2(\int_0^{\infty} h_1(\tau)d\tau)^4 + 3p^2(\int_0^{\infty} h_1(\tau)d\tau)^2 \cdot (\int_0^{\infty} h_1'(\sigma)d\sigma)^2 +
$$
$$
3p^2(\int_0^{\infty} h_1'(\tau)d\tau)^2 \cdot (\int_0^{\infty} h_1(\sigma)d\sigma)^2 + 3p^2(\int_0^{\infty} h_1'(\tau)d\tau)^4
\tag{12}
$$

The solution (13) shows these four terms are not zero, since the integral of the respective impulse functions of Gabor filters, are proved to be not zero by the Eqs. (5) and (6). Thus, the orthogonality ratio here is 0%.

Under the stimulus condition $(b)$ in Fig. 4, the correlation between outputs $y(t)$ and $z(t)$ between the energy model units in Fig. 6, is computed as follows,

$$
3p^2\left(\int_0^{\infty} h_1(\tau)d\tau\right)^4 + \left\{pp'\left(\int_0^{\infty} h_1(\tau)d\tau\right)^2 \cdot \left(\int_0^{\infty} h_1'(\sigma)d\sigma^2\right) + 0 + 0\right\} +
$$
$$
3p^2\left(\int_0^{\infty} h_1'(\tau)d\tau\right)^2 \cdot \left(\int_0^{\infty} h_1(\sigma)d\sigma\right)^2 + \left\{p'^2\left(\int_0^{\infty} h_1'(\tau)d\tau\right)^4 + 0 + 0\right\},
\tag{13}
$$

where the orthogonality ratio is $(1/12) \times 4 = (1/3) \simeq 33\%$, since the number of 0 is 4.

Under the stimulus condition $(c)$ in Fig. 4, the correlation between outputs $y(t)$ and $z(t)$ between the energy model units in Fig. 6, is computed similarly.

Then, the orthogonality ration is $(1/12) \times 6 = (1/2) = 50\%$.

Under the stimulus condition $(d)$ in Fig. 4, the correlation between outputs $y(t)$ and $z(t)$ between the energy model units in Fig. 6, is computed similarly.

Then, the orthogonality ratio is $(1/12) \times 4 = (1/3) \simeq 33\%$.

Under the stimulus condition $(e)$ in Fig. 4, the correlation between outputs $y(t)$ and $z(t)$ between the energy model units in Fig. 6, is computed similarly.

Then, the orthogonality ratio here is 0%.

### 4.3 Comparison of Orthogonal Properties Between the Asymmetric Network and the Energy Model

We can compare the orthogonal properties between the asymmetric network and the conventional energy model by the computed values in the previous section. Under the different stimulus conditions as shown in Fig. 4, the orthogonality ratios are summarized as shown in Fig. 7.

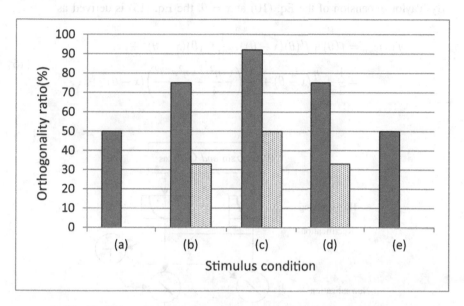

**Fig. 7.** Comparison of orthogonality ratio under stimulus conditions

In Fig. 7, the filled bar shows the orthogonality ratio value by the asymmetric neural networks in Fig. 3, while the dotted bar shows by the symmetric network called energy model in Fig. 6 under the same stimulus conditions. The asymmetric network shows higher orthogonality ratio, compared with the conventional symmetric network.

## 5 Application of Asymmetric Networks to Bio-inspired Neural Networks

Figure 8 is a connected network model of V1 followed by MT [8], in which V1 is the front part of the total network, while MT is the rear part of it. Figure 8 is transformed to the approximated one as follows.

## 5.1  Combinations of Orthogonal Pairs to Generate Independent Subspaces

The half-wave rectification in Fig. 8 is approximated in the following equation.

$$f(x) = \frac{1}{1 + e^{-\eta(x-\theta)}} \tag{14}$$

By Taylor expansion of the Eq. (10) at $x = \theta$, the Eq. (15) is derived as

$$
\begin{aligned}
f(x)_{x=\theta} &= f(\theta) + f'(\theta)(x - \theta) + \frac{1}{2!}f''(\theta)(x - \theta)^2 + \dots \\
&= \frac{1}{2} + \frac{\eta}{4}(x - \theta) + \frac{1}{2!}\left(-\frac{\eta^2}{4} + \frac{\eta^2 e^{-\eta\theta}}{2}\right)(x - \theta)^2 + \dots
\end{aligned}
\tag{15}
$$

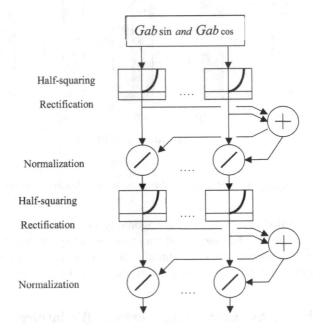

**Fig. 8.** Model of neural network of brain cortex V1 followed by MT [7]

In Fig. 9, the nonlinear terms, $x^2, x^3, x^4, \dots$ are generated from the Eq. (15). Thus, the combinations of Gabor function pairs ($G_{ab\,sin}$, $G_{ab\,cos}$) are generated according to the Eq. (15) as shown in Fig. 8, in which the transformed network consists of two layers. The characteristics of the extended asymmetric network have pathways with higher order nonlinearities. In Fig. 9, linear notation $L$ shows term, $Ae^{-\frac{t'^2}{2\sigma^2 \xi^2}} sin(t')$ or $Ae^{-\frac{t'^2}{2\sigma^2 \xi^2}} cos(t')$, while $S$ shows their doubles. Selective independent sub-spaces are generated in the layers in Fig. 9. The combination pairs of the Gabor filters are shown in (16) by approximated computations. Under the same stimulus condition in Fig. 4, the orthogonality is computed between items connecting solid lines in Fig. 10. Here, the connecting line shows over 50% orthogonality ratio in Fig. 10.

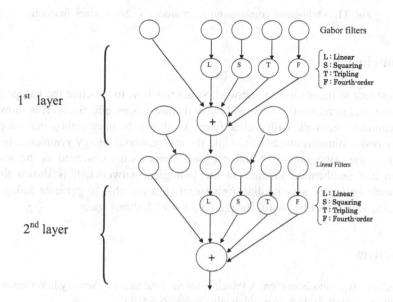

**Fig. 9.** A transformed network model for the layered network for one pathway in Fig. 6.

$$Ae^{-\frac{t'^2}{2\sigma^2\zeta^2}}sin(t') \quad Ae^{-\frac{t'^2}{2\sigma^2\zeta^2}}cos(t') \quad A^2(e^{-\frac{t'^2}{2\sigma^2\zeta^2}})^2cos^2(t') \quad A^3(e^{-\frac{t'^2}{2\sigma^2\zeta^2}})^3cos^3(t')... \quad (16)$$

**Fig. 10.** Orthogonal combination pairs among Gabor sine and cosines functions

Since $Ae^{-\frac{t'^2}{2\sigma^2\zeta^2}}sin(t')$ and $Ae^{-\frac{t'^2}{2\sigma^2\zeta^2}}cos(t')$ are orthogonal, thus they have independent relations which are shown in the solid line in Fig. 10. Similarly, $Ae^{-\frac{t'^2}{2\sigma^2\zeta^2}}sin(t')$ and $A^2(e^{-\frac{t'^2}{2\sigma^2\zeta^2}})^2cos^2(t')$ are orthogonal, thus they have independent relation. Then, similar independent relations hold between the left $cos(t')$ and the right $sin(t')$, $sin^3(t')$.. terms, respectively in the Eq. (17) in Fig. 11, which are shown in the solid line.

Similar independent relations hold between $sin^3(t')$ and $cos(t')$, $cos^2(t')$, $cos^3(t')$... terms, respectively. Thus, selective independent subspaces are generated by the combination pairs of sine and cosine terms in Fig. 9. In the 2$^{nd}$ layer in Fig. 9, combination pairs are increased by those generated in the 1$^{st}$ layer and created newly in the 2$^{nd}$ layer.

$$\underline{Ae^{-\frac{t'^2}{2\sigma^2\varsigma^2}}cos(t')} \; \underline{Ae^{-\frac{t'^2}{2\sigma^2\varsigma^2}}sin(t')} \; \underline{A^3(e^{-\frac{t'^2}{2\sigma^2\varsigma^2}})^3 sin^3(t')} \; \underline{A^5(e^{-\frac{t'^2}{2\sigma^2\varsigma^2}})^5 sin^5(t')...} \qquad (17)$$

**Fig. 11.** Orthogonal combination pairs among Gabor cosines functions

## 6  Conclusion

It is important to make clear the network structures how to generate the independence and orthogonality relations, which will make feature spaces, effectively. It is shown that the asymmetric network with Gabor filters is shown to have orthogonal properties strongly under stimulus conditions, while the conventional energy symmetric network is weak in the orthogonality. Nonlinear characteristics are observed as the squaring function and rectification function in the biological networks. It is shown that the asymmetric networks with nonlinear characteristics are able to generate independent subspaces, which will be useful for the creation of features spaces.

## References

1.  Reichard, W.: Autocorrelation, A Principle for the Evaluation of Sensory Information by the Central Nervous System. Rosenblith edn. Wiley, NY (1961)
2.  Adelson, E.H., Bergen, J.R.: Spatiotemporal energy models for the perception of motion. J. Optical Soc. Am. A **2**, 284–299 (1985)
3.  Pan, H., Jiang, H.: Learning convolutional neural networks using hybrid orthogonal projection and estimation. In: ACML 2017, Proceedings of the Machine Learning Research, vol. 77, 1–16 (2017)
4.  Bansal, N., Chen, X., Wang, Z.: Can we gain more from orthogonality regularization in training deep CNNs? In: 32nd International Conference on Neural Information Processing System, NIPS 2018, pp. 4266–4276 (2018)
5.  Widrow, B., Greenblatt, A., Kim, Y., Park, D.: The *No-Prop* algorithm: a new learning algorithm for multilayer neural networks. Neural Netw. **37**, 182–188 (2013)
6.  Huang, L., Liu, X., Lang, B., Yu, A.W., Wang, Y., Li, B.: Orthogonal weight normalization: solution to optimization over multiple dependent stiefel manifolds in deep neural networks. In: 32nd AAAI Conference on AI, AAAI-18, pp. 3271–3278 (2018)
7.  Shi, W., Gong, Y., Cheng, D., Tao, X., Zheng, N.: Entropy orthogonality based deep discriminative feature learning for object recognition. Pattern Recogn. **81**, 71–80 (2018)
8.  Simoncelli, E.P., Heeger, D.J.: A model of neuronal responses in visual area MT. Vision. Res. **38**, 743–761 (1996)
9.  Marmarelis, P.Z., Marmarelis, V.Z.: Analysis of Physiological Systems – The White Noise Approach. Plenum Press, New York (1978)
10. Sakuranaga, M., Naka, K.-I.: Signal Transmission in the Catfish Retina.III. Transmission to Type-C Cell. J. Neurophysiol. **53**(2), 411–428 (1985)
11. Naka, K.-I., Sakai, H.M., Ishii, N.: Generation of transformation of second order nonlinearity in catfish retina. Ann. Biomed. Eng. **16**, 53–64 (1988)

12. Ishii, N., Deguchi, T., Kawaguchi, M., Sasaki, H.: Motion detection in asymmetric neural networks. In: Cheng, L., Liu, Q., Ronzhin, A. (eds.) ISNN 2016. LNCS, vol. 9719, pp. 409–417. Springer, Cham (2016). https://doi.org/10.1007/978-3-319-40663-3_47
13. Ishii, N., Deguchi, T., Kawaguchi, M., Sasaki, H.: Distinctive features of asymmetric neural networks with gabor filters. In: de Cos Juez, F., et al. (eds.) HAIS 2018. LNCS, vol. 10870, pp. 185–196. Springer, Cham (2018). https://doi.org/10.1007/978-3-319-92639-1_16

# Deep CNN-Based Recognition of JSL Finger Spelling

Nam Tu Nguyen[2], Shinji Sako[2], and Bogdan Kwolek[1(✉)]

[1] AGH University of Science and Technology, 30 Mickiewicza Av.,
30-059 Krakow, Poland
bkw@agh.edu.pl
[2] Frontier Research Institute for Information Science, Nagoya Institute
of Technology, Gokiso-cho, Showa-ku, Nagoya 466-8555, Japan
sako@msp.nitech.ac.jp
http://home.agh.edu.pl/~bkw/contact.html

**Abstract.** In this paper, we present a framework for recognition of static finger spelling in Japanese Sign Language on RGB images. The finger spelled signs were recognized by an ensemble consisting of a ResNet-based convolutional neural network and two ResNet quaternion convolutional neural networks. A 3D articulated hand model has been used to generate synthetic finger spellings and to extend a dataset consisting of real hand gestures. Twelve different gesture realizations were prepared for each of 41 signs. Ten images have been rendered for each realization through interpolations between the starting and end poses. Experimental results demonstrate that owing to sufficient amount of training data a high recognition rate can be attained on images from a single RGB camera. Results achieved by the ResNet quaternion convolutional neural network are better than results obtained by the ResNet CNN. The best recognition results were achieved by the ensemble. The JSL-rend dataset is available for download.

## 1 Introduction

Hand detection, tracking and recognition of hand gestures are important research topics with a large potential for application in human-machine-communication, virtual reality [1], entertainment, robotics [2–4], medicine, and assistive technologies for the handicapped and the elderly [5]. Hand gesture is one of the most intuitive and flexible ways to attain user-friendly man-machine communication [6,7]. Touch-less gesture recognition is introduced in automotive user interfaces as they possess a strong potential to enhance safety and driving comfort [8]. Gesture recognition on images acquired by a single color camera is very useful, yet difficult task, due to occlusions, variations in gesture expressions, differences in hand anatomy and appearance, etc. In order to achieve robust, on-line gesture recognition many challenging sub-tasks such as detection of hands and other body parts, motion modeling and tracking, pattern recognition and classification should be advanced and improved. Despite the huge efforts of many research

© Springer Nature Switzerland AG 2019
H. Pérez García et al. (Eds.): HAIS 2019, LNAI 11734, pp. 602–613, 2019.
https://doi.org/10.1007/978-3-030-29859-3_51

teams [7,9], there are still challenges to be addressed for achieving recognition performance required for real-life applications.

In the last decade, several approaches to recognition of static gestures on color images have been proposed [10]. Despite the encouraging progress in learning deep Convolutional Neural Networks (CNNs), a recent survey devoted to hand gesture recognition [10] reports only one significant approach, i.e. [11]. In [12], a CNN has been designed and trained to classify six hand gestures to control robots on the basis of colored gloves. More recently a CNN has been implemented in Theano framework and then applied on the Nao humanoid robot [13]. In recently published work [14], a CNN has been learned on one million of data samples to classify sign characters. However, only a subset of data, namely 3361 manually labeled frames into 45 classes has been made publicly available. One of the obstacles to using the deep CNNs on a larger scale is lack of properly aligned datasets of sufficient size as well as shortage of robust real-time hand detectors.

Finger spelling is a form, and frequently an integral part of sign language, where each sign corresponds to a word of the alphabet. Tabata and Kuroda [15] proposed a Stringlove system to recognize hand shapes for finger spelling in Japanese Sign Language (JSL). The system is based on a custom-made glove that is equipped with sensors to capture finger features. The glove is build on nine contact sensors and 24 inductcoders that jointly estimate several features like: thumb and wrist rotations, adduction/abduction angles of fingers, the joint flexion/extension of fingers and the contact positions among fingertips of the fingers. In [16], a modified kind of the shape matrix being capable of capturing salience of the finger spelling postures on the basis of precise sampling of regions and contours has been proposed. In [17], recognition of JSL finger spellings was realized on the basis of embedding calculated by multiple Siamese CNNs. A dataset consisting of real images for recognition of real images has been introduced in discusses work. It contains 5311 training images and 579 test images of size 64 × 64.

In this work, we present a framework for recognition of static finger spelling in Japanese Sign Language on RGB images. The finger spelled signs were recognized by an ensemble consisting of a ResNet-based convolutional neural network and two ResNet quaternion convolutional neural networks. A 3D articulated hand model has been utilized to render realistic finger spellings and to extend the JSL dataset for training a deep network. Twelve different gesture realizations were prepared for each of 41 signs. Ten images have been rendered for each realization through interpolations between the starting and end poses. The JSL-rend dataset is available for download.

## 2    Relevant Work

In [13], a multichannel convolutional neural network for hand posture recognition has been presented. The model employed a cubic kernel to enhance the features for the classification. The method has been evaluated on Nao humanoid robot. In [18], a glove was used to provide contour representation of a gesture. A neural network has been trained on a dataset with 100 images per gesture. A 90% classification accuracy has been reported in the discussed work. In a recently proposed algorithm [19] for estimation of hand orientation from 2D monocular images a staged probabilistic regressor (SPORE) has been proposed. It has been demonstrated experimentally that simultaneously learning hand orientation and pose significantly improves the pose classification performance on 2D monocular images. In [20] an algorithm for detection and extraction of shape for static fingerspelling recognition on the basis boundary tracing and chain code has been proposed. The system has been designed for ASL language recognition. On images of size $320 \times 240$ acquired by a webcam and some image collection obtained from freely available resources the recognition accuracy was 97.75% and 96.48% for alphabet characters and numbers, respectively.

## 3 Japanese Sign Language

Japanese Sign Language, also known by the acronym JSL, is a visual sign language in Japan. As other sign languages, the JSL comprises words, or signs, and the grammar with which they are bonded together. The Japanese Sign Language syllabary is a system of manual kana utilized as part of the JSL. In general, fingerspelling is used mostly for foreign words and last names. The JSL finger spelling is performed by five fingers of the hand and direction it points. For example, the signs na, ni, ha are all expressed with the first two fingers of the hand extended straight, but for the sign na the fingers point down, for ni across the body, and for ha toward the partner or audience. The signs for te and ho are both made with open flat hand, but in te the palm faces the viewer, and in ho it faces away. These and many other factors make recognition of JSL fingerspelling on the basis of a single camera a difficult problem.

Most of finger spellings are expressed through static postures, but some of them are dynamic postures. In addition, dullness, half dullness, and long sound are represented by dynamic postures. In this study, we only focused on static finger spellings. There are 41 static finger spellings in JSL. Figure 3 depicts the considered static hand signs. In the following columns and rows (from left to right and from top to bottom), the words that make up the Hiragana alphabet are visualized.

## 4 Methods

At the beginning we explain fingerspelling modeling and rendering. In the next subsection we present our 3D-model based dataset for JSL recognition. Finally, we discuss the selected network architecture.

### 4.1 Fingerspelling Modeling and Rendering

A 3D hand model has been prepared in Blender software. The articulated hand model consists of 21 bones and has five control bones, see Fig. 1. The blue lines illustrate the control bones. The bones in the skeleton form a structure of rigid bodies that are connected together by joints with one or more degrees of freedom. The model has 37 degrees of freedom (DOF), the 3D mesh contains 528 vertices and is composed of 1036 triangles. The 26 element skeleton (armature) is bound to such a 3D mesh. The root joint is located in the wrist. The model has been exported to MD5 data format to perform animations in external programs.

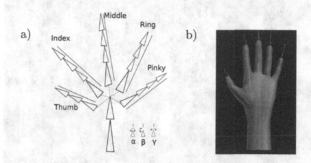

**Fig. 1.** 26 DOF skinned hand model. Kinematic tree (a), digital skeleton bound to the 3D mesh (b). (Color figure online)

The 3D articulated hand model has been used to generate synthetic finger spellings and to extend our dataset consisting of real hand gestures. Twelve different gesture realizations were prepared for each of 41 signs. Ten images have been rendered for each realization through interpolations between the starting and end poses. Figure 2 depicts the rendered images for the sign 'a'. For each starting gesture a final gesture has been created and ten interpolated images were rendered among them.

**Fig. 2.** Example realizations of JSL gesture 'a'.

## 4.2   3D-Model Based Dataset for JSL Recognition

The dataset consists of 5109 color images of size 64 × 64. For each JSL finger-spelling there are twelve different realizations. This means that gestures differ in hand postures to express different realizations of the gesture by different persons.

For each realization of the gesture we modeled starting and final posture and then interpolated hand postures between them. The number of images generated on the basis of the interpolation is equal to eight. The dataset has been stored in .mat files and can be easy imported to python. The whole JSL-rend dataset is freely available at: http://home.agh.edu.pl/~bkw/data/hais.

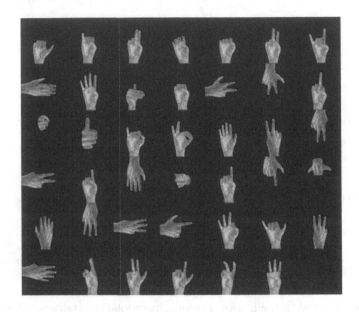

**Fig. 3.** Examples of rendered Hiragana signs.

### 4.3   ResNet Convolutional Neural Network

Network depth is of crucial importance to increase the recognition performance, but deeper networks are more difficult to train. However, increasing network depth by simply stacking layers sequentially might lead to difficulties in training. Deep networks are hard to train because of the vanishing gradient problem. As the gradient is back-propagated to earlier layers, subsequent multiplications may make the gradient infinitively small. Consequently, when the network is deeper, its performance gets saturated or even starts degrading rapidly. [21] introduced residual networks (ResNets), which provide an important contribution to training very deep networks. While deeper networks have been found to get better generalization, especially on large datasets, large network depths lead to training difficulties, even with batch normalization and the correct initialization, and generalization begins to level off. The residual learning framework simplifies the training of such networks, and enables them to be substantially deeper, which leads to improved performance. The residual networks are much deeper in comparison to their ordinary counterparts, yet they need a similar number of parameters. The general idea of the residual network is use blocks that re-route the input, and add to the concept learned from the previous layer. The constituent building unit of the ResNet architecture is the ResNet block. A deeper network can be built by simply repeating this module, i.e the smaller network. A desired underlying mapping $H(x)$ can be approximated by a few stacked nonlinear layers, so it can also be achieved through underlying mapping $F(x) = H(x) - x$. As a result, it is possible to reformulate it as $H(x) = F(x) + x$, which consists of the Residual Function $F(x)$ and input $x$. The connection of the input to the output is called a skip connection or identity mapping. The central idea is that if multiple nonlinear layers can approximate the complicated function $H(x)$, then it is possible to approximate the residual function $F(x)$. Therefore the stacked layers are not used to fit $H(x)$, instead these layers approximate the residual function $F(x)$.

### 4.4   Quaternion Convolutional Neural Network

Convolutional neural networks have demonstrated excellent abilities to capture the high-level relations that take place between neighbor features, such as shape and edges on an image. In order to enhance internal dependencies within the features a quaternion convolutional neural network (QCNN) has been proposed recently [22]. Let $\gamma_{ab}^l$ and $S_{ab}^l$ denote the quaternion output and the preactivation quaternion output at layer $l$ and at the indexes $(a, b)$ of the feature map, and $w$ be a quaternion-valued weight filter map of size $K \times K$. The convolution can be expressed as:

$$\gamma_{ab}^l = \alpha(S_{ab}^l) \tag{1}$$

where

$$S_{ab}^l = \sum_{c=0}^{K-1} \sum_{d=0}^{K-1} w^l \otimes \gamma_{(a+c)(b+d)}^{l-1} \tag{2}$$

where $\alpha$ stands for quaternion split activation function [23] defined in the following manner:

$$\alpha(Q) = f(r) + f(x)\mathbf{i} + f(y)\mathbf{j} + f(z)\mathbf{k} \tag{3}$$

where $f$ is related to any standard activation function. A derivation of the back-propagation algorithm for quaternion neural networks has been presented in [24].

Recently, in [25] a quaternion convolutional neural network for color image processing has been introduced. A color image is represented in the quaternion domain as a quaternion matrix. While the conventional real-valued convolution is only able to execute scaling transformation on the input, the quaternion convolution provides the scaling and the rotation of input in the color space, which carries out more structural representation of color information [25]. Since the QCNN enforces an implicit regularizer on the architecture of network, more complicated relationships across different channels are achieved in the training. Thus, better learning results with fewer parameters compared with real-valued CNNs can be accomplished.

## 4.5   Ensemble of CNNs

Constructing ensembles of models consists in combining predictions from multiple statistical models to form one final prediction. Essentially, ensembles tend to yield better results when there is a significant diversity among the models [26]. There are many different types of ensembles, and stacking is one of them. Stacking involves training a learning algorithm to combine the predictions of several other learning algorithms. At the beginning all of the other algorithms are trained using the available data. Afterwards, a combiner algorithm is trained to make a final prediction using all the predictions of the other algorithms. The simplest form of stacking involves taking an average of outputs of models in the ensemble. Since averaging doesn't need any parameters, there is no need to train such an ensemble and only its models should be trained in advance.

# 5   Experimental Results and Discussion

Experimental evaluations have been performed on JSL finger spelling dataset [17] and our JSL-rend dataset. All experiments were carried out on color RGB images of size $64 \times 64$. Altogether 10420 images for training and 579 test images were collected and employed in the evaluations consisting in recognition of 41 JSL static hand gestures. The performance has been assessed in a person independent scenario, wherein gestures performed by some subject have been used for testing, whereas gestures performed by remaining subjects have been employed for training of the models. The images rendered on the basis of the 3D-model were concatenated with the training subset of the JSL dataset. Since the JSL dataset is somewhat unbalanced, this way we balanced the dataset as well as extended its size.

At the beginning of the experiments we investigated various approaches to rendering of the images representing the JSL gestures. We exported the models representing the gesture to md5 data format and then used our parser in order to import the mesh and the animation data for OpenGL-based rendering. The mesh data and animation data in the discussed format are separated in distinct files. One of the advantages of the md5 data format is that data is stored in ASCII files and are human readable. The rotations are represented by quaternions. Through modifying the values of the parameters stored in the plain text files it is possible to configure the skeleton of the model into the required poses, and then render the model. In our approach, after parsing the skeleton and animation data, in every frame the 3D hand has been rotated about randomly generated angles to simulate observing the hand from different camera views. Afterwards, we rendered the hands in Blender without light effects. Finally, in order to obtain more photorealistic images we illuminated the hands using virtual lights, see Fig. 4.

After rendering the images the JSL-rend dataset has been created and then used in the experiments. It turned out that owing to extending the JSL dataset about the rendered images a more advanced convolutional neural networks in

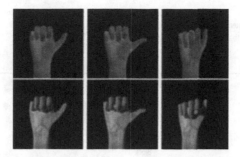

**Fig. 4.** Example images of 01_a sign, no lighting (top row), with lighting (bottom row).

comparison to networks used in [17] can be trained without the risk of under-fitting/overfitting. We experimented with various neural networks both trained from scratch and fine-tuned deep CNNs. At this stage of experiments we focused on pre-trained VGG-16 CNNs. The last fully connected layers, which are viewed as classification layers, were reset and a smaller learning rate has been applied to the pretrained layers. As the VGG model expects the input images of size $224 \times 224 \times 3$, the images were resized to the above mentioned size. The evaluations of the fine-tuned models were performed both on gray and RGB images.

Next, we implemented a ResNet based convolutional neural network consisting of three ResNet-blocks, see Fig. 5. The neural network has been trained on RGB images of size $64 \times 64 \times 3$. Each model was trained using the Adam optimizer ($lr = 0.001$, beta_1 $= 0.9$, beta_2 $= 0.999$, epsilon $=$ 1e-08) and categorical cross-entropy loss, with a small learning rate. The learning rate was scheduled to be reduced after 20, 30, 40, 50 epochs. The values of the hyper-parameters were selected empirically. The number of the trainable parameters is equal to 282 953.

Afterwards, we implemented a QCNN, and then in the ResNet we substituted the convolutional blocks by the quaternion-based convolutional blocks. The neural network has been trained on RGB images of size $64 \times 64 \times 3$ using the same Adam optimizer as well as parameters. The number of trainable parameters is equal to 577 945. Finally, an ensemble consisting of one ResNet and two independently trained QCNNs has been constructed. The output is determined by voting.

Table 1 presents the classification performance that has been obtained on test part of JSL dataset using ResNet neural network that has been learned on training subset of the JSL dataset. During learning an-online data augmentation has been executed. As we can notice, considerable improvement of classification accuracy in comparison to classification accuracy in [17] has been achieved.

Afterwards, we evaluated the classification performance on the test part of JSL dataset using ResNet neural network that has been trained on concatenated JSL dataset (training part) and introduced in this work the JSL-rend dataset. As we can observe, almost 3% improvement in classification accuracy

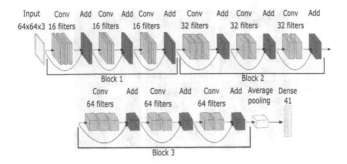

**Fig. 5.** Flowchart of the ResNet used for JSL fingerspelling classification.

**Table 1.** Classification performance in performer independent experiment.

|  | Accuracy | Precision | Recall | F1-score |
|---|---|---|---|---|
| CNN - JSL [17] | 0.753 | 0.792 | 0.754 | 0.730 |
| CNN - JSL | 0.881 | 0.893 | 0.884 | 0.865 |
| CNN - JSL + JSL-rend | 0.910 | 0.931 | 0.912 | 0.900 |
| QCNN - JSL + JSL-rend | 0.921 | 0.931 | 0.922 | 0.914 |
| Ensemble - JSL + JSL-rend | 0.941 | 0.947 | 0.941 | 0.931 |

has been obtained owing to use of the JSL-rend dataset. The classification accuracy achieved by the quaternion ResNet is better about 1% in comparison to ResNet classification accuracy. The ensemble improves the classification performance more than 1%. Figure 6 depicts the confusion matrix of the final classifier.

As we can observe, the hand gesture 04_e has the lowest classification performance. All four gestures belonging to this class were classified as 11_sa gestures. Figure 7 depicts representative images from both classes. As we can notice the hand shapes in the considered gestures are quite similar, which may explain the reason for the incorrect predictions of the classifier. The next poorly classified class is 06_ka. Out of fifteen samples in this class, ten of them were wrongly classified as 41_ra and one sample was incorrectly classified as 10_ko. As we can observe, the hand shapes in classes 06_ka and 41_ra are pretty alike.

One of the major reasons of insufficient classification performance for a few classes are strong inter-class similarities. We investigated several approaches to improve the recognition performance, including rendering additional images for the classes with lower classification ratios, ensembles of classifiers, synthesis of images on the basis of Adversarial Generative Models (GANs). Initial experimental findings show that in order to achieve better distinguishing between similar classes the ResNet neural network should be trained on larger images. It appears that the images with resolution 128 × 128 will be a reasonable choice having on regard that additional data, both real and rendered should be collected.

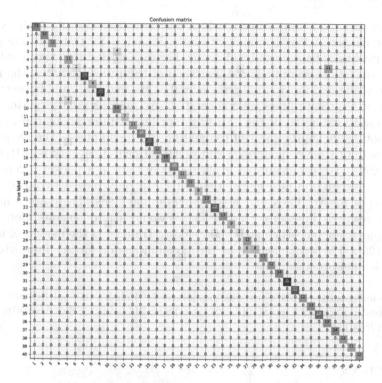

**Fig. 6.** Confusion matrix - each row represents the real class while each column represents the predicted class of the gestures.

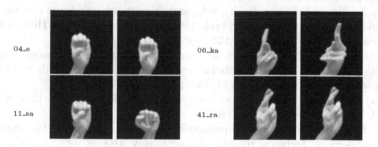

**Fig. 7.** Example inter-class similarities, 04_e – 11_sa and 06_ka – 41_ra.

The neural network has been trained on TitanX GPU with epoch size set to 64 and number of epochs set to 100. The neural networks were implemented in Python using TensorFlow/Keras frameworks.

## 6   Conclusions

In this paper we demonstrated a framework for recognition of static finger spellings on color RGB images. The recognition of hand gestures is performed by a convolutional neural network, which has been trained using both real and synthetic images. A few thousands of synthetic images for training were generated on the basis of a skinned hand model. For each 41 gestures, twelve different gesture realizations were prepared. For each of the twelve gestures, the model has been configured in an initial pose as well as the end pose, and between such poses 10 intermediate poses were generated. We demonstrated experimentally that owing to extension of the JSL fingerspelling dataset about the rendered data a noticeable improvement of recognition can be obtained. The recognition performance of ResNet quaternion convolutional neural network is better than ResNet CNN. We demonstrated experimentally that in a person independent scenario a recognition rate higher than 94% can be achieved by an ensemble consisting of ResNet CNN and ResNet quaternion CNN. There are a number of future research directions. First of all, we would like to generate more training samples using our articulated hand model. Training of the models of images of size $128 \times 128$ would be one of the possibilities for further improving the recognition performance.

**Acknowledgments.** This work was supported by Polish National Science Center (NCN) under a research grant 2017/27/B/ST6/01743 and JSPS KAKENHI under a grant 17H06114.

# References

1. Sagayam, M., Hemanth, J.: Hand posture and gesture recognition techniques for virtual reality applications: a survey. Virtual Reality **21**(2), 91–107 (2017)
2. Chen, F., Zhong, Q., Cannella, F., Sekiyama, K., Fukuda, T.: Hand gesture modeling and recognition for human and robot interactive assembly using Hidden Markov Models. Int. J. Adv. Rob. Syst. **12**(4), 48 (2015)
3. Raj, M.D., Gogul, I., Thangaraja, M., Kumar, V.: Static gesture recognition based precise positioning of 5-DOF robotic arm using FPGA. In: Trends in Industrial Measurement and Automation (TIMA), pp. 1–6 (2017)
4. Liu, H., Wang, L.: Gesture recognition for human-robot collaboration: a review. Int. J. Ind. Ergon. **68**, 355–367 (2018)
5. Patil, S., et al.: GesturePod: programmable gesture recognition for augmenting assistive devices, Technical report, Microsoft, May 2018
6. Rautaray, S., Agrawal, A.: Vision based hand gesture recognition for human computer interaction: a survey. Artif. Intell. Rev. **43**(1), 1–54 (2015)
7. Al-Shamayleh, A.S., Ahmad, R., Abushariah, M., Alam, K.A., Jomhari, N.: A systematic literature review on vision based gesture recognition techniques. Multimedia Tools Appl. **77**(21), 28121–28184 (2018)
8. Ohn-Bar, E., Trivedi, M.: Hand gesture recognition in real time for automotive interfaces: a multimodal vision-based approach and evaluations. IEEE Trans. Intell. Transp. Syst. **15**(6), 2368–2377 (2014)
9. Pisharady, P., Saerbeck, M.: Recent methods and databases in vision-based hand gesture recognition. Comput. Vis. Image Underst. **141**, 152–165 (2015)
10. Oyedotun, O., Khashman, A.: Deep learning in vision-based static hand gesture recognition. Neural Comput. Appl., 1–11 (2016)

11. Tompson, J., Stein, M., LeCun, Y., Perlin, K.: Real-time continuous pose recovery of human hands using convolutional networks. ACM Trans. Graph. **33**(5) (2014)

12. Nagi, J., Ducatelle, F., et al.: Max-pooling convolutional neural networks for vision-based hand gesture recognition. In: IEEE ICSIP, pp. 342–347 (2011)

13. Barros, P., Magg, S., Weber, C., Wermter, S.: A multichannel convolutional neural network for hand posture recognition. In: Wermter, S., et al. (eds.) ICANN 2014. LNCS, vol. 8681, pp. 403–410. Springer, Cham (2014). https://doi.org/10.1007/978-3-319-11179-7_51

14. Koller, O., Ney, H., Bowden, R.: Deep hand: how to train a CNN on 1 million hand images when your data is continuous and weakly labelled. In: IEEE Conference on Computer Vision and Pattern Recognition, pp. 3793–3802 (2016)

15. Tabata, Y., Kuroda, T.: Finger spelling recognition using distinctive features of hand shape. In: International Conference on Disability, Virtual Reality and Associated Technologies with Art Abilitation, pp. 287–292 (2008)

16. Kane, L., Khanna, P.: A framework for live and cross platform fingerspelling recognition using modified shape matrix variants on depth silhouettes. Comput. Vis. Image Underst. **141**, 138–151 (2015)

17. Kwolek, B., Sako, S.: Learning siamese features for finger spelling recognition. In: Blanc-Talon, J., Penne, R., Philips, W., Popescu, D., Scheunders, P. (eds.) ACIVS 2017. LNCS, vol. 10617, pp. 225–236. Springer, Cham (2017). https://doi.org/10.1007/978-3-319-70353-4_20

18. Rosalina, L.Y., Hadisukmana, N., Wahyu, R.B., Roestam, R., Wahyu, Y.: Implementation of real-time static hand gesture recognition using artificial neural network. In: CAIPT, pp. 1–6 (2017)

19. Asad, M., Slabaugh, G.: SPORE: staged probabilistic regression for hand orientation inference. Comput. Vis. Image Underst. **161**, 114–129 (2017)

20. Dawod, A.Y., Nordin, M.J., Abdullah, J.: Static fingerspelling recognition based on boundary tracing algorithm and chain code. In: International Conference on Intelligent Systems, Metaheuristics & Swarm Intelligence, pp. 104–109. ACM (2018)

21. He, K., Zhang, X., Ren, S., Sun, J.: Deep residual learning for image recognition. In: IEEE Conference on Computer Vision and Pattern Recognition (CVPR), pp. 770–778 (2016)

22. Parcollet, T., et al.: Quaternion convolutional neural networks for end-to-end automatic speech recognition. In: Interspeech, ISCA, pp. 22–26 (2018)

23. Popa, C.A.: Learning algorithms for quaternion-valued neural networks. Neural Process. Lett. **47**(3), 949–973 (2018)

24. Nitta, T.: A quaternary version of the back-propagation algorithm. In: Proceedings of International Conference on Neural Networks, vol. 5, pp. 2753–2756 (1995)

25. Zhu, X., Xu, Y., Xu, H., Chen, C.: Quaternion convolutional neural networks. In: Ferrari, V., Hebert, M., Sminchisescu, C., Weiss, Y. (eds.) ECCV 2018. LNCS, vol. 11212, pp. 645–661. Springer, Cham (2018). https://doi.org/10.1007/978-3-030-01237-3_39

26. Opitz, D., Maclin, R.: Popular ensemble methods: an empirical study. J. Artif. Int. Res. **11**(1), 169–198 (1999)

# Algorithm for Constructing a Classifier Team Using a Modified PCA (Principal Component Analysis) in the Task of Diagnosis of Acute Lymphocytic Leukaemia Type B-CLL

Mariusz Topolski[1] and Katarzyna Topolska[2(✉)]

[1] Department of Systems and Computer Networks, Faculty of Electronics,
Wroclaw University of Science and Technology, Wybrzeze Wyspianskiego 27,
50-370 Wroclaw, Poland
mariusz.topolski@pwr.edu.pl
[2] WSB Universities, Fabryczna 29/31, 53-609 Wroclaw, Poland
Katarzyna.topolska@wsb.wroclaw.pl

**Abstract.** Systems of data recognition and data classification are getting more and more developed. There appear newer algorithms that solve more difficult and complex decision problems. Very good results are obtained using sets of classifiers. The authors in their research focused on certain data characteristics. The characteristics concerns recognition of classes of objects whose features can be grouped. Clusters created in this manner can contribute to better recognition of certain decision classes. One such example is a diagnosis of forecast in the case of acute lymphocytic chronic leukaemia B-CLL type. In this document, the authors present a modified selection method of features of the PCA object. The modification concerns the rotation of objects in relation to decision classes. In addition to grouping similar features using Varimax rotation, a procedure for grouping patients in these PCA groups was developed. Within each PCA, two classifiers - strong and weak ones were built. In the research part, the developed method was compared to the one-stage recognition algorithms known from the literature. The obtained results have a significant contribution to medical diagnostics. They allow to develop a procedure for treatment of B-CLL lymphocytic leukaemia. Making an appropriate diagnosis allows to increase a patient's survival chance by implementing appropriate treatment.

**Keywords:** Analysis of major components · Classifiers ·
Lymphocytic leukaemia

## 1 Introduction

Contemporary problems of recognizing objects are often based on very complex and large volumes of data or on very few data. The few data relates mainly to the situation when implementation of an experiment is too time-consuming or we do not have sufficient research material. As the machine learning progresses, algorithms are developed to solve increasingly complex decision problems. One of such difficult issues of classification is face recognition. Despite great efforts, we still cannot find an

© Springer Nature Switzerland AG 2019
H. Pérez García et al. (Eds.): HAIS 2019, LNAI 11734, pp. 614–624, 2019.
https://doi.org/10.1007/978-3-030-29859-3_52

algorithm that would unpretentiously deal with such a task. One of the methods aiming at increasing efficiency of algorithms is to combine classifiers into teams. It turns out that the right team selection can bring many benefits. Construction of classifiers may be carried out using various methods, including: AdaBoost, Bagging, and Selection. In a situation where the decision-making system is built of at least several classifiers, one should choose a classifier that suits best the features of an object being recognized. An interesting solution is to build strong classifiers that are able to teach weaker classifiers to interpret data. Currently, in classification tasks, it is very popular to use advanced exploration methods. One of these methods is PCA (principle component analysis). It is often used to reduce a multidimensional vector of features. In the data classification task, it can be used to group similar features. Then one can assign data cases (records) to a given principal component group. Recognizing such single clusters at the assumption of interdependence with other clusters may contribute to solving the problems of certain medical data.

One such algorithm is AdaBoost based on the so-called Adaptive strengthening. An algorithm for this property that in subsequent interactions trains k-weaker classifiers on a set of translations with weights [1]. Complex recognition methods use ensemble classifiers [2, 3]. In other studies authors propose the novel classification method, employing a classifier selection approach, which can update its model when new data arrives [4].

One such example is acute lymphocytic leukaemia B-CLL. The B-CLL problem does not relate to its diagnosis but to development of patient prognosis survival procedures. Here medicine is constantly looking for a solution. In the next section, the authors briefly present a problem of understanding the prognosis of survival of patients with acute lymphocytic leukaemia.

The aim of the research is development of a method facilitating prognosis of lymphocytic leukaemia. In the task of classification, one should develop a method of selection of features which will contribute significantly to improvement of accuracy of assessment of a prognosis of a patient's survival. Continuous works and research aim at determination of factors leading to a better prognosis, which in turn may contribute to implementation of a correct treatment procedure to save lives of many patients with B-CLL. Application of the developed method may be applied in statistical packages e.g. medical STATISTICA in a set of advanced methods of selection and data classification.

The development of advanced computational methods gives more and more opportunities to solve various complex problems.

## 2 B-CLL Acute Lymphocytic Leukaemia

Cancers that are connected to B lymphocytes are the most numerous group of lymphoproliferative diseases. One of the diseases of this group is B-cell chronic lymphocytic leukaemia (B-CLL). Chronic B-cell lymphocytic leukaemia is a lymphoproliferative group of accumulation nature. In its clinical course, lymphocytes accumulate in the peripheral blood [5–7]. In clinical trials in the task of recognition of lymphocytic B-cell leukaemia, the presence of a monoclonal population of lymphocytes B in the peripheral blood is determined. Doctors recognize leukaemia based on the first results of monoclonal population of lymphocytes >5000/μl and B-CLL cells in the bone marrow >30%. These cells are of immune-phenotypic nature [5–8].

Chronic lymphocytic leukaemia has a varied course and evaluation of a patient's prognosis is not easy to determine by an oncologist. Often, a course of the disease is asymptomatic. Evaluation of prognosis in the case of B-CLL is extremely difficult. The reason for this may be an asymptomatic course, and in some cases, in some patients, even in a short period of time there occur lymphadenopathy and organo-nomeny [9–11]. Contemporary medicine is facing the problem of forecasting a further course of the disease. Researchers are looking for factors that will be able to make such a prognosis. Current research results show that B-CLL prognosis is most affected by the number of peripheral lymphocytes, bone marrow depletion rate, percentage of peripheral blood lymphoid cells, and time of lymphocyte doubling [12]. Performed tests allow to distinguish factors foreseeing further course of the disease process in patients with B-cell chronic lymphocytic leukaemia. These factors are:

- period of clinical advancement according to Binet,
- period of clinical advancement according to Rai,
- bone marrow infiltration in the assessment of trephine biopsy,
- bone marrow infiltration in cytological assessment,
- leucocytosis
- time to double the number of leukocytes
- serous markers,
- cytogenetic aberrations,
- expansion CD38 on leukemic cells,
- status of the altered gene for the immunoglobulin heavy chain (IgVH),
- expression of ZAP70 on leukemic cells [8].

In the diagnosis and prognosis of chronic lymphocytic b-cells leukaemia some of the above features may form certain groups that are sufficient to prognosticate further development of the disease. Dividing the features into similar ones allows to reduce the multidimensional vector of these features to these discriminating most the prognosis. It should also be noted that even if one manages to group features into certain clusters in order to build a classifier, this will not necessarily be the key for all patients. Therefore, it is important to assign not only features, but also patients into these groups. As a last resort, it is possible to build a cluster, in which a link of such a cluster is a single classifier. Description of the model supporting prognosis of patients with B-cell chronic lymphocytic leukaemia is presented in the next chapter.

## 3    Algorithm for Building a Team of Classifiers Using a Modified PCA Main Component Analysis

This section presents a model for building a set of classifiers using PCA (principal component analysis). In the further part we will call this model: "**Model PCA**". The PCA is not, in this case, to reduce features of an object, but to indicate connections between them. It should be noted that application of PCA makes sense when the following conditions are met:

- normality of distribution of measured variables,
- sample size of at least 50 observations.
- elimination of outliers,
- missing data should be replaced, for example, by an average of a feature in the decision class [13, 14].

Let $x \in X$ be a $d$-dimensional vector of variables (features) that have been measured. In the task of classifying objects, in particular medical data, features have a different field of value. To bring them to the common measuring scale $Z$, we use standard normalization $X \rightarrow N(0; 1)$:

$$Z = \frac{X - m}{\delta} \qquad (1)$$

Where: $m$ is an average value and $\delta$ is a standard deviation.

We apply normalization (1) to all quantitative features of the object. The data can now be presented in the form of normalized variables:

$$Z = \begin{bmatrix} z_{11} & z_{12} & \cdots & z_{1d} \\ z_{21} & z_{22} & \cdots & z_{2d} \\ \vdots & \vdots & \ddots & \vdots \\ z_{n1} & z_{n2} & \cdots & z_{nd} \end{bmatrix}. \qquad (2)$$

The formal record of the PCA model is:

$$PC_i = w_{i1}Z_1 + w_{i2}Z_2 + \ldots + w_{ik}Z_k, \qquad (3)$$

where: j is the decision class of the object

$$\sum_{j=1}^{k} w_{ij}^2 = 1 \qquad (4)$$

Equation (4) presents weights that are assigned to subsequent variables in the process of creating the principal component

We assume that the explored factors are independent and have a standardized normal distribution (1) and fulfil the conditions:

$$w_j' w_j^{-1} = 1; \quad w_j' w_j = 0 \qquad (5)$$

In order to calculate the coefficients, one should assume a criterion of maximizing variance by choosing the value of the vector w. This selection should be made using the Lagrange multipliers method, which will lead to the solution of the own matrix $S$ vector.

$$S(R) = \sum_{i=i}^{N} \left( z^{(i)} - \bar{z} \right) \left( z^{(i)} - \bar{z} \right)^{T}, \tag{6}$$

Where R - is the correlation matrix (covariance) of N output variables. Next, we determine eigenvectors and the eigenvalues of the S matrix:

$$S(R) = A \wedge A^{T}, \tag{7}$$

where $A$ is the matrix of eigenvectors, and $\wedge$ is a diagonal matrix, on the diagonal of which the eigenvalues of the matrix are placed.

The determinant equation of the $S(R)$ matrix can be represented as follows:

$$|R - \lambda I| = 0 \tag{8}$$

The next step of the algorithm is to organize the eigenvalues of the $S(R)$ covariance matrix in a decreasing order:

$$\lambda_1^S \geq \lambda_2^S \geq \ldots \geq \lambda_k^S \geq 0 \tag{9}$$

Then a new subset of features is determined in the new space, which is unbuttoned by the orthogonal principal components. New variable groups are determined based on the maximization of observation of object features with simultaneous minimization of information loss.

Therefore, from the matrix of own vectors A, we choose as many PCA factors for which the eigenvalue of the factor is >1:

$$Rw_i = \lambda_i w_i \tag{10}$$

where $w_i$ is the own vector of the correlation matrix (covariance)

The constituent values of the vector (10) will be the values of the coefficients of orthogonal variables whose linear combinations generate new principal components.

Thus, we managed to divide $d$ the dimensional vector of input features into $k$ quantitative variables that are in the strongest relationship with each other. For the construction of the p set of $\psi_{ijp}$ classifiers, the cases most strongly related to the $PC_i$ factor should be chosen. For this purpose, we can use the modified square function of cosines:

$$\bigvee_j \cos_j^2(\alpha) = \wedge + \sin^2(\alpha) \oplus \sin_j^2(\alpha), \tag{11}$$

where: $\alpha$ - is the rotation angle

We rotate factors by assigning variables to them according to the decision class, including the occurrence of all classes.

To build a given classifier $\Psi_p$ we first determine the affiliation of objects for each class $j$ to a given factor $PC_i$ based on:

$$\max_j \left[ \cos^2(\alpha) \right] \rightarrow PC_i \ i = 1, 2, \ldots, k \tag{12}$$

Then, for each $PC_i$ we sort all cases from the largest value of $\cos^2(\alpha)$. Strong classifiers $\psi_{ijp}$ are those for which values $\cos^2(\alpha) \geq 0,7$. The number of these classifiers is equal to $2ij$. In order for a given object to be able to join a group of objects that create strong classifiers, the average value of coherent features classified to the PCi for a given object class $j$ must meet the criterion:

$$P(t) = P \left( \frac{\frac{\sum_{t=1}^{q_j} z_{jt}}{q_j} - z_{jo}}{\delta_q} \right) > 0,05 \tag{13}$$

where: $p$-is the significance level of $t$-Student statistics for a single sample, $q_j$ – the number of features for a given class $j$ in a given $PC_i$, $z_{jt}$-the value of standardized features (1) in a given $PC_i$. Therefore, an object is attached to a given classifier when the average value of all $PC_i$ features determined by the $t$-Student for one sample does not differ from the average value of $PC_i$ features for which there are strong dependencies $\cos^2(\alpha) \geq 0,7$.

For each $i$-th component of the main and $j$-th decision-making class, $p$ classifiers are created within this component. The team of classifiers can be expressed as follows

$$\bigvee_i \bigvee_j \psi_{ijp} = \left[ \psi_p(PC_i, A_{ij}), \psi'_p(PC_i, A_{ij}) \right], \tag{14}$$

where: $A_{ij}$ is a set of cases of a given class j for the $i$-th main component

Objects that within the given $PC_i$ have not been classified to a strong classifier have been included in weak classifiers $\psi'_{ijp}$.

There are $2ij$ all possible classifiers. The discrimination power of a given classifier is an average value $\overline{\cos_d^2(\alpha)}$ of all its objects.

When recognizing objects in the first stage, the average value of the group of its measured features is compared with the average of each classifier $\psi_{ijp}$ using the $t$-Student test statistics (13).

All classifiers pass into the second classification stage, which in the Eq. (13) obtained $P(t) > 0,05$. This means that the average result of the features of a recognized object does not differ significantly from the average result of the entire group of objects within a given classifier. In other words, we determine the consistency of the object recognized with a given classifier. Finally, the diagnosis is made on the basis of the class determined by the strong classifier $\psi_{ijp}$ with the maximal significance value *max* *[P(t)]*, assuming $P(t) > 0,05$. If none of the strong classifiers meets the criterion

$P(t) < 0,05$, we perform the same operations for the set of weak classifiers. If for all classifiers the average value of the features of the recognized object $P(t) < 0,05$, then from all strong classifiers we also choose from $max[P(t)]$. The choice of a given classifier determines the decision class transferred by this classifier.

The following is an exemplary pseudocode of the PCA task implementation

------------------------------------------------

**Stage of learning**

**Enter dataset**

>N=239 data of real patients with B-CLL

**Normalise data**

>According to the distribution Z: N (0;1)

**Perform PCA procedure with rotation**

$$\bigvee_j \cos^2_j(\alpha) = \wedge + \sin^2(\alpha) \oplus \sin^2_j(\alpha)$$

>Rotate until achievement of maximalisation of variance in $j$ classes

**Classify cases to a given** $PCi$ **factor**

$$\max_j \left[\cos^2(\alpha)\right] \rightarrow PC_i \quad i = 1,2,...,k$$

**Build** $2ij$ **of classifiers within each factor**

>**Within** $PCi$ build classifiers strong for lambda >0.7

>Calculate average standard deviations of features for each classifier

>Label j to each classifier

**Stage of testing**

**Enter features of an** $X$ **object under recognition**

**Apply a method of triple cross validation**

**Perform the following procedure for each testing object**

>**Compare an average value of features of (an object) under recognition from the test set with the average of features of the classifiers from the learning set**

>>Calculate weightiness $p$ for the test t-Student

>>Select a classifier with the highest $p$ value.

>**Conclusion**

>>Class with the highest size in a given classifier.

------------------------------------------------

# 4   Test Results

In order to verify the method of building classifiers based on the PCA method proposed in item 3, the data showing prognosis of B-CLL type lymphocytic leukaemia were used. The data refers to 239 patients with B-CLL leukaemia. The patients were subjected to homogeneous treatment with chemotherapy. Patients underwent follow-up examinations and the following results were reported: z1 - bone marrow infiltration in cytological assessment, z2 - leucocytosis, z3 - percentage of pro-lymphocytes in peripheral blood, z4 - doubling time of lymphocyte count, z5 - β2-microglobulin concentration, z6 – tyrosine kinase activity of lymphocytes, z7 - CD23 expression on leukemic cells, z8 - CD38 expression on leukemic cells, z9 - ZAP70 expression on leukemic cells. Based on these results, patients were classified into three classes j = [0-favorable prognosis, 1-no favourable prognosis 2-very unfavourable prognosis).

In the research, the test results were compared with the algorithms known from literature. References to these algorithms are given in the Table 2. In the classification task the method of triple cross validation was applied.

In the task of model building on the basis of the Kaiser criterion three principal components of the PCA were identified. The eigenvalues and the percentage of explained variance by these factors are presented in Table 1.

**Table 1.** Own values of the PCA method

| PCAi | Own values | % of total variance | Accumulation of own value | Accumulated percentage |
|------|------------|---------------------|---------------------------|------------------------|
| PCA1 | 3,06 | 38,27 | 3,06 | 38,27 |
| PCA2 | 2,74 | 34,29 | 5,80 | 72,55 |
| PCA3 | 1,69 | 21,14 | 7,50 | 93,69 |

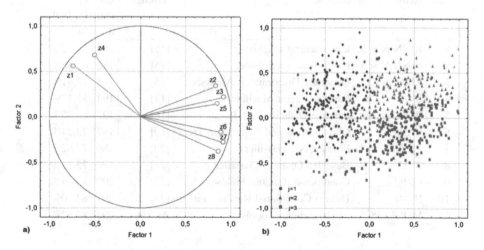

**Fig. 1.** Projection of (a) factors and (b) cases (objects), on the plane

Three principal components were distinguished in the learning process. PCA1 component = {z1, z4}, PCA2 = {z2, z3, z5} and PCA3 = {z6, z7, z8}. According to (13) 18 classifiers were identified. Then a series of tests was carried out comparing the quality of the method's classification with the one-step recognition algorithms known from the literature. The results of the classification are presented in Table 2.

The obtained results indicate that the use of methods for selecting object features may give better results for some data than classical recognition methods (Fig. 1).

The result i.e. 84,52% of classification correctness is affected by the characteristics of B-CELL data. This means that the prognosis of a patient's survival is influenced by all the features, but some of them have the greatest impact. Grouping of features and creation of a set of classifiers is an important element of modern recognition systems. By analysing certain features locally, we can get more accurate classifications. Applying the author's method of rotation of factors as a sum of orthogonal general rotation and in classes, it was possible to obtain very good results of selection of features within the decision classes. In comparison to PCA method known from literature it turned out that it was possible to match features to the classes which represented them in the strongest manner. This improvement of results is affected mainly by a reduction of features within classifiers which contributes to decrease a classification error. For comparison, a tradition PCA was carried out and this enabled us to obtain classifications by 5% worse than in the proposed method. After performing the Chi-square test, we may significantly ($p < 0,05$) conclude that frequency of the correct classifications is significantly higher in the proposed model than in traditional PCA, K-NN-3 and 5, RDA, SVM, ID3, CART, FK-NN and BOC.

**Table 2.** Results of the B-CLL type lymphocytic leukaemia prognosis

| Ref. | Abbr. | Method | Reference | Correct classification | |
|------|-------|--------|-----------|---|---|
| | | | | n | % |
| 1 | K-NN N = 3 | k = 3-nearest neighbour | [15] | 186 | 77,82% |
| 2 | K-NN N = 5 | k = 5-nearest neighbour | [15] | 166 | 69,46% |
| 3 | RDA | Quadratic Discriminant Analysis | [16] | 185 | 77,41% |
| 4 | SVM | Support Vector Machine | [17] | 194 | 81,17% |
| 5 | ID3 | Iterative Dichotomiser 3 | [18] | 193 | 80,75% |
| 6 | CART | Classification and Regression Trees | [19] | 187 | 78,24% |
| 7 | FK-NN | Fuzzy k-NN classifiers | [20] | 186 | 77,82% |
| 8 | BOC | Bayesian Optimal Classifier | [21] | 180 | 75,31% |
| 9 | CRF | Conditional Random Fields | [22] | 200 | 83,68% |
| 10 | HCRF | Hidden Conditional Random Fields | [23] | 201 | 84,10% |
| 11 | MODELPCA | Model Article (Topolski) | Article | 202 | 84,52% |

# 5 Conclusion

The article focuses on an important medical problem. It concerns the prognosis of patients with acute chronic B-CLL lymphocytic leukaemia. The diagnosis of B-CLL is not difficult for physicians. The problem arises when a patient's prognosis of survival is to be estimated. At a time of contemporary research, we already know which factors can be the basis for forecasting survival. However, it is still unknown how to combine these features to get the best possible measurement information. Knowing the different course of the disease and the survival rate of patients with B-CLL, we can seek to respond to this problem. For this purpose, the authors developed a concept for construction of a set of classifiers based on the modified method of PCA. This modification applies to Varimax rotation in relation to decision classes. Such rotation allows to increase the probability of class recognition by the best selection of discriminating variables. The right assumption was made that by grouping features and assigning appropriate groups of patients to these groups, we can create classifiers locally acting better. Experience shows that weak classifiers are located within the boundaries of decision classes. The method used gives comparable results to the methods such as Conditional Random Fields and Hidden Conditional Random Fields. The results obtained may contribute to orientating researchers to the problems of B-CLL prognosis. Development of methods supporting the estimation of prognosis by doctors gives a possibility of more precise and quick orientation of a patient for further treatment. Further directions of the research will cover methods of selection of features of an object using the fusion of multidimensional analysis of correspondence with PCA. It will make it possible to take into consideration quality and quantity variables in one model. The author is working on a concept of moving the axis of rotation of coordinate factors of centre of gravity of many features in classes. The applied algorithm may be applicable in the module of the programme STATISTICA medicine in selection and classification of date which will enable researchers to test directly a variety of medical problems, and in particular the ones regarding research on features of prognosis of acute lymphocytic leukaemia.

**Acknowledgement.** This work was supported by the statutory funds of the Department of Systems and Computer Networks, Faculty of Electronics, Wroclaw University of Science and Technology.

# References

1. Burduk, R.: Integration base classifiers based on their decision boundary. In: Rutkowski, L., Korytkowski, M., Scherer, R., Tadeusiewicz, R., Zadeh, L.A., Zurada, J.M. (eds.) ICAISC 2017. LNCS (LNAI), vol. 10246, pp. 13–20. Springer, Cham (2017). https://doi.org/10.1007/978-3-319-59060-8_2
2. Woźniak, M., Ksieniewicz, P., Cyganek, B., Kasprzak, A., Walkowiak, K.: Active learning classification of drifted streaming data. Procedia Comput. Sci. **80**, 1724–1733 (2014)
3. Krawczyk, B., Ksieniewicz, P., Woźniak, M.: Hyperspectral image analysis based on color channels and ensemble classifier. In: Polycarpou, M., de Carvalho, A.C.P.L.F., Pan, J.-S., Woźniak, M., Quintian, H., Corchado, E. (eds.) HAIS 2014. LNCS (LNAI), vol. 8480, pp. 274–284. Springer, Cham (2014). https://doi.org/10.1007/978-3-319-07617-1_25
4. Zyblewski, P., Ksieniewicz, P., Woźniak, M.: Classifier selection for highly imbalanced data streams with *Minority Driven Ensemble*. In: Rutkowski, L., Scherer, R., Korytkowski, M., Pedrycz, W., Tadeusiewicz, R., Zurada, J.M. (eds.) ICAISC 2019. LNCS (LNAI), vol. 11508, pp. 626–635. Springer, Cham (2019). https://doi.org/10.1007/978-3-030-20912-4_57
5. Kay, N., Hamblin, T., Jelinek, D., et al.: Chronic lymphocytic leukemia. American Society of Hematology, Hematology, pp. 193–213 (2002)
6. Dmoszyńska, A., Robak, T.: Podstawy hematologii. Wydawnictwo Czelej, Lublin, wyd 2 (2008)

7. Hallek, M., Cheson, B., Catovsky, D., Caligaris-Cappio, F., Dighiero, G.: Guidelines for the diagnosis and treatment of chronic lymphocytic leukemia: a report from the International Workshop on Chronic Lymphocytic Leukemia (IWCLL) updating the National Cancer Institute-Working Group (NCI-WG) 1996 guidelines, vol. 111, pp. 5446–5456 (2008)

8. Monserrat, E., Gine, E., Bosch, F.: Redefining prognostic elements in chronic lymphocytic leukemia. Hematol J. 4(suppl. 3), 180–182 (2003)

9. Hamblin, T.J.: CLL: How many diseases? Hematol J. 4(suppl. 3), 183–186 (2003)

10. Rai, K.R., Chiorazzi, N.: Determining the clinical course and outcome in chronic lymphocytic leukemia. N. Engl. J. Med. 348, 1797–1799 (2003)

11. Bosch, F., Villamor, N.: ZAP-70 expression in chronic lymphocytic leukemia: a new parameter for an old disease. Hematologica 88, 724–726 (2003)

12. Brugiatelli, M., Mannina, D., Neri, S., et al.: Recent update of prognosis and staging of chronic lymphocytic leukemia. Hematol J. 88(suppl. 10), 30–31 (2003)

13. Grabiński, T.: Metody taksonometrii. Akademia Ekonomiczna, Kraków (1992)

14. Stanisz, A.: Przystępny kurs statystyki z zastosowaniem Statistica PL na przykładach z medycyny. T. 3: Analizy wielowymiarowe. StatSoft, Kraków (2007)

15. Fix, E., Hodges, J.L.: Discriminatory analysis. Nonparametric discrimination: consistency properties, Report Number 4, Project Number 21-49-004, 1951, Reprinted in International Statistical Review, 57, pp. 238–247 (1989)

16. Fukunaga, K.: Introduction to Statistical Pattern Recognition, 2nd edn. Academic Press, New York (1990)

17. Vapnik, V.: The Nature of Statistical Learning Theory. Springer, New York (1995). https://doi.org/10.1007/978-1-4757-3264-1

18. Quinlan, J.R.: Discovering rules by induction from large collections of examples. In: Expert Systems in the Micro Electronic Age, pp. 168–201. Edinburgh University Press (1979)

19. Breiman, L., Friedman, J.H., Olshen, R.A., Stone, C.J.: Classification and Regression Trees, Wadsworth, Belmont (1984)

20. Keller, J.M., Gray, M.R., Givens, J.A.: A fuzzy K-nearest neighbor algorithm. IEEE Trans. Syst. Man Cybern. 15(4), 580–585 (1985)

21. Devijver, P.A., Kittler, J.: Pattern Recognition: A Statistical Approach. Prentice-Hall, London (1982)

22. Sutton, C., McCallum, A.: An introduction to conditional random fields for relational learning. In: Getoor, L., Taskar, B. (eds.) Introduction to Statistical Relational Learning, pp. 93–128. MIT Press, Cambridge (2006)

23. Zhang, J., Gong, S.: Action categorization with modified hidden conditional random field. Pattern Recogn. 43, 197–203 (2010)

# Road Lane Landmark Extraction: A State-of-the-art Review

Asier Izquierdo[1], Jose Manuel Lopez-Guede[2,4(✉)], and Manuel Graña[3,4]

[1] Airestudio Geoinformation Technologies Scoop, Albert Einstein Kalea, 44, E6,
Oficina 8, 01510 Vitoria-Gasteiz, Spain
[2] Department of Systems Engineering and Automatic Control,
Faculty of Engineering of Vitoria, Basque Country University (UPV/EHU),
Nieves Cano 12, 01006 Vitoria-Gasteiz, Spain
jm.lopez@ehu.es
[3] Department of Computer Science and Artificial Intelligence, Faculty of Informatics,
Basque Country University (UPV/EHU), Paseo Manuel de Lardizabal 1, 20018
Donostia-San Sebastian, Spain
[4] Computational Intelligence Group, Basque Country University (UPV/EHU),
Donostia-San Sebastian, Spain

**Abstract.** In this paper we present a state-of-the-art review about road lane landmark extraction. Automatic lane landmark extraction has been studied during the last decade for different practical applications. The purpose of this paper is to gather and discuss methodologies of road lane landmark extraction based on signals from different sensors in order to automate the extraction of horizontal road surface lane signs and get an accurate map of road lane landmarks. Specific algorithms for each kind of sensors are analyzed, describing their basic ideas, and discussing their pros and cons.

## 1 Introduction

In the last few years, the need to elaborate and update road map databases has rapidly increased. These databases are used for different purposes such as advanced driver assistance systems (ADAS), autonomous car navigation systems, roadway inventory, and any application based on real-time localization. Whatever the required application, an accurate model of the road is needed. Such an accurate model should contain information about horizontal and vertical road signs. In this paper, we focus on horizontal road signs, such as lane lines, crosswalks and arrow marks. Methodologies to extract horizontal lane image landmarks are heavily dependent on the sensor used to collect the road information. Depending on the sensor signal, different algorithms are used. Currently, it is possible to classify them into three big groups: high resolution optical image, Light Detection and Ranging (LiDAR) data, and a combination of optical images and LiDAR data.

H. Pérez García et al. (Eds.): HAIS 2019, LNAI 11734, pp. 625–635, 2019.
https://doi.org/10.1007/978-3-030-29859-3_53

## 1.1 Background

In this section some points are going to explain briefly in order to introduce a background in road lane mark extraction. To begin with, we discuss general issues detected in road lane mark segmentation systems [2,13,22]. Regardless of the algorithm or the sensor used, the main classification problems are the following ones:

1. False positives: signal features wrongly classified as road lane marks.
2. Image brightness issues: when the image is too bright, it is difficult to differentiate between different features.
3. Lane landmark occlusions created by nearby vehicles.
4. Poor visibility: Inability to detect road features accurately due to visibility conditions, such as heacy rain or fog.
5. Shadows may create misleading edges and texture on the road.

**Table 1.** Algorithms summary

|                    | Imagery | LIDAR | Imagery + LIDAR |
|--------------------|---------|-------|-----------------|
| Template matching  | [6]     |       |                 |
| Hough transform    | [1]     |       |                 |
| SIFT               | [15]    |       |                 |
| SURF               | [3]     |       |                 |
| NURBS              |         | [21]  |                 |
| Otsu thresholding  |         | [10]  |                 |
| PCA                |         | [5]   |                 |
| RANSAC             |         | [18]  |                 |

In order to be able to segment road lane mark information, different algorithms have been proposed depending on the sensor used. The main goal of the proposed algorithms is to extract road lane marks from a given road in an efficient way, solving noise and uncertainty issues. The most commonly used algorithms are referred in Table 1 distinguishing the typology of the sensor used.

## 1.2 Paper Structure and Contents

In summary, the main contribution of this paper is to explain the different methods that are used in the literature to extract horizontal road signs depending on the type of sensor used. Section 2 is focused on optical image sensors. Section 3 explains sensors based on information from Light Detection and Ranging data including different algorithms to gather horizontal road signs from captured LIDAR point cloud. Section 4 introduces methods that combine images and LiDAR data. Finally, in Sect. 5, as a conclusion, future works are proposed in order to extend the actual methodologies and improve the horizontal road sign detection.

## 2    Image Sensors

Image sensors are commonly used by autonomous car navigation systems [12,14]. Segmentation methods usually work on single images, so that the landmark extraction is repeated *de novo* for each image every $s$ seconds, the time interval chosen between images. A CCD sensor is installed on the vehicle to obtain high quality images. Some authors think that the best way to capture the road surface and its lane marks is using vision based approaches [13], since other technologies such as LiDAR based sensor are not capable to detect features in a flat plane accurately, such as road painted marks.

Using this type of imaging sensor to capture road landmarks can suffer different problems:

– Applying a threshold on brightness, may lead to false positives due to occlusion by nearby vehicles or brightness variations in the same trajectory.
– Another inevitable problems are occluded landmarks by other vehicles on the road.
– Other problems may occur, such as different lane signs, changed width of the marks, image clarity issues (cast shadow on the image, saturated images, etc.), poor visibility conditions due to fog, heavy rain or reflections on wet roads at night time, or shadows in the road due to sun position in the sky, as shown in Fig. 1 [2].

**Fig. 1.** Shadows in the road due to sun position

Many algorithms can be applied in order to extract horizontal lane marks. The most popular algorithms are the following ones:

- Template matching: It is a technique in digital image processing for finding small parts of an image which match a template image [6].
- Hough Transform: The purpose of the technique is to find imperfect instances of objects within a certain class of shapes by a voting procedure. This voting procedure is carried out in a parameter space, from which object candidates are obtained as local maximum in a so-called accumulator space that is explicitly constructed by the algorithm for computing the Hough Transform [1]. Hough transform has been notoriously used for line detection.
- SIFT: The Scale-Invariant Feature Transform (SIFT) is a feature detection algorithm in computer vision to detect and describe local features in images [15].
- SURF: Speeded-Up Robust Features (SURF) is a feature descriptor that extracts keypoints from different regions of a given image in order to find similarities between different images [3].

The main issue of template matching is the need for precise templates and the need of overcome distortions and noise. Being a correlation based algorithm, it can be robust to some kind of noises, but it is very sensitive to spatial distortions such as scale or perspective transformations. The Hough transform is quite robust but limited to specific features, such as lines. Some generalization has been proposed to detect objects under occlusion which can be useful in our problem given an adequate dataset of examples. So called image feature extraction algorithms, such as SIFT and SURF, are quite robust against deformations and occlusion, because they extract salient image points that characterize the objects in the image. In fact, they are under patent and therefore they can not be used freely in industrial applications. Alternative open and free algorithms are proposed in the literature.

## 3   LiDAR Based Sensors

As remarked in Sect. 2, vision based sensor methodologies are susceptible to shadow, weather conditions, and brightness, so some authors proposed road lane mapping methods using LiDAR based sensors [16]. These methods are based on Mobile Mapping Systems (MMS) which use LiDAR technology, with scanners capable of collecting up to $10^6$ measurements per second. The data acquired by MMSs are:

- Range and intensity data from LiDAR scanners.
- Visual data (RGB) from panoramic images.
- Positioning data:
  - Absolute positioning by Differential Global Positioning Systems (DGPS) which are enhancements to the Global Positioning System (GPS) providing improved location accuracy, in the range of operations of each system, from the 10 ms nominal GPS accuracy to about 10 cm in the best implementations.

**Fig. 2.** Intensity point cloud

**Fig. 3.** RGB point cloud

- Inertial data from Inertial Measurement Unit (IMU).
- Vehicle speed by odometer.

The data are fused by proprietary software of MMSs. The output usually consists of:

- A trajectory file, which contains information in terms of latitude, longitude, projected coordinates, altitude, sensor intrinsic rotations' angles (pitch, raw, yawn) and panoramic images information.
- Georeferenced panoramic images obtained from spherical image based sensor(s).
- Georeferenced point cloud with RGB and intensity data, as shown in Figs. 3 and 2.

Some studies [4,20] propose to generate lane level maps by driving along the centerline of the road and then by analyzing the trajectory obtained by

Global Navigation Satellite System (GNSS). However, this approach is neither efficient nor accurate. Other authors use only georreferenced point cloud and trajectory file, in spite of the fact that MMS systems images are available [16,22]. Their methodology consists of a set of processing modules that allow to obtain a digital image from the point cloud. This process is necessary since lane marks will be extracted from a nadir point of view, obtained by image processing. Different workflows are proposed:

- In order to obtain a digital image from a point cloud, [16] proposed the following schema to extract objects that lie on the road surface:
  1. Re-project point cloud coordinates from geographical to planar coordinates.
  2. Geometric filtering.
  3. Interpolation on a regular grid, where the generation of a raster 2D image is generated, as shown in Fig. 4.
  4. Apply a thresholding to obtain a binary image.
  5. Reduce noise applying morphological operations.
  6. Isolate objects labeling and calculating morphological indicators.
  7. Peak detector for lane line detection, including line validation and attributes generation.
  8. Apply a template matching for arrow marks.
  9. Detect and extract crosswalks applying morphological indicators.

- On the other hand, [22] proposes a different workflow to improve some problems such as time-consuming processing of point clouds (noise, unorganized points, big amount of points, etc.), incontinence in the reflective intensity values unevenness caused by dust of the road, intensity of lane marks or vehicle speed amongst others. As [16] proposed, 2D image is generated from the point cloud, but proposing different methods which include the following steps:
  1. After the point cloud is gridded, unexpected changes in the normal vector are used in order to place the curb grids on both sides of the trajectory. Road edges are placed between the curb grids and used to segment the road points. So to facilitate the following lane extraction, the inconstancy of the reflective intensity of the road points are corrected. Lane-marking extraction: After 3D road points are mapped into a 2D image, a self-adaptive thresholding method is developed to extract lane markings from the 2D image.
  2. Lane mapping: In order to obtain the 3D lane lines, it is needed to cluster and render points belonging to the same lane line type. To complement missing lines caused by occlusion, a global post-processing steep that integrates all local extraction is used.

Since lane signs are generally more reflective than the background surface, the intensity value can be useful to find lane-marking classified points. Some authors use a single-global threshold to extract lane signs [21]. Road boundaries are detected using a two-steep method based on local road shape. Road sides and main axis are fitted by arc-length parametrized NURBS allowing

**Fig. 4.** Left: RGB point cloud. Right: generated 2D image

to compute roads with a curvature at the desired resolution. Using the reflectivity information provided by LiDAR sensors, road markings are extracted, as shown in Fig. 5.

- The Otsu thresholding method is proposed by [10] in order to find an intensity threshold value which maximizes the intracluster variance of road markings and road-surface LiDAR clusters, optimizing the segmentation of point clouds into asphalt and road marks.

## 4   Fusion of MMS Data and Image

Only a few papers talk about combining images and point clouds provided by MMSs to extract lane marks. According to [22] the following difficulties or challenges exist using MMS-based lane mapping methods;

- Time-consuming processes due to unorganized points or noisy classified points.
- Inconsistency of the reflective intensity measured point clouds due to different scanning ranges, incidence angles, surface proprieties, etc.
- Interferences caused because of the unevenness caused by the dirt or dust on the road.
- Because of instrument settings, velocity of the vehicle and quality of the MMS system.
- Those lane-marks points surveyed, in particularly those which are further from the vehicle become sparse and difficult to recognize.
- Work with image texture and reflective intensity of the point cloud.

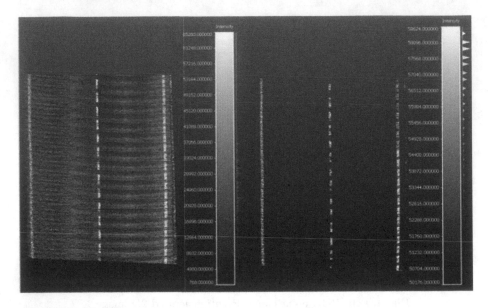

**Fig. 5.** Left: Point cloud represented as intensity values. Right: Lane marks extraction using intensity values

Generally, MMS-based lane mapping methods are divided in two steps. In a first stage, the road surface is extracted; in the following step the road lane-marks are extracted. To extract the road surface, many segmentation methods have been proposed, divided in two groups:

– Planar-surface-based methods [17] generally use fitting algorithms. Most of them use the normal vectors of points which are usually estimated using RANdom SAmple Consensus (RANSAC) [25], Principal Components Analysis (PCA) [5], Robust and Diagnostic Principal Components Analysis or Hough transform [8].
– Edge-based methods [26], in a first process, linear road edges are detected and fitted using those edges in order to segment road surface [23].

Preliminary studies on the fusion of MMS images and point clouds have been carried out to detect tree species [24], pedestrian [11,19] or facades [7,9], but not many studies have accomplished to extract lane marks from image and point cloud fusion.

A validation method proposed by [22] fuses image texture saliency with the geometrical distinction of the lane-marking point cloud to improve the robustness of lane mapping. The proposed methodology is shown in Fig. 6, where the process is divided in three sub-processes:

1. Road surface extraction, based on the Normal vector using the RANdom SAmple Consensus (RANSAC) [18] algorithm, that is an iterative method to estimate parameters of a mathematical model from a set of observed data that contains outliers.

2. Lane marking extraction based on a 2D image obtained from the 3D point cloud so as to obtain elevation values. The 2D image correlates with the 3D point cloud.

3. Lane mapping: Lane markings in a local section are clustered and fitted into lane lines. Those point are clustered and refined according to the typology of the lane line belong to; continuous line, broken line, etc. To fill in the missing lanes caused by local occlusion, a global post-processing is adopted. In order to improve the robustness of lane mapping, textural saliency analysis is proposed using images obtained from MMS sensor. It is used to validate candidate lane lanes.

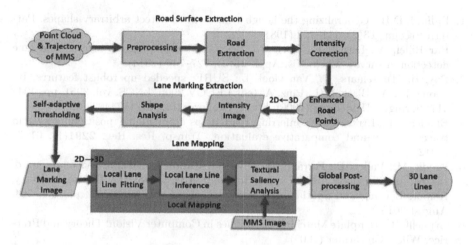

**Fig. 6.** Point cloud and image fusion proposal [22]

## 5   Conclusions

In this paper, different methodologies using different sensors to extract road's lane marks accurately have been presented. As it is explained in Sect. 4, a methodology that combines image and LiDAR is to be tackled since shortcomings of image based methods should be improved using LiDAR information and vice versa, leveraging advantages of MMS in order to use the information given from all sensors, and not only use partial information as it is commonly done. Currently, there is an effort to produce detailed road labeling in navigation systems, that are able to produce detailed and accurate localization of road features. This is also important for road catalog and maintenance. In this regard, actual data capture campaigns are being carried which use both LiDAR and image sensors. Camera calibration and fusion of multiple views are therefore of big importance to obtain accurate visual information for posterior processing in order to get signaling attributes (stop, speed limits, direction and other kind of signals). Attempts to apply modern deep learning techniques may be of paramount impact, due to the intrinsic flexibility and robustness of such techniques. However, data collection and labeling pose strong impediments for its systematic application.

**Acknowledgments.** The work in this paper has been partially supported by Airestudio Geoinformation Technologies Scoop and Basque Government's BIKAINTEK grant. The work has also been supported by FEDER funds for the MINECO project TIN2017-85827-P, and projects KK-2018/00071, KK-2018/00082 of the Elkartek 2018 funding program of the Basque Government.

# References

1. Ballard, D.H.: Generalizing the hough transform to detect arbitrary shapes. Pattern Recogn. **13**(2), 111–122 (1981)
2. Bar Hillel, A., Lerner, R., Levi, D., Raz, G.: Recent progress in road and lane detection: a survey. Mach. Vis. Appl. **25**(3), 727–745 (2014)
3. Bay, H., Tuytelaars, T., Van Gool, L.: SURF: speeded up robust features. In: Leonardis, A., Bischof, H., Pinz, A. (eds.) ECCV 2006. LNCS, vol. 3951, pp. 404–417. Springer, Heidelberg (2006). https://doi.org/10.1007/11744023_32
4. Biagioni, J., Eriksson, J.: Inferring road maps from global positioning system traces: survey and comparative evaluation. Transp. Res. Rec. **2291**(1), 61–71 (2012)
5. Brédif, M., Vallet, B., Ferrand, B.: Distributed dimensonality-based rendering of LIDAR point clouds. ISPRS - International Archives of the Photogrammetry, Remote Sensing and Spatial Information Sciences, vol. XL-3/W3, pp. 559–564, August 2015
6. Brunelli, R.: Template Matching Techniques in Computer Vision: Theory and Practice. Wiley, Chichester (2009)
7. Hammoudi, K., Dornaika, F., Paparoditis, N.: Generating virtual 3D model of urban street facades by fusing terrestrial multi-source data. In: 2011 7th International Conference on Intelligent Environments, pp. 330–333 (2011)
8. Hammoudi, K., Dornaika, F., Soheilian, B., Paparoditis, N.: Extracting outlined planar clusters of street facades from 3D point clouds. In: Proceedings of the 2010 Seventh Canadian Conference on Computer and Robot Vision (CRV 2010), pp. 122–129: 2010 Seventh Canadian Conference on Computer and Robot Vision (CRV 2010), 31 May–2 June 2010, ON, Canada, Ottawa (2010)
9. Hammoudi, K., Dornaika, F., Soheilian, B., Vallet, B., Paparoditis, N.: Generating occlusion-free textures for virtual 3d model of urban facades by fusing image and laser street data. In: IEEE Virtual Reality Conference 2012 Proceedings, IEEE Visualization and Graph Technical Communication, IEEE Computer Society (2012)
10. Hata, A., Wolf, D.: Road marking detection using LIDAR reflective intensity data and its application to vehicle localization. In: 17th International IEEE Conference on Intelligent Transportation Systems (ITSC), pp. 584–589 (2014)
11. Jun, W., Wu, T., Zheng, Z.: LIDAR and vision based pedestrian detection and tracking system. In: Proceedings of 2015 IEEE International Conference on Progress in Informatcs and Computing (IEEE PIC), pp. 118–122 (2015)
12. Levinson, J., et al.: Towards fully autonomous driving: systems and algorithms. In: 2011 IEEE Intelligent Vehicles Symposium (IV), pp. 163–168, June 2011

13. Lim, K., Hong, Y., Ki, M., Choi, Y., Byun, H.: Vision-based recognition of road regulation for intelligent vehicle. In: 2018 IEEE Intelligent Vehicles Symposium (IV), pp. 1418–1425 (2018)
14. Liu, X., Deng, Z.: Segmentation of drivable road using deep fully convolutional residual network with pyramid pooling. Cognitive Comput. 10(2), 272–281 (2018)
15. Lowe, D.G., et al.: Object recognition from local scale-invariant features. In: ICCV 1999, no. 2, pp. 1150–1157 (1999)
16. Mancini, A., Frontoni, E., Zingaretti, P.: Automatic road object extraction from mobile mapping systems. In: Proceedings of 2012 IEEE/ASME 8th IEEE/ASME International Conference on Mechatronic and Embedded Systems and Applications, pp. 281–286, July 2012
17. Nguyen, H.L., Belton, D., Helmholz, P.: Planar surface detection for sparse and heterogeneous mobile laser scanning point clouds. ISPRS J. Photogrammetry Remote Sens. 151, 141–161 (2019)
18. Patil, H., Deshmukh, S.S.: Homography estimation using RANSAC. IJREAT Int. J. Res. Eng. Adv. Technol. 1(3), 1–4 (2013)
19. Premebida, C., Ludwig, O., Nunes, U.: Lidar and vision-based pedestrian detection system. J. Field Robot. 26(9), 696–711 (2009)
20. Rogers, S.: Creating and evaluating highly accurate maps with probe vehicles. In: ITSC 2000, 2000 IEEE Intelligent Transportation Systems. Proceedings (Cat. No. 00TH8493), pp. 125–130 (2000)
21. Smadja, L., Ninot, J., Gavrilovic, T.: Road extraction and environment interpretation from LIDAR sensors. In: Paparoditis, N., Pierrot Deseilligny, M., Mallet, E., Tournaire, O. (eds.) PCV 2010 - Photogrammetric computer vision and image analysis, PT I, Volume 38 of International Archives of the Photogrammetry Remote Sensing and Spatial Information Sciences, pp. 281–286. ISPRS Tech Commiss (2010)
22. Wan, R., Huang, Y., Xie, R., Ma, P.: Combined lane mapping using a mobile mapping system. Remote Sens. 11(3), 305 (2019)
23. Wang, H., et al.: Automatic road extraction from mobile laser scanning data. In: 2012 International Conference on Computer Vision in Remote Sensing, pp. 136–139 (2012)
24. Wu, J., Yao, W., Polewski, P.: Mapping individual tree species and vitality along urban road corridors with LiDAR and imaging sensors: point density versus view perspective. Remote Sens. 10(9), 1403 (2018)
25. Zeng, X., Araki, S., Kakizaki, K.: An improved extraction method of individual building wall points from mobile mapping system data. In: 2016 Sixth International Conference on Innovative Computing Technology (INTECH), IEEE (2016)
26. Ziou, D., Tabbone, S.: Edge detection techniques - an overview. Int. J. Pattern Recogn. Image Anal. 8, 537–559 (1998)

# CAPAS: A Context-Aware System Architecture for Physical Activities Monitoring

Paulo Ferreira[1], Leandro O. Freitas[1(✉)], Pedro Rangel Henriques[1],
Paulo Novais[1], and Juan Pavón[2]

[1] ALGORITMI Centre, University of Minho, Braga, Portugal
pauloaferreira@gmail.com, leanfrts@gmail.com, {prh,pjon}@di.uminho.pt
[2] Facultad de Informática, University Complutense Madrid, 28040 Madrid, Spain
jpavon@fdi.ucm.es

**Abstract.** Attribute grammars are widely used by compiler-generators since it allows complete specifications of static semantics. They can also be applied to other fields of research, for instance, to human activities recognition. This paper aims to present CAPAS, a Context-aware system Architecture to monitor Physical ActivitieS. One of the components that is present in the architecture is the attribute grammar which is filled after the prediction is made according to the data gathered from the user through the sensors. According to some predefined rules, the physical activity is validated after an analysis on the attribute grammar, if it meets those requirements. Besides that it proposes an attribute grammar itself which should be able to be incorporated in a system in order to validate the performed physical activity.

**Keywords:** Attribute Grammar · Intelligent environment · Activity recognition

## 1 Introduction

Nowadays it is a well known fact that practicing physical activities can prevent a lot of diseases both in the short and long term. However, getting injured is relatively easy if the activity is not being executed in the right way. So, activity monitoring systems are of the vital importance.

Attribute Grammars are such an important part of a program once it defines its semantic. Semantic is what gives a program meaning. Attribute grammars were first introduced in [1] allowing compilers to process programming languages in a much easier way, since the parts from the compiler which dealt with semantics were becoming very complex considering the evolution of programming languages and computers.

Besides the consolidation of it in programming languages it is possible to apply this type of formalism to any kind of application. This exact thought was

© Springer Nature Switzerland AG 2019
H. Pérez García et al. (Eds.): HAIS 2019, LNAI 11734, pp. 636–647, 2019.
https://doi.org/10.1007/978-3-030-29859-3_54

what led to the development of this paper. In other words, applying an attribute grammar to a context-aware system which recognises physical activities and gives suggestions according to the context. The system acts both before and during the physical activity. For example, it can suggest the user to go for a run outside if the weather presents good conditions and during the physical activity it suggests a change on the activity if something is wrong. By using this kind of grammars in such application it is possible to get a more well-defined system in what concerts the validation of a physical activity. A physical activity is validated only if all of its sub activities are well executed.

The purpose of this work is to apply an attribute grammar to an human activity recognition platform when it comes to validate a physical activity performed by the user. The use of an attribute grammar allows to define semantic rules which should be satisfied to validate the activity. It defines what can be accepted or not. At first the Abstract Syntax Tree (AST) should be filled accordingly to the analysis of raw data. Then the system should evaluate the AST in order to validate the physical activity. The physical activity is composed of several activities. If at least one of the activities which compose the physical activity is not validated after the analysis of the grammar, a warning should be given to the user saying that the activity is not being properly executed.

The document is structured as follows. Section 2 presents concepts related to context, context-aware computing and attribute grammars. In Sect. 3 it is described the proposal that uses an attribute grammar for the recognition of human activity. After that, Sect. 4 presents the CAPAS architecture for context-aware systems, where the grammar will be applied. Following, Sect. 5 presents the discussion where some case studies are presented and explained. Finally, final considerations and future directions are presented in Sect. 6.

## 2   Background

### 2.1   Physical Activity Monitorization

The current development in technology allows to detect and recognise physical activities in a really easy way. Nowadays, it is possible to recognise activities such as running, lifting weights, push-ups, cycling and many others using different sensors like accelerometers, gyroscopes and heart rate [2]. With that in mind it is possible to monitor physical activity in order to prevent injuries, by analysing the patterns of an user on a certain activity for example.

Currently there are some works which focus on activity and physical activity monitoring. For instance, the work [3] reviews many of the used sensors in order to evaluate whether they are efficient or not. Besides that it describes the most used physiological parameters for this purpose. For example body temperature, heart rate and blood pressure. Also, it addresses some challenges when it comes to development of an application for activity monitoring. For instance, the fact that wearables need to be light in order to be comfortable when the user uses it. Energy is something that needs to be considered as well, the device's battery should last a reasonable amount of time. This work is quite helpful since it reviews a lot of sensors and helps when it comes to choose the proper sensors. Besides that, it mentions important factors which should be taken into account when developing this application as the mentioned problems, for instance.

Another interesting work is [4], where it was presented a smartphone based architecture with the goal to monitor patients that suffer from the disease Parkinson. That application detects the freezing of gait (FOG). Which causes a paralysis on the patient's muscles where the risk of falling highly increases. So the main goal is to detect it and warn someone whenever it happens. Even though it monitors the activity of the user as in the presented work. Its goal is not to give suggestions.

There is another related work [5] which monitors elderly and gives them physical activities suggestions according to their performance. It first starts by evaluating the execution of the activity. If the user performed it correctly it goes into a new activity, if he/she didn't perform it correctly the system keeps in the same activity until it is correctly executed. This is the most similar work to the one on this paper so many things are common between both works. However, the difference is that it uses a different approach when it comes to detect the activity. Instead of using wearables it uses cameras. Video-based activity recognition systems are widely studied in the literature [6–8]. However, this technique is not always the most used because it generates some distrust in the protection of the user's privacy. This is the reason why in this work, those recognition systems that use cameras to identify specific patterns are not taken into account.

What differentiates this work from previous studies is that with the attribute grammar semantic rules can be added to the model. For instance, the work [9] predicts activities with the same classification algorithm, which is support vector machine. However, a further validation of the workout with be harder. The semantic rules added by the attribute grammar, such as, a workout is composed of several activities turns the process of validation easier. For instance, in order to validate a workout, every single activity that composes that workout needs to be well executed. Therefore, just by analysing the abstract syntax tree it is possible to check if the exercises are well executed or not.

## 2.2    Context Awareness

Context awareness uses data from environments to adapt applications and services according to the changes detected there. The main goal of this process is to improve human assistance. Thus, this section aims to describes the concepts of context and context-aware computing that was adopted for the development of the research.

**Context.** As pointed in [10], there are several definitions about the term context in the literature. Some consider it as the environment or situation, others consider it to be the user's environment while others consider it as the environment of the application. According to [11], context is divided in two categories which are operational and conceptual. The former helps to understand issues and challenges that comes with each data acquisition technique. The latter helps to better understand the connection between contexts.

Another way to look at context is by the perspectives of active context. Which can be described as discovered context and passive context where the application presents the context to the user instantly or it can also store it for the user to retrieve later as defined in [12].

However, according to [10] context-aware applications look at the *who's*, *where's*, *when's* and *what's* of entities and according to that information it is possible to know why a situation is occurring. In spite of being important variables of context, they are not enough and there are certain types of context more important than others such as **location**, **identity**, **activity** and **time**. With the former ones (who, where, when and what) the application can only know information about location and identity. However, in order to characterise a situation, activity and time information are needed. These context types allow to answer the questions of *who, what, when,* and *where*. Besides that, it also acts as indices into other sources of contextual information.

For the development of this paper it was considered the definition of [10] which says that context is any information that can be used to describe the current situation of an entity. Where an entity can be a person, place, object or anything that can be relevant to the interaction between an user and an interface. As mentioned in [10], context consists only of implicit information.

**Context-Aware Computing.** Context awareness was first introduced by [13], where it was defined as "the ability of a mobile user's applications to discover and react to changes in the environment they are situated in" [13]. It refers to the identification of the current state of entities and how they influence the system. However to better comprehend context awareness, it is needed a better understanding on what context is. So this section discusses a definition of context and some other aspects about context awareness.

According to [12], context-aware computing is a mobile computing paradigm where applications discover user's contextual information such as location, time of the day and so on. Then, it takes advantage of that information in order to perform a certain activity. The work [12] fits context in two categories the active and the passive context. Those categories lead context-aware computing to be defined in two perspectives, active context awareness which automatically adapts to discovered context by changing the application's behaviour and passive context awareness which presents the new or updated context to an user or stores it in order to use it later.

Considering that, a context-aware application needs to use the context it is inserted in, in order to provide information or automatically execute a service, e.g., playing a song when the user arrives at a certain place or as presented in the case study [14], a message is displayed on his smart phone, for instance when it is time to take medicine. According to [13] a context-aware application adapts itself to the context. Another interesting and more general definition given at [10] says that a context-aware application uses context to not only adapt its behaviour according to it but it also uses context to display relevant information. For instance an application that only shows the temperature of a room, it is not modifying its own behaviour but it still is a context-aware application while the definition given by Schilit and Teimer [13] gives the idea that only reactive applications, like an application that turns on the air conditioner when the room is too hot, would be a context-aware application.

This application is able to give suggestions on the go due to the context-aware computing paradigm. It analyses the context the user is inserted in, fills the attribute grammar and gives the user a suggestion.

However, there are problems with context-aware applications. As pointed in [15], *Evaluation* is one of those problems. The process of evaluation on a context-aware system fail to take into account many context factors. The use of attribute grammars can reduce this problem. They allow to define a set of rules in order for the context to be validated.

## 2.3  Attribute Grammar

According to [16], the semantics of a context-free grammar is defined by an attribute grammar. It was previously used by compiler developers in order to define the static semantics of a programming language.

Nowadays, the components used in the semantics analysis phase by compiler-generators is generated automatically by attribute grammars which take it from the user's specification. According to [16], usually in the process of compilation there are three phases. The first one refers to the lexical part, which converts a stream of characters into a stream of terminal symbols. The second part is the syntactical phase, which converts those terminal symbols into an attributed syntax tree. The last phase is the semantic analysis which receives an attributed syntax tree as input in order to be examined and the obtained output is the evaluated attributed syntax tree.

Still according to [16], an attribute grammar is an extension of the notion of context-free grammars through two different types of attributes. The inherited attributes and synthesised attributes. The former are used in a top-down approach where it specifies the flow of information from a node while the latter characterise information in a bottom-up approach. The proposed attribute grammar in this work is composed of both inherited and synthesised attributes.

An attribute grammar is composed of non-terminal symbols, terminal symbols and productions. Productions are what define the rules of the grammar. It defines what can be accepted or not by the attribute grammar. Semantic rules express the relations between attribute values of different nodes [16]. Another important part of an attribute grammar is the testing for cycles since it is impossible to evaluate attributes which depend on themselves. The semantic rules need to be well defined. They need to lead to definitions of all attributes at all nodes of all derivation trees [1]. An attribute grammar does not contain a cycle if there is no attribute A which depends on B if B depends on A.

## 3    Attribute Grammar Proposal for Physical Activity Monitoring

This paper intends to present a formal validation of data related to human activities recognition. One of its goals is to present a solution for the process of validation of a physical activity performed by an user. It intends to recognise the performed activity and give instant feedback to the user whether it is being well executed or not. The data related to the activities is validated through an Attribute Grammar. Therefore the application will have three phases, the analysis of the raw data, then the grammar should verify the values accordingly to the previously analysed data and then the validation of the data present in the attribute grammar. Following it is presented the productions of the attribute grammar.

**p1:** *physicalActivity* → *id, type, activities, location, begin, duration*
**p2:** *activities* → *activity*+
**p3:** *name* → *name1, begin*
**p4:** *activity* → *name, duration, begin, execution*
**p5:** *duration* → *CT, begin*
**p6:** *execution* → *TRUE|reason*

The grammar considers that each physical activity is composed of an identification code (*id*) which identifies the physical activity, a *type* which can be for example upper if the physical activity has a focus on the upper part of the body. The *location* where it is being performed such as a park or a gymnasium, its *beginning*, which is equivalent to the beginning of the first activity, its *duration* and a set of *activities* [**p1**]. Regarding the *duration* of the whole physical activity, it results from the sum of each sub activity's duration. Besides that, it is a synthesised attribute since it depends on the value of attribute duration of every single activity.

First of all the *name* of the activity results from the combination of the name of the activity and its beginning [**p3**]. This attribute grammar considers that a physical activity should be composed of a set of *activities* since that it is the natural evolution of a physical activity. Usually a physical activity is composed of two or more *activities* [**p2**]. While an activity is composed of its *name, duration, beginning* and *execution* [**p4**].

The duration of a sub *activity* results from the current time and the *beginning* of it [**p5**]. The value of the symbol *execution* can vary accordingly to the user's execution of certain activity [**p6**]. If the user executes the activity correctly, it should hold the value TRUE. If not it should have the *reason* why it is not correctly executed. A user performs an activity correctly when the required parameters are met which were defined during the implementation of the application. For instance during a run, if the heart rate bpm is over 85% it can be considered the wrong way, therefore this should have the reason why it is a failed execution.

The production *execution* in the attribute grammar is particularly useful on applications which evaluate the correct execution of a physical exercise. If a user is not performing it correctly the system should not validate it and then it should recommend a change in the way the user is executing it. Apart from that, it can also be possible to infer some health problem with the user or even prevent it since a wrong movement can lead into a bad injury and being able to analyse it and correct it can prevent them. The system can detect an health problem by analysing the pattern throughout the time. For instance, if the user has been performing an activity correctly and there is one day in which the system detects some anomaly it might be possible that something is wrong.

The validation of the attribute grammar should be done following some requirements. The first one is that every activity present on the physical activity should have the value of the attribute execution set to TRUE. Meaning that it was correctly executed by the user. Then it analyses the rest of the data such as location, duration and all the other parameters. If those values are between a predefined range the attribute grammar should be validated.

## 4   Context-Aware System Architecture for Physical ActivitieS (CAPAS)

Following, in Fig. 1 it is possible to see the CAPAS architecture for context-aware systems that aim to monitor physical activities.

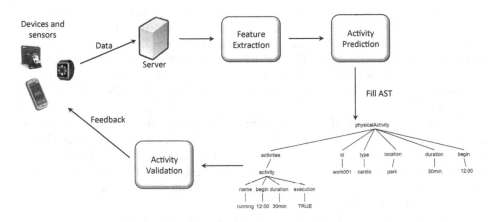

**Fig. 1.** The system architecture CAPAS

First, data is gathered from the user through sensors. There is all sorts of data gathered from many types of sensors such as accelerometer, gyroscope, heart rate and many more which might be useful to detect the activity. Then, the data is sent to the server through appropriated communication protocols. After that, data will go through the pre-processing phase. It is where data is taken care of. For example, null values and outliers are deleted and where other anomalies are corrected.

The next phase is the feature extraction phase. It is the most important step for the classification of the activity. The chosen features allow to differentiate the activities from each other. For example, mean, standard deviation, fourier transform and many others which might be useful for a specific activity.

Following the feature extraction step, there is the prediction of the activity, where according to the extracted features a prediction will be given. It uses a supervised approach to classify the activity which means that the system learned using labeled data [2]. Support Vector Machine is the used algorithm since it has an high accuracy on human activity recognition and it has been widely used for this kind of tasks for the past years [2]. Next, the Abstract Syntax Tree (AST) is filled according to the prediction made and the data gathered from the sensors. For the given example, which is running, the system first starts to generate an identification for the activity. Then the production *activities*, as only one activity was detected, a new production is made. First it starts to fill the *name* of it according to the activity which was predicted by the model. The *duration* is calculated as the difference between the current time and the beginning of it and for last the *execution*. It is calculated according to how the exercise was performed. It is possible to know if the exercise is correctly executed or not since for every exercise which is recognised by the model, it is trained with the most common errors in each of them. The parameter *type* is filled according to the category the exercise is inserted in. There is a pre defined database with all the exercises which can be recognised by the model and its respective category.

The last step is the validation of the physical activity. The system analyses the AST and according to the defined requirements it gives the user feedback about the activity. In order for a physical activity to be validated, every sub activity that composes the physical activity should be evaluated. For example, in a physical activity composed of two sub activities such as bicep curl and chest press, both should hold the value true in the parameter *execution*. Otherwise, it is not validated and the user is advised to change the way it is being executed.

## 5   Discussion

This section presents three case studies aiming to discuss the partial results as well as its contributions. The first case study addresses the validation of a cardio physical activity, the second one regards a physical activity which focus on the upper body and the last one is a football match. Following in Fig. 2 it is presented the Abstract Syntax Tree (AST) of an attribute grammar when the system recognises the activity of *running*. The values presented in the tree are filled accordingly to data acquired from heart rate and accelerometer sensors, since the recognised activity is *running* and those sensors are usually the chosen ones for this kind of activity [17]. After an analysis is finished on the raw data gathered from them, the system fills the attributes with the values in order to validate the physical activity.

However, for the physical activity to be validated the sub activities also need to be validated. This is done by analysing the production execution. In Fig. 2 it is

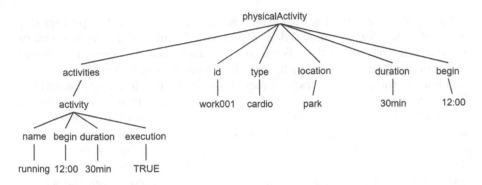

**Fig. 2.** Attribute grammar for a cardio physical activity

possible to see that it holds the value TRUE since according to the gathered data from several sensors the user is executing it correctly considering the system's requirements whether it be an heart rate beat per minute threshold, a wrong movement or any other pre defined parameter. What defines the correctness of the activity is implemented directly on the system.

Then, the validation of the whole structure is done after an analysis of the values. If they all meet the pre established requirements then it is validated. The validation means that the exercise was well executed so there is no need to warn the user. After the analysis on the sub activities that compose the physical activity it should check the other parameters and see if it corresponds to the expected ones. This way it is possible to detect any problem with the user and then it should be able to warn the user saying he is not performing it correctly.

Regarding the *location* it is particularly useful since it can be helpful when the system is recognising the activity. For example, if the user is in a park it might be possible that he/she is running. The system can also be used to monitor the user while he is at the gym performing a physical activity. For instance, Fig. 3 presents the Abstract Syntax Tree for a physical activity which focus on the upper body.

In this case, it is possible to see that this physical activity is composed of two activities, a bicep curl and a chest press. However, a physical activity can be composed by as many activities as the ones the user does.

Regarding the first activity, everything is as expected. However, the execution of the second activity holds the value *rightArmPosition*. Indicating that the user's right arm is not correctly positioned. Therefore the system will not validate the performed activity and will give a warn to the user saying that he did not perform it correctly so he/she needs to change the way. The sensors that were used to detect this activity were the gyroscope and accelerometer. The former was used to detect the position of the arms while the latter to detect the movement.

With the information that this exercise is not being correctly executed, an injury can be prevented, as already mentioned before or even have an hint that

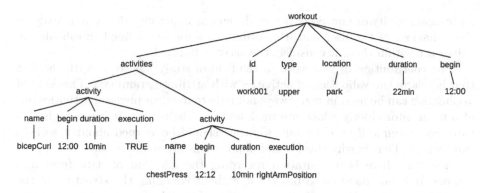

**Fig. 3.** Attribute grammar for an upper body physical activity

something is going on with the user speaking in terms of health. Also it can be used to monitor more general physical activities such as a football game. Figure 4 presents an AST for a football game.

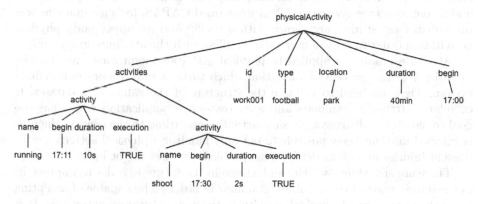

**Fig. 4.** Attribute grammar for football recognition

Even though a football game is composed of several activities, for this example were only considered two activities, shoot and running. Since it is hard to define what a correct way is, when it comes to shoot the ball for example. More activities were not considered because the AST would be too long to be represented here.

Considering the two activities represented on the tree, they should be validated since they are considered to be executed as the right way. The validation of the activity should be done after an analysis of the tree. If the parameters are as expected it should validate it. An activity is considered validated when it is being correctly executed by the user. Besides that, the conclusion of an activity should be precise since the type of suggestions change if the user is currently in a physical activity or not. So to detect the end of an activity the system analyses the behaviour of the user and if according to some pre-defined thresholds the values are within them it considers the activity finished.

The contribution of this work to the field of study has to do with the fact that it relates the validation of activities with attribute grammars. This kind of application can be used in many ways not only to monitor physical activities but also to monitor elderly which are weak and need help when it comes to health care. For example if an elder falls someone should be warned about it so help can be sent. That is why this kind of applications are really useful.

However, there is a limitation regarding the addition of data from new sensors. It is not possible to process it without changing the structure of the attribute grammar. Since all the parameters should have a pre-defined source from where it's data should be collected.

## 6   Final Considerations

This paper described an attribute grammar developed to be used to validate systems for physical activity monitoring. Its productions were described in details and a context-aware system architecture, named CAPAS, for such domains was presented. Case studies are shown with a cardio and an upper body physical activities in order to let the user more familiar with the attribute grammar.

Attribute grammars applied to physical activities monitoring provides one more layer in the process of validation which turns it into a more well defined system. They are used to validate the structure of the values of a dataset. It correlates attribute grammars and a context-aware application which has the goal to monitor and suggest physical activities according to the context the user is inserted in. Being now possible to not only monitor a physical activity but to prevent injuries as well as detect some problem the user might have.

There are still things which can be done in the future in order to improve it. For instance, expand the attribute grammar in order to be capable of accepting more than one type of physical activity in the same abstract syntax tree. It is quite usual for a person to do a specific physical activity for upper or lower body and complement it with a cardio physical activity. However, the current version of this attribute grammar treats it as two different physical activities.

Detection of health problems with the user could also be tackled. If the parameters of an exercise is different considering the user's pattern , the application could predict what is wrong with the user. Which would need to have new productions to describe the problem that happened and what it might be when talking about the human body. Future directions of research includes to analyse the efficiency of this proposal. At this point, it is not possible to do such thing.

**Acknowledgements.** "This work has been supported by FCT – Fundação para a Ciência e Tecnologia within the Project Scope: UID/CEC/00319/2019."

# References

1. Knuth, D.E.: Semantics of context-free languages. Math. Syst. Theor. **2**, 127–145 (1968)
2. Lara, O.D., Labrador, M.A.: A survey on human activity recognition using wearable sensors. IEEE Comm. Surv. Tutorials **15**(3), 1192–1209 (2013)
3. Mukhopadhyay, S.C.: Wearable sensors for human activity monitoring: a review. IEEE Sens. J. **15**(3), 1321–1330 (2015)
4. Capecci, M., Pepa, L., Verdini, F., Ceravolo, M.: A smartphone-based architecture to detect and quantify freezing of gait in Parkinson's disease. Gait Posture **50**, 08 (2016)
5. Angelo Costa, M.C., Martin, E.M., Pharos, V.J.: Physical assistant robot system. Sensors **18**(8), 2633 (2018)
6. Xia, L., Aggarwal, J.K.: Spatio-temporal depth cuboid similarity feature for activity recognition using depth camera. In: Proceedings of the IEEE Conference on Computer Vision and Pattern Recognition, pp. 2834–2841 (2013)
7. Ming, Y., Ruan, Q., Hauptmann, A.G.: Activity recognition from RGB-D camera with 3D local spatio-temporal features. In: 2012 IEEE International Conference on Multimedia and Expo, pp. 344–349. IEEE (2012)
8. Jalal, A., Kamal, S., Kim, D.: Shape and motion features approach for activity tracking and recognition from kinect video camera. In: 2015 IEEE 29th International Conference on Advanced Information Networking and Applications Workshops, pp. 445–450. IEEE (2015)
9. He, Z.-Y., Jin, L.-W.: Activity recognition from acceleration data using ar model representation and SVM. In: 2008 International Conference on Machine Learning and Cybernetics, vol. 4, pp. 2245–2250. IEEE (2008)
10. Abowd, G.D., Dey, A.K., Brown, P.J., Davies, N., Smith, M., Steggles, P.: Towards a better understanding of context and context-awareness. In: Gellersen, H.-W. (ed.) HUC 1999. LNCS, vol. 1707, pp. 304–307. Springer, Heidelberg (1999). https://doi.org/10.1007/3-540-48157-5_29
11. Alegre, U., Augusto, J.C., Clark, T.: Engineering context-aware systems and applications. J. Syst. Softw. **117**(C), 55–83 (2016)
12. Musumba, G.W., Nyongesa, H.O.N.: Context awareness in mobile computing a review. Int. J. Mach. Learn. Appl. **2**(1), 5 (2016)
13. Schilit, B.N., Theimer, M.M.: Disseminating active map information to mobile hosts. IEEE Network **8**(5), 22–32 (1994)
14. Freitas, L.O., Henriques, P.R., Novais, P.: Uncertainty in context-aware systems: a case study for intelligent environments. WorldCIST (2018)
15. Yujie, Z., Licai, W.: Some challenges for context-aware recommender systems. In: 2010 5th International Conference on Computer Science Education, August 2010, pp. 362–365 (2010)
16. Karol, S.: An introduction to attribute grammars. Department of Computer Science. Technische Universitat Dresden, Germany (2006)
17. Hassan, M.M., Uddin, M.Z., Mohamed, A., Almogren, A.: A robust human activity recognition system using smartphone sensors and deep learning. Future Gener. Comput. Syst. **81**, 307–313 (2018)

# Anomaly Detection Using Gaussian Mixture Probability Model to Implement Intrusion Detection System

Roberto Blanco[1,2], Pedro Malagón[1,2(✉)], Samira Briongos[1,2], and José M. Moya[1,2]

[1] LSI-Universidad Politecnica de Madrid, Madrid, Spain
[2] CCS-Center for Computational Simulation, Madrid, Spain
{r.bandres,malagon,samirabriongos,josem}@die.upm.es

**Abstract.** Network intrusion detection systems (NIDS) detect attacks or anomalous network traffic patterns in order to avoid cybersecurity issues. Anomaly detection algorithms are used to identify unusual behavior or outliers in the network traffic in order to generate alarms. Traditionally, Gaussian Mixture Models (GMMs) have been used for probabilistic-based anomaly detection NIDS. We propose to use multiple simple GMMs to model each individual feature, and an asymmetric voting scheme that aggregates the individual anomaly detectors to provide. We test our approach using the NSL dataset. We construct the normal behavior models using only the samples labelled as normal in this dataset and evaluate our proposal using the official NSL testing set. As a result, we obtain a F1-score over 0.9, outperforming other supervised and unsupervised proposals.

**Keywords:** Intrusion Detection · Gaussian Mixture Model · Voting

## 1 Introduction

In addition to general security concerns, service providers have to deal with attacks to their infrastructures, which can affect their service availability, their clients or industrial privacy, integrity or reliability of their solutions. Moreover, the irruption of the Internet of Things has lead to an exponential growth of the number of devices connected to the Internet. The challenges related to protect services, networks and devices are drastically increasing in complexity.

Rule-based protection mechanisms, such as firewalls, are not as effective as Intrusion Detection Systems (IDS) [5] when dealing with new security threats and complex systems. Intrusion Detection Systems are based on the assumption that an attack or an intrusion will change the pattern of resource usage or network flow. Traditionally, IDS are classified as signature-based or anomaly-based [1] depending on how they face detection. Signature-based detectors check if the collected samples match with known attacks, whereas anomaly-based detectors

© Springer Nature Switzerland AG 2019
H. Pérez García et al. (Eds.): HAIS 2019, LNAI 11734, pp. 648–659, 2019.
https://doi.org/10.1007/978-3-030-29859-3_55

build statistical models that characterize normal behavior and look for abnormal patterns. In practice, both approaches require to monitor network packets or to collect representative samples of the system they want to protect.

Training and evaluating a NIDS requires a comprehensive dataset that is representative of real traffic packets passing through a firewall. This dataset must contain normal and abnormal samples. Indeed, each sample of these datasets usually contains multiple features and its corresponding label. There are multiple datasets available for research purposes [15, 17] which are commonly used to train IDS and test their performance, efficiency and accuracy.

The IDS classifies the data into categories using different methods. Multiple machine learning algorithms have been proposed for implementing the classifier of the IDS, including both supervised and unsupervised algorithms. Supervised algorithms are capable of detecting known varieties of attacks. However, new or undocumented attacks may go undetected. For this reason, it is commonly suggested to implement IDS based on anomaly detection algorithms. Besides, building a labelled dataset to model the new Internet of Things applications, including normal traffic and attacks, can be more expensive (or even impossible) than generating one with only normal patterns.

Our proposal is, consequently, based on anomaly detection [8]. We use Gaussian Mixture Models (GMM) [20] to model normal behavior. We propose a set of classifiers, which evaluate the individual probability of each of the features of a sample, to be considered normal according to GMM. This information is then used as the input to a voting based aggregation method which decides if the sample is normal or abnormal. This method does not require any anomalous sample during the training phase.

We use the NSL-KDD [17] dataset for our experiments. Our approach obtains an F1-score over 0.9 using a test set with completely new traces which are not related to the training set, neither normal nor attacks. We compare our solution to existing supervised and unsupervised methods, and the voting GMM outperforms most of the considered algorithms.

## 2 Related Work

The main goal of an anomaly detection algorithm is to filter out outliers. This task is critical in many disciplines, including medical diagnosis, fault prediction, fraud detection or network intrusion detection [10].

According to Domingues et al. [6] anomaly detection algorithms are divided into different families. Probabilistic methods fit the behavior of the system in a set of known functions (Gaussian with GMM [20] or other generic functions in Kernel Density Estimators, KDE [12]). Distance-based algorithms, such as the Local Outlier Factor (LOF) [4], are applied to Gaussian models or clusters of neighbors (using K-means or K-nearest neighbors). Neural networks constitute another family in which the most common type of network used for anomaly detection is known as Self-Organizing-Maps (SOM) [21]. Finally, domain-based algorithms, such as one-class Support Vector Machines (SVM) [22] have been used to establish an irregular multi-dimensional boundary to the normal data.

Network intrusion detection systems (NIDS) have been evaluated with multiple and well-known datasets. The KDD99 dataset [11] was used in [13] considering new attacks. The NSL-KDD dataset was evaluated in [19] and the UNSW-NB15 dataset in [16]. NSL-KDD includes a separate official testing set, with traces of attacks and normal traffic not present in the training set, which makes it ideal for testing anomaly detection algorithms.

GMM based algorithms have been proposed to implement NIDS in [2]. In [14] fuzzy logic was used to implement clustering using GMM. In [3], they present a method that uses a lower dimensional space and adapts to changes in time. A majority voting scheme is used in [9] with votes in time windows to reduce noise in outlier detection. These algorithms, in order to be applied in real environments, need to be fast and select realistic features. For example, Dromard et al. [7] use a clustering anomaly detection algorithm to meet real time constraints.

## 3   Materials and Methods

### 3.1   Dataset Description

NSL-KDD dataset [17] is a refined version of the KDD cup 99 dataset (a well known benchmark for the research of Intrusion Detection techniques). It contains the essential records of its predecessor balancing the proportion of normal versus attack traces, and excluding redundant records. Each record is composed of 41 attributes unfolding four different types of features of the flow, and its assigned label which classifies it as an attack or as normal. These features include basic characteristics of each network connection vector such as the duration or the number of bytes transferred, content related features like the number of "root" accesses, contextual time related traffic features such as the number of connections to the same destination, and host based traffic features like the number of connections to the same port number. The whole amount of records covers one normal class and four attack classes grouped as denial of service (DoS), surveillance (Probe), unauthorized access to local super user (R2L) and unauthorized access from a remote machine (U2R).

### 3.2   Data Preprocessing

NSL-KDD dataset contains numeric and categorical features. The most convenient method for managing categorical features when feeding them to machine learning algorithms is the one hot encoding conversion. However, in this dataset there are only three categorical features (protocol, service and flag) that are not independent from each other. We have removed the service and flag and we have only selected tcp traces for our experiments, as this is the most relevant and abundant protocol. Moreover, our intention is to build an anomaly detection model that could be applied in a router node of the network, so we have also removed the content related features which the router should not be able to reach. After this process, the number of data features has been reduced to 24 and we have a train dataset including 53600 normal and 49040 attack records and a test dataset with 7842 normal and 10971 attack traces. It is important to mention that the test dataset includes attacks that have not being included in any entry of the training set.

## 3.3    Normalization

Since the range of values of the raw data varies widely, normalization is a must step for some machine learning algorithms. It allows to calculate distances between points using the Euclidean distance or even accelerates the convergence of many optimization algorithms such as gradient descent. We use feature scaling to adjust all column feature values into the range [0,1] and avoid large variations in data.

$$X' = \frac{X - X_{min}}{X_{max} - X_{min}}$$

## 3.4    Principal Component Analysis

Principal component analysis (PCA) is a statistical procedure that uses an orthogonal transformation to convert a set of observations of possibly correlated variables into a set of linearly uncorrelated variables called principal components. This transformation is defined in such a way that the first principal component has the largest possible variance (that is, accounts for as much of the variability in the data as possible), and each succeeding component has, in turn, the highest possible variance under the constraint that it is orthogonal to the preceding components. This technique is mostly used to reduce the dimensionality of the data while preserving the maximum amount of information among the features. The problem to solve becomes much simpler, as it deals with less features and the solution is still good enough. In this work, we are mainly interested in the capability of PCA to obtain uncorrelated features. Our original space has not too many dimensions, for this reason we do not care about dimensionality reduction. We propose to use the PCA technique in order to make a transformation that allows to consider each generated feature as if it was independent from the others. We know it is not really true but it is a better approximation if we make the assumption after the PCA. We have explored both approaches.

## 3.5    Feature Gaussian Mixture Probability Model

A Gaussian mixture model is a probabilistic model that assumes all the data points are generated from a mixture of a finite number of Gaussian distributions with unknown parameters (Fig. 1 left). For a given set of data we can apply an expectation-maximization statistical iterative algorithm and obtain which points come from which Gaussian latent component. The algorithm provides a classification of the points and the latent components, which are useful to our approach. Our goal is to implement an anomaly detection algorithm by modeling the normal behavior of a system. Assuming that every feature of our system normal traffic follows a Gaussian mixture distribution, we are able to obtain its latent components; i.e., we can estimate the mean and the variance of every Gaussian in the mixture. We consider an algorithm to distinguish normal from anomalous traffic using the obtained normal model. The simple Gaussian mixture model only gives us the probability of each sample to belong to every latent component of the mixture, but this is only useful when classifying and in our problem we don't know what an anomaly is and we should not use any attack record in the model building step. Therefore, we cannot use explicitly this model to detect any behavior different from the normal one. This is the reason why we need to obtain the latent components, because with them we can compute a probability of occurrence. With the assumption of the normal traffic in our system following a Gaussian mixture distribution we can obtain for a given traffic vector the latent component that each one of the vector features belongs with the highest probability. As we have characterized the latent components we can then compute the probability of occurrence for this traffic vector values following the corresponding latent components as if it was the worst case, that means that we compute the area under the normal curve for all the possible values with an absolute value greater than the analyzed value. For example, if we analyze the probability of occurrence for a value that match the mean of the latent component, this would be 1; on the other hand, if we consider the probability of occurrence for a value that match the mean plus the variance of the latent component, its probability wouldl be $1 - (0.3413 + 0.3413) = 0.3174$ (Fig. 1 right).

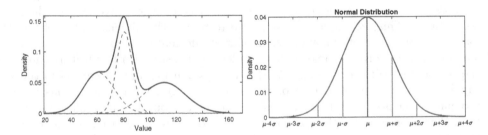

**Fig. 1.** Gaussian Mixture and Gaussian probabilities.

### 3.6   Probability Voting Scheme

The main point in this paper is the Feature Gaussian Mixture Probability Model that is supposed to obtain the occurrence probability of a certain value of a feature in a given traffic vector. This is in fact a probabilistic statistical model in which we expect that the normal values have a higher probability than the anomaly ones. In order to make decisions we need a method that aggregates all the features probabilities and applies a certain threshold so we can classify the entire traffic vector as normal or anomalous. Taking in account this idea we propose a simple voting scheme that evaluates each feature probability independently and then estimates the nature of the traffic vector based on the number of independent positive evaluations. Our method needs two hyperparameters, one for establishing the individual feature probability threshold and the other as the minimum number of anomalous features to consider the whole vector as an attack. We have called the first one as $\alpha$ and it is the percentage error we can afford when classifying normal features. For every normal feature in our training dataset, we compute the occurrence probability. The value of alpha represents the percentage of the training normal feature probabilities that will be considered as anomalous for the model, so the decision threshold will be set as the maximum occurrence probability in the $1 - \alpha$ remaining percentage. For simplicity this percentages are normalized from 0 to 1. We have called the second one as consensus and it is just the number of positive (feature probability larger than the threshold) evaluations needed to consider the whole traffic vector as anomalous.

### 3.7   Other Machine Learning Algorithms

In order to compare the proposed methods, other state-of-art algorithms are introduced.

**K-Means** is an unsupervised learning algorithm that is mostly used for clustering. Given a set of data vectors with the same number of features (dimensionality d) and a number of desired clusters c, the algorithm is able to seek and find the optimum c points in the space d that minimize the sum of squared distances of the whole set of data vectors to its closest point. At the end this means that the algorithm can organize the data in c groups or clusters making use of its underlying structure. We use the algorithm to solve a binary classification problem. Therefore, we could set the number of desired clusters to two. However, normal traffic can be distributed in more than one cluster. We apply the algorithm several times, varying the number of considered clusters, and then define each cluster as normal or anomalous looking at the proportion of normal and attack records that it contains. In our problem, K-Means is an appropriate method for building up a classifier due to its unsupervised nature. However, as we are trying to detect anomalies and we should not know how anomalies are before we want to detect them, K-Means in its standard way can only be applied for a reactive model and not for a predictive one. But we can use K-Means in another way as well, instead of try to make different clusters with the whole set of data and then try to identify which clusters are for each class we can only give the algorithm the normal class data that we want to model. This produces different clusters for only what we know it is normal traffic. Once we have these clusters, we measure the distances from the normal records to the centroids of the method and then the distances from the attacks to the same centroids. Ideally, we should obtain larger distances for the attacks than the distances for the normal traces, as we have built the algorithm to minimize the distances to the normal traces. This is what we called in this paper the K-Means distance method. It is a pure anomaly detection algorithm because it is unsupervised and only needs one class to model the normal scenario.

Anomaly detection is very similar to novelty detection, which detects a sample that is different to an initial set of data. Considering novelty detection algorithms, there is a well known method that is a variation of the **Support Vector Machine** algorithm with the objective of obtaining a membership decision boundary for only one class of data. As SVMs are max-margin methods, this algorithm does not model a probability distribution of the data. It only finds a function that is positive for regions with high density of points and negative for small densities. In this work, both SVM approaches are included, so the classifier has been called SVM-2 and the novelty detector has been named as SVM-1.

A **Decision Tree** is a flowchart-like structure in which each internal node represents a "test" or a decision on an attribute. Each branch represents the outcome of the test, and each leaf node represents a class label. The full paths from root to leaf represent classification rules. We can train a decision tree structure with input train data and a desired class or label output in a supervised learning framework in order to adapt it to our problem. This simple algorithm is very convenient to compare with in this paper because we propose a new voting scheme algorithm which solution is in fact very similar to a decision tree structure.

A **Multilayer Perceptron** (MLP) is a type of feedforward artificial neural network. It is typically composed by three different layers. The first layer is called the input layer and it is fed with the numerical values that we want to be the input of the network. The second layer is called the hidden layer and it usually contains several nodes or neurons. Each neuron is fully connected to all of the input values of the first layer and it applies a non-linear activation function to a linear weighted combination of the inputs. The last layer is the output layer and it has the same number of nodes as values we want to estimate. When the network is used in a classification problem the output layer is supposed to give an approximated probability for every class we want to distinguish. MLP utilizes a supervised learning technique called backpropagation for training so it is a supervised algorithm. Moreover, the nonlinear activation functions of the layers make it able to distinguish data that is not linearly separable. Unlike the Decision tree algorithm this is a parametric method, and once optimized it offers a complex mathematical function that approximates the solution to the problem.

## 4    Experimental Setup

We conduct a set of experiments using the same original dataset described in the proposal. We perform three different transformation techniques to the data in order to cover all the possibilities. We contemplate all the combinations so at the end we obtain eight different datasets that are just numerical transformed versions of the original one. Figure 2 shows the full decision diagram of data transformations applied to the original dataset to obtain each of the used input datasets. The normalization, principal component parameters and the Feature Gaussian Mixture Probability Model latent components for each feature are computed only with the information of the normal records in the training set. Once adjusted, the three techniques are applied to our training and testing dataset without changing any configuration.

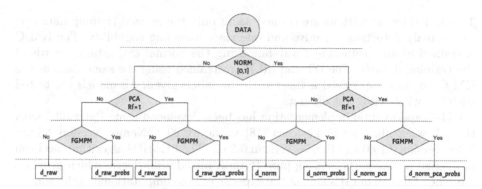

**Fig. 2.** Data transformation diagram

The eight generated datasets are:

- **d_raw**: The original NSL dataset without any transformation of the numerical values.
- **d_raw_probs**: We apply the FGMPM to the original NSL dataset values and change each feature value for the occurrence probability of each feature in the normal model.
- **d_raw_pca**: The uncorrelated version of the original NSL dataset with the same number of features.
- **d_raw_pca_probs**:We apply the FGMPM to the uncorrelated version of the original dataset and obtain the occurrence probabilities for this uncorrelated values of the features.
- **d_norm**: The original NSL dataset with the normal training values normalized to the range [0–1] and the remaining values normalized according to the previous scaler.
- **d_norm_probs**: We apply the FGPM to the normalized version of the dataset.
- **d_norm_pca**: The uncorrelated version of the normalized dataset.
- **d_norm_pca_probs**: The occurrence probabilities of the uncorrelated features of the normalized dataset.

For every mentioned dataset we build up six different models with the following algorithms:

- **Voting**: Our proposed voting scheme method for anomaly detection that can only be applied to the probability datasets.
- **KM-D**: The well known K-Means algorithm using the anomaly detection approach with the squared euclidean distances.
- **SVM**: A one class SVM for novelty detection.
- **KM-C**: K-Means algorithm in its standard clustering approach.
- **DT**: A default decision tree classifier.
- **MLP**: A simple multilayer perceptron with a hidden layer of 100 neurons and an output layer with 2 cells: attack or non-attack.

The first three algorithms are trained using only the normal training data due its anomaly detection objective and one class modeling capability. The KM-C is trained in an unsupervised way but using the normal and attack records of the training dataset. The DT and MLP are trained using the same data as the KM-C but in a supervised manner with the labels given. All models are tested with the whole NSL test dataset.

The experimental implementation has been developed using Python3.5 with the following libraries: Scikit-learn [18] version 0.20.2, Numpy version 1.13.0, Scipy version 0.19.0 and Pandas version 0.20.2. The FGMPM algorithm has been developed by the authors using pure Python mixed scikit-learn, which provides more flexibility in our research at the expense of less computational performance.

## 5    Results

We first introduce the set of metrics used to evaluate the performance of the proposed models with state of the art algorithms. Although we are facing an anomaly detection problem, our test can be considered as simple binary classification experiments: normal and attack traffic vectors. Therefore, we use the four basic metrics of the binary confusion matrix: True Positives (TP) and True Negatives (TN) for correctly classified records, and False Negatives (FN) and False Positives (FP) for the misclassified samples. These four values lead us to more interesting metrics in anomaly detection:

- *Sensitivity*: Positive detection rate.
- *Positive Predictive Value (PPV)*: True positives vs predicted positives rate.
- *Negative Predictive Value (NPV)*: True negatives vs predicted negatives rate.
- *F1 Score (F1)*: Harmonic mean between PPV and Sensitivity.
- *B*: Attack percentage in the whole testing dataset.
- *Intrusion Detection Capacity (CAP)*: A more complex and sensitive metric that relates the PPV and NPV with B and gives a very accurate idea of the complete performance of the model.

We select the three most interesting metrics for the anomaly detection systems for the evaluation: Sensitivity, in order to compare the anomaly detection rate, the F1-Score, as a measure of the test's accuracy, and Intrusion detection capacity (CAP), which best reflects the effectiveness of the models.

Table 1 shows the values obtained on the three selected metrics with every algorithm on every generated dataset. The detection methods are sorted, beginning with those who require less information to be trained. We highlight in red the experiments in which the algorithm does not converge to a valid solution. Also we have placed '-' where our proposed voting scheme makes no sense because the dataset is not composed by probabilities.

The SVM and the K-Means algorithms, in its classical approach, generally do not converge using not normalized data. It is an expected result, as both of them rely on the distances among the data. The K-Means algorithm, in its anomaly detection variant, does not converge well for the probabilities obtained after the PCA transformation. In general, supervised learning algorithms perform worse

**Table 1.** CAP, F1-Score and Sensitivity. E1 stands for d_norm, E2 for d_norm_probs, E3 for d_norm_pca, E4 for d_norm_pca_probs, E5 for d_raw, E6 for d_raw_probs, E7 for d_raw_pca and finally E8 for d_raw_pca_probs. The best results are highlighted in bold, whereas the experiments in which convergence was not achieved are in italic

|  |  | E1 | **E2** | E3 | **E4** | E5 | **E6** | E7 | **E8** |
|---|---|---|---|---|---|---|---|---|---|
| CAP | **Voting** | - | 0,4714 | - | **0,4972** | - | 0,4797 | - | **0,4958** |
|  | KM-D | **0,4502** | 0,3897 | **0,4502** | *0,0306* | 0,2155 | 0,4236 | 0,2155 | *0,0405* |
|  | SVM-1 | 0,3536 | 0,2011 | 0,3536 | 0,3536 | *0,0022* | 0,3067 | *0,0023* | 0,171 |
|  | KM-C | 0,42 | **0,5127** | 0,4215 | 0,4097 | *0,0005* | **0,5113** | *0,0005* | 0,4772 |
|  | DT | 0,3801 | 0,3456 | 0,3396 | 0,3347 | **0,3934** | 0,3659 | **0,408** | 0,3305 |
|  | SVM-2 | 0,3144 | 0,3557 | 0,3144 | 0,3144 | 0,3271 | 0,3366 | 0,2873 | 0,3216 |
|  | MLP | 0,3505 | 0,305 | 0,3572 | 0,3318 | 0,3401 | 0,3027 | 0,3136 | 0,3333 |
| F1 | **Voting** | - | 0,8703 | - | **0,8838** | - | 0,8715 | - | **0,9061** |
|  | KM-D | 0,8558 | 0,8499 | 0,8558 | *0,0163* | 0,7169 | 0,8475 | 0,7169 | *0,4066* |
|  | SVM-1 | 0,7729 | 0,6255 | 0,7729 | 0,7729 | *0,7372* | 0,7303 | *0,7373* | 0,6177 |
|  | KM-C | **0,8631** | **0,8976** | **0,8636** | 0,8826 | *0,7369* | **0,8981** | *0,7369* | 0,9054 |
|  | DT | 0,7766 | 0,7504 | 0,7429 | 0,7458 | **0,7877** | 0,7681 | **0,8015** | 0,7364 |
|  | SVM-2 | 0,7171 | 0,7565 | 0,7171 | 0,7171 | 0,7568 | 0,7382 | 0,72 | 0,7678 |
|  | MLP | 0,7513 | 0,7088 | 0,7575 | 0,8151 | 0,7796 | 0,7025 | 0,7972 | 0,7967 |
| Sensitivity | **Voting** | - | 0,7952 | - | 0,8184 | - | 0,7944 | - | 0,9032 |
|  | KM-D | 0,7692 | 0,7865 | 0,7692 | *0,0088* | 0,5946 | 0,7616 | 0,5946 | *0,2715* |
|  | SVM-1 | 0,6397 | 0,4651 | 0,6397 | 0,6397 | *0,9934* | 0,5839 | *0,9938* | 0,4635 |
|  | KM-C | **0,8046** | **0,8512** | **0,8048** | **0,8913** | *0,9999* | **0,8539** | *0,9999* | **0,9495** |
|  | DT | 0,6392 | 0,6055 | 0,5954 | 0,6008 | 0,6545 | 0,6287 | 0,6746 | 0,5875 |
|  | SVM-2 | 0,5628 | 0,6128 | 0,5628 | 0,5628 | 0,6203 | 0,589 | 0,5733 | 0,5836 |
|  | MLP | 0,6059 | 0,553 | 0,6139 | 0,7323 | **0,6555** | 0,5447 | **0,7018** | 0,6911 |

than unsupervised algorithms in our experiment. This is because of the fact that the testing dataset has different attacks than the training dataset, so the supervised algorithms cannot generalize as good as unsupervised ones. Although supervised algorithms seems to be always worse, they have a very high specificity, higher than unsupervised algorithms. The best result in the table is achieved by the K-Means clusters for the d_raw_pca_probs dataset. However, our voting scheme has a higher CAP than KM-C with this data because the overall performance of the model is better, although the KM-C has a higher anomaly detection rate. KM-C and our Voting scheme are the best algorithms here, but we have to consider that KM-C needs attacks in its training phase and does not provide a strictly normal model. On the other hand, our voting scheme offers a good normal model but it always needs the occurrence probabilities to be computed. Regarding the two hyperparameters of the Voting scheme, the best performance is achieved for a value of alpha equal to 0,013 and a consensus of 5. It has been noticed that an alpha increase can bee compensated with a consensus decrease and vice-versa in order to achieve a good performance. The KM-D algorithm seems to be a good alternative when we cannot normalize the data nor compute these probabilites.

# 6    Conclusions

Considering anomaly detection, unsupervised models are better suited to the real scenario, with unknown or untagged attacks or anomalies in datasets. We evaluate the impact of different preprocessing on the anomaly detection performance of different algorithms. We consider normalization, PCA and the probabilities of normal features with GMM. Moreover, we propose a Voting scheme algorithm. We train and test our voting scheme and multiple known unsupervised algorithms. The best results are obtained using KM-C and our Voting scheme, although the latter requires less information than the former. Considering the preprocessing, normalized data usually leads to a better performance. However, using the probabilities of normal features with GMM in NIDS with NSL-KDD, not normalized data generates more accurate probabilites and more sensitive detection algorithms. The PCA slightly improves the sensitivity of the anomaly detection algorithms, while it seems to have less effect with supervised algorithms. Finally, we have proved that using the occurrence probabilities improves the performance of the anomaly detection models and, specially, it allows the usage of a simple voting scheme to achieve a very good detector with F1-scores over 0.88 and CAP over 0.49, better than other more complex algorithms evaluated.

**Acknowledgements.** This work was supported by the Spanish Ministry of Economy and Competitiveness under contracts TIN-2015-65277-R, AYA2015-65973-C3-3-R and RTC-2016-5434-8.

# References

1. Axelsson, S.: Intrusion detection systems: a survey and taxonomy. Chalmers University of Technology, Tech. rep. (2000)
2. Bahrololum, M., Khaleghi, M.: Anomaly intrusion detection system using Gaussian mixture model. In: 2008 Third International Conference on Convergence and Hybrid Information Technology, November 2008, vol. 1, pp. 1162–1167. https://doi.org/10.1109/ICCIT.2008.17
3. Barkan, O., Averbuch, A.: Robust mixture models for anomaly detection. In: 2016 IEEE 26th International Workshop on Machine Learning for Signal Processing (MLSP), September 2016, pp. 1–6. https://doi.org/10.1109/MLSP.2016.7738885
4. Breunig, M.M., Kriegel, H., Ng, R.T., Sander, J.: LOF: identifying density-based local outliers. In: Chen, W., Naughton, J.F., Bernstein, P.A. (eds.) Proceedings of the 2000 ACM SIGMOD International Conference on Management of Data, 16–18 May 2000, Dallas, Texas, USA, pp. 93–104. ACM (2000). https://doi.org/10.1145/342009.335388
5. Denning, D.E.: An intrusion-detection model. IEEE Trans. Softw. Eng. **13**(2), 222–232 (1987). https://doi.org/10.1109/TSE.1987.232894
6. Domingues, R., Filippone, M., Michiardi, P., Zouaoui, J.: A comparative evaluation of outlier detection algorithms: experiments and analyses. Pattern Recogn. **74**, 406–421 (2018)

7. Dromard, J., Roudière, G., Owezarski, P.: Online and scalable unsupervised network anomaly detection method. IEEE Trans. Netw. Serv. Manage. **14**(1), 34–47 (2017). https://doi.org/10.1109/TNSM.2016.2627340
8. Heady, R., Luger, G., Maccabe, A., Servilla, M.: The architecture of a network level intrusion detection system. Tech. rep., Los Alamos National Lab., NM, United States, New Mexico University, Albuquerque (1990)
9. Hock, D., Kappes, M.: A self-learning network anomaly detection system using majority voting. In: Dowland, P., Furnell, S., Ghita, B.V. (eds.) Proceedings Tenth International Network Conference, INC 2014, Plymouth, UK, 8–10 July 2014, pp. 59–69. Plymouth University (2014). http://www.cscan.org/openaccess/?paperid=225
10. Hodge, V.J., Austin, J.: A survey of outlier detection methodologies. Artif. Intell. Rev. **22**(2), 85–126 (2004). https://doi.org/10.1007/s10462-004-4304-y
11. Kdd cup 1999, October 2007. http://kdd.ics.uci.edu/databases/kddcup99/kddcup99.html
12. Kim, J., Scott, C.D.: Robust kernel density estimation. J. Mach. Learn. Res. **13**(1), 2529–2565 (2012). http://dl.acm.org/citation.cfm?id=2503308.2503323
13. Kukielka, P., Kotulski, Z.: Analysis of neural networks usage for detection of a new attack in IDS. Ann. UMCS Inf. **10**(1), 51–59 (2010)
14. Liu, D., Lung, C., Lambadaris, I., Seddigh, N.: Network traffic anomaly detection using clustering techniques and performance comparison. In: 2013 26th IEEE Canadian Conference on Electrical and Computer Engineering (CCECE), May 2013, pp. 1–4. https://doi.org/10.1109/CCECE.2013.6567739
15. Moustafa, N., Slay, J.: UNSW-NB15: a comprehensive data set for network intrusion detection systems (UNSW-NB15 network data set). In: Military Communications and Information Systems Conference (MilCIS), pp. 1–6. IEEE Stream (2015)
16. Moustafa, N., Slay, J.: The evaluation of network anomaly detection systems: statistical analysis of the UNSW-NB15 data set and the comparison with the KDD99 data set. Inf. Secur. J. A Global Perspect. **25**(1–13), 1–14 (2016)
17. NSL-KDD data set for network-based intrusion detection systems, March 2009. http://nsl.cs.unb.ca/NSL-KDD/
18. Pedregosa, F., et al.: Scikit-learn: machine learning in Python. J. Mach. Learn. Res. **12**, 2825–2830 (2011)
19. Revathi, S., Malathi, A.: A detailed analysis on NSL-KDD dataset using various machine learning techniques for intrusion detection. Int. J. Eng. Res. Tech. **2**(12), 1848–1853 (2013)
20. Reynolds, D.D.: Gaussian Mixture Models. In: Li, S.Z., Jain, A. (eds.) Encyclopedia of Biometrics. Springer, Boston (2009). https://doi.org/10.1007/978-0-387-73003-5
21. Shahreza, M.L., Moazzami, D., Moshiri, B., Delavar, M.: Anomaly detection using a self-organizing map and particle swarm optimization. Scientia Iranica **18**(6), 1460–1468 (2011). https://doi.org/10.1016/j.scient.2011.08.025
22. Zhang, R., Zhang, S., Muthuraman, S., Jiang, J.: One class support vector machine for anomaly detection in the communication network performance data. In: Proceedings of the 5th Conference on Applied Electromagnetics, Wireless and Optical Communications, pp. 31–37. ELECTROSCIENCE'07, World Scientific and Engineering Academy and Society (WSEAS), Stevens Point (2007)

# Combining *Random Subspace* Approach with SMOTE Oversampling for Imbalanced Data Classification

Paweł Ksieniewicz[✉] [iD]

Wrocław University of Science and Technology, Wrocław, Poland
pawel.ksieniewicz@pwr.edu.pl

**Abstract.** Following work tries to utilize a hybrid approach of combining *Random Subspace* method and SMOTE oversampling to solve a problem of imbalanced data classification. Paper contains a proposition of the ensemble diversified using Random Subspace approach, trained with a set oversampled in the context of each reduced subset of features. Algorithm was evaluated on the basis of the computer experiments carried out on the benchmark datasets and three different base classifiers.

**Keywords:** Imbalanced classification · SMOTE · Random Subspace · Classifier ensembles

## 1 Introduction

A major part of the pattern recognition problems presents the task of classification, in which we train a model capable of assigning new, unknown objects to predefined groups on the basis of a knowledge extracted from a set of labeled patterns [5]. Most classical classification algorithms assume an equal percentage of each class and encounters a problem when the proportions between them are strongly disturbed, tending to favor prediction of the more common one. Data about this characteristic is called *imbalanced data* [22]. Most of the real problems, such as diagnosis of diseases, SPAM-detection or fraud recognition, require detection of events far from normal and therefore are not balanced, which makes necessary to modify the pattern recognition models for their needs [10].

In order to eliminate this problem and to construct a model capable of classifying imbalanced data, three approaches are most commonly used [13]. The first of these are methods of data pre-processing, in which we do not modify the learning process, but we introduce changes in the training set itself. The simplest examples are *random undersampling*, in which we take into training set the full minority class and a random subset of the same size from majority class and *random oversampling*, where we use the full majority class and randomly selected objects of the minority class of the same cardinality, regardless of the repetition of the patterns [6].

© Springer Nature Switzerland AG 2019
H. Pérez García et al. (Eds.): HAIS 2019, LNAI 11734, pp. 660–673, 2019.
https://doi.org/10.1007/978-3-030-29859-3_56

More complex solutions of this type are *oversampling* algorithms, which instead of repeating existing minority class patterns, generate new synthetic objects based on the information contained in their distribution. The most common of them are ADASYN [9] and SMOTE [3], developed into a multitude that also takes into account the distribution of the majority class of varieties such as *Borderline*-SMOTE [8], *Safe-Level*-SMOTE [2] or LN-SMOTE [18].

Another approach is to use *inbuilt mechanisms* into the classifier learning process itself. The most commonly used is the one-class classification, insensitive to the distribution of classes of the problem [14] and the cost-sensitive classification that takes into account the loss-function asymmetric in favor of the minority class [10].

The final group of considered approaches are hybrid solutions that combine preprocessing methods with classifier ensembles. The modifications of *Bagging* and *Boosting* are the most popular [20], but there are also methods based on combining a team of classifiers built on the basis of various methods of oversampling or random split undersampling [15].

An important factor that we must take into consideration during the experiments on imbalanced data is also the metric used to assess the quality of constructed models [12]. Typical accuracy, in a strongly imbalanced problem, gives us results being far from the truth, showing, for example, 90% accuracy when wrongly classifying the entire minority class occurring in one in ten samples. In binary classification problems, therefore, the most-used are taking into account the proportions of the measurement classes F-measure or geometric mean score. In multi-class problems, where the above measures can not be calculated, we use the most often the balanced accuracy score.

Following paper attempts to propose a new hybrid method, based on classifier ensembles built in accordance with the *Random Subspace* [26] principle common for multidimensional data and the SMOTE algorithm used for oversampling of objects in the subspaces of each of the member classifiers [24]. Previous studies have already used a combination of these methods, but *oversampling* is performed there before application of *Random Subspace*, which may not properly use the profits achieved by finding a subspace that allows effective determination of the decision boundary [11].

The main contributions of this work are:

- Proposition of the method of joint use of weighted classifier ensemble obtained by the *Random Subspace* method with the use of SMOTE oversampling.
- Implementation of the proposed method in varieties taking into account the separate use of each of its elements.
- Experimental evaluation of the impact of SMOTE, *Random Subspace* and weighing the ensemble fuser on the quality of imbalanced data classification.

## 2    Method Design

*Random Subspace.* The construction of the classifier ensemble gives us two basic difficulties [25]. The first is to provide a diverse pool of classifiers that allow to perform independent, parallel prediction. We can achieve this by using different classification algorithms or various subsets of the training set. The proposed method uses the second approach, where each member of the committee is built on a random subspace of the training set.

SMOTE. Before training each member of the ensemble, minority oversampling is performed using the basic version of the SMOTE algorithm. Training takes place using the base classifier chosen by the experimenter, which must, however, be a probabilistic classifier, or at least have probabilistic interpretation.

*Fuser.* The second difficulty before the effective construction of the classifier ensemble is the construction of its *fuser* – the function responsible for making decisions on behalf of the ensemble based on the opinions of its members [16]. Among the concepts used, according to names popular in implementations, there are *hard fusers* – based on voting principles and *soft fusers* – based on support accumulation. The use of probabilistic base classifiers allows in this case the use of a method with accumulation of support, additionally enriched by weighing. The weights of member classifiers in the proposed method will be their quality measured with the *f-measure* metric determined for the training data. Utilized F-score is determined as ratio between duplicated product of precision and recall relative to its sum [21].

The full processing scheme of the method proposed in this paper is presented in Fig. 1. The available training set is divided into a random subsets of features using the *Random Subspace* method and minority class in all the subspaces is independently oversampled using the SMOTE method. The structure of classifiers constructed in this way allows for prediction by support accumulation weighted with *f-measure* obtained on a training set.

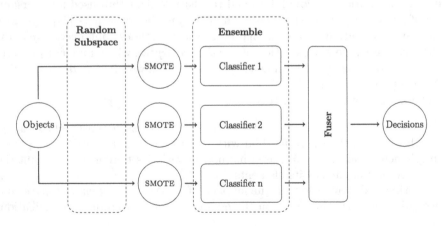

**Fig. 1.** Random subspace approach to build classifier ensemble

# 3   Experiments Set-Up

The experimental evaluation was carried out on the basis of 30 datasets with various imbalance ratio, available in the KEEL *data repository* [1]. The application of the *Random Subspace* method was based on arbitrarily set 30 subspaces using three features each [23]. As base classifiers, three standard methods with probabilistic interpretation were adopted:

1. Gaussian Naive Bayes
2. Logistic Regression
3. Support Vector Machine

The method was programmed in accordance with the programming interface of the *scikit-learn* environment [19], so it was possible to use the basic classifiers implemented in it. The used implementation of the SMOTE algorithm was the version available in the *imbalanced-learn* library [17].

The study identified and tested six approaches for each base classifier:

1. Base method.
2. SMOTE oversampling
3. RS – *Random Subspace*
4. RS+SMOTE – Combined *Random Subspace* and SMOTE
5. WRS – *Random Subspace* weighted by *f-score* obtained on a training set
6. WRS+SMOTE – Combined *Random Subspace* and SMOTE weighted by *f-score* obtained on a training set.

In the course of the experiment stratified 5-fold crossvalidation was used [4], measuring quality of all predictions using the *f-score* method. Tests analyzing statistical dependence were carried out using the Wilcoxon test [7]. The full code of the proposed solution as well as the experiments themselves was made available through the public Git repository[1].

# 4   Experimental Evaluation

The detailed results of the experiments, including the averaged f-measure for all folds of cross-validation along with the evaluation of the statistical dependence of the analyzed solutions for each of the data sets are presented in Tables 1, 2 and 3 respectively for all three base classifiers.

Table 4 contains the ranks determined by ranked Wilcoxon test for each of the analyzed methods and base classifiers. Based on them, assuming the levels of significance .9 and .95, a summary was prepared, counting the cases of advantage of each solution over the others and stored in Table 5.

As may be observed from the obtained results, which was also in line with expectations, regardless of the underlying classifier, using only SMOTE, without

---

[1] https://github.com/w4k2/wrssmote.

**Table 1.** Classification results on *Gaussian Naive Bayes*

| Dataset | Base method | SMOTE | RS | RS+SMOTE | WRS | WRS+SMOTE |
|---|---|---|---|---|---|---|
| *australian* | 0.719 | 0.718 | 0.755 | 0.795 | **0.859** | **0.863** |
|  | — | — | — | [1,2,3] | [1,2,3,4] | [1,2,3,4] |
| *glass-0-1-2-3-vs-4-5-6* | **0.709** | **0.718** | 0.694 | **0.744** | 0.680 | **0.737** |
|  | — | — | — | — | — | — |
| *glass-0-1-4-6-vs-2* | 0.219 | 0.231 | 0.292 | 0.270 | 0.266 | 0.226 |
|  | — |  |  |  |  |  |
| *glass-0-1-5-vs-2* | 0.218 | 0.161 | 0.016 | 0.198 | 0.218 | 0.176 |
|  | [3] | [3] | — | [3] | — | [3] |
| *glass-0-1-6-vs-2* | 0.199 | 0.190 | 0.192 | 0.228 | 0.155 | 0.193 |
|  |  |  |  |  |  |  |
| *glass-0-1-6-vs-5* | 0.760 | 0.760 | 0.636 | 0.557 | 0.717 | 0.648 |
|  | — | — | — | — | — | — |
| *glass-0-4-vs-5* | 0.960 | 0.893 | 0.727 | 0.727 | 0.893 | 0.893 |
|  |  |  |  |  |  |  |
| *glass-0-6-vs-5* | 0.893 | 0.893 | 0.733 | 0.670 | 0.796 | 0.664 |
|  | — | — |  |  |  | — |
| *glass0* | 0.642 | 0.639 | 0.633 | 0.644 | 0.644 | 0.649 |
|  | — | — |  |  |  |  |
| *glass1* | 0.604 | 0.638 | 0.504 | 0.626 | 0.603 | 0.626 |
|  | — | — |  |  |  |  |
| *glass2* | 0.189 | 0.193 | 0.177 | 0.181 | 0.202 | 0.203 |
|  | — | — |  | — | — | — |
| *glass4* | 0.237 | 0.509 | 0.200 | 0.580 | 0.200 | 0.567 |
|  | — | — |  | — | — | — |
| *glass5* | 0.768 | 0.768 | 0.591 | 0.498 | 0.659 | 0.590 |
|  | — | — |  | — | — | — |
| *glass6* | 0.772 | 0.786 | 0.815 | 0.826 | 0.800 | 0.811 |
|  | — | — |  | — | — | — |
| *heart* | 0.802 | 0.808 | 0.797 | 0.800 | 0.827 | 0.808 |
|  | — | — | — | — | — | — |
| *hepatitis* | 0.719 | **0.851** | 0.649 | 0.588 | **0.917** | **0.893** |
|  | — | — | — | — | [1,4] | [4] |

**Table 1.** (*continued*)

| Dataset | Base method | SMOTE | RS | RS+SMOTE | WRS | WRS+SMOTE |
|---|---|---|---|---|---|---|
| page-blocks-1-3-vs-4 | **0.493** | **0.540** | **0.470** | **0.498** | **0.435** | **0.471** |
|  | — | — | — | — | — | — |
| pima | **0.621** | **0.665** | 0.523 | **0.657** | 0.597 | **0.664** |
|  | 3 | 3,5 | — | 3,5 | 3 | 3,5 |
| shuttle-c0-vs-c4 | 0.980 | 0.980 | **0.975** | **0.981** | **0.996** | **1.000** |
|  |  |  |  |  |  | 1,2 |
| vowel0 | **0.709** | **0.592** | 0.057 | **0.568** | **0.426** | **0.569** |
|  | 3 | 3 | — | 3 | 3 | 3 |
| wisconsin | **0.943** | **0.944** | **0.959** | **0.957** | **0.959** | **0.959** |
|  |  |  |  |  |  |  |
| yeast-0-2-5-6-vs-3-7-8-9 | **0.262** | **0.488** | 0.023 | **0.377** | 0.113 | **0.471** |
|  | 3 | 3,5 | — | 3 | — | 3,5 |
| yeast-0-2-5-7-9-vs-3-6-8 | 0.201 | 0.173 | **0.619** | 0.195 | **0.706** | **0.650** |
|  | — | — | 1,2,4 | — | 1,2,4 | 1,2,4 |
| yeast-0-3-5-9-vs-7-8 | **0.269** | **0.216** | **0.244** | **0.186** | **0.151** | **0.080** |
|  | — | 6 | — | — | — | — |
| yeast-0-5-6-7-9-vs-4 | 0.175 | 0.180 | **0.386** | 0.193 | **0.429** | **0.445** |
|  | — | — | 1,2 | — | — | 1,2,4 |
| yeast-2-vs-4 | 0.297 | 0.270 | **0.574** | 0.312 | **0.743** | **0.671** |
|  | — | — | 1,2,4 | — | 1,2,4 | 1,2,4 |
| yeast-2-vs-8 | 0.254 | 0.175 | **0.683** | 0.109 | **0.658** | **0.658** |
|  | — | — | 1,2,4 | — | 2,4 | 2,4 |
| yeast1 | 0.457 | 0.458 | **0.557** | 0.484 | 0.489 | **0.557** |
|  | — | — | 1,2,4,5 | 1,2 | 1,2 | 1,2,4,5 |
| yeast3 | 0.236 | 0.252 | **0.642** | 0.243 | **0.608** | **0.686** |
|  | — | — | 1,2,4 | — | 1,2,4 | 1,2,4 |
| yeast5 | 0.154 | 0.192 | 0.181 | 0.107 | **0.582** | **0.485** |
|  | 4 | 4 | 4 | — | 1,2,3,4 | 1,2,3,4 |

using Random Subspace, positively influences the quality of classification against the training of the classifier just with the original data set. At the same time, using only *Random Subspace* leads to very poor results, even worse than the base method.

**Table 2.** Classification results on *Logistic Regression*

| Dataset | Base method | SMOTE | RS | RS+SMOTE | WRS | WRS+SMOTE |
|---|---|---|---|---|---|---|
| *australian* | **0.842** | **0.846** | 0.841 | **0.859** | **0.859** | **0.853** |
| | — | — | — | — | — | — |
| *glass-0-1-2-3-vs-4-5-6* | **0.775** | **0.807** | 0.547 | **0.747** | 0.583 | **0.720** |
| | — | — | — | — | — | — |
| *glass-0-1-4-6-vs-2* | 0.000 | **0.258** | 0.000 | **0.290** | 0.000 | **0.286** |
| | — | 1,3,5 | — | — | — | — |
| *glass-0-1-5-vs-2* | 0.000 | **0.184** | 0.000 | **0.172** | 0.000 | **0.181** |
| | — | 1,3,5 | — | 1,3,5 | — | 1,3,5 |
| *glass-0-1-6-vs-2* | 0.000 | **0.176** | 0.000 | **0.196** | 0.000 | **0.282** |
| | — | 1,3,5 | — | 1,3,5 | — | — |
| *glass-0-1-6-vs-5* | 0.149 | **0.451** | 0.000 | **0.519** | 0.000 | **0.519** |
| | — | 3,5 | — | 3,5 | — | 3,5 |
| *glass-0-4-vs-5* | 0.367 | **0.720** | 0.000 | **0.829** | 0.162 | **0.829** |
| | — | 3,5 | — | 3,5 | — | 3,5 |
| *glass-0-6-vs-5* | 0.347 | **0.584** | 0.000 | **0.638** | 0.000 | **0.657** |
| | 3,5 | 3,5 | — | 3,5 | — | 3,5 |
| *glass0* | 0.516 | **0.678** | 0.237 | **0.675** | 0.369 | **0.678** |
| | 3 | 3,5 | — | 3,5 | — | 3,5 |
| *glass1* | 0.243 | **0.568** | 0.000 | **0.553** | 0.133 | **0.538** |
| | 3,5 | 1,3,5 | — | 1,3,5 | 3 | 1,3,5 |
| *glass2* | 0.000 | **0.167** | 0.000 | **0.202** | 0.000 | **0.195** |
| | — | 1,3,5 | — | 1,3,5 | — | 1,3,5 |
| *glass4* | 0.167 | **0.578** | 0.000 | **0.591** | 0.200 | **0.566** |
| | — | 1,3 | — | 1,3 | — | 1,3 |
| *glass5* | 0.149 | **0.510** | 0.000 | **0.465** | 0.000 | **0.428** |
| | — | 3,5 | — | 3,5 | — | 3,5 |
| *glass6* | **0.759** | **0.832** | 0.570 | **0.825** | 0.742 | **0.825** |
| | — | — | — | — | — | — |

**Table 2.** (*continued*)

| Dataset | Base method | SMOTE | RS | RS+SMOTE | WRS | WRS+SMOTE |
|---|---|---|---|---|---|---|
| *heart* | **0.829** | **0.817** | **0.804** | 0.794 | **0.829** | 0.794 |
|  | 4,6 | — | — | — | — | — |
| *hepatitis* | **0.919** | **0.875** | **0.912** | **0.904** | **0.912** | **0.904** |
|  | — | — | — | — | — | — |
| *page-blocks-1-3-vs-4* | **0.560** | **0.683** | **0.421** | **0.537** | **0.442** | **0.501** |
|  | — | — | — | — | — | — |
| *pima* | 0.618 | **0.684** | 0.432 | **0.653** | 0.560 | **0.665** |
|  | 3,5 | 1,3,5 | — | 3,5 | 3 | 3,5 |
| *shuttle-c0-vs-c4* | **0.996** | 1.000 | **0.996** | **0.996** | **0.996** | **0.996** |
|  | — | — | — | — | — | — |
| *vowel0* | **0.579** | **0.652** | 0.087 | **0.637** | **0.550** | **0.628** |
|  | 3 | 3 | — | 3 | 3 | 3 |
| *wisconsin* | **0.947** | **0.956** | **0.950** | **0.959** | **0.945** | **0.959** |
|  | — | — | — | — | — | — |
| *yeast-0-2-5-6-vs-3-7-8-9* | 0.019 | **0.461** | 0.000 | **0.423** | 0.000 | **0.420** |
|  | — | 1,3,5 | — | 1,3,5 | — | 1,3,5 |
| *yeast-0-2-5-7-9-vs-3-6-8* | 0.236 | **0.587** | 0.000 | **0.563** | 0.000 | **0.575** |
|  | 3,5 | 1,3,5 | — | 3,5 | — | 1,3,5 |
| *yeast-0-3-5-9-vs-7-8* | 0.067 | **0.275** | 0.000 | **0.244** | 0.067 | **0.261** |
|  | — | 1,3,5 | — | 1,3,5 | — | 1,3,5 |
| *yeast-0-5-6-7-9-vs-4* | 0.000 | **0.468** | 0.000 | **0.466** | 0.000 | **0.488** |
|  | — | 1,3,5 | — | 1,3,5 | — | 1,3,5 |
| *yeast-2-vs-4* | 0.170 | **0.690** | 0.000 | **0.664** | 0.103 | **0.680** |
|  | — | 1,3,5 | — | 1,3,5 | — | 1,3,5 |
| *yeast-2-vs-8* | 0.080 | **0.546** | 0.000 | **0.480** | 0.080 | **0.658** |
|  | — | 1,3,5 | — | 1,3,5 | — | 1,3,5 |
| *yeast1* | 0.329 | **0.584** | 0.062 | **0.565** | 0.241 | **0.593** |
|  | 3 | 1,3,5 | — | 1,3,5 | 3 | 1,3,5 |
| *yeast3* | 0.109 | **0.671** | 0.000 | **0.681** | 0.056 | **0.686** |
|  | 3 | 1,3,5 | — | 1,3,5 | — | 1,3,5 |
| *yeast5* | 0.000 | **0.480** | 0.000 | **0.461** | 0.000 | **0.462** |
|  | — | 1,3,5 | — | 1,3,5 | — | 1,3,5 |

**Table 3.** Classification results on *Support Vector Machines*

| Dataset | Base method | SMOTE | RS | RS+SMOTE | WRS | WRS+SMOTE |
|---|---|---|---|---|---|---|
| *australian* | 0.000 — | 0.247 — | 0.693 [1,2] | **0.795** [1,2,3,5] | 0.675 [1,2] | **0.813** [1,2,3,5] |
| *glass-0-1-2-3-vs-4-5-6* | **0.741** — | **0.864** — | 0.674 — | 0.835 — | 0.666 — | 0.842 — |
| *glass-0-1-4-6-vs-2* | 0.000 — | **0.296** [1,3,5] | 0.000 — | **0.366** [1,3,5] | 0.000 — | **0.375** [1,3,5] |
| *glass-0-1-5-vs-2* | 0.000 — | **0.264** [1,3,5] | 0.000 — | **0.377** [1,3,5] | 0.000 — | **0.389** [1,3,5] |
| *glass-0-1-6-vs-2* | 0.000 — | **0.286** [1,3,5] | 0.000 — | **0.364** [1,3,5] | 0.000 — | **0.380** [1,3,5] |
| *glass-0-1-6-vs-5* | 0.000 — | **0.544** [1,3,5] | 0.000 — | **0.492** [1,3,5] | 0.000 — | **0.466** [1,3,5] |
| *glass-0-4-vs-5* | **0.431** [3] | **0.762** [3] | 0.031 — | **0.629** [3] | 0.693 [3] | 0.629 [3] |
| *glass-0-6-vs-5* | **0.467** [3,5] | **0.800** [3,5] | 0.000 — | **0.700** [3,5] | 0.000 — | **0.700** [3,5] |
| *glass0* | 0.460 [3] | **0.699** [1,3,5] | 0.198 — | **0.731** [1,3,5] | 0.455 [3] | **0.738** [1,3,5] |
| *glass1* | **0.578** [3] | **0.603** [3] | 0.184 — | **0.596** [3] | **0.618** [3] | **0.588** [3] |
| *glass2* | 0.000 — | **0.247** [1,3,5] | 0.000 — | **0.287** [1,3,5] | 0.000 — | **0.309** [1,3,5] |
| *glass4* | **0.671** [3] | **0.772** [3] | 0.000 — | **0.762** [3] | **0.593** [3] | **0.781** [3] |
| *glass5* | 0.000 — | **0.403** [1,3,5] | 0.000 — | **0.537** [1,3,5] | 0.000 — | **0.472** [1,3,5] |
| *glass6* | **0.824** — | **0.809** — | **0.827** — | **0.847** — | **0.827** — | **0.847** — |
| *heart* | 0.000 — | 0.063 [1] | **0.764** [1,2,5] | **0.802** [1,2,5] | 0.657 [1,2] | **0.714** [1,2] |
| *hepatitis* | **0.912** — | **0.912** — | **0.912** — | **0.927** — | **0.912** — | **0.906** — |

**Table 3.** (*continued*)

| Dataset | Base method | SMOTE | RS | RS+SMOTE | WRS | WRS+SMOTE |
|---|---|---|---|---|---|---|
| *page-blocks-1-3-vs-4* | 0.000 | 0.037 | 0.114 | **0.478** | 0.057 | **0.398** |
| | — | — | — | 1,2,3,5 | — | 1,2,3,5 |
| *pima* | 0.000 | 0.015 | 0.000 | **0.400** | 0.000 | **0.368** |
| | — | — | — | 1,2,3,5 | — | 1,2,3,5 |
| *shuttle-c0-vs-c4* | 0.190 | 0.606 | **0.987** | 0.983 | **0.987** | 0.987 |
| | — | 1 | 1,2 | 1,2 | 1,2 | 1,2 |
| *vowel0* | **0.547** | **0.566** | 0.208 | **0.706** | **0.600** | **0.678** |
| | — | 3 | — | 3 | 3 | 3 |
| *wisconsin* | **0.933** | **0.936** | 0.961 | **0.968** | 0.961 | **0.968** |
| | — | — | — | — | — | — |
| *yeast-0-2-5-6-vs-3-7-8-9* | 0.000 | **0.493** | 0.000 | **0.420** | 0.118 | **0.402** |
| | — | 1,3,5 | — | 1,3,5 | — | 1,3,5 |
| *yeast-0-2-5-7-9-vs-3-6-8* | 0.000 | **0.652** | 0.053 | **0.559** | 0.167 | **0.579** |
| | — | 1,3,5 | — | 1,3,5 | — | 1,3,5 |
| *yeast-0-3-5-9-vs-7-8* | 0.000 | **0.198** | 0.000 | **0.239** | 0.114 | **0.241** |
| | — | 1,3 | — | 1,3 | — | 1,3 |
| *yeast-0-5-6-7-9-vs-4* | 0.000 | **0.472** | 0.000 | **0.495** | 0.000 | **0.497** |
| | — | 1,3,5 | — | 1,3,5 | — | 1,3,5 |
| *yeast-2-vs-4* | 0.000 | **0.697** | 0.463 | **0.664** | 0.589 | **0.682** |
| | — | 1 | 1 | 1 | 1 | 1 |
| *yeast-2-vs-8* | **0.613** | **0.658** | 0.000 | **0.487** | **0.658** | **0.658** |
| | 3 | 3 | — | 3 | 3 | 3 |
| *yeast1* | 0.073 | **0.573** | 0.097 | **0.574** | 0.345 | **0.584** |
| | — | 1,3,5 | — | 1,3,5 | 1,3 | 1,3,5 |
| *yeast3* | 0.000 | 0.681 | 0.000 | **0.721** | **0.759** | **0.701** |
| | — | 1,3 | — | 1,3 | 1,2,3 | 1,3 |
| *yeast5* | 0.000 | **0.457** | 0.000 | **0.522** | 0.000 | **0.523** |
| | — | 1,3,5 | — | 1,3,5 | — | 1,3,5 |

Enhancing the Random Subspace method with weighing, which – thanks to the f-measure – is realized in the context of the imbalanced problem, makes it to be more effective than the base method, although it still does not affect the quality of classification as positively as SMOTE.

**Table 4.** Ranks computed by the Wilcoxon test

| | Base method | SMOTE | RS | RS+SMOTE | WRS | WRS+SMOTE |
|---|---|---|---|---|---|---|
| **GNB** | - | 191.0 | 258.0 | 245.0 | 196.0 | 132.0 |
| SMOTE | 274.0 | - | 273.0 | 279.0 | 194.0 | 137.0 |
| RS | 207.0 | 192.0 | - | 240.0 | 102.5 | 75.0 |
| RS+SMOTE | 220.0 | 186.0 | 195.0 | - | 153.0 | 100.0 |
| WRS | 269.0 | 241.0 | 362.5 | 312.0 | - | 203.5 |
| WRS+SMOTE | 333.0 | 298.0 | 360.0 | 335.0 | 231.5 | - |
| **LR** | - | 10.0 | 416.5 | 18.0 | 389.0 | 22.0 |
| SMOTE | 455.0 | - | 460.0 | 283.0 | 453.0 | 235.0 |
| RS | 18.5 | 5.0 | - | 4.0 | 52.0 | 4.0 |
| RS+SMOTE | 417.0 | 182.0 | 431.0 | - | 429.0 | 187.0 |
| WRS | 46.0 | 12.0 | 383.0 | 6.0 | - | 7.0 |
| WRS+SMOTE | 413.0 | 230.0 | 431.0 | 248.0 | 428.0 | - |
| **SVM** | - | 2.0 | 213.0 | 7.0 | 101.5 | 1.0 |
| SMOTE | 433.0 | - | 371.0 | 137.0 | 348.5 | 114.0 |
| RS | 222.0 | 64.0 | - | 1.0 | 114.0 | 5.0 |
| RS+SMOTE | 458.0 | 328.0 | 464.0 | - | 431.0 | 234.0 |
| WRS | 363.5 | 116.5 | 321.0 | 34.0 | - | 26.5 |
| WRS+SMOTE | 464.0 | 321.0 | 430.0 | 231.0 | 438.5 | - |

The extension of the basic form of *Random Subspace* with SMOTE leads to slightly better results than using only SMOTE, which suggests that the positive influence on the quality of classification by both methods may be independent and it may be complemented. This is confirmed by the use of the full proposal, based on the weighted Random Subspace from SMOTE, whose quality is definitely the best in the competition and except one case (dataset *heart* with the Linear Regression) there are no situations where it is not among the best approaches in the considered pool.

The *Random Subspace* method is particularly popular in the case of mul-tidimensional data, being a solution to the significant problem of the curse of dimensionality. Lowering the number of features analyzed by each model, using the same cardinality of patterns, reduces the decision space, and thus compacts the samples in space. On the other hand, the role of SMOTE is to equalize the density of the occurrence of patterns in the problem by compacting the objects of the minority class. Using both methods together, controlling the influence of each subspace on the final prediction of the ensemble by weighing it with a

**Table 5.** Summary of the Wilcoxon test. •= the method in the row improves the method of the column. ○= the method in the column improves the method of the row. Upper diagonal of level significance $\alpha = 0.9$, Lower diagonal level of significance $\alpha = 0.95$

| | Base method | SMOTE | RS | RS+SMOTE | WRS | WRS+SMOTE |
|---|---|---|---|---|---|---|
| **GNB** | - | | | | | ○ |
| SMOTE | | - | | | | ○ |
| RS | | | - | | ○ | ○ |
| RS+SMOTE | | | | - | | ○ |
| WRS | | • | | | - | |
| WRS+SMOTE | • | | • | • | | - |
| **LR** | - | ○ | • | ○ | • | ○ |
| SMOTE | • | - | • | | • | |
| RS | ○ | ○ | - | ○ | ○ | ○ |
| RS+SMOTE | • | | • | - | • | |
| WRS | ○ | ○ | • | ○ | - | ○ |
| WRS+SMOTE | • | | • | | • | - |
| **SVM** | - | ○ | | ○ | ○ | ○ |
| SMOTE | • | - | • | ○ | • | ○ |
| RS | | ○ | - | ○ | ○ | ○ |
| RS+SMOTE | • | • | • | - | • | |
| WRS | • | ○ | • | ○ | - | ○ |
| WRS+SMOTE | • | • | • | | • | - |
| $\alpha = .9$ | 2 | 6 | 0 | 7 | 4 | 11 |
| $\alpha = .95$ | 2 | 6 | 0 | 7 | 4 | 10 |

measure adequate to the imbalanced problem may lead to close to the optimal placement of training patterns in the classification space, and thus lead to a model with a high discriminatory ability, as shown by carried out experiments.

## 5 Summary

This paper proposes the use of a classifier ensemble diversified using the *Random Subspace* approach and trained on sets oversampled independently in each subspace using the SMOTE algorithm. The experiments performed for its needs are testing the quality of such solution with relation to both concepts present in it, using 30 imbalanced data sets and three probabilistic base classifiers.

As shown by the results of experiments, it is a promising approach, able to effectively use the advantages of both methods leading to a better solution than each of them individually. Research shows that the *Random Subspace* method, although in itself, does not allow to improve the prediction of imbalanced data, in the weighted option may positively affect the achieved quality of classification. Combining it with the creation of synthetic patterns in the subspace areas of the problem gives an effective solution with high usefulness in the processing of this type of problems.

**Acknowledgements.** This work was supported by the Polish National Science Center under the grant no. UMO- 2015/19/B/ST6/01597 and by the statutory fund of the Faculty of Electronics, Wroclaw University of Science and Technology.

# References

1. Alcalá-Fdez, J., et al.: Keel data-mining software tool: data set repository, integration of algorithms and experimental analysis framework. J. Mult.-Valued Log. Soft Comput. **17**, 255–287 (2011)
2. Bunkhumpornpat, C., Sinapiromsaran, K., Lursinsap, C.: Safe-Level-SMOTE: safe-level-synthetic minority over-sampling technique for handling the class imbalanced problem. In: Theeramunkong, T., Kijsirikul, B., Cercone, N., Ho, T.-B. (eds.) PAKDD 2009. LNCS (LNAI), vol. 5476, pp. 475–482. Springer, Heidelberg (2009). https://doi.org/10.1007/978-3-642-01307-2_43
3. Chawla, N.V., Bowyer, K.W., Hall, L.O., Kegelmeyer, W.P.: SMOTE: synthetic minority over-sampling technique. J. Artif. Intell. Res. **16**, 321–357 (2002)
4. Diamantidis, N., Karlis, D., Giakoumakis, E.A.: Unsupervised stratification of cross-validation for accuracy estimation. Artif. Intell. **116**(1–2), 1–16 (2000)
5. Dietterich, T.G.: Ensemble methods in machine learning. In: Kittler, J., Roli, F. (eds.) MCS 2000. LNCS, vol. 1857, pp. 1–15. Springer, Heidelberg (2000). https://doi.org/10.1007/3-540-45014-9_1
6. García, S., Herrera, F.: Evolutionary undersampling for classification with imbalanced datasets: proposals and taxonomy. Evol. Comput. **17**(3), 275–306 (2009)
7. Gehan, E.A.: A generalized Wilcoxon test for comparing arbitrarily singly-censored samples. Biometrika **52**(1–2), 203–224 (1965)
8. Han, H., Wang, W.-Y., Mao, B.-H.: Borderline-SMOTE: a new over-sampling method in imbalanced data sets learning. In: Huang, D.-S., Zhang, X.-P., Huang, G.-B. (eds.) ICIC 2005. LNCS, vol. 3644, pp. 878–887. Springer, Heidelberg (2005). https://doi.org/10.1007/11538059_91
9. He, H., Bai, Y., Garcia, E.A., Li, S.: ADASYN: adaptive synthetic sampling approach for imbalanced learning. In: 2008 IEEE International Joint Conference on Neural Networks (IEEE World Congress on Computational Intelligence). IEEE, June 2008
10. He, H., Garcia, E.A.: Learning from imbalanced data. IEEE Trans. Knowl. Data Eng. **9**, 1263–1284 (2008)
11. Huang, H.Y., Lin, Y.J., Chen, Y.S., Lu, H.Y.: Imbalanced data classification using random subspace method and SMOTE. In: The 6th International Conference on Soft Computing and Intelligent Systems, and the 13th International Symposium on Advanced Intelligence Systems. IEEE, November 2012

12. Jeni, L.A., Cohn, J.F., De La Torre, F.: Facing imbalanced data-recommendations for the use of performance metrics. In: 2013 Humaine Association Conference on Affective Computing and Intelligent Interaction, pp. 245–251. IEEE (2013)

13. Krawczyk, B.: Learning from imbalanced data: open challenges and future directions. Prog. Artif. Intell. **5**(4), 221–232 (2016)

14. Krawczyk, B., Woźniak, M., Herrera, F.: On the usefulness of one-class classifier ensembles for decomposition of multi-class problems. Pattern Recogn. **48**(12), 3969–3982 (2015)

15. Ksieniewicz, P.: Undersampled majority class ensemble for highly imbalanced binary classification. In: Torgo, L., Matwin, S., Japkowicz, N., Krawczyk, B., Moniz, N., Branco, P. (eds.) Proceedings of the Second International Workshop on Learning with Imbalanced Domains: Theory and Applications. Proceedings of Machine Learning Research, PMLR, ECML-PKDD, Dublin, Ireland, vol. 94, pp. 82–94, 10 September 2018

16. Kuncheva, L.I.: Combining Pattern Classifiers: Methods and Algorithms. Wiley, Hoboken (2004)

17. Lemaître, G., Nogueira, F., Aridas, C.K.: Imbalanced-learn: a Python toolbox to tackle the curse of imbalanced datasets in machine learning. J. Mach. Learn. Res. **18**(1), 559–563 (2017)

18. Maciejewski, T., Stefanowski, J.: Local neighbourhood extension of SMOTE for mining imbalanced data. In: 2011 IEEE Symposium on Computational Intelligence and Data Mining (CIDM). IEEE, April 2011

19. Pedregosa, F., et al.: Scikit-learn: machine learning in Python. J. Mach. Learn. Res. **12**(Oct), 2825–2830 (2011)

20. Quinlan, J.R., et al.: Bagging, boosting, and C4. 5. In: AAAI/IAAI, vol. 1, pp. 725–730 (1996)

21. Sasaki, Y., et al.: The truth of the F-measure. Teach Tutor Mater **1**(5), 1–5 (2007)

22. Sun, Y., Wong, A.K., Kamel, M.S.: Classification of imbalanced data: a review. Int. J. Pattern Recogn. Artif. Intell. **23**(04), 687–719 (2009)

23. Topolski, M.: Multidimensional MCA correspondence model supporting intelligent transport management. Arch. Transp. Syst. Telemat. **11**, 52–56 (2018)

24. Topolski, M.: Algorithm of multidimensional analysis of main features of PCA with blurry observation of facility features detection of carcinoma cells multiple myeloma. In: Burduk, R., Kurzynski, M., Wozniak, M. (eds.) CORES 2019. AISC, vol. 977, pp. 286–294. Springer, Cham (2020). https://doi.org/10.1007/978-3-030-19738-4_29

25. Wozniak, M.: Hybrid Classifiers: Methods of Data, Knowledge, and Classifier Combination. Studies in Computational Intelligence, vol. 519. Springer, Heidelberg (2013). https://doi.org/10.1007/978-3-642-40997-4

26. Yu, G., Zhang, G., Domeniconi, C., Yu, Z., You, J.: Semi-supervised classification based on random subspace dimensionality reduction. Pattern Recogn. **45**(3), 1119–1135 (2012)

# Optimization of the Master Production Scheduling in a Textile Industry Using Genetic Algorithm

Leandro L. Lorente-Leyva[1]([⊠]), Jefferson R. Murillo-Valle[1],
Yakcleem Montero-Santos[1], Israel D. Herrera-Granda[1],
Erick P. Herrera-Granda[1], Paul D. Rosero-Montalvo[1],
Diego H. Peluffo-Ordóñez[2,3], and Xiomara P. Blanco-Valencia[3]

[1] Facultad de Ingeniería en Ciencias Aplicadas, Universidad Técnica del Norte,
Av. 17 de Julio, 5-21, y Gral. José María Cordova, Ibarra, Ecuador
lllorente@utn.edu.ec
[2] Escuela de Ciencias Matemáticas y Tecnología Informática,
Yachay Tech, Hacienda San José s/n, San Miguel de Urcuquí, Ecuador
[3] SDAS Research Group, Ibarra, Ecuador
http://www.sdas-group.com/

**Abstract.** In a competitive environment, an industry's success is directly related to the level of optimization of its processes, how production is planned and developed. In this area, the master production scheduling (MPS) is the key action for success. The object of study arises from the need to optimize the medium-term production planning system in a textile company, through genetic algorithms. This research begins with the analysis of the constraints, mainly determined by the installed capacity and the number of workers. The aggregate production planning is carried out for the T-shirts families. Due to such complexity, the application of bioinspired optimization techniques demonstrates their best performance, before industries that normally employ exact and simple methods that provide an empirical MPS but can compromise efficiency and costs. The products are then disaggregated for each of the items in which the MPS is determined, based on the analysis of the demand forecast, and the orders made by customers. From this, with the use of genetic algorithms, the MPS is optimized to carry out production planning, with an improvement of up to 96% of the level of service provided.

**Keywords:** Master Production Scheduling · Optimization · Textile industry · Genetic algorithm · Production planning · Forecasting

## 1 Introduction

The textile industries have chosen to achieve an improvement in the production planning process, so they have used several classic methods to determine the amount of production sufficient to meet the needs of customers and thus achieve good business development, but despite this process can compromise the efficiency of production and the costs of manufacturing the products. Therefore, in an environment of global

© Springer Nature Switzerland AG 2019
H. Pérez García et al. (Eds.): HAIS 2019, LNAI 11734, pp. 674–685, 2019.
https://doi.org/10.1007/978-3-030-29859-3_57

competition, the success of a company is directly related to the level of optimization of its processes in general, but, particularly, how to plan and executes the production.

Different methods, techniques and mathematical models have been developed to treat master production schedule optimization also known as MPS. According to Higgins and Browne [1], Slack et al. [2] the MPS is the most important activity in planning and controlling production. In 2012 Wu et al. use the ant colony algorithm for the optimization of the MPS, which better achieved the load capacity of the machinery [3].

Different authors deal in a general way with everything related to the MPS [4] and the application of methods for its optimization [5]; and as in most industries around the world, the creation of an MPS considers the conflicting objectives, such as the maximization of service levels, efficient use of resources and minimization of inventory levels [6–8].

According to Golmohammadi [5] metaheuristics refer to the design of fundamental types of heuristic procedures for solving an optimization problem. According to Jonsson and Kjellsdotter in 2015 [6] these procedures are a kind of approximate methods that are designed to solve difficult problems of combinatorial optimization, in which classical heuristics are neither effective nor efficient. They provide a general framework for creating new hybrid algorithms combining different concepts derived from artificial intelligence, biological evolution and statistical mechanics [3, 4].

In 2013 Alba et al. [9] mention that these techniques obtain solutions that comply with the required quality and the delay times imposed in the industrial field, in addition, they allow to study generic classes of problems instead of instances of particular problems. In general, the techniques that work best in solving complex, real-world problems use metaheuristics. Their fields of application range from combinatorial optimization, bioinformatics, telecommunications to economics, software engineering, etc., which need fast solutions with high quality [10–13].

In 2009 Soares and Vieira developed a multi-objective optimization method for MPS problems based on genetic algorithm (GA), obtaining satisfactory results in the minimization of inventory levels, the inventory that is below the safety stock level, as well as the identification of needs that are not met and estimation of overtime required [14].

Currently the applications of GA in the optimization of problems arising in production have diversified. With the aim of achieving better control and planning of the same, as well as minimizing losses. In 2019 Luo et al. propose an improved GA to solve the manufacturing efficiency problem in the manufacturing industry [15]. Furthermore, Pinto and Nagano discusses the use of GA to solve a specific variation of the order batching and sequencing problem, where they provide satisfactory quality solutions to any optimized selection picking instance [16]. Goli et al. addresses a robust multi-objective multi-period aggregate production planning problem based on different scenarios under uncertain seasonal demand. Finally, they evaluate the performance of the non-dominated genetic classification algorithm II (NSGA-II) designed to solve this problem [17].

Lin et al. in 2019, propose an algorithm based on a depth-first search with an expanding technique to meet the demand and develop a two-stage approach based on the NSGA-II in a real case of t-shirt production [18]. Ben-Ammar et al. also optimize of multi-period supply planning under stochastic lead times and a dynamic demand, developing a GA to determine planned delivery times and the level of safety stock, while minimizing the expected total cost [19].

## 2 Materials and Methods

### Master Production Scheduling in Textile Industry

The Master Production Schedule (MPS), is a plan which details the quantity of final products to be produced, in a time determined by the customer and the entrepreneur, taking into account the production capacity of the company and what it needs to meet demand and comply with the production plan. This program establishes which items are to be produced and when: It disaggregates the aggregate production plan. While the aggregate production plan is defined in very broad terms, such as product families and the MPS is established in terms of specific products [20–22].

Because production planning works with batch size, lead times, safety inventories previously established without providing help for these configurations, the MPS does not generate alternative plans versus non-viable plans. In addition, it does not consider the availability of materials in the work centers, times of supplies depending on the production batch and there is an excess of workers performing the same task.

In Imbabura, Ecuador most of the companies are manufacturing, so they have decided to implement the MPS to have a better development in the production area, but there are new inconveniences that affect the production planning, so it is necessary to carefully analyze this and determine the most relevant problems to carry out the optimization.

There are several companies in Imbabura that have an MPS, use the initial inventory and sales forecast for a particular product, a planner can calculate the amount of production needed per period to meet the anticipated demand of customers. This calculation becomes more complex when it comes to a multi-product, where forecasting errors and capacity constraints can lead to greater uncertainty in the planning process.

But these companies still do not have an optimized MPS, which makes them go through big problems of profitability and generate failures in their processes, because the current plan is not stable and efficient.

The development of this research will be done by generating a model to optimize the MPS with the application of genetic algorithm. For this purpose, historical data will be taken from industries in the textile area, which already have a Master Production Scheduling. This will seek greater stability of the MPS, generate alternatives to non-viable plans and make companies more profitable.

### Mathematical Model

It is a system where all the options can be simulated by means of mathematical equations, whose variables are previously established according to what we want to demonstrate. It allows to obtain results based on previous experiences or statistics. It is used in demand or sales forecasts, in inventory control. Every mathematical model has an error when compared with reality, since it will always be a calculation and external factors that do not allow accuracy.

The MPS acts as a bridge between strategic planning and production scheduling because it harmonizes the market and demand with the company's internal resources. As the number of products, the number of resources, and number of periods (production cells, production lines, assembly lines, machines).

In 2017 Abu et al. [11] mention production and planning problems generally include conflicting objectives, for example, reducing inventory and maximizing service levels. Because of these complexities, the use of heuristics or metaheuristics is highly recommended in solving these problems.

**Objective Function and Restrictions**

The MPS problem can be mathematically modeled as a mixed integer program as follows [11, 13, 20]:

K: Total quantity of different products (SKU)
R: Total quantity of different productive resources
P: Total number of planning periods
$TH_p$: Available time at each period $p$.
TH: Total planning horizon
$OH_k$: Initial available inventory, at the first scheduling period
$GR_{kp}$: Gross requirements for product $k$ at period $p$
$BS_{kp}$: Standard lot size for product $k$ at period $p$
$NR_{kp}$: Net requirements for product k at period p, considering infinity capacity
$SS_{kp}$: Safety inventory level for product $k$ at period $p$.
$UR_{kr}$: Production rate for product $k$ at resource $r$ (units per hour).
$AC_{rp}$: Avalaible capacity, in hours, at resource $r$ at period $p$.

According to Abu et al. 2017 [11], Soares and Vieira, 2009 [13] and Ribas 2003 [20] objective function and its restrictions can be defined as:

$$Minimize = C_1 * AIL + C_2 * RNM + C_3 * BSS + C_4 * OC \tag{1}$$

Where:

The overall average inventory level, considering all products and the entire planning horizon (AIL) is:

$$AIL = \sum_{K=1}^{K} \left( \frac{\sum_{P=1}^{P} AIL_{kp}}{TH} \right) \tag{2}$$

The average requirements not met for all products and periods (RNM) is:

$$RNM = \frac{\sum_{K=1}^{K} \sum_{P=1}^{P} RNM_{kp}}{TH} \tag{3}$$

The average quantity bellow safety inventory level, considering all products and periods (BSS) is:

$$BSS = \frac{\sum_{K=1}^{K} \sum_{P=1}^{P} BSS_{kp}}{TH} \tag{4}$$

The average over capacity needed, considering all resource and the entire planning horizon (OC) is:

$$OC = \sum_{r=1}^{R} \sum_{p=1}^{P} OC_{rp} \qquad (5)$$

Subject to:

$$TH = \sum_{r=1}^{P} TH_p \qquad (6)$$

$$BI_{kp} = \left\{ \begin{array}{c} OH_k \quad se(p = 1) \\ . \\ EL_{k(p-1)} \quad se(p > 1) \end{array} \right\} \qquad (7)$$

$$AIL_{kp} = \frac{(EI_{kp} + BI_{kp}) \; x \; TH_P}{2} \qquad (8)$$

$$EI_{kp} = \max \left[0, \left((MPS_{kp} + BI_{kp}) - GR_{kp}\right)\right] \qquad (9)$$

$$MPST_{kp} = \sum_{r=1}^{R} MPS_{kpr} \qquad (10)$$

$$MPS_{kpr} = BN_{kpr} \; x \; BS_{kpr} \qquad (11)$$

$$RNM_{kp} = \max \left[0, \left(GR_{KP} - (MPST_{kp} + BI_{KP})\right)\right] \qquad (12)$$

$$BSS_{kp} = \max \left[0, \left(SS_{kp} - EI_{kp}\right)\right] \qquad (13)$$

$$CUH_{rp} = \sum_{k=!}^{K} \frac{(BS_{kp}BN_{kp})}{UR_{kr}} \qquad (14)$$

$$CUH_{rp} = \; \leq AC_{rp} \qquad (15)$$

$$OC_{rp} = max \left[0, \left(\frac{CUH_{rp}}{AC_{rp}} - 1\right)\right] \qquad (16)$$

As this is a minimisation problem, a small Z brings the solution closer to the optimal solution. Therefore, Z cannot be used directly as a potential fitness function. Nevertheless, a change can be made to this. To consider the function as a measure of individual fitness, the function is expressed as follows [11]:

$$fitness = [1/1 + Z_n] \qquad (17)$$

where:

$$Z_n = C_1 * AIL/AIL_{max} + C_2 * RNM/RNM_{max} + C_3 * BSS/BSS_{max} + C_4 * OC \qquad (18)$$

## Decision Variables for the MPS

BNkpr: Quantity of standard lot sizes needed for the production of the product $k$ at resource $r$, at period $p$

MPSkpr: Total quantity to be manufactured of the product $k$ at resource $r$, at period p

MPSTkp: Total quantity to be manufactured of the product $k$ at resource $r$, at period $p$; (considering all available resources r)

CUHrp: Capacity used from the resource $r$ at period $p$

GRkp: Gross requirements for product k at period p

RMkp: Total requirements met for product $k$ at period $p$

RMkpr: Total requirements met for product $k$ at period $p$, at resource $r$.

RNMkp: Requirements not met for product $k$ at period $p$

SLkp: Service level, relation of the requirements met RMkp, and the gross requirements for product $k$ at period $p$

AILkp: Average inventory level generated for product $k$ at period $p$

## Genetic Algorithm for MPS Optimization

The Genetic Algorithm (GA) makes the natural selection on the group of solutions of the problem to solve. They are based on the creation of successive generations of individuals representing possible solutions to the problem. The code of a solution is interpreted as the chromosome of the individual composed of a certain number of genes to which certain chromosomes correspond. The most common coding of the chromosomes that make up the solutions is through binary chains. A fitness function must be generated for each problem specifically. It must be able to punish bad solutions and reward good ones, so that good solutions are the ones that spread the fastest.

In 2017 Abu et al. [11] and Soares and Vieira [13] suggest the following configuration parameters of the genetic algorithm for MPS optimization in any type of company:

- Population size: $\geq 100$
- Crossover rate: $0,6 \leq Pc \leq 0,99$
- Mutation rate: $0,001 \leq Pm \leq 0,01$
- Stopping criteria: Based on proposed convergence method
- Elitism strategy: Yes, maintaining only the best individual from the current generation.
- Crossover method: Two points
- Selection method: Tournament - using groups of three
- Control the population diversity: No need for MPS problem
- Method applied on generation of the initial population: Heuristic

Considering these suggested parameters, we proceed to perform the optimization of MPS for the textile industry and achieve the improvement of key aspects of production depending on the level of service performed.

## Current MPS

The company has an MPS which was developed with the application of classic tools and methods, starting from this current MPS, applies the mathematical modeling proposed by [11, 13, 20] to perform the optimization. In such a way that the obtained results serve for the improvement in the planning and decision making. In Table 1, the data of the current MPS of the company are shown:

**Table 1.** Current MPS

|           | Sublimated uniforms | Sport T-shirts | Sublimated T-shirts | Polo T-shirts |
|-----------|--------------------|----------------|--------------------|---------------|
| January   | 151                | 503            | 218                | 285           |
| February  | 100                | 382            | 249                | 316           |
| March     | 151                | 503            | 218                | 285           |
| April     | 100                | 382            | 249                | 316           |
| May       | 151                | 503            | 218                | 285           |
| June      | 100                | 382            | 249                | 316           |
| July      | 151                | 503            | 218                | 285           |
| August    | 100                | 382            | 249                | 316           |
| September | 151                | 503            | 218                | 285           |
| October   | 100                | 382            | 249                | 316           |
| November  | 151                | 503            | 218                | 285           |
| December  | 100                | 382            | 249                | 316           |

Once the objective function with all its restrictions has been established and defined, each of the values of the decision variables is determined (Table 2).

**Table 2.** Time available in each period p

| THp | January  | 284 | May     | 262 | September | 249 |
|-----|----------|-----|---------|-----|-----------|-----|
|     | February | 224 | June    | 262 | October   | 287 |
|     | March    | 267 | July    | 274 | November  | 249 |
|     | April    | 262 | August  | 274 | December  | 223 |

The total planning horizon (TH) is determined by means of a sum of the company's available times in each period (Table 3).

Within the decision variables is $k = 4$, being the total of products produced by the company, the total number of machines $r = 11$ and total of planning periods $p = 12$. All these data were provided from the forecast that the company made for last year.

The safety stock (SS) for product $k$ in period $p$ is determined using the annual demand and the standard deviation of demand, as well as the duration of the year and the number of work shifts, as shown below:

$$SS_1 = \left(\frac{2103}{365} * 1\right) + (1,96 * 69,52) = 142,02 \text{ Sublimated Uniforms} \quad (19)$$

$$SS_2 = \left(\frac{5176}{365} * 1\right) + (1,96 * 127,10) = 263,29 \text{ Sport T-shirts} \quad (20)$$

**Table 3.** Initial available inventory

| OHkp | Sublimated uniforms | 83 |
|------|---------------------|-----|
|      | Sport T-shirts      | 324 |
|      | Sublimated T-shirts | 224 |
|      | Polo T-shirts       | 177 |

$$SS_3 = \left(\frac{2776}{365} * 1\right) + (1,96 * 60,92) = 127,01 \text{ Sublimated T - shirts} \qquad (21)$$

$$SS_4 = \left(\frac{2855}{365} * 1\right) + (1,96 * 68,68) = 142,43 \text{ Polo  T - shirts} \qquad (22)$$

The level of service (LS) provided current is determined in relation to the requirements met RMkp, and the gross requirements of product $k$ in period $p$ (SLkp) as indicated below:

$$LS = \prod\left(1 - \frac{\sum Failures}{\sum Planned}\right) \qquad (23)$$

$$LS = \prod\left(1 - \frac{RNM_{kp}}{MPS_{kpr}}\right)$$

$$LS = \prod\left(1 - \frac{3856}{13224}\right) = 70,84\%$$

## 3   Results and Discussion

### Optimization of Master Production Scheduling

In order to optimise MPS, the demand forecast for 2019 is developed, with which the new planning will be carried out and will serve as the basis for the company's production planning. The values of demand in 2016, 2017 and 2018 of the textile industriy under study were taken as a reference. For the MPS optimization through GA, we used a HP computer with Intel Core i5-4210U 2.4 GHz processor, 8 GB DDR3 RAM. In Fig. 1 the behavior of the demand forecast in relation to the previous years can be clearly evidenced.

Thus, to estimate AILmax, RNMmax, and BSSmax, a pre-processing step is needed. During this phase the GA operates by its general logic, however, with high mutation and crossover rates to generate the maximum possible values that can be found during the actual execution of the GA. That is, an estimation of the maximum values and a higher value can be found during the actual execution of the GA [11, 13]. If this happens, the corresponding maximum values are updated. In the Table 4 presents the optimization of the MPS with the GA application.

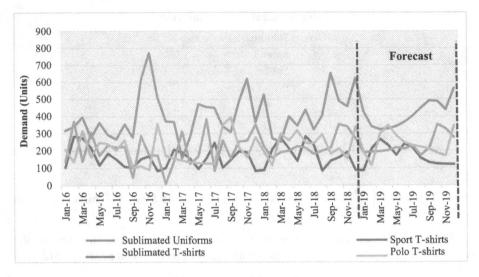

**Fig. 1.** Demand forecasting

**Table 4.** Optimization of the MPS with GA

| Optimized MPS | | | | |
|---|---|---|---|---|
| | Sublimated uniforms | Sport T-shirts | Sublimated T-shirts | Polo T-shirts |
| January | 89 | 526 | 224 | 196 |
| February | 212 | 341 | 193 | 116 |
| March | 273 | 323 | 196 | 300 |
| April | 232 | 399 | 202 | 347 |
| May | 174 | 344 | 226 | 318 |
| June | 283 | 437 | 216 | 248 |
| July | 231 | 401 | 229 | 243 |
| August | 157 | 447 | 199 | 294 |
| September | 142 | 652 | 217 | 209 |
| October | 161 | 489 | 354 | 213 |
| November | 192 | 457 | 340 | 171 |
| December | 120 | 623 | 276 | 345 |

With the development of the demand forecast a better production planning is achieved and the fulfillment of the monthly requirements of the products to be manufactured, as shown in Fig. 2.

As shown in Table 5, the number of unfulfilled requirements for product $k$ in period $p$ by optimizing the MPS through GA decreased overall and the service level achieves satisfactory results of 96.3%, as observed in Eq. (24).

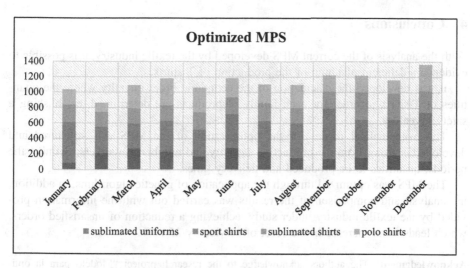

**Fig. 2.** MPS optimization

**Table 5.** Requirements not met for product $k$ at period $p$, after optimization

| RNM$_{kp}$ | Sublimated Uniforms (units) | | | | | | | | | | | |
|---|---|---|---|---|---|---|---|---|---|---|---|---|
| E | F | M | A | M | J | J | A | S | O | N | D |
| 0 | 0 | 0 | 0 | 0 | 0 | 0 | 0 | 0 | 0 | 0 | 0 | 0 |
| **Sport T-shirts (units)** | | | | | | | | | | | | |
| E | F | M | A | M | J | J | A | S | O | N | D |
| 0 | 0 | 0 | 0 | 0 | 0 | 0 | 0 | 0 | 0 | 0 | 0 | 0 |
| **Sublimated T-shirts (units)** | | | | | | | | | | | | |
| E | F | M | A | M | J | J | A | S | O | N | D |
| 218 | 0 | 0 | 0 | 0 | 0 | 0 | 0 | 0 | 0 | 0 | 0 | 0 |
| **Polo T-shirts (units)** | | | | | | | | | | | | |
| E | F | M | A | M | J | J | A | S | O | N | D |
| 285 | 0 | 0 | 0 | 0 | 0 | 0 | 0 | 0 | 0 | 0 | 0 | 0 |

$$LS = \prod \left( 1 - \frac{\sum Failures}{\sum Planned} \right) \tag{24}$$

$$LS = \prod \left( 1 - \frac{RNM_{kpoptimized}}{MPS_{kproptimized}} \right)$$

$$LS = \prod \left( 1 - \frac{503}{13557} \right) = 96,3\%$$

## 4    Conclusions

With the analysis of the current MPS developed by the textile industry, it is possible to evidence a deficient planning of the production of T-shirts.

It was developed the calculation of the economic order quantity where the quantities of T-shirts to order for each family of products are determined, establishing a safety inventory.

The mathematical model for the optimization of the MPS was structured and developed, starting from the productive capacity and number of workers; defining the periods, quantities to be produced and delivery times.

The MPS was optimized through the application of genetic algorithms, in addition, the analysis and comparison of the results was carried out with the information provided by the textile industry under study, achieving a reduction of unsatisfied orders, which leads to an increase in the level of service by 25.86%.

**Acknowledgment.** The authors acknowledge to the research project "Modelo para la optimización del Master Production Scheduling en entornos inciertos aplicando técnicas metaheurísticas" supported by Agreement HCD Nro. UTN-FICA-2017-0640 by Facultad de Ingeniería en Ciencias Aplicadas from Universidad Técnica del Norte. As well, authors thank the valuable support given by the SDAS Research Group (www.sdas-group.com).

## References

1. Higgins, P., Browne, J.: Master production scheduling: a concurrent planning approach. Prod. Plan. Control **3**(1), 2–18 (1992)
2. Slack, N., Chambers, S., Johnston, R.: Operations Management, 4th edn. Pearson, Upper Saddle River (2004)
3. Wu, Z., Zhang, C., Zhu, X.: An ant colony algorithm for Master production scheduling optimization. In: Proceedings of the 2012 IEEE 16th International Conference on Computer Supported Cooperative Work in Design (ISCAS), pp. 775–779 (2012). https://doi.org/10.1109/CSCWD.2012.6221908
4. Díaz-Madroñero, M., Mula, J., Peidro, D.: A review of discrete-time optimization models for tactical production planning. Int. J. Prod. Res. **52**(17), 5171–5207 (2014). https://doi.org/10.1080/00207543.2014.899721
5. Golmohammadi, D.: A study of scheduling under the theory of constraints. Int. J. Prod. Econ. **165**, 38–50 (2015). https://doi.org/10.1016/j.ijpe.2015.03.015, Art. no. 6034
6. Jonsson, P., Kjellsdotter Ivert, L.: Improving performance with sophisticated master production scheduling. Int. J. Prod. Econ. **168**, 118–130 (2015). https://doi.org/10.1016/j.ijpe.2015.06.012
7. Korbaa, O., Yim, P., Gentina, J-C.: Solving transient scheduling problem for cyclic production using timed Petri nets and constraint programming. In: European Control Conference, ECC 1999 - Conference Proceedings, pp. 3938–3945 (2015). https://doi.org/10.23919/ECC.1999.7099947, Art. no. 7099947
8. Akhoondi, F., Lotfi, M.M.: A heuristic algorithm for master production scheduling problem with controllable processing times and scenario-based demands. Int. J. Prod. Res. **54**(12), 3659–3676 (2016). https://doi.org/10.1080/00207543.2015.1125032

9. Alba, E., Luque, G., Nesmachnow, S.: Parallel metaheuristics: recent advances and new trends. Int. Trans. Oper. Res. **20**, 1–48 (2013)
10. Abedini, A., Li, W., Ye, H.: An optimization model for operating room scheduling to reduce blocking across the perioperative process. Procedia Manufact. **10**, 60–70 (2017). https://doi.org/10.1016/j.promfg.2017.07.022
11. Abu, M., Abbas, I., AlSattar, H., Khaddar, A-G., Atiya, B.: Solution for multi-objective optimisation master production scheduling problems based on swarm intelligence algorithms. J. Comput. Theor. Nanosci. **14**(11), 5184–5194 (2017). https://doi.org/10.1166/jctn.2017.6729
12. Lorente, L., et al.: Applying lean manufacturing in the production process of rolling doors: a case study. J. Eng. Appl. Sci. **13**(7), 1774–1781 (2018). https://doi.org/10.3923/jeasci.2018.1774.1781
13. Soares, M., Vieira, G.: A new multi-objective optimization method for master production scheduling problems based on genetic algorithm. Int. J. Adv. Manuf. Technol. **41**, 549–567 (2009). https://doi.org/10.1007/s00170-008-1481-x
14. Lorente-Leyva, L.L., et al.: Developments on solutions of the normalized-cut-clustering problem without eigenvectors. In: Huang, T., Lv, J., Sun, C., Tuzikov, Alexander V. (eds.) ISNN 2018. LNCS, vol. 10878, pp. 318–328. Springer, Cham (2018). https://doi.org/10.1007/978-3-319-92537-0_37
15. Luo, T., Li, G., Yu, N.: Research on manufacturing productivity based on improved genetic algorithms under internet information technology. Concurrency Comput. **31**(10), e4859 (2019). https://doi.org/10.1002/cpe.4859
16. Pinto, A.R.F., Nagano, M.S.: An approach for the solution to order batching and sequencing in picking systems. Prod. Eng. Res. Devel. **13**(3–4), 325–341 (2019). https://doi.org/10.1007/s11740-019-00904-4
17. Goli, A., Tirkolaee, E.B., Malmir, B., Bian, G.B., Sangaiah, A.K.: A multi-objective invasive weed optimization algorithm for robust aggregate production planning under uncertain seasonal demand. Computing **101**(6), 499–529 (2019). https://doi.org/10.1007/s00607-018-00692-2
18. Lin, Y.K., Chang, P.C., Yeng, L.C.L., Huang, S.F.: Bi-objective optimization for a multistate job-shop production network using NSGA-II and TOPSIS. J. Manufact. Syst. **52**, 43–54 (2019). https://doi.org/10.1016/j.jmsy.2019.05.004
19. Ben-Ammar, O., Bettayeb, B., Dolgui, A.: Optimization of multi-period supply planning under stochastic lead times and a dynamic demand. Int. J. Prod. Econ. **218**, 106–117 (2019). https://doi.org/10.1016/j.ijpe.2019.05.003
20. Ribas, P.C.: Análise do uso de têmpera simulada na otimização do planejamento mestre da produção. Pontifícia Universidade Católica de Paraná, Curitiba (2003)
21. Wang, B., Guan, Z., Ullah, S., Xu, X., He, Z.: Simultaneous order scheduling and mixed-model sequencing in assemble-to-order production environment: a multi-objective hybrid artificial bee colony algorithm. J. Intell. Manuf. **28**(2), 419–436 (2017). https://doi.org/10.1007/s10845-014-0988-2
22. Muñoz, E., Capón-García, E., Muñoz, M., Montoya, P.: Decision-support platform for industrial recipe management. In: Mejia, J., Muñoz, M., Rocha, Á., Quiñonez, Y., Calvo-Manzano, J. (eds.) CIMPS 2017, vol. 688, pp. 198–206. Springer, Cham (2018). https://doi.org/10.1007/978-3-319-69341-5_18

# Urban Pollution Environmental Monitoring System Using IoT Devices and Data Visualization: A Case Study

Paul D. Rosero-Montalvo[1,2,3,5](✉), Vivian F. López-Batista[1,5], Diego H. Peluffo-Ordóñez[2,3,5], Leandro L. Lorente-Leyva[2,5], and X. P. Blanco-Valencia[4,5]

[1] Universidad de Salamanca, Salamanca, Spain
[2] Universidad Técnica del Norte, Ibarra, Ecuador
pdrosero@utn.edu.ec
[3] Instituto Tecnológico Superior 17 de Julio, Urcuquí, Ecuador
[4] Yachay Tech University, Urcuquí, Ecuador
[5] SDAS Research Group, Urcuquí, Ecuador

**Abstract.** This work presents a new approach to the Internet of Things (IoT) between sensor nodes and data analysis with visualization platform with the purpose to acquire urban pollution data. The main objective is to determine the degree of contamination in Ibarra city in real time. To do this, for one hand, thirteen IoT devices have been implemented. For another hand, a Prototype Selection and Data Balance algorithms comparison in relation to the classifier k-Nearest Neighbourhood is made. With this, the system has an adequate training set to achieve the highest classification performance. As a final result, the system presents a visualization platform that estimates the pollution condition with more than 90% accuracy.

**Keywords:** Intelligent system · Environmental science computing · Environmental monitoring · Data analysis

## 1 Introduction

The world climate conditions are very complex systems, whose it an alteration in any place has an impact on the entire planet earth. This climate change raises the global temperature and it may cause an increase in ocean waters (20 cm right now) [1]. Unfortunately, with the increase of industries and the extermination of green areas, it hopes the next century the ocean waters will increase very rapidly [2]. As a result, some cities near the coast will face flooding in 2050. In some cases, these cities will need an embankment to conserve their life. In one hand, the global temperature variation in different areas causes aggressive rains in replacement of a balanced rain [3]. Another hand, other cities of the world have been exposed to long periods of drought causing famine, animal deaths, human

© Springer Nature Switzerland AG 2019
H. Pérez García et al. (Eds.): HAIS 2019, LNAI 11734, pp. 686–696, 2019.
https://doi.org/10.1007/978-3-030-29859-3_58

disease, among others. Because of that, studies conducted in recent years have found the highest global temperature in comparison with other decades.

One of the principal reasons for global warming is air pollution due to the effects of industrialization, urbanization and individual mobility in vehicles, which has become a great risk to health in all countries of the world [4]. The World Health Organization (WHO) estimates that one out of every eight premature deaths is due to the effects of air pollutant particles [5]. In addition, every year there are around 3 million people who die from air pollution [6].

In order to counteract climate change, many institutions have invested in intelligent data acquisition systems to monitor environmental conditions to implement some strategies and their effect on air pollution. For do that, the Internet of Things (IoT) is fundamental, because it is in charge of developing electronic devices that collect data and exchange it with the common goal of technological advancement to improve the human conditions. In addition, to these tasks, it must have sensors to convert environmental signals to electric ones. As a result, the electronic device with sensors becomes a node of IoT cite Saha2018. However, the amount of data can be acquired with noise for many reasons like not linearity of the electronics elements, electronic device wearing, among others. For these reasons, it must go through a cleaning and selection process to choose the ideal ones that represent the phenomenon studied. As a result, measurements and monitoring of air quality are necessary for the analysis of heterogeneous environments with different emission sources like urban areas [7]. Studies such as [8–11] have developed data collection and monitoring systems. However, there are open problems, such as the appropriate decision making in remote systems, the consumption of batteries of electronic devices, adequate selection of data, among others.

The present system is installed in Ibarra-Ecuador. It can classify between 3 cases: (i) high contamination levels, (ii) Normally gases presence and (iii) No emissions. To do that, 13 sensor nodes are ubicated on the city that sends data about environmental conditions through a 4G cellphone network to an IoT server. The server allows the visualization of information on fundamental environmental parameters that must be considered when protecting the health of people. Also, the user can visualize the city situation by traffic light colors. Due to this, a prototypes selection criteria and classification algorithms are implemented in each sensor node for improving the decision task.

The rest of the document is structured as follows: Section 2 presents the design and development of the electronic system. Section 3 presents the data analysis and the implementation of the decision algorithm. Section 4 shows the results obtained in the measurement of environmental conditions. Finally, the conclusions and recommendations are presented in Sect. 5.

## 2 Design and Development of Electronic System

The sensor node has an Arduino UNO with PCB board with sensors MQ-7 (Carbon Monoxide), MQ-135 (Carbon Dioxide), ML8511 (UV Rays), DTH11

(Temperature and Humidity). One of the main characteristics of an embedded system must be adaptability. That is, they are able to emulate some processing skills that the human brain performs. The same that implies in some way, the ability to make decisions, to learn from external stimuli, to adapt to changes or the possibility of executing intelligent mathematical algorithms. Implicitly, it is based on a computational paradigm that receives or processes data to achieve a task assigned [12]. Under this concept, it is expected that the sensors can provide the best information as possible [13].

Related in the environmental monitoring field. The sensors that can acquire data of the most harmful and detectable gases for health caused by vehicular traffic were selected. This means, on the one hand, nitrogen oxides (NOx) which is a generic term that refers to a group of highly reactive gases such as nitric oxide (NO) and nitrogen dioxide (NO2) containing nitrogen and oxygen in various proportions [14]. The main sources of NOx are diesel buses, power plants and other industrial, commercial and domestic sources that burn fuels [1]. In the atmosphere, nitrogen oxides can contribute to the formation of photochemical ozone (smog or polluting mist) and have health consequences. If prolonged or continuous exposure occurs, the nervous system and cardiovascular system may be affected, leading to neurological and cardiac alterations. On the other hand, carbon monoxide (CO) [15] Its main source is the transportation sector due to the incomplete combustion of gas, oil, gasoline, and coal. The domestic appliances that burn fossil fuels such as stoves, stoves or heaters, among others. Your health conditions are mental confusion, vertigo, headache, nausea, weakness, and loss of consciousness. It also contributes to global warming and can cause acid rain. Both gases also act as precursors of ozone formation which potentially aggravates climatic conditions [16,17].

Finally, the system has a UV sensor to determine the maximum radiation rates and a temperature and relative humidity sensor to know its status during the day. The Fig. 1 presets the electronic connection and his PCB design. The developed system is under free hardware and software components. Its plate design is in the form of a mountable device for the Arduino (shield). In this way it can be connected and disconnected in an easy way. For the correct functioning of the sensors, a burning stage must be carried out. Finally, the system has a UV sensor to determine the maximum radiation rates and the temperature and relative humidity sensor to know its status during the day. The figure ref img1 presents the electronic connection and its PCB design.

The developed system is under free hardware and software components. Its plate design has the form of a mounting device for the Arduino (shield). In this way, it can be connected and disconnected in an easy way. For the correct functioning of the sensors, each one must be exposed to fire. In this sense, the sensor can be exposed to different environmental conditions so that the acquisition of data can be stabilized. Subsequently, the internal resistance must be configured to select the gas to be measured. This is because the harmful gas sensors have the capacity to perceive other gases. Finally, each node only acquires data every 30 min, stored it in external memory (MicroSD) as a backup, then the system sends it through the mobile network. In Figs. 1 and 2 the system developed and ready to go into operation is shown.

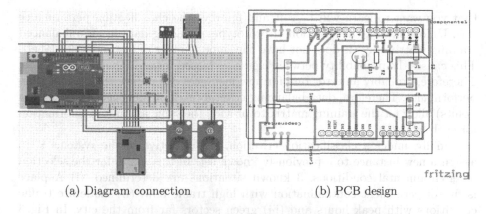

(a) Diagram connection              (b) PCB design

**Fig. 1.** Electronic system and connection design

**Fig. 2.** Electronic system developed

# 3  Data Analysis

In this section presents a data analysis in two stages. The first one, a comparison with Prototype Selection and Data Balance algorithms to find new small data sets. The second one, with these training sets, is tested with k-Nearest Neighbourhood to find the best classifier performance.

## 3.1  Data Analysis Scheme

Sensor nodes were located in the different sectors of Ibarra city. The information collected was sent to a repository in the cloud. However, this information may be subject to different errors in reading, shipping, among others. For this reason, some machine learning criteria allow choosing the ideal data sets that represent the large volume of information acquired [18]. As a result, prototype selection algorithms focus on eliminating redundant data that does not provide information to the classifier. In this way, the decision borders of each of the categories that you want to learn will be soft and improve the classification performance [17]. Unfortunately, the selection of prototypes reducing data in an unbalanced training matrix way (it does not have the same number of instances per label). This causes the probability between classes is biased towards the majority class. Therefore, probably an error rate for new instances can reduce the classification performance. Therefore, data balancing criteria (each cluster has the same data points) allow for the training matrix to provide the same amount of information for each label [19].

Finally, under a classification criterion, the objective of the systems is to assign a new instance to a previously known set. In this sense, for the selection of environmental conditions, 3 known situations are determined. (i) A place is full of permanent contamination with high traffic flow. (ii) moderate traffic conditions with peak hours and (iii) green sectors far from the city. In Fig. 3 the proposed data analysis scheme that allows the system to make the correct decision is presented.

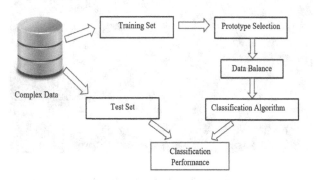

**Fig. 3.** Data analysis scheme

## 3.2   Data Analysis Criterion

Considering prototype selection criteria (PS) and data balancing (DB) are performed outside the electronic system (in a high-performance computer) different algorithms can be tested. Due to this, the most representative of each criterion is analyzed. On the one hand in PS, the algorithms to be used are All-k Edited Nearest Neighbors (AENN), Condensed Nearest Neighbor (CNN), Edited Nearest Neighbor (ENN), Reduced Nearest Neighbor (RNN), Decremental Reduction Optimization Procedures 3 (DROP3). Each one represents different ways to eliminate data (i): edition, (ii) condensations and (iii) hybrid. On the other hand, in DB the algorithms can be Kenard-Stone, Dúplex, ShenkWset and Naes [20]. All of them are implemented in different **R** software libraries. About classification algorithms, according to [20] and due to limitations of computational resources of the embedded system used (Arduino) it is advisable to use the k-Nearest Neighborhood (k-NN) classifier. For this reason, a comparison of instances removed by the PS algorithms and their execution time is performed. These data are presented in Table 1. The size of the training matrix is 1200 data points for each sensor.

**Table 1.** Prototype selection comparison

| Algorithm | Remov. inst. | % Remov. inst. | Time ejec. |
|-----------|--------------|----------------|------------|
| AENN | 11 | 1.015 | 3.51 s |
| CNN | 1042 | 96.481 | 2.66 s |
| ENN | 10 | 0.925 | 3.15 s |
| RNN | 1042 | 96.481 | 2.66 s |
| DROP3 | 1033 | 95.678 | 21.45 s |

Data matrix reduced by the PS algorithms, the next step is testing DB algorithms in comparison with classification algorithm k-NN to find the best performance. As a result of each database shows in Table 2.

## 4   Results

CNN database has been chosen to maintain a high classification performance. In each node has been implemented a classification algorithm (k-NN) and CNN. Because of this, the own system can improve your training set. Therefore, the node sensor sends only a cleaned lecture with the decision label trough 4G network to IoT network and it going to store in local micro SD. As a final IoT scheme, thirteen nodes that acquire data were implemented. In the beginning, the first readings were systematically to achieve an optimal location of the nodes around the city. Each node was installed at least with 2 km distance of another one. As a result, it was possible to determine the ranges of affectation in relation to the decision taken by each node. With this, it was possible to show graphically the sectors with the highest concentration of gases. At Fig. 4 shows a borderline classifier label decision in two dimensions.

**Table 2.** k-NN performance with data balance matrix

| Algorithm | Data balance | K-NN performance | | |
|-----------|--------------|------|------|-------|
| | | CNN | RNN | DROP3 |
| KENSTONE | 210 | 0.911 | 0.911 | 0.937 |
| DUPLEX | 210 | 0.914 | 0.914 | 0.948 |
| SHENKWEST | 210 | 0.322 | 0.327 | 0.28 |
| NAES | 210 | 0.837 | 0.911 | 0.948 |

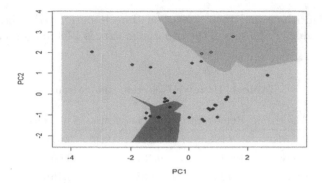

**Fig. 4.** Borderline classifier label decision, green: no pollution, blue: allowed range, red: caution (Color figure online)

For the correct visualization of information, IoT server was installed a Processing (visualization software). Each node sends data with their location, the visualization platform changes in relation to all of them. When the user selecting in each node, the data that is being monitored at that moment is displayed and the pollution state in colors (green: no pollution, yellow: allowed range, red: caution). For reasons of subsequent analysis, the maximum and minimum values are also stored with their time and date of all the variables to be measured. The city map and the air quality decision borders are shown in Fig. 5 Consequently, in Fig. 6 indicates the individual interface of each node. In this way, you can extract information tables with different reports in relation to the need for the study as shown in Table 3.

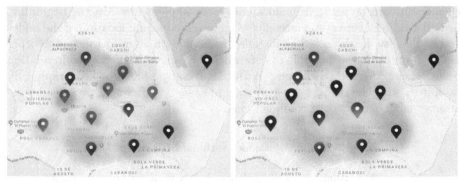

(a) Diagram connection              (b) Nodes sensing data at midday

**Fig. 5.** Nodes sensing data at night (Color figure online)

**Fig. 6.** Data visualization for each node in Ibarra city

**Table 3.** Sensor node in Bolivar street, Ibarra

| Bolivar street | V. Max | V. Min |
| --- | --- | --- |
| CO2 | 53.93 | 51.19 |
| CO | 24 | 22 |
| Temperature | 30°C | 17°C |
| Humidity | 68 % | 40 % |
| UV | 9.7 | 0 |

With the coordinates of the sensor nodes, it can be analyzed within a platform such as Google maps. Of this way, It can have a greater specificity of the environments your pollution status. Figure 7 shows the decision edges with greater precision in relation to traffic light color (red: high-level pollution, yellow: normal pollution and green: no pollution). Finally, Fig. 8 indicates a node with a rechargeable battery monitoring the environment in a green area.

**Fig. 7.** Google maps with environmental analysis (Color figure online)

**Fig. 8.** Node acquiring data

As a remarkable result, the system has been tested between some specific air pollution applications. The visualization platform has more exact values because these It only shows an estimated average. In this sense, in real conditions, nodes can classify the air pollution conditions with 92% accuracy among the nodes.

## 5    Conclusions and Future Works

The present environmental monitoring project allowed to know the places of pollution inside the city. With this, different strategies can be planned to mitigate these problems. With respect to data analysis, the proposed proposal achieved the objective of providing the correct information to the classifier and allowed it to make an adequate decision in each node. To do this, technologies such as the IoT allow an agile collection of data and interconnect a large number of devices for the extraction of knowledge. In addition, this analysis must have an interface that allows the user to know the actions taken by nodes and be able to perform different types of analysis with the stored databases.

With regard to the analyses carried out, it was possible to deduce that the concentration of the greatest amount of harmful gases could affect up to a kilometer around where there is no emission of this type. Another point to consider is that the gases in the area of vehicular traffic by schedules gradually decrease after 25 or 30 min. Finally, the mist in cold places allows a faster dissipation by the condensation of water.

As future work, it is expected to have the IoT system actively and have nodes that can be mobile and perform a more comprehensive analysis.

**Acknowledgment.** This work is supported by the Smart Data Analysis Systems - SDAS group. http://sdas-group.com/.

# References

1. Saha, A.K., et al.: A raspberry Pi controlled cloud based air and sound pollution monitoring system with temperature and humidity sensing. In: 2018 IEEE 8th Annual Computing and Communication Workshop and Conference, CCWC 2018, vol. 2018, January 2018
2. Wang, D., Duan, E., Guo, Y., Sun, B., Bai, T.: Numerical simulation of the effect of over-fire air on NOx formation in furnace. In: 2013 International Conference on Materials for Renewable Energy and Environment, pp. 780–783. IEEE, August 2013. http://ieeexplore.ieee.org/lpdocs/epic03/wrapper.htm?arnumber=6893790
3. Guariso, G., Volta, M. (eds.): Air Quality Integrated Assessment. SAST. Springer, Cham (2017). https://doi.org/10.1007/978-3-319-33349-6
4. Sujatha, K., Bhavani, N.P.G., Ponmagal, R.S.: Impact of NOx emissions on climate and monitoring using smart sensor technology. In: 2017 International Conference on Communication and Signal Processing (ICCSP), pp. 0853–0856. IEEE, April 2017. http://ieeexplore.ieee.org/document/8286488/
5. Bashir Shaban, K., Kadri, A., Rezk, E.: Urban air pollution monitoring system with forecasting models. IEEE Sens. J. **16**(8), 2598–2606 (2016). http://ieeexplore.ieee.org/document/7370876/
6. Maraj, A., Berzati, S., Efendiu, I., Shala, A., Dermaku, J., Melekoglu, E.: Sensing platform development for air quality measurements and analysis. In: 2017 South Eastern European Design Automation, Computer Engineering, Computer Networks and Social Media Conference (SEEDA-CECNSM), pp. 1–5. IEEE, September 2017. http://ieeexplore.ieee.org/document/8088233/
7. Lin, Y.-L., Kyung, C.-M., Yasuura, H., Liu, Y. (eds.): Smart Sensors and Systems. Springer, Cham (2015). https://doi.org/10.1007/978-3-319-14711-6
8. Fioccola, G.B., Sommese, R., Tufano, I., Canonico, R., Ventre, G.: Polluino: an efficient cloud-based management of IoT devices for air quality monitoring. In: 2016 IEEE 2nd International Forum on Research and Technologies for Society and Industry Leveraging a better tomorrow (RTSI), pp. 1–6. IEEE, September 2016. http://ieeexplore.ieee.org/document/7740617/
9. Wang, W., De, S., Zhou, Y., Huang, X., Moessner, K.: Distributed sensor data computing in smart city applications. In: 2017 IEEE 18th International Symposium on a World of Wireless, Mobile and Multimedia Networks (WoWMoM), pp. 1–5. IEEE, June 2017. http://ieeexplore.ieee.org/document/7974338/
10. Kafli, N., Isa, K.: Internet of Things (IoT) for measuring and monitoring sensors data of water surface platform. In: 2017 IEEE 7th International Conference on Underwater System Technology: Theory and Applications (USYS), pp. 1–6. IEEE, December 2017. http://ieeexplore.ieee.org/document/8309441/
11. Kumar, S., Jasuja, A.: Air quality monitoring system based on IoT using Raspberry Pi. In: 2017 International Conference on Computing, Communication and Automation (ICCCA), pp. 1341–1346. IEEE, May 2017. http://ieeexplore.ieee.org/document/8230005/
12. Rosero-Montalvo, P.D., et al.: Intelligence in embedded systems: overview and applications. In: Arai, K., Bhatia, R., Kapoor, S. (eds.) FTC 2018. AISC, vol. 880, pp. 874–883. Springer, Cham (2019). https://doi.org/10.1007/978-3-030-02686-8_65
13. Chiu, S.-W., Hao, H.-C., Yang, C.-M., Yao, D.-J., Tang, K.-T.: Handheld gas sensing system. In: Lin, Y.-L., Kyung, C.-M., Yasuura, H., Liu, Y. (eds.) Smart Sensors and Systems, pp. 155–190. Springer, Cham (2015). https://doi.org/10.1007/978-3-319-14711-6_8

14. Bae, H.: Basic principle and practical implementation of near-infrared spectroscopy (NIRS). In: Lin, Y.-L., Kyung, C.-M., Yasuura, H., Liu, Y. (eds.) Smart Sensors and Systems, pp. 281–302. Springer, Cham (2015). https://doi.org/10.1007/978-3-319-14711-6_12

15. Peng, L., Danni, F., Shengqian, J., Mingjie, W.: A movable indoor air quality monitoring system. In: 2017 2nd International Conference on Cybernetics, Robotics and Control (CRC), pp. 126–129. IEEE, July 2017. http://ieeexplore.ieee.org/document/8328320/

16. Air Quality Expert Group: air quality and climate change: a UK perspective. http://webarchive.nationalarchives.gov.uk/20130403220722/archive.defra.gov.uk/environment/quality/air/airquality/publications/airqual-climatechange/documents/fullreport.pdf

17. Rosero-Montalvo, P.D., et al.: Air quality monitoring intelligent system using machine learning techniques. In: 2018 International Conference on Information Systems and Computer Science (INCISCOS), pp. 75–80. IEEE, November 2018. https://ieeexplore.ieee.org/document/8564511/

18. Rosero-Montalvo, P., et al.: Prototype reduction algorithms comparison in nearest neighbor classification for sensor data: empirical study. In: 2017 IEEE Second Ecuador Technical Chapters Meeting (ETCM), pp. 1–5. IEEE, October 2017. http://ieeexplore.ieee.org/document/8247530/

19. Rosero-Montalvo, P., et al.: Neighborhood criterion analysis for prototype selection applied in WSN data. In: 2017 International Conference on Information Systems and Computer Science (INCISCOS), pp. 128–132. IEEE, November 2017. http://ieeexplore.ieee.org/document/8328096/

20. Rosero-Montalvo, P.D., Peluffo-Ordóñez, D.H., López Batista, V.F., Serrano, J., Rosero, E.A.: Intelligent system for identification of wheelchair user's posture using machine learning techniques. IEEE Sens. J. **19**(5), 1936–1942 (2019)

# Crowd-Powered Systems to Diminish the Effects of Semantic Drift

Saulo D. S. Pedro[✉] and Estevam R. Hruschka Jr.

Universidade Federal de São Carlos, UFSCar., São Carlos, Brazil
saulods.pedro@gmail.com

**Abstract.** Internet and social Web made possible the acquisition of information to feed a growing number of Machine Learning (ML) applications and, in addition, brought light to the use of crowdsourcing approaches, commonly applied to problems that are easy for humans but difficult for computers to solve, building the crowd-powered systems. In this work, we consider the issue of semantic drift in a bootstrap learning algorithm and propose the novel idea of a crowd-powered approach to diminish the effects of such issue. To put this idea to test we built a hybrid version of the Coupled Pattern Learner (CPL), a bootstrap learning algorithm that extract contextual patterns from an unstructured text, and SSCrowd, a component that allows conversation between learning systems and Web users, in an attempt to actively and autonomously look for human supervision by asking people to take part into the knowledge acquisition process, thus using the intelligence of the crowd to improve the learning capabilities of CPL. We take advantage of the ease that humans have to understand language in unstructured text, and we show the results of using a hybrid crowd-powered approach to diminish the effects of semantic drift.

**Keywords:** Crowdsourcing · Crowd-powered systems · Semantic drift

## 1 Introduction

Many applications in Machine Learning depend on large unstructured data sets that are expensive to preprocess to be ready for mining. Sometimes, even when the data is structured, it might be inaccurate or not useful. In such scenarios, the assistance of an oracle could be helpful, that is, someone with extensive knowledge of the problem could look at the data and select the best from it to build the data set [2,21]. Approaches like these, have been used to improve ML tasks by filtering irrelevant data, thus reducing the cost of expensive tasks such as labeling.

In an Active Learning task, this oracle could be a human and if the volume of the data is too big, it may not be possible for the oracle to provide enough assistance in reasonable time. Such a situation encourages the discussion of having an automatic agent assisting in ML, proactive in taking ownership of demands

© Springer Nature Switzerland AG 2019
H. Pérez García et al. (Eds.): HAIS 2019, LNAI 11734, pp. 697–709, 2019.
https://doi.org/10.1007/978-3-030-29859-3_59

like validation, revision and feeding. Scenarios where a problem can be easily resolved by human experts (an oracle) and poorly resolved by machines, motivates *human-in-the-loop* approaches [1] that consider adding humans to take part into the knowledge acquisition process of a Machine Learning system. To provide information with enough accuracy, in such *human-in-the-loop* approaches, agents would have to access human generated content and extract information from it automatically, whenever the system needs it. One possibility, would be to build an agent capable of identifying its own needs for human-supervision, and based on those needs, proactively start conversation with people in social network environments, asking them to assist the intelligent machine.

Crowdsourcing can be seen as a process to complete a task with the contribution of a large set of people, and it is often used to gather resources for collaborative systems such as Wikipedia[1] and for work market places such as Amazon's Mechanical Turk (AMT)[2]. Encouraged by large data sets that need human attention, crowdsourcing has been used as input of computational systems. Such solutions can be called crowd-powered systems [3], and include algorithms that gather and apply the intelligence of humans. The power of the crowd has been applied in ML to boost projects in research fields such as image recognition [23], word processing [4], sentiment analysis [7], product classification [22] and labeling images [12]. Crowdsourcing has been widely used to annotate labels for data sets in ML tasks. However, more than provide labels, it has been used as part of the intelligent system itself, such as in [26], where sentences translated by people are combined to provide a translation that outperforms the accuracy of professional translation.

In Machine Learning, a great part of the communication management between people and machines has been done through AMT. However, to achieve more accurate annotations or to provide a larger control of how humans participate in the learning process, some intelligent systems have their own platform to connect with humans [14]. An interesting work on knowledge acquisition through the crowd is the Curious Cat [6], an agent that talks to people through an application to collect knowledge. This work also brings approaches to resolve the issues of working with malicious content generated from humans. The authors look for redundant content created by other humans and perform a consistency check with a large KB like CyC [15]. In addition, the agent learns context from the users creating a channel of communication that targets a single user allowing for example, to ask specific questions for users that are more likely to be able to answer.

In this work, we explore the properties of crowd-powered systems to improve the performance of a bootstrap learning algorithm. These algorithms were initially proposed to extract named entity categories from text [20], but were also used in tasks such as web page classification [5] and word sense disambiguation [25]. They can learn from a large amount of unlabeled data and a small set of labeled data, called "seeds", in a semi-supervised learning task. The seeds are

---

[1] wikipedia.org.
[2] mturk.com.

used to train an initial model that labels part of the unlabeled data, and then use this model to train a new model that uses the seeds and the recently self-labeled examples. The process is iterative, which allows the amount of labeled data to expand with minimal supervision. However, the ambiguity of language introduce errors in the iterative process which can lead to semantic drift, a situation where wrongfully labeled data propagates through the learning iterations [11].

To address the problem of semantic drift, the work in [16] identified that the choice of seeds affects the propagation of errors and proposed a bagging of random seeds that reduce the semantic drift in the task of extracting semantic lexicons from text. In bootstrap learning algorithms, a common approach to diminish semantic drift, is to couple different functions that perform the same task, which allows the functions to supervise each other through the integration of their knowledge [20, 24]. Another important example of coupling functions to reduce semantic drift is the Coupled Pattern Learner(CPL), which is a model that learns to extract predicates from an unstructured text in a bootstrap learning fashion. CPL relies only on a few seeds of those predicates for training and uses its previously learned predicates to learn new predicates [9].

In this work, we propose the novel idea of using a crowd-powered approach to reduce the effects of semantic drift in a bootstrap learning algorithm. We put this idea to test by having humans to tell CPL how good its predicates are. To do that we built a hybrid version of CPL with the SSCrowd component, an agent that allows a learning system to post questions and read answers from users on social networks. This approach keeps the idea of building machines that resolve their questions like humans do, that is, talking to other humans, as attempted in [19]. The SSCrowd component has been put to test on Yahoo! Answers and Twitter [17, 18]. Section 2.1 details the problem of semantic drifting for CPL, Sect. 2.2 describes how SSCrowd uses Twitter to allow learning systems to ask questions to humans and Sect. 2.3 details our hybrid approach to improve CPL. In Sect. 3 we show the results of this strategy.

## 2    Learning Through Conversation with Humans

The knowledge representation of CPL is a hierarchical structure of predicates divided into categories and relations. A category can be seen as a class of instances. For example, John Travolta is an actor because *"John Travolta"* is an instance of the category *actor*. In addition, John Travolta starred in the movie Pulp Fiction because the pair *(John Travolta, Pulp Fiction)* is an instance of the relation *actorstarredinmovie*. This relation must be satisfied by the constraint that the in the pair *(X, Y)*, X must be an instance of the category *actor* and Y must be an instance of the category *movie*.

CPL access unstructured text data sets to learn contextual patterns that are good extractors of predicates. For example, the sentences *"X and other actors"* and *"X agreed to star in Y"* are contextual patterns that could be used to respectively extract instances of the category *actor* and pairs *(actor, movie)* for the relation *actorstarredinmovie*. To learn contextual patterns, CPL performs two distinct operations:

(i) Part-of-speech tags found in the sentences that match $X$ and $Y$, are said to co-occur with the contextual patterns and turn into candidate instances of the category or relation extracted by the pattern. The candidates are filtered and ranked to be promoted into actual instances.

(ii) CPL then re-access the data set looking for candidate contextual patterns that co-occurs with the promoted instances. These candidates are also filtered and ranked to be promoted into actual contextual patterns.

CPL performs the operations (i) and (ii) described above iteratively. The contextual patterns learned in operation (ii) are reused to extract more instances in operation (i), which closes a loop that allows CPL to learn in a bootstrap learning fashion. Although this model can learn predicates for categories and relations, in this work, we are only addressing categories due to space constraints. The steps to promote candidate patterns and instances are described in Sect. 2.4

## 2.1   Using Human Generated Content to Avoid Semantic Drifting

To use contextual patterns and instances promoted in previous iterations to keep learning could allow CPL to perform tasks such as Named Entity Resolution. However, since language can be ambiguous and under-constrained, it can also lead to semantic drifting. In [9], this problem is approached by learning predicates for multiple categories simultaneously and penalizing predicates that co-occurs with predicates from mutual exclusive categories. For example, John Travolta is an actor, and *"cities such as X"* is a contextual pattern that extracts cities. Since actors cannot be cities, the categories are mutually exclusive and therefore any candidate instance of *actors* that co-occurs with *"cities such as X"* are violating the coupling and will be penalized, hence, being less likely to be promoted.

The semantic drift in CPL raises when the instances learned in previous iterations for a given category $c$ yield contextual patterns that are extractors of a wider range of categories than $c$. For example, consider that for the category *food*, CPL has learned the instances *carrots*, *onions* and *tomatoes*, which yielded the contextual pattern *"asian species of X"*. This contextual pattern extracts those instances, but it could also extract *bears* and *monkeys*, which are instances of the category *animals*.

In this work, we propose the novel idea of using a *human-in-the-loop* approach to reduce the effects of semantic drift in a bootstrap learning algorithm. We put our idea to test through the improvement of CPL performance by asking humans to evaluate CPL's candidate contextual patterns and reward those that are more likely to extract only instances from the pattern's category. Such evaluation task is hard for machines to do, due to the ambiguity of language, but it is an easier task for humans.

## 2.2   SSCrowd

SSCrowd (Self-Supervisor Agent Based on the Wisdom of Crowds)[19], is a component designed to take advantage of the wisdom of crowds and put to test the idea of using human collaboration to assist in Machine Learning tasks. Its steps could be defined as follows:

1. receive information from the learning system that needs to be assessed by humans
2. translate the information into a human understandable question
3. post the question to a Web community and wait for answers
4. find consensus in the answers and return it to the learning system

This component was designed to perform a proactive and autonomous task that allows self-supervision on demand to learning systems, and it was first tested to validate the knowledge acquired by NELL (Never Ending Language Learner) a system that aims to learn forever by taking advantage of the self-supervision provided by coupling different models with a shared objective [10].

SSCrowd can also be seen as an implementation of a Conversing Learning(CL) agent [18]. CL focuses on enabling an intelligent system to proactively start and keep conversations with Web users, actively and autonomously looking for human supervision, by asking people (Web users) to take part into the knowledge acquisition process, thus using the intelligence from the crowd to improve the agent's learning capabilities. [17,19]

## 2.3 Enabling CPL to Work with SSCrowd

The scenario of an evaluation task that is hard for machines and easy for humans, motivated us to use a Conversing Learning setup as an extra step to the CPL pipeline. Algorithm 1 shows the adapted version of the CPL algorithm originally proposed in [9]. In our version, at the end of the promotion step, we take all the candidate contextual patterns that were promoted and build a question to be

---

**Algorithm 1:** CPL Algorithm with SSCrowd

**input** : seed instances and seed patterns for each category $c \in C$
**output:** instances and patterns for each category $c \in C$
for $i=1,2,...,\infty$ do
    for $c \in C$ do
        EXTRACT new candidate instances/contextual patterns using recently promoted patterns/instances
        FILTER candidate contextual patterns that violate coupling
        FILTER candidate instances with SSCrowd validated contextual patterns
        RANK candidate contextual patterns
        RANK candidate instances with SSCrowd validated contextual patterns
        PROMOTE top candidate instances/contextual patterns
        ASK Twitter users to validate candidate contextual patterns
    end
end

---

posted on Twitter through the SSCrowd algorithm. The consolidated answers from users creates a score that we use to modify how the filter and the rank step work for candidate instances. We could also use SSCrowd to validate candidate instances, as it has been done in [19], however CPL is more sensitive to contextual patterns since there are less of them, and they can yield more instances than instances could yield patterns.

**Gathering the Opinion from the Crowd:** We translated CPL needs of evaluation of contextual patterns by questioning people whether they believe a noun fits a given contextual pattern. Figure 1 shows an example with the category *sportsgame* and the contextual pattern *"team lost in X"* asked in Twitter. We have tried to place the question in a natural manner, as human speech is. The question template and the options for answers are always the same, and we expected people to either answer with representative letters (A, B, C, D) or the sentence or both.

In this work, we do not aim to micro-read the answers of the users, therefore we needed a way to motivate people to give a straightforward answer. To do that, we took the idea of *driven feedback*, that is, to modify the question in a way that it yields answers that are easier for machines to understand, thus reducing efforts in understanding language. This idea was mentioned in [18], and we implemented it by using polls to capture the intention from the users. By the time we performed the experiments, the Twitter API would not allow polls, so we have used images instead. The feedback for CPL is the consensus through majority vote. Ties were decided in favor of worse contextual patterns. We then use the consensual votes to modify the filter and the rank step. The consensual votes determine $sscrowd(p)$ for the pattern $p$ as follows:

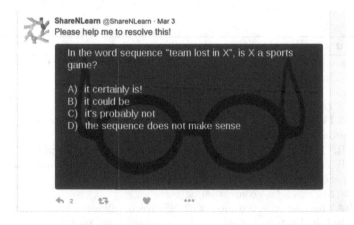

**Fig. 1.** Example of a question asked in Twitter

$$sscrowd(p) = \begin{cases} 0 & \text{if majority vote is C or D} \\ 1 & \text{if majority vote is B} \\ 2 & \text{if majority vote is A} \end{cases}$$

With this setup, the contextual patterns that receive an "A" answer are the best extractors, followed by "B". "C and D" are the answers for bad extractors.

## 2.4   Execution Steps for Modified CPL

**Extraction Step:** At its first iteration, CPL uses a few labeled seed examples to start its execution. Those seeds are automatically promoted into instances and patterns. In the following iterations, CPL will use previously promoted instances and patterns to learn new instances and new patterns. During the extraction step, it finds instances and contextual patterns that co-occur with each other to build a list of candidate instances and candidate contextual patterns.

**Filter Step:** In this step, for each category, some candidate instances will be rejected if the number of times they co-occur with promoted patterns $p \in P$ is not greater than $\tau$ times the number of times they co-occur with patterns from mutually exclusive categories $p \in \bar{P}$. In our version, CPL uses the votes gathered from SSCrowd to improve filtering. To do that, we added $sscrowd(p)$ to this step in a way that instances that co-occur with good extractors are less likely to be rejected and instances that co-occur with bad extractors are rejected. The $\tau$ parameter controls the softness of the filtering. The boolean value for rejection in the filter step is calculated as follows:

$$reject(i) = \sum_{p \in P} count(i, p) \times sscrowd(p) < \sum_{p \in \bar{P}} count(i, p) \times \tau \qquad (1)$$

In Eq. 1, $count(i, p)$ is the number of times the instance $i$ co-ocurrs with the pattern $p$. Candidate patterns are filtered in the same manner, using promoted instances instead of promoted contextual patterns. However, they would not use any form of assistance from SSCrowd, since, in this work, SSCrowd was not used to evaluate instances.

**Rank Step:** During the rank step, category instances are ranked higher when they co-occur with most of the contextual patterns. $sscrowd(p)$ was added in the rank step to rank higher candidate instances that co-occurs with the best extractors, our approach to rank candidate instances is calculated as follows:

$$rank(i) = \sum_{p \in P} pair(i, p) \times sscrowd(p) \qquad (2)$$

where,

$$pair(i,p) = \begin{cases} 0 & \text{if } i \text{ do not co-occurs with } p \\ 1 & \text{if } i \text{ co-occurs with } p \end{cases}$$

Candidate patterns are ranked differently, based on a score that favors candidates that co-occurs with most promoted instances. For candidate patterns, we use the same rank setup as described in [9].

**Promotion Step:** This step promotes the instances and contextual patterns that are ranked higher based on a threshold parameter. In our work we promoted the top 3 contextual patterns and the top 20 instances.

## 3   Experiments

To show that a crowd-powered approach can be used to reduce semantic drift in a bootstrap learning algorithm like CPL, in this section, through our experiments, we want to answer the following question: How the evaluation of contextual patterns through talking to people would affect the promotions and precision of the category instances learned by CPL?

### 3.1   Experimental Setup

CPL was first implemented using a data set from 200 million web pages [9]. In our experiments we used the English Wikipedia, which, by the time this article was written, had 6 million pages. The Wikipedia articles were tokenized and parsed with spaCy[3] to extract noun phrases. We then created a data set of unique noun phrase and contextual pattern co-occurrence counts, and filtered out those that occurred only once. This process yielded a 70 million unique pairs of noun phrases and contextual patterns.

As shown in Algorithm 1, at each iteration CPL calls for SSCrowd's assistance. CPL sends to SSCrowd a contextual pattern and a category name. Then, SSCrowd builds the question and post it on Twitter. To give time for people to answer, the questions were posted every 2 h, which means that a queue of questions is formed, which delays the execution of CPL. The answers were gathered after 24 h of posting. Answers received after that window were not considered. This is reasonable time, since questions receive most of the answers immediately, mostly because on Twitter, older messages are quickly replaced by new ones in the user feed.

The experiments were performed over a set of 42 categories based on the original experiments described in CPL's first implementation [9] plus the addition of the categories *disease* and *fruit*. At each iteration CPL would learn, for each category, at most 3 contextual patterns and 20 instances. After the end of

---

[3] https://spacy.io/.

each iteration CPL waits for one day before closing the window for answers. The answers are gathered and a consensus is determined, as described in Sect. 2.3. For the experiments described in this work, we posted 375 questions, one for each contextual pattern promoted by CPL using two different Twitter accounts. In both accounts, people would know that they were taking part in a scientific experiment, however, no instructions were given, so we could capture most of the Web users intentions while answering the questions. The questions received a total 1372 answers, from which 1248 were useful (some answers were too complex to understand). With this setup, after 10 iterations, CPL would not learn any new instances or patterns, therefore, we have ended the execution at this point.

We ran two sets of experiments, the first with CPL without considering the $sscrowd(p)$ parameter. We will reference this experiment as "CPL". The second experiment is the combination of CPL and SSCrowd as described in Eqs. 1 and 2. We will reference this experiment as "CPL + SSCrowd". For both experiments, we used $\tau = 3$ to enforce filtering. As the work in [9], we also did not perform deep sensitivity analysis for $\tau$.

After the execution of both experiments, we have manually evaluated all the instances from CPL. The result of this evaluation is shown in Table 1.

## 3.2 Experimental Evaluation

The first thing we have noticed is that CPL itself does not perform very well with a small corpus for a considerably large number of categories. More data allows CPL to grow its coupling constraints, thus learning more accurate instances. In Table 1, we show the promotions and precision of the category instances learned by "CPL" and "CPL + SSCrowd", where both were trained using Wikipedia as corpus and compare them with the results described in [9], where the same "CPL" algorithm was trained with 200 million Web pages as corpus. This data set is a preliminary version of the ClueWeb09 data set [8]. In the table, we refer to these results as "CPL CW". For each category, the precision measure is the total of category instances correctly identified divided by the sum of categories instances found by CPL. The promotion measure is the number of category instances that made through the promotion step defined in Sect. 2.4.

As shown in Fig. 2, CPL + SSCrowd would not promote as many instances as CPL did, however, it will assist in diminishing the effects of semantic drift. Over 10 iterations, CPL learned 3474 instances with a precision of 47% and CPL + SSCrowd learned 2552 instances with a precision of 65%, also improving the precision of 37 out of the 44 categories tested as shown in Table 1. For both algorithms, most of the instances are learned in the first iterations, where the precision is higher, boosted by the gold standard quality of the seed instances. As iterations move forward, CPL was less likely to accurately find new instances. Because people can easily understand language and flag bad extractors, their opinion would block a great part of the noise of bad contextual patterns thus, learning fewer instances, but more accurately. Table 2 shows the kind of assistance that comes

**Table 1.** Promotions (#) and estimated precision (%) for all categories

|  | Estimated Precision | | | Promotions | | |
|---|---|---|---|---|---|---|
|  | CPL CW | CPL | CPL SSCrowd | CPL CW | CPL | CPL SSCrowd |
| Academic field | 70 | 68 | 91 | 46 | 16 | 12 |
| Actor | 100 | 58 | 75 | 199 | 74 | 88 |
| Animal | 80 | 100 | 100 | 741 | 36 | 36 |
| Athlete | 87 | 46 | 80 | 132 | 128 | 66 |
| Award trophy tournament | 57 | 25 | 54 | 86 | 60 | 35 |
| Board game | 80 | 31 | 47 | 10 | 57 | 38 |
| Body part | 77 | 45 | 52 | 176 | 51 | 44 |
| Building | 33 | 55 | 81 | 597 | 106 | 58 |
| Celebrity | 100 | 55 | 66 | 347 | 115 | 84 |
| Ceo | 33 | 58 | 62 | 3 | 34 | 37 |
| City | 97 | 72 | 83 | 1000 | 85 | 97 |
| Clothing | 97 | 52 | 89 | 83 | 95 | 57 |
| Coach | 93 | 44 | 56 | 188 | 87 | 64 |
| Company | 97 | 39 | 56 | 1000 | 107 | 75 |
| Conference | 93 | 30 | 50 | 95 | 81 | 65 |
| Country | 57 | 56 | 72 | 1000 | 89 | 76 |
| Disease | N/a | 66 | 80 | N/a | 69 | 55 |
| Economic sector | 60 | 16 | 21 | 1000 | 105 | 123 |
| Emotion | 77 | 65 | 65 | 483 | 55 | 55 |
| Food | 90 | 54 | 100 | 811 | 112 | 95 |
| Fruit | N/a | 26 | 54 | N/a | 114 | 57 |
| Furniture | 100 | 47 | 79 | 55 | 55 | 34 |
| Geometric shape | 77 | 26 | 25 | 43 | 56 | 58 |
| Hobby | 77 | 31 | 43 | 357 | 57 | 46 |
| Kitchen item | 73 | 20 | 61 | 11 | 77 | 31 |
| Mammal | 83 | 53 | 87 | 224 | 64 | 40 |
| Movie | 97 | 35 | 66 | 718 | 74 | 39 |
| Newspaper | 90 | 80 | 80 | 179 | 20 | 20 |
| Politician | 80 | 28 | 42 | 178 | 121 | 77 |
| Product | 90 | 34 | 93 | 1000 | 129 | 45 |
| Profession | 73 | 22 | 33 | 916 | 155 | 109 |
| Professional organization | 93 | 23 | 23 | 104 | 52 | 52 |
| Reptile | 95 | 93 | 93 | 19 | 30 | 30 |
| Room | 64 | 15 | 27 | 25 | 132 | 83 |
| Scientist | 97 | 34 | 56 | 83 | 129 | 75 |
| Sport | 77 | 45 | 52 | 283 | 71 | 61 |
| Sports equipment | 20 | 22 | 100 | 58 | 80 | 17 |
| Sports league | 100 | 88 | 88 | 11 | 35 | 36 |
| Sports team | 90 | 63 | 100 | 301 | 55 | 35 |
| Stadium | 93 | 24 | 28 | 102 | 81 | 78 |
| State or province | 77 | 92 | 85 | 202 | 56 | 68 |
| Tool | 40 | 27 | 41 | 561 | 115 | 77 |
| University | 93 | 47 | 64 | 1000 | 80 | 62 |
| Vehicle | 67 | 44 | 50 | 460 | 74 | 62 |
| Mean | 75 | 47 | 65 | 338 | 78 | 58 |

from Twitter users. One distinct aspect is the pattern *"large amount of X"* being flagged as bad extractor. Although it could extract *foods*, it is too generic and it was rejected.

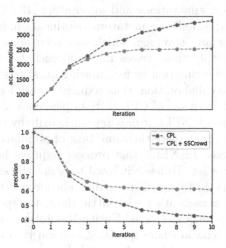

**Fig. 2.** Accumulated promotions and precision per iteration for CPL and CPL + SSCrowd

**Table 2.** Sample of consensual votes for contextual patterns by Twitter users

|       | A                            | B                    | C and D                 |
|-------|------------------------------|----------------------|-------------------------|
| Food  | Herbs such as $X$            | Served with raw $X$  | Large amount of $X$     |
| City  | $X$ important cities like $X$ | Player born in $X$   | $X$ and other countries |
| Actor | Foreign actor including $X$  | Neo played by $X$    | Is forced to by $X$     |

## 4   Conclusion

In this work, we introduced the novel idea of applying a crowd-powered approach to reduce the effects of semantic drift in a bootstrap learning algorithm. We put our idea to test through an attempt to reduce the semantic drift in the CPL, a bootstrap learning algorithm that extracts instances of categories and relations from unstructured text. The crowd is added to the loop of learning through SSCrowd, an algorithm that takes questions from learning systems and asks them on Twitter. We built a modified version of CPL that uses inputs from humans through SSCrowd. Our results show that the ease of humans to understand ambiguity in language is an advantage to address the problem of semantic drift. The supervision of humans help to prevent CPL from learning wrongfully instances, thus reducing the semantic drift.

Gathering the crowd to assist ML tasks raises some challenges, such as, discover how to keep the crowd motivated, figure how to manage communication channels like Amazon's Mechanical Turk and how to ensure quality by avoiding noisy collaborations and malicious people [6]. These issues are usually addressed by identifying the best annotations and annotators [13]. One possible way to reach such annotation, as well as annotators identification, is to study the trade-off between consensus and coverage, which has been done in [7]. In this work we do not explore deeply these issues and approach the problem of annotation/annotator quality estimation by formulating questions that encourage people to provide specific collaboration, thus reducing noisy feedback.

One of the most known use of CPL is its implementation as a tinker toy for NELL [10]. In this model, NELL avoids semantic drift by receiving supervision of multiple tinker toys that share the same task of populating an ontology with instances for categories. In NELL this process requires heavy computational power and a large data set. To use SSCrowd to assist the extraction contextual patterns and category instances like we did, encourages the idea of building applications for text classification with little data, for specific domains. CPL could be used in tasks such as Named Entity Resolution, only relying on a few seed instances. In a scenario where specialist opinion is available and text data sets are scarce, SSCrowd could be of great help to connect the machine that needs assistance to the specialists that can provide it.

# References

1. Amershi, S., Cakmak, M., Knox, W.B., Kulesza, T.: Power to the people: the role of humans in interactive machine learning. AI Magazine **35**(4), 105–120 (2014)
2. Balcan, M.-F., Urner, R.: Active learning-modern learning theory. In: Kao, M.-Y. (ed.) Encyclopedia of Algorithms, pp. 8–13. Springer, New York (2016)
3. Bernstein, M.S.: Crowd-powered systems. KI-Künstliche Intelligenz **27**(1), 69–73 (2013)
4. Bernstein, M.S., et al.: Soylent: a word processor with a crowd inside. In: Proceedings of the 23nd Annual ACM Symposium on User Interface Software and Technology, pp. 313–322. ACM (2010)
5. Blum, A., Mitchell, T.: Combining labeled and unlabeled data with co-training. In: Proceedings of the Eleventh Annual Conference on Computational Learning Theory, pp. 92–100. ACM (1998)
6. Bradeško, L., Starc, J., Mladenic, D., Grobelnik, M., Witbrock, M.: Curious cat conversational crowd based and context aware knowledge acquisition chat bot. In: 2016 IEEE 8th International Conference on Intelligent Systems (IS), pp. 239–252. IEEE (2016)
7. Brew, A., Greene, D., Cunningham, P.: Using crowdsourcing and active learning to track sentiment in online media. In: ECAI, pp. 145–150 (2010)
8. Callan, J., Hoy, M., Yoo, C., Zhao, L.: Clueweb09 data set (2009)
9. Carlson, A.: Coupled semi-supervised learning. Tech. rep., Machine Learning Department, Carnegie Mellon University (2010)
10. Carlson, A., Betteridge, J., Kisiel, B., Settles, B., Hruschka Jr, E.R., Mitchell, T.M.: Toward an architecture for never-ending language learning. In: AAAI, vol. 5, p. 3 (2010)

11. Curran, J.R., Murphy, T., Scholz, B.: Minimising semantic drift with mutual exclusion bootstrapping. In: Proceedings of the 10th Conference of the Pacific Association for Computational Linguistics, vol. 6, pp. 172–180. Citeseer (2007)
12. Kamar, E., Hacker, S., Horvitz, E.: Combining human and machine intelligence in large-scale crowdsourcing. In: Proceedings of the 11th International Conference on Autonomous Agents and Multiagent Systems, vol. 1, pp. 467–474. International Foundation for Autonomous Agents and Multiagent Systems (2012)
13. Karger, D.R., Oh, S., Shah, D.: Iterative learning for reliable crowdsourcing systems. In: Advances in neural information processing systems, pp. 1953–1961 (2011)
14. Lasecki, W.S., Wesley, R., Nichols, J., Kulkarni, A., Allen, J.F., Bigham, J.P.: Chorus: a crowd-powered conversational assistant. In: Proceedings of the 26th Annual ACM Symposium on User Interface Software and Technology, pp. 151–162. ACM (2013)
15. Lenat, D.B.: CYC: a large-scale investment in knowledge infrastructure. Commun. ACM **38**(11), 33–38 (1995)
16. McIntosh , T., Curran, J.R.: Reducing semantic drift with bagging and distributional similarity. In: Proceedings of the Joint Conference of the 47th Annual Meeting of the ACL and the 4th International Joint Conference on Natural Language Processing of the AFNLP, pp. 396–404 (2009)
17. Pedro, S.D.S., Appel, A.P., Hruschka Jr, E.R.: Autonomously reviewing and validating the knowledge base of a never-ending learning system. In: Proceedings of the 22nd International Conference on World Wide Web, pp. 1195–1204. ACM (2013)
18. Pedro, S.D.S., Hruschka, E.R.: Conversing learning: active learning and active social interaction for human supervision in never-ending learning systems. In: Pavón, J., Duque-Méndez, N.D., Fuentes-Fernández, R. (eds.) IBERAMIA 2012. LNCS (LNAI), vol. 7637, pp. 231–240. Springer, Heidelberg (2012). https://doi.org/10.1007/978-3-642-34654-5_24
19. Pedro, S.D.S., Hruschka Jr, E.R.: Collective intelligence as a source for machine learning self-supervision. In: Proceedings of the 4th International Workshop on Web Intelligence & Communities in conjunction with WWW 2012, p. 5. ACM (2012)
20. Riloff, E., Jones, R., et al.: Learning dictionaries for information extraction by multi-level bootstrapping. In: AAAI/IAAI, pp. 474–479 (1999)
21. Settles, B.: Active learning literature survey. University of Wisconsin, Madison **52**(55–66), 11 (2010)
22. Sun, C., Rampalli, N., Yang, F., Doan, A.H.: Chimera: Large-scale classification using machine learning, rules, and crowdsourcing. Proc. VLDB Endowment **7**(13), 1529–1540 (2014)
23. Von Ahn, L., Maurer, B., McMillen, C., Abraham, D., Blum, M.: reCAPTCHA: human-based character recognition via web security measures. Science **321**(5895), 1465–1468 (2008)
24. Yangarber, R.: Counter-training in discovery of semantic patterns. In: Proceedings of the 41st Annual Meeting on Association for Computational Linguistics, vol. 1, pp. 343–350. Association for Computational Linguistics (2003)
25. Yarowsky, D.: Unsupervised word sense disambiguation rivaling supervised methods. In: 33rd Annual Meeting of the Association for Computational Linguistics (1995)
26. Zaidan, O.F., Burch, C.C.: Crowdsourcing translation: professional quality from non-professionals. In: Proceedings of the 49th Annual Meeting of the Association for Computational Linguistics: Human Language Technologies, vol. 1, pp. 1220–1229. Association for Computational Linguistics (2011)

# Prediction of Student Performance Through an Intelligent Hybrid Model

Héctor Alaiz-Moretón[1]([✉])(iD), José Antonio López Vázquez[2](iD),
Héctor Quintián[2](iD), José-Luis Casteleiro-Roca[2](iD), Esteban Jove[2](iD),
and José Luis Calvo-Rolle[2](iD)

[1] Department of Electrical and Systems Engineering, University of León,
Escuela de Ingenierías, Campus de Vegazana, 24071 León, Spain
`hector.moreton@unileon.es`
[2] Department of Industrial Engineering, University of A Coruña,
Avda. 19 de febrero s/n, 15405 Ferrol, A Coruña, Spain

**Abstract.** The present work addresses the problem of low academic performance in engineering degree students. Models capable of predicting academic performance are generated through the application of several intelligent regression techniques to a dataset containing the official academic records of students of the engineering degree in the University of A Coruña. The global model, specifically the hybrid model based on K-means clustering, can predict the grade subject based on previous courses. In addition, an LDA (Linear Discriminant Analysis) has been implemented in order to identify the important features and visualize the classification clearly. Thus, the developed model makes it possible to estimate the academic performance of each student as well as the most important variables associated with it.

**Keywords:** Clustering · Ranking features · Regression ·
Academic performance · Visualization · Projection Techniques

## 1 Introduction

The main concern is the academic performance of students under the European Higher Education Area (EHEA) system. The studies tracking is legally regulated and mandatory in official university degrees [39]. As a result of this regulation, educational institutions pay special attention to improving quality indicators, in terms of academic results and performance [16].

These education indicators include: exam absence and failure rates, the number of attempts needed to pass an exam or the Grade Point Average (GPA). However, regardless the education parameter, a low academic performance trend is kept over the years.

Focusing on technical engineering degrees, studies like [15] show that only 11% of students obtain the graduate after the years determined by the educational curriculum. It is also shown that the abandonment rate is around 70%, the

© Springer Nature Switzerland AG 2019
H. Pérez García et al. (Eds.): HAIS 2019, LNAI 11734, pp. 710–721, 2019.
https://doi.org/10.1007/978-3-030-29859-3_60

average time needed to finish the degree is 5.41 years, and the performance rate remains close to 56%. According to a more recent report drafted by the Spanish University System (SUE), the academic performance of technical degree students is about 60% [27].

In these circumstances, higher education institutions must face the task of improving academic performance. The use of tools to guide or assist the students in their degree may be an effective solution.

Hence, the prediction of academic performance may be the first step to guide a student on their academic path. To this end, the use of different techniques, such as traditional statistics algorithms, advanced datamining, decision trees, artificial neural networks or Bayesian networks are considered [10,21,26,31,40].

This work aims to predict, at the individual level, the academic performance of electrical engineering students. More concretely, a grade is predicted for a specific subject for the second and third years. Given the high non-linearity of the initial dataset, clustering and intelligent regression techniques are applied. In this work, Multilayer Perceptron, Extra Tree Regressor and Support Vector Machine Regressor are the techniques utilized.

Also, the possibility of creating hybrid models by grouping the dataset with the K-means clustering algorithm is considered and then, the regression techniques are trained and tested for each cluster, leading to successful results [7,12,13,17,19,20,23,24,30].

This paper is structured as follows: the case study is briefly described in the next section. The prediction and clustering methods are introduced in Sect. 3. The experiments and their results are described in Sect. 4 and the conclusions and suggestions for future work are presented in Sect. 5.

## 2    Case Study

The present work employs intelligent techniques to predict the number of attempts that a student will have to make at an exam before passing the subject. A general description of the case study is presented in this section.

### 2.1    Dataset Description

These degrees are taught at the Polytechnic University College (EUP) of the University of A Coruña (UDC) and they last three year with a total amount of 25 subjects: the 9 subjects are done annually in the first and 7 subjects in third year. The total amount of subjects corresponds to 236 academic credits.

The following information is available about each student:

- The access method: the means by which a student enrolled in the university. Two main cases are considered; secondary school or associate degree.
- The admission grade: grade obtained by the students in their previous studies.
- For each subject:
  • The number of attempts required to pass the exam.
  • The obtained grade.

## 2.2  Preliminary Analysis and Preprocessing

In the beginning, the academic data of students enrolled in the EUP from 1996/1997 to 2008/2009 were considered. This dataset consisted of 2.736 students. With this information, a preliminary statistical analysis was carried out, reaching the following conclusions:

- On average, students enrol for the first time at the age of 20.86.
- The students graduate at an average age of 26.40.
- An average of 5.96 years are required to finish the studies.
- The 99.4% of the students are Galician.
- The 91.65% of the enrolled students come from secondary education.
- The 20% of the students are female.

However, the students who did not finish the degree are also discarded, as well as the ones that enrolled before 2001, due to the lack of numeric grades on their academic records.

After the final preprocessing, the dataset used to develop the regressive models is composed of the academic records of 225 electrical engineering graduates.

## 3  Methods

### 3.1  Preprocessing

A prepossessing procedure has been implemented, it consisting in deleting incorrect data samples, to later apply a standardization over the dataset. The applied standardization criterion has been the MaxMin Scaler [32], presented in Eq. (1):

$$\frac{X_j - minimum(X)}{maximum(X) - minimum(X)} \tag{1}$$

### 3.2  K-means and Elbow Method for Clustering

K-means is a clustering method utilized to classify a set of data into a number of groups. This is based on a first approach, where cluster centers are distributed randomly, then, to improve the similarity between groups, always trying to keep the clusters as far as possible from each other. K-means uses Euclidean Distance to measure the distance between groups with the following formula Eq. 2:

$$\sqrt{(X_1 - X_2)^2 + (Y_1 - Y_2)^2} \tag{2}$$

Moreover, the Elbow method calculates the optimal number of groups that a dataset can be divided into, according to the % of variance explained as a function of the number of groups. The Elbow method can be combined with K-means to address the classification problem [9,22,28].

### 3.3    Linear Discriminant Analysis

This is a generalization of Fisher's linear discriminant [25], the main goal of this method is to characterize or separate two or more classes of objects or events like a linear combination of features. In addition, it has been utilized as a two-dimension screening method [25] in this paper.

### 3.4    Regression Methods

**Support Vector Machines:** Support Vector Machines (SVMs) has a modality in order to get robust forecasts, named SVR. This technique is employed to choose a hyperplane regressor with the best tuning for the training dataset [3,8,11].

**Extra Trees Regressor:** This ensemble method is based on a meta estimator for regression purposes. It is based on a set of dataset sub-samples that are fitted with a number of randomized decision trees to improve accuracy and control over-fitting issues [1,18,29,33].

**Multi-layer Perceptron:** This Artificial Neural Network, for supervised learning purposes, is able to learn thanks to function: $Fun(\cdot) : R^n \to R0$. In this paper this technique has been implemented by means of Python Scikit-Learn [4,5]. The training process has been oriented to back-propagation architecture with a single neuron in the output layer with linear activation function inside.

### 3.5    Tuning Hyper-parameter Techniques

With the aim of creating a flux of transformation of raw input data, the *Pipeline* tool from *Python Scikit-Learn* is used [37]. This tool makes it easier to define a sequence of steps for pre-processing, training and validation purposes.

Pipeline manages all steps together within a *Ten-fold Grid Search Cross Validation* process. In this way, the cross validation is able to train the algorithms described previously with different parameters, choosing a combination of the best ones to obtain the best regression model [2,6,14,36].

## 4    Experiments and Results

### 4.1    Cluster and Features Importance

Once the normalization process (Sect. 3.1) has been applied, the clustering algorithm K-means, combined with the Elbow method, has been implemented to find the optimal number of clusters located in the data set. Thus, the number of inspected clusters ranged from 2 to 10. The result is three clusters, where a turning point of the Fig. 1 is located. It indicates that is the maximum average of squares sum between 2 and 4 clusters.

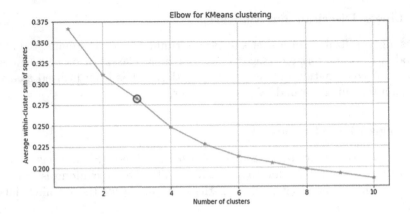

**Fig. 1.** Elbow: optimal number of clusters

When the optimal number of clusters for the normalized dataset is identified by means of the Elbow method, K-means in combination with Linear Discriminant Analysis is implemented in 3 clusters. Thereby, the output of K-means is utilized like supervised learning in the Linear Discriminant Analysis, in order to extract a 2-D visual representation. In the Fig. 2, the visual dataset projection with the 3 well-delimited clusters.

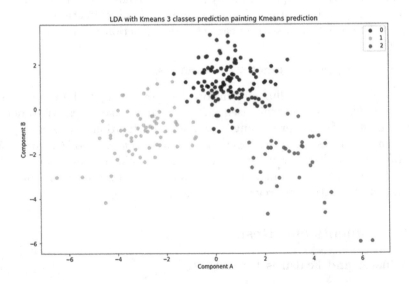

**Fig. 2.** LDA projection with 3 clusters

PCA (Principal Component Analysis) [41] could not be implemented due to large number of features included in the study, a total of 56, and the low number of cases 225.

Additionally, the LDA algorithm has the capacity to rank feature importance. Figures 3 and 4 show features ranked by their importance for component A (horizontal) and component B (vertical).

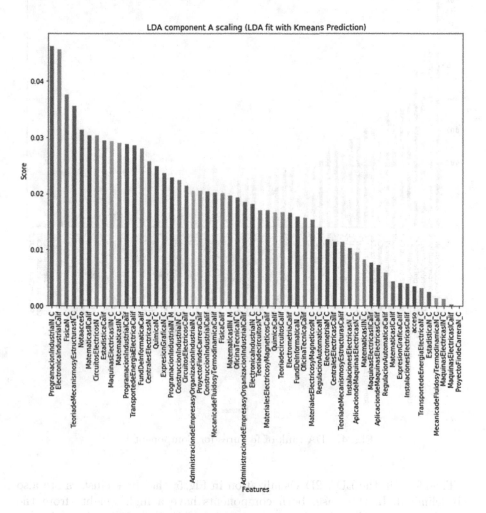

**Fig. 3.** LDA rank of features for Component A

To validate the proper working of the combination between K-means and LDA, a new experiment was developed combining K-means (3 clusters) choosing only the top four features in the rank of LDA for the component A. This component has been selected because the 3 clusters are better delimited along the horizontal axis (component A) than the vertical axis (components B) (see Fig. 2). The top four features have been selected because there is a considerable difference in the importance of the coefficients of the fourth and the fifth features in component A (Fig. 3).

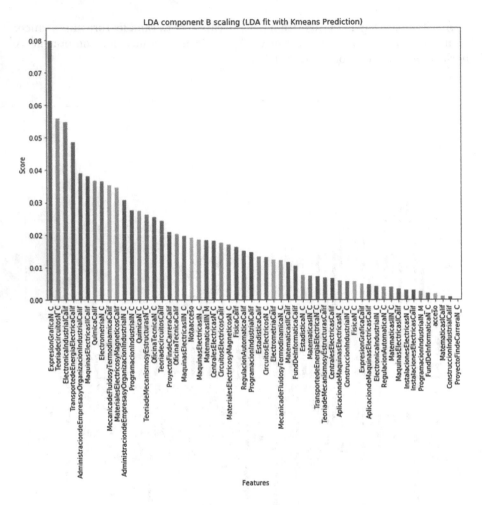

**Fig. 4.** LDA rank of features for Component B

Therefore, in the LDA 2D visualization in Fig. 5, the three clusters are also well-delimited. In this case, both components have a high weight (from the importance point of view) when it is intended to delimited the three clusters correctly.

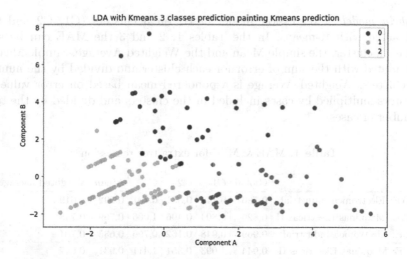

**Fig. 5.** LDA projection with 3 clusters choosing four features of top importance

## 4.2 Regression

The final objectives of this work have been oriented to predicting two different grades for the subjects considered in the previous experiments.

The subject of second year is "Industrial Electronics", second most important subject according to with Component A of LDA ranking features. The third-year subject that has been chosen is "Electric Machines II" because it is the first subject in the feature ranking of Component A. Therefore, the input variables predicted for each subject have been:

- Access qualification.
- Access modality.
- Number of attempts in subjects from previous courses.
- Grades in subjects from previous years.

To obtain a robust regression model, the technique based on Cross-validation has been applied to get the best combination of parameters for the best regression model. The error measure chosen for Grid Search as Cross-Validation has been the *Mean Absolute Error* (*MAE*) (Eq. 3). The hyper-parameters according to the three implemented machine learning techniques are explained in depth at [34,35] and [38].

$$MAE = \frac{1}{n} \sum_{i=1}^{n} |Yobs_i - Ypred_i| \qquad (3)$$

The final achieved results are considered form two different perspectives.

- *Global model*: regression model that includes all cases of the dataset (Tables 1, 2 and 3).

- *Hybrid model:* a regression model based on 3 clusters (C1, C2 and C3) extracted with K-means. In the Tables 1, 2 and 3 the MAE can be seen for each cluster, the simple Mean and the Weighted Average. Simple Mean is calculated with the sum of error for each cluster and divided by the number of clusters. Weighted Average is a pondered mean based on error value for clusters multiplied by cases included in the clusters and divided by the total number of cases.

**Table 1.** MAE & MSE for extra tree regression

|  | Global | $C1$ | $C2$ | $C3$ | $Mean$ | Weighted average |
|---|---|---|---|---|---|---|
| MAE: Electronica Industrial | 0,766 | 0,742 | 0,331 | 0,701 | 0,592 | 0,619 |
| MAE: Maquinas Electricas II | 0,826 | 0,804 | 0,496 | 1,006 | 0,769 | 0,748 |
| MSE: Electronica Industrial | 0,969 | 0,948 | 0,162 | 0,796 | 0,635 | 0,701 |
| MSE: Maquinas Electricas II | 0,944 | 1,093 | 0,351 | 1,411 | 0,931 | 0,647 |

**Table 2.** MAE & MSE for SVR

|  | Global | $C1$ | $C2$ | $C3$ | $Mean$ | Weighted average |
|---|---|---|---|---|---|---|
| MAE: Electronica Industrial | 0,766 | 0,779 | 0,287 | 0,854 | 0,640 | 0,651 |
| MAE: Maquinas Electricas II | 0,716 | 0,641 | 0,425 | 0,742 | 0,603 | 0,595 |
| MSE: Electronica Industrial | 1,076 | 1,07 | 0,141 | 1.071 | 0,763 | 0,809 |
| MSE: Maquinas Electricas II | 0,909 | 0,849 | 0,24 | 0,874 | 0,68 | 0,505 |

**Table 3.** MAE & MSE for MLP

|  | Global | $C1$ | $C2$ | $C3$ | $Mean$ | Weighted average |
|---|---|---|---|---|---|---|
| MAE: Electronica Industrial | 0,809 | 0,782 | 0,259 | 1,357 | 0,799 | 0,723 |
| MAE: Maquinas Electricas II | 0,973 | 0,735 | 0,803 | 1,060 | 0,23 | 0,765 |
| MSE: Electronica Industrial | 1,107 | 1,024 | 0,105 | 2,225 | 1,118 | 0,949 |
| MSE: Maquinas Electricas II | 1,292 | 0,883 | 0,902 | 1,413 | 0,863 | 0,892 |

## 5 Conclusions and Futures Works

A robust procedure has been developed to predict student performance in second- and third-year subjects for students doing the electrical engineering Bachelors degree. This procedure is based on a hybrid model composed of three clusters, where SVR has been applied as a regression technique. The optimal number of clusters has been obtained using the Elbow method. Moreover, to visualize the classification output as well as the feature ranking, K-means, combined with LDA has been implemented. Very good results have been achieved in general terms. With the proposal it could be possible pay special attention on some specific subjects that had more weigth in the academic students record. Even when few data were available, the students were given useful tips. It would be interesting to consider different data knowledge extraction and educational techniques and tools in future works.

**Acknowledgements.** This work is partially supported by:
- Junta de Castilla y Leon - Consejería de Educacion. Project: LE078G18. UXXI2018/000149. U-220.
- Powered by NVIDIA GPU Grant Program.

# References

1. Alaiz-Moreton, H., Aveleira-Mata, J., Ondicol-Garcia, J., Muñoz-Castañeda, A.L., García, I., Benavides, C.: Multiclass classification procedure for detecting attacks on MQTT-IoT protocol. Complexity **2019** (2019)
2. Alaiz-Moreton, H., Fernández-Robles, L., Alfonso-Cendón, J., Castejón-Limas, M., Sánchez-González, L., Pérez, H.: Data mining techniques for the estimation of variables in health-related noisy data. In: Pérez García, H., Alfonso-Cendón, J., Sánchez González, L., Quintián, H., Corchado, E. (eds.) SOCO/CISIS/ICEUTE -2017. AISC, vol. 649, pp. 482–491. Springer, Cham (2018). https://doi.org/10.1007/978-3-319-67180-2_47
3. Bengio, Y., Goodfellow, I.J., Courville, A.: Deep learning. Nature **521**(7553), 436–444 (2015)
4. Buitinck, L., et al.: API design for machine learning software: experiences from the scikit-learn project. In: ECML PKDD Workshop: Languages for Data Mining and Machine Learning, pp. 108–122 (2013)
5. Calvo-Rolle, J.L., Quintian-Pardo, H., Corchado, E., del Carmen Meizoso-López, M., García, R.F.: Simplified method based on an intelligent model to obtain the extinction angle of the current for a single-phase half wave controlled rectifier with resistive and inductive load. J. Appl. Logic **13**(1), 37–47 (2015)
6. Castej, M., et al.: Coupling the PAELLA algorithm to predictive models. In: Pérez García, H., Alfonso-Cendón, J., Sánchez González, L., Quintián, H., Corchado, E. (eds.) SOCO/CISIS/ICEUTE -2017. AISC, vol. 649, pp. 505–512. Springer, Cham (2018). https://doi.org/10.1007/978-3-319-67180-2_49
7. Casteleiro-Roca, J.L., Barragán, A.J., Segura, F., Calvo-Rolle, J.L., Andújar, J.M.: Fuel cell output current prediction with a hybrid intelligent system. Complexity **2019**, 10 (2019)
8. Casteleiro-Roca, J.L., Calvo-Rolle, J.L., Meizoso-López, M.C., Piñón-Pazos, A., Rodríguez-Gómez, B.A.: Bio-inspired model of ground temperature behavior on the horizontal geothermal exchanger of an installation based on a heat pump. Neurocomputing **150**, 90–98 (2015)
9. Casteleiro-Roca, J.L., Calvo-Rolle, J.L., Méndez Pérez, J.A., Roqueñí Gutiérrez, N., de Cos Juez, F.J.: Hybrid intelligent system to perform fault detection on BIS sensor during surgeries. Sensors **17**(1), 179 (2017)
10. Casteleiro-Roca, J.L., Jove, E., Gonzalez-Cava, J.M., Méndez Pérez, J.A., Calvo-Rolle, J.L., Blanco Alvarez, F.: Hybrid model for the ANI index prediction using remifentanil drug and EMG signal. Neural Comput. Appl. 1–10 (2018). https://doi.org/10.1007/s00521-018-3605-z

11. Casteleiro-Roca, J.L., Jove, E., Sánchez-Lasheras, F., Méndez-Pérez, J.A., Calvo-Rolle, J.L., de Cos Juez, F.J.: Power cell SOC modelling for intelligent virtual sensor implementation. J. Sens. **2017**, 1–10 (2017)
12. Casteleiro-Roca, J.L., Pérez, J.A.M., Piñón-Pazos, A.J., Calvo-Rolle, J.L., Corchado, E.: Modeling the electromyogram (EMG) of patients undergoing anesthesia during surgery (2015). https://doi.org/10.1007/978-3-319-19719-7_24
13. Casteleiro-Roca, J.L., Perez, J.A.M., Piñón-Pazos, A.J., Calvo-Rolle, J.L., Corchado, E.: Intelligent model for electromyogram (EMG) signal prediction during anesthesia. J. Multiple-Valued Logic Soft Comput. **32**, 205–220 (2019)
14. Duan, K., Keerthi, S.S., Poo, A.N.: Evaluation of simple performance measures for tuning SVM hyperparameters. Neurocomputing **51**, 41–59 (2003)
15. Espina, A.: La formación técnica postsecundaria y la competitividad de la economía española. Reis, pp. 69–115 (1997)
16. Ferreira, F.H.G., Gignoux, J.: The measurement of educational inequality: achievement and opportunity. World Bank Econ. Rev. **28**(2), 210–246 (2014). https://doi.org/10.1093/wber/lht004
17. Garcia, R.F., Rolle, J.L.C., Castelo, J.P., Gomez, M.R.: On the monitoring task of solar thermal fluid transfer systems using nn based models and rule based techniques. Eng. Appl. Artif. Intell. **27**, 129–136 (2014). https://doi.org/10.1016/j.engappai.2013.06.011. http://www.sciencedirect.com/science/article/pii/S0952197613001127
18. Geurts, P., Ernst, D., Wehenkel, L.: Extremely randomized trees. Mach. Learn. **63**(1), 3–42 (2006)
19. Gonzalez-Cava, J.M., Reboso, J.A., Casteleiro-Roca, J.L., Calvo-Rolle, J.L., Méndez Pérez, J.A.: A novel fuzzy algorithm to introduce new variables in the drug supply decision-making process in medicine. Complexity **2018**, 15 (2018)
20. González Gutiérrez, C., Sánchez Rodríguez, M.L., Fernández Díaz, R.Á., Calvo Rolle, J.L., Roqueñí Gutiérrez, N., Javier de Cos Juez, F.: Rapid tomographic reconstruction through GPU-based adaptive optics. Logic J. IGPL **27**(2), 214–226 (2018)
21. Jove, E., et al.: Modelling the hypnotic patient response in general anaesthesia using intelligent models. Logic J. IGPL **27**(2), 189–201 (2018)
22. Kodinariya, T.M., Makwana, P.R.: Review on determining number of cluster in k-means clustering. Int. J. **1**(6), 90–95 (2013)
23. Manuel Vilar-Martinez, X., Aurelio Montero-Sousa, J., Luis Calvo-Rolle, J., Luis Casteleiro-Roca, J.: Expert system development to assist on the verification of "tacan" system performance. Dyna **89**(1), 112–121 (2014)
24. Marrero, A., Méndez, J., Reboso, J., Martín, I., Calvo, J.: Adaptive fuzzy modeling of the hypnotic process in anesthesia. J. Clin. Monit. Comput. **31**(2), 319–330 (2017)
25. Mika, S., Ratsch, G., Weston, J., Scholkopf, B., Mullers, K.R.: Fisher discriminant analysis with kernels. In: Neural Networks for Signal Processing IX: Proceedings of the 1999 IEEE Signal Processing Society Workshop (cat. no. 98th8468), pp. 41–48. IEEE (1999)
26. Palmer, S.: Modelling engineering student academic performance using academic analytics. Int. J. Eng. Educ. **29**(1), 132–138 (2013)
27. Pérez, C.: El rendimiento académico en las universidades españolas. In: Conferencia de Rectores de las Universidades Españolas, Madrid (2012)

28. Quintián, Héctor, Casteleiro-Roca, José-Luis, Perez-Castelo, Francisco Javier, Calvo-Rolle, José Luis, Corchado, Emilio: Hybrid intelligent model for fault detection of a lithium iron phosphate power cell used in electric vehicles. In: Martínez-Álvarez, Francisco, Troncoso, Alicia, Quintián, Héctor, Corchado, Emilio (eds.) HAIS 2016. LNCS (LNAI), vol. 9648, pp. 751–762. Springer, Cham (2016). https://doi.org/10.1007/978-3-319-32034-2_63

29. Quintián, H., Corchado, E.: Beta scale invariant map. Eng. Appl. Artif. Intell. **59**, 218–235 (2017)

30. Quintian Pardo, H., Calvo Rolle, J.L., Fontenla Romero, O.: Application of a low cost commercial robot in tasks of tracking of objects. Dyna **79**(175), 24–33 (2012)

31. Romero, C., Ventura, S.: Educational data mining a review of the state of the art. IEEE Trans. Syst. Man Cybern. Part C Appl. Rev. **40**(6), 601–618 (2010)

32. Scikit-learn: Min max scaler (2018). http://scikit-learn.org/stable/modules/generated/sklearn.preprocessing.MinMaxScaler.html

33. Sunter, D., Berkeley, P., Kammen, D.: City-integrated photovoltaics sustainably satisfy urban transportation energy needs. WIT Trans. Ecol. Environ. **204**, 559–567 (2016)

34. Extra trees regressor (2019). http://scikit-learn.org/stable/modules/generated/sklearn.ensemble.ExtraTreesRegressor.html. Accessed 22 Apr 2019

35. Random forest regressor (2019). http://scikit-learn.org/stable/modules/generated/sklearn.ensemble.RandomForestRegressor.html. Accessed 22 Apr 2019

36. Grid search cross validation (2019). http://scikit-learn.org/stable/modules/generated/sklearn.model_selection.GridSearchCV.html. Accessed 22 Apr 2019

37. Pipeline (2019). http://scikit-learn.org/stable/modules/generated/sklearn.pipeline.Pipeline.html. Accessed 22 Apr 2019

38. Svr (2019). http://scikit-learn.org/stable/modules/generated/sklearn.svm.SVR.html. Accessed 22 Apr 2019

39. http://www.ehea.info/. Accessed 19 Mar 2017 (2017)

40. Vega Vega, R., Quintián, H., Calvo-Rolle, J.L., Herrero, Á., Corchado, E.: Gaining deep knowledge of android malware families through dimensionality reduction techniques. Logic J. IGPL **27**(2), 160–176 (2018)

41. Wold, S., Esbensen, K., Geladi, P.: Principal component analysis. Chemometr. Intell. Lab. Syst. **2**(1–3), 37–52 (1987)

# A Hybrid Automatic Classification Model for Skin Tumour Images

Svetlana Simić[1], Svetislav D. Simić[2], Zorana Banković[3],
Milana Ivkov-Simić[1], José R. Villar[4], and Dragan Simić[2(✉)]

[1] Faculty of Medicine, University of Novi Sad,
Hajduk Veljkova 1-9, 21000 Novi Sad, Serbia
{svetlana.simic,milana.ivkov-simic}@mf.uns.ac.rs
[2] Faculty of Technical Sciences, University of Novi Sad,
Trg Dositeja Obradovića 6, 21000 Novi Sad, Serbia
{simicsvetislav,dsimic}@uns.ac.rs, dsimic@eunet.rs
[3] Frontiers Media SA, Paseo de Castellana 77, Madrid, Spain
zbankovic@gmail.com
[4] University of Oviedo, Campus de Llamaquique, 33005 Oviedo, Spain
villarjose@uniovi.es

**Abstract.** In medical practice early accurate detection of all types of skin tumours is essential to guide appropriate management and improve patients' survival. The most important is to differentiate between malignant skin tumours and benign lesions. The aim of this research is classification of skin tumours by analyzing medical skin tumour dermoscopy images. This paper is focused on a new strategy based on hybrid model which combines mathematics and artificial techniques to define strategy to automatic classification for skin tumour images. The proposed hybrid system is tested on well-known *HAM10000 data set*, and experimental results are compared with similar researches.

**Keywords:** Automatic classification · Dermoscopy images ·
Support Vector Machine · *K*-Nearest Neighbors · Multilayer perceptron

## 1 Introduction

Skin has an important immunologic role, and for that reason the correlation between skin tumours development and immunologic mechanisms are intensely studied. Non-immunologic risk factors like individual predisposition, sun and environmental exposure are all contributing to the skin neoplasia incidence. Early accurate detection of all skin tumour types is essential for patients' morbidity or survival.

Generally, in medical practice, the most important skin tumours, could be classified either as malignant tumours or benign lesions. The aim of this research is classification of skin tumours on malignant tumours and benign lesions for skin tumour dermoscopy images. In this research, only malignant melanomas will be analyzed. Melanomas are the most malignant skin tumours. They grow in melanocytes, the cells responsible for pigmentation. This type of skin cancer is rapidly increasing, but its related mortality rate is increasing more modestly. The critical factor in assessment of a patient

© Springer Nature Switzerland AG 2019
H. Pérez García et al. (Eds.): HAIS 2019, LNAI 11734, pp. 722–733, 2019.
https://doi.org/10.1007/978-3-030-29859-3_61

prognosis in skin cancer is early diagnosis. More than 60,000 people in the United States (US) were diagnosed with invasive melanoma in 2000, and more than 8,000 died of the disease [1]. In 2011, an estimated 70,230 adults in the US were diagnosed with melanoma. It is estimated that 8,790 deaths from melanoma will occur within a year. Almost 10,130 deaths from melanoma will appear over this (2019) year [2]. Its frequency is rising in many other countries, for example, 10 cases were reported each year in Algeria [3].

Currently, experienced dermatologists use dermoscopy analysis and observation based on dermatology criteria to pose a working diagnosis of a type of a skin tumour. Availably of an objective system able to classify moles helps physicians-dermatologists in diagnosis and early detection of melanoma. In development of such a system it is not necessary to classify every mole but to achieve such a precision that all malignant tumours are classified as dangerous, and none is overlooked. The evaluation of the clinical exams' accuracy showed that without use of a dermatoscope, dermatologists can detect only 65–80% of cases of melanoma. The dermoscopy increases the diagnostic accuracy rate for about 10–27% [4].

This paper presents hybrid automatic classification model for skin tumour images. The proposed classification system for skin tumour images is based on golden rule based on dermatology criteria, named *ABCDE* rule: *Asymmetry, Border, Colour, Dimension,* and *Evolution.* The proposed hybrid automatic classification model uses dermatology criteria *ABC – Asymmetry, Border,* and *Colour,* in this research. The *Principal Component Analysis* (PCA) has been used in analysis of *Asymmetry;* for *Border irregularity* – analysis of *edge* – outlining of a convex polynomial is measured; and for *Colour* – analysis of *colours* – standard deviation for all three red-green-blue (RGB) channels on pixels which are included on skin tumour images are measured. In general, three artificial intelligence techniques are used for classification, particularly: *Support Vector Machine* (SVM), *k-nearest neighbors* (*k*-NN) and artificial neural network – *Multilayer perceptron* (ANN-MLP). The implemented model is tested on the part of the skin images of *HAM10000 Data Set.* Also, this paper continuous the authors' previous research in different applications on medical domain presented in [5–8].

The rest of the paper is organized in the following way: Sect. 2 provides an overview of the basic idea on image classification and related work. Section 3 presents modeling the automatic classification for skin tumour. The Preliminary experimental results testing with well-known *HAM10000 Data Set* are presented in Sect. 4. Section 5 provides conclusions and some points for future work.

## 2   Image Classification and Related Work

Classification is a process of assigning a new item or observation to its proper place in an established set of categories. Classification is used mostly as a supervised learning method, and the goal of classification is predictive. General references regarding data classification are presented in [9], and the very good contemporary hybrid classification techniques can be found in the textbook [10].

## 2.1 Related Work in Image Skin Tumour Classification

In the past three decades, many approaches have been proposed to solve skin tumour image classification problem to help physicians to make decision regarding this particular illness and future patient treatments. In the paper [1] the discrete wavelet transforms (DWT) for feature extraction is used, and skin images have been reduced using PCA. In the classification stage two classifiers have been developed. The first classifier is based on feed forward back propagation artificial neural network (FP-ANN) and the second classifier is based on k-nearest neighbor (k-NN). The features hence derived are used to train a neural network based binary classifier, which can automatically infer whether the image is that of a normal skin or a pathological skin, suffering from skin tumour. According to the classification results of both classifiers, in *decision* phase, normal or abnormal skin lesion is defined.

Several dermatoscopic rules for automatic detection of melanoma in order to generate new high-level features – ABCD rule, 7-point checklist, Menzies method and CASH (*colour, architecture, symmetry, homogeneity*) algorithm – allowing semantic analysis are shown in [4]. The extracted features are based on shape, colour and texture features. 206 images of skin lesions are used. They were all extracted from the two online public databases *Dermatology Information System* and *DermQuest*, 119 are melanomas, and 87 are not melanoma. A neural network classifier is used for decision making, but it is not noted which type of ANN is used.

The paper [11] presents a unified method for histopathology image representation learning, visual analysis interpretation, and automatic classification of skin histopathology images as either having basal cell carcinoma or not. The novel approach is inspired by ideas from image feature representation learning and deep learning (DL) and yields a DL architecture that combines an autoencoder learning layer, a convolutional layer. A softmax classifier for cancer detection, a generalization of a logistic regression classifier [12], and visual analysis interpretation are used.

Smartphone applications are readily accessible and potentially offer an instant risk assessment of the likelihood of malignancy so that the people could immediately seek further medical attention from a clinician for more detailed assessment of the lesion. There is, however, a risk that melanomas might be missed and treatment delayed if the application reassures the user that their lesion is low risk [13].

## 3    Modeling the Automatic Classification for Skin Tumour

The proposed flow-diagram for modeling the automatic classification for skin tumour image is presented in Fig. 1. It presents one of the standard steps: (1) *Preprocessing*; (2) *Segmentation*; (3) *Analysis* in three sub-steps – (a) *Analysis of asymmetry*; (b) *Analysis of edge*; (c) *Analysis of Colours*; (4) Extraction the *Set of Features*; (5) *Classification* with three classifiers – (a) SVM; (b) k-NN; (c) ANN-MLP. Therefore, before such an examination, it is necessary to start by preprocessing and segmenting the skin tumour image.

Preprocessing is the first step in skin tumour images. Many skin features may have impact on digital images like hair and colour, and other features such as brightness, and

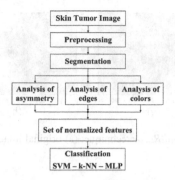

**Fig. 1.** Flow-diagram of the proposed hybrid techniques

the type of the scanner or digital camera. In the preprocessing step, the border detection procedure, colour space transformation, contrast enhancement, and artifact removal are treated. The idea is that if there is a transaction on edge detection of a source noised image, it can be located with other additional edges due to the presence of noise. Therefore, filtering the noised image is necessary.

### 3.1 Preprocessing and Segmentation - Filtering the Noised Image

Some images include artefacts, mostly hair; these artefacts can be misleading for the segmentation algorithm. The *DullRazor* technique, an artefact removal preprocessing technique, deals well with hair and other artefacts. It tends to erase the details of the image by making the pigmented network unclear, and separate only skin tumor.

| **Algorithm 1:** *The DullRazor algorithm for* ***Filtering the Noised Image*** |
|---|
| ***Step 1:***   *Dilate then erode the image to remove the small details* |
| ***Step 2:***   *Calculate the difference between the obtained image and the original one* |
| ***Step 3:***   *Dilate then erode the mask of difference, to remove noise* |
| ***Step 4:***   *Create a Boolean mask containing the location of the artefacts* |
| ***Step 5:***   *From the original image, replace the pixels covered by the mask by the pixels corresponding to the original image* |

The basic steps of the DullRazor algorithm proposed in [14], are summarized by the pseudo code shown in Algorithm 1. The *DullRazor* technique has fully addressed the problem of human hairs including the imaged lesions by designing an automatic segmentation program to differentiate skin lesions from the normal healthy skin. This technique performed well with most of the images, and with those where hairs, especially dark thick hairs, cover part of the lesions as shown on Fig. 2.

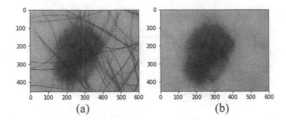

**Fig. 2.** Hair removal by the *DullRazor* technique: (a) original image, (b) image after removal of hair

## 3.2 Analysis of Asymmetry

*Asymmetry* is an essential parameter in differentiating malignant tumours from benign lesions. It is generally evaluated by dermatologists through observation by comparing the two halves of the lesion according to the principal axis. The usage method proposed in [15] calculates the index of symmetry by the differences between the areas defined with main axes and areas defined with main axes rotated by the 180 axes, compared to the centre of gravity of the skin tumour [16].

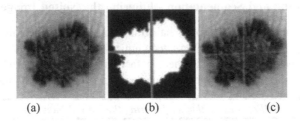

**Fig. 3.** Calculating the symmetry following the two principal axes: (a) image after filtering, (b) binary mask, (c) detection axis of inertia

Therefore, it can be concluded that asymmetry is a quantifiable property and that parameter can be used for discriminating and characterizing the melanomas. The main axes is defined by classical technique PCA and one of the most common methods used to achieve reduction without losing too much information. The central symmetry can be determined by a rotation of 180° around the centre of gravity which is presented in Fig. 3. The axial symmetry around the principal and the secondary axis of inertia are considered. It is focused on two formulations: *minimizing the residual sum of squares*, and *maximizing the variance captured* which is in detail discussed in [17].

## 3.3 Analysis of Edges

*Analysis of Edges* (*Border*) is measured by circumscribed convex polygon around the contours of skin tumour and skin tumour comparisons surface contours and polygons. Circumscribed convex polygon is designed usage of convex hull algorithm. The completely convex hull algorithm is in detail described in [18]. The segmentation of an image into a complex of edges is a useful prerequisite for object identification. However, although many low-level processing methods can be applied for this purpose, the problem is to decide which object boundary each pixel in an image falls within and which high level constraints are necessary (Fig. 4).

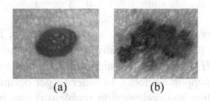

(a)                    (b)

**Fig. 4.** Extraction *Border* from skin tumours images: (a) benign, (b) malignant [15]

### 3.4   Analysis of Colours

The pigmentation of skin tumours can be characterized by several colours – five to six colours may be present in a malignant lesion. The skin tumour **Colour** – *Analysis of Colours* – is measured by mean of standard deviation for all three red-green-blue (RGB) channels on pixels which are included in skin tumour images. Description of applied algorithm is summarized by the pseudo code and is shown in Algorithm 3.

| **Algorithm 2:** *The Analysis of Colours* |
| --- |
| **Step 1:**  Select only pixels which are included on skin tumour image |
| **Step 2:**  For every selected pixel calculate *red-green-blue* color components |
| **Step 3:**  For every pixels calculate *variance* for every RGB channels: *Var*Red, *Var*Green, *Var*Blue; where *standard deviation* is the square root of the *variance* |
| **Step 4:**  The entire image mean *Var*RGB value is computed with the following equation  *Var*RGB = (1/3)·(*Var*Red + *Var*Green + *Var*Blue) |

The last equation computes the entire image mean of variance for all three (RGB) channels.

### 3.5   Classification – Support Vector Machine

Support Vector Machine (SVM) [19] is mathematical computational unit just like neural network that constructs hyperplanes defining decision boundary of classification. Main intuition behind contusions of SVM is to maximize separation between classified labels. SVM is applicable for binary classification and multi-class classification. Original SVM was stated for linear classification but non-linear classification can be obtained by using kernel function. Kernel function maps low dimension data into high dimensional space where a linear separation of data is possible. SVM derived from statistical learning and uses supervised learning model for training. On solving it gets converted into quadratic programming problem solving quadratic equation with linear constraints. There are infinite numbers of possible decision boundaries but SVM selects the best decision boundary. Associated hyperplane is mathematically formulated by following:

$$X_i * w + b \geq 1 \quad if \ y_i = +1$$
$$X_i * w + b \leq 1 \quad if \ y_i = -1$$

where, $x_i$ = feature vector, $y_i$ = class label, $w$ = weights, $b$ = Threshold/bias value, $d$ + = the shortest distance to the closest positive point, $d-$ = the shortest distance to the closest negative point. $H1$ and $H2$ are "supporting hyperplane". The points on the planes $H1$ and $H2$ are "support vectors". In order to get optimum results, the margin has to be maximized. Thus, the optimization problem $\psi$ can be defined by: $\psi = min$ $(1/2 \ w^T w)$; subjected to $y_i (w^T x_i + b) \geq 1, i = 1, 2, \ldots m$ ($m$ - total number of training instances). The quadratic optimization problem of SVM is solved using *Sequential Minimal Optimization* algorithm, in which two parametric values are solved at an instant keeping rest constant. The values are updated at every iteration till the convergence is reached.

### 3.6    Classification – *k-Nearest Neighbors*

One of the most straightforward instance-based learning algorithms is the nearest neighbor algorithm – *k-Nearest Neighbors* (*k*–NN). The *k*–NN is based on the principle that the instances within a dataset will generally exist in close proximity to other instances that have similar properties [20]. If the instances are tagged with a classification label, then the value of the label of an unclassified instance can be determined by observing the class of its nearest neighbors. The *k*-NN locates the *k* nearest instances to the query instance and determines its class by identifying the single most frequent class label.

| **Algorithm 3:** *The k-Nearest Neighbors classification* | |
|---|---|
| *Step 1:* | Compute the distance between *t* and each instance in *I* |
| *Step 2:* | Sort the distances in increasing numerical order and pick the first *k* elements |
| *Step 3:* | Compute and return the most frequent class in the **k nearest neighbors**, optionally weighting each instance's class by the inverse of its distance to *t* |

Description of the algorithm summarized by the pseudo code is shown in Algorithm 3. The training phase for *k*-NN consists of simply storing all known instances and their class labels. A tabular representation can be used, or a specialized structure such as a *kd*-tree. When it is necessary to tune the value of *k* and/or perform feature selection, n-fold cross-validation can be used on the training dataset. The testing phase for a new instance *t*, is given a known set *I*.

### 3.7    Classification – Artificial Neural Network - Multilayer Perceptron

A *Multilayer perceptron* can be viewed as a logistic regression classifier where the input is first transformed using a learnt non-linear transformation φ. A *Multilayer perceptron* (MLP) (or Artificial Neural Network – ANN-MLP) with a single hidden layer can be represented graphically as in Fig. 5.

Formally, a one-hidden-layer MLP is a function $f: \mathfrak{R}^D \to \mathfrak{R}^L$, where $D$ is the size of input vector $x$ and $L$ is the size of the output vector $f(x)$, such that, in matrix notation:

$$f(x) = G(b^{(2)} + W^{(2)}(s(b^{(1)} + W^{(1)}x)))$$

with bias vectors $b^{(1)}$, $b^{(2)}$; weight matrices $W^{(1)}$, $W^{(2)}$ and activation functions $G$ and $s$.

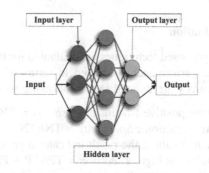

**Fig. 5.**  Artificial Neural Network – *Multilayer perceptron*

The vector $h(x) = \varphi(x) = s(b^{(1)} + W^{(1)}x)$ constitutes the hidden layer. $W^{(1)} \in \mathfrak{R}^{D \times Dh}$ is the weight matrix connecting the input vector to the hidden layer. Each column $W_{.i}^{(1)}$ represents the weights from the input units to the $i$-th hidden unit. Typical choices for $s$ include *Rectified Linear Unit* (ReLU), with $f(x) = 0$ for $x < 0$; $f(x) = 1$ for $x \geq 0$; and $f'(x) = 0$ for $x < 0$; $f'(x) = 1$ for $x \geq 0$. In this research this function is used, because it typically yields to faster training and sometimes also to better define local minima. The output vector is then obtained as: $o(x) = G(b^{(2)} + W^{(2)} h(x))$.

## 4    Experimental Results and Discussion

The proposed hybrid automatic classification system for skin tumour images which combines following techniques: DullRazor technique, PCA, convex hull algorithm, and classifiers SVM, *k*–NN, and ANN-MLP was further on, in our research, tested on tumour images from well-known repository *HAM10000 Data Set* [21]. *HAM10000 Data Set* is collection of 10015 dermatoscopic images from different populations, acquired and stored by different modalities.

## 4.1  Data Set

This research puts stress on recognition of the first three characteristics of an ABCD(E) method: *Asymmetry*, *Border* and *Colour* (ABC), considering that this data set did not contain clinical dimensions of moles on the images, nor any other data that could help determine the size of the mole, it was not possible to follow the evolution procedure. From content of repository *HAM10000 Data Set* which consists of 10015 dermato-scopic images, the first images that are eliminated are images with vignette, because vignette makes skin image very asymmetric, and it is difficult to recognize differences between skin tumour and vignette. Then, according *HAM10000* metadata, approximately the same number of benign and malignant skin tumours, 971 cases of benign skin tumour (*Positives*) and 1163 cases of malignant skin tumours (*Negatives*) are selected, which means total of 2134 skin tumour images are used in this research.

## 4.2  Performance Evaluation

The performances of the proposed techniques are evaluated for the skin tumour images. The proposed techniques performance evaluated in terms of sensitivity, specificity, precision, recall, and accuracy are defined in the following way:

1. *Sensitivity = Recall* (true positive fraction): *Sensitivity* = TP/(TP + FN)
2. *Specificity* (true negative fraction): *Specificity* = TN/(TN + FP)
3. *Precision* (*false predictive value*): the result indicates a good measure to determine the costs of *False Positive* is high: *Precision* = TP/(TP + FP)
4. *Accuracy*: the probability that the diagnostic test is performed correctly:

$$Accuracy = (TP + TN)/(TP + TN + FP + FN)$$

5. *F1 Score*: the success rate of classification:

$$F1\ Score = 2 * Precision * Recall/(Precision + Recall)$$

Where, TP (*True Positives*) – Correctly classified positive cases; TN (*True Negative*) – Correctly classified negative cases; FP (*False Positives*) – Incorrectly classified negative cases; FN (*False Negative*) – Incorrectly classified positive cases.

## 4.3  Experimental Results

Training the model and classifying skin tumours in categories is based on three characteristics – (a) *Symmetry*, (b) *Border regularity*, (c) *Colour deviation mean*; which are measured and normalized according to the description of the data themselves. The experimental results are presented in Table 1. Three different artificial techniques are used for classification: (a) SVM, (b) *k*-NN and (c) ANN–MLP classifiers.

The data set is divided in two classes, 70% for training and 30% for testing classifiers. Therefore, there are 1493 training instances and 641 testing instances. Every three-classifier experiment is repeated one hundred (100) times.

**Table 1.** Experimental results for Symmetry, Border regularity, Colour deviation mean for *Positives* and *Negatives* skin tumours

|  | Positives | Negatives |
|---|---|---|
| Symmetry mean | 0.752 | 0.891 |
| Border regularity mean | 0.838 | 0.937 |
| Colour deviation mean | 0.457 | 0.346 |

Experimental clinical diagnostics performance *Sensitivity, Specificity, Precision, Accuracy,* and *F1 Score* for all three classifiers SVM, *k*-NN, and ANN-MLP are presented in Table 2. The best performance for *Specificity, Precision, Accuracy,* and *F1 Score* has SVM classifier, but it must be noticed that ANN-MLP classifier has near close performance with SVM classifier. Maybe some correction in hidden layer can be significantly improved in ANN-MLP performance. The experimental results for different skin images used by SVM classifier are presented in Table 3.

**Table 2.** Compare clinical diagnostics performance for three classifiers

| Performance | | Classifier | | | Performance | Classifier | | |
|---|---|---|---|---|---|---|---|---|
| | | SVM | k-NN | ANN-MLP | | SVM | k-NN | ANN-MLP |
| Sensitivity | *min* | 0.5970 | 0.6760 | 0.6640 | Specificity | 0.8670 | 0.8220 | 0.8340 |
| | *mean* | 0.6623 | 0.7156 | 0.7001 | | 0.9056 | 0.8586 | 0.8653 |
| | *max* | 0.7460 | 0.7660 | **0.7800** | | **0.9380** | 0.8980 | 0.9050 |
| Precision | *min* | 0.7890 | 0.7660 | 0.7490 | | | | |
| | *mean* | **0.8547** | 0.8080 | 0.8148 | | | | |
| | *max* | 0.9020 | 0.8600 | 0.8490 | | | | |
| Accuracy | *min* | 0.7660 | 0.7780 | 0.7660 | F1 Score | 0.7090 | 0.7400 | 0.7200 |
| | *mean* | 0.7947 | 0.7935 | 0.7896 | | 0.7456 | 0.7586 | 0.7528 |
| | *max* | **0.8480** | 0.8080 | 0.8240 | | **0.7960** | 0.7820 | 0.7930 |

## 4.4  Discussion

These experimental results could be compared with some other skin tumour images analysis. In research presented in [3] *Sensitivity* is 67.5%, *Specificity* is 80.5%, correct classifications *Accuracy* is 74%, but another data set which consists of 180 skin tumour images is used there. In our research best experimental results for *Sensitivity* is 78%, *Specificity* is 93.8%, *Precision* is 85.5%, *Accuracy* is 84.8%, and *F1 Score* is 79.6%. All performances in our research are better than in the mentioned research. But, on the other side, paper [1] presents system which got 100% for *Sensitivity*, 95% for *Specificity*, and 97.5% for *Accuracy*, but the data set used there consists of only 40 images (20 normal and 20 abnormal). The number of skin images in our research is 2134 which is at least ten times more in comparison to the other research.

**Table 3.** Comparison of the original type defined in HAM10000 metadata of skin tumours and predicted type with Support Vector Machine classifier (T – True; F – False)

| Image | Original | Prediction | T/F | Image | Original | Prediction | T/F |
|-------|----------|------------|-----|-------|----------|------------|-----|
| ISIC_0025072 | 0 | 0 | T | ISIC_0032072 | 1 | 1 | T |
| ISIC_0027411 | *1* | *0* | *F* | ISIC_0033055 | 1 | 1 | T |
| ISIC_0028279 | 0 | 0 | T | ISIC_0030550 | 0 | 0 | T |
| ISIC_0028848 | 0 | 0 | T | ISIC_0034222 | 1 | 1 | T |
| ISIC_0028878 | 1 | 1 | T | ISIC_0034316 | 1 | 1 | T |
| **Successful rate** | | | | | | | **90%** |

# 5  Conclusion and Future Work

The aim of this paper is to propose the new hybrid strategy for automatic classification of skin tumour images. First, the algorithms are employed for the extraction of the set of features based on *ABC* dermatology criteria – *Asymmetry*, *Border*, and *Colour*. Different mathematical and artificial intelligence techniques are combined: DullRazor technique, PCA, convex hull algorithm, and three classifiers SVM, *k*–NN, and ANN-MLP. The system is tested on 2143 selected tumour images from repository *HAM10000 Data Set* which consists of 10015 dermatoscopic images.

Preliminary experimental results encourage the further research by the authors because the proposed strategy on experimental data set has: *Sensitivity* of 78%, *Specificity* of 93.8%, *Precision* of 85.5%, *Accuracy* of 84.8%, and *F1 Score* of 79.6%. But the improvements are possible in many aspects. Our future research will focus on: (1) improving segmentation and preprocessing, as well as analyzing the system that could lead to improving the robustness of the solution (2) certain corrections in ANN-MLP classifier which will improve its performance; (3) broadening the data set with new classes which exist in the meta data set (4) creating new hybrid model combined with novel artificial intelligence techniques which will efficiently solve real-world skin tumours data sets.

# References

1. Elgamal, M.: Automatic skin cancer images classification. Int. J. Adv. Comput. Sci. Appl. 4(3), 287–294 (2013)
2. Swetter, S.M., Tsao, H., Bichakjian, C.K., Curiel-Lewandrowski, C.: Guidelines of care for the management of primary cutaneous melanoma. J. Am. Acad. Dermatol. 80(1), 208–250 (2019)
3. Messadi, M., Bessaid, A., Taleb-Ahmed, A.: Extraction of specific parameters for skin tumour classification. J. Med. Eng. Technol. 33(4), 288–295 (2009)
4. Abbes, W., Sellami, D.: High-level features for automatic skin lesions neural network based classification. In: IEEE IPAS 2016: International Image Processing, Application and Systems Conference, Hammamet, Tunisia (2016). https://doi.org/10.1109/ipas.2016.7880148

5. Krawczyk, B., Simić, D., Simić, S., Woźniak, M.: Automatic diagnosis of primary headaches by machine learning methods. Open Med. **8**(2), 157–165 (2013)
6. Simić, S., Banković, Z., Simić, D., Simić, S.D.: A hybrid clustering approach for diagnosing medical diseases. In: de Cos Juez, F., et al. (eds) Hybrid Artificial Intelligent Systems. HAIS 2018. LNCS, vol. 10870, pp. 741–775. Springer, Cham. https://doi.org/10.1007/978-3-319-92639-1_62
7. Simić, S., Banković, Z., Simić, D., Simić, Svetislav D.: Different approaches of data and attribute selection on headache disorder. In: Yin, H., Camacho, D., Novais, P., Tallón-Ballesteros, Antonio J. (eds.) IDEAL 2018. LNCS, vol. 11315, pp. 241–249. Springer, Cham (2018). https://doi.org/10.1007/978-3-030-03496-2_27
8. Simić, S., Milutinović, D., Sekulić, S., Simić, D., Simić, S.D., Đorđević, J.: A hybrid case-based reasoning approach to detecting the optimal solution in nurse scheduling problem. Logic J. IGPL (2018). https://doi.org/10.1093/jigpal/jzy047, https://academic.oup.com/jigpal/advance-article/doi/10.1093/jigpal/jzy047/5107037
9. Aggarwal, C.C.: Data Classification: Algorithms and Applications. Chapman and Hall/CRC, Boca Raton (2014)
10. Wozniak, M.: Hybrid Classifiers: Methods of Data, Knowledge, and Classifier Combination. Springer, Heidelberg (2016). https://doi.org/10.1007/978-3-642-40997-4
11. Cruz-Roa, A.A., Arevalo Ovalle, J.E., Madabhushi, A., González Osorio, F.A.: A deep learning architecture for image representation, visual interpretability and automated basal-cell carcinoma cancer detection. In: Mori, K., Sakuma, I., Sato, Y., Barillot, C., Navab, N. (eds.) MICCAI 2013. LNCS, vol. 8150, pp. 403–410. Springer, Heidelberg (2013). https://doi.org/10.1007/978-3-642-40763-5_50
12. Krizhevsky, A., Sutskever, I., Hinton, G.E.: ImageNet classification with deep convolutional neural networks. Commun. ACM **60**(6), 84–90 (2017)
13. Chuchu, N., et al.: Smartphone applications for triaging adults with skin lesions that are suspicious for melanoma. Cochrane Database Syst. Rev. **12** (2018). https://doi.org/10.1002/14651858.cd013192. Art. No.: CD013192
14. Lee, T., Gallagher, R., Coldman, A., McLean, D.: Dullrazor®: a software approach to hair removal from images. Comput. Biol. Med. **21**(6), 533–543 (1997)
15. Andreassi, L., et al.: Digital dermoscopy analysis for the differentiation of atypical nevi and early melanoma. Arch. Dermatol. **135**, 1459–1465 (1999)
16. Tran, N.M., Burdejová, P., Osipenko, M., Härdle, W.K.: Principal Component Analysis in an Asymmetric Norm. SFB 649 Discussion Paper 2016–040 (2016). http://sfb649.wiwi.hu-berlin.de/papers/pdf/SFB649DP2016-040.pdf
17. Jolliffe, I.: Principal Component Analysis, 2nd edn. Springer, New York (2002). https://doi.org/10.1007/b98835
18. Aussenhofer, M., Dann, S., Langi, Z., Toth, G.: An algorithm to find maximum area polygons circumscribed about a convex polygon. Discrete Appl. Math. **255**, 98–108 (2019). https://doi.org/10.1016/j.dam.2018.08.017
19. Cortes, C., Vapnik, V.: Support-vector networks. Mach. Learn. **20**(3), 273–297 (1995). https://doi.org/10.1007/BF00994018
20. Cover, T., Hart, P.: Nearest neighbor pattern classification. IEEE Trans. Inf. Theory **13**(1), 21–27 (1967)
21. https://dataverse.harvard.edu/dataset.xhtml?persistentId=doi:10.7910/DVN/DBW86T

# Texture Descriptors for Automatic Estimation of Workpiece Quality in Milling

Manuel Castejón-Limas[iD], Lidia Sánchez-González[(✉)][iD], Javier Díez-González,
Laura Fernández-Robles[iD], Virginia Riego, and Hilde Pérez[iD]

Departamento de Ingenierías Mecánica, Informática y Aeroespacial,
Universidad de León, 24071 León, Spain
{manuel.castejon,lidia.sanchez,jdieg,l.fernandez,hilde.perez}@unileon.es

**Abstract.** Milling workpiece present a regular pattern when they are correctly machined. However, if some problems occur, the pattern is not so homogeneous and, consequently, its quality is reduced. This paper proposes a method based on the use of texture descriptors in order to detect workpiece wear in milling automatically. Images are captured by using a boroscope connected to a camera and the whole inner surface of the workpiece is analysed. Then texture features are computed from the coocurrence for each image. Next, feature vectors are classified by 4 different approaches, Decision Trees, K Neighbors, Naïve Bayes and a Multilayer Perceptron. Linear discriminant analysis reduces the number of features from 6 to 2 without loosing accuracy. A hit rate of 91.8% is achieved with Decision Trees what fulfils the industrial requirements.

**Keywords:** Quality estimation · Milling machined parts ·
Haralick descriptors · Wear detection

## 1 Introduction

Industry 4.0 is assumed as the next industrial revolution that is already happening in factories. The final goal is to use collaborative robots in order to help human beings in their daily routines. Currently, one of the procedures that are carried out by the operators is the visual inspection of the machined workpiece. Collaborative robots are aimed at helping the operator in the decision-making process, in order to discard out of specifications parts. This paper presents an automated process to determine the workpiece surface quality by means of computer vision techniques.

Some research has already been done as for example, defects of injection parts have been analyzed by following an inspection model in order to correct the process parameter values [3]. In deep drilling operations Bayesian networks are employed to predict surface roughness of manufactured steel components

© Springer Nature Switzerland AG 2019
H. Pérez García et al. (Eds.): HAIS 2019, LNAI 11734, pp. 734–744, 2019.
https://doi.org/10.1007/978-3-030-29859-3_62

like moulds or dies [2]. [8] presents a hybrid chromosome genetic algorithm to detect surface defects considering different features based on geometry, shape or texture, among others.

More specifically, in manufacturing processes like milling, there are works that have used digital image processing in order to assess tool condition [4, 6, 12] or to automatically determine the tool tip position [10].

Texture provides intuitively properties about the smoothness, roughness and regularity of a region [1]. Texture also gives information about the spatial order of colours and the intensity or luminance of an image. Conceptually, the main feature of a texture is the repetition of a basic pattern named texel. Texture analysis is a complex task due to the weak definition of texture. Basically, there are two approaches, one based on structural features (texture is a repetition or distribution of patterns or texels) and statistical features (in this case, texture is a quantitative measure of the pixel distribution in a region). Using real images, textures do not present a regular geometric structure and texels hardly can be segmented manually. In such cases, texture can be better described by means of statistical models.

In [9], entropy - a statistical texture descriptor based on the Gray Level Co-ocurrence Matrix (GLCM) - is computed in order to assess tool wear and correlate it with the workpiece surface texture.

This work proposes a method based on the texture features computed from the GLCM in order to detect if a workpiece presents wear so as to assess its quality.

The paper is structured as follows. Section 2 explains the inspection method and how the texture features are computed. In Sect. 3 the experiments carried out to validate the method are presented. Finally, Sect. 4 gathers the achieved conclusions and future works.

## 2   Inspection Method

The proposed method is based on an industrial boroscope connected to a microscope camera and using a white LED based lighting system. That vision system allows us to acquire digital images from the mechanized pieces in order to automatically analyze them to assess their quality. This procedure comprises the stage described in the following sections. A scheme of the whole system is shown in Fig. 1.

Images have been captured by a microscope camera connected to an industrial boroscope. The lighting system is based on LED lights with brightness adjustable. The lightning adjusting has been done manually by visualizing the images in the Dino Capture software [5] provided by the camera. It has also been used to calibrate the system.

Images are RGB images of $2592 \times 1944$ pixels with a resolution of 300 ppp. The four manufactured parts considered are presented in Fig. 2. For each part, a set of images have been captured along its external and internal surface, covering the whole part. Some examples of the acquired images are shown in Fig. 3.

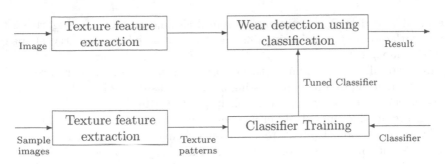

Fig. 1. Scheme of the proposed method based on texture descriptors. Given an image, texture features are computed and the trained classifier detects if defects appear

Fig. 2. Parts considered in experiments.

Fig. 3. Samples of acquired images.

For the acquired images, a contrast-limited adaptive histogram equalisation (CLAHE) process [11,13] is carried out in order to improve the contrast quality of the images and enhance them.

Otsu method is applied to the images so as to convert them to black and white images. This image processing is shown in Fig. 4.

Original                    Clahe                    Otsu

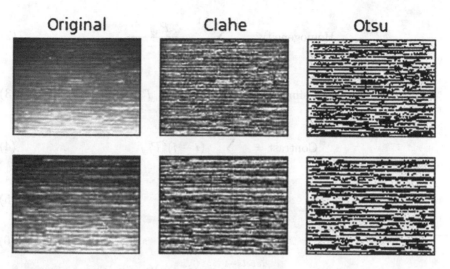

**Fig. 4.** Original image converted to gray levels (left), enhanced image (middle) and binary image (right).

Certain descriptors to measure quantitatively the texture are considered. The robustness of several classifiers are also analysed for texture pattern detection despite the fact that the provided information is inconsistent, weak or noisy. In the first step, a set of measures which characterizes texture for each pixel is computed considering a certain neighbourhood. To classify pixels of an image, different classifiers are trained with a set of texture patterns for each texture class. Two classes are considered: one with no defects and another formed by wear workpiece.

Second order statistics are considered since they comprise both intensity distribution and pixel position. Second order statistics are computed from a coocurrence matrix. Let $P$ a position operator and an image $I$ with $L$ grey levels, the coocurrence matrix is defined as a $L \times L$ matrix whose items $c_{ij}$ are the number of pairs of pixels $i$ and $j$ which satisfy the spatial condition determined by the operator $P$. A common position operator can be defined by the polar coordinates $(r, \theta)$ where $r$ is the module of a shift and $\theta$ its angle. Usually, the considered values of $\theta$ are $0°$, $45°$, $90°$ and $135°$. Using another notation, the position operator would be $(dr, dc)$ where $dr$ is the vertical shift (rows) and $dc$ the horizontal shift (columns). Both expressions are the same since $(dr, dc) = (-r * sen\theta, r * cos\theta)$. So, for a given shift vector $d = (dc, dr)$ and an image $I$, the coocurrence matrix $C_d$ for the image $I$ can be defined as follows:

$$C_d(i, j) = |(r, c)/I(r, c) = i \cap I(r + dr, j + dc) = j| \qquad (1)$$

The normalised coocurrence matrix gives the probability of a pair of points $x$ and $y$ which satisfy $P$ have grey levels $i$ and $j$ respectively. The sum of all the items of such matrix is one. Coocurrence matrices provide information about textures, but to compare textures some features derived from these matrices have to be computed. These features describe texture in a more sound way. The most known statistics are the following [7]:

$$\text{Homogeneity} = \sum_{i,j=0}^{levels-1} \frac{P_{i,j}}{1+(i-j)^2} \tag{2}$$

$$\text{Dissimilarity} = \sum_{i,j=0}^{levels-1} |i-j| \, P_{i,j} \tag{3}$$

$$\text{Contrast} = \sum_{i,j=0}^{levels-1} (i-j)^2 P_{i,j} \tag{4}$$

$$\text{ASM} = \sum_{i,j=0}^{levels-1} P_{i,j}^2 \tag{5}$$

$$\text{Energy} = \sqrt{ASM} \tag{6}$$

$$\text{Correlation} = \sum_{i,j=0}^{levels-1} P_{i,j} \frac{(i-\mu_i)(j-\mu_j)}{\sigma_i \sigma_j} \tag{7}$$

where

$$\mu_i = \sum_i i \sum_j P_{i,j} \tag{8}$$

$$\mu_j = \sum_j j \sum_i P{i,j} \tag{9}$$

$$\sigma_i = \sum_i (i-\mu_i)^2 \sum_j P_{i,j} \tag{10}$$

$$\sigma_j = \sum_j (j-\mu_j)^2 \sum_i P_{i,j} \tag{11}$$

The above descriptors characterize the content of the coocurrence matrices. For instance, energy is a measure of its randomness, achieving its minimum value when all the items in the matrix are the same. Contrast has a value quite low when the great values of the matrix are next to the main diagonal (differences $(i-j)$ are smaller there). Homogeneity has the opposite effect than the contrast.

Spread items in the coocurrence matrix influence badly to estimate the joint probability distribution. To deal with that, the number of the image grey levels is reduced. Feature calculations are also accelerated. About the used orientation to compute the coocurrence matrices, the 0° orientation is only considered since no meaningful improvements were obtained using the common four orientations (0°, 45°, 90° and 135°).

# 3   Experimental Results

A set of images acquired from 4 different parts have been considered for experiments. Analysing the whole inner surface of the parts, 587 images were captured. 437 of the images presented some wear and the remaining 152 kept a high quality with no wear. Firstly, images are preprocessed as it is explained in previous section, by enhancing the contrast and then are converted into binary images. After that, the texture descriptors are computed for each image.

The complete dataset is split into a training set, formed by the 70% of the images, and a test set, formed by the remaining 30%, which will be eventually used as hold out set to evaluate the performance of the chosen model.

The image collection is processed and Homogeneity, Dissimilarity, Contrast, ASM, Energy, and Correlation texture descriptors are computed. Owing to their definition, texture descriptors such as those computed are frequently highly correlated. Figure 5 shows the absolute value of the correlation amongst these descriptors.

For the specific dataset under study, the variables are distributed into two distinct groups, as the correlation matrix displayed in Fig. 5 suggests. The first group of descriptors comprises Homogeneity, Dissimilarity, Contrast, and

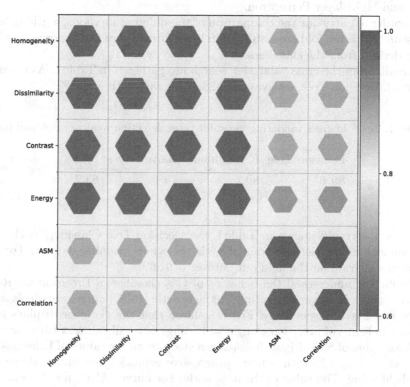

**Fig. 5.** Correlation (absolute value) showing two blocks of similar texture descriptors

Energy; the second, ASM and Correlation. Given the colinearities within groups of descriptors, the information contained in the whole set of texture descriptors is still preserved when reducing the choice of descriptors to a collection of representatives of each of the blocks suggested by the correlation matrix. For this reason, Homogeneity and ASM are chosen as descriptors of interest.

Figure 6 shows the pairs plot of this reduced set of descriptors along with a representation of the empirical approximation of their probability density functions. The figure suggests that the two classes present in the dataset display distributions sharing a significant overlapping area.

A different representation of the dataset can be obtained by using a linear discriminant analysis (LDA) estimator. LDA provides a projection, of the dataset onto a reduced dimensionality space, that maximizes both the compactness of the clusters present in the dataset, and the distance between clusters. Following this approach in this binary classification case, the dataset is projected onto a new axis (LDA 1). Figure 7 shows the violin plot of the transformed dataset. On this axis, the two classes are better separated and it is clear that the information contained in the descriptors is capable of discriminating between the two classes. Nevertheless, the non negligible overlapping area still present demands for alternative classification techniques. Thus, a collection of classification techniques is applied to the dataset; specifically: K-Nearest Neighbors, Naive Bayes, Decision Tree, and Multi-layer Perceptron.

In order to carry out the comparison of the different methods, a grid exhaustive search with ten fold cross validation is performed over the hyperparameters' space derived from the classifiers.

Results from the cross validation search are gathered in Table 1. As it can be observed, Decision Tree achieves the higher hit rate with a 89.6%.

**Table 1.** Ten-fold cross validating mean hit rates detecting wear in machined parts

| Decision Tree | K Neighbors | Naive Bayes | MLP |
|---|---|---|---|
| **89.6** | 89.2 | 75.1 | 63.7 |

Given the results shown in Table 1, the Decision Tree Classifier, as the best performance model, is now trained with the whole training dataset. For this model the hit rate on the hold out dataset is 91.8%.

To further understand the behavior of this classifier, a Precision vs. Recall analysis, and a Receiver Operating Characteristic (ROC) curve, are displayed in Fig. 8, and Fig. 9, respectively. Figure 8 shows that the classifier displays great stability in terms of precision along a wide range of recall values, with an average precision value of 89%. Figure 9 displays a steep trend for values of False positive rate below circa 0.2, with a more progressive evolution for values above that threshold value. The value of the area under the curve (AUC) for this classifier is 0.94.

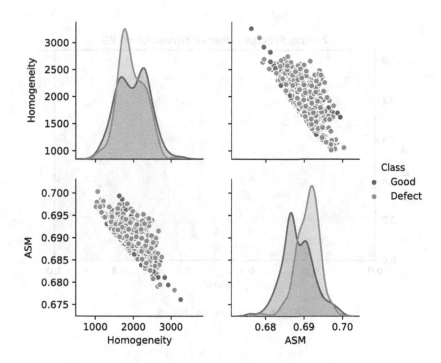

**Fig. 6.** Pairs plot showing the reduced set of features

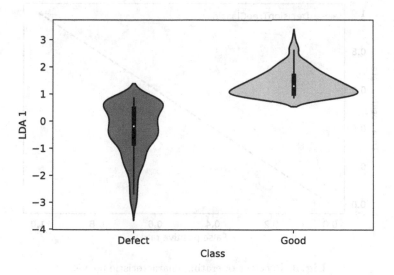

**Fig. 7.** Violin plot showing the LDA space

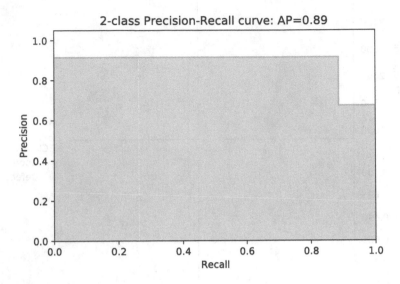

**Fig. 8.** Precision vs. Recall

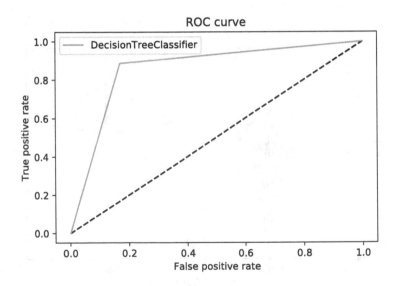

**Fig. 9.** Receiver operating characteristic curve

# 4   Conclusions

In this paper a method based on texture descriptors is proposed in order to detect heterogeneity in the machined surface of the workpiece which, at certain level, could reject the part. Applying linear discriminant analysis, the initial 6 features computed from the coocurrence matrix are reduced to 2, Homogeneity and ASM. With these features, 4 different classifiers are employed: Decision Trees, K Neighbors, Naive Bayes and a Multilayer Perceptron. Results show the good performance of the method since a hit rate of 91.8% is achieved with the Decision Tree. That result satisfied the industrial requirements.

**Acknowledgements.** We gratefully acknowledge the financial support of Spanish Ministry of Economy, Industry and Competitiveness, through grant DPI2016-79960-C3-2-P.

# References

1. Arivazhagan, S., Ganesan, L.: Texture classification using wavelet transform. Pattern Recogn. Lett. **24**(9–10), 1513–1521 (2003)
2. Bustillo, A., Correa, M.: Using artificial intelligence to predict surface roughness in deep drilling of steel components. J. Intell. Manufact. **23**(5), 1893–1902 (2012). https://doi.org/10.1007/s10845-011-0506-8
3. Chaves, M.L., Vizán, A., Márquez, J.J., Ríos, J.: Inspection model and correlation functions to assist in the correction of qualitative defects of injected parts. Polym. Eng. Sci. **50**(6), 1268–1279 (2010). https://doi.org/10.1002/pen.21647, https://onlinelibrary.wiley.com/doi/abs/10.1002/pen.21647
4. Dai, Y., Zhu, K.: A machine vision system for micro-milling tool condition monitoring. Precis. Eng. **52**, 183–191 (2018). https://doi.org/10.1016/j.precisioneng.2017.12.006, http://www.sciencedirect.com/science/article/pii/S0141635917302817
5. Dunwell Tech Inc.: Dinocapture 2.0: microscope imaging software (2019). https://www.dinolite.us/dinocapture
6. Dutta, S., Pal, S., Mukhopadhyay, S., Sen, R.: Application of digital image processing in tool condition monitoring: a review. CIRP J. Manufact. Sci. Technol. **6**(3), 212–232 (2013). https://doi.org/10.1016/j.cirpj.2013.02.005, http://www.sciencedirect.com/science/article/pii/S1755581713000072
7. Haralick, R., Shanmugan, K., Dinstein, I.: Texture features for image classification. IEEE Syst. Man Cybern. **3**(6), 610–621 (1973)
8. Hu, H., Liu, Y., Liu, M., Nie, L.: Surface defect classification in large-scale strip steel image collection via hybrid chromosome genetic algorithm. Neurocomputing **181**, 86–95 (2016). https://doi.org/10.1016/j.neucom.2015.05.134, http://www.sciencedirect.com/science/article/pii/S0925231215018482. Big Data Driven Intelligent Transportation Systems
9. Li, L., An, Q.: An in-depth study of tool wear monitoring technique based on image segmentation and texture analysis. Measurement **79**, 44–52 (2016). https://doi.org/10.1016/j.measurement.2015.10.029, http://www.sciencedirect.com/science/article/pii/S0263224115005631

10. López-Estrada, L., Fajardo-Pruna, M., Sánchez-González, L., Pérez, H., Fernández-Robles, L., Vizán, A.: Design and implementation of a stereo vision system on an innovative 6DOF single-edge machining device for tool tip localization and path correction. Sensors 18(9) (2018). https://doi.org/10.3390/s18093132, http://www.mdpi.com/1424-8220/18/9/3132
11. Park, G.H., Cho, H.H., Choi, M.R.: A contrast enhancement method using dynamic range separate histogram equalization. IEEE Trans. Consum. Electron. 54(4), 1981–1987 (2008)
12. Szydłowski, M., Powałka, B., Matuszak, M., Kochmański, P.: Machine vision micro-milling tool wear inspection by image reconstruction and light reflectance. Precis. Eng. 44, 236–244 (2016). https://doi.org/10.1016/j.precisioneng.2016.01.003, http://www.sciencedirect.com/science/article/pii/S0141635916000052
13. Zuiderveld, K.: Contrast limited adaptive histogram equalization. In: Heckbert, P.S. (ed.) Graphics Gems IV, pp. 474–485. Academic Press Professional Inc., San Diego, CA, USA (1994). http://dl.acm.org/citation.cfm?id=180895.180940

# Surface Defect Modelling Using Co-occurrence Matrix and Fast Fourier Transformation

Iker Pastor-López$^{(\boxtimes)}$, Borja Sanz, José Gaviria de la Puerta, and Pablo G. Bringas

University of Deusto, Avenida de las Universidades 24, 48007 Bilbao, Spain
{iker.pastor,borja.sanz,jgaviria,pablo.garcia.bringas}@deusto.es

**Abstract.** There are several industries that supplies key elements to other industries where they are critical. Hence, foundry castings are subject to very strict safety controls to assure the quality of the manufactured castings. In the last years, the use of computer vision technologies to control the surface quality. In particular, we have focused our work on inclusions, cold laps and misruns. We propose a new methodology that detects and categorises imperfections on the surface. To this end, we compared several features extracted from the images to highlight the regions of the casting that may be affected and, then, we applied several machine-learning techniques to classify the regions. Despite Deep Learning techniques have a very good performance in this problems, they need a huge dataset to get this results. In this case, due to the size of the dataset (which is a real problem in a real environment), we have use traditional machine learning techniques. Our experiments shows that this method obtains high precision rates, in general, and our best results are a 96,64% of accuracy and 0.9763 of area under ROC curve.

**Keywords:** Defect characterization · Machine learning · Feature engineering · Image segmentation

## 1 Introduction

Computer vision for surface quality control have gained a lot of interest from several industries to automate the inspection processes, and to increase the quality of the resulting products. In some industries, especially when the inspection is dangerous or critical, computer vision can replace or complement human inspection reducing errors and danger [17].

Thus, real-time visual inspection is crucial to improve the results of a manual inspection. In recent years, the advances in technology have allowed the development of new methods based on computer vision to perform this recurring tasks in a better and more safety way than traditional human inspection [31]. Although persons can accomplish some of these tasks better than machines [21], in some

© Springer Nature Switzerland AG 2019
H. Pérez García et al. (Eds.): HAIS 2019, LNAI 11734, pp. 745–757, 2019.
https://doi.org/10.1007/978-3-030-29859-3_63

situations they are slower an get easily fatigued. Moreover, human operators require specific learning skills and capabilities that normally take a long time to acquire.

In the literature, there are many examples of these applications in different industries like timber [15], textile [29] or metallurgical [20]. In most of them, inspection performed by operators depends on human factor, machine vision has become an autonomous way to perform these task reducing the errors. Concretely, in this paper we will focus on iron foundries, which produce iron castings. This output products are used as input for important industries like automotive, aeronautic, naval or weaponry. In some of these industries, they play a key role, so they have to overcome several quality and safety processes.

In the last years, there have been a great evolution in machine learning area, using algorithms that are known as *Deep Learning* (DL) [18], that is a sub-field of *Artificial Neural Networks* (ANNs). A standard neural network (NN) is a collection of a simple connected processors called neurons. Each neuron produce a sequence of real-valued activations. The neurons that obtain the inputs form the environment are called input neurons, and other neurons get activated through weighted connections from previously active neurons. The output neurons shows the final results. DL algorithms uses the same bases, but includes different hidden layers, and in the last years have been developed different architectures in this kind of networks [12].

These techniques also have been applied to the industrial inspection [33]. The *Optical Quality Control* (OQC) ensure that the product is visually free of imperfections or defects. In this way, it is common practice in the industry create a new set of features manually engineered when a problem arise. Deep Learning techniques allow us to avoid this manual step and make the features engineering process automatically. In order to achieve this, it is necessary to get a big labelled data-set. Unfortunately, in the industrial process there are several issues that makes things difficult to get this dataset. For example, the size of typical data-set to train these system may serve as an example. The IRIS data-set [9] it is a traditional data-set that is include in the all traditional machine learning courses. It is composed by 150 instances with 4 attributes each one of them. On the other hand, MINST [19] data-set it is the equivalent data-set in the deep learning area. It is composed by around 70.000 instances of $20 \times 20$ images (e.g., $20 \times 20 \times 3$ channels (Red, Green and Blue), 1.200 features by instances. In summary, the first data-set contains 600 elements and the second one around 84 millions of elements.

Thus, nowadays the manual feature engineering process are still use in the industry. As the Sect. 2 shows, our data-set it is not big enough to apply deep learning techniques. In the last years, there have been several authors that try to develop new techniques that avoid the use of this manual feature engineering process to the small data-sets [16]. Our research is focused in this "small data" area.

We organize the rest of the paper as follows. First, we start with a complete description to prepare the dataset that is used for surface defect categorisation. Secondly, we describe the machine learning algorithms that we use for the experimentation. Finally, we present the results obtained with the combination of different groups of variables and algorithms, as well as the conclusions of our work.

## 2 Dataset Preparation

### 2.1 Data Gathering

To acquire the data from the surface of the castings, we have developed a functional machine vision system (see Fig. 1) [23] composed of: (i) a laser-based triangulation camera with 3D technology, (ii) a robotic arm and (iii) a computer with several data processing capabilities [24].

**Fig. 1.** Proposed prototype for data gathering.

To start with the data acquisition, the casting is put over a black painted table and adjusted on a special silicon mould. We use this color to decrease unwished reflections of the laser thus minimizes the noise. Besides, the mould ensures that all the scanned castings are in same position. Then, making a linear movement with the robotic arm, the laser is projected over the surface of the casting, and with this projection, a set of height points are calculated. The precision of our system is of 0.2 mm. Then, we remove the points related with the working table in order to reduce the noise. As a result of the process, we obtain a matrix with the height values of the casting in each point.

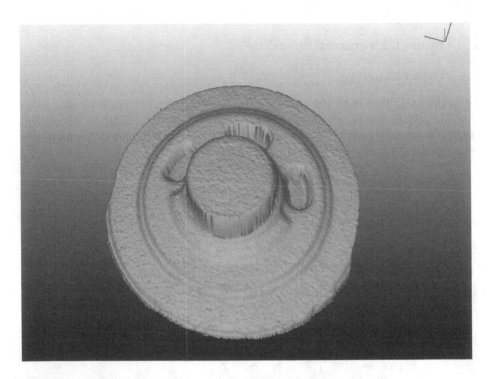

**Fig. 2.** The height matrix represented in 3 dimensions.

In addition, we have develop some software tools, that allow us to perform some operations over the castings data. Specifically, in Fig. 2, we can see a three-dimensional viewer in which we can zoom in and zoom out, and mark regions.

### 2.2    Data Representation

Using this height matrix, we have created other three representations (see Fig. 3) of the data. With these representations we want to have more complete information about the castings surface and different aspects of the data.

- **Grey-scale Height Map** [30]: In this representation we have converted each value of the height matrix into a range between 0 and 255. As a result, we obtain a grey-scale image with different levels of grey.
- **Normals map:** It does not show data about the heights, but also the direction of the surface in each point. The resulting vectors for each point, have three component, one per dimension $(x, y, z)$. Then, we codify each one with RGB color code $(R = x, G = y, B = z)$, resulting a image with a vectors data [24].

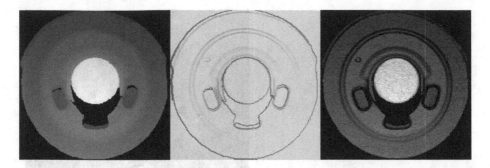

**Fig. 3.** Generated representations of the casting surface. The first one is the grey-scale height map. The second one, normals map codified as RGB image. The third one, the merge of the normals map and grey scale height map, in which we can see a more complete visual information of the casting surface.

- **Merge of the normals map and grey-scale height map:** The last generated representation, merges the data of the grey-scale height map and the normals vector one. In particular, we follow the next process:
  1. We calculate the cosine distance between each normal vector and a model vector defined as $(0, 0, 1)$. This model vector represents a flat surface. High value for this distance implies a pixel with a casting edge or surface defect.
  2. The resulting distances map with the distances in each point, are converted to a range between 0 and 255, in order to transform the distances matrix into a grey-scale map.
  3. We use substract filter between grey-scale height map and the resulting grey-scale map in the previous step.

  With this image, we represent the values of the height, and the normals vector in each point. In this way, we obtain a more complete representation of the casting surface data (see the third image in Fig. 3).

### 2.3 Regions of Interest Extraction Process

Regions of interest are parts of a image in which there is a relevant information for a particular purpose [4]. We apply this concept, in order to focus only on the areas of the images where there may be a defect or a edge of a casting [6]. To this end, we use a segmentation method, based on previously scanned castings without defects [24].

As we can see in Fig. 4, after applying our segmentation method, we obtain different areas in the image. Some of this areas represent not only edges in castings or tolerable surface imperfections, but also defects like cold lap and inclusions in this example. Finally, we extract the minimum rectangular area that contains the selected region.

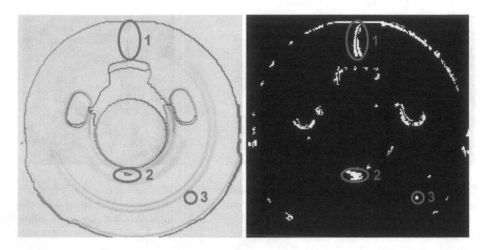

**Fig. 4.** The result of extracting regions of interest. In the first image we can see a normal map in RGB representation of a casting with 3 surface defects: (1) represents a cold-lap, and (2, 3) are inclusions with different sizes. In the second image, we show the regions of interest extracted from the previous representation after we apply the segmentation process.

### 2.4 Extracted Features for Defect Categorisation

To perform a categorisation of the defects thought machine learning techniques, we need to transform the resulting images in vectors of variables. To this end, we have used different features extracted from castings images, getting as much information as possible to represent them.

**Fast Fourier Transformation (FFT) Based Features.** Fourier transformation is a mathematical function that has been broadly used in several areas such as signal processing, telecommunications or image processing [2].

Our method utilizes the Fast Fourier Transformation (FFT) that is an implementation of the Discrete Fourier Transformation (DFT). The transformation discomposes the input image in its real and imaginary components, obtaining in this way a representation of the image in the frequency domain. The number of resultant frequencies is the same as the number of pixels of the input image. Formally, the transformation is performed by:

$$F(x,y) = \sum_{m=0}^{M-1} \sum_{n=0}^{N-1} f(m,n) \cdot e^{-i2\pi(x\frac{m}{M}+y\frac{n}{N})} \tag{1}$$

where $F(x,y)$ is the value in the Fourier domain, corresponding to the coordinates $x, y$ and $M$ and $N$ are the dimensions of the image, and $f(m,n)$ is the value of the pixel in the input image for the coordinates $m, n$.

To transform back to the spatial domain, we can compute the following:

$$f(m,n) = \frac{1}{M \cdot N} \cdot \sum_{x=0}^{M-1} \sum_{y=0}^{N-1} F(x,y) \cdot e^{i2\pi(x\frac{m}{M}+y\frac{n}{N})} \tag{2}$$

where $F(x,y)$ is the value in the coordinates $x,y$ in the transformed frequency domain space.

This transformation has been applied to the potentially faulty regions using different representations. Then, we have extracted the following features in the Fourier domain:

- **Features extracted from the FFT transformed image from the grayscale height map:** We obtain 2 histograms from the transformation with the different levels of gray. In the first one, our method uses every possible value of gray, whereas in the second one, our method discards the black values ($R = 0$, $G = 0$, and $B = 0$). The extracted features are the maximum value, minimum value, median, average, standard deviation and entropy.
- **Features extracted from the FFT transformed image from the normal vector matrices codified in RGB:** We start by transforming this information to grayscale. Hereafter, as we did in the height map representation, 2 histograms from the transformation with the different levels of gray were obtained. One, utilizing every possible value of gray, and the other, discarding the black values in the transformed image. Maximum value, minimum value, median, average, standard deviation and entropy were extracted.

**Co-occurrence Matrix Based Features.** COM (co-occurrence Matrix) is a method that can be applied to measure the textures in an image [1,14,22]. In particular, these matrices are computed as follows:

$$CM_{\Delta x, \Delta y}(i,j) = \sum_{m=1}^{k} \sum_{n=1}^{l} \begin{cases} 1, \text{ If } I(m,n) = i \text{ and} \\ \quad I(m+\Delta x, n+\Delta y) = j \\ 0, \text{ otherwise} \end{cases} \tag{3}$$

where $I$ is an images of size $k \times l$ and $\Delta x$ and $\Delta y$ denote the offset. Our method extracts the well-known Haralick textural features [13]: normalization, contrast, dissimilarity, homogeneity, angular second moment, energy, maximum probability, entropy, mean, variance, standard deviation, and correlation.

## 3   Machine Learning Algorithms

Boosted by the technological giants (Google, Facebook or Amazon, among others), there is a lot of research conducted in *machine learning* [8]. Traditionally, these algorithms are categorized with regards to the availability of labeled instances in the training data-set. The algorithms learn using this previous knowledge and can create new predictions. In this paper, we have use several machine learning algorithms that use a train data-set to get the knowledge.

### 3.1   K-Nearest Neighbors

K-Nearest Neighbors (KNN) [10] is a simple supervised machine learning method. This is a traditional non parametric technique that classifies new samples, based on distance between the sample vector and the class of the $k$ closest vectors in the training space. This algorithm is very $k$ dependant, and different values of this parameter will causes change in the performance (e.g., using a big $k$ value, the number of neighbours will increase the classification time and will influence in the accuracy of the result).

This algorithm does not have model training stage, and it just compare the distance between several instances. Traditionally, the metric to evaluate the distances is euclidean distance and the aggregation metric to evaluate the class of the instances use to be simple vote [34].

### 3.2   Bayesian Networks

Bayesian Networks [25], which are based on the Bayes Theorem, are defined as directed acyclic graph models (DAG) for multivariate analysis. This model can help to know the statistical dependencies between system variables. Each node represents problem variables that can be either a premise or a conclusion and each link represents conditional dependencies between such variables. They have an associated probability distribution function.

Moreover, the probability function illustrates the strength of these relationships in the graph [5]. The most important capability of Bayesian Networks is their ability to determine the probability that a certain hypothesis is true (e.g., the probability of an executable to be malware [26]) given a historical data-set.

### 3.3   Support Vector Machines (SVM)

SVM algorithms first map the input vector into a higher dimension space, and then algorithms divide the n-dimensional space representation of the data into two regions using a *hyperplane*. In binary classification, the main goal of the algorithm is to maximize the margin between those two regions or classes. The margin is defined by the farthest distance between the examples of the two classes and computed based on the distance between the closest instances of both classes, which are called supporting vectors [32].

### 3.4   Decision Trees

Decision Tree classifies a sample through a sequence of decisions, in which the current decision helps to make the subsequent decision. In this algorithms, nodes represent conditions regarding the variables of a problem, whereas final nodes represent the ultimate decision of the algorithm [27]. Thus, it can be represented graphically as trees.

To train the models, we use Random Forest, a combination of weak classifiers (i.e. ensemble) of different randomly-built decision trees [3], and J48, the WEKA [11] implementation of the C4.5 algorithm [28].

## 4    Empirical Validation

In order to evaluate the performance of the our detector, we have used a collected dataset form a foundry specialized in the automotive sector. More specifically, the foundry is focused in the creation of pieces in safety and precision components for this industry.

To create the dataset, we collected 645 foundry castings using the segmentation system previously described. We used 176 correct casting to train the model, and 469 for testing. With this seed, we create a dataset composed by 5785 segments to train the machine-learning models.

We have focused in the detection of 3 different defects: (1) inclusion, (2) cold lap and (3) misrun. We also have included a new category, called 'Correct', that represents the segments that are corrects, even thought the method has marked them as faulty. The number of samples in each category are indicated in Table 1.

Table 1. Number of samples for each category.

| Category | Number of samples |
| --- | --- |
| Inclusion | 387 |
| Cold Lap | 16 |
| Misrun | 52 |
| Correct | 5030 |

The criterion to acceptance is based on the final requirements of the customer. Due to the final destination industry, the quality standard are very restrictive. To this end, we labelled each possible segment with its defects within the castings.

The dataset was not balanced for the existing classes due to scarce data. To minimise the problems that these algorithms tend to have in this cases (scarce and unbalanced data), we applied Synthetic Minority Over-sampling TEchnique (SMOTE) [7], which is a combination of over-sampling the less populated classes.

Next, we conducted the following methodology to evaluate the precision of our methods to categorize the segments:

- **Cross validation:** it is a generally applied methodology in machine-learning evaluation. We use 10 as value of $k$ (e.g., we splitted 10 times our dataset in 10 different in a learning set (90% of dataset) and testing set (10% of the total data).
- **SMOTE:** In order to balance the dataset, we applied this method, which has been described previously.
- **Training step:** In this step, we used the algorithms described in Table 2 to find the algorithm that have the best performance.
- **Testing the model:** At last, to evaluate the results, we used Accuracy, Weighted Average True Positive Ratio (WATPR), Weighted Average False Positive Ratio (WAFPR) and Area Under ROC Curve (AUC).

**Table 2.** List of the algorithms used in our empirical validation.

| Algorithm Acronym | Description |
| --- | --- |
| BN: K2 | Bayesian Network, using a K2 kernel. (Ref: 3.2) |
| BN: TAN | Bayesian Network, using a TAN kernel. (Ref: 3.2) |
| Naïve Bayes | Bayesian Network, using a Naïve. (Ref: 3.2) |
| SVM: PK | SVM, using Polynomial Kernel (Ref: 3.3) |
| SVM: NPK | SVM Normalised Polynomial kernel (Ref: 3.3) |
| SVM: P | SVM, with a Pearson VII kernel (Ref: 3.3) |
| SVM: RBF | SVM, with Radial Basis Function kernel (Ref: 3.3) |
| KNN K=1 | K-Nearest Neighbors. K = 1. (Ref: 3.1) |
| KNN K=2 | K-Nearest Neighbors. K = 2. (Ref: 3.1) |
| KNN K=3 | K-Nearest Neighbors. K = 3. (Ref: 3.1) |
| KNN K=4 | K-Nearest Neighbors. K = 4. (Ref: 3.1) |
| KNN K=5 | K-Nearest Neighbors. K = 5. (Ref: 3.1) |
| DT: J48 | J48 Decision tree, (Ref: 3.4) |
| DT: RF N=10 | Random Forest, with 10 trees (Ref: 3.4) |
| DT: RF N=25 | Random Forest, with 25 trees (Ref: 3.4) |
| DT: RF N=50 | Random Forest, with 50 trees (Ref: 3.4) |
| DT: RF N=75 | Random Forest, with 75 trees (Ref: 3.4) |
| DT: RF N=100 | Random Forest, with 100 trees (Ref: 3.4) |

We compare the detection capabilities using the different algorithms. The results of the experiments are shown in the Table 3. In general, we can see that the results of FFT are significantly higher than COM. Specifically, the accuracy results are on average 7% better; and the WATPR and WAFPR are on 10% better. The Area Under ROC Curve, the WATPR, the WAFPR and the accuracy using the both categorization strategies have significant gap (e.g., using a bayesian network and a K2 kernel (BN: K2), the difference between the accuracy are 14.65%, between the WATPR are 0.26, between the WAFPR are 0.051 and 0.15 in the case of the AUC).

The Area Under ROC Curve and the accuracy using the both categorization strategies are very close. Some algorithms has a very low performance in this categorization strategies (e.g, Naïve Bayes with FFT only get 30,99% of accuracy, worst than flip a coin). As usual, we get the best results using KNN algorithm, with 5 as the value of $K$. Concretely, we obtain more than 84% of accuracy, 0.7159 of WATPR, 0.1433 of WAFPR and 0.8694 of Area under ROC Curve with the FFT categorization method.

**Table 3.** Results of the categorization using FFT, COM.

| Classifier | FFT | | | | COM | | | |
|---|---|---|---|---|---|---|---|---|
| | Acc. | WATPR | WAFPR | AUC | Acc. | ATPR | AFPR | AUC |
| BN: K2 | 0.7592 | 0.8422 | 0.2469 | 0.8715 | 0.6127 | 0.6200 | 0.3877 | 0.7140 |
| BN: TAN | 0.8617 | 0.6784 | 0.1248 | 0.8821 | 0.7967 | 0.4148 | 0.1758 | 0.7342 |
| Naïve Bayes | 0.3099 | 0.9774 | 0.7364 | 0.8306 | 0.4276 | 0.8793 | 0.6000 | 0.7275 |
| SVM: PK | 0.6780 | 0.8987 | 0.3333 | 0.8568 | 0.5498 | 0.7301 | 0.4608 | 0.7628 |
| SVM: NPK | 0.6882 | 0.8938 | 0.3225 | 0.8671 | 0.5831 | 0.6704 | 0.4221 | 0.7594 |
| SVM: P | 0.7589 | 0.8554 | 0.2469 | 0.8859 | 0.6733 | 0.6129 | 0.3229 | 0.7742 |
| SVM: RBF | 0.6753 | 0.8993 | 0.3351 | 0.8499 | 0.5348 | 0.7604 | 0.4781 | 0.7315 |
| KNN K=1 | 0.8612 | 0.5702 | 0.1182 | 0.7553 | 0.7918 | 0.4458 | 0.1836 | 0.6875 |
| KNN K=2 | 0.8877 | 0.4922 | 0.0834 | 0.8224 | 0.8408 | 0.3538 | 0.1233 | 0.7233 |
| KNN K=3 | 0.8576 | 0.6631 | 0.1286 | 0.8519 | 0.7792 | 0.4986 | 0.2011 | 0.7432 |
| KNN K=4 | 0.8686 | 0.6353 | 0.1149 | 0.8635 | 0.8005 | 0.4560 | 0.1751 | 0.7500 |
| KNN K=5 | 0.8474 | 0.7159 | 0.1433 | 0.8694 | 0.7628 | 0.5246 | 0.2204 | 0.7566 |
| DT: J48 | 0.8598 | 0.5315 | 0.1165 | 0.7329 | 0.7571 | 0.4374 | 0.2209 | 0.6628 |
| DT: RF N=10 | 0.8914 | 0.5420 | 0.0832 | 0.8754 | 0.8323 | 0.4102 | 0.1375 | 0.7569 |
| DT: RF N=25 | 0.8930 | 0.5789 | 0.0840 | 0.8864 | 0.8319 | 0.4294 | 0.1395 | 0.7697 |
| DT: RF N=50 | 0.8949 | 0.5840 | 0.0824 | 0.8907 | 0.8339 | 0.4233 | 0.1368 | 0.7723 |
| DT: RF N=75 | 0.8949 | 0.5907 | 0.0829 | 0.8928 | 0.8345 | 0.4250 | 0.1364 | 0.7739 |
| DT: RF N=100 | 0.8947 | 0.5876 | 0.0829 | 0.8942 | 0.8349 | 0.4250 | 0.1361 | 0.7747 |

## 5   Discussion and Conclusions

In conclusion, in this first experimentation we can conclude that the best strategy is to use the FTT as categorization method, comparing with by COM algorithms, and we discourage the use of these characterization method in these type of problems. Regarding to the machine learning algorithms, we can conclude that, despite the KNN obtained the best results, SVM algorithms in general have better performance when face new instances, so in this case will be the best option to include in a prototype of the system, due to over-fitting problems of decision trees aforementioned.

## References

1. Bodnarova, A., Williams, J., Bennamoun, M., Kubik, K.: Optimal textural features for flaw detection in textile materials. In: IEEE Region 10 Annual Conference. Speech and Image Technologies for Computing and Telecommunications, TENCON 1997, Proceedings of IEEE, vol. 1, pp. 307–310. IEEE (1997)
2. Bracewell, R.: The Fourier Transform and its Applications (1999)
3. Breiman, L.: Random forests. Mach. Learn. **45**(1), 5–32 (2001)

4. Brinkmann, R.: The Art and Science of Digital Compositing: Techniques for Visual Effects, Animation and Motion Graphics. Morgan Kaufmann, Boston (2008)
5. Castillo, E., Gutierrez, J.M., Hadi, A.S.: Expert Systems and Probabilistic Network Models. Springer Science & Business Media, New York (2012). https://doi.org/10.1007/978-1-4612-2270-5
6. Castleman, K.: Digital image processing. Second (1996)
7. Chawla, N.V., Bowyer, K.W., Hall, L.O., Kegelmeyer, W.P.: SMOTE: synthetic minority over-sampling technique. J. Artif. Intell. Res. **16**, 321–357 (2002)
8. Christopher, M.B.: PAttern Recognition and Machine Learning. Springer, New York (2016)
9. Fisher, R.A.: The use of multiple measurements in taxonomic problems. Ann. Eugenics **7**(2), 179–188 (1936)
10. Fix, E., Hodges Jr., J.L.: Discriminatory analysis-nonparametric discrimination: consistency properties. California University Berkeley, Technical report (1951)
11. Garner, S.R., et al.: WEKA: the waikato environment for knowledge analysis. In: Proceedings of the New Zealand Computer Science Research Students Conference, pp. 57–64. Citeseer (1995)
12. Goodfellow, I., Bengio, Y., Courville, A., Bengio, Y.: Deep Learning, vol. 1. MIT Press, Cambridge (2016)
13. Haralick, R., Shanmugam, K., Dinstein, I.: Textural features for image classification. IEEE Trans. Syst. Man Cybern. **3**(6), 610–621 (1973)
14. Iivarinen, J., Rauhamaa, J., Visa, A.: Unsupervised segmentation of surface defects. In: Proceedings of the 13th International Conference on Pattern Recognition, vol. 4, pp. 356–360. IEEE (1996)
15. Kamal, K., Qayyum, R., Mathavan, S., Zafar, T.: Wood defects classification using laws texture energy measures and supervised learning approach. Adv. Eng. Inform. **34**, 125–135 (2017)
16. Kitchin, R., Lauriault, T.P.: Small data in the era of big data. Geo J. **80**(4), 463–475 (2015)
17. Kopardekar, P., Mital, A., Anand, S.: Manual, hybrid and automated inspection literature and current research. Integr. Manuf. Syst. **4**(1), 18–29 (1993)
18. LeCun, Y., Bengio, Y., Hinton, G.: Deep learning. Nature **521**(7553), 436 (2015)
19. LeCun, Y., Bottou, L., Bengio, Y., Haffner, P.: Gradient-based learning applied to document recognition. Proc. IEEE **86**(11), 2278–2324 (1998)
20. Mery, D., Arteta, C.: Automatic defect recognition in x-ray testing using computer vision. In: 2017 IEEE Winter Conference on Applications of Computer Vision (WACV), pp. 1026–1035. IEEE (2017)
21. Mital, A., Govindaraju, M., Subramani, B.: A comparison between manual and hybrid methods in parts inspection. Integr. Manuf. Syst. **9**(6), 344–349 (1998)
22. Monadjemi, A.: Towards efficient texture classification and abnormality detection. Ph.D. thesis, University of Bristol (2004)
23. Neogi, N., Mohanta, D.K., Dutta, P.K.: Review of vision-based steel surface inspection systems. EURASIP J. Image Video Process. **2014**(1), 50 (2014)
24. Pastor-López, I., Santos, I., Santamaría-Ibirika, A., Salazar, M., de-la Pena-Sordo, J., Bringas, P.G.: Machine-learning-based surface defect detection and categorisation in high-precision foundry. In: 2012 7th IEEE Conference on Industrial Electronics and Applications (ICIEA), pp. 1359–1364. IEEE (2012)
25. Pearl, J.: Bayesian networks: a model of self-activated memory for evidential reasoning. In: Proceedings of the 7th Conference of the Cognitive Science Society (1985)

26. de la Puerta, J.G., Sanz, B., Santos, I., Bringas, P.G.: Using dalvik opcodes for malware detection on android. In: Onieva, E., Santos, I., Osaba, E., Quintián, H., Corchado, E. (eds.) HAIS 2015. LNCS (LNAI), vol. 9121, pp. 416–426. Springer, Cham (2015). https://doi.org/10.1007/978-3-319-19644-2_35
27. Quinlan, J.R.: Induction of decision trees. Mach. Learn. **1**(1), 81–106 (1986)
28. Quinlan, J.R.: C4.5: Programs for Machine Learning. Elsevier, San Francisco (2014)
29. Siegmund, D., Samartzidis, T., Fu, B., Braun, A., Kuijper, A.: Fiber defect detection of inhomogeneous voluminous textiles. In: Carrasco-Ochoa, J.A., Martínez-Trinidad, J.F., Olvera-López, J.A. (eds.) MCPR 2017. LNCS, vol. 10267, pp. 278–287. Springer, Cham (2017). https://doi.org/10.1007/978-3-319-59226-8_27
30. vom Stein, D.: Automatic visual 3-D inspection of castings. Foundry Trade J. **180**(3641), 24–27 (2007)
31. Tout, K., Retraint, F., Cogranne, R.: Automatic vision system for wheel surface inspection and monitoring. In: ASNT Annual Conference, pp. 207–216 (2017)
32. Vapnik, V.: The Nature of Statistical Learning Theory. Springer Science & Business Media, New York (2013). https://doi.org/10.1007/978-1-4757-3264-1
33. Weimer, D., Scholz-Reiter, B., Shpitalni, M.: Design of deep convolutional neural network architectures for automated feature extraction in industrial inspection. CIRP Ann. **65**(1), 417–420 (2016)
34. Zhang, M.L., Zhou, Z.H.: ML-KNN: a lazy learning approach to multi-label learning. Pattern Recogn. **40**(7), 2038–2048 (2007)

# Reinforcement Learning Experiments Running Efficiently over Widly Heterogeneous Computer Farms

Borja Fernandez-Gauna[⊠], Xabier Larrucea, and Manuel Graña

Computational Intelligence Group, University of the Basque Country, UPV/EHU, Leioa, Spain
borja.fernandez@ehu.eus

**Abstract.** Researchers working with Reinforcement Learning typically face issues that severely hinder the efficiency of their research workflow. These issues include high computational requirements, numerous hyperparameters that must be set manually, and the high probability of failing a lot of times before success. In this paper, we present some of the challenges our research has faced and the way we have tackled successfully them in an innovative software platform. We provide some benchmarking results that show the improvements introduced by the new platform.

## 1 Introduction

We present the features of an open source software platform [4] for efficient execution of Reinforcement Learning (RL) experiments over wildly heterogeneous computer farms. Our research group has been working for quite some time on control applications of Reinforcement Learning (RL) methods [3]. The aim of RL [6,9] is learning a control policy that maximizes the control goal, which is expressed as a reward signal. In on-line learning settings, agents interact with the environment in a closed-loop: every time-step the agent observes the state, selects an action, and then, observes the reward and the new state of the system. By repeating this process, the agent aims to learn a parameterized policy that maximizes the expected long-term accumulated reward. This process usually requires a lot of computation invested in simulated learning processes and is simply unfeasible to apply it directly to real-world environments. Furthermore, experimenting with RL involves setting many hyper-parameters that parameterize the behavior of the experiment elements: the learning algorithm, the environment, the reward function, and so on. Some of these hyper-parameters are often very difficult to set and add an extra level of complexity to the practical application of RL, requiring a huge number of trials before the process is successful.

A bit of history: We first developed a software tool in $C++$ that implemented a fully configurable RL closed-loop. The process was configured via an XML file that we edited manually. This approach is rather inconvenient for several

© Springer Nature Switzerland AG 2019
H. Pérez García et al. (Eds.): HAIS 2019, LNAI 11734, pp. 758–769, 2019.
https://doi.org/10.1007/978-3-030-29859-3_64

reasons: manually editing the configuration file is an inherently error-prone process, changes in the source code (which are very frequent in research) imply very frequent changes in the configuration files, and grid search for optimal hyper-parameter settings are incredibly tedious. As a first step to run experiments with hundreds of hyper-parameter variations, we scripted the process and run the different experiments in batches. Each of the batches spawned as many experiments as CPU cores the user's machine had. We next considered distributing experiments among all our machines. At the university, though, researchers have no administrative rights on any of the computers and, thus, we had to ask for permission to the computer resources administration staff every time we wanted to update the software. Administrative delays, and the manual distribution of the configuration files among the machines was a source of frustration. Furthermore, we had to work over a wildly heterogeneous computer farm: different processor capabilities, two different Operative Systems (*Windows* and *Linux*), and two different architectures (*x86* and *x64*). *C++* sources can be compiled for virtually any computer architecture, but each architecture requires a different version of the binaries and the appropriate version of the dependencies.

In this paper, we present the software platform that we have developed to overcome these difficulties and make the most out of the hardware available to us. In Sect. 2 we formally state the goals of the platform presented in Sect. 3. Section 4 discusses how this platform improved our workflow and Sect. 5 presents our conclusions.

## 2    Problem Statement

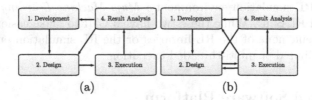

(a)                                    (b)

**Fig. 1.** (a) Conventional experimental workflow used in research. (b) Experimental workflow with early experiment failure detection.

Experimental research workflow usually consists of four different work phases depicted in Fig. 1(a): development, design, execution and result analysis. In the first phase the software is developed, then experiments are designed, executed and results are analysed. Analysis of the results may lead back to either the development phase (i.e., an algorithm or technique is added or modified) or the design phase (i.e., a hyper-parameter is modified). We have analysed the main obstacles to an efficient research workflow, identifying the following key goals to improve the experimentation cycle:

- **Efficient experiment design.** Hyper-parameter optimal value search should be a core feature of the software platform that should be managed in a systematic manner, including validation of the configurations in the experiment-design phase, and tools that allow the user to set and edit multiple values to any hyper-parameter. Independence between code and experimental design: Changes in the source code should not involve big changes in existing or new experiment configuration files.
- **Efficient execution.** The software platform should allow distributed execution over networks of wildly heterogeneous machines, prioritizing the use of machines with higher processing performance. Ideally, the results should be received in the user's computer automatically in a seamless way. Software should be automatically deployed on the remote machines to avoid depending on administrative staff for lack of administrative rights. Deploying software via the network adds a new aspect to consider: RL software may be large (some Deep Learning libraries require more than 500 Mb space on disk) and multi-platform support makes this problem even more acute. This means that the prerequisites of each experiment should be managed in an efficient way.
- **Early experiment failure detection.** Some experiments take days or weeks to finish and, in research, failed experiments are far more frequent that successful ones. Thus, it is crucial to detect failed experiments as early in the process as possible. Figure 1(b) represents our objective workflow: the user should be able to detect failed experiments during execution and go back to the design or even the development phase.

The most commonly used software packages in scientific research are Deep Neural Network libraries (*Tensor Flow* [1], *CNTK* [8]), or RL libraries that provide algorithms and environments (*pyBrain* [7], *RLPy* [5]). One level above, we note two full RL simulation environments: *Maja Machine Learning Framework* (MMLF)[1] and RL Sim[2]. While these two have a GUI to make easier the design of an experiment, none of the RL libraries or the RL simulation environments supports distributed or multi-threaded execution.

## 3   Proposed Software Platform

In this section, we describe the main features of the experimentation platform developed by our research group to address the goals listed in Sect. 2. This platform is based on the following main tools:

---

[1] http://mmlf.sourceforge.net.
[2] https://www.cs.cmu.edu/~awm/rlsim.

- The **Experiment Application** runs an instance of an experiment. A hierarchy of object classes with their respective parameters define an experimental unit[3] and its hyper-parameters. Figure 2 shows part of the class hierarchy and the parameters of each object class.
- The **Experiment Designer** features a *Graphical User Interface* (GUI) to design an experiment setting the values of the experiment application. It allows the use of multi-valued hyper-parameters.
- The **Experiment Monitor** distributes the workload among the available machines and shows the evolution of the experiment in real-time.
- The **Distributed Agent:** a service that allows any machine in the local network to receive job queries, run the tasks, report live monitoring information, and send the results to the *Experiment Monitor*.

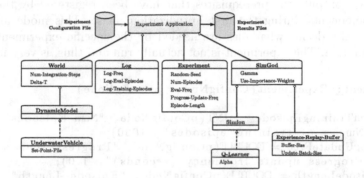

**Fig. 2.** A simplified UML diagram of the basic class hierarchy and their configurable parameters.

## 3.1 Experiment Application

This is the computational core of the platform: a data-driven application written in *C++*. It implements a fully configurable close-loop online learning process. The input of the application is an XML file with the values of all the hyper-parameters, and it outputs log files with the variables of the system, functions learned, and so on. Figure 3 shows a snippet of the code used to initialize the configurable parameters of the object class *Experiment*. The use of pre-defined hyper-parameter objects (i.e., *INT*, *DOUBLE*) allows us to easily initialize parameters automatically mapping hyper-parameters in the object class hierarchy and nodes in the XML configuration file (encapsulated by class

---

[3] We use the terms *experiment* and *experimental unit* to distinguish two different concepts. The former refers to a configuration containing multi-valued hyper-parameters that will require several executions to finish, whereas the latter reefers to each of the single-valued configuration instances produced by combining the values of an experiment.

*ConfigNode*). We enforce the inclusion of meta-data in the hyper-parameter constructors, such as user-friendly names, hyper-parameter descriptions, and default values. Figure 4 shows a XML snippet used to initialize the Experiment object class's parameters.

We divide the initialization of the application in two different phases: in the first phase the values of the hyper-parameters are initialized and each object registers its *prerequisites* (input files) and output files, whereas the second phase does all the time-consuming initialization (i.e., allocating memory buffers for the function weights). We use these two phases to define two different execution modes:

- **Regular.** This mode performs both initialization phases and runs the experiment. This is the default execution mode.
- **Prerequisites.** This second mode performs only the first initialization phase and then outputs the prerequisites that have been registered by the learning components during the first initialization phase. This mode allows to dynamically decide what files are needed by the specific experimental unit configuration. The experiment is not actually run and thus, is very fast.

```
Experiment :: Experiment ( ConfigNode* pConfigNode )
{
   m_numTrainingEpisodes = INT( pConfigNode , "Num-Episodes"
      , "Number of training episodes", 1000);
   m_progUpdateFreq = DOUBLE( pConfigNode , "Progress-Update-Freq"
      , "Progress update frequency (seconds)", 1.0);
   m_episodeLength = DOUBLE( pConfigNode , "Episode-Length"
      , "Length of an episode (seconds)", 10.0);
}
```

**Fig. 3.** Code snippet from the construction method of an *Experiment* class object. *DOUBLE* and *INT* are classes that include meta-data and initialize hyper-parameters in a uniform manner.

```
<Experiment>
        <Num-Episodes >1000</Num-Episodes>
        <Progress-Update-Freq >0.5</Progress-Update-Freq>
        <Episode-Length >200.0</Episode-Length>
</Experiment>
```

**Fig. 4.** Sample XML code used to initialize the parameters of class *Experiment*.

## 3.2   Experiment Designer

Aiming at an efficient experiment design phase, we decided to create a GUI to configure experiments because it also added the chance to target a broader audience that included non-programming users. As pointed out in Sect. 2, one of the problems for an efficient experiment design while doing research is that its code changes very often. To mitigate the coupling between the source code of the *Experiment Application* and the GUI objects used in the *Experiment Designer*, we built a custom parser that transformed the initialization code in the object classes constructors to a hyper-parameter definition file. The code in Fig. 3 produces the definition in Fig. 4. These definitions include not only the type of the parameter for validation, but also meta-data to be used in the Experiment Designer. The definition of default values as part of the meta-data in the source code allows us to reuse as much as possible old configuration files. The parsing process is automatically triggered when the source code is recompiled (Fig. 5). This updates the hyper-parameter definition file that is loaded by the *Experiment Designer* on initialization. After successfully designing an experiment and validating the values given to the hyper-parameters, the application creates the experiment configuration file used by the *Experiment Application*. The process is depicted in Fig. 6.

```
<CLASS Name="Experiment">
  <INT Name="Num-Episodes"
     Comment="Number of training episodes" Default="1000"/>
  <DOUBLE Name="Progress-Update-Freq"
     Comment="Progress update frequency (seconds)" Default="1.0"/>
  <DOUBLE Name="Episode-Length"
     Comment="Length of an episode(seconds)" Default="10.0"/>
</CLASS>
```

**Fig. 5.** Sample output of the source code parser with the definition of class *Experiment* and its parameters.

## 3.3   Distributed Agent

In order to use all the available machines in the farm, we developed a service listening for job queries. These job queries contain lists of tasks, input files, and output files that are to be sent back to the client after the execution of each task. The definition of these jobs and tasks is totally independent of the *Experiment Application* we currently use so that it can be used for any other project that may benefit from distributed computation. To avoid deployment and update issues due to our lack of administrative rights on some of the computers, job queries include the executable files and any required file for every job to be carried out. This way, we only need to ask permission to install and update the *Distributed Agent* once, instead of having to ask for permissions every time the *Experiment Application* is updated (very often in our context). Of course, this means that the application must not be installed and can be executed using only the binaries and input files sent to the remote machine. This tool is heavily inspired on the *Swarm Agent* used in *Unreal Engine 4* [2]. The description of a task may also include a request to redirect its output to the client, so that the process can be monitored remotely. *Distributed Agents* can be discovered in run-time by a UDP

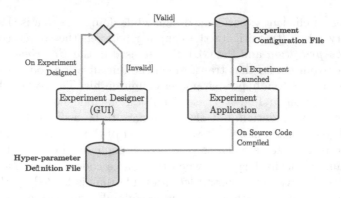

**Fig. 6.** Data-flow in the platform development cycle. Every time the experiment application is compiled, a source code parser generates a file with the definitions and parameters of the object classes in the source code. Another module reads this file on initialization to assure that the GUI matches the parameters in the source code, and then, every time an experiment is launched, it generates the input file with the experiment parameters read by the experiments' executable.

broadcast, which they respond with an description of their properties so that the client can select the agents and binaries compatible with the platform. This description is formatted as XML as shown in Fig. 7 and includes the running version of the software, the properties of the platform underneath (OS, number of CPU cores, GPU computing support) and current state (i.e., busy).

```
<Agent>
    <HerdAgentVersion >1.2.3.0 </HerdAgentVersion>
    <ProccesorId>BFEBFBFF000306D4</ProccesorId>
    <NumCPUCores>4</NumCPUCores>
    <Architecture>Win−64</Architecture>
    <ProcessorLoad >63.34</ProcessorLoad>
    <Memory>8Gb</Memory>
    <CUDA>9.3.1 </CUDA>
    <State>available </State>
</Agent>
```

**Fig. 7.** Sample description of a *Distributed Agent*'s description sent to a client that reached the agent via a UDP broadcast.

## 3.4 Experiment Monitor

This application dispatches jobs to the available *Distributed Agents* and monitorizes the progress of the experiment. Its input is an experiment configuration file output by the *Experiment Designer* that may contain multi-valued hyperparameters. Because the *Experiment Application* needs an experimental unit with single-valued hyper-parameters, multi-valued hyper-parameters must first be combined. This combination of multi-valued hyper-parameters produces a set of experimental units that are iteratively assigned to the available *Distributed Agents*. Depending on the computation capabilities of the agents and the run-time requirements of each experimental unit, more than one experimental unit may be included in the same batch assigned to the agent. For example, if a *Distributed Agent* reports 4 CPU cores, 4 experimental units that only require one CPU core to run may be batched together in the job sent to the agent. On the other hand, if an experimental unit requires using all the available CPU cores or the GPU, it will not be batched with any other experimental unit. Figure 8 shows a diagram representing the process that converts experiment configuration files to jobs ready to be sent to a *Distributed Agent*.

The calculation of the prerequisites is done by merging the platform pre-requisites and the run-time prerequisites calculated in run-time. *Platform pre-requisites* are static and depend on the Distributed Agent's platform and the application definition (Fig. 9 shows a sample of an application with two different target platforms, each with its own platform requirements). On the other hand, *run-time prerequisites* are calculated dynamically by running the *Experiment Application* in *Prerequisites* mode (see Sect. 3.1). These pre-requisites may be platform-dependent, and so, may require the description of the selected *Distributed Agent*. For example, an experiment may require all the libraries from *Microsoft's Cognitive Toolkit*, but the binaries from these libraries will be different for *Windows* and for *Linux*. The process used to calculate the prerequisites is represented in Fig. 10. The *Experiment Monitor* requests *Distributed Agents* to redirect the output of the *Experiment Applications* launched so that the Experiment Monitor can monitor the experiment. The monitoring information include the progress and the average reward obtained during evaluation episodes. This allows for an early detection of failed experiments and a faster research iteration cycle (Sect. 2).

## 4   Computational Performance Results

In this section, we provide some experimental results showing the gains introduced by the platform in our workflow. We provide two examples and compare the execution time against a single local machine. The two experiment instances, denoted as *Exp-A* and *Exp-B*, are described in Table 1[4]. In the table, $n_e$ represents the number of experimental units created from each experiment, $t_e$ the approximated execution time in seconds per experimental unit, and $s_e$ the size in *Mb* required by each experiment's prerequisites. Experimental units sprouting from *Exp-A* run on a single thread on the CPU, whereas those generated from *Exp-B* require the use of a GPU and the *Cognitive Toolkit*. For these examples, we will use the machines in Table 2 and note here that this is a very conservative comparison since it doesn't include the 75 machines in the students' labs

---

[4] *Exp-B* requires *Microsoft's Cognitive Library*, which only runs on x64 platforms. That's the reason *Exp-B* cannot run on *Windows-x32* machines.

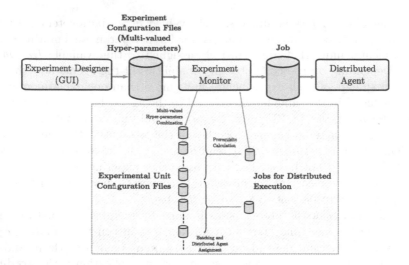

**Fig. 8.** Diagram of the process that converts experiment configuration files that potentially includes multi-valued to jobs ready for remote execution.

```
<App Name="RLSimion" FileVersion="1">
  <Version Name="Win-x64">
    <Exe>../bin/RLSimion-x64.exe</Exe>
    <Requirements>
      <Input-File>../bin/x64/msvcp140.dll</Input-File>
      <Input-File>../bin/x64/vcruntime140.dll</Input-File>
      <Architecture>Win-64</Architecture>
    </Requirements>
  </Version>
  <Version Name="Linux-x64">
    <Exe>../bin/RLSimion-linux-x64.exe</Exe>
    <Requirements>
      <Architecture>Linux-64</Architecture>
    </Requirements>
  </Version>
</App>
```

**Fig. 9.** Sample application definition file with two different target platforms, each of them with different dependencies.

for simplicity. If we assume that all the experimental units generated from an experiment take approximately the same time ($t_e$) and thus, all finish at the same time, we can approximate the total time required to execute an experiment using the formula:

$$t_{total}^d = n_r \cdot (t_e + t_t),$$

(1)

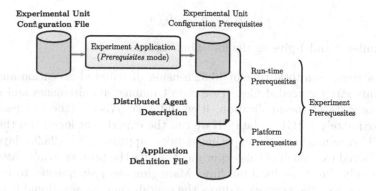

**Fig. 10.** Prerequisite calculation diagram.

**Table 1.** Characteristics of the experiments used for benchmarking purposes. $n_e$ = number of experimental units, $t_e$ = execution time in seconds per experimental unit, $s_e$ = size in $Mb$ per experiment.

| Experiment | $n_e$ | $t_e$ | $s_e$ |
|---|---|---|---|
| *Exp-A* | 3, 125 | 340 | 1.9 Mb (*Win-x32*) |
| | | | 2.3 Mb (*Win-x64*) |
| | | | 5.8 Mb (*Linux-x64*) |
| *Exp-B* | 9 | 25500 | *N/A* (*Win-x32*) |
| | | | 625 Mb(*Win-x64*) |
| | | | 576 Mb (*Linux-x64*) |

**Table 2.** The characteristics of the machines used for comparison.

| Machines | OS | Num. CPU Cores | GPU |
|---|---|---|---|
| User's machine | *Win-x64* | 4 | Yes |
| Machine #1 | *Linux-x64* | 2 | No |
| Machine #2 | *Linux-x64* | 4 | Yes |
| Machine #3 | *Win-x64* | 4 | Yes |
| Machine #4 | *Win-x64* | 4 | No |
| Remote machine #5 | *Win-x32* | 4 | No |

where $n_r$ represents the number of *rounds* (the number of jobs sent to each agent) needed to finish all the experimental units, and $t_t$ is the time needed to send the prerequisites over the network. The number of rounds can be calculated with $n_r = n_e/n_c$,where $n_c$ is the number of experimental units run concurrently in the whole system. The transmission time can be calculated $t_t = s_e/t_{Mb}$, where $t_{Mb}$ is the time needed to transmit 1 Mb over the network (we used $t_{Mb} = 0.082$ s). On the other hand, the time required to execute an experiment using only the local machine can be calculated with the formula:

$$t^l_{total} = t_e \cdot n_e/n_c. \tag{2}$$

Formulae 2 and 1 give us the following values:

- *Exp-A* requires a total time of 13.52 h using distributed execution and sending only strictly needed files. If we didn't manage requirements and sent all the contents of the application, it would take 17.41 h (the full package is approximately 1.2 Gb of data). If we run the experiment locally on the user's 4 CPU core-machine, the experiment takes approx. 678 h (28.25 days). The distributed execution of this experiment makes the process 97.5% faster than using only the user's local machine. Managing the prerequisites to minimize data sent over the network reduces the overall time an additional 0.57%.
- *Exp-B* requires a total time of 21.29 h using distributed execution and managing prerequisites. Without prerequisite management it requires 21.33 h and running only on the user's computer, about 63.75 h. This means that distributed execution is 66% faster than local execution.

## 5    Conclusions

In this paper, we have presented some of the problems we encountered while developing an efficient RL experimentation platform for our group, and the main changes that improved the efficiency of our workflow: a faster experiment design via a GUI that allows to use multi-valued hyper-parameters, distributed execution of the experiments by distributing different combinations of the multi-valued hyper-parameters to the machines available on the network, and live monitoring of the experiment for an early failure detection.

**Acknowledgements.** The work in this paper has been partially supported by FEDER funds for the MINECO project TIN2017-85827-P, and projects KK-2018/00071 and KK-2018/00082 of the Elkartek 2018 funding program of the Basque Government.

## References

1. Abadi, M., et al.: TensorFlow: a system for large-scale machine learning. In 12th USENIX Symposium on Operating Systems Design and Implementation (OSDI 2016), Savannah, GA, pp. 265–283. USENIX Association (2016)
2. EpicGames: Unreal swarm (2019)
3. Fernández-Gauna, B., Fernandez-Gamiz, U., Graña, M.: Variable speed wind turbine controller adaptation by reinforcement learning. Integr. Comput. Aided Eng. **24**(1), 27–39 (2017)
4. Gauna, B.F., Graña, M., Zimmermann, R.S.: SimionZoo: a software bundle for Reinforcement Learning applications, February 2019. https://doi.org/10.5281/zenodo.2579013
5. Geramifard, A., Dann, C., Klein, R.H., Dabney, W., How, J.P.: RLPy: a value-function-based reinforcement learning framework for education and research. J. Mach. Learn. Res. **16**, 1573–1578 (2015)

6. van Hasselt, H., Guez, A., Silver, D.: Deep reinforcement learning with double q-learning. In: Proceedings of the Thirtieth AAAI Conference on Artificial Intelligence, AAAI 2016, pp. 2094–2100. AAAI Press (2016)
7. Schaul, T., et al.: PyBrain. J. Mach. Learn. Res. **11**, 743–746 (2010)
8. Seide, F., Agarwal, A.: CNTK: Microsoft's open-source deep-learning toolkit. In: Proceedings of the 22nd ACM SIGKDD International Conference on Knowledge Discovery and Data Mining, KDD 2016, pp. 2135–2135. ACM, New York (2016)
9. Silver, D., et al.: Mastering the game of go with deep neural networks and tree search. Nature **529**, 484–503 (2016)

# Author Index

Printed in the United States
By Bookmasters